Selected Writings of Sidney J. Holt

Documenting a Lifelong Pursuit of Sustainable and Profitable Fishing

Selected Writings of Sidney J. Holt
Documenting a Lifelong Pursuit of Sustainable and Profitable Fishing

Emory D. Anderson
NOAA Fisheries (retired)
22 Indian Pond Point, P.O. Box 950
Marstons Mills, Massachusetts 02648, USA
E-mail: emoryanderson@comcast.net

and

Michael Earle
Route d'Amonines, 15
6987 Rendeux, Belgium
E-mail: michael.earle.1967@gmail.com

American Fisheries Society
Bethesda, Maryland
2021

A suggested citation format for this book follows.

Anderson, E. D., and M. Earle. Editors. 2021. Selected writings of Sidney J. Holt: documenting a life-long pursuit of sustainable and profitable fishing. American Fisheries Society, Bethesda, Maryland.

Front cover photos: Sidney J. Holt in earlier and later years, provided by Holt in 2011.

Back cover photo: Sidney Holt walking through his olive grove at his residence in Paciano, Umbria, Italy. Photo provided by David Lavigne.

Printed in the United States of America on acid-free paper.

Library of Congress Control Number 2021936589
ISBN 978-1-934874-62-2

American Fisheries Society Web site address: *www.fisheries.org*

American Fisheries Society
425 Barlow Place, Suite 110
Bethesda, Maryland 20814
USA

Table of Contents

Part III: Getting Back to First Principles

Foreword

Sidney Holt was one of a kind; a melding of intellectual brilliance, persistent focus, and international policy acumen meted out across seven decades of marine conservation advocacy. A renowned sprinter in youth, this founding father of modern fisheries science co-authored its most frequently cited text while still in his twenties. Throughout the seventy years that followed, he never slowed down.

Well into his nineties, from his sunny perch in Paciano, Umbria, Sidney continued to drive discussions, debates, strategy sessions, and tutorials with senior government officials, scientific colleagues, academic experts, and campaigners around the ocean planet. Generations of genuflecting students, select NGO leaders, and hardened "whale-huggers" alike reveled in Sidney's epic e-mails, presentations, and papers, as provocative as they were prolific. They remain so today.

Sidney had a desk and a life strewn with accrued notes and newspapers, reports by other giants ancient and new, and his own writings—including the hitherto obscure or unpublished works so carefully collected here. Fisheries scientists Emory Anderson and Michael Earle took on the leviathan challenge of selecting them. All of us at the International Fund for Animal Welfare (IFAW) are very grateful for their patient efforts and pleased to welcome the publication of this important volume.

Peer-reviewed science is a tool and template for effective wildlife conservation. Sidney's place in the pantheon is secure—across sectors. From the portals of academia to the corridors of international policymaking, he remains without peer. Close colleagues and critics alike credit Sidney Holt with saving the world's great whales from extinction. Like the end of high seas whaling and other victories hard-won, IFAW's long collaboration with Sidney is a high watermark in our continuing marine conservation work.

Our support for this unique publication completes that collaboration, preserving some of Sidney Holt's most important and provocative thinking for present-day fisheries scientists, whale biologists, and generations of wildlife conservationists yet to come. The torch the fleet-footed sprinter so long carried, he now has passed to us.

Azzedine T. Downes
President and Chief Executive Officer
International Fund for Animal Welfare (IFAW)

Preface

In the course of human endeavor throughout recorded history, many individuals have been recognized as outstanding, deserving of recognition, and worthy of honor and tribute in their chosen field of work or career path. Such people may have performed a singular act or feat or served admirably for a period of time or perhaps an entire working career. However, rarely has such an individual performed and contributed so importantly in his or her profession for such an exceptionally long time. The person being honored in this book, Sidney Joseph Holt, is one such individual who served brilliantly and with distinction in his chosen fields of fisheries and whale biology, population dynamics, and stock assessment for portions of eight decades, beginning at age 21 in 1946 and continuing until his death at age 93 on December 23, 2019.

Sidney as a student at Reading. Photo taken from Best (2010).

Sidney's[1] career in marine science was launched in 1946 when, after graduating from the University of Reading with a first-class degree in biology, he was hired as a "naturalist" at the Fisheries Laboratory in Lowestoft, England. His initial work involved going to sea, teaching himself calculus, and acquiring additional knowledge from Director Michael Graham who was "the best mentor anyone could wish for." He and his subsequent collaborator, Raymond Beverton, who was four years older, first met that spring and were told by Graham, "I want the two of you to really see if you can't put the whole of this fish population stuff on a more substantial basis." Shortly thereafter, Beverton returned to Cambridge to complete his Master's degree before returning to Lowestoft in summer 1947; Graham then said: "Well, I'll give you four years ... I can't tell you how to do it ... If you don't succeed at the end of four years, I can't protect you any longer." Their four-year collaboration culminated in the publication, in 1957, of their *magnum opus* that subsequently became the most-cited text on quantitative fish stock assessment and fondly referred to as "the bible of fisheries science": *On the Dynamics of Exploited Fish Populations* (Beverton and Holt 1957). Although the book was not published until 1957, Sidney left Lowestoft in 1950 for a position in Scotland with The Nature Conservancy and three years later, in 1953, was persuaded to join the Fisheries Division of the United Nations (UN) Food and Agriculture Organization (FAO) in Rome.

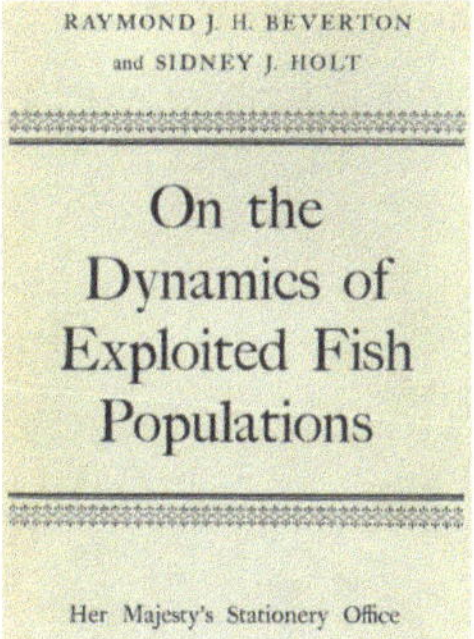

Cover of "the bible of fisheries science."

His work with FAO initially involved fisheries, including intimate involvement in discussions on management under the auspices of the UN conferences leading up to the 1958 Law of the Sea. That changed in 1960 when he was nominated to serve on a "Committee of Three Scientists" (K. Radway Allen, Douglas Chapman, and Sidney Holt) for the International Whaling Commission (IWC); a year later, the Committee was expanded to include a fourth scientist (John Gulland). The work of the Committee led to quotas on blue *Balaenoptera musculus*, fin *B.*

[1]Some of the anecdotes in this Preface are drawn from papers cited below, particularly Anderson (2002a, 2002b, 2011), Beverton and Anderson (2002), and Holt (2020). Others are personal by one or other of the co-editors.

physalus, and humpback *Megaptera novaeangliae* whales in the Antarctic, and Sidney's continued involvement in whale stock assessment work eventually led to a moratorium on whaling in 1982 (effective in 1986) and, a decade later, to the Revised Management Procedure. In the words of his former colleague Ray Beverton, "Sidney had achieved a wonderful result. He had saved the great whales in the early 1970s." His active participation in the IWC continued until 2002, over 40 years, including the "Committee of Three," the Scientific Committee, and the Commission itself either as a member, observer, adviser (France), or commissioner (Seychelles).

Sidney formally retired from the UN in 1979, but during his 25-year tenure, had served at various times as Director of the Fisheries Resources and Operations Division of FAO in Rome, FAO Liaison Officer with the Intergovernmental Oceanographic Commission (IOC) of the United Nations Educational, Scientific and Cultural Organization (UNESCO) in Paris (1969), Executive Secretary of IOC in Paris (1970–1972), and Director of UNESCO's Marine Sciences Division in Paris, the latter three positions held during a three-year secondment from FAO to UNESCO.

Following his retirement from the UN, and during his continued participation in the IWC, he also served as the first Director of the International Ocean Institute in Malta and assisted in the drafting of the UN Convention on the Law of the Sea in 1982. He devoted considerable energy to the overall conservation and protection of the great whales through a number of non-governmental organizations such as Greenpeace, World Wildlife Fund (WWF), International Fund for Animal Welfare (IFAW), and Sea Shepherd Conservation Society; for many of these he served as an adviser and author of articles.

Sidney and Ray Beverton together for the last time in 1984. Photo courtesy of Sidney Holt.

Beginning in the early 2000s and continuing until the last few years of his life, Sidney refocused his time and effort back to fisheries management problems, mainly those involving the European Union (EU) and the revision of its Common Fisheries Policy (CFP). As a result, he became engaged with the EU Parliament (primarily through the Greens/EFA Group) and wrote numerous articles pertaining to the faults of maximum sustainable yield (MSY), the folly of using it as a management objective, and the need to set appropriate target fishing mortality rates for sustainable fishing below those generally associated with MSY. Many of his writings in this context, previously unpublished in journals or books, are included in this volume.

In the final two decades of his life, Sidney received many honors and was a frequent speaker to various groups. In 2013, he was selected as "Buckland Professor" by the Frank Buckland Foundation in the UK and gave the traditional "Buckland Lecture" in London's prestigious Fishmongers' Hall. He was presented numerous awards for his contributions to science and conservation, including the Gold Medal of the WWF, the Global 500 Award of UNEP, the Royal Golden Ark of The Netherlands, the Blue Planet award from IFAW, and, in 2017, the Beverton Medal of the Fisheries Society of the British Isles.

After his death, Sidney was eulogized by many from both the fisheries and whaling disciplines, e.g., Clifton (2019), Brown (2020), Earle (2020), Engelhard and Pinnegar (2020), Kent MacKay (2020), Lavigne (2020), Pauly (2020), and The Times (2020). The *ICES Journal of Marine Science (IJMS)* announced that it would publish a special issue or themed

Sidney in the library of the Institut Océanographique de Paris in 2011. Photo by Tim Holt.

section of contributed papers in honor of Sidney. Interestingly, a timely publication by the *IJMS* in late January 2020 of an essay (Holt 2020) submitted by Sidney in 2017, but which he was unable to revise following peer review, was rewritten by Emory D. Anderson, one of the *IJMS* editors and co-editor of this book, and Andrew I. L. Payne, a former *IJMS* Editor-in-Chief, using the original peer reviews as well as modifying the original text based on their personal knowledge of Sidney. That essay, when it appeared, attracted the attention of Michael Earle, the other co-editor of this book. Earle, a retired fisheries adviser for the Greens in the European Parliament, had persuaded Sidney, in 2006, to become involved in the reform of the CFP, as mentioned above. For a decade thereafter, Sidney wrote analyses and briefings for the Greens on the pros and cons of various fisheries models as well as on the thorny question of where MSY lies and the trade-offs between a reduction in fishing effort and a reduction in fisheries yields. Having in hand all of those original papers by Sidney, which had thus far not been widely read outside the EU institutions, Earle inquired whether they could somehow be used by the *IJMS*. It was agreed that papers such as those clearly had important lessons and insights based on Sidney's lengthy and wide-ranging experience and thus warranted a wider distribution in order that current and future generations of fisheries scientists and managers could benefit from his expertise and wisdom. Since the *Journal* was not the appropriate venue for publication, it was finally decided that a fitting way to preserve his writings would be to incorporate them into a book, together with other of his writings from throughout his long career that either were never published or were published outside of mainline scientific journals or books. Because Sidney's career incorporated both fisheries and whale science and stock assessment, it was felt that material encompassing his entire career should be included. Sidney, being a prolific writer throughout his life, authored or co-authored numerous publications, reports, and unpublished essays. Anderson and Earle, as a labor of love, agreed to collaborate and co-edit the book as their way of honoring this great man who had contributed so much to the world of marine science.

This book is a collection of 62 selected writings authored by Sidney (sometimes in collaboration with one or more co-authors), with each article constituting a chapter. The chapters are divided into three parts, each depicting a major sector of his working and writing career: 1) "Launching a legacy" (writings during his years at Lowestoft and early years at FAO; 1940s–1950s); 2) "Saving the great whales" (1960s–2000); and 3) "Getting back to first principles" (2006–2016). The majority of the articles in the second part were written for the IWC, while most in the third part concern the objectives of management in the context of the CFP, often put into a unique historical perspective. The book also contains a chronological bibliography of Sidney's publications and writings comprising 465 entries.

The papers have been chosen to reflect the duration, breadth, and depth of Sidney's professional involvement in the science and practice of the management of the exploitation of marine species. What emerges is a clear, consistent, and unwavering commitment to rational, science-based management. The principles he espoused in his later years in the context of the CFP, the need for conservation, the—to some—apparently contradictory idea that maximizing catches would almost inevitably be unprofitable, and the need to ensure a solid basis behind the modeling, can already be seen in the FAO work included here. Sidney remained true to his original beliefs and commitments, all the while evolving as a scientist and adapting his work to the needs of the period in which he was working.

In many of the articles in this book, readers will discover that Sidney often told the same or similar stories or repeated the same arguments or opinions on a specific topic. This is because the articles, or the lectures on which the articles were based, were made to different audiences and at different times. Many such stories pertained to the origin of the MSY concept by U.S. researchers in the post-World War II period in the Pacific that he abhorred throughout his career as an appropriate basis for fisheries management.

Sidney communing with nature. Photo by David Lavigne.

Sidney fervently believed in the value of studying the "gray literature," both as a window into the historical developments of his disciplines and for the detailed information contained therein (see Chapter 41). This volume seeks to put a selection of Sidney's own gray literature into wider circulation.

Every attempt was made to retain the original text of each article, while making punctuation and spelling changes where needed or correcting obvious errors of any type. Scientific names of species have been added; Sidney rarely included them. In many articles, especially those in the third part, Sidney made extensive use of footnotes for including bibliographic information of papers; in such cases, we opted to cite those papers in a traditional list of references. For all papers that had been published elsewhere, permission was obtained from the publisher or owner of the article (e.g., IWC, ICNAF/NAFO, FAO). In some cases, such permission is noted in a footnote on the first page of the article.

Both editors of this book (Anderson and Earle) had personal relationships with Sidney. Earle, a Canadian living in Belgium, first met Sidney during the late 1980s when Greenpeace (for whom Earle worked at the time) was initiating—at Sidney's behest—its global campaign to rein in overfishing. They remained in contact over the years, occasionally working together on projects such as a protocol for the scientific evaluation of culling marine mammals, included as Chapter 33 (see also Earle 2020), until 2006, when Earle sent him a plea for help in putting the EU's CFP on the right track. That began a decade of intense collaboration, in which Earle was privileged to have Sidney as a personal tutor on the various scientific and political aspects of fisheries management.

Anderson first met Sidney during his first year of graduate school (1965–1966 academic year): "I first met Sidney that school year; he was one of the 'visiting firemen' brought in from around the world by my major professor to present seminars to the fisheries graduate students. Coming from FAO in Rome and with his hair combed forward along the sides, he looked to me like Julius Caesar!" His next interaction with Sidney was not until 2000 when Sidney presented a paper (Holt 2002) (Chapter 34 in this book) at the ICES Symposium "100 Years of Science under ICES" held in Helsinki. "True to form throughout the four-day symposium, Sidney sat attentively in the front row and asked questions of nearly every presenter." Anderson (2011) noted a humorous incident involving Sidney and Katherine (Kathy) Richardson, when the latter presented an invited keynote paper about the theory of plankton production and links to fish recruitment (Richardson 2002). Not knowing each other, they had chatted while walking together from their hotel to a reception the first evening of the symposium, both talking about their respective scientific interests, hers being plankton and his being whales in the Antarctic. Based on his attire, physical appearance, and interest in whales, Kathy had him pegged as an NGO observer. It wasn't until later at the reception when someone told her that she had walked in with Sidney Holt, which caused her considerable embarrassment. Interestingly, they became friends and corresponded for some time after the symposium.

About ten years after the 2000 ICES History Symposium, Anderson re-established contact with Sidney while gathering information and photos for his 2011 article in the *ICES Insight* magazine about the lives of Sidney and his former colleague Ray Beverton entitled "Standing on the shoulders of giants" (Anderson 2011). His last encounter with Sidney was over breakfast with him and his son Tim in 2012 at the ICES Annual Science Conference in Bergen, Norway, where Sidney presented a paper (Holt 2012; Chapter 49 in this book).

During the preparation of this volume, we contacted numerous individuals who had known Sidney, and organizations with which he had been associated. Almost without fail, our requests and queries were met with alacrity and enthusiasm. Sidney has clearly profoundly touched and influenced a wide range of people, and the goodwill we encountered in this project does his memory proud.

We are indebted to the American Fisheries Society for publishing this book and to Aaron Lerner (Director of Publications) and Kurt West (Book Production Editor) for their assistance in this process. We also thank the International Fund for Animal Welfare, an organization that Sidney worked with for many years, for their generosity in assisting with the funding of this book. We thank many individuals who assisted in various ways in providing access to articles or otherwise offering advice and counsel in the preparation of this book: Gun and Gudrun Akyuz, Cherry Allison, Kirsten Anderson, Colin Bannister, Manuel Barrange, Karl Bours, Howard Browman, Doug Butterworth, Steven Cadrin, Julie Creek, Mark Dickie-Collas, Greg Donovan, Stella Duff, Carmel Finley, Ranier Froese, Frances Gulland, David Lavigne, Rebecca Lent, Sebastian Losada, Pamela Mace, Michael Mainelli, Judy Meadmore, Alexis Pacey, Vassili Papastavrou, Daniel Pauly, Tony Pitcher, Andre Punt, Patrick Ramage, Robert Rocha, Brian Rothschild, and Mark Tandy.

Emory D. Anderson and Michael Earle
Editors

References

Anderson, E. D. Editor. 2002a. The Raymond J. H. Beverton lectures at Woods Hole, Massachusetts. Three lectures on fisheries science given May 2–3, 1994. NOAA Technical Memorandum NMFS-F/SPO-54.

Anderson, E. D. Editor. 2002b. 100 years of science under ICES. A symposium held in Helsinki, 1–4 August 2000. ICES Marine Science Symposia 215.

Anderson, E. D. 2011. Standing on the shoulders of giants. ICES Insight 48:4–12.

Best, S. (Producer and Director). 2010. Sidney Holt: In His Own Words Part 1. Video by Animal Alliance of Canada. Available: https://www.youtube.com/watch?v=np2Eep99ohg. (April 2021).

Beverton, R. J. H., and E. D. Anderson. 2002. Reflections on 100 years of fisheries research. ICES Marine Science Symposia 215:453–463.

Beverton, R. J. H., and S. J. Holt. 1957. On the dynamics of exploited fish populations. Her Majesty's Stationery Office, UK.

Brown, P. 2020. Sidney Holt obituary. Fisheries scientist who fought for nearly 60 years to save great whales from extinction. The Guardian, 8 January 2020. Available: https://www.theguardian.com/environment/2020/jan/08/sidney-holt-obituary. (April 2021).

Clifton, M. 2019. Sidney Holt, 93, achieved goal that great whales would outlive him. Animals 24-7, December 28, 2019. Available: https://www.animals24-7.org/2019/12/28/sidney-holt-93-achieved-goal-that-great-whales-would-outlive-him/. (April 2021).

Earle, M. 2020. Obituary Sidney Holt. A towering intellect. Samudra Report No. 82:16–18. Available: https://www.icsf.net/images/samudra/pdf/english/issue_82/276_ICSF_Samudra%2082_February_2020.pdf. (April 2021).

Engelhard, G. H., and J. K. Pinnegar. 2020. Sidney J. Holt, eminent fisheries scientist and conservationist (28 February 1926–22 December 2019). Fish and Fisheries 21(3):doi: 10.1111/faf.12456.

Holt, S. J. 2002. ICES involvement in whaling and whale conservation, and implications of IWC actions. ICES Marine Science Symposia 215:464–473.

Holt, S. J. 2012. The some good and mostly bad about maximum sustainable yield as a management target. ICES Document CM 2012/L:02.

Holt, S. J. 2020. Becoming a marine scientist: helped by a daily quota of three lumps of coal. ICES Journal of Marine Science, doi:10.1093/icesjms/fsaa009.

Kent MacKay, B. 2020. The great whales lose a champion. Sidney Holt 1926–2019. Available: https://www.bornfreeusa.org/2020/01/13/the-great-whales-lose-a-champion-sidney-holt-1926-2019/. (April 2021).

Lavigne, D. M. 2020. Sidney J. Holt (1926–2019): In Memorium. Ocean Yearbook 35:xi–xvi.

Pauly, D. 2020. Sidney Holt (1926–2019). Science 367(6479):744.

Richardson, K. 2002. Linking plankton and fish production throughout the history of ICES. ICES Marine Science Symposia 215:156–163.

The Times. 2020. Sidney Holt obituary. British marine scientist who outmanoeuvered Japanese whalers in the 1970s and 1980s and almost managed to ‘save the whale’. The Times, February 13, 2020. Available: https://www.thetimes.co.uk/article/sidney-holt-obituary-f68zh6f7l. (April 2021).

Symbols and Abbreviations

The following symbols and abbreviations may be found in this book without definition. Also undefined are standard mathematical and statistical symbols given in most dictionaries.

A	ampere
AC	alternating current
Bq	becquerel
C	coulomb
°C	degrees Celsius
cal	calorie
cd	candela
cm	centimeter
Co.	Company
Corp.	Corporation
cov	covariance
DC	direct current; District of Columbia
D	dextro (as a prefix)
d	day
d	dextrorotatory
df	degrees of freedom
dL	deciliter
E	east
E	expected value
e	base of natural logarithm (2.71828…)
e.g.	(exempli gratia) for example
eq	equivalent
et al.	(et alii) and others
etc.	et cetera
eV	electron volt
F	filial generation; Farad
°F	degrees Fahrenheit
fc	footcandle (0.0929 lx)
ft	foot (30.5 cm)
ft^3/s	cubic feet per second (0.0283 m^3/s)
g	gram
G	giga (10^9, as a prefix)
gal	gallon (3.79 L)
Gy	gray
h	hour
ha	hectare (2.47 acres)
hp	horsepower (746 W)
Hz	hertz
in	inch (2.54 cm)
Inc.	Incorporated
i.e.	(id est) that is
IU	international unit
J	joule
K	Kelvin (degrees above absolute zero)
k	kilo (10^3, as a prefix)
kg	kilogram
km	kilometer
l	levorotatory
L	levo (as a prefix)
L	liter (0.264 gal, 1.06 qt)
lb	pound (0.454 kg, 454g)
lm	lumen
log	logarithm
Ltd.	Limited
M	mega (10^6, as a prefix); molar (as a suffix or by itself)
m	meter (as a suffix or by itself); milli (10^{-3}, as a prefix)
mi	mile (1.61 km)
min	minute
mol	mole
N	normal (for chemistry); north (for geography); newton
N	sample size

NS	not significant
n	ploidy; nanno (10^{-9}, as a prefix)
o	ortho (as a chemical prefix)
oz	ounce (28.4 g)
P	probability
p	para (as a chemical prefix)
p	pico (10^{-12}, as a prefix)
Pa	pascal
pH	negative log of hydrogen ion activity
ppm	parts per million
qt	quart (0.946 L)
R	multiple correlation or regression coefficient
r	simple correlation or regression coefficient
rad	radian
S	siemens (for electrical conductance); south (for geography)
SD	standard deviation
SE	standard error
s	second
T	tesla
tris	tris(hydroxymethyl)-aminomethane (a buffer)
UK	United Kingdom
U.S.	United States (adjective)
USA	United States of America (noun)
V	volt
V, Var	variance (population)
var	variance (sample)
W	watt (for power); west (for geography)
Wb	weber
yd	yard (0.914 m, 91.4 cm)
α	probability of type I error (false rejection of null hypothesis)
β	probability of type II error (false acceptance of null hypothesis)
Ω	ohm
μ	micro (10^{-6}, as a prefix)
′	minute (angular)
″	second (angular)
°	degree (temperature as a prefix, angular as a suffix)
%	per cent (per hundred)
‰	per mille (per thousand)

Part I

Launching a Legacy

Foreword to
Part I: Launching a Legacy

Among the writings of Sidney J. Holt, the most familiar, of course, is the book he co-authored with Raymond J. H. Beverton: *On the Dynamics of Exploited Fish Populations* (Beverton and Holt 1957). However, his writing skills began to be honed early in his time at the Lowestoft Fisheries Laboratory. His first two publications (as a co-author) emerged in 1947, the first being Hulme et al. (1947), a short note on the calculation of the rate of yield, which was published in *Nature* in May of that year, which alluded to further details and its application (perhaps hinting at what would come later from the joint work of Sidney and Ray), and the second, Margetts and Holt (1947), an analysis of the effect of the cessation of fishing on the English North Sea trawl fisheries during World War II, was presented in October 1947 at an International Council for the Exploration of the Sea (ICES) Special Scientific Meeting on "The Effect of the War on Stocks of Commercial Food Fishes."

Ray Beverton (left) and Sidney Holt (right) in 1949. Photo courtesy Centre for Environment, Fisheries and Aquaculture Science, Lowestoft.

During his relatively short tenure at Lowestoft (1946–1950), Sidney directed most of his writing efforts towards work on Beverton and Holt (1957). According to Sidney, he "did the majority of the mathematics and calculating, and contributed most to the first parts of the book" (Anderson 2011). Nevertheless, several other papers were produced (e.g., Holt 1949a, 1949b; see Bibliography) before he left.

Between the time he left Lowestoft in 1950 and joined the Fisheries Division of the United Nations Food and Agriculture Organization (FAO) in Rome in late 1953, other publications appeared (Holt 1951, 1955), the latter of which was based on his own research on wood ants *Formica rufa* while employed by The Nature Conservancy in Scotland, the results of which were referenced in Beverton and Holt (1957) as an example of the method of examining dispersal data to identify the effects of environmental factors.

Part 1 of this book is relatively short, with only four chapters. These writings stem entirely from material authored or co-authored by Sidney during the first four years of his employment with FAO and were prepared for presentation at various international conferences, with three of the four articles being for the International Technical Conference on the Conservation of the Living Resources of the Sea held in Rome, April 18–May 10, 1955. The fourth paper was for the International Commission for the Northwest Atlantic Fisheries (ICNAF) Symposium on Some Problems for Biological Fishery Survey and Techniques for their Solution held at Biarritz,

France, March 1–10, 1956. To some extent, they reflected the work program of FAO's Fisheries Division at that time which was directed towards improving international fisheries science and management, but they also conveyed Sidney's emerging views about his concerns for maximum sustainable yield (MSY) and its application for fisheries management. As readers examine these chapters, they will note his scrutiny of researchers of that time from the USA whom he thought were viewing MSY as implying that taking a maximum catch from a fish stock is a desirable and practicable objective for regulation. Sidney felt that this was not necessarily true and, therefore, this and subsequent writings of his argued thus. In Chapter 4, he introduces the ratio of mortality to growth rates (*M/K*) which allows broad generalizations of the dynamics of fish populations, much used in subsequent work by him and others. Sidney shows the utility of this ratio in many of the chapters in Part 3.

Sidney received a PhD from Reading in 1959 based on his work in Beverton and Holt (1957). Photo taken from Best (2010).

Many of the chapters in the final section of this book (Getting Back to First Principles) include Sidney's own analyses and anecdotes about this period of his professional life and the impact it has had on fisheries law and the science of fisheries management, especially Chapters 41, 42, 49, 53, 57, and 59.

Emory D. Anderson and Michael Earle
Editors

References

Anderson, E. D. Editor. 2002. The Raymond J. H. Beverton lectures at Woods Hole, Massachusetts. Three lectures on fisheries science given May 2–3, 1994. NOAA Technical Memorandum NMFSF/SPO-54.

Anderson, E. D. 2011. Standing on the shoulders of giants. ICES Insight 48:4–12.

Best, S. (Producer and Director). 2010. Sidney Holt: In His Own Words Part 1. Video by Animal Alliance of Canada. Available: https://www.youtube.com/watch?v=np2Eep99ohg. (March 2021).

Beverton, R. J. H., and S. J. Holt. 1957. On the dynamics of exploited fish populations. Her Majesty's Stationary Office, UK.

Holt, S. J. 1949a. Biodynamic assessments of fish populations. Challenger Society 3:10.

Holt, S. J. 1949b. The fishing intensity on the plaice stock of the southern North Sea. Annales Biologiques Conseil Permanent International pour l'Exploration de la Mar 4:111.

Holt, S. J. 1951. Review of W. F. Thompson. "The Effect of Fishing on Stocks of Halibut in the Pacific," and Anon. "Regulation and Investigation of the Pacific Halibut Fishery in 1947 and 1948." Journal du Conseil Permanent International pour l'Exploration de la Mar 17(3):320–322.

Holt, S. J. 1955. On the foraging activity of the wood ant. Journal of Animal Ecology 24(1):1–34.

Holt, S. J. 2020. Becoming a marine scientist: helped by a daily quota of three lumps of coal. ICES Journal of Marine Science, doi:10.1093/icesjms/fsaa009.

Hulme, H. R, R. J. H. Beverton, and S. J. Holt. 1947. Population studies in fisheries biology. Nature 159:714–715.

Margetts, A. R., and S. J. Holt. 1947. The effect of the 1939–1945 war on the English North Sea Trawl Fisheries. Rapports et Procès-Verbaux des Réunions du Conseil Permanent International pour l'Exploration de la Mar 122:26–46.

Chapter 1

Some Observations on the Concept of "Maximum Equilibrium Catch" in Relation to Conservation Problems[1]

The term "equilibrium" has been used in fishery theory in several ways. Frequently, the meaning is not clearly indicated (e.g., "the equilibrium between catch and the amount of fishing"), but in general, reference is implied to stability either of a fish stock or of a fishing industry. It is clear that a stock may be stabilized at any level of abundance within wide limits, and that the possibilities within a given economy of stabilizing the level of operation and production of a particular industry are equally wide. In both cases, external constraints—on the one hand, for example, the amount and kind of fishing and, on the other, the economic conditions outside the industry determining costs and prices—may determine the particular equilibrium attained; management of a fishery may involve adjustment of any of these factors.

"Maximum equilibrium catch" may therefore mean either the greatest sustained yield that a stock can give or the greatest catch that can be taken whilst allowing economic equilibrium, according to whether the fish stock or the fishing industry is the system, the dynamics of which are under consideration. It has frequently been assumed that attainment of the former is economically possible and desirable, or at least not economically undesirable, but this assumption needs critical examination.

"Equilibrium" in a unit fish stock is here understood to mean a condition in which decrements occurring in the population through the capture of some fish and "natural" death of others are, on average, equal to the increments due to supplementation of numbers of fish by recruitment and to increase in size of individual fish by growth. Such a condition is better referred to as a steady-state.

It is a characteristic, though not an invariable, property of animal populations or communities that they are able to attain a steady-state with a wide range of external conditions, and it has, therefore, been fruitful in the field of population dynamics to consider first the population as an "open" system in steady-state and then to examine the consequences of departures from this state.

[1]Editors' note: to the best of the editors' knowledge, this manuscript was never published. The version here was produced from a scanned version of an original typed copy with handwritten notes indicating UN Conf. Fish., April May 1954 Rome, misc. manuscripts and clippings, Dr. Finn (referring to Dr. D. B. Finn, Director of the FAO Fisheries Division, 1946–1964), and marked as fourth draft. We are not aware of a UN conference on fisheries held in 1954, but there was one in 1955 (International Technical Conference on the Conservation of the Living Resources of the Sea, Rome, 18 April–10 May 1955). At the time of this writing, Holt was employed by the Food and Agriculture Organization of the United Nations in its Fisheries Biology Branch. Permission to reproduce this article has been given by FAO. The views expressed in this publication are those of the author and do not necessarily reflect the views or policies of the Food and Agriculture Organization of the United Nations.

Trends towards or from particular steady-states may or may not be reversible. Reversibility would, for a fish stock, mean that if a change in the intensity of fishing caused a movement from one steady-state to another, then a further change back to the original intensity would result eventually in the stock regaining the original steady-state. There are indications that this is generally true for several marine fish stocks. The most conclusive evidence has come from studies of the effects of temporary cessation during wartime of fishing on depleted stocks.

Russell (1931) has given us a symbolic statement of the relation between stock size and the various increments and decrements occurring in the stock. His axiomatic equation, in fact, describes a fish stock as an open system, and it can give simple expression to the steady-state condition. It cannot, as it stands, be used to assess the state of the stocks, and the needs for corrective measures, since all the quantities symbolized in it cannot be directly measured. However, it represents a starting point for a working formulation of the sustained catch problem. We shall return to this formulation later.

The first variable, the relation of which with steady catch we need to discover, is the "amount of fishing." The most direct approach, if a long series of commercial fishery statistics is available, is to pick out several periods during each of which the intensity of fishing was fairly constant for long enough to make reasonable the assumption of equilibrium. The average annual catch for each period is then plotted against the corresponding fishing effort. Graham (1951) has remarked that the statistics alone are usually insufficient to give a coherent graph, and he gives examples in support of this from what are generally considered to be well-studied fisheries.

Some further assumptions are, therefore, necessary concerning the general shape of the required curve relating the equilibrium catch to the fishing effort. A device that has been employed is to fit the Verhulst–Pearl (Verhulst 1845; Pearl 1922) logistic curve to the available data. Graham (1935) was able to do this using observations of the effect of war-caused changes in the amount of fishing on the catch and the catch per unit effort from North Sea trawl fisheries. Although the validity of his assumption that the total growth of all demersal fish populations in the area can be represented by the logistic function cannot be tested directly, it has been accepted as a not-unreasonable first approximation. Since the growth curve is certainly sigmoid—whether or not it is a symmetrical logistic—the stock will have a maximum rate of growth at some intermediate stock size when the inflexion is reached; the argument is that a maximum equilibrium catch will be obtained when the stocks are reduced by fishing to a level corresponding with that inflexion.

A difficulty in developing further the above method is that no account can be taken of changes in the age composition of the stock that inevitably result from changes in fishing intensity; equilibrium catch is considered only as a function of a gross measure of population size.

Schaefer (1954) has recently attempted to make fuller use of series of annual catch and effort statistics to investigate the behavior of fish populations when they are being exploited. His method is essentially the analysis of the fluctuations in the series for the purpose of predicting steady-states. Thus, the growth potential, and hence the steady-state yield from a stock of a size equal to the average stock size in a particular year, is estimated as the catch in that year plus the difference between the stock size at the beginning of the year and at the end of it.

Schaefer plotted figures for "equilibrium catch" obtained in this way against catch per unit effort for the Pacific Halibut *Hippoglossus stenolepis* and Pacific Pilchard *Sardinops sagax caerulea* stocks. His figures indicate that, for both species, over the observed range, the equilibrium catch increases with increasing stock size. It is possible to fit a logistic function to data treated in this way, but the scatter of points is found to be so great that no useful estimate can be made of the stock size that would give the "maximum equilibrium catch." In fact, the data give almost no sign even of the existence of a maximum. They could, however, be taken as evidence that, for the situations examined, a reduction in fishing would result in an increased average yield, provided it is also shown that the stock abundance is largely controlled by the fishing intensity.

Perhaps the main objection to the above method is that no knowledge is obtained of the extent to which the year-to-year changes in stock size are, on the one hand, the result of mortality and growth within the stock and, on the other, due to the recruitment of new individuals into the stock. Now, the magnitude of recruitment in a particular year can hardly be regarded, in general, as a simple function of the stock size in that same year. It is, for all species but those having a lifespan of less than one year, a function of the stock size during an earlier period when the new year class was spawned. Such considerations remind us that man rarely, if ever, exploits the whole of a population of fish, but rather selects only individuals above a certain size and age that constitute the available stock, and frequently only a readily accessible fraction of these.

To illustrate the errors of interpretation that may arise if these facts are neglected, it is sufficient to consider the case where fishing is very intense, but highly selective, only the larger fish being caught. In this situation, the catch per unit effort may be an index of the size of the exploited stock, but this bears no relation to the total size of the population. Yet, it is the growth of fish before recruitment, which, in this case, determines the steady yield. To give a practical example: analysis of the North Sea Plaice *Pleuronectes platessa* stock shows that, even where fish having a possible lifespan of over 20 years become liable to capture, on average, soon after they are three years old, rather more than one-half of the stock growth from which the yield is derived occurs before recruitment.

For further development of conservation theory, this distinction between total, fishable, and fished stocks must be fully recognized and the relations between them taken into account. Fundamentally, the fished stock from which the catch comes cannot be treated as a self-reproducing unit having, at any given time, a potential for growth, which is a function mainly of its size at that time; this potential is essentially a complex function of stock size and composition at the time and of the size and composition of the total population during some previous period.

The above discussion leads us to consider another approach to the maximum yield problem. It has frequently been pointed out that there is a best age at which to harvest a particular crop. For cultivated stocks—whether these are livestock or fish—this best age may be that for which the net gain is at a maximum and after which the cost of, for example, continued feeding, may be greater than the value of the additional growth. For wild fish stocks, the best age is considered to be that at which a particular year class reaches its greatest total weight, after which the decrements by natural mortality are greater than the summed growth increments of the survivors. But, in marine fisheries, it is not possible to capture all the fish at a particular age, so that, in practice, it is best to start catching fish when they are rather younger; how much younger depends upon how intense is the fishing—the harder you can fish, the longer you can safely leave the fish to grow before you try to catch them.

Allen (1953), in discussing the best legal size limits for fish, has shown that for a given intensity of fishing, the weight of an individual of the size at which it is best to start fishing is equal to the average weight of the fish in the catch multiplied by the proportion of fish reaching that size and age that are eventually caught. However, like Russell's equation, this formula defines a steady-state situation, after the event, as it were, but does not provide a means of *a priori* calculation and prediction.

To unify these two approaches to conservation problems (i.e., consideration of the best fraction of the stock to remove annually and of the best age at which to catch the fish) it is necessary to examine the analytical methods developed by Baranov (1918) and several other workers since (e.g., Thompson and Bell 1934). These depend essentially upon the construction of mathematical models of exploited population containing coefficients which describe the factors—migration and recruitment, growth and mortality—causing the decrements and increments in a fish stock mentioned above, and the relations between these factors.

The various models put forward (e.g., Ricker 1944, 1946; Tester 1953; Tanaka 1954) have differed from each other principally in the shape of the curve used to describe growth of the individual fish, and the properties exhibited by the models depend, to a considerable extent, upon the accuracy with which the particular growth function reproduces the general characteristics of fish growth. When the growth function used has been a realistic one, these models have frequently shown a peak in the curve of steady yield against fishing intensity. When, however, the rate of natural mortality is high in relation to the overall growth rate, or when the dependence of growth rate of individuals on the population density is taken into account, then the curve may have no peak, but may, instead, approach an asymptote. Again, when the size at which fish are first liable to capture by the gear is increased, then the maximum is found to be obtained at a higher fishing intensity, and the peak of the yield curve becomes less pronounced. Eventually a size is reached for which the peak disappears and the yield curve becomes asymptotic.

Beverton and Holt (see Beverton 1953) have studied the effect on these models of varying together both the age at which capture begins and the magnitude of the fishing mortality. They have plotted contours of equilibrium yield with these two variables as coordinates. The picture so obtained is most usefully summarized by a eumetric yield curve in which steady yield is again plotted against fishing intensity, but where, for each intensity, the selective properties of the gear are adjusted so that the greatest yield is obtained, giving conditions satisfying Allen's equation mentioned above.

In the present context, the most important characteristic of eumetric yield curves for single species is that they are always asymptotic. The highest equilibrium yield is obtained only with an infinitely high fishing intensity, i.e., when all fish could be caught as soon as they reach a certain best age. If several species are considered together, a maximum may appear again, but it is clear that the attainment of a maximum equilibrium catch with some finite fishing intensity cannot be regarded as a universal or even a general characteristic of the relation between fish stocks and fishing activity, in situations where there is any possibility of adjusting the selectivity of the gear. Even where a maximum may exist, it is as likely as not to occur at an extremely high fishing intensity.

Almost all analyses have, so far, been concerned with trawl fisheries or with gear, such as seines, having generally similar selective properties. However, preliminary examination of models for gillnet and line fisheries indicate that if appropriate gear adjustments are possible—by changing the mesh or hook size, type of bait, etc.—the equilibrium yield curve for which selectivity and fishing intensity are varied simultaneously, according to the principles outlined above, would have similar properties.

What, then, are the implications of these findings? Apparently, no single biological criterion can be used as a guide to where it would be desirable for a fishery to operate. Thus, the maximum possible yield, i.e., the asymptote of the eumetic yield curve, can be attained only with an infinitely high fishing intensity and hence at a correspondingly high cost; it is, therefore, a totally unreal objective for resource use. It seems that some sacrifice of yield must be made if the cost is to be such that fishing is practicable.

The importance of the economic conditions under which the yield is obtained has been stressed by Graham (1943) in saying that, "... the benefit of efficient exploitation lies more in the economy of effort than in increase of yield or preservation of future stocks, though both of these purposes may also be served." This dichotomy of regulative function has been discussed in detail by Herrington (1943) and Nesbit (1943) who use the terms <u>conservation</u> to mean regulation designed purely to increase the yield, and <u>economic management</u> for regulation that is concerned with the economic and social conditions in the fishing industry. Herrington regarded these as independent objectives, with economic management to be undertaken or not, according to policy. Nesbit, on the other hand, while agreeing with Herrington in regarding conservation as the primary objective, believed that both should always be considered. Now, both Herrington and Nesbit were supposing that a maximum catch can be taken and that this is a practicable objective for regulation. It is shown above that this is not necessarily true, so that the whole question needs re-examination.

It might be thought that eumetry itself could be taken alone as the objective of regulation, and it is instructive to consider for a moment what would happen if this were done. Regulative action would be limited to ensuring that the selective properties of the gear were always eumetric with whatever fishing intensity was being exerted. The latter would not be regulated and economic incentive may cause it to increase until a point upon the eumetric yield curve was reached at which the profit margin was so low that the incentive no longer existed. This steady-state would differ from that reached in the absence of regulation of any kind only in that the yield would be rather larger because the mesh size would be eumetric—it would, in fact, be the largest that could be obtained by an economically independent industry. While such a procedure could result in a rather larger amount of fish being caught than in an unregulated fishery, economic conditions in the fishing industry would, for all practical purposes, remain unchanged. Conversely, it is well enough realized that the greatest catch per unit effort and, roughly, the greatest rate of profit, is obtained by one ship alone or a very few ships exploiting the stocks, but, of course, the total yield in such circumstances is small.

Thus, we conclude that not only is it impossible to maximize both yield and economic return together, but if one is maximized, the other is automatically reduced, if not minimized. It is difficult to visualize a situation in which, once recognized, either extreme would be acceptable as an objective for conservation, either to the industry or to society generally. It seems then that it is misleading to regard conservation and economic management as two separate and independent functions of regulation, especially as to do so apparently leads either to the former being regarded as obligatory and the latter voluntary, or at least to the former being assigned the greater weight on principle. Rather, it is suggested that intervention intended to benefit a fishery should be conceived on a broad enough basis to embrace biological, economic, and social factors on, a priori, equal terms. It should have as its general objective the adjustment of these factors so that, in each particular case, an acceptable balance is achieved between the benefits on the one hand to the producer, in the form of profit to the fishing industry and a good living for fishermen, and on the other to the consumer as a large and steady supply of fish at a reasonable price.

To see what may be implied—from the purely technical point of view—by an attempt to give practical effect to the above approach, we may again examine the concept of eumetric fishing. It is evident that theoretical curves for trawl fisheries, in which steady yields are plotted against eumetric fishing intensity, are only meaningful insofar as the size of mesh may, in practice, be freely varied, at least over a certain range. Not quite so obvious, but equally important, is the condition that the cost involved in changing the selectivity is negligible compared with that of changing the intensity of fishing. In fact, economic considerations have been taken into account even in choosing the amount of fishing as the significant variable in which we are primarily interested (we may add here that this is also true for the choice of yield in weight as the significant dependent variable, it being, in reality, used as a simple index of "value").

In some situations, a considerable proportion of human effort and hence of cost may be expended in activity other than the fishing itself. For example, an increase in hook size to adjust the selectivity of a line fishery may mean a great increase in the amount of bait required. Again, if, as sometimes suggested, flatfish in the North Sea were transplanted on a commercial scale to better feeding grounds, realistic assessment of conservation needs would involve evaluation of the relative amounts of effort taken up by this activity and in fishing; and so on. The main criterion by which the desirability of such an action is judged is whether the predicted difference between the additional value of the catch and the cost of carrying out the action is sufficiently large.

Then again, variation in the amount of fishing on a resource consisting of populations of a number of different species of fish would be accompanied by variation in the species composition of the catch. It was mentioned earlier that, in such a situation, a maximum may appear even in the eumetric yield curve, but to obtain it may mean the deliberate loss of nearly all of the species that escape more easily through large meshes, in order to make bigger catches of the others. But the various species may have very different unit prices on the market, and there may be other reasons for preserving variety in the catch.

Clearly, criteria for conservation need to specify both the qualitative characteristics of the desired yield and its quantitative composition—for example, the relative amounts of smaller and larger fish—as well as the total magnitude.

Bearing all the above factors in mind, we may define conservation generally as "*the planning and control of human utilization of a natural resource intended to ensure yields of such quality and quantity, and over such a period, as are best suited to the needs of the communities which exploit the resource and the nations of which those communities are parts*" and then see if more specific statements can be made for the particular situations with which we are faced.

Firstly, conservation measures may be aimed at checking or reversing a change in a resource which otherwise would proceed irreversibly. Examples of such changes may be (i) on the land—the erosion of soil and the change in vegetative composition of pastures through overgrazing and (ii) in the sea—the reduction of a species towards extinction. So far, there is positive evidence of this only for marine mammals. It is technically and economically possible for whaling operations to continue even when the stock is reduced to a level at which the net reproduction rate is critically dependent on the population size and the danger of extinction, therefore, arises. We may also expect developments giving large increases in the ability of fishing gear to exert a force of mortality on a fish population, at little extra cost (for example, sensory aids such as echosounders and general improvements in searching methods), to lead to a similar situation.

Recent calculations suggest that another change in trawling efficiency as great as that associated with the change from sailing to steam and motorized fishing vessels might, theoretically, and if not accompanied by some compensating increases in gear selectivity, lead to the virtual extinction of certain species for which a relation between spawning stock and egg production and subsequent recruitment can already be detected (see discussion of North Sea Haddock *Melanogrammus aeglefinus* population, Beverton and Holt 1955). Where such a possibility is found to exist, we may say, on purely biological grounds, that fishing should not be sustained at an intensity exceeding a particular critical value. However, although more work needs to be done in developing methods for determining such critical points, it is important to note that such an intensity of fishing could not normally be maintained for long since the catch per unit effort would drop to an extremely low level before the critical stock size was reached.

Secondly, conservation may be aimed at securing a particular steady level of stock abundance and yield. Although here, the system to which the steady-state refers is the resource itself, the desired level is determined by reference to an equilibrium desired for the fishing industry as a whole, in which a certain difference between input and output (net gain or total profit) is to be sustained. This latter may be maximized or balanced in relation to that being obtained in industries other than fishing, or it may be decided on social and political grounds that some other criterion is to be used. In any case, it is necessary to formulate clearly the criteria for use of the resource in such a way that the biological implications can be investigated and the results presented in a useful form.

Lastly, measures may be proposed not so much to obtain a particular steady-state in the industry as to reduce the fluctuations in catch. The task here is to counteract the features of the industry which may tend to cause violent fluctuations in production, and the first problem is to locate and identify these features. They may or may not be connected with changes in the resource. For example, Scott Gordon (1955) has found that of seven fisheries for which correlation analysis was carried out, in four of them the economic factors tested could account for more than 50% of the variation in production; in the remaining three, fluctuation could have been largely associated with stock changes. When responsibility for fluctuations can, in part, be traced to the fish stock, it remains necessary to distinguish changes due directly to the fishing itself from changes in abundance arising through other causes and from variations in availability and accessibility, which may be associated with changes in the physical environment.

It should be mentioned here that the problem of fluctuation in yield or stock size cannot be considered entirely separately from the question of maintaining a particular average level for these quantities. For example, an increase in fishing intensity, which reduces the general level of the stock will also change its age composition. The direction of this change is such that fluctuations in annual recruitment become increasingly reflected in the catch.

In all the above cases, once the particular problem is identified, appropriate investigations might be carried out and proposals for corrective measures made, but these will not necessarily refer, even indirectly, to maximizing either the gross productivity of the resources or the yield from selected stocks.

However, this is not to say that attainment of a maximum yield may not be a useful aim in certain situations. It may be that technical factors make eumetric fishing impossible and the biological characteristics of the stock are such that the curve of steady yield against fishing effort has a maximum, that attainment of this maximum is economically possible, and that it is socially desirable. Even where these conditions are not all met, the principle of maximizing yield might be adopted as a temporary expedient. It might, for example, be decided to adjust the selectivity of the gear to that which would give the maximum yield if the fishing effort did not change. By this means, considerable increases in yield might be obtained. These would be accompanied initially by increases in catch per unit effort, making fishing more profitable, and this might be expected to lead to increase in the amount of fishing until the net return was reduced to its former level. We could, however, safely say that the economic state of the industry would not at the end be any worse than before; the total catch would be higher and, at least, the principle of encouraging scientific management of the resource will have been introduced.

In other situations, reference to maximizing the yield could be most misleading. Suppose that the gear selectivity cannot be varied and the curve of steady yield from a resource against fishing intensity is rather flat-topped, as may often by the case. Then, the fishing intensity may become stabilized for economic reasons at a level somewhat less than that giving the maximum yield. Some measure of conservation might be needed, though not in the direction of maximum yield, but away from it, since it may be that by decreasing fishing, the profitability may be enormously increased with but a negligible loss in yield.

These examples have been given to illustrate the need for understanding of the economic and social factors as well as the biological ones that determine our attitude to resource conservation. Sometimes, it may be possible, even when an economic analysis has not or cannot be made, to say that biological knowledge indicates that a particular action is unlikely to be harmful and will most probably result in considerable benefits. Even then, it would be most unwise to attach too much importance to a particular criterion of wise resource use, be it "maximum steady yield" or any other.

References

Allen, K. R. 1953. A method for computing the optimum size-limit for a fishery. Nature 172:210.

Baranov, F. I. 1918. On the question of the biological basis of fisheries. Nauchnyi Issledovatelskii Ikhtiologischeskii Institut Isvestia 1:81–128.

Beverton, R. J. H. 1953. Some observations on the principles of fishery regulation. Journal du Conseil Permanent International pour l'Exploration de la Mer 19:56–68.

Beverton, R. J. H., and S. J. Holt. 1955. Fishery Investigations, London, Series 3. (in press).[2]

Gordon, H. S. 1955. Economic factors in catch fluctuations. Journal of the Fisheries Research Board of Canada 12(1):85–92.

Graham, M. 1935. Modern theory of exploiting a fishery, and application to North Sea trawling. Journal du Conseil Permanent International pour l'Exploration de la Mer 10(3):64–274.

Graham, M. 1943. The fish gate. Faber and Faber, London.

Graham, M. 1951. Overfishing. Pages 20–24 *in* Proceedings of the UN Scientific Conference on the Conservation and Utilization of Resources: Volume 7, Wildlife and Fish Resources. United Nations, Department of Economic Affairs, New York.

Herrington, W. C. 1943. Some methods of fishery management and their usefulness in a management program. Pages 3–22 *in* W. H. Herrington, and R. A. Nesbit, editors. Fishery management. U.S. Fish and Wildlife Service, Special Scientific Report 18.

Nesbit, R. A. 1943. Biological and economic problems of fishery management. Pages 23–53 *in* W. H. Herrington, and R. A. Nesbit, editors. Fishery management. U.S. Fish and Wildlife Service, Special Scientific Report 18.

[2]Editors' note: the "in press" designation refers to the eventual 1957 publication of "On the dynamics of exploited fish populations" co-authored by Raymond Beverton and Sidney Holt.

Pearl, R. 1922. The biology of death. Lippincott, Philadelphia.

Ricker, W. E. 1944. Further notes on fishing mortality and effort. Copeia 1944(1):23–44.

Ricker, W. E. 1946. Production and utilization of fish populations. Ecological Monographs 16:373–391.

Russell, E. S. 1931. Some theoretical considerations on the "overfishing" problem. Journal du Conseil Permanent International pour l'Exploration de la Mer 6:3–20.

Schaefer, M. B. 1954. Fisheries dynamics and the concept of maximum equilibrium catch. Proceedings of the Gulf and Caribbean Fisheries Institute 6:53–64.

Tanaka, S. 1954. The effect of restriction of fishing effort on the yield. Bulletin of the Japanese Society of Scientific Fisheries 20(7):599–603.

Tester, A. L. 1953. Theoretical yields at various rates of natural and fishing mortality in stabilized fisheries. Transactions of the American Fisheries Society 82:115–122.

Thompson, W. F., and F. H. Bell. 1934. Biological statistics of the Pacific halibut fishery. (2) Effect of changes in intensity upon total yield and yield per unit of gear. Report of the International Fisheries [Pacific Halibut] Commission 8.

Verhulst, P.-F. 1845. Recherches mathématiques sur la loi d'accroissement de la population. Nouveaux Mémoires de l'Académie Royale des Sciences et Belles-Lettres de Bruxelles 18:1–42.

Chapter 2

Classification of International Conservation Problems[1]

Introduction

The purpose of the present paper is to present a brief survey of resources and areas in respect of which no treaty, convention, or agreement has been promulgated and to consider the information available concerning these resources[2] in order to know whether a need for conservation is indicated. It is also necessary to learn if there are sufficient data to indicate the type of action that should be taken. Where the information is found to be deficient in this regard, it is necessary to provide a basis for discussing what research ought to be undertaken to supplement existing information so that a plan for conservation may be formulated. In the following pages, an attempt is made to examine the conditions under which these purposes may be fulfilled and to lay the basis for a task that cannot, at this point, be carried far beyond the preliminary stage.

There can be no doubt that a survey of available information is desirable at this time, and that some conclusions may be drawn from such a survey, though the conclusion must often be that still more information is required. While any program of exploitation requires more and more information, it seems desirable to stipulate at the outset that such a survey of the fishery problems of what might be called the "non-treaty" resources and areas must not be taken to indicate that all these problems are of a nature requiring action. A brief discussion is given below of the place of conservation measures in the entire plan of exploitation since, in attempting to identify "conservation problems," it is necessary to be clear about what is meant by conservation and what are the problems.

A major difficulty in identifying conservation problems, for which the governments concerned have not developed or agreed upon measures or procedures by which they can be handled, is that the already existing agreements are, in their investigational aspect, so exten-

---

[1]Editors' note: this paper was co-authored by Geoffrey Kesteven and Holt and presented at the International Technical Conference on the Conservation of the Living Resources of the Sea, Rome, 18 April–10 May 1955 as UN Doc. No. A/Conf.10/6 under the title "Identification of Types of International Fisheries Conservation Problems for the Handling of which the Governments Concerned have not yet Developed or Agreed upon Measures or Procedures." It was then included in International Technical Conference on the Conservation of the Living Resources of the Sea (1956). At the time of this writing, Kesteven and Holt were employed by the Food and Agriculture Organization of the United Nations in its Fisheries Biology Branch. Permission to reproduce this article has been given by FAO. The views expressed in this publication are those of the authors and do not necessarily reflect the views or policies of the Food and Agriculture Organization of the United Nations.

[2]The term "resources" here relates to individual species or to groups of species or the general resources of the seas in a particular area.

sive and relate to such a large part of the world and its fishery resources. Responsibility for the North Atlantic and adjacent waters is shared by the permanent commission under the Convention for the Regulation of the Meshes of Fishing Nets and the Size Limits of Fish (1946), the International Council for the Exploration of the Sea, and the International Commission for the Northwest Atlantic Fisheries. The North Pacific is covered by the International North Pacific Fisheries Commission. The entire Indo-Pacific area, whose limits are not yet precisely defined, is under the Indo-Pacific Fisheries Council. The Mediterranean is covered by the General Fisheries Council for the Mediterranean and by the International Commission for the Scientific Exploration of the Mediterranean Sea. A commission has been proposed for Latin America; a Caribbean commission has certain responsibilities in respect of that area. The Commission for Technical Cooperation in Africa South of the Sahara has certain responsibilities in its area, and the Permanent Commission for the Exploitation and Conservation of the Maritime Resources of the South Pacific in a large sector of the Pacific. There are commissions for all Antarctic whale stocks and for halibut, tuna, salmon, and seals. In consequence, apart from such areas as the Red Sea and the Persian Gulf, and some uncertainty of jurisdiction in the western Indian Ocean, the eastern Pacific Ocean and the broad stretches of the South Atlantic, there is, as to area alone, a coverage of a large part of the globe.

If, therefore, the establishment of a fisheries agreement for an area, when entered into by various governments, were mistakenly considered a solution of conservation problems—in the sense that machinery thus exists for gathering information, assessing it, and recommending action—there would remain only few areas in respect of which further solutions were yet to be proposed; and even for these, there are signs that recommendations for action are already contemplated.

However, the establishment of an agreement, such as that under which the Indo-Pacific Fisheries Council is constituted, can scarcely be held to solve the problems of the fisheries in that region, whether these problems are identified as requiring conservation or as any other kind. Such a step is to be regarded as the preliminary administrative move to make possible the types of international collaboration that will accelerate the gathering of information concerning the resources, improve methods of research, and interpretation of data and, eventually, lead to such discussion of the results as may permit realistic conclusions about the fishery and appropriate action. Even in the case of certain other agreements concerned with the regulation of particular resources, as, for example, the convention establishing the International Pacific Halibut Commission, it cannot be held that problems of conserving these resources have been wholly solved by the adoption of a treaty, or even by subsequent work in research and management.

In consequence, it seems to be an unavoidable conclusion that either the treaty areas must be excluded from consideration in this paper, which thereby would be reduced to relatively negligible proportions, or that the majority, if not all, of the resources of these areas must be included within its scope. Moreover, it is important to reject the possible implication that the existence of various treaties and agreements means that the situations to which they refer are definable either as involving conservation problems or as partial solutions of these problems.

Conservation Measures and Conservation Problems

Conservation measures involve control of fishing operations in order that the resources and the catch should continue to possess desired quantitative and qualitative characteristics—

especially in the light of other forces affecting such resources. It will be recognized that in this definition, the emphasis is upon restraints placed on fishing operations. The purpose of such restraints, however, is to bring about or to maintain certain biologically possible conditions in the resource for reasons based on economic and social grounds.

The term "conservation problem" may mean one or more of the following: (a) whether there is need for conservational measures in the sense defined above; (b) where it is established that there is a need for such measures, what kind of measures should be adopted; and (c) the measures having been decided upon, how they can be implemented. Other supplementary problems might be listed, such as the arrangements that should be made to carry out investigations for solving each of the foregoing problems.

It should again be emphasized that both the formulation and the solution of any of these problems, while involving scientific and technical considerations, also include decisions as to the end to be achieved, which involve economic and social judgments. There is no single isolated conservation problem; further, a particular fishery presents conservation problems only in consequence of the factors that determine its situation, including not only biological, but also economic and political elements.

Conservation in an Exploitation Program

If the general history of fisheries is shown in graphic form (Figure 1), with the catch plotted over a period of time, the curve showing its growth has very often a sigmoid shape. When the resource is first discovered, the catch, of course, is small. Thereafter, as experience grows and skill develops and the number of fishing units rises, the catch increases at an accelerating rate, but at some point, this rate of increase slows down and, though the catch continues to increase, it does so at a progressively slower rate until the curve begins to flatten out and the catch becomes more or less stabilized. This situation is represented in the accompanying diagram on which three principal stages in the development of a fishing industry have been marked. This diagram, of course, is extremely generalized, and the history of no single fishery would precisely follow this pattern. However, in principle, it is true that for each individual fishery, there is a preliminary exploratory phase, a developmental phase, and one or more phases of relative stability. It is obvious that, in the first two stages, the exploitation program is concerned with increasing the operations in respect both of their efficiency and of the total catch. For fisheries in these stages, there is no immediate question of imposing conservation measures, though it is conceivable that in some situations, it might be possible to foresee changes which will presumably take place and to carry out sufficient investigations to be able to propose such measures for the management of the industry as would lead to its most effective development, and yet so to control these operations that, in effect, there would be conservation.

Generally speaking, however, conservation measures become necessary only towards the end of a developmental stage or in a stabilized stage of a fishery. Nevertheless, demonstration that a fishery is at either of these stages does not necessarily mean that deliberate restraint upon fishing operations, such as may be described as a conservation measure, is, in fact, required. There must, at times, be doubt as to whether the fishery has actually completed a developmental stage, and there are numerous examples of fisheries which appear to have reached a stabilized condition, but which have suddenly expanded far beyond that level. It is, therefore, again important to emphasize that a fishery's history reflects the play of many

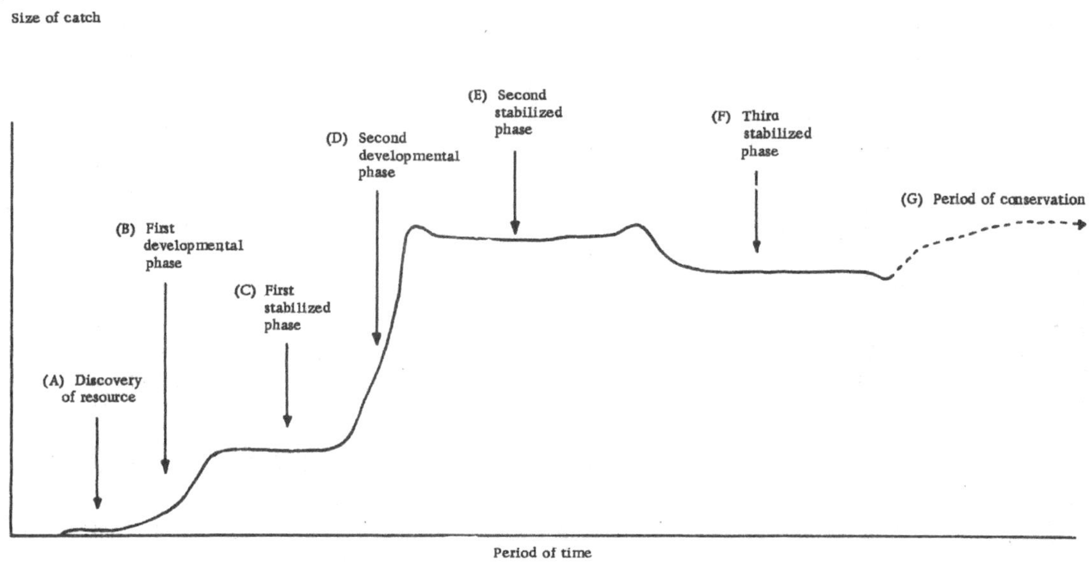

Figure 1. Hypothetical general curve of historical development of a fishery. Note: After discovery of a resource (A), fishing activity may increase (B), but expansion may be limited (C) by economic and technical factors. Renewed and rapid expansion (D) may be permitted by such factors as the invention of better fishing gear or the opening up of new markets for the catch. Production reaches a peak, but then settles down to another steady level (E), higher than the first, but fishing is not necessarily more profitable. It may be that new technical improvements which increase efficiency or reduce costs will encourage further rise in production, but this may only be temporary. The catch settles down to a stable level (F), determined by the capacity of the stock to increase under the particular conditions of exploitation. This level may be no higher, and may perhaps be lower, than the previous one. In this situation, control of the kind and amount of fishing may result in sustained catch (G) equal to the earlier levels, or even higher, obtained with less effort. The problem of evaluation is to distinguish these different kinds of limitation on the catch.

factors—economic and social factors not least among them. A fishery history may show stabilization of levels of catch other than the maximum the stock can sustain. This, in fact, is only a special aspect of a more general rule—that none of the features of the historical curve of development may be taken, without adjustment (for which other kinds of information are necessary), as direct evidence concerning the magnitude and productivity of the stocks from which the catch is taken.

Interpretation of the history of a fishery can be achieved only after there has been a biological analysis of the fish stocks—for this purpose also, certain technical and economic data are necessary—and measurement of their properties, combined with an analysis of the technical, economic, and social factors involved. After such a study has been made, and it is shown that the catch has approached the level above which no amount of fishing, in practical terms, can be expected to increase it, then that knowledge is sufficient to determine with confidence the nature of the effect which exploitation is producing on the stocks. Some measures for controlling the exploitation may then be proposed, either to maintain certain established features with the stocks, to produce other features, or to restore still others. This is the true conservation situation.

Precautionary Conservation

From time to time, research workers, administrators, and members of the industry have the impression, from certain relatively isolated pieces of evidence, that a fishery has reached its final stage and that the effect of fishing operations is such that it is necessary to exercise some restraint upon them. In these situations, resort is had to certain general principles, such as the desirability of ensuring that each fish has the opportunity of spawning at least once, and restrictions are imposed upon fishing in order to observe these principles. In these cases—aside from the fact that the need for conservation is not truly established—there is no certainty that the measures applied will have the effects that are sought or even that the effects themselves are truly desirable.

Of course, such precautionary conservation will not necessarily prove a failure, but, on the other hand, those who impose such measures cannot determine that they are not causing an important and unnecessary economic loss to the industry. Since, however, there appears to be no logical basis for the adoption of such measures and since it is also impossible to predict when such measures may appear desirable, it is not possible in this paper to consider as true conservation problems the situations where precautionary conservation is being or might be practiced; certainly no such situations can be anticipated.

Recapitulation

Each fishery normally passes through at least three stages in its history. In the first of these stages, which is exploratory, and generally in the second also, there is no occasion for conservational action. Such action can be considered only in relation to fisheries in or approaching an asymptotic stage of their history. To prove that a fishery is at such a stage, it is necessary to analyze evidence concerning the biological, technical, and human elements, and the various factors influencing these. If it can shown that fishing operations are likely to continue with an intensity endangering the future of the fishery as a whole, measures to restrain these operations should be designed in such a way that they will produce effects on the stocks, which will be of economic and social advantage to the countries concerned.[3]

In a schedule of fisheries, conservational problems are present only where the existence of some such characteristics has been recognized and is considered undesirable for one reason or another. Among fisheries for which the conservation problem is recognized, it may be one or more of the three types listed earlier: (a) whether there is need for conservation measures; (b) what kind of measures should be adopted; and (c) how the measures should be implemented.

Schedule of World Fisheries

The schedule at the end of this paper presents a list of the unit fisheries of the world, identified by their common names, with a certain amount of grouping to permit treatment in terms of the foregoing definitions (Figure 2 provides a key to the numbered areas). So far as

---

[3]The idea of "endangering" denotes a wide range of possible undesirable consequences. Graham (1956), presented at the same conference as this paper, mentions declines in total catch, in catch per unit effort, and in average size of fish. To this list may be added others, such as undesirable changes in the composition of the species and other variations tending to reduce the value of the catch.

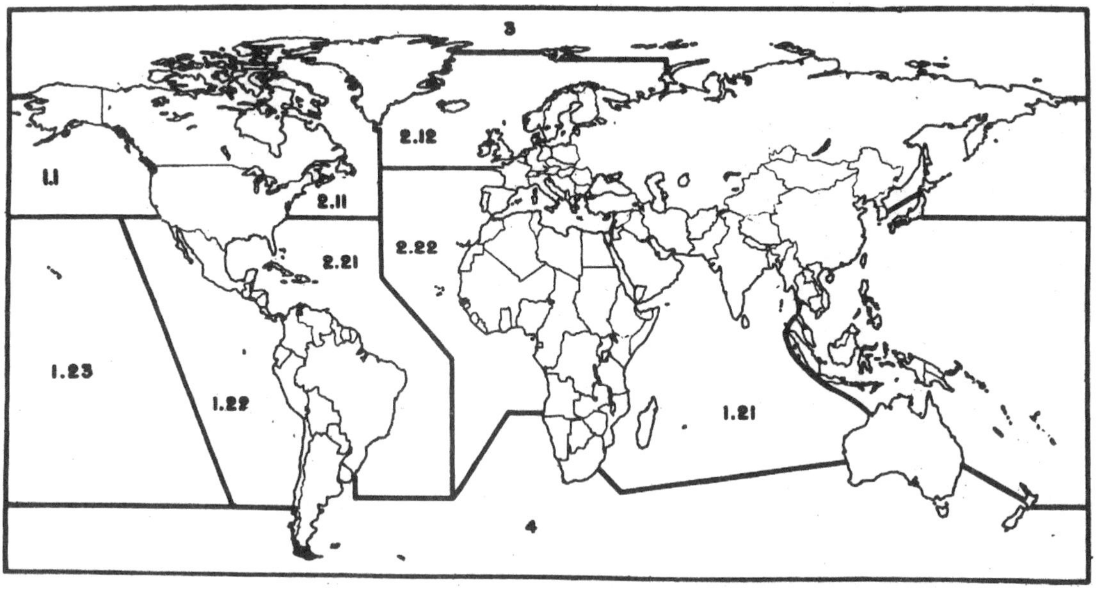

Figure 2. Map showing division of oceans for classification of unit fisheries. Note: The boundaries shown represent a compromise among several criteria, including oceanographic and zoogeographical factors.

possible, the schedule indicates, for each of the unit fisheries, the species being exploited, the location of the fishery, the gear which is used, and the countries participating in the fishery.

In preparing this schedule, the terms "renewable resources," "management unit," and "unit fishery" had to be defined. It is essential to regard a renewable resource broadly, including not merely the size groups of fish of economic value, nor only the entire population of the species which is of present economic interest, but the community of associated living organisms to which this economic species belongs. The importance of this viewpoint derives from the fact that the entire association is affected by fishing operations and, conversely, that fishing operations themselves are, in large measure, dependent upon forces operating within the entire biological system. The magnitude of the stocks which are of immediate interest and are actually being fished is determined by the magnitude of the entire population of which these stocks are part, and the magnitude of that population is, in turn, dependent upon the community within which the species lives. The resource, therefore, must be defined as a dynamic system of which one part, the population of the species, is of interest to man who, by his fishing operations, removes a part of this population.

In his paper, "The scientific basis for a conservation programme," M. B. Schaefer (1956) has indicated that the unit for management purposes is the separate population of a species. The importance of this unit, its characteristics, and the method of examining and measuring them are described in his paper; the relations of this unit to its environment are discussed by J. B. Tait (1956) in his paper on the "Role of the environment in the biology of economically valuable stocks."

The term "unit fishery" is less amenable to definition. It is best understood in terms of the common names of fisheries, such as trawl fishery, herring fishery, whale fishery, and so forth. In some cases, the unit fishery is based upon a single unit stock, and it is of great importance to be able to confirm that the area of the fishery is coextensive with that of the stock. In other cases, the unit fishery is based upon a particular community, for example, in a fishery exploiting demersal stocks, where the gear used may take two or more species simultaneously. Sometimes the fishery is thought of in terms of particular areas, or of particular types or groups of craft or gear, but in many of such cases, it is a subunit rather than a true unit that is being discussed, since the fishing exploits only a part of the population. From the biological point of view, all fishing upon a particular unit stock must be included within the same unit fishery.

In compiling this list of unit fisheries, a quick review was made of available literature concerning them. This literature is, of course, exceedingly voluminous, and some fisheries are more fully described than others. For example, literature on the resources of the North Sea and the North Atlantic is most extensive; literature on other resources, such as the Flyingfish *Exocoetus volitans* of the Celebes, is sparse. From this literature, some attempt was made to summarize available information concerning resources being exploited by each fishery. It was at once apparent that this would take more time than was available and would result in a very bulky document. Moreover, authoritative reduction of these data, from the point of view of determining their reliability and usefulness for formulating conservation programs, could scarcely be made in a short time from a single office, even if the compilation were complete. In fact, preparation of this compilation is the substantive part of the program on which the Biology Branch of the Fisheries Division of the Food and Agriculture Organization of the United Nations has now embarked. It is not expected that it will be completed for several years; it is hoped that regional fishery commissions and councils will assist on the project.

Thus, without further studies of the character indicated above, identification cannot be achieved of the type of international fisheries conservation problems for which the governments concerned have not yet developed or agreed upon measures or procedures.

A tentative list of the most important unit fisheries of the world is given below (Table 1). The information, except perhaps for treaty arrangements, is almost certainly incomplete, but it may provide a basis for discussion and for the compilation of a more comprehensive account.

Table 1. Schedule of world fisheries presented as a list of the unit fisheries of the world by area and resources, gear and methods of fishing, participating countries, and treaty arrangements. The abbreviations for the treaty arrangements used in this list are shown in a note at the end of the table.

Area and resources	Gear and method of fishing	Participating countries	Treaty arrangements
1. WORLD-WIDE			
Whales	Pelagic whaling	Denmark, Japan, Netherland, Norway, Panama, USSR, UK	International Whaling Commission, South Pacific Commission
	Shore-based whaling	Argentina, Australia, Brazil, Canada, Chile, France (for French Equatorial Africa), New Zealand, Peru, Portugal, Union of South Africa	International Whaling Commission, South Pacific Commission
Tunas, other oceanic pelagic species (swordfish, spearfish, sailfish, etc.):			
Central and South American tunas	Lines, purse seines	Chile, Costa Rica, Ecuador, Mexico, Peru, USA	Tuna Commission
Mediterranean tunas	Traps, lines, purse seines	Egypt, France (French territories?), Greece, Israel, Italy, Libya, Spain, Turkey	GFCM, International Commission for the Scientific Exploration of the Mediterranean Sea

Table 1. Continued.

Area and resources	Gear and method of fishing	Participating countries	Treaty arrangements
1. WORLD-WIDE			
Pacific tunas	Lines	Australia, China: Taiwan, Japan, Pacific territories, Philippines, USA	Indo-Pacific Fisheries Council
Tunas in other waters, not specifically referred to in treaty agreements			
2. INDO-PACIFIC			
2.1 North Pacific			
Salmon (five species, *Oncorhynchus*)	Traps, purse seines, gillnets, lines	Canada, Japan, Korea, USSR, USA	Salmon Commission, North Pacific Commission
Sardine (*Sardina* or *Sardinops melanosticta*, *Etrumeus micropus*, *Engraulis japonicus*)	Nets	Japan, Korea, USSR	North Pacific Commission
Herring (*Clupea pallasi*)	Purse seines	Canada, USA	North Pacific Commission
Mackerel (*Scomber japonicus*), Jack (*Trachurus japonicus*)	Lines, nets	Japan	North Pacific Commission
Saury (mackerel), Pike (*Cololabis saira*)	Dipnets, lines	Japan	North Pacific Commission
King Crab (*Paralithodes camtschatica*)	Tangle nets, trawls	Canada, Japan, USSR, USA	North Pacific Commission
Halibut (*Hippoglossus stenolepis*)	Lines	Canada, USA	Halibut Commission, North Pacific Commission
Groundfish, American coast: Pleuronectids, gadoids (*Sebastes*)	Trawls, some lines	Canada, USA	North Pacific Commission
Shrimp (*Penaeus carinatus*), Yellow Sea	Trawl nets	China, Japan	Indo-Pacific Fisheries Council
Groundfish, East China Sea, Yellow Sea (*Pagrosomus major, Paralichthys olivaceus, Nibea argentata*)	Trawls	China, Japan, Korea	Indo-Pacific Fisheries Council
Fur seal (*Callorhinus ursinus*)	Pelagic and island fishing	Canada, Japan, USSR, USA	Fur seal agreement, 1942
2.2 Tropical Indo-Pacific			
2.21. Indian Ocean:			
Reef fish (percomorphs: Lutjanidae, Carangidae, Sciaenidae)	Lines	Madagascar, Mauritius, Réunion, Seychelles	Indo-Pacific Fisheries Council
Stolephorus species	Traps	Most Asian countries	Indo-Pacific Fisheries Council
Groundfish (percomorphs: Lutjanidae, Carangidae, Sciaenidae)	Trawls, lines	Ceylon, India, Japan, Pakistan	Indo-Pacific Fisheries Council
Indian Mackerel (*Rastrelliger kanagurta*)	Traps, surrounding nets	India, Indochina, Malaya, Thailand	Indo-Pacific Fisheries Council
Indian Oil Sardine (*Sardinella longiceps*)	Surrounding nets	Southern Arabia, India, Pakistan	Indo-Pacific Fisheries Council
Clupeids, Persian Gulf (*Sardinella, Engraulis* species)	Various nets	Arabian countries, Iran	Indo-Pacific Fisheries Council
Groundfish, Persian Gulf (percomorphs: Lutjanidae, Carangidae, Sciaenidae)	...	Arabian countries, Iran	Indo-Pacific Fisheries Council

Table 1. Continued.

Area and resources	Gear and method of fishing	Participating countries	Treaty arrangements
2.2 Tropical Indo-Pacific			
2.21. Indian Ocean:			
Clupeids, Red Sea	Various nets	Arabian countries, Egypt, Ethiopia, Iran	
Groundfish, Red Sea	...	Arabian countries, Egypt, Ethiopia, Iran	
Reef fish (percomorphs)	Traps, lines	Australia	
Rock lobster (*Panulirus* species)	Traps, lines	Australia	
2.22 Eastern Pacific:			
Rock lobster (*Jasus* species)	...	Chile	
Sardine (*Sardinops sagax*)[4]	Purse seines	Chile, Ecuador, Peru	
Shrimp (peneids)	Trawls	Costa Rica, Mexico,Panama	
Groundfish, hake (*Merluccius*), pleuronectids	Trawls, lines, gillnets	Chile	
2.23 Central Pacific:			
Reef fish (percomorphs: Lutjanidae, Carangidae, Sciaenidae)	Lines, traps	All countries	Indo-Pacific Fisheries Council, South Pacific Commission
Flying fish (Exocoetidae)	...	Indonesia	Indo-Pacific Fisheries Council
Mackerel ("Kanbong, plathu," *Rastrelliger kanagurta*)	Surrounding nets, traps	Malaya, Thailand	Indo-Pacific Fisheries Council
Groundfish (Sciaenidae, Foriniidae, *Stromateus*)	Trawls	China, Japan, Korea, USSR	Indo-Pacific Fisheries Council
Pearl oyster	Diving	Australia, Japan	
2.1–2.22 Transition region:			
Pilchard (*Sardinops caerulea*), anchovies (*Engraulis mordax*, *Anchoviella* species), mackerel, horse mackerel (*Pneumatophorus diego*, *Trachurus symmetricus*)	Purse seines	USA	
3. ATLANTIC			
3.11 Northwest Atlantic			
Herring (*Clupea harengus*), other clupeids	Pelagic gear, weirs, seines, traps	Canada, USA	ICNAF
Mackerel (*Scomber scombrus*), chub (*Pneumatophorus colias*)	Purse seines	Canada, USA	ICNAF
Lobster (*Homarus americanus*)	Traps	Canada, France, Spain, UK, USA	ICNAF
Groundfish on banks (gadoids, pleuronectids, *Sebastes*)	Trawls, lines	Canada, France, Spain, UK, USA	ICNAF
Shrimp, prawn (*Nephrops*, etc.)	Trawls	Denmark	ICNAF

[4]Also basis for guano.

Table 1. Continued.

Area and resources	Gear and method of fishing	Participating countries	Treaty arrangements
3.11–3.12 Transition region:			
Groundfish (gadoids, pleuronectids)	Trawls, lines	Denmark, Iceland, Norway, Portugal, UK	ICNAF, ICES, Overfishing convention, 1946
3.12 Northeast Atlantic			
Groundfish (gadoids, pleuronectids)	Trawls, demersal seines	European countries	ICES, Overfishing convention, 1946
Herring (*Clupea harengus*)	Pelagic and demersal gear (trawls, driftnets, ringnets)	European countries, USSR	ICES, Overfishing convention, 1946
3.12–3.22 Transition region:			
Hake (*Merluccius merluccius*)	Trawls	European countries	ICES
3.11–3.21 Transition region:			
Menhaden (*Brevoortia tyrannus*)	Purse seines	USA	
3.21 Western central Atlantic:			
Shrimp (*Penaeus* species)	Trawls	Cuba, Mexico, USA	
Reef fish	Lines, traps	Caribbean countries	
Groundfish	Trawls	Argentina, Brazil, Uruguay	
Reef percomorphs	Lines, seines, traps	Brazil	
3.22 Eastern central Atlantic and Mediterranean:			
Groundfish, many species	Trawls	Mediterranean countries	GFCM, ICES, International Commission for the Scientific Exploration of the Mediterranean Sea
Sardine (*Sardinella aurita, Sardina pilchardus*)	Surrounding nets	France, Greece, Italy, Morocco, Portugal, Spain, Yugoslavia	GFCM, ICES, International Commission for the Scientific Exploration of the Mediterranean Sea
Anchovy, horse mackerel (*Engraulis encreaicholus, Trachurus trachurus*), Black Sea	Ringnets	Turkey, USSR	
Dolphin (*Delphinus delphis*), Black Sea	Ringnets, shooting	Turkey, USSR	
Groundfish (percomorphs)	Lines, trawls	French, Portuguese, and UK territories in Africa and southern Africa	ICES, Commission for Technical Cooperation in Africa South of the Sahara
4. ARCTIC			
Herring (*Clupea harengus*)	Seines	USSR	
Groundfish (pleuronectids)	Trawls	USSR	
5. SOUTHERN SEAS			
5.1 Temperate waters:			
Pilchard (*Sardinops ocellata*), Maasbanker (*Trachurus trachurus*)	Purse seines	Union of South Africa	
Groundfish (*Merluccius capensis*), pleuronectids (*Austroglossus* species), etc.	Trawls	Union of South Africa	
Snoek (*Thyrsites atun*)	Jiglines	South West Africa, Union of South Africa	

Table 1. Continued.

Area and resources	Gear and method of fishing	Participating countries	Treaty arrangements
4. ARCTIC			
Herring (*Clupea harengus*)	Seines	USSR	
Groundfish (pleuronectids)	Trawls	USSR	
5. SOUTHERN SEAS			
5.1 Temperate waters:			
Percomorphs (*Atractoscion aequidens, Johnius hololepidotus, Argyrozona argyrozona*)	Handlines	Union of South Africa	
Sardine (*Sardinops neopilchardus*)	Nets	Australia	
Groundfish (*Neoplatycephalus richardsoni*)	Trawls	Australia	
Barracouta (*Leionura atun*)	Jiglines	Australia	
Reef fish (*Chrysophrys guttatus*)	Lines, traps	Australia	
Shark	Longlines	Australia	
Groundfish (percomorphs, pleuronectids)	Trawls	New Zealand	
Groundfish	Trawls	Argentina	
5.2 Antarctic:			
Fur seals (*Arctophalus pusillus*)	...	Union of South Africa	

Note: In addition to the above, there are various diadromous fishes which are not, at present, caught on the high seas, but they spend an important part of their lives at sea. These include:

3.11–3.21 Transition region:			
Shad (*Alosa sapiddissima*)	...	North America	
2.21 Indian Ocean			
Indian shad (*Hilsa* fish, *Hilsa ilisha*)	...	India	
Hilsa herring (*Hilsa reevesii*)	...	China, Korea	

Note: Fur seal agreement, 1942—Provisional fur seal agreement between Canada and the USA, 1942; GFCM—General Fisheries Council for the Mediterranean; Halibut Commission—International Pacific Halibut Commission; ICES—International Council for the Exploration of the Sea; ICNAF—International Commission for the Northwest Atlantic Fisheries; North Pacific Commission—International North Pacific Fisheries Commission; Overfishing convention, 1946—Convention for the Regulation of the Meshes of Fishing Nets and the Size Limits of Fish (Overfishing conference, 1946); Salmon Commission—International Pacific Salmon Fisheries Commission; South Pacific Commission—Permanent Commission for the Exploitation and Conservation of the Maritime Resources of the South Pacific; Tuna Commission—Inter-American Tropical Tuna Commission.

References

Graham, M. 1956. Concepts of conservation. Pages 1–13 *in* Papers Presented at the International Technical Conference on the Conservation of the Living Resources of the Sea, Rome, 18 April to 10 May 1955. United Nations Publication, Sales No: 1956.II.B.1. UN, New York.

International Technical Conference on the Conservation of the Living Resources of the Sea. 1956. Papers Presented at the International Technical Conference on the Conservation of the Living Resources of the Sea, Rome, 18 April to 10 May 1955. United Nations Publication, Sales No: 1956. II.B.1. UN, New York.

Schaefer, M. B. 1956. The scientific basis for a conservation program. Pages 15–55 *in* Papers Presented at the International Technical Conference on the Conservation of the Living Resources of the Sea, Rome, 18 April to 10 May 1955. United Nations Publication, Sales No: 1956.II.B.1. UN, New York.

Tait, I. B. 1956. Role of the environment in the biology of economically valuable stocks. Papers Presented at the International Technical Conference on the Conservation of the Living Resources of the Sea, Rome, 18 April to 10 May 1955: 62. United Nations Publication, Sales No: 1956.II.B.1. UN, New York.

Chapter 3

Observations on the Concept of Maximum Sustainable Yield as a Criterion for the Wise Exploitation of the Living Resources of the Sea[1]

There are considerable difficulties in the application of the maximum sustainable yield concept, as formulated in the papers by Dr. Schaefer (1956) and other contributors to this meeting, to fisheries in which it is possible for man to control the sizes or ages at which fish are captured by the fishing gear.

Consider the life history of a particular brood of fish spawned in one year. By various agents of mortality, the numbers of this brood are caused to decrease from year to year, but at first, this loss is more than offset by the growth in size of the surviving fish. From the beginning, the brood increases in total weight and at an accelerating rate. However, after a time, growth of the individual fish begins to slow down and eventually a time is reached when the total weight of the brood reaches a maximum, and thereafter it begins to decline. The maximum sustainable yield from the stock will usually be obtained when the whole of each brood can be caught at this optimum age. With our present methods of fishing, this can only be achieved by fishing very intensely though highly selectively. Indeed, in marine fisheries, it is not possible to capture all the fish at this particular optimum age, so that, in practice, it is best to start catching them when they are a little younger; how much younger depends upon how intense is the fishing—the harder and more effectively we can fish, the longer we can safely leave the fish to grow before we try to catch them. Under these conditions, the stock is not reduced to the level of the inflexion of the normal population growth curve, but rather is maintained at a level only slightly below its maximum possible abundance. Only the number in the age groups being intensely fished will be substantially reduced.

A similar argument may be applied at the other extreme where, as in the case of the Flyingfish *Exocoetus volitans* of the Celebes, it is the eggs of the fish which are, at present, the products most useful to man. This case is not, however, of such general interest as the one mentioned above.

---

[1]Editors' note: written by Sidney Holt for the International Technical Conference on the Conservation of the Living Resources of the Sea, Rome, 18 April–10 May 1955, although the paper does not appear in the publication *Papers Presented at the International Technical Conference on the Conservation of the Living Resources of the Sea, Rome, 18 April to 10 May 1955* (International Technical Conference on the Conservation of the Living Resources of the Sea 1956). At the time of this writing, Holt was employed by the Food and Agriculture Organization of the United Nations in its Fisheries Biology Branch. Permission to reproduce this article has been given by FAO. The views expressed in this publication are those of the author and do not necessarily reflect the views or policies of the Food and Agriculture Organization of the United Nations.

Now, we know that there is a tendency in an unregulated fishery for the intensity of fishing to increase to a point at which fishing is no longer profitable, and at that point the yield may be less than the maximum obtainable on a sustained basis by that kind of fishing, but it is equally true that in an unregulated fishery, using a type of gear the selectivity of which can be modified as by a change in the size of mesh, the fisherman, unless restrained, tends to adopt the smallest mesh that he can practically work. The "sigmoid curve theory of fishing" is inadequate to deal with such situations.

My first point, then, is this: that there is no single maximum sustainable yield, but many, depending not only on the amount, but equally on the kind, of fishing, specifically its selectivity. And, we must remember that fishing may be selective not only because of the mechanical properties of the gear, but through the tendency in some species for the age or size classes to be segregated in different shoals and also through the differing behavior of fish of different size or age towards certain kinds of gear.

Whilst it is true that there are situations where a certain reduction in fishing can lead to increases in yield and also to increases in catch per unit effort, it is also true that technical developments may make possible a more selective fishing, and that these, if correctly applied, may lead to even greater sustainable increases in yield without any reduction in effort; or increases in effort may continue to give higher yields and without reduction in the catch per unit effort. Only in cases where technical developments have proceeded far enough for selectivity to be varied at will can we usefully speak of the maximum sustainable yield, and then we must recognize that to approach this, regulation of the intensity of fishing and of the kind of fishing are equally necessary and, indeed, complementary.

For each kind of fishing activity, the maximum sustainable yield, or, in fact, any particular sustainable yield, is dependent not only upon the level at which the stock is maintained, but upon the productivity of the ecosystem in which that stock is an element. This productivity is not a constant quantity and neither is the proportion of it passing into the fish stocks in which we are immediately interested. Dr. Herrington (1956) has indicated one of the ways that this proportion may be changed—that is, by redistribution of a population over its feeding grounds—and Dr. Zenkovich (1956) has described another way in which the fraction of the primary productivity of the seas which is turned into useful form may be increased. There are certainly other ways as well. This is my second reason for suggesting that we are not accurately formulating our modern knowledge of fishery biology by speaking of "maximum yield."

My third objection to this term as a criterion for the conservation of the living resources of the seas is that it is an entirely theoretical concept, about the determination of which biologists and others are still arguing and will undoubtedly continue to argue. Should we set up such a criterion as the basis of recommendations from the conference concerning the regime of the high seas? I think not. There are other more homely concepts more suitable for our present purpose. I think we all understand the idea of "getting larger" or "getting smaller" better than the idea of attaining a maximum. I think we are agreed that there are fisheries situations where the sustained yield cannot be increased unless we rationalize our fishing activities; unless we all agreed not to compete for the fish to the detriment of everyone, but instead learn to control both the amount and kind of our fishing in such a way that we can continue to increase our exploitation of the seas' resources and continue to increase the efficiency with which we do this. We know there are limits to the sustained yield we can obtain with uncontrolled fishing. But, as far as we know, there are virtually no limits to the yield we can gain by rational, controlled exploitation, and it would be wrong if, in our formulation, we implied that there are limits and, thus, perhaps sometime in the future, to use the words of Michael Graham at the UNSCCUR (1951) meeting "prevent something being done when it needs to be done." The ultimate limit is, for all practical purposes, set by man's technical ability to take and use the resources, his biological knowledge as to the means by which this is to be done in a way ensuring that the increase can be sustained in the future, and by his success in reaching international understanding as to these means and agreement to adopt them.

My proposal is then that instead of speaking, as in the annotations to Item 9 of the provisional agenda (A/CONF.10/2),[2] of the maximum sustainable yield, we should speak of controlling man's activities so as to produce the greatest possible sustainable continuing increases in the yield of products in the form most useful to man. Then, we can apply ourselves to the task of defining the situations in which the enterprises of the fisherman must be not restrained, but turned in other directions that will be more productive both for him and for the community at large. A beginning has been made in this by the formulation of the modern theory of fishing, given by Michael Graham (1956). Dr. Kesteven and I have tried, in our graph of the generalized history of a fishery (Kesteven and Holt 1956), to indicate the various situations to be considered.

References

Graham, M. 1956. A first approximation to the modern theory of fishing. Pages 56–60 *in* Papers Presented at the International Technical Conference on the Conservation of the Living Resources of the Sea, Rome, 18 April to 10 May 1955. United Nations Publication, Sales No: 1956.II.B.1. UN, New York.

Herrington, W. C. 1956. Comments on the principle of abstention. Pages 344–349 *in* Papers Presented at the International Technical Conference on the Conservation of the Living Resources of the Sea, Rome, 18 April to 10 May 1955, United Nations Publication, Sales No: 1956.II.B.1. UN, New York.

International Technical Conference on the Conservation of the Living Resources of the Sea. 1956. Papers Presented at the International Technical Conference on the Conservation of the Living Resources of the Sea, Rome, 18 April to 10 May 1955. United Nations Publication, Sales No: 1956.II.B.1. UN, New York.

Kesteven, G. L., and S. J. Holt. 1956. Classification of international conservation problems. Pages 350–371 *in* Papers Presented at the International Technical Conference on the Conservation of the Living Resources of the Sea, Rome, 18 April to 10 May 1955. United Nations Publication, Sales No: 1956.II.B.1. UN, New York.

Schaefer, M. B. 1956. The scientific basis for a conservation program. Pages 15–55 *in* Papers Presented at the International Technical Conference on the Conservation of the Living Resources of the Sea, Rome, 18 April to 10 May 1955. United Nations Publication, Sales No: 1956.II.B.1. UN, New York.

UNSCCUR (United Nations Scientific Conference on the Conservation and Utilization of Resources). 1951. Proceedings of the United Nations Scientific Conference on the Conservation and Utilization of Resources, 17 August–6 September 1949, Lake Success, New York, Volume VII, Wildlife and Fish Resources. United Nations, New York.

Zenkovich, L. 1956. Biological appraisal of the ocean, and the problem of transoceanic acclimatization. Pages 127–144 *in* Papers Presented at the International Technical Conference on the Conservation of the Living Resources of the Sea, Rome, 18 April to 10 May 1955. United Nations Publication, Sales No: 1956.II.B.1. UN, New York.

[2]https://digitallibrary.un.org/record/3849057?ln=en

Chapter 4

The Evaluation of Fisheries Resources by the Dynamic Analysis of Stocks, and Notes on the Time Factors Involved[1]

Introduction

To devise means of reducing the time required to assess fish stocks, it may be fruitful and, in view of the present active development of several methods of analysis (see e.g., Schaefer 1956) opportune, to re-examine the aims of this work, and the definitions of the concepts involved. The aims are, I believe, to develop methods of evaluating fishery resources and then to determine, for a given time and situation, what action is appropriate to increase the economic production from those resources.

Although arguments have been advanced for defining fishery resources as whole ecosystems, rather than as populations of commercially valuable fish (Kesteven and Holt 1955), it is an important characteristic of present methods that we begin by analysis of particular parts of those ecosystems. The unit fish stock is, for this purpose, selected as a part of the resource susceptible to separate treatment. The stock is thus considered as a partial system, and the procedure is to attempt to link this with the other elements of the ecosystem—that is, the environment of the stock—considering each relation in turn.

The fundamental attributes of the unit fish stock that we seek to describe are:

a) its identity as an independent population, its general location, and the bounds of its distribution;
b) the behavior exhibited by the individuals or schools composing it, e.g., changes of habitat at certain stages in the life history, migrations within the general area of distribution, local movements in response to various stimuli, and other forms of activity; and
c) its magnitude.

The time factor might be considered in studying all three attributes, but in this note, I wish to draw attention only to what seem to be important aspects of the last mentioned, since I believe this is the one for which the reduction of time in analysis can most profitably be discussed.

---

[1]Editors' note: this was originally presented as a contributed paper at an ICNAF Symposium on Some Problems for Biological Fishery Survey and Techniques for their Solution held at Biarritz, France, 1–10 March 1956 and was then published as Holt (1958). It is included here, with minor editorial changes, with the permission of the NAFO Secretariat.

The appropriate measure of magnitude of a unit fish stock to use when considering it as an element in a resource system is not obvious. The chief requirements of a suitable measure are comparability of values as between different stocks, and a consistent relation to the catch obtainable from the stock under well-defined conditions.

Clearly, population size itself (standing crop or biomass) is not adequate, since the standing crops of different populations of the same size may give completely different catches. The catch, being the integral of a rate of activity over a certain time-period, is more closely related to the production (turnover) of living organic material, but whereas this might provide a rough index of resource potential, it is not entirely suitable when we study the fish populations themselves in detail. For example, production is a quantity that itself varies with the fishing activity, as also does the specific productivity (fractional turnover rate, or the rate of production per gram of standing crop).

A measure having many advocates and, to a certain extent, established by being written into several international treaties, is the "maximum equilibrium (or sustainable or sustained) catch."

The term "equilibrium" has been used in fishery theory in several ways, but, in general, there is an implication of stability either of a fish stock or of a fishing industry. Now, a stock may be stabilized at any level of size within wide limits, and the possibilities within a given economy of stabilizing the level of operation and production of a particular industry are equally wide. In both cases, external constraints—on the one hand, for example, the amount and kind of fishing, and on the other, the economic conditions outside the industry setting costs and prices—may determine the particular equilibrium attained.

"Maximum equilibrium catch" may, therefore, mean either the greatest sustained catch that a stock can give or the greatest catch that can be taken whilst allowing economic equilibrium, according to whether the fish stock or the fishing industry is the system the dynamics of which are under consideration. It has frequently been assumed that it is economically possible and desirable, or at least not economically undesirable, to take the greatest amount that the stock can sustain, but this proposition needs critical examination.

Equilibrium of a unit fish stock is here understood to mean a condition in which the weight of fish removed from the population, by capture of some and "natural" death of others, is, on average, equal to the weight added through supplementation of numbers of fish by recruitment and increase in size of individual fish by growth. Such a condition is better referred to as a steady-state.

It is a characteristic, though not an invariable, property of animal populations or communities that they are able to attain steady-states within a wide range of external conditions, and it has been usual in the field of population dynamics, as with analogous physical problems, to consider first the population as an "open" system in steady-state and then to examine the consequences of departures from this state.

Trends to or from particular steady-states may or may not be reversible. Reversibility would, for a fish stock, mean that if a change in the intensity of fishing caused a movement away from a steady-state, then a further change back to the original intensity would result eventually in the stock regaining the original steady-state. There are indications that this is generally true for several marine fish stocks, the most conclusive evidence being the effects of temporary cessation during wartime of fishing on depleted stocks.

Russell (1931) has given us a symbolic statement of the relation between stock size and the various additions to and removals from the stock. His axiomatic equation, in fact, describes a fish stock as an open system and it can give simple expression to the steady-state condition. It has been recognized that while this cannot, as it stands, be used to assess the state of the stocks and the needs for corrective measures, it nevertheless represents a starting point for a working formulation of the sustained catch problem.

The most important variable, the relation of which with steady catch we need to discover, is the "amount of fishing." The most direct method, if a long series of commercial fishery statistics is available, is to pick out several periods during each of which the intensity of fishing and the catch were fairly constant for long enough to make possible the assumption that a steady-state was set up during each period. The average annual catch for each period is then plotted against the corresponding fishing effort. Graham (1951) has remarked that the statistics alone are usually insufficient to give a coherent graph, and he gives examples in support of this from what are generally considered to be well-studied fisheries.

In an attempt to interpret the limited data, some further assumptions may then be made concerning the general shape of the required curve relating equilibrium catch to fishing effort. A device that has been employed is to fit the Verhulst–Pearl logistic curve to the available data. Graham (1935) was able to do this using observations of the relation of war-caused changes in the amount of fishing to changes in the catch and the catch per unit effort from North Sea trawl fisheries. Although the validity of his assumption, that the total growth of all demersal fish populations in the area can be represented by the logistic function, cannot be tested directly, it has been accepted as a not unreasonable first approximation. Since the natural growth curve is certainly sigmoid—whether or not it is a symmetrical logistic—the stock will have a maximum rate of increment at some intermediate stock size when the inflexion is reached; the argument is that a maximum equilibrium catch will be obtained when the stocks are, by fishing, kept at the level of that inflexion.

One difficulty in developing this method further is that—equilibrium catch being considered only as a function of a gross measure of population size—no account can be taken of changes in the age composition of the stock that inevitably result from changes in fishing intensity.

Schaefer (1953) has recently attempted to make fuller use of series of annual catch and effort statistics to investigate the behavior of fish populations when they are being exploited. His method is essentially the analysis of the fluctuations in the series for the purpose of predicting steady-states. The growth potential, and hence the steady catch that could be taken from a stock of a size equal to the average stock size in a particular year, is estimated as the algebraic sum of the actual catch in that year, and the difference between the stock size at the beginning of the year and at the end of it. Schaefer plotted figures for "equilibrium catch" obtained in this way against catch per unit effort for the Pacific Halibut *Hippoglossus stenolepis* and Pacific Pilchard *Sardinops sagax caerulea* stocks. His figures indicate that, for both species, the "equilibrium catch" increases with increasing stock size over the observed range. It is possible to fit a logistic function to data treated in this way, but the scatter of points is so great that no useful estimate can be made of the stock size that would give the "maximum equilibrium catch." In fact, the data give almost no sign even of the existence of a peak in the curve. They could, however, be taken as evidence that, for the situations examined, a reduction in fishing would result in an increased average catch, provided it were also shown that the stock abundance is largely controlled by the fishing intensity.

A fundamental objection to the above method is that no knowledge is obtained of the extent to which the year-to-year changes in stock size are, on the one hand, the result of mortality and growth within the stock and, on the other, due to the recruitment of new individuals into the stock. The recruitment in a particular year can hardly be regarded, in general, as a simple function of the stock size in that same year. It is, for all species but those having a lifespan of less than one year, a function of the stock size during an earlier period, at the beginning of which the new year class was spawned.

To illustrate the errors of interpretation that may arise if these facts are neglected, it is sufficient to consider the case where fishing is very intense, but highly selective, only the larger fish being caught. In this situation, the catch per unit effort may be an index of the size of the part of the stock being exploited, but this bears no simple relation to the total size of the population. Yet, it is the growth of fish before recruitment which, in this case, determines the steady yield. Analysis of the North Sea Plaice *Pleuronectes platessa* stock shows that even where these fish, having a possible lifespan of over 20 years, become liable to capture, on average, soon after they are three years old, rather more than one-half of the stock growth from which the catch is derived occurs before recruitment.

Such considerations remind us that man rarely, if ever, exploits the whole of a population of fish, but rather selects only individuals within a certain range of size and age which constitute the available stock, and frequently only a readily accessible fraction of these. For further development of fishing theory, this distinction among total, fishable, and fished populations must be fully recognized and the relations among them taken into account. Fundamentally, the fished stock from which the catch comes cannot be treated as a self-reproducing unit having, at any given time, a potential for growth, which is a function mainly of its size at that time; this potential is essentially a complex function of stock size and composition at the time and of the size and composition of the total population during some previous period.

The above discussion leads us to consider another approach to the maximum catch problem. It has frequently been pointed out that there is a best age at which to harvest a particular crop. For cultivated stocks, this best age may be that for which the net gain is at a maximum and after which the cost of, for example, continued feeding, may be greater than the value of the additional growth. For wild fish stocks, the best age is considered to be that at which a particular year class reaches its greatest total weight, after which the losses by natural mortality are greater than the summed growth increments of the survivors. In marine fisheries, it is not possible to capture all the fish at a particular age, so that, in practice, it is best to start catching fish when they are rather younger; how much younger depends upon how intense is the fishing—the harder and more effective the fishing, the longer the fish can be left to grow before one tries to catch them.

Allen (1953) has discussed the best legal size limit of fish for any particular rate of fishing. From his theorem, we may deduce that the weight of an individual of the size at which it is best to start fishing is equal to the average weight of the fish in the catch multiplied by the ratio of the fishing mortality rate to the total mortality rate in the stock. However, like Russell's equation, this formula defines a steady-state situation, *a posteriori*, but does not provide a means of prediction.

To unify these two approaches to conservation problems, i.e., consideration of the best fraction of the stock to remove annually and of the best age at which to catch the fish, the analytical methods developed by Baranov (1918) and several other workers since are necessary. These depend essentially upon the construction of mathematical models containing coefficients which describe the factors of migration and recruitment, growth, and mortality, and causing the removals from and additions to a fish stock mentioned above, and the relations between these factors. The various models put forward (e.g., Ricker 1944; Tester 1953; Tanaka 1954; among others) have differed from each other principally in the shape of the curve used to describe growth of the individual fish, and the characteristics of each model depend to a considerable extent upon the accuracy with which the particular growth function used reproduces the general features of fish growth. When it has been a realistic one, these models have frequently shown a peak in the curve of steady yield against fishing intensity. When, however, the rate of natural mortality is high in relation to the overall growth rate, or when the dependence of growth rate of individuals on the population density is taken into account, then the curve may have no peak, but may instead approach an asymptote. Again, when the size at which fish are first liable to capture is increased, then the maximum is found to occur at a higher fishing intensity and the peak of the curve becomes less pronounced. Eventually a size is reached for which the peak disappears and the curve becomes asymptotic.

A clue as to a suitable measure of stock magnitude is found by considering the nature of fishing activity from the biological point of view. It has two characteristic properties—amount and selectivity—especially with regard to the sizes and ages of fish taken, but also to species, sex, and other attributes. We have seen that much work on the theory of fishing has considered only the former characteristic, though sometimes, especially when workers have been concerned with the management of fisheries by mesh regulation and the imposition of size limits, the selectivity of the gear for certain size classes of fish has been the main consideration.

In fact, both characteristics have always to be treated simultaneously in any acceptable theory of fishing. Doing so leads inevitably to a three-dimensional representation of the relation of steady catch to the fishing activity, but this may be shown in a two-dimensional diagram by the device of plotting contours of catch against the two axes of amount of fishing and selectivity of fishing. One such representation is the steady-catch isopleth diagram (Beverton 1953; there referred to as a yield-isopleth diagram), which is obtained by calculation from an equation containing parameters representing the recruitment, growth, and mortality rates of fish in a stock, and which refers to a fishery based on a stock with an annual recruitment, which is effectively independent of the size of the spawning population, a constant natural mortality that can be represented by an exponential curve, and an invariant growth pattern of the individual fish conforming to von Bertalanffy's equation.

In that diagram were plotted not values of characteristics of the fishing activity itself, but the equivalent biological parameters, so that there are illustrated the inherent properties of the stock itself, as exhibited under the influence of a particular kind of fishing operation, trawling in the example given. It is found that for trawling operations, the most useful parameters are the instantaneous coefficient of fishing mortality F (which is conveniently proportional to the fishing intensity expressed in standard terms) and the age at first liability to capture by the gear t_c, which is primarily a simple function of the size of mesh in the trawl codend.

Similar procedures can be followed for dealing with hook and line and other types of fishery, but there is a further difficulty in reconciling the alternative catch-isopleth diagrams that might be drawn for a particular stock with the possibility of being exploited by different types of gear for which the relations of selectivity to size of fish are different. With a trawl or other "bag" net such as a seine, the fishing mortality coefficient is, to a first approximation, constant for all fish above a certain age at first capture, this age being determined either by the pattern of recruitment or by the size of mesh. Rollefsen's (1953) and other data indicate, however, that for hook and line gear, the instantaneous fishing mortality coefficient may vary in a curvilinear manner (possibly exponentially) with age of fish. The same is undoubtedly true of gill- and trammelnets. In general, the mortality coefficient is some function—but probably usually a rather simple function—of the size or age of fish; each kind of fishing unit has a characteristic function and, for trawls, this is of the simplest possible kind. It is possible to define an amount and selectivity of one kind of fishing (say trawling) that will result in the same sustained catch and have roughly the same effect on the stock magnitude as any given amount and selectivity of another kind of fishing, such as lining with a particular size of hook. An isopleth diagram referring to a certain kind of method can, therefore, be transformed to an equivalent one for some standard fishing method, and it is convenient to take trawling as the standard.

In this way, standard exploitation conditions can, for the purposes of resource appraisal, be defined in biological terms so that the steady catch may be used in deriving a basic measure of magnitude.

The procedures of drafting an isopleth diagram and subsequent operations could be simplified in several ways, including some standardization of methods of observation and presentation of data, compilation, and critical review of data on special subjects such as growth of fish, mathematical approximation, and short cuts in computation and plotting.

Other summary curves indicating certain aspects of the contour pattern may be drawn on isopleth diagrams of steady catch. The most important is the eumetric fishing curve, showing the best selectivity of the gear for each particular fishing mortality and from which may be deduced the eumetric catch curve giving the maximum steady catch as a function of the amount of eumetric fishing. For the complementary line, shown dotted in Figure 2, adjustment of selectivity is considered to be limited for technical or other reasons, and for each age at first liability to capture, the best fishing mortality is imposed. On this curve, lie all points of "maximum sustainable catch" obtained when only the variation of amount of fishing is considered as a factor governing catch.

The horizontal line in these diagrams marks the age of recruitment. Below this line, fish are not accessible to the kind of gear under consideration and could only be caught, if at all, by adopting entirely different methods.

In the present context, the most important characteristic of eumetric catch curves is that they are asymptotic, the greatest steady catch being obtained only with infinitely intense, but very selective, fishing, that is, when all fish can be caught as soon as they reach a certain best age. I propose to call this asymptotic steady catch the potential yield. This definition differs to some extent from previous usage, "catch" and "yield" in marine fisheries having usually been treated as synonymous. However, it is useful, as we shall see, to distinguish the two quantities, and this is in keeping with the distinction made between "harvest" and "yield" in agricultural and forestry science. The potential yield is, in fact, that quantity which could be taken by ideal cropping methods; in marine fisheries, this includes the use of gear of optimum selectivity and highly efficient capture operations. The amount by which catch falls short of potential yield is a measure of the absolute efficiency with which the harvesting operation is being conducted; this means efficiency of organization and control, of course, as well as technical efficiency. I suggest that the assessment of potential yield in the above sense is the correct focus for work on evaluation of fishery resources, though this must, of course, be linked with biological production at different levels in the ecosystem of which the fish stock is part and with the practical requirements of exploitation at the present time. Having measured potential yield, we may, on the one hand, examine the factors limiting the catch in relation to the potential yield and, on the other hand, the means of increasing the potential yield itself in relation to the rate of organic production by the stock by, for example, transplantation, predator control, and other forms of intervention in the resource system.

Using the simple equation given by Beverton for a trawl fishery, we may derive expressions for two quantities which are sufficient to define a eumetric catch curve. One of these is the initial slope of the curve, which might be called the responsiveness of the stock, and it is given by:

$$Q = RW_\infty \sum_{n=0}^{n=3} \{1 - \exp[-(M + nK)]\} \exp(-nKt_r)\, a_n/(M + nK)$$

where R is annual recruitment, W_∞ and K are parameters of von Bertalanffy's equation for growth in weight, M is the instantaneous coefficient of natural mortality, t_r is the age of fish at recruitment, L is the fishable lifespan, and a_0, a_1, a_2, and a_3 are +1, –3, +3, and –1, respectively. When L is large and t_r small, as may often be assumed, this approximates to:

$$Q \approx 6RW_\infty K^3 / \operatorname{Prod.}_{n=0}^{n=3} (M + nk)$$

The other quantity is the asymptote, the potential yield, and it is:

$$Y = 27RW_\infty K^3 M^{M/K} \exp(Mt_r)/(M + 3K)^{3+M/K}$$

Rearrangement show that the potential yield per recruit is, when expressed as a fraction of the asymptotic weight W_∞, a simple function of the ratio $K/M = i$, thus:

$$y = Y \times \exp(-Mt_r)/RW_\infty = 3i^3/(3i + 1)^{3+1/i}$$

This function is plotted in Figure 1 on logarithmic scale to show the relative effects on the potential yield of proportionate changes in the ratio K/M. It will be seen that as this ratio increases (by increase in K or decrease in M), the yield can be improved relatively more effectively by increases in W_∞ or R. Such considerations might assist in evaluating alternatives of, say, transplantation and predator control.

The weight of the individual fish at age t_y, when the brood attains its maximum aggregate weight, is given by:

$$\mathrm{Wt}_y = W_\infty[K/(K - M/3)]^3$$

The relation between the quantities Y and Q can be represented by the empirical equation:

$$C_u = Y[1 - \exp(-2QF/Y)]/[1 + \exp(-2vQF/Y)]$$

where C_u is the eumetric steady catch and v is a constant lying between 0 and 1.

The equations for Y and Q contain only parameters referring to the recruitment, growth, and mortality of fish in the stock, all of which factors are measurable. For this population model, these parameters have been assumed to be constant, and if, for example, the growth rate were strongly dependent on the stock density, the deductions might be invalid. This first approximation might, however, be taken as a working basis from which to develop more comprehensive methods of estimation, and it seems to help in sorting out concepts, identifying the necessary types of data, and linking widely different kinds of observation. The advantage of using terminology and methods that are applicable equally to marine and to inland resources,

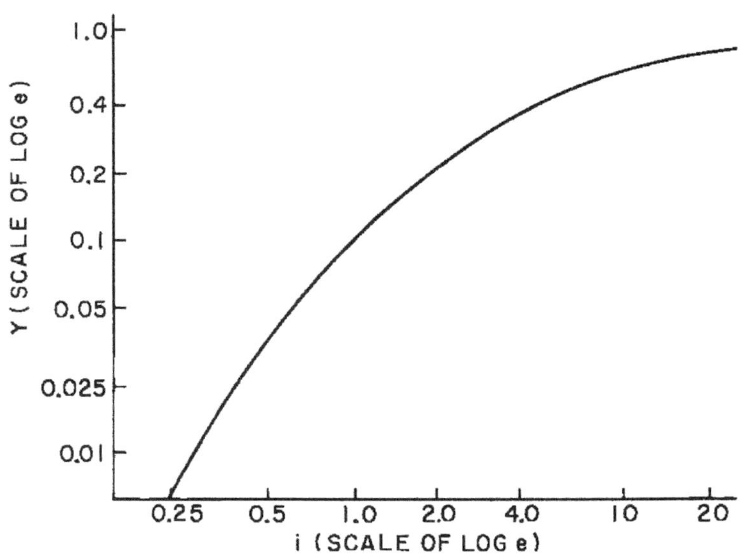

Figure 1. Potential yield per recruit per gram asymptotic weight (*y*) as a function of the ratio (*i*) of the growth parameter *K* to the natural mortality coefficient *M*.

and to cultural practices as well as to simple exploitation of wild stocks, might also be borne in mind.

We now consider the relations between potential yield and organic production and harvest. The annual production of the fishable stock is equal, under steady-state conditions, to the total weight of fish dying in each year (in absolute terms), and the ratio of potential yield to this production is given by:

$$\sum_{n=0}^{n=3} a_n\,[M/(M+3K)]^{n+(M/K)} \Big/ \sum_{n=0}^{n=3} a_n\,[M/(M+3K)]^{n+(M/K)}\, e^{-(M+nK)t_r}[1-M/(M+nK)\,e^{Mt_r}]$$

Such an expression relates yield to production at this particular trophic level and, through transfer factors, to production at other levels, so that ecological data can be integrated with the detailed studies of commercially exploited fish populations.

Turning now to the significance of the potential yield in putting the magnitude of the catch into perspective, we may again consider the isopleth diagram for steady catch. A simplification can be made by selecting the contour line passing through the observed value of the steady catch, which means that the diagram shall be drawn with reference to a given set of data defining an observed initial situation. On one side of this line lies improvement of catch, on the other side, decrease.

The other principal characteristics of the catch that have been identified as significant in relation to conservation problems, and to the exploitation program generally (see report of the UN Technical Conference on the Conservation of the Living Resources of the Sea, Rome, 1955) (ITCCLRS 1956), are the catch per unit effort and certain indications of "quality," notably the average size or age of the fish and, in certain cases, the species composition of the catch. Isopleths of these quantities expressed in biological terms, e.g., the ratio *C/F* and the average weight of fish in the catch ($\overline{W}$), can also be drawn on the same diagram, again choosing for preference the particular isopleths passing through the reference point. The diagram is, by their intersection, cut into segments in some of which there is universal improvement in all catch characteristics, and in others, universal detriment; whilst in other segments, one characteristic is improved to the detriment of one or both other characteristics.

An example is given of this procedure (Figure 2), the material being taken from Beverton and Holt (1956) and referring to the Plaice stock of the North Sea. The curve for average weight of fish, used as an index of quality, almost coincides in this case with the lower part of the curve for amount of catch. The eumetric line and its complement are shown to indicate relief on the diagram, and the potential yield is marked in.

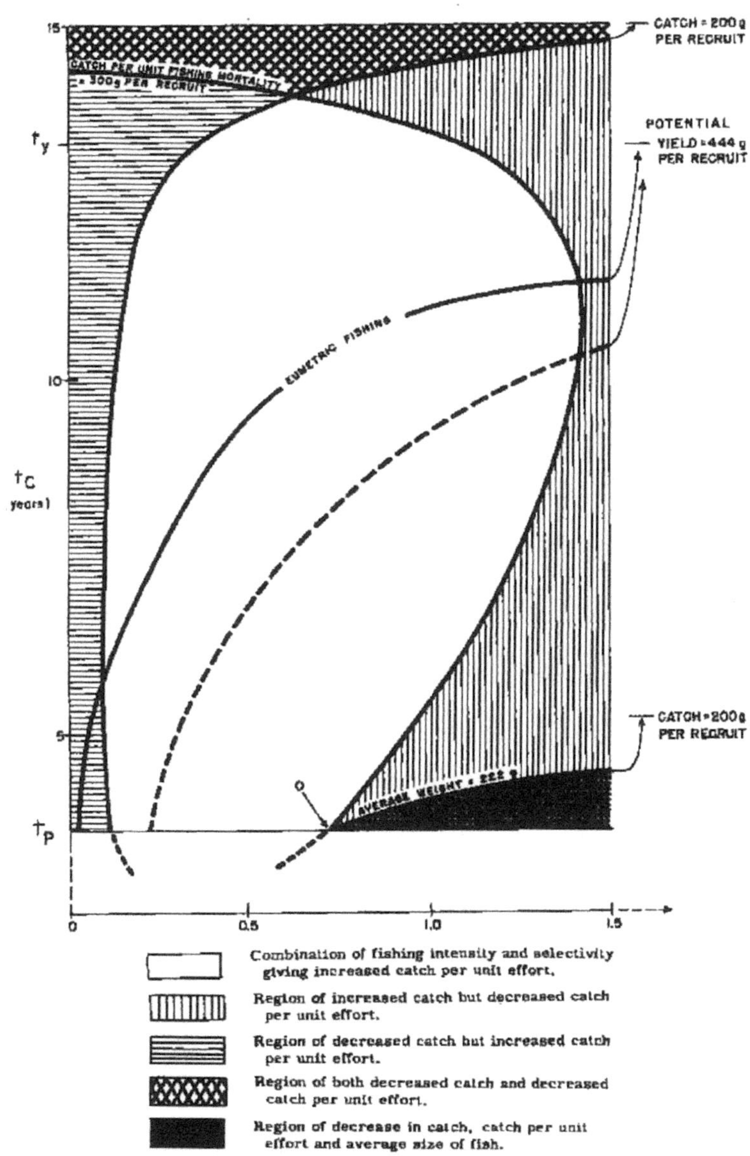

Figure 2. Evaluation diagram for North Sea Plaice *Pleuronectes platessa*.

Such diagrams can provide the basis for technical and economic evaluations by suitable transformation of the coordinates. Thus, F is proportional to standard fishing intensity, and catch and effort can be converted to value and cost to obtain curves for net economic gain. Diagrams for different stocks may be compared to show conditions under which returns from more valuable species begin to decline in relation to less valuable species, and diagrams for different types of fishing can be compared by superposing them directly or after transformation into cost-value diagrams. They thus serve to reduce to simple terms the essential information about the magnitude of the exploited stocks that is obtainable from the records of fishing operations and catch and from special research projects.

Similar presentations for mixed populations of two or more species are complicated but not different in principle.

Consideration of the situation where more than one species is taken by the same gear provides an example of this procedure and, incidentally, illustrates the practical value of distinguishing potential yield from catch. Figure 3 shows a combination diagram for both Plaice and Haddock in the North Sea. Since the relation between codend mesh size and t_c is different for the two species, the scale of the selectivity axis has for both been transformed to mesh size. The fishing mortality coefficient is expressed in relative instead of absolute terms, so that the two reference points of observed state O coincide at unity. In this case, the catch and

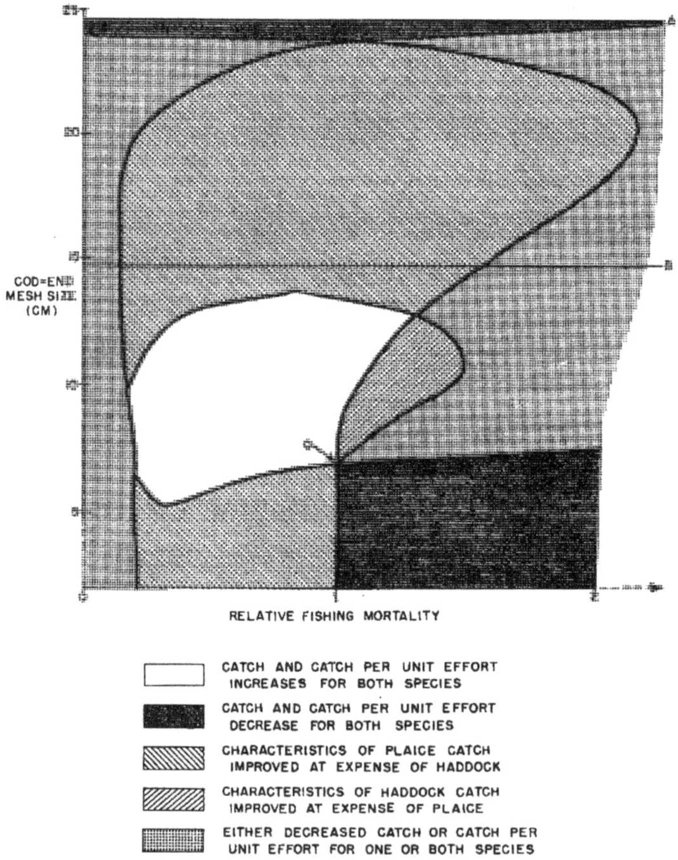

Figure 3. Combined evaluation diagram for North Sea Plaice and Haddock.

catch per unit effort lines only have been plotted, and they cut the diagram into several sectors showing different degrees of advantage or disadvantage. Not all possible combinations are distinguished here, certain groups of segments having been blended to show the general pattern more clearly. The significant feature is that although each stock has a characteristic potential yield, in practice, it would not be sound policy to attempt to manage the fishery to obtain these yields—this could only be done if species-selective gears could be used. This conclusion would indeed be even more obvious if we had plotted curves for, say, Sole *Solea solea*, a valuable fish which escapes more easily through large meshes. However, it will be seen that there is a considerable area (white on the diagram) within which there is scope to improve both characteristics of the catches of both species. The extent by which the catches obtained under such conditions fail to reach the combined potential yield of the two stocks can provide an objective index by which could be judged the value of special innovations, either in species-selectivity of gear or in the tactics of fishing so as to catch preferentially one species or another at times or places when they are segregated.

I suggest that, given the identification of a unit stock, we might restate the principal aims of studies in stock dynamics as: (i) the measurement of potential yield, as a standard for comparison with catches to characterize the general status of a fishery; and (ii) the determination of the topography of an evaluation diagram in the region of the point characterizing the present situation of an industry.

The leading questions then become "Is the catch a large or small proportion of the potential yield of that stock?" and "In what direction, and at approximately what rate, will the important characteristics of the catch change with changes (in either direction) of the amount and/or the selectivity of fishing?" To answer them, standard methods of assessment need to be worked out, comparable, for example, with the systems for routine assessment of forest increment. At present, much time is absorbed in trying to devise means of dispensing with such empirical measurements by using only time-series of catches and of fishing effort and a minimum of other information. It will surely be agreed that catch and effort data, no matter with what precision they are obtained, are not sufficient alone. The various methods seem to differ mainly in the kinds of other experience brought to bear on the problem and in the manner in which this is done. This experience may be expressed in the form of broad theories, such as that of logistic growth, based on studies of populations other than the fish stock in question; or it may consist of physiological or other observations than biodemographic ones on the stock in question or similar stocks. In any case, the time taken to make a reliable evaluation will vary inversely with the success gained in applying other kinds of data to this purpose. For example, suppose it is necessary to know, with a certain precision, the growth rate of fish in a stock. Von Bertalanffy's equation might be fitted to age–weight data and K and W_∞ calculated. But, if at least the order of magnitude of K is known from physiological experiments on that species, then fewer observations would be needed to find W_∞ with the same precision. If the physiological experiments were backed up by comparative studies of related species, the situation would be even better. Much depends upon the efficiency with which past experience is accumulated and digested and statistical population studies developed in close relation with other kinds of biological work.

Not only are data on catch and effort insufficient by themselves, but they may not even be the most important of the array of data required to assess resources. There seems no reason why the potential yield of a stock could not be measured before it is exploited. To find the relation between catch and potential yield, effort data are not indispensable, but to fix points and gradients on evaluation diagrams, it seems essential. If the methods of processing catch and effort data to compute particular parameters are the best (i.e., statistically the most efficient) that can be devised, then since we cannot increase the rate at which these kinds of data are produced, we can only try to reduce the sources of variability in them by seeking relevant information of a different kind. By the same reckoning, we may improve the testing of, for example, the effects of regulation or predictions of future events by paying attention not so much to fluctuations in catches as to changes in quantities such as mortality rates.

The prediction problem is essentially the estimation of steady-state conditions from data referring usually to a limited period of time and to transient situations. There are at least two ways of attempting this. Suppose it is required to predict the change in growth rate of fish that may be expected with change in population density caused by some sustained adjustment in the amount or kind of fishing. The fluctuations in growth rate occurring over a long run of years might be correlated with fluctuations in relative stock size, and then the presumption made that for steady-state situations and different average stock levels, the same regression lines will apply. Predictions could be made in a much shorter time from controlled experiments on food utilization of that particular species of fish, but the research effort required for this might well be greater. Some field test would have to be applied, and this would include measurement of available food. By appropriately combining the two approaches, the time required to reach a conclusion on which action might reasonably be based could be minimized.

In the above paragraph, a distinction has been made between total time and research time. As a guide for discussion, the research can be considered in three phases:

a) collection and tabulation of data,
b) calculation of population parameters, and
c) computation and presentation of evaluations in the form of diagrams, tables, etc.

The time factor in (a) has been discussed by Gulland (1955) in terms of sampling methods to reduce labor and bias and to increase accuracy. The relative extents to which effort can be reduced further by improved instrumentation, by international collaboration in reducing overlapping work, or filling in gaps in knowledge, and by standardization of practices also demand examination.

In phase (b), the various parameters can be considered in turn.

Mortality rates: The requirements in measuring these were discussed in detail in papers presented to the ICES Special Meeting on Sampling, 1954. One of the chief problems is standardization of fishing intensity measures such that catch per unit intensity is a valid index of stock density. With refinement of procedures of measurement, the concept of catch per unit effort is losing the economic significance it once had. Definitions of effort should be stabilized to recognize this, and Ricker's term "effective effort" used for that which is directly and simply related to the induced fishing mortality.

The time required for natural mortality estimation might be reduced by discovering the causes of mortality and their incidence, e.g., by measurement of morbidity rates and using the results to reduce the variance in regressions of total mortality (from age composition) on fishing intensity.

Growth: If there could be agreement to present growth data in a standard form, then much labor in analysis could be avoided. Von Bertalanffy's equation gives, apparently, a means of separating density-dependent and density-independent processes, of estimating parameters from data varying from year to year, of comparing stocks, and of making extensive use of physiological and ecological data. A systematic analysis of the available growth data to permit comparison between related stocks could reduce substantially the time required in future investigations.

Recruitment: The urgent need is for a method of examining the dependence of the number of recruits on the size of the stock which spawned them, other than by regression over a period of many years of varying stock size, because the apparent fluctuations in survival of pre-recruits are so great. If the sources of variance can be reduced by bringing in other kinds of related information, such as some index of relative favorability of the environment in different years, the time-period could be shortened (see Kesteven 1947).

In phase (c), it would seem to be more economical of effort, and perhaps more useful, to think in terms of slopes on response surfaces rather than curves of catch under varying conditions. This seems to be proving a very useful approach to the analogous problem of determining optimum conditions for industrial chemical reactions (Box 1954). Means of estimating yield and slopes on the "reaction surface", by special experiments comparable with Allen's method for determining eumetric points, might, therefore, be worked out.

Lastly, standardization of terms and symbols would simplify work and should not, at this stage, be difficult to achieve.

References

Allen, K. R. 1953. A method of computing the optimum size limit for a fishery. Nature 172:210.

Baranav, F. I. 1918. On the question of the biological basis of fisheries. Nauchnyi Issledovatelskii Ikhtiologischeskii Institut Isvestia 1:81–128.

Beverton, R. J. H. 1953. Some observations on the principles of fishery regulation. Journal du Conseil Permanent International pour l'Exploration de la Mer 19:56–68.

Beverton, R. J. H., and S. J. Holt. 1956. A review of methods for estimating mortality rates in exploited fish populations, with special reference to sources of bias in catch sampling. Rapports et Procès-Verbaux des Réunions du Conseil Permanent International pour l'Exploration de la Mer 140:67–83.

Box, G. E. P. 1954. The exploration and exploitation of response surfaces: some general considerations and examples. Biometrics 10:16–60.

Graham, M. 1935. Modern theory of exploiting a fishery, and application to North Sea trawling. Journal du Conseil Permanent International pour l'Exploration de la Mer 10(3):264–274.

Graham, M. 1951. Overfishing. Proceedings of the UN Scientific Conference on the Conservation and Utilization of Resources: Volume 7, Wildlife and Fish Resources 20–24.

Gulland, J. A. 1955. Estimation of growth and mortality in commercial fish populations. Fishery Investigations, London, Series 2 18(9).

Holt, S. J. 1958. The evaluation of fisheries resources by the dynamic analysis of stocks, and notes on the time factors involved. Pages 77–95 *in* Some problems for biological fishery survey and techniques for their solution (A Symposium held at Biarritz, France, 1–10 March 1956). ICNAF Special Publication No. 1.

International Technical Conference on the Conservation of the Living Resources of the Sea (ITC-CLRS). 1956. Papers Presented at the International Technical Conference on the Conservation of the Living Resources of the Sea: Rome, 18 April to 10 May 1955. United Nations Publication, Sales No: 1956.II.B.1. UN, New York.

Kesteven, G. L. 1947. Population studies in fisheries biology. Nature 159:10–12.

Kesteven, G. L., and S. J. Holt. 1955. A note on the fisheries resources of the North West Atlantic. FAO Fisheries Papers 7.

Ricker, W. E. 1944. Further notes on fishing mortality and effort. Copeia 1:23–44.

Rollefsen, G. 1953. The selectivity of different fishing gear used in Lofoten. Journal du Conseil Permanent International pour l'Exploration de la Mer 19:191–194.

Russell, E. S. 1931. Some theoretical considerations on the "overfishing" problem. Journal du Conseil Permanent International pour l'Exploration de la Mer 6:3–20.

Schaefer, M. B. 1953. Fisheries dynamics and the concept of maximum equilibrium catch. Proceedings of the Gulf and Caribbean Fisheries Institute 6:53–64.

Schaefer, M. B. 1956. The scientific basis for a conservation program. Pages 15–55 *in* Papers Presented at the International Technical Conference on the Conservation of the Living Resources of the Sea, Rome, 18 April to 10 May 1955. United Nations Publication, Sales No: 1956.II.B.1. UN, New York.

Tanaka, S. 1954. The effect of restriction of fishing effort on the yield. Bulletin of the Japanese Society of Scientific Fisheries 20(7):599–603.

Tester, A. L. 1953. Theoretical yields at various rates of natural and fishing mortality in stabilized fisheries. Transactions of the American Fisheries Society 82:115–122.

Part II

Saving the Great Whales

Foreword to
Part II: Saving the Great Whales

In 1960, Sidney's life changed forever when he was appointed as a member of the Committee of Three (K. Radway Allen, Douglas Chapman, and Sidney Holt) with the International Whaling Commission (IWC), also known as the "Three Wise Men" (a year later, the Committee was expanded to four with the addition of John Gulland). According to Sidney's own account (Holt 2011), this came about due to an inability of the members of the IWC to agree on action in the face of declining blue whale *Balaenoptera musculus* catches. The suggestion to form the Committee of Three came officially from the United Kingdom delegation, but it had been quietly promoted for some time by Donald Finn, Director of the United Nations Food and Agriculture Organization (FAO) Fisheries Division, and the American fisheries biologist William Herrington (both of whose names appear in the first and final parts of this book, respectively). The idea met with approval among some IWC delegations, as well as members of the IWC's Scientific Committee (SC); there were concerns not only about the independence of some government-appointed members of the SC, but also that the SC lacked the expertise in quantitative assessment techniques that had been developed in fisheries primarily, but not only, by Sidney (with Ray Beverton). It was felt that these techniques should be applied to the whale fishery.

FAO Biology Branch staff in 1960. Left to right: Bill Beckman, Sidney Holt, Pablo de la Fuente, unknown, Geoffrey Kesteven, Taivo Laevestu, unknown, and Erdogan Akyüz. Front: Horatio Rosa. Photo provided by Sidney Holt.

The final report of the Committee of Three (Chapter 6) provides a devastatingly clear analysis of the failures of an international management body, in this case the IWC, in adequately conserving the whale resources it was responsible for protecting. Lessons can be learned from the thoroughness and rigor of the analyses in the report and the clarity and inexorability of the presentation of its conclusions. This is not "old hat" science, but remains relevant for many management institutions even today.

Sidney often pointed out (see, for example, Chapter 42) that it is not evident today, when global whale catches are limited to a few thousand animals annually, just how important the whaling industry was in the 1950s in terms of production and economics. The 1950 catch of whales comprised a far-from-negligible portion of the total catch of marine species. Global whale catches in 1950 amounted to 2.5 million tonnes, whereas the FAO estimate of global fish catch for that year was only 20 million tonnes.[1] Antarctic catches had been declining for

[1]We are grateful to Robert Rocha for supplying data on whale catches and weights. See also Rocha et al. (2015).

some time—the global catch of whales in 1937, for example, had been 3.6 million tonnes. Many countries thus had an interest in rebuilding the industry and so were motivated to seek external scientific help.

Sidney wearing his "Holt Faction" cap, a label given him and three others by the Japanese delegation to the IWC. Photo by Tim Holt.

The United Nations Conference on the Human Environment held in Stockholm in June 1972 marked a turning point in the development of international environmental politics and law as well as the regulation of whaling. Among the many recommendations from that conference was one to strengthen the IWC, increase international research efforts, and call for a ten-year moratorium on commercial whaling. Sidney met whale activist Joan McIntyre at that conference; she was an advocate of the moratorium and kindled new thinking in his mind about whales and whaling. One of the major results of the Stockholm conference was the creation of the United Nations Environment Programme (UNEP), within which a Marine Mammals Project was established that Sidney was placed in charge of during 1973–1978.

In that capacity, Sidney was instrumental in the creation and organization of the Working Party on Marine Mammals of the FAO Advisory Committee on Marine Resources Research. That endeavor, beginning in 1972 at the FAO's Committee on Fisheries, brought together many of the pre-eminent marine mammal scientists of the era in a series of workshops over several years and culminated in the four-volume series *Mammals in the Seas*, published between 1978 and 1982. Two papers by Sidney in Volume 1 are included here as Chapters 21 and 22.

According to Sidney (Best 2010), a major turning point for him relative to his thinking about whales and whaling came in 1975 when he was invited to speak at a National Whale Symposium held at Indiana University in Bloomington, Indiana. This gathering of nearly 1,000 participants was organized by a university undergraduate student, John Goodman, and was a multidisciplinary public conference devoted to the study and celebration of whales, dolphins, and porpoises. Eminent environmentalists, biologists, musicians, and scholars were invited to speak about these mammals and their relationships to humanity. Preserving the endangered whales and public policy initiatives undertaken on their behalf were major themes of the conference. Attending the conference and, among other things, hearing for the first time the evocative sound of humpback whales *Megaptera novaeangliae* singing, recorded by biologist Roger Payne, introduced to him another way of looking at whales instead of through the lens of population dynamics. It had a lasting impact on Sidney's thinking, influenced his attitude towards whales, and, over the subsequent several years, he became convinced that the thing to do was not to regulate whaling, but to stop it! When he retired from FAO in 1979, his mission of actively promoting the cessation of commercial whaling began in earnest.

The impact of the Committee of Three's findings, the subsequent development, implementation, and failure of the New Management Procedure (NMP), and the path that led to the Revised Management Procedure (RMP) have been written about extensively elsewhere, including by Sidney (see, for example, Chapter 27 as well as his paper at the 2000 International Council for the Exploration of the Sea History Symposium—Holt 2002, here included as Chapter 34), and are beyond the scope of these introductory remarks. Rather, an attempt has been made to illustrate the scientific controversies and advances that occurred in the IWC from the 1960s through to the early 1990s via Sidney's own writings included in this part of the book. A brief chronology of relevant developments in the IWC is included below to help readers unfamiliar with the history of the regulation of whaling.

The chapters here include a broad range of articles by Sidney, including the final reports of the Committees of Three and Four, papers on specific questions such as whether to manage the whale fishery by numbers or by weight (Chapter 11), the applicability of MSY as a management objective (Chapters 16 and 21) for whales (with implications for the same question with respect to fish), and a number of longer review articles on subjects of particular interest. Because there were so many articles written by Sidney on whale issues, only a representative few could be used in order to illustrate his diversity of interest and propensity to address any problem that aroused either his anger or his disagreement with positions taken by other whale scientists.

One critically important issue from this period is, unfortunately, underrepresented in this selection, and that is the development of the RMP during the late 1980s and early 1990s. Sidney was intimately involved in this, beginning with his "recruitment" of William de la Mare from Australia to the whaling issue and the latter's 1985 presentation to the SC (de la Mare 1986) which demonstrated, very clearly, that the NMP had not been sufficiently well specified to deal with problems relating to, among others, various uncertainties; these meant that the NMP could still not guarantee that a stock would not experience heavy depletion. Though Sidney continued to produce papers for the SC on numerous matters, most of his involvement in the RMP was as an active member of the Scientific Committee's Sub-Committee on the Revised Management Procedure, which had been assigned the task of developing a catch limit algorithm (CLA) to form the basis of the RMP—he did not attempt to develop a candidate procedure himself. As he later noted (Chapter 39), "My role was to think of plausible scenarios that might cause the CLAs to fail to perform as desired; I was singularly unsuccessful." His submissions to the IWC on the conclusion of the selection process, included here, provide thought-provoking insights into how to prioritize management objectives and the respective roles of scientists and politicians in management. The short note (Chapter 26) on sensitivity of procedures to discrepancies in catches serves as a timely lesson on the frailties and vulnerabilities of human institutions, in this case an international management body. In Chapters 29 and 30, he discusses selection of the best procedure.

Certain of these chapters comprise discussions of issues which were of critical political and/or scientific importance at the time; though they may appear rather dated, they contain valuable insights into the politics and science of management (such as Chapter 13 on the utility of a quota for bottlenose whales *Hyperoodon ampullatus*). Others (Chapters 15 and 22) are detailed and meticulous descriptions of the properties and assumptions of the various models being used in stock assessment. One of Sidney's primary messages throughout this period was that the behavior of models, and their built-in assumptions—both explicit and hidden—must ceaselessly be reviewed, thoroughly understood, and constantly verified on the basis of the available data. The SC had the time, talent, and political motivation to explore these assumptions in detail, for there were those who wanted to kill more whales and those who wanted to kill fewer, and States in the plenary which would provide political support to both factions—hence, no serious view could be summarily dismissed. The (usually veiled) ferocity of the mutual examinations of models by the various camps in the SC doubtless led to better science and management.

As with his FAO period, Sidney's writings in the next part (Getting back to first principles) provide extensive accounts of the trials and controversies of the IWC and his thoughts about their relevance for fisheries. As he points out, it was only when the adoption of the moratorium on whaling, which came into effect in 1986, relieved the SC of the need to devote extensive effort to basic stock assessment, that it was free to give thought to entirely new ways of managing whaling. Another important factor was the constructively competitive nature of the process, with five teams of scientists vying to produce the "best" result, while each giving different emphases to one or other of the objectives, as decided by the politicians (IWC Plenary). The procedure eventually adopted by the IWC and referred to as the C-procedure was the one developed by Justin Cooke (Cooke 1996; see also Cooke 1999).

It was the extraordinarily intense international political attention focused on the IWC that led to the moratorium and fostered the developments described above. That particular condition, and the degree of scientific talent it attracted, was probably unique; while other fisheries scientific bodies (RFMOs and others) spend time analyzing models, assumptions, and biases in the data, it is doubtful that they approach the prolonged and intense scrutiny of the IWC Scientific Committee at that time.

The RMP, even though it remains unimplemented by the IWC to this day—for reasons having nothing to do with the Procedure itself—was a revolutionary development in fisheries management and is being increasingly adopted in various fora, often under the name of Management Strategy Evaluation. The fisheries expert (Sidney) coming to the whaling issue led to a major development in the science of resource exploitation management which was then adopted by the fisheries world—a fitting return to source.

Emory D. Anderson and Michael Earle
Editors

IWC Historical Timeline

1946	International Convention for the Regulation of Whaling establishes the IWC
1960	Agreement to form Committee of Three
1964	Committee of Four terminates work
1974	Resolution on management principles adopted
1975	New Management Procedure (NMP) adopted
1979	Indian Ocean Sanctuary established
1982	Moratorium decided
1985	De la Mare paper presented to SC (published in 1986)
1986	Moratorium legally in effect
1986	Scientific Committee began work on Revised Management Procedure (RMP)
1990	Planned completion of comprehensive assessment
1991	Adoption of C-procedure under RMP (IWC 1992)

Limericks by William de la Mare, 2004 (Lavigne 2006)

There once was a Briton named Holt
Whose studies gave whalers a jolt
He said "MSY
Is pie in the sky
And if you don't believe that you're a dolt"

One participant was Sidney
He says sustainable reaches infidney
He drew lots of curves
That unsettled our nerves
But he looked ever so clever, did'n he?

Artist: Marcy Saddy, London, ON, from a Tim Holt photograph, commissioned by David Lavigne.

References

Best, S. (Producer and Director). 2010. Sidney Holt: In His Own Words Part 1. Video by Animal Alliance of Canada. Available: https://www.youtube.com/watch?v=np2Eep99ohg. (March 2021).

Cooke, J. G. 1996. The International Whaling Commission's Revised Management Procedure as an example of a new approach to fishery management. Pages 647–657 *in* A. Schytte Blix, L. Walløe, and Ø. Ulltang, editors. Whales, seals, fish and man. Elsevier Science, Amsterdam.

Cooke, J. G. 1999. Improvement of fishery-management advice through simulation testing of harvest algorithms. Pages 797–810 *in* A. I. L. Payne, editor. Confronting uncertainty in the evaluation and implementation of fisheries-management systems. ICES Journal of Marine Science 56(6).

de la Mare, W. K. 1986. Simulation studies on management procedures. Reports of the International Whaling Commission 36:429–450.

Holt, S. J. 2002. ICES involvement in whaling and whale conservation, and implications of IWC actions. ICES Marine Science Symposia 215:464–473.

Holt, S. J. 2011. Save the whale! Memoirs of a whale hugger. Trolley Books, London.

International Whaling Commission. 1992. Chairman's Report of the Forty-Third Meeting, Appendix 4. Reports of the International Whaling Commission 42:47–48.

Lavigne, D. M. 2006. Gaining ground: in pursuit of ecological sustainability. International Fund for Animal Welfare, Guelph, Canada and the University of Limerick, Limerick, Ireland.

Rocha, R. C., Jr., P. J. Clapham, and Y. V. Ivashchenko. 2015. Emptying the oceans: a summary of industrial whaling catches in the 20th century. Marine Fisheries Review 76(4):37–48. doi: dx.doi.org/10.7755/MFR.76.4.3

Chapter 5

Establishing the Committee of Three[1]

Introduction

At its Twelfth Meeting held in London in June 1960, the International Whaling Commission requested its Chairman, in consultation with the Vice-Chairman and the Chairman of the Scientific Committee "to obtain the services of three scientists qualified in population dynamics or other appropriate science and drawn from countries not engaged in pelagic whaling in the Antarctic *to examine in the light of the advice and data available to the Commission the scientific evidence and to the extent justified by this evidence report as a committee to the Commission within one year of their appointment on the condition of the Antarctic whale stocks, on the level of sustainable yield that can be supported by these stocks and on any conservation measure that would increase this sustainable yield*." The Commission declared its intention, in setting up this special group of scientists, to bring, not later than 31 July 1964 "the Antarctic catch limit into line with the scientific findings, having regard to the provisions of paragraph 2 of Article V of the Convention." These scientists were asked to work with the then *Ad Hoc* Scientific Committee of the Commission, and to assist it in its allotted task to

(i) review the data available and make a statement on their deficiencies.

(ii) study the manipulative and analytical methods employed in whale research and propose ways of improvements and standardization; or if such proposals would appear premature, to indicate, where possible, research work leading to development of this kind.

(iii) study the data available on selected problems, making the fullest use of developed methods of population dynamics research, and seek to arrive at conclusions on these data.

(iv) review the methods at present in use for measurement and calibration of the efforts expended in capturing whales with a view to proposing ways in which these methods might be improved so as to provide more effective and reliable data and if necessary to propose details of a project of research in these matters.

(v) consider what specific collections and observations could usefully be made as a routine in whaling factories. For example, comprehensive collections of ear plugs for age determination and certain records of whales observed are desirable.

---

[1]Editors' note: this was taken from Appendix V "Reports of the Committee of Three Scientists on the Special Scientific Investigation of the Antarctic Whale Stocks," which was presented at the Fifteenth Meeting of the IWC held in London in July 1963 and published in Reports of the International Whaling Commission 14:32–106, 1964.

By March 1961, the following three scientists had been appointed to the special group:

- Dr. D. G. Chapman, University of Washington, Seattle, USA (subsequently elected by the group as its Chairman).
- Mr. K. R. Allen, Marine Department, New Zealand.
- Mr. S. J. Holt, Biology Branch, Fisheries Division, Food and Agriculture Organization of the United Nations.

The special group, hereafter referred to as the "Committee of Three," met several times and reported as follows:

Rome, 24 April–5 May 1961

This meeting was held simultaneously with the "Workshop" of the *Ad Hoc* Scientific Committee. A first interim report was prepared (6 May 1961) and circulated at the Thirteenth Meeting of the Commission (London, 19–23 June 1961) as Document IWC/13/9. It has since been published in the Thirteenth Report of the Commission, 1962, pp. 36–38.

The report of the Rome meeting of the *Ad Hoc* Scientific Committee is also pertinent, as it reviews the available data, sets out considerations on the measurement of fishing effort, on age determination and on marking, and gives the *pro formas* of data tables which were to be the basis for the work of the Committee of Three. It has also been published in the Thirteenth Report of the Commission, pp. 38–70.

Both reports from the Rome meetings were considered by the *Ad Hoc* Scientific Committee at its meeting in London, 13–16 June 1961, immediately preceding the Thirteenth Meeting of the Commission. This Committee's report (16 June 1961) has been published in an Appendix of the Twelfth Report of the Commission, 1961, pp. 22–31; and Appendix, Programme of Calculations, pp. 31–32.

Seattle, 6–19 December 1962

At this time, the Committee of Three met jointly with the Scientific Committee of the Commission, under the chairmanship of Dr. Chapman. Previously, Mr. Holt had met with the Scientific Committee at its meeting in London in July 1962 at the time of the Commission's Fourteenth Meeting, as a result of which additional data compilations and analyses had been put in hand.

Immediately after the Seattle meeting, the Committee of Three submitted to the Commission a Second Interim Report, dated 14 January 1963. This was circulated to Contracting Governments in Document IWC/15/7. In the Second Interim Report were given provisional conclusions concerning the sustainable catches in the 1962/63 season, and maximum sustainable yields. Subsequently, further analyses were made and further data processed, including those for the 1961/62 Antarctic season, and the results of these, as well as of the Seattle meeting, were embodied in the Final Report of the Committee of Three, circulated as Document IWC/15/9 and Addendum. Certain parts of Section 2 of the "Final Report" (dealing with analysis of data) and its Appendices are identical with Appendices of the Provisional Report (dated January 1963) of the Joint Meeting in Seattle of the Scientific Committee and the Committee of Three Scientists, circulated as Document IWC/15/6.

London, 24–27 June 1963

At this time, the Committee of Three met separately and also with the Commission's Scientific Committee, immediately prior to the Fifteenth Meeting of the Commission. Since preparing its Final Report, the Committee of Three had discussed certain problems with members of the Scientific Committee, had studied the discussion of the results of the Seattle meeting by the Scientific Subcommittee in London, April 1963, as reported in Document IWC/15/8, and had attended the discussion of the Final Report by the Scientific Committee and had the opportunity there to answer questions put to them. Furthermore, the statistical data for the 1962/63 Antarctic season had become available, and it was possible to reappraise the situation and present estimates of the sustainable yields of the stocks as at mid-1963. As a result, a supplementary report of the Committee of Three was prepared and circulated as Document IWC/15/14.

The Supplementary Report was considered together with the Final Report and with the Report of the Meeting of the Scientific Committee, 25–27 June 1963 (Document IWC/15/3) by the Commission at its Fifteenth Meeting in London in July 1963. The Commission authorized the publication of the reports of the Committee of Three and made arrangements for its work to continue until the Sixteenth Meeting of the Commission in 1964. At the same time, the group was augmented to a Committee of Four Scientists by the addition of Mr. J. A. Gulland, Fisheries Laboratory, Lowestoft, UK. It should be acknowledged here that Mr. Gulland had given most valuable assistance to the Committee of Three at the time of the Seattle meeting and thereafter, and his contribution to the analyses and calculations on which the Final and the Supplementary Reports were based was greatly appreciated.

The present Report comprises the texts of the Second Interim Report, the Final Report with its Appendices, and the Supplementary Report. These have been published essentially in their original form with only minor corrections and amendments, and after some editing. The Committee of Three wish to emphasize that the Report represents the work not only of the Committee members, but also of members of the Commission's Scientific Committee, both collectively and individually. In this connection, special mention should be made of analyses of the effects of weather on fishing effort made by Dr. Kesteven (Australia) and Mr. Gulland, and by Norwegian and Japanese scientists, and to the analysis of humpback whale *Megaptera novaeangliae* stocks by Mr. Chittleborough (Australia). Summaries of such analyses are presented in this Report because they provide essential links in the arguments on which our conclusions are based; these are, however, not full detailed accounts of the studies made, which may be published elsewhere by the scientists and national groups concerned.

Chapter 6

Special Committee of Three Scientists Final Report[1]

Section 1: Summary

A. Introduction

The Committee of Three made reports to the Commission following its first meeting in Rome and its second meeting in Seattle. These reports indicated the action taken at those meetings, the results of the studies made, and further plans. Since the Seattle meeting, further analyses have been made and further data processed by machine.

These further analyses have in no way contradicted, but only confirmed, the drastic need for action that was pointed out in the Second Interim Report. While this report is labeled final and is the definitive report as required by the Commission in its request to the Committee, it does not preclude the Committee providing supplementary information for the Commission when the Committee meets in June and has the 1962/63 season's data available to it for analysis.

In any case, it should be emphasized that while this may be the final report of the Committee of Three, it should not be the final study of Antarctic whale stocks. The analyses presented here are the Committee's best estimate of the situation based on current data. As new data become available, new and revised analyses must be made by the Commission's own scientists, which will serve to refine and sharpen these results.

B. Recommendations

The analysis given in Section II of this reports leads to the following best present estimate of the state of the Antarctic whale stocks:

Species	Catch in 1961/62	Sustainable catch in 1962/63 (would not change stock level)	Maximum sustainable yield when stocks have been permitted to rebuild	Years to reach the level of maximum sustainable yield if no catch after 1961/62
Blue (other than pygmy blue)	700	Less than 200	6,000	50 or more
Humpback Group IV and V[1]	810	Less than 100	Less than 1,000	50 or more
Fin	28,000	Less than 7,000	20,000	5 or more
Pygmy blue	460	-	Probably a few hundred	-

[1] Includes catches in temperate waters in 1961 and refers to total sustainable yields.

---

[1]Editors' note: this was presented at the Fifteenth Meeting of the IWC held in London in July 1963 as Document IWC/15/9 and was published as Chapman et al. (1964).

It is clear from this assessment that the blue whale *Balaenoptera musculus* and the Group IV and V humpbacks *Megaptera novaeangliae* are in serious danger of extermination unless adequate protective measures are taken immediately. Fin whale *B. physalus* stocks have been seriously depleted and are far below the levels of maximum sustainable yield.

Our recommendations, therefore, are:

1) That there be a complete cessation of catching of blue and humpback whales (Group IV and V) for a considerable number of years.
2) That the quota on fin whale catches be reduced to 7,000 or less. The greater the reduction in the present quota, the more rapidly will fin whale stocks rebuild to the level of maximum sustainable productivity.

Supplementary recommendations are:

3) The "blue whale unit" be eliminated and separate quotas established for each species.
4) Population analyses be continued, based on the pooled biological data of all the countries participating in Antarctic whaling.

The reasons for recommendation 3) were outlined in the Second Interim Report, but we repeat them here for emphasis. "While the concept of the 'blue whale unit' may have had some administrative convenience and given some apparent flexibility to the operation of the quota system in the past, the maximum sustainable yield can only eventually be obtained by taking the maximum sustainable yield for each of these species considered separately. If, even after stocks of all species had built up to the optimum level, the total catch taken in a certain season were equal to the maximum sustainable total catch, but the species were not taken in the correct proportions, then the stocks of one or two of them would be reduced below the optimum level and the others would be allowed to rise above the optimum, with a consequent reduction in the sustainable catch of each and all of them. While, as at present, all stocks are below the optimum level, any incorrect distribution of the current total sustainable yield would involve overexploitation of certain species leading to further reduction of their stocks or, at best, to deferment of recovery to the optimum level. In either case, this would result in a loss to the future total catch of that species which could never be recovered."

Section II: Analysis

A. Introduction

Since most of the analysis was done at the Seattle meeting and reported on in the appendices to the report of that meeting (IWC/15/6), this report necessarily repeats much of that material. Some parts have, however, been omitted as of secondary importance for the current analyses and some have been revised in the light of newer machine analyses. Some appendices of that report not included in the main body of this report are added as appendices here to facilitate reference to them. [*Editors' note: in the interest of space, these appendices are not included, which include those referenced in the next several paragraphs. Interested readers can find those in the Fourteenth Report of the Commission (https://archive.iwc.int/pages/view.php?ref=1814&k=).*]

Some information is available from marking and since it pertains to both fin and blue whales, this analysis is given first. Besides this, three general methods have been used: an analysis based on age–length keys (Beverton and Holt 1957 method: IWC/15/6/Appendix 1—now Appendix 1 of this report), an analysis based on the decline in catch per unit of effort (DeLury 1947 method), and an analysis based on estimation of the available "surplus" stock for each season or period of years (Schaefer 1954 method). Since most of the analyses relied heavily on measures of catch per unit of effort as indices of the abundance of the exploited stocks, calculations were made of the effect of weather on catcher efficiency (IWC/15/6/Appendices 3 and 4—now attached as Appendices 2 and 3 of this report) and of the increases in catcher efficiency over the years (IWC/15/6/Appendix 5—attached as Appendix 4 of this report).

That weather plays an important role in catcher success is well known. The studies of Gulland and Kesteven succeeded in measuring the effect of weather quantitatively. Further studies indicated that the weather index varies greatly from expedition to expedition and month to month. While such an index could greatly improve the analysis, time and cost prohibit its use in the present study. Such an index should be used in further studies.

The study of Hylen (Appendix 4) indicated that catcher efficiency is approximately proportional to tonnage. Machine analyses of the same question proved unsatisfactory because of variations in efficiency between countries, due to weather, and other causes. This is one of the problems that needs much future work.

Nevertheless, the tagging analysis confirms that efficiency of the catchers has been increasing so that the abundance is, in fact, declining faster than the catch per catcher days' work. However, analyses have been included based on several measures of catch per unit of effort, and these all lead to the same general conclusions.

B. Review of Theory of Fishing

In the early years of most fisheries, the catch is small in relation to the total number in the stock. This was true in Antarctic whaling. As the exploitation continues and intensifies, this proportion rises and the effect of exploitation in reducing the stock numbers becomes evident in a change in the catch per unit effort. This decline in stock is not, in itself, evidence of overexploitation, though it does indicate that catching (i.e., whaling) is becoming a major factor determining natural stock size. Whales in a stock of a particular size and composition have a certain capacity for reproduction and a certain rate of mortality. The difference between these, that is, the excess of reproduction and subsequent recruitment to the exploitable stock, over the natural deaths is the measure of the "surplus" population or, in other words, the catch which could be taken from that stock without either causing it to decline or allowing it to grow. This is what in the report we term the *sustainable yield* at any given time. The sustainable yield is zero in an unexploited stock. In such a stock, which is neither growing nor decreasing, effective reproduction and natural deaths must be balancing each other. As the stock is reduced by whaling, the rate of recruitment r (defined as the ratio of effective reproduction to stock size) must increase or the rate of natural mortality M (the ratio of natural deaths to stock size) must decrease or both; this process results in an excess of recruitment over natural mortality which can be taken as sustainable catch.

The instantaneous fishing mortality coefficient F is equal to the ratio of catch C over a certain period (say, a season) to the average total stock P during that period. Then, in any season, $C = F \times P$, but the fishing mortality to give the sustainable catch is $F_s = r - M$ and C_s

$= P(r - M)$. If, in any season, F exceeds F_s, the catch will be greater than the catch a stock of that size can sustain and the stock will decline. If the stock does decline, only an increasing amount of effort, to increase F proportionately will result in the previous catch level being maintained, and that only at the expense of a still further reduced stock. When the stock is very small, the sustainable catch is also very small. At some intermediate size, the sustainable catch is at a maximum. Our analyses have been essentially to estimate values of F, r, M, and P and to use these to determine for each stock:

a) the catch which the stock could, at its 1962/63 level, sustain without increasing or decreasing;
b) the maximum sustainable yield it might produce if permitted to rebuild;
c) the minimum time it would take for the stock to grow to the size at which it could sustain the maximum yield. (This would occur if whaling were temporarily suspended. If whaling continued, though at a reduced rate, such that catches each season were kept below the levels sustainable at that time, the stock could eventually attain the optimum size, but this would take longer.)

The precision and range of data required to make assessment b) are greater than those to make assessment a), but as the optimum level is approached, it may be estimated with progressively increasing confidence. The assessment given below for maximum sustainable yields should, therefore, be considered only as approximate estimates, while those relating to the present sustainable catches are more firm.

C. Marking Analysis

Total Mortality Between 1930s and 1950s

During the seasons 1957/58 to 1960/61, marks were returned from both pre-war and post-war markings. The chances of recovery of marks and fishing mortality during these seasons are presumably the same for both groups of marks. The difference in the percentages of the numbers originally marked which are returned during these four most recent seasons will, therefore, be due to the mortality of the pre-war marked whales before the time of post-war marking. The comparison of the percentages returned from pre- and post-war marking experiments has been done for three groups of fin whales—those marked in Area II, in Area III, and in all other areas. The data are given in Table 1. The ratio of the two percentages will be equal to e^{-Zt}, where t is the time between the mean dates of pre-war and post-war marking, and Z is the average annual total mortality coefficient during this period. Taking an average of t of about 20 years (from 1935 to 1955), average values of Z have been calculated and are shown in the table.

The average annual pelagic catch during the period in question was 14,300, so that if F was 0.11, the mean stock size was at least 130,000. The mean stock size index in this period (from Table 9) was 1.35, 1.00 being the index for 1950/51–1956/57, so that the stock size is estimated in the latter period to have been at least 130/1.35 or 96. If M was about 0.5, stock size in 1950/51–1956/57 was $96[0.11/(0.11–0.05)] = 176{,}000$.

Table 1. Comparison of the percentage returns in 1957 to 1960 of whales marked before and after the war.

Area		Pre-war	Post-war	Ratio of %	*Zt*	*Z*
II	N - number marked	2,205	262			
	n - number returned	16	8			
	% returned	0.726	3.053	0.238	1.43	0.071
III	N - number marked	749[1]	300			
	n - number returned	21	31			
	% returned	2.80	10.33	0.271	1.30	0.065
Other areas	N - number marked	289	276			
	n - number returned	3	54			
	% returned	0.038	19.56	0.053	2.94	0.147
Total	N - number marked	3,243	838			
	n - number returned	40	93			
	% returned	1.233	11.10	0.111	2.20	0.110

[1] Marking in 1934/35 and 1935/36 seasons only.

Comparison of Rate of Returns from Pre- and Post-War Marking

Returns, expressed as a percentage of the number marked returned during each season, will differ from experiment to experiment for the following reasons:

a) different fishing effort, in crude terms (e.g., number of catcher days)
b) different efficiency of the units of fishing effort (e.g., bigger and faster catchers)
c) changes in the preference for the particular species
d) changes in the proportion of marks in captured whales which are detected and returned.

a), b), and c) are real and reflect real differences in fishing mortality; a) can be eliminated by expressing the returns not as percentages, but as percentages per unit effort.

The available data are set out in Table 2. For fin whales, only returns in the area of marking and effort in the area of marking have been considered. The table gives the returns in seasons 1 and 2 (i.e., omitting the actual season of marking and taking the two following seasons). For blue whales, there are not enough returns to treat each area separately, and total returns and total effort have been used. Because of scarcity of returns, data for seasons 0 and 1 have been used instead of seasons 1 and 2, but in case season 0 returns are abnormal, data for the two years are shown separately.

Table 2. Rate of return of marked whales per unit effort.

	Fins Area II		Fins Area III		Blue (all areas)			
					Season 0		Season 1	
Parameter	**Pre-war**	**Post-war**	**Pre-war**	**Post-war**	**Pre-war**	**Post-war**	**Pre-war**	**Post-war**
Number marked	2,171	267	749	394	693	249	691	249
Number returned	45	19	6	10	22	26	9	13
% returned	2.1	7.0	0.8	2.5	3.2	10.4	1.3	5.2
% per unit effort	1.28	8.17	0.46	2.84	1.72	6.38	0.77	3.15
Mean year of return	1937	1956	1936	1957	1936	1957	1937	1958
Ratios of returns per unit effort	6.37		6.15		3.71		4.09	

The table shows that for blue whales, percentage returns per unit effort have increased fourfold, while for fin whales by more than sixfold. Changes in catcher efficiency and chance of recovery will be common to both species, and the difference is presumably due to the increased attention paid to fin whales. Assuming that the attention paid to blue whales has not decreased, then the increased attention paid to fins has increased the fishing mortality by a factor of:

$$(6.37 + 6.15)/(3.71 + 4.09) = 12.52/7.80$$

Changes in catcher efficiency and in the chances of recovery cannot be distinguished by this analysis, but the combined effect is an increase of fourfold.

Total Mortality and Fishing Mortality of Blue Whales

Of 3,243 fin whales marked pre-war (about 1937), 122 were returned in 1946/47 or later seasons. For blue whales, these numbers are 693 and 7, respectively. If the number returned are the same proportion of the number of marked whales remaining in 1947, then we have:

$$(3{,}243 \times 7)/(122 \times 693) = e^{-(Z_b - Z_f)t}$$

where t is time elapsed from 1937 to 1947, i.e., 10 years, and subscripts b and f refer to blue and fin whales, respectively. Hence, $Z_b - Z_f = 0.14$. Since the marking data indicate Z_f averaged 0.11 for years 1935 to 1955, Z_f for seasons 1936/37 to 1946/47 is < 0.10 because there was little whaling during the war years (in terms of catches, the six war years were equivalent to about one year of normal activity). So, $Z_b = 0.25$ for the years 1937–47. We now write

$$10(Z'_b - Z'_f) = 10(M'_b - M'f) + 5(F_b - F_f)$$

where F is the fishing mortality in a "normal" year, Z' is the corresponding total mortality, and M is the natural mortality. If we assume $M_b = M_f$, we find $F_b + F_f = 0.28$. But $M_f < 0.11$. If $M_f = 0.11$, then $F_f = 0$ and $F_b = 0.28$, at least. More reasonably, $M_f = 0.05$ so that $F_f = 0.06$ and $F_b = 0.33$, that is the average fishing mortality in the period 1937–47, excluding the war years, was 0.33 for blue whales.

D. Fin Whale Analysis

1(a) Estimation of Mortality Rates and Population Size from Catch Data

For the purpose of this analysis, each whaling Area (I to VI) was divided into two zones (north and south of 60°S). The age distribution in each zone in each month and for each sex was estimated from the length distribution using a combined age–length key for the whole Area and all months in the season. The age–length key was determined from age readings of ear plug laminations assuming two laminations per year. The validity of this assumption is reasonably established by comparison with other methods for the younger age classes. Because the assumption may fail for older animals and because length is a less satisfactory criterion of age for older animals, methods of analysis have been used that do not depend upon the exact breakdown of age data of older animals. Mortality coefficients were first estimated as the natural logarithm of the ratio of the catches per catcher day of whales 5 years old and older in a particular month and zone in one year to the catches per catcher day of whales 4 years old and older in the corresponding month and zone in the previous year. A probably better estimate of mortality was then obtained by adding to this first estimate a factor equal to the logarithm of the ratios of the mean catcher tonnages in the two seasons. This gave up to six estimates (two zones each for January, February, and March) in each area of the mortality between successive seasons. These estimates varied considerably; and to reduce the variance, the mean estimate for each area was calculated. Because the estimates in any one year appear to be highly correlated, this averaging still leaves more considerable variance, and in Tables 3a and 3b, the mean mortalities in 4-year periods are given for sexes separately and combined.

These figures have not all the same precision; those for Areas V and VI especially are based on few and variable estimates, and the most reliable figures are probably those for Areas II and III.

To break down these total mortality estimates into the two components, fishing mortality and natural mortality, it is customary to plot total mortality against effort. This requires a reasonably satisfactory measure of trends in effort throughout the whole period, and the corrections for changes in the efficiency of the effort are much more crucial than in the annual figures where changes in efficiency, selection, etc. will usually be small between any two consecutive seasons. For this reason, a method that does not require such assumptions is first used.

This method is based upon an independent estimate of M, the natural mortality rate. This can be obtained from the catch curves in seasons at the beginning of exploitation. Heincke's method of estimation has been applied to the catch curves for 1931/32, 1932/33, and 1945/46, i.e., seasons following a relatively small fin whale catch. It should be noted that there had

Table 3a. Estimated mortality rates of male, female, and both sexes combined fin whales: effort measured as CDW.

Period	Area I	Area II	Area III	Area IV	Area V	Area VI	Average all areas
Males							
1931/32–1933/34	-	0.026	0.636	0.808	-	-	0.247
1934/35–1937/38	-	0.102	0.088	0.189	-	-	0.068
1945/46–1948/49	-	0.048	0.014	0.033	0.126	-	0.024
1949/50–1952/53	-	0.049	0.114	0.092	0.179	0.196	0.070
1953/54–1956/57	0.772	0.380	0.132	0.296	0.130	0.078	0.298
1957/58–1960/61	0.414	0.237	0.339	0.533	0.548	0.486	0.426
Females							
1931/32–1933/34	-	0.126	0.310	0.345	-	-	0.260
1934/35–1937/38	-	0.062	0.211	0.195	-	-	0.156
1945/46–1948/49	-	0.104	0.130	0.115	0.112	-	0.063
1949/50–1952/53	-	0.033	0.024	0.141	0.209	0.258	0.031
1953/54–1956/57	0.473	0.078	0.202	0.208	0.044	0.181	0.198
1957/58–1960/61	0.075	0.227	0.362	0.460	0.462	0.422	0.335
Both sexes combined							
1931/32–1933/34	-	0.050	0.163	0.133	-	-	0.007
1934/35–1937/38	-	0.020	0.150	0.003	-	-	0.044
1945/46–1948/49	-	0.076	0.058	0.074	0.119	-	0.044
1949/50–1952/53	-	0.008	0.045	0.116	0.104	0.277	0.050
1953/54–1956/57	0.622	0.229	0.167	0.252	0.087	0.130	0.248
1957/58–1960/61	0.244	0.212	0.350	0.496	0.505	0.454	0.380

Table 3b. Estimated mortality rates of male, female, and both sexes combined fin whales: effort corrected for tonnage.

Period	Area I	Area II	Area III	Area IV	Area V	Area VI	Average all areas
Males							
1931/32–1933/34	-	0.039	0.649	0.093	-	-	0.260
1934/35–1937/38	-	0.062	0.128	0.149	-	-	0.028
1945/46–1948/49	-	0.026	0.060	0.107	0.200	-	0.098
1949/50–1952/53	-	0.089	0.074	0.132	0.139	0.156	0.030
1953/54–1956/57	0.806	0.414	0.166	0.330	0.164	0.112	0.332
1957/58–1960/61	0.450	0.273	0.375	0.569	0.584	0.522	0.462
Females							
1931/32–1933/34	-	0.113	0.297	0.332	-	-	0.247
1934/35–1937/38	-	0.102	0.251	0.235	-	-	0.196
1945/46–1948/49	-	0.030	0.204	0.189	0.186	-	0.137
1949/50–1952/53	-	0.007	0.064	0.181	0.011	0.218	0.009
1953/54–1956/57	0.507	0.112	0.216	0.242	0.078	0.215	0.232
1957/58–1960/61	0.111	0.263	0.398	0.496	0.498	0.458	0.371
Both sexes combined							
1931/32–1933/34	-	0.037	0.176	0.120	-	-	0.006
1934/35–1937/38	-	0.020	0.190	0.043	-	-	0.084
1945/46–1948/49	-	0.002	0.132	0.148	0.193	-	0.118
1949/50–1952/53	-	0.048	0.005	0.156	0.064	0.187	0.010
1953/54–1956/57	0.656	0.263	0.201	0.286	0.121	0.164	0.282
1957/58–1960/61	0.280	0.268	0.386	0.532	0.541	0.490	0.416

been some catch of fin whales prior to 1931/32, but mostly at land stations. These catches should, however, have not affected recruitment to any degree so that the catch curve estimates represent an upper bound to the natural mortality rate. Heincke's method (cf. Ricker 1958, p. 42) has been used because it is insensitive to inaccuracies in the older age groups. The results are shown in Table 4.

The 1945/46 catch curve is from a population that had been exposed to extensive catching through 1938/39, but the total wartime catches approximated one year's catch in the 1930s.

Table 4. Mortality estimates from catch curves: fin whales (Heincke's method).

Season	5 and up	6 and up	7 and up	Mean
		Male		
1931/32	0.048	0.049	0.045	0.047
1932/33	0.055	0.066	0.057	0.059
1945/46	0.067	0.076	0.059	0.067
Mean	0.057	0.064	0.054	0.058
		Female		
1931/32	0.048	0.066	0.051	0.055
1932/33	0.082	0.081	0.054	0.072
1945/46	0.104	0.104	0.060	0.089

It seems reasonable, on the basis of these data, to accept 0.05 (the 1931/32 average for males and females) as the best estimate of natural mortality, with 0.07 as a maximum estimate. Such estimates are in agreement with the results obtained from the analysis of tag recoveries (Section IIC).

Another method of estimating natural mortality is available, viz. from the ratios of catches of animals 5 and more years old in the 1938/39 season to catches of the same year classes at 12 and more years of age in 1945/46, or to catches at 13 and older in 1946/47. Estimates obtained in this way for females are shown in Table 5.

Table 5. Estimates of total mortality for female fin whales over the wartime period.

		1938/39–1945/46	1938/39–1946/47
All areas and months	Northern series	0.19	0.05
	Southern series	0.19	0.18
All areas and both series	January	0.17	0.06
	February	0.26	0.29
	March	0.10	0.10
All months, both series	Area II	0.17	0.18
	Area III	0.22	0.13
	Area IV	-	0.11
All areas and months, both series		0.19	0.14

These estimates reflect not only natural mortality, but also the element of fishing mortality that did occur during this time-span. A more serious difficulty is that they depend on the estimates of catches in the age classes 12 and over or 13 and over, respectively. For this reason, these estimates are not considered to be as reliable as those obtained from the catch curves.

While the best treatment would be one which deals separately with each population, because at the present time the separate stocks are not clearly defined and because the estimates are so variable, it is simpler to use the average mortality rate and apply it to the whole stock. Subtracting M, estimated at 0.05, from the total mortality estimate of 0.380 yields a fishing mortality rate (1957/58–1960/61 average) of 0.330. Since the average catch in these four seasons was 26,082, the resulting estimated total population (again an average for the four seasons) is 26,082/0.330 or 79,000. It should be emphasized that this is the estimated stock available to the fishery.

If the mortality rates are estimated using CDW (corrected for tonnage) as the measure of effort, then the resulting population estimate is 71,000.

1(b) Recruitment of Fin Whales

An approximation to the relation between the number of recruits and the number of the parent stock from which they are derived can be obtained by plotting the catch per unit of effort of the youngest fully recruited age group against the catch per unit of effort of the parents. A problem arises as to which age group should be selected for this purpose. A preliminary comparison was made of the rate of catch of 2, 3, and 4-year olds. It appears that 4-year olds are the youngest fully recruited age group. Since many whales are taken at ages 1, 2, and 3, an adjustment has been made to the index obtained from 4-year-old catches. The catch-per-unit-effort figures have been corrected for catcher tonnage and for partial effort as in Section IID1(a).

Table 6 shows stock indices for 4-year olds and the corresponding parent stock (females). Table 7 shows the summarized data used in the analysis and particularly the rate of recruitment. The female data are also shown in graphical form in Figure 1.

The monthly means and uncorrected overall means are the means over the series of the average catch per CDW. The corrected means have been divided by catcher tonnage/1000 and corrected for partial effort as in Section IID2.

It must be pointed out that the parent stock indices in these tables refer to the 5-year olds and older, not to the total stock being fishing upon. There has been some tendency for the proportion of the 5-year olds and older in the total catch to decrease. This is seen from Figure 2.

Table 6. Stock indices for 4-year olds (female) and corresponding parent stock (female)—5-year olds and older.

Year of capture	Parent stock	January	February	March	Mean	Corrected mean
1961/62		0.059	0.060	0.050	0.056	0.086
	1957/58	0.576	0.452	0.473	0.500	0..877
1960/61		0.086	0.072	0.054	0.070	0.110
	1956/57	0.689	0.616	0.369	0.558	1.018
1959/60		0.082	0.081	0.057	0.073	0.115
	1955/56	0.600	0.509	0.527	0.545	1.062
1958/59		0.112	0.082	0.104	0.099	0.165
	1954/55	0.539	0.445	0.425	0.470	0.919
1957/58		0.112	0.084	0.096	0.097	0.170
	1953/54	0.507	0.464	0.516	0.496	0.996
1956/57		0.088	0.078	0.058	0.075	0.138
	1952/53	0.397	0.367	0.384	0.383	0.783
1955/56		0.103	0.075	0.066	0.081	0.158
	1951/52	0.384	0.422	0.343	0.383	0.963
1954/55		0.114	0.091	0.068	0.091	0.178
	1950/51	0.317	0.298	0.218	0.278	0.725
1953/54		0.093	0.080	0.076	0.083	0.167
	1949/50	0.349	0.392	0.295	0.345	1.073
1952/53		0.084	0.068	0.053	0.068	0.151
	1948/49	0.361	0.369	0.302	0.344	1.113
1951/52		0.058	0.053	0.041	0.051	0.121
	1947/48	0.378	0.384	0.379	0.380	1.344
1950/51		0.040	0.038	0.033	0.037	0.112
	1946/47	0.349	0.339	0.305	0.331	1.534
1949/50		0.041	0.047	0.043	0.044	0.138
	1945/46	0.282	0.308	0.302	0.297	1.233
1938/39		0.034	0.043	0.046	0.041	0.241
	1934/35	0.195	0.265	0.250	0.237	2.135
1937/38		0.050	0.056	0.049	0.052	0.265
	1933/34	0.260	0.126	0.264	0.283	4.188
1936/37		0.030	0.050	0.042	0.041	0.314
	1932/33	0.104	0.264	0.182	0.183	3.444
1935/36		0.036	0.042	0.030	0.036	0.396
	1931/32	0.048	0.196	0.157	0.134	3.129

Table 7. Stock indices for parent stock in a given season four years later.

Season	Index of parent stock (female)	Index of recruitment four years later (female)	Four-year-old recruitment rate
1957/58	0.877	0.096	0.109
1956/57	1.018	0.110	0.108
1955/56	1.062	0.115	0.108
1954/55	0.919	0.165	0.180
1953/54	0.996	0.170	0.171
1952/53	0.783	0.138	0.176
1951/52	0.963	0.158	0.164
1950/51	0.725	0.178	0.246
1949/50	1.073	0.167	0.156
1948/49	1.113	0.151	0.136
1947/48	1.344	0.121	0.090
1946/47	1.534	0.112	0.073
1945/46	1.233	0.138	0.104
1934/35	2.135	0.241	0.113
1933/34	4.188	0.265	0.063
1932/33	3.444	0.314	0.091
1931/32	3.129	0.396	0.127

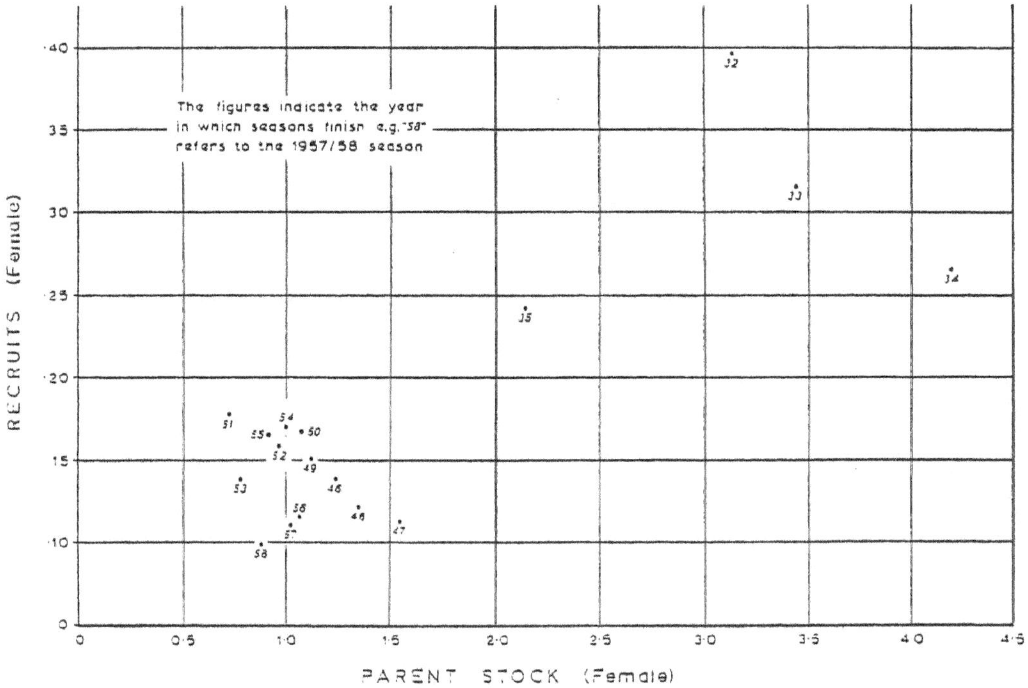

Figure 1. Graphical representation of the relationship between recruits and parent stock of female fin whales.

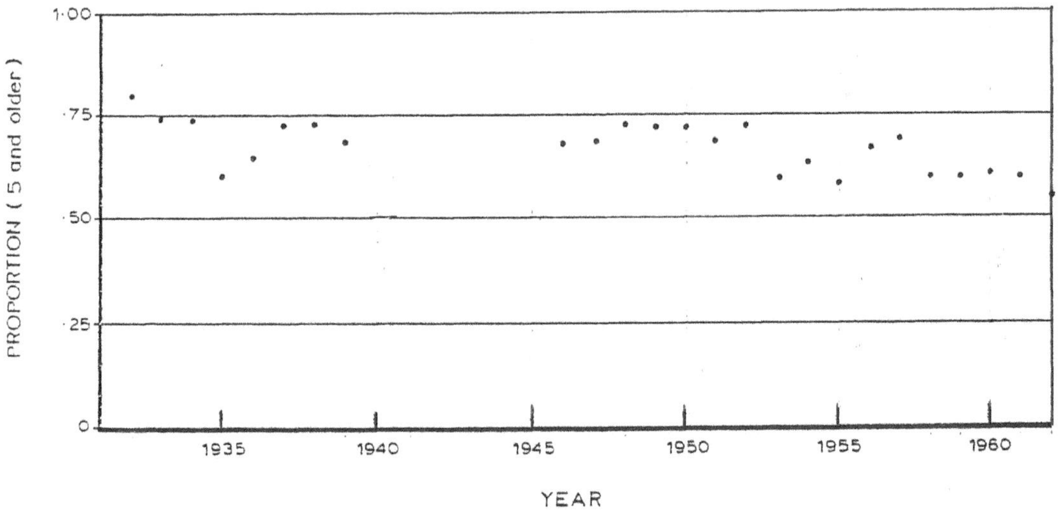

Figure 2. Proportion of 5-year olds and older fin whales in total catch, 1931/32–1961/62.

Referring to Table 7, the pre-war stock indices seem to be aberrant. Partly, this is due to the aspect just noted, but this does not account for all of the discrepancy. It is probable that the corrections made in the measurement of effort are not satisfactory in the pre-war data. For this reason and because a large correction must be made for the 2- and 3-year-old catches, analyses have been based primarily on recent data and only simple methods have been used.

The year classes 1949–57, which are measured as recruits in the seasons 1952/53 through 1961/62, were produced by parent stocks of approximately the same size. The average 4-year-old recruitment rate of these ten year classes is determined to be 0.155.

To obtain the true recruitment rate, i.e., to adjust for catches at earlier ages, again a simple averaging approach is used. Since a good estimate of the average population size is necessary, and this is available for the 1957/58 to 1960/61 seasons [cf Section IID1(a) above], the adjustment will be based on data from these years. The required basic data are shown in Table 8.

From Section IID1(a), the estimated average population available to the fishery in 1957/58 to 1960/61 was 79,000. Of this, half are estimated to be female and 59.3% mature (i.e., 5 and older). Hence, the number of 5+ females is estimated to be (79,000) (0.50) (0.593) or 23,400. Now, this is equivalent to the average index value (catch per unit of effort) 0.756, so that the relationship between population in numbers (N) and population as measured by catch per unit effort (I) is $N = 0.310(10^5)I$. Now, the average recruitment index for this time-period is 0.140, which is, therefore, equivalent to a recruitment (female) of 4,340.

To find an adjustment factor, it is necessary to calculate the recruitment at age 4 (in numbers) together with the earlier catches. These earlier catches would have been subject to additional natural mortality if not caught, and this must be allowed for. The estimated total recruitment is:

Average 4-year-old recruitment (estimated)	4,340
Average 3-year-old catch less 5% mortality	1,604
Average 2-year-old catch less 2 years mortality at 5% per year	1,484
Average 1-year-old catch less 3 years mortality at 5% per year	353
Total	7,781

Now, total recruitment is 7,781/4,334 or 1.79 times the 4-year-old recruitment; consequently, the rate calculated above can be adjusted by multiplying by 1.79. Multiplying 0.155 by 1.79 yields 0.277 as the estimated recruitment rate applying to the population of 5-year olds and older in recent years.

In further sections, it will be convenient to apply a recruitment rate to the whole stock available to the fishery. Since the 5-year old and older animals form, on the average, about 60% of the available stock, the recruitment rate for the whole stock is easily obtained by

Table 8. Data for adjustment of recruitment estimate due to 2-year-old catch.

	1957/58	1958/59	1959/60	1960/61	Average
Catch at age 1 (estimated)	423.1	435.2	370.8	417.1	411.6
Catch at age 2 (estimated)	1,692.7	1,794.0	1,481.8	1,610.3	1,644.7
Catch at age 3 (estimated)	1,799.4	1,763.3	1,517.5	1,672.3	1,688.1
Catch at age 4 (estimated)	1,455.2	1,401.2	1,220.5	1,404.4	1,370.3
Catch per unit of effort of mature females (as in Table 2)	0.877	0.852	0.694	0.600	0.681
Catch per unit of effort of 4-year-old females (from Table 2)	0.170	0.165	0.115	0.110	0.126
Ratio of 5-year-old female catch to total female catch	0.569	0.586	0.617	0.600	0.593

multiplying the rates determined above by 0.6. Hence, the estimated recruitment rate for the whole stock is 0.166.

This estimate of r is based upon the population estimate using CDW as a measure of effort. If the estimate of 71,000, based on tonnage adjustment, is used, similar calculations give an estimate for r for the whole stock as 0.175.

2. Estimates of Fin Whale Stocks—DeLury Method

Catches of fin whales expressed as catch per unit effort for all areas show, after an initial rise, a fairly steady period and then a fall, as in the following table, which shows catch per unit of effort both for the situation where effort is measured in catcher day's work and where it is corrected for changes in catcher efficiency by dividing by the annual mean tonnage of the catchers/1,000.

Season	Catch (1000s)	C[1] E	C[2] E	Season	Catch (1000s)	C[1] E	C[2] E	Accumulated catch (1000s)
1949/50	17.9	0.99	2.33	1955/56	25.1	1.69	3.29	25.1
1950/51	17.3	0.91	2.00	1956/57	25.5	1.65	3.03	50.6
1951/52	20.3	1.20	2.54	1957/58	25.1	1.51	2.68	75.7
1952/53	21.0	1.23	2.49	1958/59	25.7	1.58	2.64	101.4
1953/54	24.7	1.58	3.17	1959/60	26.3	1.24	1.96	127.7
1954/55	25.6	1.54	3.01	1960/61	27.3	1.14	1.78	155.0
				1961/62	26.4	0.88	1.35	181.4

[1]Effort measured in catcher's day's work (CDW).
[2]Effort corrected by dividing by mean tonnage/1,000.

For the second period (1955/56) onward, catch per unit effort shows an almost linear relation with accumulated catch. A crude application of the DeLury method might be made, but this would neglect recruitment and natural mortality, which are clearly not negligible here.

The following method provides a basis for allowing for these. This method assumes:

1) that over the period concerned, recruitment has been proportional to the stock three years earlier,
2) that over the period, natural mortality in each year has been a fixed proportion of the initial stock,
3) that the stock was constant for the five years ending in 1955/56,
4) that the stock in any season is proportional to catch per unit effort as corrected; it ignores the effect of any change in selection from blue to fin whales.

On this basis, the stock in 1961/62 can be considered to be the stock in 1955/56 + recruitment in the interval – natural mortality in the interval – catch. This can be written as:

$$S_6 = S_0 + r(3S_0 + S_1 + S_2 + S_3) - M(S_0 + S_1 + S_2 + S_3 + S_4 + S_5) - (C_0 + C_1 + C_2 + C_3 + C_4 + C_5)$$

where X = catch per unit effort, r = recruitment as a proportion of parent year class, M = annual natural mortality rate, C = catch, S = stock, and subscripts 0–6 represent seasons 1955/56 to 1961/62. Dividing by S_0, we get: $S_6/S_0 = 1 + r[3 + (\Sigma_1^3 S)/S] - M(\Sigma_0^5 S/S_0) - \Sigma_0^5 C/S_0$ or $X_6/X_0 = 1 + r[3 + (\Sigma_1^3 X)/X_0] - M(\Sigma_0^5 X/X_0) - \Sigma_0^5 C/S_0$ and hence

$$S_0 = \sum_0^5 C / [1 + (X_6/X_0)] + [3 + (\sum_1^3 X / X_0)] - (M \sum_0^5 X/X_0$$

By inserting values of X and C, we get:

1) $S_0 = 155.0/[0.480 + 5.22(r - M) + 0.59r]$ if effort is measured by catcher day's work.

2) $S_0 = 155.0/[0.590 + 4.67(r - M) + 0.87r]$ if effort is corrected for tonnage as indicated.

Inserting various possible values of $(r - M)$ and r in these, we get estimates of S_0, the stock in 1955/56 as set out in the tables (in thousands).

DeLury estimate of S_0 (stock size 1955/56); effort – CDW

	r					
r* – *M	**0.100**	**0.125**	**0.150**	**0.175**	**0.200**	**0.025**
0.050	194	190	187	184	180	177
0.075	166	164	161	159	157	154
0.100	146	144	142	140	138	136
0.125		128	127	125	124	122
0.150			115	113	112	111
0.175				104	102	102
0.200					94	94

DeLury estimate of S_0 (stock size 1955/56); effort – CDW/(mean tonnage/1000)

	r					
r* – *M	**0.100**	**0.125**	**0.150**	**0.175**	**0.200**	**0.025**
0.050	171	167	163	159	155	152
0.075	149	147	144	140	138	136
0.100	139	132	130	128	126	124
0.125		120	118	117	115	113
0.150			109	107	106	104
0.175				100	98	97
0.200					91	90

In Section IID1(b), the recruitment rate was estimated to be 0.169 for the whole herd. Putting $r = 0.169$, $M = 0.05$ in the first formula above (i.e., when effort is measured by CDW) yields $S_0 = 129{,}000$. Since the average catch per unit of effort was 1.69 in 1955/56, while the 1957/58 to 1960/61 average was 1.3725, a population of 133,000 in 1955/56 corresponds to a 1957/58 to 1960/61 average of 105,000.

Alternatively, if the second formula is used (i.e., effort is corrected for tonnage), $r = 0.175$, $M = 0.05$ yields $S_0 = 117{,}000$. The catch per unit effort, with effort corrected for tonnage in 1955/56 was 3.29, while the 1957/58 to 1960/61 average was 2.265. This implies an estimate of 81,000 for the four-season period.

3. Fin Whale Stock—Estimation of Sustainable Yield by Schaefer's Method

The method uses data for catch and catch per unit effort in each season and an estimate of fishing mortality on stock size at some time during the period under consideration. Total Antarctic and pelagic catches and corresponding pelagic catches per catcher-ton-day's work were analyzed. In earlier years, fin whale abundances are progressively underestimated because of the preference for blue whales when the latter were relatively more abundant. This effect is related in part to a northward shift in later years in the latitude of fishing and can be reduced by calculating catches per unit effort in averages of those for the different series of squares treated separately. It was found that an equivalent correction could be made by multiplying the overall pelagic catch per unit effort in each season by the ratio of all species to fin whales in the catch in that season for seasons up to 1953/54, after which time the blue whales had become so scarce that there was effectively no selection against fin whales, and the total:fin ratio stayed at about 0.9 except for disturbances caused by rather large catches of humpback or sei whales *B. borealis* in certain recent years. Table 9, column A shows the series of catches and stock sizes analyzed; the latter are expressed as fractions or multiples of the average stock in the period 1951/52 to 1955/56, when it was relatively unchanging. (Values in parentheses for certain seasons were derived from catch per catcher-ton, suitably adjusted, instead of catch per catcher-ton-day.)

For each season, a sustainable yield, or "surplus," was calculated in the usual way as the sum of the catch in that season and half the difference between the stock size in the preceding season and the following season. This is an estimate of the total catch which could have been taken in that season without, on the one hand, causing a decline in stock or, on the other hand, permitting a rise in it.

Table 10 summarizes the results for three post-war five-year periods, and the pre-war period for which there are data, giving the average stock size index in each period and an estimate of it in the preceding period five years earlier, with related sustainable yields for three hypotheses as to the average stock size in the period 1950/51 to 1956/57 and the corresponding average fishing mortality coefficient F.

The harvestable "surplus" in any season is produced by recruitment during it and since the preceding season, and this is a function of the stock about five years earlier. The results show this function follows the expected pattern. In the 1930s, when the rate of exploitation of fin whales was rather light and the stock was large and near to its unexploited level, the sustainable yield was low, with hardly any excess of recruitment over natural mortality. The situation was little changed in the first years after the war, though the stock was rather smaller than in the 1930s, not having, apparently, recovered much from a decline which began in the later 1930s. In the second five-year period after the war, the stock was reduced a little more, catches were greater, and the sustainable yield was also greater, indicating a greater excess of recruitment over mortality, probably mainly the result of an increase in the pregnancy rate.

In the third post-war five-year period, the stock was reduced still further and the sustainable yield declined sharply. The annual catches were greater than ever before and considerably in excess of the sustainable yield of the reduced and declining stock at that time.

In Figures 3–5, the sustainable yield has been plotted against the stock in the previous period for each of the three hypotheses as to the stock size in the period 1950/51 to 1956/57. Stock sizes, in thousands of fin whales, of 180, 135, and 90 in this period correspond with 349, 262, and 175, respectively, in 1932/33 to 1937/38. For comparison, the value from sightings, thought to be an underestimate, was 165, with a possible range of 107–225. On this evidence, the second or third values are probably nearer the truth than the first.

On the graphs have also been plotted diagonal lines representing various values of excess of recruitment over natural mortality, expressed as instantaneous rates $r - M = 0.1$, 0.2, and, in

Table 9. Fin whale analysis by Schaefer's method: basic data.

Season	Annual catch (thousands)		Annual indices of stock size	
	Total Antarctic	**Pelagic**	**A**	**B**
1919/20–1923/24	3.8	2.0	-	-
1924/25–1928/29	5.9	3.2	-	-
1929/30	11.5	8.1	1.39 }	-
1930/31	10.0	8.6		-
1931/32	2.9	1.1	1.67	0.45
1932/33	5.2	4.4	1.78	0.57
1933/34	7.2	5.5	1.96	0.86
1934/35	12.5	11.7	1.91	1.10
1935/36	9.7	9.2	2.08	0.83
1936/37	14.4	13.3	1.91	1.12
1937/38	28.0	26.4	1.84	1.45
1938/39	20.8	19.5	1.32	0.97
1939/40	18.7	17.8	(1.48) }	-
1940/41	7.8	7.1		-
1941/42	1.2	0.0		-
1942/43	0.8	0.0		-
1943/44	1.2	0.5	(1.36) }	-
1944/45	1.7	0.7		-
1945/46	9.2	7.7	1.42	1.27
1946/47	14.5	12.9	1.37	1.01
1947/48	21.1	18.9	1.24	1.12
1948/49	19.1	17.1	9.97	0.87
1949/50	20.1	17.9	1.02	0.80
1950/51	19.5	17.3	0.90	0.82
1951/52	22.5	20.3	1.01 }	0.83 }
1952/53	22.9	21.0	0.92 }	0.87 }
1953/54	27.7	24.7	1.05 } 1.00	
				1.07 }
1954/55	28.6	25.6	1.00 }	1.01 }
1955/56	28.0	25.1	1.02 }	1.22 }
1956/57	27.8	25.5	1.00 }	1.02
1957/58	27.5	25.1	0.89	0.89
1958/59	27.1	25.7	0.88	0.75
1959/60	27.6	26.3	0.65	0.59
1960/61	28.8	27.3	0.59	0.57
1961/62	7	26.4	0.45	0.41

Table 10. Fin whale analysis by Schaefer's method: results under different hypotheses.

	Hypotheses			A	B	C
	If stock (thousands) in 1950/51–1956/57 =			**180**	**135**	**90**
	***F* in 1950/51–1956/57 =**			**0.14**	**0.19**	**0.28**
Seasons	**Mean catch (thousands)**	**Mean stock index**		***"Surplus" (thousands) (sustainable yield)***		
		Current period	**Five years earlier**			
1932/33–1937/38	12.8	1.94	1.80	2.5	5.9	7.8
1945/46–1949/50	16.8	1.20	1.40	1.2	6.4	8.9
1950/51–1954/55	24.2	0.98	1.20	26.2	25.5	25.2
1955/56–1959/60	27.6	0.89	0.98	13.4	18.2	20.7
1960/61	28.8	0.59	0.99	10.4	16.6	19.6

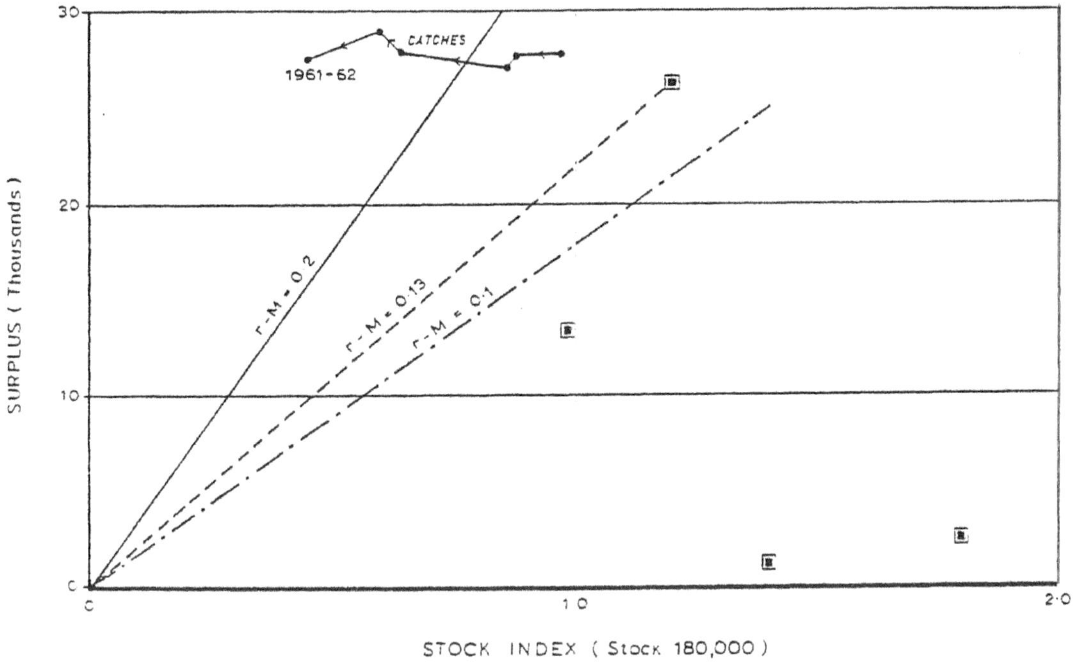

Figure 3. Fin whale. Schaefer method (hypothesis A).

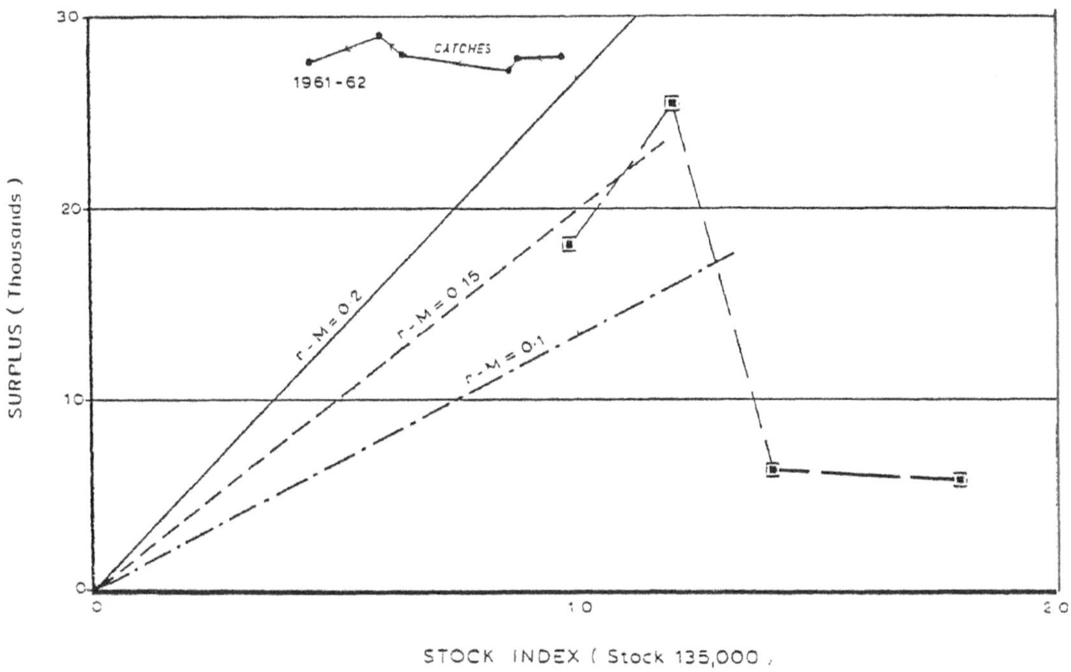

Figure 4. Fin whale. Schaefer method (hypothesis B).

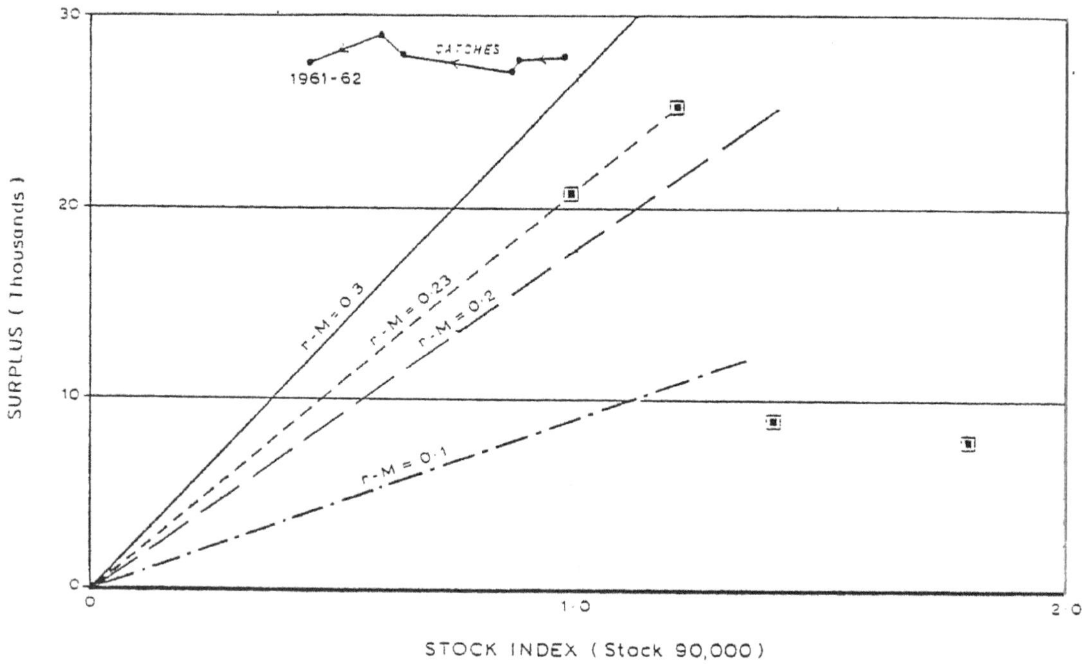

Figure 5. Fin whale. Schaefer method (hypothesis C).

Table 11. Sustainable yields in recent years.

Season	Catch	Sustainable yield in that season not more than (thousands)					
1961/62	28.0	10	16	9	12	9	11
1960/61	28.8	13	21	12	15	12	16
1959/60	27.6	14	-	13	17	13	17
1958/59	27.1	19	-	18	24	18	24
1957/58	27.5	19	-	18	24	19	24
Stock in 1950/51–1956/57		180		135		90	
$r - M$ not greater than		0.13	0.20	0.15	0.20	0.23	0.30
Years for recovery (minimum) (see text)		7	4 or 5	6	4 or 5	4	3

one case, 0.3. The ascending left-hand limb of a curve relating sustainable yield to stock must be below and to the right of whichever such diagonal is the correct one for the very reduced stock. Examination of sex ratios, pregnancy rates, and ratios of mature to total stock show the greatest rate of reproduction cannot be greater than 20–25%, and $r - M$ is, therefore, almost certainly less than 0.2 and, if M is not negligible, may even be less than 0.1.

The curve of sustainable yield against stock must, of course, start at the origin (or, if reproduction were not possible at all below a certain small, but finite, stock size, a little to the right along the "stock" axis), and it appears to rise steadily towards a maximum of not more than about 25,000. Then, the right limb descends rather sharply to cut the "stock" axis at a point corresponding with an exploited stock about double the 1950/51 to 1956/57 level.

On the graphs are also plotted the actual catches in each of the recent seasons 1956/57 to 1961/62. For all three hypotheses, these catches clearly greatly exceed the catches which the stock in each of these years could sustain. Table 11 above gives for comparison, and for each hypothesis, estimates of the greatest catches in each of the first five seasons which the stock *could* have sustained.

The last row of Table 11 gives estimates of the number of years it would take to reach a stock size near to that which would give the maximum sustainable yield of about 20–25,000 starting from the level of 1961/62, if there were no more fin whales caught until that time had elapsed. If there were any caught after 1961/62, even if these were less than the stock could sustain in the year in which they were caught, the stock would take longer to reach the level for maximum sustainable yield. Thus, in a new analysis, the catches of the 1962/63 season must be taken into account in estimating permissible catches in 1963/64 and subsequent seasons.

Analysis by areas and series

To determine whether the above conclusions were substantially biased by the procedures and to obtain indices of stock size in each year, calculations were made from catches for catcher-ton-day in each area and latitudinal series. No different trends were found in the different areas and so data for these were combined. There were larger discrepancies between series, especially in the earlier years. It appeared that the catch per unit effort was lower in the northern than in the southern series in these years, relative to the catches in later years in each series. It was suggested that this was because the small catchers at that time could not work effectively in the relatively bad weather of the northern waters and, therefore, did better nearer to the shelter of the ice edge. The catch per unit effort in the southern series was, therefore, considered as an index of stock size, but even here there is evidently an underestimation of the fin whale abundance in the earlier years when blue whales were abundant and preferentially sought. The results of applying Schaefer's method to the southern series data tabulated in Table 9, column B are given in Table 12 below. They do not lead to conclusions different in substance from those obtained by the previous analysis.

4. Summary of Fin Whale Analysis—Maximum Sustainable Yield

The analyses of the earlier sections show that if effort is corrected for tonnage, the two different methods—catch analysis and DeLury method—yield population estimates that are quite close (71–81,000 as the 1957/58–1960/61 average). The agreement is not so good for the estimates obtained when effort is measured by CDW (79 vs. 105,000). These estimates, in turn, imply the population of fin whales available to the fishery in the early 1950s was from 100 to 140,000.

Table 12. Application of Schaefer's method to fin whale total catches, and abundance estimates derived from catch per unit effort in southern series.

	1950/51–1956/57			Stock (thousands)		
				180	135	90
	Mean catch	Mean stock index		"Surplus" (thousands)		
Period	(thousands)	Current period	Five years earlier	(sustainable yield)		
1932/33–1949/50 (excluding war years)	15.0	1.00	1.10	19.3	18.3	17.1
1950/51–1954/55	24.2	0.92	1.00	33.4	31.1	28.2
1955/56–1959/60	27.6	0.89	0.92	8.3	13.2	18.0
1960/61	28.8	0.57	0.89	12.6	16.7	20.7

If the more reliable estimates, 71–81,000 or an average of 76,000 are accepted, then it may be estimated that the 1961/62 population was 1.35/2.265 or 0.60 of this. The ratio 1.35/2.265 is obtained by comparing the catch per unit of effort (corrected for tonnage) in 1961/62 with the average of the previous four seasons. The resulting estimate of the 1961/62 stock is 46,000. If the net recruitment rate ($r - M$) is 0.119, a stock of 46,000 would provide a sustainable yield of 5,500.

If the somewhat less pessimistic population estimates based on CDW as an effort measure are used, the average of the two estimates for the 1957/58 to 1960/61 period is 92,000. The corresponding estimate for 1961/62 is (0.64) (92,000) or 59,000. If $r - M = 0.125$, a stock of 60,000 could sustain a catch of 7,400.

The estimates of the maximum sustainable yield are much less firm at this time. Additional analysis is necessary to measure populations and recruitments more accurately in the pre-war and early post-war years. The Schaefer analysis suggests the maximum sustainable yield is about 25,000. This probably is too high since almost immediately after catches began to exceed 25,000, catch per unit of effort began to decline.

Another estimate can be obtained from the analysis of Section IID2 (DeLury). From this analysis, it was estimated that the population in 1955/56 was 117–129,000. With a net recruitment rate ($r - M$) of 0.119 to 0.125, this implies a sustainable yield of 14–16,000.

The best value that can now be adopted is the mean of these two estimates, roughly 20,000, though it is clear that further analysis will be necessary as further data are obtained, and such further analysis may well modify this figure by as much as 3 or 4,000 in either direction.

While this is true, it is also clear that the present sustainable yield, even viewed most optimistically, is much below the maximum sustainable yield and the level of recent seasons' catches.

E. Blue Whale Analysis

1. Estimates of Blue Whale Stock—DeLury Method

Tabulations were made of the catch of blue whales in each season 1953/54 to 1960/61 in each area at latitudes higher than 59°S—to exclude catches of pygmy whales *B. m. brevicauda*—and the corresponding numbers per catcher day. The catch per catcher day shows, for each area, a declining trend when plotted for each season against the cumulative catch up to and including that season. The plotted points are rather erratic, however, and so data for all areas were combined, with the results given in the table below.

Season	Catch per catcher day	Cumulative catch
1953/54	0.256 (0.137)	2,527
1954/55	0.207 (0.107)	4,581
1955/56	0.208 (0.107)	6,190
1956/57	0.202 (0.098)	7,613
1957/58	0.232 (0.106)	9,272
1958/59	0.142 (0.059)	10,272
1959/60	0.114 (0.045)	11,060
1960/61	0.081 (0.031)	11,650
1961/62	0.086	12,284

(in parentheses are given the catch per catcher day divided by the mean horsepower of catchers in each season, in thousands of HP)

For seasons 1953/54 to 1957/58, there was a slow decline in catch per catcher day and a much faster decline in 1957/58 to 1960/61. The slow initial decline may be an indication that recruitment in those years is continuing to compensate in part for mortality. By extrapolating the rate of decline in the later seasons, we find that if the difference between natural mortality and recruitment in those years is negligible in relation to the fishing mortality, the cumulative catch for zero catch per catcher day is about 13–14,000. This is one upper estimate of the initial blue whale stock, i.e., in 1953/54.

During this period, the average size and power of catchers were steadily increasing. If the increase in efficiency is about proportional to increase in size and power, we can correct the figures for catch per catch day to allow for the increase in efficiency. This leads to an upper estimate of 1953/54 stock still at about 13,000 blue whales. It is conceivable that the apparent rate of stock decline in recent years has been exaggerated by a change during those years in the probability that a blue whale, once sighted in the area in question and during the blue whale season, will be taken. Although over a longer period, there has been a change in "preference" towards fin whales rather than blue, this has been manifest mainly in a change in the latitude of whaling, which has been taken care of by restricting the present analysis to the 600 and 700 series of squares. It does not appear, furthermore, that a blue whale, *once seen*, was any less likely to be taken in, say, the 1960s than in the mid-1950s. Thus, although in theory, the DeLury estimate could lead to an underestimate of the stock for such a reason, in practice, it is unlikely that such an effect is appreciable in this case. It should be noted also that an error due to such a cause in the blue whale estimate will result in a similar, but opposite, error in estimates of fin whale stock by the same method.

If we ignore the earlier seasons and plot only cumulative catch from 1957/58 against catch per catcher day, we estimate the stock in that season as 5,500 and, hence, by ratios of catch per catcher days, as 6,000 in 1953/54. A similar estimate of stock in 1956/57 is about 7–8,000, and hence about 9–10,000 in 1953/54.

According to Mackintosh and Brown (1956), an independent estimate, from sightings, of average blue whale stock size in 1933/34 to 1938/39 is about 15% of 220,000 (the total Antarctic whale stock) plus or minus about 50%—33,000 (limits 21,000–51,000). The mean catch of blue whales per catcher day in that period was 0.84 (ranging from 0.53 up to 1.22 in different seasons) as compared with 0.26 in 1953/54. If this represents the order of difference in stock size, we obtain estimates for stock in 1953/54 as about 10,000 (limits 7,000–15,000). The average tonnage and horsepower of catchers were both, however, by 1953/54 nearly double what they were during the period 1933/34–1938/39. If catcher efficiency increased about proportionally, we get revised estimates of 1953/54 stock size as 5,000 (limits 3,500–7,500). The values obtained by Antarctic sightings are believed to be underestimates of total stock as it is now believed that the proportion of whales outside the Antarctic is greater than was presumed at the time the estimates were published. It is possible also that older whales are disturbed by and move away from the vessel from which sightings were made. This belief is consistent with the rather high numbers in Antarctic catches during the period in comparison with the stock estimates. On the other hand, the sightings include small whales not included in estimates of exploited stock given by analysis of catches.

At this stage, we can only say that the data from sightings are not inconsistent with the DeLury method estimates. An absolute upper estimate of 1953/54 stock is about 14–15,000 blue whales, and a likely value is about 10,000 or less. F_b in 1953/54 was, therefore, certainly not less than nearly 0.2, and a probable value is 0.25 or more. A catch per catch day (catcher tonnage about 500, HP about 1,900) of 0.1 thus corresponds with a stock of 4,000 whales or less, and 1,000 such catcher days generate $F_b = 0.025$ or more.

The fishing intensity in the pre-war period 1938–40 was about 25,0000 catcher days per year, with catchers averaging about 1,100 HP and 300 tons. An estimate of F_b in those years is, therefore, $0.025 \times 25 \times 1{,}100/1{,}900 = 0.36$. This is in good agreement with estimates obtained from marking data.

2. Application of Schaefer's Method to Blue Whales

The population size was estimated from the catch per catcher day divided by the average gross tonnage of the catcher. So as to include as much of the early data as possible, the tonnages of catchers before 1933, when the first published data are available, were estimated by extrapolating backwards, taking the average tonnage in 1925 as 200 (i.e., an increase of about 3% per year between 1925 and 1933, which is similar to the observed increase between 1933 and 1939). Several alternative hypotheses were made concerning the relation between this catch per unit and stock size, but only a limited range gave reasonable answers, as explained below. The values of stock size corresponding to a catch of 1 whale per 1,000 catcher ton-days, which were used were 15, 20, 30, and 40,000 whales. Using these figures, the same calculation as for fin whales was carried out to give the average annual surplus (i.e., the sustainable catch) during each five-year period. The mean value for the ten-year period covering the war years was also calculated. The stock size in each period has been expressed as a proportion of the stock in the "basic" season 1953/54. The results are given in Table 13.

The column for $F = 0.16$ gives obviously wrong surplus stock in the last period—they imply that the decline in the stock in these years was greater than the numbers being caught. The stock in 1953/54 was, therefore, less than 12,600 (this corresponds in essentials to the estimate by the DeLury method). The column for $F = 0.48$ gives unreasonable results for the opposite reasons—the rate of increase is too high; thus in the 1931–36 period, a stock of $4.2 \times 8.32 = 35{,}000$ whales is presumed to be able to increase at a rate of 13,600 = 39% per year, which is higher than could be achieved even if every calf survived to maturity. Outer limits for the stock in 1953/54 are, therefore, 4,200–12,600.

In Table 13, the entries for the first five-year period are probably erroneous. During this period, the catch per unit effort was increased and, in particular, between the 1930/31 season, when 29,410 whales were caught, and the 1931/32 season, the catch per 1,000 ton days rose

Table 13. Sustainable yields of blue whales for different hypotheses.

	Stock size in 1953/54 (thousands)			**4.2**	**6.3**	**8.4**	**12.6**		**8.4**
	Fishing mortality in 1951–56			**0.48**	**0.33**	**0.24**	**0.16**		**0.24**
		Index of annual stock						**1920s adjusted**	
Seasons	**Catches**	**Current period**	**Five years earlier**	**Surplus stock (1000s)**				**Index of stock**	**Surplus**
1926/27–1930/31	15.0	8.32	(10)	16.4	17.2	17.9	19.4	15.80	7.2
1931/32–1935/36	15.4	10.29	8.32	13.6	12.7	11.9	10.2	10.29	11.9
1936/37–1945/46	6.5	4.13	7.43	4.7	3.8	2.9	1.1	4.13	2.9
1946/47–1950/51	7.4	2.65	(3.50)	5.7	4.7	3.9	2.0	2.65	3.9
1951/52–1955/56	3.1	1.03	2.65	2.3	1.8	1.5	0.6	1.03	1.5
1956/57–1960/61	1.2	0.46	1.03	0.8	0.6	0.3	0.2	0.46	0.1
	Stock size in 1953/54 (thousands)			4.2	6.3	8.4	12.6		
	Present sustainable yield			210	120	60	Nil		

from 3.71 to 4.83, even though the 1930/31 catch was at least 30% of the stock. Between these two seasons, the numbers of catchers fell from 200 to 33, and it is likely that these 33 were the more efficient vessels. The observed increase in catch per unit effort was due to increased catcher efficiency rather than to increased stock. The observed catches per unit effort were, therefore, increased by the following factors:

Season	1925/26	1926/27	1927/28	1928/29	1929/30	1930/31
Factor	2.0	1.9	1.8	1.7	1.6	1.5

This is equivalent to assuming that the efficiency doubled between 1925/26 to 1931/32. This may be an overestimate, but will result in an underestimate of the sustainable yield (just as the uncorrected figures give an overestimate of the sustainable yield). These adjusted figures are given in the last column of Table 13. This column probably approaches most closely the real relation between stock and sustainable yield, and in Figure 6 the corresponding data for individual years (with the sustainable yields smoothed by three) have been plotted. As full recruitment probably occurs at 5 years old, the sustainable yield has been plotted against the stock five years earlier. As well as the individual years, the mean surplus stock is fitted closely by the proportional line $Y = 0.1P$, i.e., the stock is capable of increasing at 10% per year. A curve has been fitted by eye through all the points, and this suggests that a maximum sustainable yield of a little over 6,000 blue whales could be taken if the stock were allowed to reach a level of around 100,000–125,000 whales. Also shown in the figure are the catches since 1950/51, plotted against the stock in the year they were taken. This shows that the catches have been consistently several times greater than the sustainable yield, causing the observed rapid decline in both stock and catch.

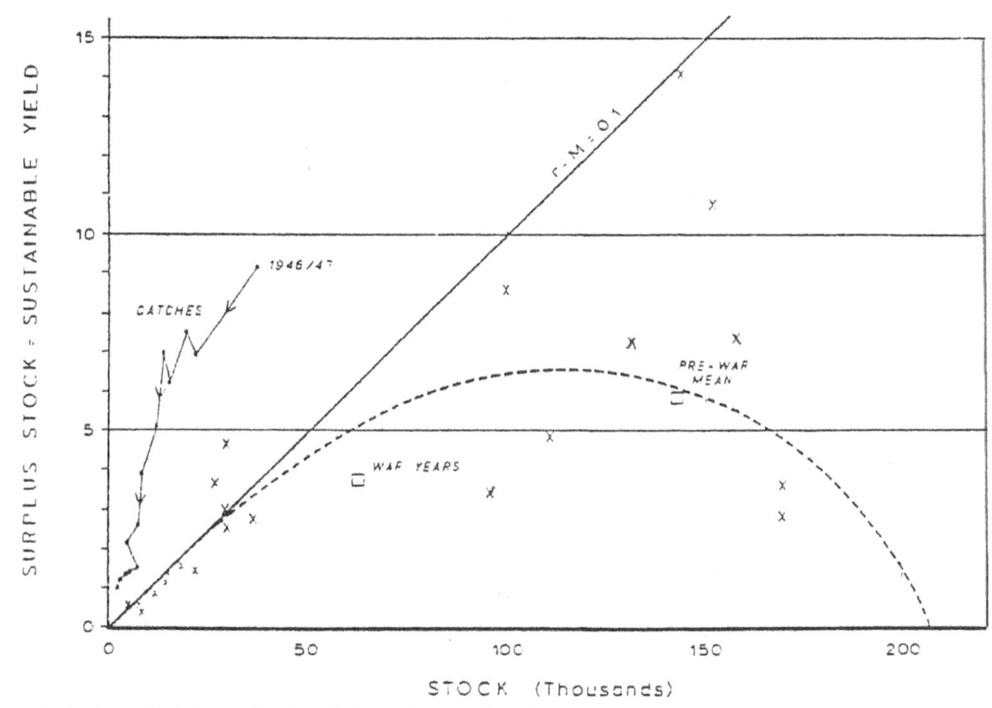

Figure 6. Antarctic blue whales. Schaefer method.

The present stock is so low that not only is the sustainable yield at this level very low (between 0 and 200 whales), but the total catch which would be obtainable by unrestricted whaling in the next few years is not much higher. The stock (omitting pygmies) in 1961/62 was about 22% of that in 1953/54; between 930 and 2,790 whales in all. Even allowing for some recruitment, whaling at the present rate, if continued unchecked, is unlikely to yield a total of more than double these figures, i.e., 2,500–5,000 whales, before blue whales become virtually extinct.

Alternatively, the catching of blue whales might be stopped at the end of the 1962/63 season and the stock allowed to increase to a level permitting a sustainable yield of 5,000 whales, i.e., a population of tenfold the 1953/54 level. By 1963, the stock will have been reduced below even the present level, probably to between 650 and 1,950 whales—which is a level at which there must be a distinct risk of complete extinction—so that the increase required is by a factor of around 70. At a rate of increase of 10% per year, this will require about 65 years. This delay period will be increased by around five years for each year that blue whale catching is continued.

F. Humpback Analysis

1. Humpbacks: Group IV stock (Western Australia)

Mortality rates

The length distribution of whales in Western Australian catches in each year from 1949 to 1962 was converted to age distribution using age–length keys. The keys were derived from ear plug readings for males (assuming two laminations per year) and ovary counts for females (based on the observed ratio of 1.1 ovulations to two ear plug laminations); different keys were applied for the periods 1949–54, 1955–58, and 1959–61.

Corresponding estimates of catches per unit effort were obtained from ships' logs—the same four catchers at Carnarvon each year—for the same seasonal period (25 June to 26 August), effort being measured as catcher steaming hours plus a weighted component for spotter aircraft flying hours. These catches per unit effort were applied to the age distributions to give relative abundance estimates for each age group in each year.

In the season of 1955 and later, the quota was reduced, and this resulted in a more vigorous management policy making the unit fishing time more effective. This increased efficiency was apparently the cause of a very low apparent mortality from 1954 to 1955 (as estimated from the ratios of catches per unit effort). This mortality has been omitted from the subsequent analysis, and the effort units in 1954 and earlier years reduced by a factor of 0.7 estimated by assuming that the real mortality in 1954/55 was the mean of the 1953/54 and 1955/56 mortalities.

From the catches per unit effort of each age, total mortality coefficients (Z) were calculated for each pair of years and year class (i.e., pairs of age groups) for each sex.

Figures for annual Australian effort were obtained by multiplying the effort of the four catchers by the ratio of the total Australian catch to the catch by those catchers for each winter season. The Antarctic (Area IV) effort was obtained by dividing the catch in each summer by the mean catch per unit effort in the two adjacent winters. Finally, the total effort responsible for the estimate of Z from a particular pair of years was estimated as the mean Australian effort in those two years, plus the Antarctic effort in the intervening summer.

The estimates of Z were then plotted against the estimated effort for each sex separately. Both gave highly significant regression lines, with intercepts on the y-axis, corresponding to natural mortality M of 0.19 (females) and 0.20 (males). In both plots, the point for the pair of years 1958/59 is anomalous. The effort value is far higher than for any other point, but the mortality, while high, is not as high as might be expected from the regression. The high effort values derive from very large Antarctic catches during the summer 1958/59; if some of these whales did not, in fact, belong to the Group IV population (cf. Chittleborough 1959), then the effort as computed would be too high; if, in fact, 40% of these catches were from outside Group IV, then the 1958/59 point would fit closely on the regression line through the other points. There, therefore, appears some justification in omitting the 1958/59 point, and the regressions fitted through the other points give intercepts of 0.09 (females) and 0.06 (males). These are probably closer to the true value of natural mortality than the previous estimates of 0.2. (These regressions are shown in Figure 9.)

Another approach to the estimation of natural mortality in humpback whales can be made from the age distribution of the virgin stock. This cannot be done in the Group IV population because of its previous history of exploitation. However, there was little exploitation of the Group V population before 1952, and the total mortality in this stock, as estimated from the age distribution of the males in 1952 catches, was 0.18, and this can be taken as nearly all natural mortality.

The total mortality in the most recent seasons has ranged from 0.85 to 1.1 (males) and 0.54 to 0.80 (females). Using the above estimates of M, these correspond to fishing mortalities of 0.65–0.9 (males) and 0.34–0.60 (females). The difference between males and females is due to the fact that all mature males can be taken by the fishery, while females with calves (35–40% of the total number of mature females) cannot be taken.

Size of population

These estimates of F, fishing mortality, can be used directly to estimate, from the catches, the average population size during each recent season. These calculations are given in Table 14.

Another independent estimate of stock size may be obtained from the DeLury method of plotting catch per unit effort against accumulated catch. This gives a very good relation for the data since 1955, with an intercept on the x-axis of 9,800 + 1,100. Because of the error introduced by ignoring recruitments and natural mortality, this estimate is biased, and the true value of the catchable stock in 1955 is less than this. The plot can also give estimates of the stock in most recent years of about 2,500 (both sexes) which, as expected, is rather greater than those in Table 14. The plot is shown in Figure 7.

Table 14. Population size of Group IV humpbacks estimated from mortality rates.

	Males					Females				
Year	**1958**	**1959**	**1960**	**1961**	**1962**	**1958**	**1959**	**1960**	**1961**	**1962**
Australian catch	508	331	271	306	304	458	369	274	272	239
Antarctic catch	645	28	3	24		458	38	1	32	
Mean catch	1,065	329	291	329		1,181	360	274	287	
Z	1.09	0.88	0.85	0.90		0.80	0.67	0.80	0.54	
F (if M = 0.2)	0.89	0.68	0.65	0.70		0.60	0.47	0.60	0.34	
Stock	1,200	485	450	470		1,970	770	460	840	
F (if M = 0.1)	0.99	0.78	0.75	0.80		0.70	0.57	0.70	0.44	
Stock	1,075	420	390	410		1,690	630	190	650	

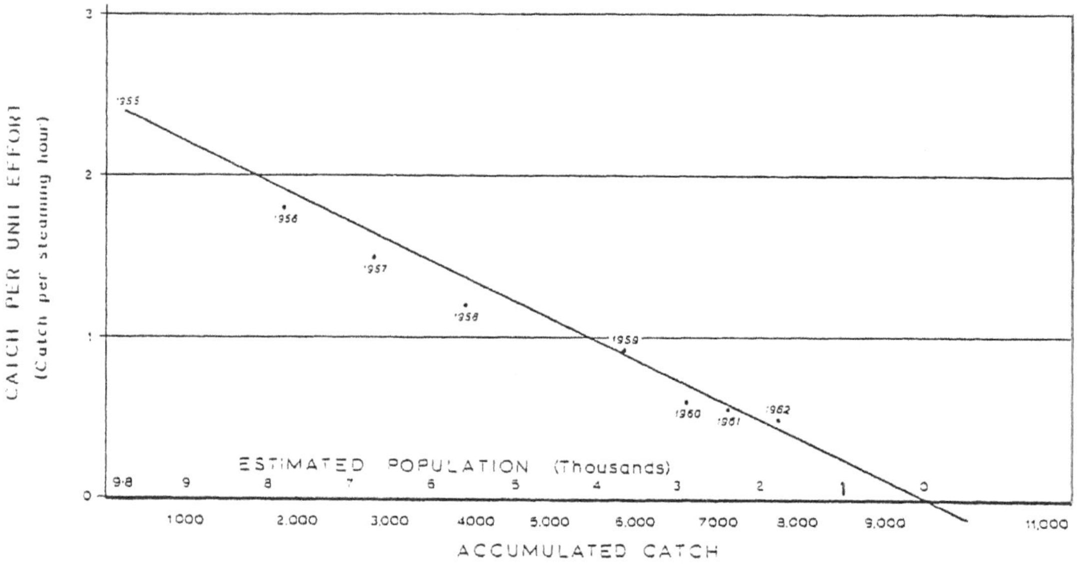

Figure 7. DeLury method on Group IV population of humpback whales.

A third independent estimate can be derived from data of sightings in 1933–39 (Mackintosh and Brown 1956). This gave a total baleen whale stock of 230–340,000. Of these, Mackintosh (1942) estimated that 10% were humpbacks, so that estimated humpback population in the 1930s was from 22 to 34,000, which has to be apportioned between the five distinct populations. A fair measure of the relative abundance of the 1930s of these groups is as follows (cf. Mackintosh 1952):

Group I	Group II	Group III	Group IV	Group V
1	1	2	3	3

so that Group IV included 30% of the total Antarctic population, and the estimated size of this population in the 1930s was 6,600–10,200 whales, the larger figure being probably the more likely.

All those estimates have been put together in Table 15. This also includes data on the best estimate of catch per unit effort, and by comparing catch per unit effort and population estimates, based on mortality rates in recent years, the former can be expressed in absolute terms for all available years. If $M = 0.1$, the relation is that a catch per unit effort of 0.1 is equivalent to a population of 2,000 whales.

Unfortunately, the periods covered by the sighting records and the other data do not overlap, so they cannot be compared directly. The history of whaling on this stock is of few catches before 1934, very heavy catches (12,000 in all) from 1934 to 1939, and no whaling from 1939 to 1949. Thus, compared with the 1950 stock, the stock in 1939 was probably considerably less, and the population in 1934 fewer than 12,000 more than that in 1939. That is, taking the population to be 9,400 in 1950, the 1939 population was somewhere in the range of 3–8,000 and the 1934 population in the range of 11–16,000. The sighting estimate of 10,000 humpbacks during the period 1933–39 fits well inside this range.

Table 15. Estimate of population size of Group IV humpbacks by various methods.

Year	Estimated population (mortality rates $M = 0.1$)	Catch per effort	Estimated population (cpe)	Estimated population (DeLury)	Estimated population (sightings)
1930s	-	-	-	-	10 200
1950	-	0.47	9,400	-	-
1951	-	0.42	8,400	-	-
1952	-	0.35	7,000	-	-
1953	-	0.35	7,000	-	-
1954	-	0.35	7,000	-	-
1955	-	0.24	4,800	9,800	-
1956	-	0.18	3,600	7,800	-
1957	-	0.15	3,000	6,700	-
1958	-	0.12	2,400	5,600	-
	2,765				
1959		0.09	1,800	3,700	-
	1,050				
1960		0.06	1,200	2,900	-
	780				
1961		0.055	1,100	2,400	-
	1,060				
1962		0.05	1,000	1,800	-

Stock and recruitment

The abundance of adult females in each year from 1949 onwards has been estimated as the sum of the catches per unit effort of female whales above 5 years of age. Age of recruitment to the fishery was taken as 4 years old. At this age, whales are not fully recruited, so that the catches per unit effort of 3-year-old females cannot be used directly, and an indirect method had to be used. This was done for each year class by taking the catches per unit effort at the fully recruited ages and converting these to actual numbers in the population using the relation that a cpe (catch per effort) of 0.1 is equivalent to a population of 2,000 whales. The numbers of 3-year-old females were then estimated by adding the known catches of females in the year class and the estimated deaths by natural mortality (assuming $M = 0.1$, but this value is not a critical one).

The number of recruits at 3 years old in each year can then be plotted against the number of females in the parent generation, i.e., the number four years earlier (that is, the pregnant females in 1950 gave birth in 1951 and these are 3-year-old recruits in 1954). The result is shown in Figure 8, and the points lie very close to proportional lines, with R (recruits) = 0.12 × stock.

Another estimate can be made from pregnancy data. Thirty-seven percent of adult female humpbacks in catches are pregnant, so that if the sex ratio at birth is 0.5, 100 females will give birth to 18.5 female calves. The number of 3-year female recruits is, therefore, $18.5\ e^{-3M}$, where M is the average natural mortality in the first three years. This juvenile mortality is almost certainly, if anything, greater than the adult natural mortality; assuming it is 0.15, then 100 mature females will produce 11.8 3-year-old females per year; i.e., $R = 0.118 \times$ stock. The figure of twelve 3-year-old females per 1,000 mature females is, therefore, a fairly reliable estimate. The pregnancy data can also be used to set an upper level to this recruitment rate. If the juvenile M averaged 0.1, then 100 mature females would produce 13.7 females

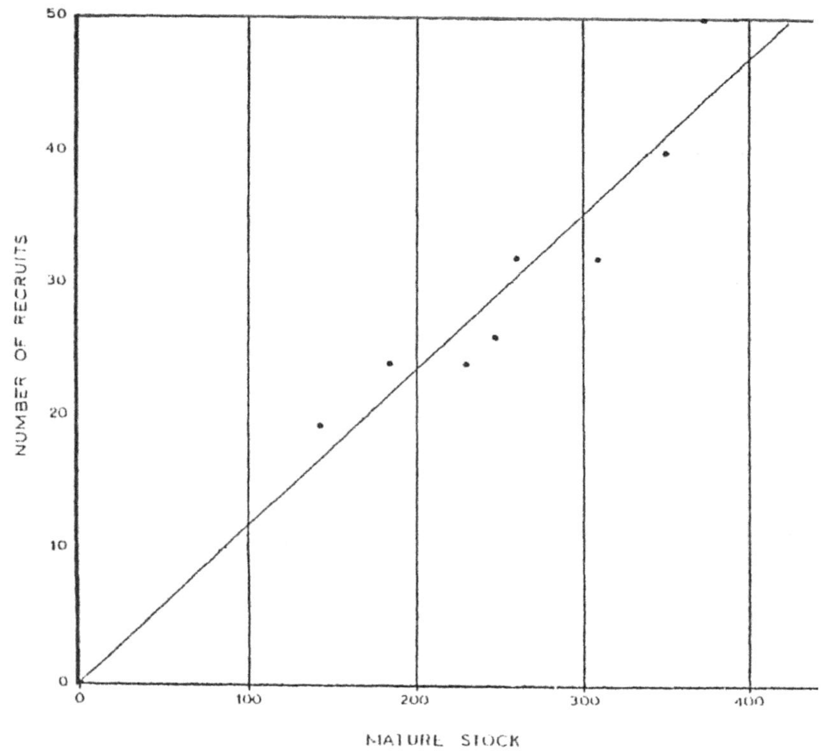

Figure 8. Stock–recruit relationship for Group IV humpback whales.

at age 3. Even if *M* were negligible, which is unlikely, the production of 3-year-old females could not exceed 18.5 per 100 mature females in the stock. The pregnancy rate itself, taken as 37%, has been established within close limits and the percent value is most unlikely to be more than 40%, which gives an extreme limit to the recruitment rate (taking $M = 0$) of 20 per 100 mature females.

Though whales become available to the catchers at 3 years old, the females are not mature until, on the average, they are 5 years old. The annual numbers of mature females from 100 mature females is, therefore, $12 \times e^{-2M} = 10$ (if there is no fishing).

Assessments

The net rate of change of the population is the difference between *r*, the recruitment coefficient, and the sum of *F*, the fishing mortality coefficient, and *M*, the natural mortality coefficient. The population will be in balance, neither increasing nor decreasing, if $F = r - M$ and the catch taken is $F \times MP = (r - M)P$, where *P* is the population size. Either or both *M* and *r* will change with population size, *M* increasing and/or *r* decreasing as the population increases, and when the population is at its maximum, the unfished state, $M = r$. The exact relation of *M* and *r* is difficult to determine, but some reasonable assumptions can be made, and hence the form of the curve $c = (r - M)P$ giving the steady yield determined, and in particular the position and height of the maximum estimated.

At the present stock abundance, *r* has been estimated as 0.10, and reasonable limits may be taken as 0.12 and 0.08. *M* is not so well known, the best estimate combining the two regressions for male and female whales is 0.07, though the data are consistent with values of from 0.05 to 0.10. A range of values of $r - M$ can now be given as follows:

Best estimate = 0.10–0.7 = 0.3
High estimate = 0.12–0.05 = 0.07
Low estimate = 0.08–0.10 = –0.02

The negative "low" estimate shows that the humpback population may be very delicately balanced and unable to stand even the smallest amount of catching.

The present total population is around 1,000 whales; of these, not more than half are mature (of 73 females caught in 1962, only 47% were mature). Taking the mature population as 500, the quotas that can be taken from this population without causing a further decline are:

Best estimate = 500 × 0.03 = 15
High estimate = 500 × 0.07 = 35
Low estimate = 0

Therefore, probably 15 whales could be taken, and the stock would maintain itself at the present level. Almost certainly if more than 35 whales were taken, the stock would decline.

The observed relation between stock and recruitment suggests that there is no reduction in r and, therefore, probably also in $r - M$ for stock sizes up to those observed in 1949. This

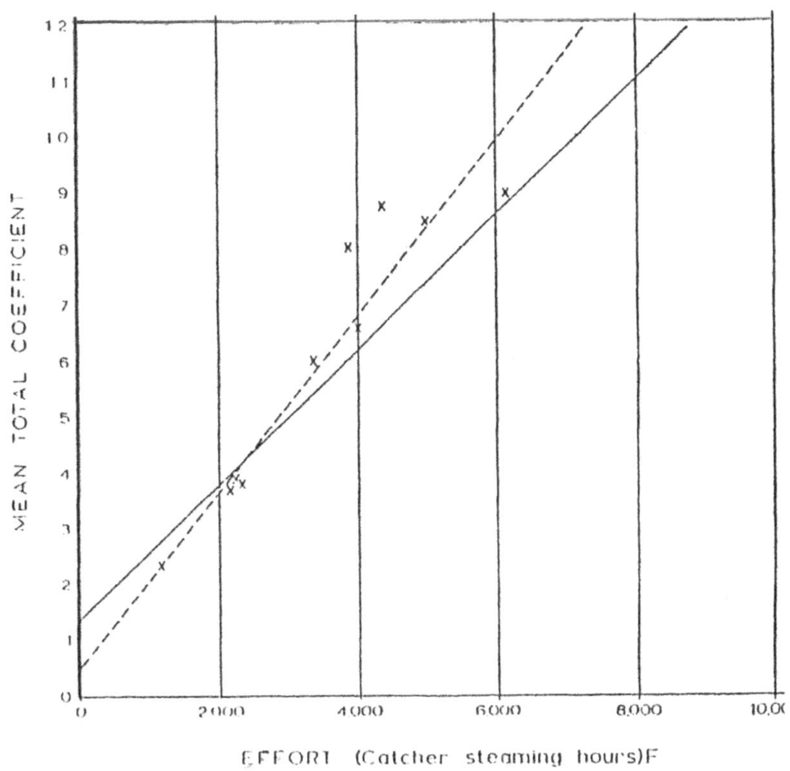

Figure 9. Mortality and effort. Group IV male humpback whales.

was 10,000 whales, of which some 85% were mature; the sustainable yield at this population level may be estimated as follows:

Best estimate = 8,500 × 0.03 = 255
High estimate = 8,500 × 0.07 = 595
Low estimate = 0.

(The "low" estimate is clearly too pessimistic, for if the stock was built up again to the 1949 level, it could stand some catches being removed, but the sustainable catch might be very low, say 50–100 whales.)

The size of the stock of humpbacks in the unfished state is not known precisely, but before 1949, the catches taken from it were 5800 whales in 1912–1916, another 3,400 whales in 1925–1928, and 12,800 in 1934–1939; assuming that the stock recovered to some extent between these periods of whaling, the unfished population was probably between 15 and 20,000. In this presumably stable population $r - M = 0$, so that as the stock increases from 10 to 20,000, there must be a critical reduction in r, or an increase in M, or both. The maximum sustainable catch, therefore, probably comes at a stock level of around 10,000 whales, and the maximum sustainable catch is probably around 250 whales, though it might be as low as 50–100 or as high as 600.

These catches have been based on mature stock; little difference will be made if catches are taken of both mature and immature whales. Catches of immature whales will reduce the net recruitment rate to the mature part of the stock, and hence the sustainable catch of mature whales. Allowing for the fact that some of the immature whales caught might have died from natural causes before reaching maturity, the sustainable catch in numbers would be very slightly increased by taking both mature and immature whales; taking into account the growth of the individual whales, the sustainable catch in weight will be slightly less.

The time taken for the stock to build up to the optimum level can be derived from the potential net rate of increase in $r - M$. If whaling on the Group IV stock was stopped now, i.e., no more whales were taken after the 1962 season, there would be initially a rather rapid increase in the mature stock because of recruitment from the rather larger mature stocks in 1958 and 1959 (which now form the relatively abundant immature stock). The mature stock in 1964 would, therefore, be about 900 (allowing for some reduction due to natural mortality from the 1,000 whales now in the stock). These 900 would then increase at a rate of $r - M$, and reach the 10,000 level after T years, where $e^{(r-M)T} = 10,000/900$, i.e., $(r - M)T = 2.41$. Using the best estimate of $r - M(0.03)$, $T = 80$ years. Using the "high" estimate of $r - M(0.07)$, $T = 34$ years. This delay period is critically dependent on the level to which the stock has been reduced before whaling is stopped. If whaling had stopped in 1959, when the stock was about 1,800, then the mature stock after two years would rise to about 1,620, and $e^{(r-M)T} = 10,000/1,620$, i.e., $(r - M)$ $T = 1.82$. $T = 61$ (best estimate) or 26 (high estimate of $r - M$).

Conversely, if whaling continued at the present level, the mature stock will decline at the rate of about $F - (r - M)$, which is about 0.7. That is, for each year whaling continues at the present level (in terms of fishing mortality rate, the numbers taken would decline proportionally to the stock) increases the delay until the stock can recover to its optimum level at 0.7/0.07 = 10 years, and probably 0.7/0.03 = 23 years.

In terms of catches lost at the optimum level, these delays are equivalent to 10 × 595 or 5,950 whales, or 255 × 23 = 5,865 whales; the actual estimate is, therefore, very little affected by the precise estimate of $r - M$; a long delay period, due to a low value of $r - M$, corresponds to a lower catch at the optimum stock level. This figure of the loss from the potential catch of 6,000 whales may be compared with the 1962 catch of 600 whales and the probable future annual catches at the present level of fishing mortality of 400 or less in 1963, and progressively less in later years.

Summary

The natural mortality of Group IV humpbacks was estimated from the regression of total mortality on effort, as from 0.07 up to 0.20. An alternative estimate from the age distribution in the unfished Group V stock was 0.18. The corresponding estimates of fishing mortality are 0.85–1.1 (males) and 0.34–0.60 (females).

The stock abundance estimated from these mortalities was 10,000 in 1949 and 1,000 in 1962; these agree reasonably well with the sighting data from the 1930s and are about 50% of the estimates using the DeLury method; the latter method is known to give an overestimate. The recruitment rate is close to twelve 3-year-old whales per 100 mature whales.

The present stock will certainly be reduced by taking 35 whales per year, and probably can only sustain an annual catch of 15 whales. The maximum sustained catch will be taken from a population of around 10,000 whales (the stock in 1949), and at this population level, the sustainable catch is around 250 whales per year (range 50–600). If whaling stopped at once, it would take probably 80 years for the stock to build up (possible as little as 34 years). Every year in which the present effort continues delays reaching the optimum level by 23 years (possible on ten years), and results in a loss of potential catch of 6,000 whales.

2. Humpback Whales—Group V Stocks

Segregation during breeding

Although individuals mingle freely in Antarctic Area V while feeding each summer, the Group V population is widely dispersed during the breeding season. From the results of whale marking, Dawbin (1959) produced evidence that this population was segregated to some extent into a number of breeding units. If whaling was carried out only in temperature and tropical regions, it might be necessary to consider each as a separate unit stock. However, as it is impossible to separate these units within commercial catches from Antarctic Area V, and as similar trends have been recorded during recent years in catches from New Zealand and the east coast of Australia, the Group V humpbacks will be considered as one stock for present purposes.

Mortality rates, east coast of Australia

The same basic methods were used as for the Group IV stock. A single age–length key for each sex was used for the whole period 1952–61 for females based on ovaries collected during each year and ear plugs collected during 1956–61.

Because of the use of tow boats, the basic unit of fishing effort was the number of hours (i.e., omitting tow boat time and time spent traveling to and from the whaling grounds) of two standard catchers at Tangalooma. Catches per unit effort for each year class were thus obtained each year and mortality rates estimated. Total effort was obtained as before, including Antarctic fishing and also that at New Zealand and Norfolk Island. These mortality and effort data are shown in Table 16. (As in Group IV, the mortalities in 1954/55 have been omitted.)

At Tangalooma (east coast of Australia), the frequency of sightings of humpback whales from the catchers was 88.2 per 100 hunting hours in 1961 and 38.6 per 100 hunting hours in 1962, both values being obtained from the same period (12 June to 6 August). This is equivalent to a total mortality coefficient of 0.98, which is in accordance with the even higher values of total mortality obtained from the 1961/62 catch-per-unit-effort results, which refer to the catchable portion of the stock (i.e., only those over 35 feet in length).

Mortality rates, New Zealand

Using the length frequency of catches of male and female humpback whales from New Zealand (records from International Whaling Statistics) and applying age–length keys derived from the large Australian material, which closely agreed with the New Zealand material of 1949–62, estimates were made of the age distribution within New Zealand catches each year from 1955 to 1962.

There are no direct measurements of effort available for these catches, which were obtained throughout the whole of the northward migration, searching being conducted from an elevated position on the shore during daylight hours. The processing capacity of the factory increased steadily from 1955 to 1960, and over the same period the efficiency of the lookouts was also increased each year (more powerful glasses, more experienced and skillful observers, etc.). The increasing number of whales sighted and captured each year from 1955 to 1960 is believed to be due to better searching and greater plant efficiency rather than to an increase in the stock.

The catch per unit effort of each age was calculated assuming that the total catch per unit effort had been constant from 1955 to 1960 (i.e., the stock did not increase), and that the effort in 1961 was the same as that in 1960. In 1962, two sighting lookouts were operating and approximately half of the sightings were made from the new lookout, so that the effort expended in 1962 was taken to be double that of 1960.

Total mortality coefficients were then estimated between pairs of years from the ratio of abundance of age groups $t + 1$ in one year to that of age group t in the preceding year, using only those age groups which were fully recruited. The mean values are shown below (Table 17).

Table 16. Total effort for Group V humpbacks and total mortality coefficients (from catches on the east coast of Australia).

Year	1952/53	1953/54	1955/56	1956/57	1957/58	1958/59	1959/60	1960/61	1961/62
Female effort	513	304	565	493	462	1,108	1,648	1,698	1,372
Mortality	0.32	0.54	0.17	0.23	0.01	0.14	0.51	1.56	1.75
Male effort	754	689	955	953	942	1,619	1,985	2,526	2,253
Mortality	0.25	0.44	0.26	0.20	0.01	0.19	0.48	1.56	1.58

Table 17. Total mortality coefficients calculated from New Zealand catches.

Year	1955/56	1956/57	1957/58	1958/59	1959/60	1960/61	1961/62
Males	0.19	0.14	1.02	0.22	0.81	1.60	1.77
Females	0.36	0.16	0.31	0.14	0.09	1.62	2.15
Mean	0.58	0.15	0.67	0.18	0.45	1.61	1.96

Total mortality can also be estimated from the records of sightings of humpback whales at Cook Strait, New Zealand, in the last three years. During the northward migration of 1960, a total of 480 humpback whales were sighted; in 1961, the total sighted from the same observation point was 88 humpbacks, while in 1962, only 12 humpbacks were seen passing the observation point. From the ratio of sightings in 1961 to that of 1960, the total mortality coefficient was 1.71, and the ratio of sightings in 1962 to that of 1961 gave a mortality coefficient of 1.97. These values are in very close agreement with the mortality coefficients calculated from New Zealand catches.

The results from both catches and sightings from two widely separated localities (east coast of Australia and New Zealand) show a very sharp rise in mortality with the Group V stock from 1960 to 1962. The evidence of some segregation into separate breeding units within this stock during the winter (breeding season) indicates that the similar and high mortalities recorded at the same time in the different localities were not due entirely to the fishing in those regions, but had a common origin. Recent catches reported from Antarctic Area V (included in the total effort recorded in Table 16) have not been sufficiently large to account for these very high mortality coefficients.

There are three alternatives possible:

a) The Group V stock of humpbacks has recently changed its long established migration paths. This hypothesis is contradicted by widely scattered and independent observers throughout the South Pacific islands, on ships, and on the Australian coast, all reporting a great reduction in the number of humpback whales during the winter seasons of 1961/62.

b) Increased natural mortality (of epidemic proportions) due to disease, parasitic infections, or starvation. This also has no supporting evidence, as the few humpbacks captured were in good health and had a normal thickness of blubber and oil yield.

c) Substantial catches of humpback whales being taken from this stock in addition to the available records in a region where the humpback whales from New Zealand and the east coast of Australia can both be fished simultaneously, i.e., Antarctic Area V. Additional mortality of little more than 4,000 humpbacks in 1960/61 and 1,000 humpbacks in 1961/62, from feeding concentrations in Antarctic Area V, would (when added to all other known catches) explain the recent abrupt increase in total mortality within this stock.

Population estimates

The method of DeLury was used to obtain estimates of the upper limits of the size of the Group V humpback stock. The catch per unit effort measured each year on the east coast of Australia and the total recorded catches from this population are given in Table 18.

Table 18. Catch per unit effort and total catches for the Group V stock of humpback whales.

Year	1952	1953	1954	1955	1956	1957	1958	1959	1960	1961	1962
Catch per unit effort	0.97	0.96	0.76	0.78	0.70	0.71	0.75	0.74	0.52	0.23	0.07
Total catch temperate regions	722	809	898	832	1,033	1,025	1,023	1,278	1,341	981	209
Antarctic catch	146	506	0	1,097	194	0	0	885	931	293	0
Cumulative catch	868	2,181	3,079	4,908	6,235	7,260	8,283	10,446	12,718	13,992	14,204

Within the Group V population, the catch per unit effort has been measured with the greatest precision at east Australia. These data should, therefore, give the most reliable estimate available of abundance in this population. When this catch per unit of effort is plotted against cumulative catch, as recorded from *all* known catches in each year from the Group V population, there was little change apparent in the size of the population from 1952 to 1959. From the regression fitted to the results from 1959 to 1962, the upper limit of the population size would have been 4,600 humpback whales in 1959. If the regression is taken only through the values for 1959/60, the intercept on the x-axis gives a 1959 population size of up to 7,600 humpbacks. The second regression would require an unexplained mortality of about 2,300 humpbacks in 1961 and about 2,600 humpbacks in 1962. Over the past two years, this would require a total additional population loss from the Group V stock of approximately 4,900 humpbacks, which is very close to the figure of about 5,000 additional humpbacks necessary to account for the present high mortality coefficients. These data are shown in Figure 10.

Mean population size between pairs of years can be estimated from the total mortality coefficients and total catches if the total mortality can be separated into its components of natural and fishing mortality. In the section on the Group IV humpbacks, it has been shown that natural mortality must be less than 0.12. The value of 0.1 applied to the Group IV population will be used here as the coefficient of natural mortality. If natural mortality was less than 0.1, the estimates of the present size of the Group V stock would be reduced very slightly, e.g., for a value as low as 0.05, the estimate for 1961/62 population size would be reduced from 380 to 320.

The mean population size during various periods is given in Table 19, the population estimate being obtained by dividing the annual catch by the fishing mortality.

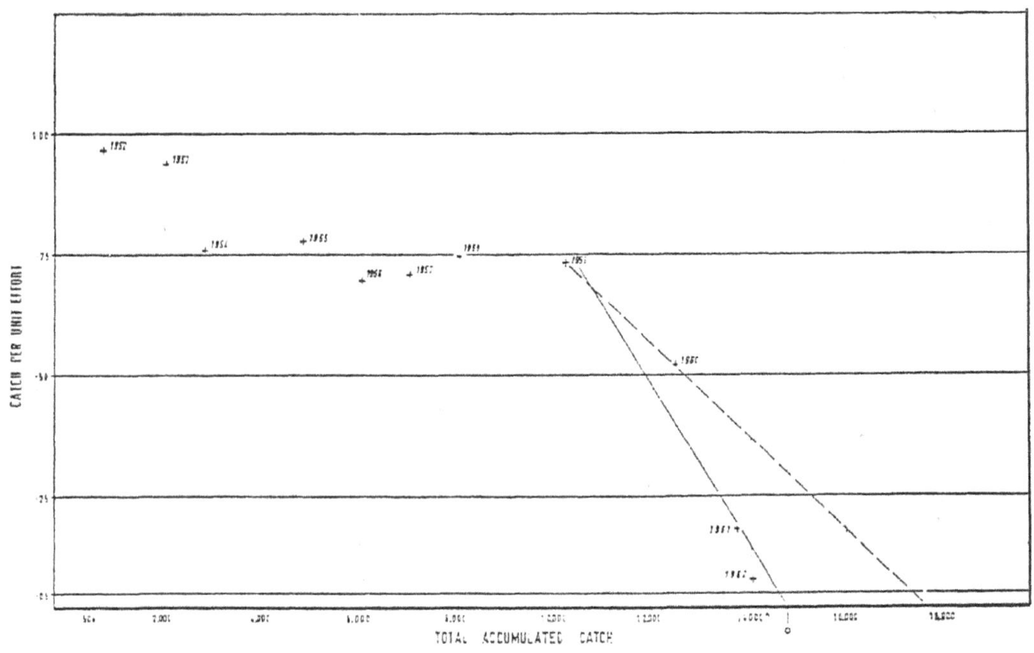

Figure 10. Group V humpback whale population—catch per unit effort and cumulative catch.

Table 19. Estimates of population size for Group V humpbacks.

Period	1952–59	1959/60	1960/61	1961/62
Average total mortality (*Z*)	0.224	0.50	1.56	1.67
Fishing mortality (*F*) (*M* = 0.1)	0.114	0.40	1.46	1.57
Annual catch (*C*)	1,329	2,241	1,456	595
Mean population size (*C*/*F*)	11,700	5,600	1,000	380

For 1960/61 and 1961/62, further estimates of mean population size were made by including additional mortality of 3,500 humpbacks in 1960/61 and of 1,500 in 1961/62. The revised estimates of mean population size would then be:

Annual catch	1,129	2,241	4,954	2,095
Mean population size	11,700	5,600	3,600	1,334

If the available catch reports were complete, this table shows that the present population is substantially less than the estimate obtained by making due allowance for the unexplained mortality.

The upper and lower estimates of the catchable stock in 1961/62 are, therefore, approximately 1,400 and 400, respectively.

In order to obtain the present size of the catchable stock, these estimates should be reduced by at least half of the catch in 1962. The present size of the catchable stock is then from 1,300 to 300 humpbacks.

Stock assessment

Pregnancy rates recorded from Cook Strait from 1948 to 1962 were almost identical with those recorded in recent years from Albancy (west coast of Australia), so a similar value for recruitment rate applied to both breeding stocks. If the rate of recruitment ranges from 0.10 to 0.12 and natural mortality during juvenile years ranged from 0.07 to 0.05, the rate of increase of the unfished population ($r - M$) would be from 0.03 to 0.07.

The catch of humpbacks from the east coast of Australia in 1962 contained 50% of mature individuals. If the present catchable stock contains 1,300 humpbacks, almost 650 of these are mature. The maximum catch from these is the difference between the recruits which these would produce and the loss due to natural mortality ($r - M$). If this ranges from 3 to 7%, the maximum sustainable yield would be 20–45 individuals.

If the present size of the catchable stock is 300, there would be 150 mature individuals and the maximum possible catch would be from 5 to 10 humpbacks per year, if the stock is to remain at its present low level.

On this evidence, the maximum catch which can be taken each year from the whole of the Group V population, in order to prevent any further decrease in this population, would range from 5 to 10 humpbacks per year, if recent catch reports are reliable, and from 20 to 65 per year if there has been some other mortality in addition to that reported.

If we accept the upper limit of the estimates of present catchable stock size (1,300 humpbacks, of which 650 are mature), and a maximum rate of natural increase of 7% per annum, the period of complete protection necessary in order to take selected levels of sustained catch can be estimated.

An annual catch of 300 humpbacks would require a mature stock of at least 6,000 humpbacks. If there was no fishing mortality, the minimum time required for an adult stock of 650 humpbacks to increase to 6,000 is given by: $6.50e^{0.07t} = 6{,}000$, i.e., $t = 32$ years.

Similarly, annual catches of 200 or 100 humpbacks could be sustained after 26 and 16 years, respectively.

If the present size of the catchable stock is only 300 humpbacks, the minimum period of protection required in order to take an annual catch of even 100 humpbacks would be 27 years.

Summary

Applying a natural mortality of 0.1, estimates of fishing mortality in the Group V stock ranged from a mean of 0.12 in earlier years (1952–59) to 1.57 in 1961/62. The sharp increase in mortality during recent years is not consistent with recorded catches of humpbacks from this stock.

The stock abundance estimated from these mortalities was a mean of 10,700 between 1952 and 1959, 5,600 in 1959/60, 1,500 in 1960/61, and 600 in 1961/62. If there has been additional catch in excess of that recorded in recent years, the mean stock size in 1960/61 would have been 3,600 and that in 1961/62 1,300 humpbacks. The latter represents the upper limit of the catchable stock size at that time.

The maximum annual catch which could now be taken from this stock is not more than 45 humpbacks and may be less than 20 per year.

If the size of the present stock is taken as the upper limit of 1,300 humpbacks, and the maximum possible rate of increase ($r - M$) as 7% per year, the minimum periods of protection required to take sustained annual catches of 100, 200, and 300 humpbacks from this population would be 16, 26, and 32 years, respectively.

G1. Pygmy Blue Whales

Because whaling on the stock of pygmy blue whales has only been operating for two seasons, no very firm estimates can be made concerning the effect of whaling on the stock or the sustainable yield from it. From the rather restricted geographical range, it is probable that it is a small stock. This is supported by the statistics of the last two seasons, taking the catches between 30 and 80°E and north of 60°S as being on the pygmy blue whale stock.

Season	Catcher days	Pygmy blue whales	Catch of blues/day
1960/61	1,908	1,126	0.59
1961/62	1,822	403	0.22

The data suggest that the 1960/61 catch of 1,126 whales greatly reduced the pygmy blue whale stock, and by the DeLury method would yield an estimate of a population at the beginning of the 1960/61 season of less than 2,000 whales. This estimate should be made more precise when the data for the 1962/63 season are available, but it is likely that the present stock is only a few hundred whales.

G2. Pygmy Blue Whale—Mortality Rate

Ichihara (1961) gives the age distribution of catches of male and female pygmy blue whales caught by Japanese whalers near Kerguelen from 17 to 26 March 1960. This was the first year of exploitation of this stock, at least since pre-war years, and 311 whales were taken. Age determination was based on 130 ear plugs, 86 of which were taken from whales in groups above the age at which they were fully recruited or were unselected because of the imposition of size limits.

The catch curves give estimates of Z about 0.22 for males and 0.34 for females for ages 10 and above and 12 and above, respectively, assuming two laminations formed per year. If this was an independent unexploited stock, as it is believed to have been, these values are estimates of M. They are rather high in comparison with estimates of natural mortalities for other exploited whale stocks, even taking account of a possible decrease in M with decrease in abundance. It is worth noting that if at the ages in question fewer laminations are formed per year—perhaps only one—the Z (and M) values are reduced proportionally to 0.11 for males and 0.14 for females.

Summary table of statistics.

Season	Catch	Total	% females	% females mature	% mature females pregnant
1959/60		311	57	83	27
1960/61	C	1,000			
1961/62	C	400			

Section III: General Remarks

The analyses summarized here depend on biological and effort data collected over many years by many groups which have numerous possibilities of error. In particular, the question of the randomness and representativeness of the samples has been raised. Also, the methods of ageing have not been fully validated nor is the best method of measuring effort known, particularly as it applies to earlier seasons and to fin whales. Further, the recently developed method of reading ages from ear plugs has served as the basis for the age–length keys for the fin whale data from 1931/32 onwards.

In regard to this, it should be re-emphasized that the analysis has been based primarily on recent data and on age determinations that require accuracy only for younger ages (e.g., Sections IID1 and 2 and consequently the conclusions in IID4). The analysis of recent data minimizes the problems associated with changes in catcher efficiency and in selection for species by the catchers.

As the Scientific Sub-Committee has also pointed out, in such extensive calculations, many of which are done by an electronic computer, it is difficult to check for errors in the calculations. What is of importance is not the results in detail, but the overall picture on which recommendations are based. As further suggested by the Scientific Sub-Committee, the agreement from four independent sources leaves no doubt that the stocks have been overexploited and a program of conservation should be initiated if the industry is to be maintained on a continuing basis. But, it is also true that the analyses made should be checked, that refinements in the methods be considered and used, and that as new data become available, additional studies be undertaken on a continuing basis.

The analyses have also neglected some biological factors, e.g., they have been based primarily on data from January, February, and March catches. Some components of the stock may be available only prior to 1 January. Also, some components of the stock may winter in tropical or temperate waters. Groups of whales that remain for several years outside the Antarctic clearly are not available to the Antarctic whaling industry and do not come within the purview of this study. Groups of whales that migrate erratically could modify the results of the analysis slightly. It is clear from the broad agreement between different methods and from the recent rapid decline in catch per unit of effort that the proportion of such anomalous migrations must be quite small. If they do exist, they may help the stocks to rebuild back slightly faster than indicated, *if the stocks are given protection by reducing quotas below the sustainable yield.*

References

Beverton, R. J. H., and S. J. Holt. 1957. On the dynamics of exploited fish populations. Her Majesty's Stationery Office, UK.

Chapman, D. G., K. R. Allen, and S. J. Holt. 1964. Special Committee of Three Scientists Final Report. Reports of the International Whaling Commission 14:39–92.

Chittleborough, R. G. 1959. Intermingling of two populations of humpback whales. Norsk Hvalfangst-tidende 48:510–521.

Darwin, W. H. 1959. New Zealand and South Pacific whale marking and recoveries to the end of 1958. Norsk Hvalfangsttidende 48:213–238.

DeLury, D. B. 1947. On the estimation of biological populations. Biometrics 3(4):145–167.

Ichihara, T. 1961. Blue whales in the waters around Kerguelen Island. Norsk Hvalfangsttidende 50:1–20.

Mackintosh, N. A. 1942. Southern stocks of whalebone whales. Discovery Reports 22:197–300.

Mackintosh, N. A. 1952. Observations on whales from ships. Marine Observer 22:87–90.

Mackintosh, N. A., and S. G. Brown. 1956. Preliminary estimates of the southern population of the larger baleen whales. Norsk Hvalfangsttidende 45:469–480.

Ricker, W. E. 1958. Handbook of computations for biological statistics of fish populations. Bulletin of the Fisheries Research Board of Canada 119.

Schaefer, M. B. 1954. Some aspects of the dynamic of populations important to the management of the commercial marine fisheries. Bulletin of the Inter-American Tropical Tuna Commission 1:25–56.

Chapter 7

Measures of Abundance of Antarctic Whale Stocks[1]

Introduction

In this paper, we give a brief account of the work done during the past two years under the auspices of the International Whaling Commission, jointly by its Scientific Committee and by a Special Committee of three scientists (subsequently four)—Dr. D. G. Chapman, Mr. K. R. Allen, and the authors—formed by the Commission in 1961 to advise it on the status of the stocks of blue *Balaenoptera musculus*, fin *B. physalus*, and humpback whales *Megaptera novaeangliae* in the Antarctic, their sustainable yields, and the appropriate conservation measures to be adopted. The Twelfth and Thirteenth Annual Reports of the Commission contain accounts of preliminary work, and the final report is now being prepared for publication in the Fourteenth Annual Report. Other members of the Committee, of course, also participated in the work we here outline.

Whaling has been conducted for many years, during about four months of each Antarctic summer, by pelagic expeditions from five countries: Netherlands, Norway, Japan, United Kingdom, and USSR (as well as by a few land-based whaling stations in the Southern Hemisphere). Each sends one or more factory mother ships accompanied by a number of small, fast, catcher boats. The total catch has, in most years, been limited to a total maximum number of "blue whale units"; these are calculated on the basis of 1 blue whale = 2 fins = 6 sei *B. borealis* = 2½ humpbacks. Further regulation is effected by opening and closing dates for season of capturing each species, and by a northern latitudinal limit to pelagic whaling. In addition, in recent seasons, each participating country has taken an agreed fraction of the total catch.

Estimates of stock sizes have been made by most of the methods known to fisheries biologists, as follows:

1) DeLury estimates from relating catch per unit effort to accumulated total catch over a period of rapidly declining stocks, when most of the mortality has been due to catching (DeLury 1947).
2) Estimates of total mortality rate from age compositions, splitting this between catching ("fishing") and natural mortality by regression on catching effort, and using data for total catches.
3) Marking, to give catching mortality, and again using data for total catch.

[1]Editors' note: written by Sidney J. Holt (FAO, Rome) and John A. Gulland (Fisheries Laboratory, Lowestoft, England) and published in 1964 in Rapports et Procès-Verbaux des Réunions du Conseil Permanent International pour l'Exploration de la Mer 155:147–151. Permission to reproduce this article, with minor editing, has been granted by ICES.

In addition, some data were available for direct estimation by sightings of whales from a research ship. All these methods gave more or less consistent results. One of the consequences of the assessments has been a complete ban by the Commission on the capture, in future seasons, of blue whales (except for a small population of "pygmy" blue whales *B. m. brevicauda*). This poses a problem in estimating the abundance of the recovering blue whale stock in the coming years when there will no longer be catch data from the catchers continuing to hunt for other species.

Catch-per-unit-effort Data

The abundance estimates from methods other than by sightings require a knowledge of catch and catching effort. Detailed data on catches are available virtually from the beginning of the industry; every whale caught is reported, with its length, sex, and, for females, maturity or whether pregnant, to the Bureau of International Whaling Statistics, with a record of the date and area of capture. The ages of sampled whales have been determined in a variety of ways by the laboratories responsible for whaling research in each country. For these assessments, these data were all pooled, and from them age/length keys derived for each sector of the Antarctic and for each month, species, and sex, which were applied to the Bureau's length frequency data to obtain age compositions as a basis for total mortality estimation. It was a feature of this work that this part of it was done entirely by electronic computer; all past data are now on punch cards, and arrangements are being made for updating these after each season.

The measure of effort recorded in the statistics is "catcher's day's work." This is the sum, for all expeditions, of the days spent on the whaling ground by the factory ship multiplied by the number of catchers operating. The effort figures are tabulated by month and by area. The BIWS has, for each expedition, a record of the noon position of each factory ship on each day, the number of whales taken by that expedition on each day, and an inventory of the catchers and certain of their characteristics such as horsepower and tonnage. In each country, the whaling companies have further information about the catchers, their equipment, and crews, especially the gunners.

The first adjustment made to the catcher's day's work figures was to deduct the number of days of "no-catch"—that is, when no whales were taken by any of the catchers of any given expedition, for whatever reasons. This was done because it was known that sometimes catches were made faster than the factory ship could process them, and a delay had to be introduced by a "stop-catch" order to smooth the production process. "No-catch" days could, however, include times when weather was too bad for operation or, occasionally, simultaneous bad luck was experienced by all catchers.

These crude catching effort figures include the times taken searching, chasing and killing, "flagging" the dead floating whale for subsequent pick-up and tow to the mother ship (by the same catchers or by particular catchers temporarily assigned to that task), and all other operations engaged in by vessels other than the mother ship.

Actual catching time could be less than the number of days in the season, on the grounds, and on which catches were made because of (1) bad weather, (2) bunkering, (3) repairs, (4) missions for species purposes, and (5) survey missions. These time losses could be a constant proportion in each season or a variable proportion—the variations being more or less random or showing a trend from season to season. Constant losses or those with small random variation could be ignored; a trend could not be ignored, especially if it were correlated with changes in the abundance of available whales.

From what is known generally about the operations, it has been considered that the incidence of time loss by special purpose expeditions is random; bunkering, weather, and "stop-catch" time loss might have a decreasing trend, but repairs and survey expeditions an increasing trend. Thus catchers have increased in size and improved in design and can, therefore, continue to operate longer in adverse weather, especially with other improved equipment. Introduction of diesel engines has reduced bunkering, but increased repair time.

The factors discussed also could have caused a change in the number of hours per whaling day in which effective operations could be carried out. The effect of weather was, however, thought *a priori* to be the only one which might show a significant trend and was hence singled out for a special investigation, as described below, using logbook records.

With regard to the fishing power of the catchers, there have been, over the years, many changes tending to increase this, perhaps greatly. This trend has been encouraged by the catch restrictions, especially before there was agreed a quota allocation to countries. Catchers have been getting larger and have more powerful engines and higher chasing speeds. Searching has been improved by navigational aids and use of aircraft; the time taken to kill a whale has been reduced by several methods, including the use of Asdic in the chase. Because chasing and killing have become so efficient, taking at most, and usually less than, three hours for each whale, and the number of whales caught has fallen to the order of one per day's work, the efficiency of the effort is now governed mainly by the efficiency of searching. The skill of gunners is variable and very important; the best gunners usually tend to have the most skilful crews supporting them, but they also get assigned to the newer, bigger, faster and otherwise better catchers. All these factors are, therefore, correlated with the size and horsepower of the catchers, and it was considered that as a first approximation, an effort correction based upon the relation of fishing power to size or horsepower would also, therefore, include the effects of the other main factors.

Weather

Catches are greatly affected by weather conditions; catching is difficult in high winds and in poor visibility (Gulland and Kesteven 1962; Kurogane and Nemoto 1963). The weather effect was analyzed in two stages, first to express the effect of weather in quantitative terms, and then to use these data to correct past measures of effort to standard values independent of the weather. Data from expeditions of all countries were analyzed. These consisted of daily records for each expedition of catch of each species in numbers, and the noon weather (wind force on the Beaufort scale, and visibility at the position of the factory ship classified as: good, moderate, or poor). These gave very clear results, with little difference between expeditions or from one year to the next. Best catches were taken when the visibility was good, and the winds below force 3; in stronger winds the catches dropped off, slowly at first, but rapidly above force 6 (when they were about half the maximum), until in gale force winds (above force 8) catches were negligible. In moderate or poor visibility, catches were less than the maximum even in light winds, and also decreased with increasing wind. Thus, given the weather on any day, the catch or effective effort can be expressed as a proportion of the catch (or of the effort) that would have been made if the weather conditions had been "standard" (taken as wind force 3–4 with good or moderate visibility). For example, a day's catching in moderate visibility and wind force 5 is equal to 0.65 of a day's fishing in standard conditions. Theoretically, therefore, all the effort data could be corrected to standard conditions. Unfortunately, the complete records were not available, nor was there sufficient time to carry out a full analysis, and only the records of Norwegian (from 1946/47 to 1957/58) and British (from 1936/37 to 1957/58) expeditions were analyzed. These showed that there are considerable year-to-year fluctuations in the effectiveness of the average day's fishing throughout the season—between 0.65 and 0.87 of a standard day. This variation is not very important, though to remove it should improve estimates such as those of year-to-year mortality. More important, as it could lead to bias, is the fact that there are trends in both Norwegian and British data, the average day's fishing becoming less effective in recent years. This is presumably due to the known change in distribution of the fleets (as the whales, and particularly the blue whales, became scarcer) away from the relatively calmer waters close to the Antarctic continent to the stormier waters farther north. However, the trend is not very strong, amounting, in the British data, to a decrease in efficiency over the years of around 15%, and in the Norwegian data (covering a shorter period) of about 10%. These changes are probably negligible in comparison with trends in the opposite direction due to increased size and efficiency of catchers and, therefore, for simplicity and to avoid further computation, the effects of weather were ignored in the analyses made so far. A more detailed analysis could, however, take these trends into account, if only to reduce the variance of estimates of mortality and other population parameters.

Fishing Power, Catcher Size, and Horsepower

Three methods are available for studying this relation from the catch statistics and logbooks. We may make comparisons of catch per catcher day by pairs of expeditions in successive years, chosen so that one of the expeditions has the same catchers in both seasons while the other has some change in the second season—for instance, replacement of one or more of its older and smaller catchers by bigger or more powerful ones. Secondly, comparisons could be made within expeditions of catches by the different catcher boats. Thirdly, comparisons could be made of catches by different expeditions in the same area and over the same time-period, in cases where the average sizes of the catchers in the expeditions differ. Not all of these methods have, so far, been successfully applied. The corrections eventually used were based on a study by Norwegian scientists of the catch of individual catchers in 4–7 Norwegian expeditions during the seasons 1938/39, 1947/48, and 1960/61. For the last two seasons, catching times were corrected for length of day[2], wind, and visibility. The catching powers of the various catchers were calculated as the ratio of their catch per corrected catching time relative to the mean catch per corrected time of catchers of 800, 1300, and 2000 IHP [indicated horsepower] in the seasons 1938/39, 1947/48, and 1960/61, respectively.

Regressions of these indices on IHP and on tonnage were calculated. The regression coefficients for IHP differed from season to season, but those for tonnage were nearly constant ($r = 0.7$). Furthermore, although the regression slopes for both IHP and tonnage were significant, the intercepts were not significantly different from zero—indeed, in some cases, they were positive and in others negative. For the purposes of the assessments, the catching powers were taken as proportional to catcher tonnage, and in making a first correction to the effort statistics, the number of catcher's day's work (excluding days of no-catch) in each season was

[2]Searching takes place by daylight; the hours of daylight change through the season and also with the latitude of operation, which as mentioned earlier has altered in a way correlated with abundance changes.

multiplied by the mean tonnage of all catchers operating during that season. Over the period considered, there had been a nearly threefold change in the average tonnage, so the correction made a substantial difference to the annual mortality estimates and their regression on effort.

Population Estimates from Marking

Whales have been marked since the 1930s by firing numbered metal cylinders into them, these being recovered in the factory vessel during processing, often so late in the processing that the individual whales from which the marks came cannot be identified. As for many fish, not all marks in recaptured whales are recovered, and also "loss" of marks similar to the death of fish at the time of, or soon after, marking occurs, in that one cannot be certain that a mark fired at a whale has hit it and become firmly embedded. Direct estimation of fishing mortality from marking is not too easy, and until the present studies, marking analyses had been confined mainly to the study of movement and the separation of stocks. Several analyses of marking data were made for these assessments. First, the percentage recoveries in the seasons 1957/58 to 1960/61 of fin whales marked in the 1930s and in the 1950s were compared. This gave a direct estimate of the total mortality of the early marked whales between the two marking periods, independent of changes in catching efficiency or of recovery efficiency. The values of average total annual mortality obtained in three different areas were 0.071, 0.065, and 0.147. Comparisons were also made of the returns per unit effort (thousand catcher-days) in the seasons after marking, of fin whales and blue whales, for the pre- and post-war marking experiments. These can differ for the following reasons: (a) different efficiency of the unit of fishing effort, (b) changes in the preference for the particular species, and (c) changes in the proportion of marks in captured whales which are detected and returned. The data are given in Table 1.

For fin whales, the numbers returned are of those reported as captured in the seasons 1 and 2, i.e. omitting the season immediately after marking, when the distribution of marks might be abnormal. For blue whales, numbers are too few to do this, so seasons 0 and 1 were used; these are shown separately.

The table shows that the percentage returns per unit effort for fin whales has increased sixfold, and that of blue whales fourfold. This difference is presumably due to the increased attention paid to fin whales as the blue whale stock decreased. If the attention paid to blue whales has not decreased, the mortality of fin whales caused by unit effort has increased by a factor of 6/4 = 1.5. The remaining fourfold increase may be due either to increased catcher efficiency or to increased probability of detection in return of marks. Unfortunately, there is no other measure of whether the chances of recovery have changed or by how much, though

Table 1. Results of marking experiments in Antarctic whale stocks.

					Blue (all areas)			
	Fins Area II		Fins Area III		Season 0		Season 1	
Item	Pre-war	Post-war	Pre-war	Post-war	Pre-war	Post-war	Pre-war	Post-war
Number marked	2,171	267	749	394	693	249	693	249
Number returned	45	19	6	10	22	26	9	13
% returned	2.1	7.0	0.8	2.5	3.2	10.4	1.3	5.2
% per unit effort	1.28	8.17	0.46	2.84	1.72	6.38	0.77	3.15
Mean year of return	1937	1956	1936	1957	1936	1957	1937	1958
Ratios of returns per unit effort	6.37		6.15		3.71		4.09	

it seems unlikely that it has increased fourfold. The marking data, therefore, show that there has been a considerable increase in catcher efficiency, possibly by fourfold; thus the rise in efficiency of about threefold in correcting the fishing effort figures, based on tonnage data, may well be an underestimate.

Population Estimates from Sightings

The earliest data collected to give direct estimates of the absolute abundance of the Antarctic whales were the sighting records from the RRS "Discovery" during the seasons 1933/34 to 1938/39 (Mackintosh and Brown 1956). During their work in Antarctic waters, a continuous watch was kept, and the numbers of whales of each species sighted within a certain distance each side of the ship were recorded. These data were divided into zones, within each of which the numbers were expressed as numbers per unit area. By multiplying by the area of the zone, the total numbers in each were determined, and hence the total numbers of each species in the Antarctic stock. The advantage of this method is obvious—a direct measure of absolute abundance is obtained, free of all the difficulties inherent in the catch-per-unit-effort-data. There are, however, disadvantages. For instance, it is possible that not all the whales of the Antarctic stocks enter Antarctic waters every season, or some may leave before the last ones arrive. This means that any count even at this peak of the season will be an underestimate of the total stock. Also, it was not easy to be sure how far away a whale was, so that some whales that should have been included in the count were omitted, or conversely. The choice of distance can itself be important—if too short (e.g. counting only whales within half a mile of the ship), few whales would be counted, and the results would be likely to be variable, whereas at longer distance, the probability that a whale would be missed or misidentified must increase. This problem was met by estimating separately the numbers of whales seen between 0–1, 1–2, 2–3, 3–4, and 4–5 miles from the ship. In the most distant zone, only about 1% of the number of whales in the nearest zone was seen. Assuming, from data of frequency of blowing etc., that 90% of the whales in the nearest zone were seen, it was calculated that only 30% of whales within 5 miles of the ship were seen, and a factor of 1/0.3 was used to correct the number of whales recorded.

It has also been assumed that, within a zone, the distribution of the survey vessel is independent of the distribution of whales. This is probably true of the "Discovery" records and could be ensured in any special survey in future, but is most unlikely to be true for some other ships, particularly records from the whaling ships themselves. However, it is hoped, now that catching of blue whales is banned (except for one area), to use the sightings of blue whales from ships hunting fin whales to give a measure of relative abundance of blue whales.

Human errors ensure that bias is still possible, even when using sightings only as a relative measure. When whales are relatively abundant, the look-out is likely to be wide awake, familiar with the appearance of whales, and confident in his ability to see and recognize them. All whales seen are, therefore, likely to be recorded (unless whales are appearing all round the ship, and this is unlikely to occur in the Antarctic within the present century). When whales are scarce, the look-out is more likely to be asleep (though this problem can be dealt with); more important, he may be less confident in recognizing whales, and some would not be recorded because of his uncertainty. Further general studies of the collection and analysis of sighting records are being continued, particularly of the sampling problems involved: how accurate is an estimate by sightings from one vessel likely to be, or, how much survey work is required to give estimates with a specified precision?

Summary

The methods of estimating the size of the stocks of Antarctic whales used by the Special Committee advising the International Whaling Commission are reviewed. These methods include several commonly used in fishery research—DeLury estimates from relating the decline of catch per unit effort to accumulated catch, estimates from relating total mortality to effort, and from tagging data. Estimates were also made directly from the numbers of whales seen in a known area along the track of a research ship. All these methods gave reasonably consistent results. The problems in defining fishing (catching) effort in the whaling industry are discussed.

References

DeLury, D. B. 1947. On the estimation of biological populations. Biometrics 3:145–167.

Gulland, J. A., and G. L. Kesteven. 1962. The effect of weather on catches of whales. Paper IWC/15/6 App. 3 presented to the meeting of IWC Scientific Committee and Three Scientists, Seattle 6–19 December 1962. Unpublished. 7 pp. [Available from the IWC Secretariat].

International Whaling Commission. 1962, 1963, 1964. Twelfth, Thirteenth, and Fourteenth Annual Reports.

Mackintosh, N. A., and S. G. Brown. 1956. Preliminary estimates of the southern population of the larger baleen whales. Norsk Hvalfangsttidende No. 9.

Kurogane, K., and T. Nemoto. 1963. The influence of meteorological factors on the Antarctic whaling. Norsk Hvalfangsttidende 52(4):89–104.

Chapter 8

Report of the Committee of Four Scientists, June 1964[1]

At its Fifteenth Annual Meeting, the Commission asked the Committee of Four, which conducted the special scientific investigation of Antarctic whale stocks, "to continue this work for one more year."

The Committee and the governments and agencies supporting them all agreed to this extension; however, approval of the budget for the continuing investigation did not come through until April of this year, and for this reason some work that might have been undertaken has not been possible. It is fortunate, therefore, that the 1963/64 results have confirmed further the original analysis in that the predictions made by the Committee of Four have been borne out by the 1963/64 catches.

Analysis of the 1963/64 Catches

Blue and Humpback Whales

Since the Commission accepted the recommendations of this Committee with respect to a cessation of catching of humpbacks *Megaptera novaeangliae* and essentially so with respect to blue whales *Balaenoptera musculus* (except for a small area which remains open for the catching of blue whales), there are effectively no new data to analyze for these species. The Committee repeats that the Commission should organize systematic recording of sightings so that some attempt could be made to measure the response of these stocks to protection and to obtain more precise predictions when it will be feasible to resume catching them. If this can be undertaken in the course of normal whaling operations, it would usefully be put into effect as soon as possible. Any cessation of present whaling operations would, however, require the organizing of special sighting expeditions.

Fin Whales

In their Supplementary Report made to the Commission at its Fifteenth Meeting (Fourteenth Report, page 93), the Committee predicted that "if whaling continued at the 1962/63 level of effort, the catch in 1963/64 would only be 14,000 (fin) whales." In fact, the catch was 13,853 fin whales *B. physalus*, though it should be pointed out that these were taken with 91% of the effort (catch days corrected for tonnage) expended in 1962/63. Thus, the forecast made upon the basis of the previous calculations was correct to within 10% of the actual catch attained.

[1]Editors' note: this was presented at the Sixteenth Meeting of the IWC held in Sandefjord, Norway in June 1964 and published in Reports of the International Whaling Commission 15:47–63, 1965.

The catch per unit of effort (catcher's day's work corrected for tonnage) in 1963/64 was 0.96 as compared to 1.19 in 1962/63. The corresponding figures together with population estimates for the past several seasons are shown in Table 1.

It should also be noted that in the Supplementary Report, it was stated that: "the stock at the beginning of the 1963/64 season would also be about 40,000 whales." Since 32,400 is the mid-season average, if to this we add half the catch (6,900), the conclusion is that the pre-season total was 39,300. Again, the agreement is excellent.

Proceeding in the same way, it appears that at the beginning of the 1964/65 season, the stock of fin whales available to the fishery will be 35,000–36,000. Thus, with the same effort as in 1963/64, we would expect the 1964/65 fin whale catch to be in the neighborhood of 12,000, and less again in the following years.

Catches of this magnitude will continue to deplete the stock despite the fact that the recruits which are the offspring of the larger mature stocks of three or four years ago continue to be relatively numerous. The sharp decrease in stock size over the past few years means that the recruitment in the next few years will also decrease sharply.

The Committee estimate the sustainable yield from the present stock to be between 4,000 and 5,000 whales. Catches much in excess of this number will cause continued rapid depletion of the stocks. Catches below this level will allow rebuilding of the stocks to increase future productions. Such rebuilding would take place at a rate increasing with the increasing difference between the sustainable yield and the actual catch. In setting the quota for future years, the Commission must decide whether it should continue to live off (and deplete) its capital (which it could only do at the present rate for, at most, four years, and that only by progressively and substantially increasing the effort), or whether it should put something back in the "bank" to produce for the future.

Sei Whales

The catch of sei whales *B. borealis* increased this season by nearly 3,000 whales, from 5,482 in 1962/63 to 8,256 in 1963/64. This increase is, it will be noted, less than the drop in the number of other species take (fin whales, blues, and humpbacks totaled 5,883 less) or even in the number of fin whales alone, which dropped by 4,783. This would support the contention, in the Supplementary Report, that the sei whale stock is not big enough to sustain "catches that will support an industry comparable to that which has been operating in the

Table 1. Fin whale average population size 1955/56–1963/64 as calculated from catch per unit of effort (catcher's day's work corrected for tonnage).

Season	Catch per unit of effort	Population size (thousands)
1955/56	3.29	110.4
1956/57	3.03	101.7
1957/58	2.68	89.9
1958/59	2.64	88.6
1959/60	1.96	65.7
1960/61	1.78	59.7
1961/62	1.35	45.3
1962/63	1.19	40.0
1963/64	0.96	32.4

Antarctic," since it may be presumed that if sei whales were abundant, the expeditions would take them in at least equivalent numbers when the other species are scarce.

The need for consideration of the sustainable yield for this species will be further accentuated if, as the Committee recommends, the blue whale unit is abolished so that separate quotas for each species have to be considered. The small amount of available data for this species makes possible only an approximate assessment of the situation. A lower limit of the present sei whale population can be obtained on the assumption that under the conditions prevailing last season when quotas were obtained with difficulty, if at all, the ratio of sei to fin whales in the catches represents the ratio of their populations. The estimate obtained in this way is 20,000 sei whales. This estimate could be subject to correction for sei whales occupying areas in the Antarctic where whaling did not take place, and an alternative approach can be obtained from the sighting surveys of 1933–1939. At this time, the proportion of sei whales seen was very low in relation to the other species, although there were difficulties in observing and recognizing this species. The total fin whale population at that time is estimated at about 125–150,000, so that 50,000 would seem to be a high estimate of the corresponding sei whale population. It is possible that some increase and some southward movement has occurred in this species as a result of relaxation of competition from other baleen whales, but on the other hand, 26,000 sei whales have been taken in the last five years. An extreme upper limit of the present sei whale population would, therefore, seem to be 70,000. Taking, in the absence of specific data, the best estimate of $(r - M)$ for fin whales (0.12) as applying to sei whales, the present sustainable yield for this species is between 2,400 and 8,400 whales, with a probability of being nearer the lower limit. Since the actual catch in 1963/64 (8,256) almost exactly equaled the upper limit, it is likely that the sei whale population is already being reduced, and that any further increase in catches will inevitably deplete it at a rapid rate.

In order to obtain more reliable estimates of the sei whale population and its sustainable yield, it is recommended that the Commission and its Scientific Committee should encourage further collection and analysis of biological and catch data regarding this species and should take immediate steps to supplement the population estimates based on catch-per-effort data by including sei whale sightings in a system of sighting records set up.

Pygmy Blue Whales

In their Final and Supplementary Reports (Fourteenth Report, pages 40, 93), the Committee made a brief appraisal of the scanty existing data for these stocks, and accepted an analysis by Dr. Doi suggesting that the initial stock was of the order of 4,000, that 2–3,000 remained after the 1962/63 season, and that the stock then could sustain a catch of no more than 3–400 whales annually, and perhaps even less than 200 if the natural mortality was higher than that of blue and fin whales, as the data seemed to suggest.

The 1963/64 catch was only 113 whales, i.e., less than the present sustainable yield. Drs. Ichihara and Doi have submitted to the Committee an extract from a paper further analyzing this stock ("Stock assessment of pygmy blue whales [*B. m. brevicauda*] in the Antarctic"). This latest paper raises some new questions in regard to possible reduction of the size limit and consequent exploitation of younger whales which the Committee have not studied; the Committee also have reservations on some of their new estimates of present stock size and its relation to the optimum level. However, the Committee are in general agreement with their conclusion that "the stock of pygmy blue whales could stand the annual catch of 400 whales for three seasons."

The Committee recommend that the Commission should make provision for safeguards that no greater number of pygmy blue whales are taken and that the status of this stock and the permitted catch be reviewed carefully as soon as additional data are available.

Further Examination of the Effects of Different Quotas, with Reference to Fin Whales

In the Supplementary Report, the Committee showed that to obtain the maximum long-term total catch, it is necessary to stop catching completely until the stock has built up to the level giving the maximum sustainable yield, and then to crop the stock at that rate. Any other pattern of management, e.g., one that permitted whaling to recommence before the stock had reached the optimum level, or one that allowed continued whaling after the present season at some low level, would give, over the years, a total yield less, and possibly very much less, than that which would be obtained by stopping whaling completely until the stock reached its optimum level.

An important element omitted from our calculations was the effort expended in taking the catch, which, of course, largely determines the cost. When the stock had built up to its optimum level, the catch per catcher day would be several times higher than at present, i.e., about 3–4 whales per catcher day at the present level of catcher efficiency. From that stock, the sustainable yield could be taken by a fraction of the present effort—around 5,000 catcher days, which could be exerted by 3–5 expeditions, depending on the number of catchers per expedition and the length of the season. This effort is no more than would be required to take, at the present low level of stock, the small sustainable yield of 4–5,000 whales. Over a long period, not only would the greatest total catch be taken by first stopping whaling for the time required for the stock to build up to the optimum level and then taking the maximum sustainable catch each year, but the total effort needed to take this total catch would be less than that required to take smaller total catches under different patterns of management.

The calculations presented in the Supplementary Report were also based on one particular set of estimates of the population parameters (recruitment, mortality, etc.). There is still some uncertainty about the exact values of these parameters. The exact values do not affect the general results obtained—over a sufficiently long period, the best management policy will always be complete closure until the stock has reached its optimum level. They do, however, affect the length of time needed for this optimum policy to show its superiority. If the net potential rate of increase is high, the stock could, with no whaling, increase to its optimum in ten years, and, cropped at this level for another ten years, give a greater yield over the 20-year period than would be obtainable from any other management policy. If, however, the potential rate of increase is low, then the stock might take 20 years of zero catches to reach the optimum level, and better catches over the 20-year period would be taken by setting a small annual quota throughout the period. The policy of closure until the stock reached the optimum level would, of course, still be the best when considering total catches over, say, a 40- or 50-year period.

For these reasons, and because intensive whaling took place during the 1963/64 season, further calculations of the effect of different quotas were required. These calculations were made on a computer using the following values:[2]

[2]The calculations and analyses reported in this section were made by Mr. Gulland before the results of the 1963/64 season were known. The actual catches were, however, so close to the predicted value that there was no need to recalculate to give new estimates of the present condition of the stocks and of the prospects for the 1964/65 and following seasons.

Stock at the beginning of the 1963/64 season	40,000 whales
1963/64 catch	14,000 whales
Net potential rate of increase ($r - M$)	0.08, 0.10, 0.12, 0.14, 0.16, 0.18

Earlier computations to predict the effects of zero catches over the next 20 years for a wide range of values of r, M, and age at recruitment showed that the results were only affected substantially by changes in the value of $r - M$, and accordingly only $r - M$ was varied for the calculation presented here. In these calculations, we took the stock at the beginning of a season, subtracted the catch, added the recruitment (from the parent stock of earlier years), and reduced this quantity by multiplying by a factor of $(1 - M)$ to give the stock at the beginning of the next season, and so on. The catch was determined by two sets of figures. A target figure of the desired steady yield was set, which was, however, only taken if it was less than the sustainable yield, which is $(r - M)$ multiplied by stock; otherwise, an initial quota, equal to zero or 1, 2, 3, 4, 5, or 6,000 whales, was taken. Target figures of sustained yields of 5, 10, 15, and 20,000 whales per year were used. This gave $4 \times 7 = 28$ possible patterns of management.

In Table 2, the results of each possible management pattern, in terms of total catch taken (in thousands of whales), and the total effort expended (in thousands of catcher days), over the 24-year period starting with the 1964/65 season, are shown for each of the six alternative values of the net potential rate of increase. At the highest rate of increase, 24 years is a period short enough to show clearly the advantage of a complete cessation of whaling until the stock level at which the maximum sustainable yield (20,000) could be taken is reached; the total catch is, in this case, higher and, with one exception, the total effort lower than for any other pattern of management. At lower rates of increase, 24 years is not long enough for the policy of initial complete closure until the optimum level is reached to be superior in all respects. Thus, for $r - M = 0.14$, the biggest catch within 24 years is taken by complete closure until the stock can sustain a steady yield of 15,000 per year, which is taken thereafter. However, by postponing the recommencement of catching until the stock could sustain a yield of 20,000 whales (i.e., another four years of closure), the yield taken would be only 11% less than the maximum and would be taken with 36% less effort.

At still lower rates of increase, the stock might not be able to recover to its optimum level (or even to a level which would give a sustainable yield of 10,000 or 15,000 whales) within 24 years. The largest catch in this period would be taken by setting some non-zero quota, though at the cost of extremely high effort and at the risk of driving the stock to extinction. The table shows clearly that by waiting until a moderate catch can be sustained, the total effort expended in the full 24-year period will be greatly reduced.

Table 3 shows, for $r - M = 0.12$, the best present estimate of the net rate of increase, the intermediate stages after 6, 12, and 18 years. The stock at the end of each period is shown as well as the total (accumulated) catch and effort during these years. Consideration of the stock sizes shows very clearly the difference between a quota just below the present sustainable yield (4,000) which allows the stock to build up and one just above the sustainable yield (5,000) which reduces the stock, at first slowly, but then more rapidly. The difference in catch over 24 years is 25%, but requires 76% more effort, and there is nearly threefold difference in stock size at the end of the period. Clearly, if 5,000 whales continued to be taken each year, the stock would very quickly be extinct. While, in view of the present uncertainties of the values of r and M, the present sustainable yield, which constitutes the critical level, will not

Table 2. Accumulated catch (in thousands of whales) and effort (in thousands of days) after 24 years (C = catch, E = effort).

Steady yield	Initial quota													
	0		1000		2000		3000		4000		5000		6000	
	C	E	C	E	C	E	C	E	C	E	C	E	C	E
	$r - M = 0.08$ (present sustainable yield 2755)													
5,000	45	47	40	48	48	75	72	143	96	340	85	326	78	299
10,000	0	0	24	31	48	75	702	143	96	340	Extinct		Extinct	
15,000	0	0	24	31	48	75	72	143	96	340				
20,000	0	0	24	31	48	75	72	143	96	340				
	$r - M = 0.10$ (present sustainable yield 3523)													
5,000	85	111	84	111	84	115	78	116	96	192	105	340	90	305
10,000	50	31	42	38	48	60	72	108	96	192	Extinct		Extinct	
15,000	0	0	24	26	48	60	72	108	96	192				
20,000	0	0	24	26	48	60	72	108	96	192				
	$r - M = 0.12$ (present sustainable yield 4321)													
5,000	110	169	112	170	111	175	110	175	110	178	120	243	114	358
10,000	120	90	114	92	104	94	86	96	96	138	120	243	Extinct	
15,000	90	45	80	51	61	56	72	84	96	138	120	243		
20,000	40	15	24	22	48	49	72	84	96	138	120	243		
	$r - M = 0.14$ (present sustainable yield 5150)													
5,000	120	166	120	166	120	166	120	166	120	166	120	166	120	166
10,000	160	145	159	139	160	151	149	150	138	150	120	166	144	302
15,000	180	109	164	102	152	105	132	107	107	114	120	166	144	302
20,000	160	70	138	68	120	74	89	78	96	107	120	166	144	302
	$r - M = 0.16$ (present sustainable yield 6012) *													
5,000	135	135	135	135	135	135	135	135	135	135	135	135	135	135
10,000	190	174	195	206	192	198	191	208	186	206	175	198	156	207
15,000	225	145	220	147	217	156	204	153	195	164	160	157	144	193
20,000	240	116	233	123	210	116	191	120	176	131	135	135	144	193
	$r - M = 0.18$ (present sustainable yield 6904) *													
5,000	240	180	240	180	240	180	240	180	240	180	240	180	240	180
10,000	210	218	213	243	208	212	212	250	210	244	205	232	204	252
15,000	255	167	262	198	256	196	252	198	239	190	230	197	216	212
20,000	300	163	290	162	282	165	276	174	256	170	240	180	200	180

* For $r - M = 0.16$ and 0.18, the steady yield of 5,000 is less than the sustainable yield; the stock, therefore, increases until after 25 (if $r - M = 0.16$) or 15 years (if $r - M = 0.18$), the stock reaches a level at which the sustainable yield is 20,000. Thereafter, a catch of 20,000 whales is assumed to be taken.

necessarily be exactly that indicated above, the essential point, which the example illustrates, is the long-term difference in results between taking a steady catch just above the critical level and taking one just below this level.

It has not been possible to examine very deeply the economic implications of the alternative management policies described here. Clearly, it is not sufficient to consider only the total catches taken; the times when the catches are taken—the immediate future or some more distant time—are also important, but probably more important are the effort (i.e., cost) required to take the catch, and also the stock (i.e., effectively the major capital asset of the whole industry) available at any time. In economic terms, the difference between the situation in the recent past and the situation which would exist when the stock had been permitted to increase to a level giving the maximum sustainable is very great. The difference between catches in the past and the potential catch is not very great, but whereas to obtain catches at the level of present years has required about 20,000 or more catcher day's work per season (and would require progressively more effort to maintain for a few more years), the maximum sustainable yield could be taken from a stock at the optimum level by a fraction of this effort, probably around 4,000–5,000 catcher days. The total cost of catching and processing would probably

Table 3. Stock abundance and accumulated catch in thousands of whales at 6-year intervals ($r - M = 0.12$) (S = stock, C = catch, E = effort).

Time interval	Initial quota																				
	0			1,000			2,000			3,000			4,000			5,000			6,000		
	S	C	E	S	C	E	S	C	E	S	C	E	S	C	E	S	C	E	S	C	E
	Steady yield 5,000																				
6 years	42	25	39	45	22	34	43	24	39	43	24	39	42	25	42	38	30	52	33	36	63
12 years	41	55	87	44	52	80	42	54	86	42	54	86	42	53	87	35	60	107	23	72	150
18 years	45	80	124	44	82	125	43	81	128	42	82	130	42	82	133	30	90	168	7	108	304
24 years	45	110	169	44	112	170	42	111	175	43	110	175	42	110	178	23	120	243			
	Steady yield 10,000																				
6 years	63	0	0	58	6	9	53	12	18	48	18	28	43	24	40	38	30	52	33	36	63
12 years	85	10	8	83	12	15	71	24	32	59	36	52	47	48	76	35	60	107	23	72	150
18 years	89	60	53	90	54	47	83	52	54	75	54	70	52	72	109	30	90	168	7	108	304
24 years	91	120	90	90	114	92	88	104	94	87	86	96	61	96	138	23	120	243			
	Steady yield 15,000																				
6 years	63	0	0	58	6	9	53	12	18	48	18	28	43	24	40	38	30	52			
12 years	95	0	0	83	12	15	71	24	32	59	36	52	47	48	76	35	60	107			
18 years	128	15	8	120	18	19	97	36	42	75	54	70	52	72	109	30	90	168			
24 years	134	90	45	128	80	51	125	61	56	99	72	85	61	96	138	23	120	243			
	Steady yield 20,000																				
6 years	63	0	0	58	6	9	53	12	18	48	18	28	43	24	40	38	30	52			
12 years	95	0	0	83	12	15	71	24	32	59	36	52	47	48	76	35	60	107			
18 years	142	0	0	120	18	19	97	36	42	75	54	70	52	72	109	30	90	168			
24 years	181	40	16	175	24	22	137	48	49	99	72	96	61	96	138	23	120	243			

not be reduced in quite the same proportion, but clearly at higher stock levels, both catchers and processing vessels could operate much more successfully.

Catch Per Unit Effort Variation Between Expeditions and Countries

The catches per catcher day over any season vary considerably as between expeditions, and such differences are not substantially eliminated if allowance is made for the varying tonnage of catchers or differences in the areas of operation. The Committee have made a preliminary study of the trends of catch per unit effort with time, grouping the expeditions by countries. Such grouping was made because although there is continuity in the name of the expedition, each may change in several possible ways from season to season, and they felt that any differences would show up more consistently on a country basis. This study has revealed national differences and, perhaps more important from the assessment point of view, national changes which the Committee cannot, at present, explain, but which could be significant.

Figure 1 shows the average catches of fin whales per catcher day over recent years by expeditions from three of the four countries at present participating in pelagic whaling. (For the fourth country, the season-to-season variation is great, as would be expected, it having only one expedition, and results for it have not, therefore, been plotted.) The important features of these graphs are:

a) Until 1957/58, the catch levels and trends were similar for all countries.
b) After that season, divergences appear; in two countries, the trends are the same, both showing a rather rapid decline, although the catches per unit effort by one of the two

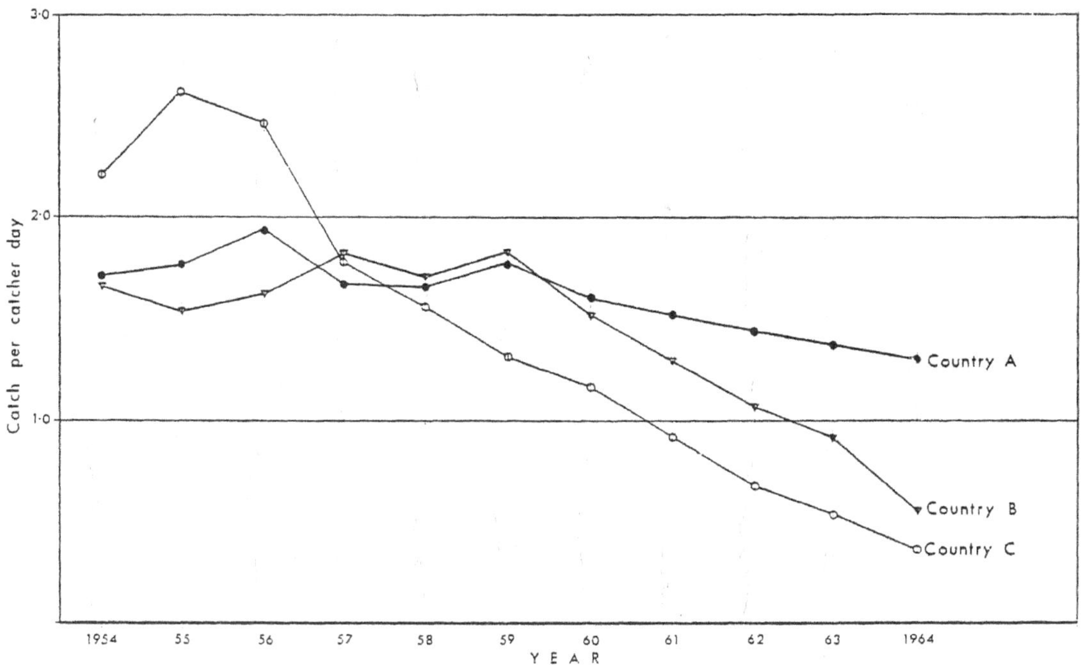

Figure 1. Trends in catches of fin whales per catcher day for expeditions of different nationalities.

are consistently lower than those by the other. For the third country, the decline is much less sharp, with the result that by the 1963/64 season, its rate of catch is double the average value for all countries.

The same features appear if catches by BWUs per catcher day are taken. These systematic differences also remain when comparisons are made between expeditions from two or more countries operating in the same area during the same month. The Committee could not verify this for all countries since, because of general differences in area of operation, there was, in some cases, little or no overlap. These differential features are, if anything, enhanced rather than diminished, if corrections are made for the differing mean tonnages of catchers as between countries.

It is difficult to see how such discrepancies could have arisen by, for example, some undetected changes in the method of recording effort—the number of catcher days per expedition per season shows little trend or variation. It may be that the expeditions of one country have achieved greater efficiency each successive season than the expeditions of the other countries. It is hardly conceivable that the inverse could be true, i.e., that the expeditions of the other countries have been progressively decreasing in efficiency.

If the expeditions of one country have been increasing markedly in efficiency and to a greater extent than is accounted for by the increase in tonnage, then inclusion of its catch-per-effort data will lead to an underestimate of the extent to which the stock has declined and will have a similar though lesser effect on the estimates derived from the overall figures which have been used by the Committee.

Not being in a position to do more than suggest hypotheses, the Committee have not attempted to take account of these differences in their assessments, but they recommend that the Scientific Committee of the Commission make further studies of these differences, with a view to identifying their causes and thus perhaps obtaining some appraisal of the changes in efficiency which may have been occurring.

Comments on Some Papers by Dr. Doi

Dr. Doi and his colleagues submitted to the Committee two papers on the fin whale stock. Some of the detailed scientific points in these papers are discussed in an appendix to this report, but the important point of agreement and disagreement between the conclusions of Dr. Doi and his colleagues and those of the Committee, as expressed in this and previous reports, are mentioned here.

First, there is agreement that the maximum sustained yield is around 20,000 whales, taken from an optimum stock of around 200,000 whales.

Secondly, there is agreement that the present population abundance is much less than that giving the maximum sustained yield.

The differences lie mainly in the estimates of the absolute numbers of whales in the present stock and in the precise shape of the curve relating adult stock and recruitment. As discussed in the appendix, there is good evidence that Dr. Doi's estimates of stock size are too high and his estimate of the optimum stock, that which will provide the maximum sustainable yield, is relatively too low. These differences mean that in Dr. Doi's report, the present sustainable yield is overestimated and the relative increase in stock, and the time required, to reach the optimum level is underestimated.

Even so, the estimate of the sustainable yield from the 1963 stock which may be made from Dr. Doi's Figures 21 and 22 is only 11,000–13,000 whales, and this is less than the 1963/64 catch, and considerably below the quota for 1963/64 (10,000 BWUs or, allowing for, say, 1,500 BWUs as blue and sei, 17,000 fin whales). The present stock is estimated as about a half to a third of the optimum. The practical implications of the analysis made by Dr. Doi and his colleagues are, therefore, the same as the Committee's recommendations, i.e.,:

(i) if the decline in fin whale stocks is to be halted, the catches must be very substantially reduced;
(ii) to reach the optimum level in the quickest time, catching should be completely stopped for a period.

Other Matters

Abolition of Blue Whale Units

In its Final Report, the Committee recommended that the blue whale unit be eliminated and separate quotas be established for each species. The Committee wish to reiterate that the rational management of the stocks now and in the future requires that each stock be treated separately. The maximum long-term utilization of the fin whale resource requires that steps be taken immediately to stop the depletion of this resource and to begin rebuilding the stock. This can be done only if the fin whale catch is substantially restricted over that taken in the past year, which can best be done if a specific fin whale quota is established.

Future Assessments

The Committee wish to reaffirm the recommendation in their Supplementary Report that population analyses be continued. A system of census by sightings is required for those stocks the catching of which is not permitted. A special detailed study is now urgently required of all existing data for sei whales, with a view to further narrowing the limits of estimates of their stock size.

Statistical Requirements

The Committee have found that the catch and effort statistics provided for them by the IBWS [International Bureau for Whaling Statistics] have been commendably prompt and satisfactory for their purposes. As regards future statistics for routine continuing assessments, the breakdown and definitions of the Rome Workshop, as amended by the following London meeting of the Commission's Scientific Committee, have proved a suitable pattern, and compilation in that form should be continued. This might, however, be an opportune time for the Commission and the IBWS jointly to review the statistical tables published by the Bureau, with a view to suggesting any changes which might make them more suitable for assessment purposes, including the possible incorporation in them of some of the special tabulations that have been made for the Committee.

Appendix to Report of Committee of Four Scientists

Comments on T. Doi's Paper "Estimation of the Size of the Fin Whale Population in the Antarctic"

The method used in this paper is based on determining an average value of "q"—the ratio between the instantaneous fishing mortality rate and the fishing effort corrected for tonnage. Briefly, Dr. Doi's method has been to determine an average value of q for each of the six areas using the mortality estimates determined by the Committee. From these, an average q is determined for each area for all seasons, pre-war and post-war combined. These are then applied to the effort in each season to obtain a population estimate by area and by season. The estimates of total population at each season are then combined with the effort data to produce an estimate of q for that season. The mean of the seasonal values of q is then used with the effort data to produce a final estimate of the population for that season. The resulting estimates are more than twice those obtained by the Committee and, therefore, it was necessary for us to examine the methods and calculations closely.

The following points need to be made:

a) Dr. Doi has taken the figures given by the Committee for Z, the *total* mortality rate, to be equal to F, the fishing mortality rate, whereas these differ by M, the natural mortality rate.
b) The formula (3) used by Dr. Doi in Section 3 is incorrect; the correct formula using Dr. Doi's notation is $C = F/(F + M)P[1 - (1 - E)e^{-M}]$, where P refers to initial population and other symbols are as defined. Formula (3) would be correct only if $M = 0$.

c) The author makes a number of assumptions; these are:
   (i) that the value of q is a constant over all seasons both pre-war and post-war;
   (ii) that the effort used has been evenly distributed over the areas; this is probably incorrect to a very great degree;
   (iii) there is no population of whales in any area which is unfished in any season.

Further, in regard to the q values actually obtained, that for Area III is very low due almost entirely to an anomalous negative estimate of fishing mortality in this area in 1932–34; the resultant population estimates for Area III are very high. For recent seasons, these populations are:

1959	159,100
1960	129,300
1961	113,400
1962	90,400 (calculated using Dr Doi's method)
1963	76,700 (calculated using Dr Doi's method)

These imply reductions of 29.8, 15.9, 23.0, and 13,700 in these four years when there were catches of 17.7, 15.8, 15.5, and 11,800 in this Area. These imply negative net recruitments in each of these years, which is inconsistent with the rest of the analysis and biologically unlikely, although some emigration to other areas would be not impossible. Thus, acceptance of the qs calculated here leads to unacceptable population estimates.

It is also possible to estimate by Dr. Doi's methods what the catch in 1963/64 would have been if the population had been at the level he estimates. To do this, the population level at the beginning of the 1963/64 season has been computed by subtracting the 1962/63 catch from the population at the beginning of that season and adding recruitment at the probable rates of $r = 0.17$ and 0.19 on the stocks of three years earlier; allowance is also made for natural mortality ($M = 0.05$). From this population estimate, expected catches have been estimated using the actual effort and Dr. Doi's estimated q. Using effort in catcher's day's work (Table 7 in the paper), the resulting estimates of expected catch in 1963/64 are 19,000 and 19,600 for the two r values, respectively. Similar estimates using effort corrected for days of no catch (Table 6 of Dr. Doi's paper) give expected catches of 17,800 and 18,400. The mean of these estimates is 18,700 as against the actual catch of 13,800. Furthermore, when the calculations of that paper are corrected (i.e., using F instead of Z and using the correct formula to get P from C), as, of course, they must be, the result is to increase the estimated population sizes and magnify the discrepancies pointed out here.

The estimates obtained by the Committee of Four were produced by two distinct methods using different data, and these gave a satisfactory level of consistency. They also enabled an estimate of the 1963/64 catch to be made, which was considerably closer to the actual catch than was obtained from Dr. Doi's method. The Committee see no reason to revise the estimates obtained in their previous reports in favor of those put forward by Dr. Doi.

Comments on "On the diagnosis of the Fin Whale Population in the Antarctic and Optimum Level of Catching," by T. Doi, S. Ohsumi, and T. Nemoto

This paper deals with two topics: first, the relation between recruitment and parent stock [(cf. Section IID1(b) of the Committee of Three's Final Report], and secondly, the level of sustainable yield. The stock–recruitment relation was studied in much the same way in both documents by calculating the abundance (in terms of catch per unit effort) of parent stock and recruits some years later, though there are differences in detail. Thus, the Japanese paper considers all recruits, rather than merely females; this causes no important difference, though by combining males and females, some of the variation in the estimates of recruitment may be reduced, but, as the Committee showed in its Supplementary Report, this introduces no systematic error. The Japanese paper also considers the stock–recruitment relation separately for each series and month. This produces diagrams in which the points are extremely scattered and often lead to biologically impossible conclusions; thus, at low values of S, the ratio of R/S, i.e., the number of recruits produced by one female each year varies from 0.33 to 3.03! It is not at all clear what is the best method of combining the various estimates to produce the relation between stock and recruitment for the Antarctic fin whale stock as a whole. The method of averaging the various estimates of theoretical parameters for each series may well cause bias and error which do not arise when considering the parent stock and recruitment obtained for whole stock. Another difference between the two documents is in the age at which recruitment is measured. The Committee of Three took 4-years old—the first age at which the whales are fully recruited—and corrected for the whales caught when 1, 2, or 3 years old. The Japanese figure is based on 3-year-old whales only and, therefore, will be an underestimate because the 3-year-old whales may not have been fully recruited in all seasons, particularly in the early years when the stock abundance was high. This error will tend to make the recruitment curve too flat and, therefore, make the maximum sustainable yield occur too far to the left, i.e., underestimate the size of the stock from which the maximum sustainable yield could be taken.

The final part of the paper by Dr. Doi and his colleagues uses the estimates of population abundance in a previous paper to express the relative sizes of stock, recruitment, and sustainable yield in actual numbers. As explained above, these estimates of population size are probably too high, so that the other estimates are also too high, though this error is in the direction opposite to that discussed in the previous paragraph.

Supplementary Note to the Report of the Committee of Four Scientists[3]

It is pointed out in the Report of the Committee of Four that the actual catch of fin whales in the Antarctic in 1963/64 and the catch predicted by the Committee agreed very closely. The whaling effort actually exerted over the season was slightly less than in 1962/63, though it was higher than in any year in the history of Antarctic whaling up to the season 1961/62. The catch was, in fact, 8% higher than expected on the basis of the actual number of catcher's day's work, corrected for the change between 1962/63 and 1963/64 in the mean tonnage of catchers. It has become clear to us, in discussions with the Scientific Committee that, however, this discrepancy could be accounted for by a difference in the weather conditions in the Antarctic in the two seasons. About half the total catch in 1963/64 was taken in Area II, and in that area, where Norwegian, Japanese, and Netherlands expeditions mainly operated, the weather was more than usually good. The conditions in Area IV (and perhaps in Area III also) where the Soviet expeditions concentrated, were not exceptionally good.

[3]Written by Sidney J. Holt and John A. Gulland.

The effect of these differences can be seen in the catches of fin whales per unit effort by area, which are tabulated below:

Area	Catch per catcher day		% decline from
	1962/63	**1963/64**	**1962/63 to 1963/64**
I	0.58	0.36	38
II	0.91	0.87	4
III	0.96	0.72	25
IV	0.62	0.31	50
V	0.51	0.35	31
VI	0.56	No data	-
All	0.83	0.68	18

In all areas for which there were data, the catch per unit effort declined, as anticipated; the expected decline was, on the basis of the Committee of Four's calculations, about 25%. In Area III, where 35% of the catches were taken, the decline was exactly this and was even greater in all other areas except Area II, where the decline was small.

Thus, our predictions were effectively in even closer agreement with the actual catches than we thought at the time of the Lowestoft meeting of the Committee of Four, and we feel that this is further confirmatory evidence for our estimates of stock size and of current sustainable yields.

Chapter 9

The Maximum Sustainable Yield in Weight, and Related Matters[1]

The operative clauses of the 1974 decision of the IWC, in which three categories of stocks are identified for management purposes, refer to maximum sustainable yield (MSY) without defining that term. The term "optimum yield" is also used, but not defined, although the resolution recognizes that management should not only be based on MSY in numbers by species, but should also take account of interspecific interactions and the weight of whales. Bearing in mind the discussion in Plenary, Technical Committee, and Scientific Committee which preceded the decision, it appears that:

a) the Commission did not intend the terms MSY to apply only to numbers, but to be more general;[2] and
b) "optimum" would include economic factors (such as the relation of input to output) and/or the maximization of a complex parameter such as the sum of the weights or values of the catches of interacting species. The wording of operative paragraph 1(a) and especially 1(b) viz, "... MSY levels <u>and then</u> [*Holt's emphasis*] at optimum ..." can imply that the "optimum" level of stock would be somewhat higher than the MSY level.

The International Convention for the Regulation of Whaling, 1946 does not refer to MSY nor, indeed, to the concept of sustainable yield. The Preamble to the Convention does, however, refer to the achievement of <u>optimum</u> levels as rapidly as possible, and Article V, paragraph 2 lays down that amendments to the Schedule should be such as are necessary, *inter alia*, "to provide for ... <u>optimum utilization</u> [*Holt's emphasis*] of the whale resources," although no definitions are given. The first explicit reference to the sustainable yield criterion for management is in the Resolution adopted by the IWC at its 12th meeting in 1960. This established the Special Committee of Three Scientists to study and report "on the level of sustainable yield that can be supported by these stocks and on any conservation measure that would increase this sustainable yield."

---

[1]Editors' note: written by Sidney Holt, presented as Document SC/SP74/Doc 12 for the special meeting of the IWC Scientific Committee held in La Jolla, California December 3–13, 1974, and published in Reports of the International Whaling Commission 26:203–214, 1976.

[2]I maintain this interpretation, despite the fact that in the first paper on this matter published in a scientific journal since the Commission meeting, Gambell (1974) discussed the MSY concept exclusively in numerical terms. See Holt (1974).

In its First Interim Report submitted to the Commission in June 1961, the Special Committee still does not refer to MSY, although it mentions the methods of analysis "to determine optimum yield." It does, however, note that "while the first assessment of the stocks and the catch that can be sustained must deal with numbers of animals, close consideration must be given to assessing the stock levels in units more clearly related to economic needs, taking into account the yield of economically valuable products and the relative cost of production, which is related to catch per unit effort." In its Supplementary Report, which was considered by the Commission in 1963, the Special Committee noted that although the question of maximum yield in terms of weight or value was not considered in its Final Report, it could be said that "for a given size limit, to the extent that older whales are bigger and more valuable than younger ones, the maximum sustainable yields in weights or values will be obtained in each case at a stock level rather greater than that level which will give the maximum sustainable yield in number." It proposed that appropriate assessments should be made as depleted stocks were permitted to increase towards their numerical optima, and observed that better data would be needed for this purpose than was available at the time.

In initially making assessments in numerical terms, the Special Committee was continuing the existing practice of the Scientific Committee. The Commission had, however, since it took over the BWU concept from the 1944 Agreement among nations interested in Antarctic whaling, been amending the Schedule in accordance with Article V(e)—"maximum catch of whales to be taken in any one season"—with reference to a criterion which was as much based on weights as on numbers. Only when the BWU was abolished (by decision of the 1973 meetings, ten years after such abolition had been recommended by the Scientific Committee and the Special Committee), did the Commission begin to regulate catches by numbers of each species and, in certain cases, of stocks.

The MSY concept is, wherever it is applied to the assessment and management of marine fisheries, usually interpreted in terms of <u>weights</u> of catches. The salmons and fur seals are exceptional in being assessed mainly numerically, in the one case because the river and estuarine catches are taken at an age and a size which is hardly subject to human choice, and in the other case because the pelt is the main product, the value of which is not proportional to the size of the animals. The concept of MSY (weight) is not needed when one is dealing with the capture of individuals whose growth has virtually been completed.

In defining the BWU for baleen whales in the Antarctic, it was recognized that the products of the four species then involved were of like kind, and that there were much greater differences in product yield per whale as between species than within species. The BWU ratios were originally determined on the basis of relative oil yield. They were not subsequently changed, notwithstanding the facts that, in later years, other products than oil became more important, that to maximize the total values, some of the oil yield had to be sacrificed (oil being left in the edible flesh), and that the oil ratios in any case did not stay constant—in particular, they changed as the average size of whales caught in one species changed more than the average size of another species. Thus, in the end, the BWU did not reflect the relative values of individuals of different species. This deficiency in BWU as a management measure was additional to the basic weakness that it permitted, and could even encourage, an imbalance in the exploitation of different species and stocks, including the possibility of continuing to take excessive number from overexploited stocks.

The numerical criterion for sustainable yields is fairly satisfactory, provided that for each species or stock, the average size of individuals caught is not substantially changed as a consequence of whaling. For the large whales, there are, thanks to the efforts of the IWC and BIWS (Bureau for International Whaling Statistics), ample data to test this. The fact that the sizes are given as lengths rather than as weights tends to obscure the degree of variation and change with respect to product value. Even so, the Scientific Committee and the BIWS have, from time to time, drawn the attention of the Commission to changes in the mean lengths of whales caught.

Particularly significant, in the present context, is the Chairman's Report of the 12th meeting, 1960, which notes "the progressive reduction in the average length of sperm whales *Physeter macrocephalus* taken in the Antarctic," as well as that in the Pacific there were "increasing catches, and large numbers of sperm whales recorded as being at or just over the minimum length." This latter observation is typical of many, in expressing concern about size changes in the context of minimum length regulations rather than as related to economic yield.

Because of variations in prices of products, and in the degree and mode of utilization of whales caught, and the international nature of the industry, it is not practicable to put forward, at this time, a strictly value criterion for MSY. The total weight of a whale catch is, however, a good index of the value of products that can be obtained from it; this is generally true even for different species of baleen whales, notwithstanding the fact that there are interspecific differences in meat and oil yield, etc., and also seasonal and area differences. Even if the weight and value are not exactly or invariably proportional, they are certainly much more closely related than are number and value. Even the sums of the cubes of the lengths give useful indices of weight (as, in fact, used by the Scientific Committee in its consideration of quotas by sex for sperm whales). There are, however, data in a series of scientific papers which permit the calculation of weights from lengths for the baleen whales and for the sperm whale. D. T. Crisp published in 1962—the year the Special Committee asked for better data—such calculations for the five species in Antarctic pelagic catches for the period 1932/33 to 1958/59. I subsequently extended his series up to 1969/70; C. Lockyer (Whale Research Unit, UK) recalculated them using all published information (1972, MS); the FAO and International Ocean Institute have now prepared comprehensive tables for each species and sex for the Antarctic (pelagic and land stations) and for the world as a whole. These tables permit a general survey of historical changes in mean size, at least by species and broad areas.

Appendix 1 summarizes the above tables; five-year averages are chosen to eliminate the rather considerable year-to-year fluctuations. These fluctuations presumably derive from several causes, including shifts in the area of whaling and also the small numbers of certain species caught in certain seasons. An overall pattern is, however, clear: in most cases, there has been a decline in mean size over the period covered, and this is particularly marked in the case of the sperm whale.

The sperm whale is overwhelmingly the most important species by weight (and probably also by value) in terms of present catches in the Antarctic and in the world (see Appendix 2), and this importance has been increasing.

The present sustainable yields in weight (thousands of tonnes) of species which are not, at present, protected are distributed roughly as follows (based on sustainable yields in numbers given by the IWC Scientific Committee for North Pacific and southern hemisphere):

Species	World	Antarctic (southern hemisphere)
Sperm		
Male	39	40
Female	9	9
Both sexes	48	49
Baleens	52	51
Fin	26	22
Sei	15	14
Minke	11	15

The sperm whale stocks (southern hemisphere and North Pacific) are considered by the Scientific Committee now to be at levels near to numerical MSY, and thus whether these are categorized as "sustained management stocks" or "protection stocks" may depend critically on the MSY definition adopted, apart from any future reassessment of their biological status.

The decline in mean size of each sex of sperm whales may appear as a result of whaling in either of two main ways or a combination of these.

a) The age range of the exploited phase is such that about two-thirds of all the potential growth in weight of the individual is within it. Thus, random capture, over the exploited age classes, which must result in a gradual reduction in the mean age of whales in the population (until a steady-state is reached with constant whaling effort), will show up as a corresponding reduction in mean size in the population, and hence in the catch.
b) To the extent that whales of different ages (and sizes) are segregated, so a shift of whaling effort from areas occupied by larger individuals towards areas occupied by smaller ones, will lead to a decline in mean size. Such a shift could be incidental to the sperm whaling operation—for example, it may be related to changes in the availability of other species hunted by the same expeditions. But, perhaps more likely, a severe reduction in the number of larger whales in areas where they are mainly to be found, as a result of previous intensive whaling, would certainly be expected to lead to a shift of effort towards the aggregations of smaller individuals. This sequence of interest arises in the same way as the sequence of interest from larger to smaller species of baleen whales because the cost in effort of catching and processing a larger individual is no more than for a smaller individual—or at least not much more—in relation to the greater product yield.

It should be noted that processes a) and b) differ little from each other in their consequences.

It could be that the decline in mean size could be related to a partial shift of whaling from one stock to another, the different stocks having different growth and/or natural mortality characteristics. This would be far more complicated to analyze, even if all the necessary data were available. However, it seems to me likely that in this case, even more stringent regulatory measures might be needed to ensure total MSY or optima. So, in looking here merely at gross figures, we would not thereby be led to excessive "conservation" in advice offered for management.

If undersized whales are occasionally "stretched" and recorded as at or over minimum legal size, one would expect this practice, if anything, to become more frequent as whales get scarcer and/or smaller. This means that the true degree of decline in mean size may be more than the BIWS length composition indicate. By accepting them at face value, we would, again, be acting "optimistically" regarding the impact of whaling on these stocks. When, however, a change is made in the legal minimum length, as it was for female sperm whales at the 1973 Commission meeting, we obviously must take special care in the interpretation of the mean length data.

The discontinuities in the decline of mean size as between the immediate pre-war (World War II) and post-war years might be evidence of some recovery of Antarctic stocks as a result of the war-time respite from whaling; the discontinuity is most clear in those species which were already heavily exploited in the 1930s. Furthermore, it will take an overexploited stock, which is then protected, considerably longer to recover the level and age/size composition at which it could sustain a maximum yield in weight than to recover to MSY level in numbers; this should temper one's view of measures needed for the wise management of stocks which are not, as yet, seriously overexploited.

Setting an objective of maximizing yields in weight does not imply that regulation must be by quotas of weight rather than of numbers. Numerical quotas and size limits could be adjusted so as to ensure that approximately the appropriate weight is taken each season from each stock. This will, as now, call for a "coarse tuning" of the whaling effort before the season opens (the decision, taken by nations and operators, as to the number of expeditions and catcher boats to employ), followed by the "fine tuning" by closing the season for each species when the appropriate catch in number has been taken. On the other hand, since lengths are measured as the whales are caught, it would, in fact, be easy to compute the estimated weight caught as the season progressed, if desired.

Sustained yield in weight is maximized by maintaining a stock at a rather high level of biomass and numbers relative to its initial state. Such a criterion should, therefore, find favor with ecologists and other persons who wish to ensure that species and stocks not only survive, but also continue to play their full role in the marine ecosystem. The stock level for MSY (weight) is nearer to the level for maximum net economic yield than is the level for MSY (number) and should, therefore, be more satisfactory to economists and governments concerned with the sustained profitable exploitation of a renewable resource. The whaling industry is itself certainly more interested in the weight of whales taken, and hence of products, than in their number. Furthermore, from the whaling industry's point of view, the catch per unit effort, in both numbers and weight, is higher at MSY (weight) than at MSY (numbers).

Since full recovery of a depleted stock to its optimal biomass is slow and would present the usual short-term difficulties for the industry, it seems imperative to ensure that no stock which is, at present, only moderately exploited is permitted to be reduced below the optimal biomass or threatened to be so reduced.

If sperm whale stocks are now near to MSY (number) levels, and the mean size has declined significantly, then they are already below their optimal biomass. The sooner action is taken to permit their recovery, the sooner they can sustain their maximum product yields. Given the age compositions, growth rates, and the usual population data, a sperm whale weight assessment model can be constructed. This may not, however, be necessary for the present purposes. From the various reports of the Scientific Committee, we can construct a time-series of stock sizes, and we now have the mean weights for each corresponding year. From these, we calculate the weight of the exploited stock and the weight of catch in each year. In a period of declining stock and declining mean sizes, the mean size in the year of a particular stock level will be greater than would be expected when a steady-state is reached. That is, the observed decline in mean size, between the initial more or less unexploited state and a later date which is still within a period of declining stock—or even at the end of such a period—is less than would eventually occur when a new steady-state is reached after the decline has been halted. So, if we treat the observed mean lengths over the past decade or two as if they related to steady-state conditions, we can be sure we are not being excessively "conservative" or "pessimistic." This is what I have done.

Some appreciation of how different may be the requirements for MSY (weight) rather than for MSY (number) can be gained by examining the numerical yield curves used by the

Scientific Committee on the basis of the model of Ohsumi and Fukuda (1972) or Allen (1973) in the region of the present situation. Consider first the southern hemisphere stocks. The present sustainable yield of females (about 1,800 animals) is only about 40% of the MSY (number). The present stock is said to be 90 or 78% (depending on the model used) of the initial stock of mature females, whereas the stock for MSY (number) would be about 57 or 46% of the initial number. This means that to attain MSY (number), one needs to reduce the present female stock by 37 or 41%.

Our tables suggest the initial mean size for southern hemisphere females was about 17 tons, and the present size about 12 tons. If the female stock were further reduced in number (to, say, the MSY level), what would the mean size become? A linear extrapolation leads to an impossibly low figure; the mean size of mature females clearly could not be less than the size of maturity (6 tons), and this would be reached with a reduction of stock by only 13 or 35% from its present level. If the females were to decline in weight by the same relative amount as did the males when they were reduced from the initial to about their MSY level (i.e. from 45 to 28 tons; see below), then the mean size of females in an MSY-level stock would be about 10.5 tons. By linear interpolation of the mean weights, one can calculate SY (weight) at intermediate stock levels; the MSY (weight) is, for this assumption, at a numerical stock level a little higher than for MSY (number). The curve of yield in weight against population size in number is, however, more flat-topped than the curve of yield in number, and its maximum is correspondingly more difficult to locate with precision.

For the present purposes, we might make the "optimistic" assumption that there would be no further decline in mean weight if the female stock were further reduced. On the other hand, if the female stock were allowed to increase (by applying lower quotas than at present), then we should presume the mean weight would increase again. Suppose the female stock were allowed to increase by 10% in number. Linear interpolation gives us a corresponding mean weight of 13.8 or 16.8 tons. In this region, the models for numerical yield are highly imprecise; they show, however, that the rate of loss of yield in number of females that this would cause cannot be balanced by the increase in mean weight. On the basis of Allen's model, it seems the SY (number) would go down by 58%; the SY (weight) would go down by 51%. Clearly, there is nothing to be gained by allowing the lightly exploited female stock to increase—unless it would lead to a possible increase in the weight of males caught.

Turning now to the males of the southern hemisphere stocks, I note first that their potential sustainable yield in number would increase beyond the present level (which is considered to be near MSY with the present sex balance) if there were some further reduction in the female stock, but this gain is very little. On the other hand, if the males were caught at a lower rate than at present, their mean weight would tend to increase from the present 28 tons (their initial weight was 45 tons). The present stock of males is 50% by number of the initial stock. The present SY (number) is about 96% of MSY, which would be obtained at a stock level 13% smaller than the present one. At that lower stock level, linear extrapolation (which is permissible in the case of the males, for which we are exploring only a small part of the yield curve near its maximum) indicates the mean weight might decline further to about 26 tons (93% of the present mean weight). Thus, at MSY (number), the SY (weight) would be about 3% less than at present.

Now suppose the male stock were permitted to increase, say, by 10 or 20%. The interpolated mean weights for these levels are, respectively, 6 and 12% higher than at present, that is, to about 29.5 and 31.5 tons. The SY (number) at these higher levels of male stocks can be estimated roughly from the slopes of the yield curves implied by the current models in the region of the present situation; they are a little more than 3 and 6% less than at present. That is, for moderate increases in the male stock, the increase in SY (weight) may be of the order of 3% for 10% increase in stock number and 5% for 20%. These estimates of percentage changes are, of course, very rough. Apart from the uncertainties in the basic models, and the approximations made here, the curve of sustained yield in weight must be flatter-topped than the corresponding curve for numbers. The direction of change required is, however, clear—it is to permit some increase in the male stock. Furthermore, even if we cannot promise much absolute increase in the SY (weight), at least the same yield as at present could be obtained with less whaling effort (and hence at lower cost). In view of the uncertainties in all present assessments and the need to be conservative rather than risk unwitting overexploitation of the resource, it seems that it would be wise to reduce the present quota of males, in the expectation that the industry would also benefit from having to catch fewer and larger whales to attain the same level of production.

At present, the mean weight of males is 2.3 times that of females. If the numbers of both sexes were allowed to increase by 10 or 20%, the weight ratio in favor of males might decline to about 2.0. But, if the males were allowed to increase and the females held steady or reduced, the ratio would rise to 2.5–2.6. The Scientific Committee's estimates of male MSY (number) are nearly twice those of the females, and recent catches and present quotas are 60% higher for males than females. We are, therefore, dealing with stocks for which under optimal management, the males contribute 4–5 times as much as the females to the total weight of catch. Estimation of the total MSY (weight) is thus increasingly sensitive to the anticipated changes in the male stock. We can say that to attain MSY requires that the number of males in the stock be higher than at present, and that females be not much fewer. To achieve this requires a reduction of the present quotas for both males and females and their eventual stabilization at less than 7,000 for males and less than 2,000 for females (present quotas being 8,000 and 5,000, respectively).

The rate at which males would be allowed to increase somewhat depends now on the interpretation of Commission's policy decisions. Regarding females, I can find no evidence which shows that there will be a substantial sustainable benefit from the reduction of females below their present level, which has an SY (number) of 1,800. I think the Scientific Committee has said, in effect, that the reduction which would certainly be caused by the present quota level of 5,000 will not be harmful and might possibly be beneficial.

The same arguments as I have made above can be followed for the North Pacific. The qualitative conclusions are roughly the same as for the southern hemisphere, but with quantitative differences because in the North Pacific, neither males nor females have been reduced, relative to their initial numbers, as far as in the southern hemisphere, and the changes in mean weight are not yet so marked. Nevertheless, it appears that the North Pacific industry would also benefit from rather lower quotas than at present.

Having in mind (i) the inevitable uncertainties in all calculation of sustainable yields [which do not preclude that some or even all the sperm whale stocks could already be somewhat below the level of MSY (number of both sexes)], (ii) the time required to rebuild any overexploited stock to attain a steady age and weight distribution, and (iii) the usual lapse of time between recommendations by the Scientific Committee and action by the Commission, I believe it is appropriate now to recommend reductions in all sperm whale quotas in order to permit some rebuilding of male stocks and to slow down or even halt the reduction in females now being caused by setting quotas at two or three times the present replacement yields.

In any consideration of the effective management of sperm whale stocks, the IWC will encounter the problem of whaling by non-Member countries. In any one year, the oil production, whether of baleen or of sperm whales, is a fair index of the weight of catch. In the year 1970/71, 3.1% of baleen oil was produced by non-Members, but 6.3% of the sperm oil; the countries concerned were, according to BIWS: Portugal, Spain, Peru, Chile, and Bahamas. In more recent years, the proportion is still higher, but cannot be determined accurately because of late or non-submission of data. In 1971/72 reporting, non-Members accounted for 6.7% of the sperm oil production. If the countries for which statistics were not published by BIWS (up to June 1974)—Chile and Bahamas—maintained their production at about the level of the previous year, the corresponding figure becomes 8%. The significance to IWC Members may be seen by comparing these catches by non-Member countries with the catches by all countries (Members and non-Members) except Japan and the Soviet Union; the factors are 33% and 35% in the two successive years.

Below is given the breakdown with respect to pelagic and land-station operations:

% by weight	**1970/71**	**1971/72**	**1972/73**
Pelagic	72.1	?	?
IWC Members	71.8	68.3	73.6
Non-Members	0.3[a]	?	?
Land stations	27.9	31.7	26.4
IWC Members	21.9	25.0	20.3
Non-Members	6.1	6.7[b]	6.1[b]
	100	100	100
Total sperm oil (tons × 10^{-6})	0.123	0.098[c]	0.105[c]

[a] Bahamas. [b] Excluding Chile. [c] Excluding Chile and Bahamas.

References

Allen, K. R. 1973. The computerized sperm whale population model. Reports of the International Whaling Commission 23:70–74.

Gambell, R. 1974. The unendangered whale. Nature 250:454–455.

Holt, S. J. 1974. Whales: conserving a resource. Nature 251:366–367.

Ohsumi, S., and Y. Fukuda. 1972. A population model and its application to the sperm whale in the North Pacific. Reports of the International Whaling Commission 22:96–110.

Appendix 1

Mean weights of whales in Antarctic and world catches (tons). Data summarized by five-year periods in which there were large enough catches to make averages meaningful.

Period	Sperm		Blue [1]	Fin [2]	Sei [3]
	Male	Female			
World					
1932/33–1936/37	26.4	-	88.5	48.5	14.4
1942/43–1946/47	26.1	16.8	73.9	45.4	14.6
1947/48–1951/52	29.8	12.7	87.1	49.1	15.7
1952/53–1956/57	26.2	13.1	82.9	48.1	16.1
1957/58–1961/62	26.9	12.2	78.3	47.9	17.4
1962/63–1966/67	23.5	13.0	63.1	43.6	17.5
1967/68–1971/72	22.8	13.7	-	43.5	16.0
1972/73	22.1	11.5	-	44.8	15.4
Antarctic					
1932/33–1936/37	47.2	-	89.0	52.0	-
1942/43–1946/47	35.9	-	82.4	48.9	19.3
1947/48–1951/52	39.3	-	88.1	51.3	19.4
1952/53–1956/57	35.9	-	84.4	49.7	19.1
1957/58–1961/62	31.6	-	79.7	49.2	19.5
1962/63–1966/67	29.8	15.4	-	48.6	18.3
1967/68–1971/72	27.6	14.1	-	48.9	17.2
1972/73	27.7	12.1	-	48.7	17.0

[1]*Balaenoptera musculus*
[2]*B. physalus*
[3]*B. borealis*

The humpback *Megaptera novaeangliae* is more variable than the other species, with a slight, but distinct, tendency to decline in size over the six 5-year periods 1932/33 to 1961/62. In all cases, the mean sizes are well above any minimum size limits in force.

Appendix 2

a) *Significance of sperm whale in recent catches*

Species	Antarctic 1972/73 pelagic		World 1972/73 and 1973	
	Tons × 10^{-3}	%	Tons × 10^{-3}	%
Sperm				
South of 40°S	**112**	**29**	373	55
Southern hemisphere	191	50	-	-
Fin	87	23	117	17
Sei and Bryde's	66	17	115	17
Minke[a]	54	14	62	9
Baleens	192	50	303	45
Total	383	100	666	100

[a] Assume mean length is 8.5 m and hence mean weight about 7 tons.

b) *Trend in importance of sperm whale*

Years	Sperm/(sperm + baleen)[a], % by weight	
	Antarctic pelagic[c]	World
1933/34–1940/41[b]	3	5
1946/47–1953/54	7	13
1954/55–1961/62	11	21
1962/63–1969/70	21	42
1970/71	24	55
1971/72	29	47
1972/73	42	60

[a] Excluding minke. [b] Averages of annual % values, eight-year periods.
[c] Including sperm catches south of 40°S.

c) I have not yet been able to locate data by which all the proportions in a) and b) above could be expressed in values rather than weights. The total Japanese 1972/73 whale catch was reported by the Economist Intelligence Unit on behalf of the Japan Whaling Association to be worth US$94.2 million. On a proportional basis, this would put the value of the world whale catch in the order of US$200 million. However, 1 ton of sperm whale is presumably worth less than 1 ton of baleen whale. The sperm meat is not much used for human consumption. The price per ton of sperm oil (on the basis of export data) may be up to 50% higher than that of whale oil; the yield of sperm oil per ton of whale is 25–30% greater than that of baleen oil. This will tend to reduce the difference in species values, but presumably not enough to offset the value of baleen meat.

The ratio sperm : sperm plus baleen in the Japanese catch is less than for the world as a whole (29% as compared with 44% in 1971/72). This difference is due mainly to the much higher ratio in the Soviet catch (57%). The total value of the world catch of all species must, therefore, be somewhat less than US$200 million, but more than US$160 million. If sperm was worth about half as much as baleen per ton, the total would have been US$175 million, of which sperm accounted for 38% and baleens for 62%, as compared with 55% and 45%, respectively, by weight. With even such a price differential, the world catch of sperm whales is worth more than half as much again as either fin or sei whales. The relative value of sperm whales will increase when efforts being made to derive edible products from the flesh are successful and widely applied.

Chapter 10

Stock Assessments of Minke Whales[1]

In Documents SC/27/1, 27/5, 27/20, and 27/27, assessment is attempted without taking account of important variations in sex ratio which are noted in SC/27/1. There are remarkable differences between the minke *Balaenoptera acutorostrata* and the other baleen whales in that the sex ratio in minke catches is rarely near to 50:50. This is to be expected from the observation in SC/26/14 that there is, at least in the North Atlantic, strong sex segregation in this species. To illustrate this phenomenon, recent statistics for the Southern Hemisphere are summarized below.

Area	Season	Numbers (and mean lengths)			
		Males	**Females**	**% females**	**Of which % pregnant**
Natal	1973–74	69 (26.2)	46 (26.3)	40	
	1972–73	125 (27.4)	48 (26.6)	28	
Brazil	1974	247 (27.8)	514 (29.1)	68	
	1973	210 (27.4)	440 (28.9)	68	
Antarctic	1974–75	1,164	1,448	55	?
	1973–74	2,958 (26.5)	4,755 (27.8)	62	61
	1972–73	2,900 (26.3)	2,843 (27.1)	50	54
	1971–72	2,925 (27.3)	2,530 (28.5)	46	63

Data for Norwegian North Atlantic catches are not all published, but the graphs in SC/26/14 show a great preponderance of females in Barents Sea catches and perhaps of males in the Lofoten catches. SC/27/Prog. Re. 3 suggests a similar phenomenon off West Greenland: of 163 whales examined in 1974, 140 (86%) were females. In 1973, Norwegian data were:

Area	% females
Barents Sea	70
East Greenland	34.2
West Greenland	85.2
All	69.5

The sex ratios cannot be accounted for in terms of differences in size of the two sexes; the mean differences are small, although they are fairly consistent from year to year, and where and when there are relatively more females, the females are bigger relative to the males, and vice versa. In the Antarctic, there are differences in the sex ratio as between months and area; changes in main areas and months of minke catching seem to account for the evident trend from season to season in that area.

The percentage pregnant among Antarctic-caught female minke whales is consistently much higher (54–63%) (see also SC/27/16) than in, for example, the fin *B. physalus* and sei

[1]Editors' note: written by Sidney Holt, presented as Document SC/27/Doc 29 at the 1975 meeting of the IWC Scientific Committee, and published in Reports of the International Whaling Committee 26:392–394, 1976.

whales *B. borealis* (40%). Annual changes in percentage pregnant correspond with changes in mean size, perhaps reflecting a change in the ratio of mature to immature animals in the catches. The annual changes in mean size are the same for males and females.

The above considerations can significantly modify conclusions about the status of minke stocks. For example, the subcommittee established by the Scientific Committee at the 1975 IWC meeting to evaluate the situation in the Southern Hemisphere concluded that the Area II stock had, by the end of the 1974/75 season, been reduced by 7% below the 1971/72 level; in absolute terms by about 2,200 whales. This change is presumed to have been the combined effect of the Brazilian catches and the Antarctic pelagic catches in Area II totaling about 4,000 whales. However, in the previous six years, Brazil reported taking 3,500 minke whales, which could have reduced an "initial stock" in 1964 by 6% by the beginning of the 1971/72 Antarctic season. Thus, the current (1975) stock size in this sector is 88% of the "initial" number. But 70% of the Brazilian catches were females. If a true initial sex ratio of 50% in the whole stock in this sector is assumed, it can readily be calculated that the reproductive capacity of the stock, in terms of the number of surviving females, has been reduced by whaling not to 88%, but to 83% of the initial capacity. Finally, it seems that because of the high pregnancy rates in catches, the impact of whaling on the rate of reproduction will be about 40% greater than that of comparable catches of fin or sei whales.

The contribution of minke whales to the Antarctic pelagic catches of baleen whales has increased in five seasons from virtually nothing to 33% by weight (Doc. SC/27/15). I have argued elsewhere that great care should now be taken not to reduce minke stocks to levels such that they cannot, with the sei whale, support an industry, in terms not only of total catch, but also of catch per unit effort, and I have suggested that to this end, the minke stocks should not be reduced to less than 3/4 of their estimated "initial" levels, at least until better assessments are available. "Better assessment" here includes account being taken of the peculiarities of sex selection and pregnancy rate in this species, as well as the relation of weight to number.

Chapter 11

Criteria for Management—Weight or Numbers? The Relevance of Whaling Effort[1]

In its report on "Classification of whale stocks" based on the report of its Special Meeting in La Jolla, December 1974, the IWC Scientific Council has described as "principles" the findings that:

a) for baleen whales so far examined, the differences between population levels for MSY (number) and for MSY (weight) are small and are negligible because they are within the limits of accuracy of estimation (Principle 7), and
b) for sperm whales *Physeter macrocephalus*, the difference is between 7 and 18% and should not be yet taken into account in management (Principle 8).

With respect to a), it is evident that the differences are much more sure than are the estimates of MSY levels. If the MSY levels are unsure, there seems all the more reason to take due account of established phenomena to gain needed protection from the harsh consequences of error in the other direction. As formulated, Principle 7 suggests that if, for example, in a particular stock the weight–number difference was even as great as 20%, but the MSY level could only be estimated to within 50% (not an unlikely situation), then the objective of management is the lower level, i.e., the MSY (number). This is surely unacceptable from the point of view both of conservation and of industrial efficiency.

Then, it is not clear to me on what scientific grounds a 7–18% difference in sperm whales is considered "negligible" especially as it is of similar magnitude to the controversial value of "Z."

I suggest that the "principles" should not, in any case, be formulated in terms of kinds of whales, i.e., "baleen whales which have been examined" and sperm whales, nor should Principle 7 then be applied to all currently exploited baleen whales, whether examined or not. The difference between weights and numbers is less in fin *Balaenoptera physalus* and sei *B. borealis* than in sperm whales, not because there is a basic difference between the two groups, but because it happens that, at least in part, because of the prevailing size limits, a greater part of the growth of sperm whales than of the two baleen species takes place within the "exploited phase" of the population. The age determinations are not adequate for certainty, but judging from the magnitude of size changes and the minimum size limit which was applied to it, the Antarctic blue whale *B. musculus* may have exhibited a greater difference than fin and sei. The same could be true for minke *B. bonaerensis*, for which there are, at present, no size limits.

---

[1]Editors' note: written by Sidney J. Holt as Document IWC SC/27/Doc 33 Revised and published in Reports of the International Whaling Commission 26:409–413, 1976.

I believe, notwithstanding the conclusions of the Scientific Committee, that immediate adoption of the MSY (weight) criterion is desirable also on other grounds. In Figure 1 are shown hypothetical curves of sustainable yield in weight (broken line) and in number (continuous line). These are expressed relative to the sustainable yield at the stock level for MSY (number) and are plotted against stock level, expressed as a fraction of initial stock number. For the purposes of illustration, the stock level for MSY (number) is taken as 0.5, and that for MSY (weight) 15% higher, i.e., at 0.575; this difference is within the range the Scientific Committee has given for the sperm whale. In such circumstances, the MSY (weight) is calculated as about 6–7% higher than the SY (weight) at the stock level of MSY (number), that is the weight of the MSY (number) catch. If this were the only benefit of adopting the weight criterion, it would perhaps not be worth pursuing the matter further. Other consequences are, however, more important.

It has been pointed out elsewhere (Holt 1975) that the percentage difference in catch per unit effort (hence, in economy of whaling) as between the weight and number maxima is considerably greater than the MSY difference. However, it can be seen from the example that a stock level which would give a catch in weight as big as that obtained from the MSY (number)—i.e., about 94% of MSY (weight)—is about 28% higher than the level for MSY (number). Furthermore, if, for safety, only 90% of the estimated MSY is taken—as has been decided with respect to number—then the catch per unit effort (by weight, and hence of products) could be 40% or more higher than from the stock at MSY (number) level. This is a potential benefit to the industry which should not be ignored if the objective is continued whaling on a sustained basis. To put it another way, by maintaining a number criterion rather than changing to a weight criterion, the industry is encouraged to exert 50% more

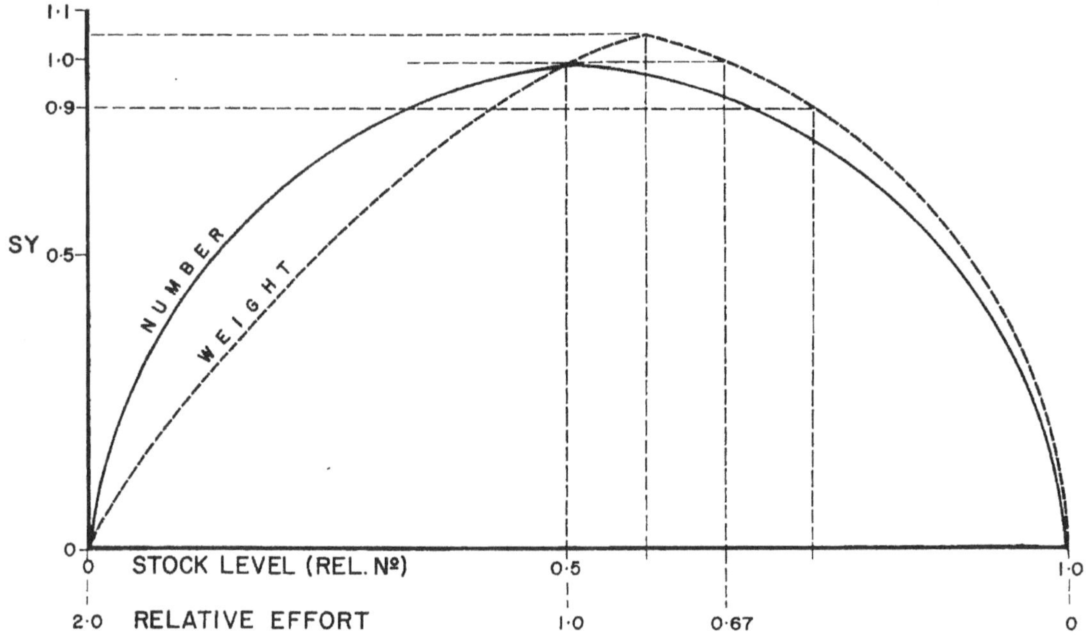

Figure 1. Hypothetical curves of sustainable yield in weight (broken line) and in number (continuous line) expressed relative to the sustainable yield at the stock level for MSY (number) and plotted against stock level, expressed as a fraction of initial stock number.

whaling effort (catcher's day's work) than is necessary to take the same yield. What this implies in terms of higher operating costs, I do not know, but it surely cannot be negligible. There are several reasons to do everything possible to avoid such a situation. Obviously, for protection stocks, there is no urgency—one can only wait until, hopefully, they recover to reach, and pass, the MSY (number) levels. For sustained management stocks which are, if anything, below rather than above MSY (number), it is a matter of whether quotas are kept level or slightly lowered to permit some increase in population. For initial management stocks and for those which, like some of the southern hemisphere sperm whale stocks, are, if anything, now somewhat above the MSY (number) level, there is a significant practical difference between the two criteria in terms of recovery times. If, in the example given, the stock were first reduced to MSY (number) level and it was subsequently decided to adopt a weight criterion, it could take 7–9 years of zero quota for the stock number to increase to the MSY (weight) level, even taking no account of time lags. Recruitment delay adds a few more years to that number. If quotas are finite, but reduced, the delay is proportionately longer. However, a stock held numerically at that new level will not instantaneously attain the equilibrium size distribution—several more years will pass before that is attained (in theory, the number of years between recruitment and reaching final size, but for practical purposes somewhat fewer). So, to permit an initial management stock to be reduced to MSY (number) level with a view to allowing it to rebuild later is irrational and detrimental to the industry. Such a policy could even prejudice the economic survival of the industry in a situation where many stocks are protected, others are exploited at low levels, and profitable whaling depends on one or a few species, such as the minke, being available at a high catch rate. Holt (1975), in considering the Antarctic fin and minke whales, argued the case for holding the minke at not less than three-quarters of its "initial number." Judging from other contributions in the present volume, it may already be too late for that, but it may not be too late to demonstrate that a sustained industry has at least as much to gain from exploitation at higher rather than lower stock levels as do the whale stocks themselves.

The above arguments in relation to weights instead of numbers apply also to other factors which tend to encourage more "conservative" exploitation. For example, the policy of taking no more than 90% of MSY (number) will either give protection against error of overestimating MSY or if such error did not occur, the loss of catch would be offset by the advantage of a higher catch per unit effort. The reason for making a special case regarding weights is that the direction of difference is known with certainty, and even its order of magnitude is better known than most parameters of whale stocks.

References

Holt, S. J. 1975. The concept of maximum sustainable yield (MSY) and its application to whaling. Document FAO/ACMRR/MM/IV/4, Arlie House Workshops on the Concepts of Management and Conservation of Wild Living Resources, Virginia.

Chapter 12

Whale Management Policy[1]

The *Ad Hoc* Group on Large Cetaceans of the ACMRR Working Party on Marine Mammals met in La Jolla, California in March 1976, immediately after the special meeting of the IWC Scientific Committee on sperm whales *Physeter macrocephalus*. Most of the participants had also attended the latter meeting. The Group discussed some of the implications of following the Commission's new management policy in cases where there is considerable uncertainty as to the identity and status of stocks. Extracts from the reports of the Group on this matter are appended; see particularly Supplement 1, paragraph 3, and General Conclusions, paragraph 2. The Group recommended that the reports' several suggestions for improvement of management in certain respects be transmitted to the IWC as well as to the Scientific Consultation, *Mammals in the Seas*, being convened by FAO in Bergen, Norway, and that I should submit a more extended explanation to both meetings. Such an explanation follows.

The new management policy provides that whale stocks be identified and then categorized according to whether they are below, at or near, or above levels from which maximum sustainable yields can be taken. That having been done, rather simple rules determine the annual quotas. The Commission itself, if only by including the term "near" in the 1974 decision which established the policy, recognized the existence of some uncertainty in the above determinations, and the Scientific Committee went further in elaborating rules to regulate the "bluntness" of the knife-edge between a Sustained Management Stock and a Protection Stock, as well as adding a safety factor whereby, in certain circumstances, only 90% of the estimated sustainable yield would be allowed to be taken. These rules, which were accepted in 1975 by the Commission, are intended to take some account of uncertainty, to allow for asymmetry in the consequences of erring on one side or the other, and to ensure an element of caution in management. They seem likely to serve those purposes when knowledge of stock dynamics is sufficient to determine fairly well the size (and composition) of a stock in relation to a curve (or surface) of equilibrium yield as a function of stock size (and composition) as calculated from a realistic model. If, however, there is a high degree of uncertainty of the model, of the state of the stock, or of both, rigid application of the policy and rules might not serve the intended purpose.

Before elaborating the reasoning behind this statement, mention should be made of the problem of stock identification. In many cases, virtually nothing is known directly about stock identity within a species; inference is made from other species inhabiting the same general region, and knowledge even about those species is far from adequate. Marking experiments

[1]Editors' note: written by Sidney J. Holt, FAO/ACMRR/UNEP Marine Mammal Project, as Document SC/28/Doc 10 for the 1976 meeting of the IWC Scientific Committee and published in Reports of the International Whaling Commission 27:133–137, 1977.

show a fairly clear separation of stocks of some baleen whales in the Southern Hemisphere; they also show that the separation is not complete. It seems that Northern and Southern Hemisphere stocks do not mix, nor do North Pacific and North Atlantic stocks.

Even when stock separation is fairly clear, practically never has it been possible to estimate the vital parameters of each stock of a species separately; most, if not all, of the parameters are assumed to have the same value in all stocks, or in a group of stocks, the only variable being the number of animals in the stock. (In memoranda to the Chairman of the Scientific Committee in 1974–1975, I drew attention to some of the consequences of such a procedure.) Practical application of the management policy is really, then, to *presumed stocks* on the basis, at best, of knowledge about the population biology of *the species*.

The situation with respect to sperm whales is especially serious. In terms of weight and value of catches, number of countries involved in exploitation, breadth of geographical distribution and, presumably, the size of present stocks, the sperm whale is by far the most important species of whale. Furthermore, it has been exploited "pelagically" over a period of time much longer than has any other species. Yet, for a number of reasons, present assessments are extremely uncertain. The effects of whaling in the 19th century and in the first half of the 20th century have not been estimated in terms of numbers, sex ratio, size and age composition, range of distribution, or social structure; the catch per unit effort is a poor index of abundance; the social structure is such that the simpler population models are not at all applicable. Apart from uncertainty as to the present stock sizes and their rates of natural mortality and reproduction, the relation between stock size and age composition in the "base year" used by the Scientific Committee—1946—and the "initial" sizes and compositions has not been evaluated. The high ages of maturity and recruitment imply long time-lags in the system which complicate both assessment and management of the whaling operations. Recovery periods, if whaling is unwittingly permitted to be too intensive for even a few years, might be very long, especially the recovery, not only of numbers, but of biomass and group structure. Further, when considering a period of exploitation of nearly two centuries, we cannot rule out the possibility of significant trends in environmental conditions affecting sperm whales in such a way that, even if there had been no whaling, the stocks would neither be constant in numbers nor have steady-state age and size compositions. Lastly, apart from the fact that they consume a lot of deep-water squid *Architeuthis dux*, practically nothing is known or can even be reasonably surmised about the dynamics of the ecosystems in which the sperm whale has a significant role. So, for this species, it seems particularly important to examine carefully the possible consequences of a more or less automatic application of the new policy and rules using central, or so-called "best," estimates of stock state. It is necessary here to examine the concept of "best estimate" and the implications of using it to establish quotas.

If there is only one estimate available, and the procedure of determining it is not such that a statistically valid measure of error can be attached to it (which is practically always the case), then it is not appropriate to call it the "best" one, certainly not in any scientific sense. Any statement concerning the likelihood of it being within a certain distance of the "true" value must be purely subjective. If several estimates are available, they are normally obtained by a variety of methods and combinations of methods. There is no scientific way of combining these to give a "best" estimate, although we may assume that the group of estimates is a sample from a probability distribution and treat them accordingly to obtain, as *central value*, a median, mode, arithmetic mean, or some other kind of average. We may assume further that the probability distribution is peaked, having a maximum somewhere between the observed extreme values, and try to estimate the location of this maximum to obtain a most probable ("best") value. But even then, we have no idea of how flat-topped the distribution is, and certainly have no reason to suppose it is like a normal curve. Thus, it is entirely possible that the most probable ("best") value is hardly more probable ("better") than other quite different values; indeed, it might be a defensible procedure to assume that all values within an observed range (perhaps even somewhat beyond it) are equally probable ("equally good"). I suggest that, on present evidence, estimates differing from the central values (which the Scientific Committee is in the process of determining) by 50% or more either way may, for practical purposes, be equally as good as each other and as the central value.

If a single general assessment is applied to give a set of central values for a group of stocks—say, the several sperm whale stocks presumed in the Southern Hemisphere—which are then used to determine quotas, the quotas are more likely than not to be *all* in error in the same direction. If the error is in the conservative direction, sperm whaling might be unduly restricted (in terms of the Commission's declared management objectives), but given the present state of knowledge and the nature of the species, this might only be noticeable after many years of monitoring as an unexpected increase in availability. If, on the other hand, the error is in the direction of "optimism," it would be possible, even likely, that the stocks would all be seriously overexploited for some years before the reduced availability could be monitored, with a danger of a catastrophic decline in recruitment a decade or so later. In such cases—and especially if the complex social structures necessary for efficient reproduction had been disrupted—even a reduction in stock to below minimal survival size could be possible. In any case, most, if not all, of the stocks would probably be similarly stressed, though not necessarily to the same degree.

However, suppose that, fortuitously, a set of quotas were applied such that there were no detectable changes in stocks over a long period. In that case, there would be practically no gain in knowledge of the population dynamics of the stocks of the species nor, if stability were attained, would it even be known if they were at or near maximum yield levels.

This is the crux of the matter: only by changing the populations can their dynamics be determined, particularly the crucial density-dependence of natural mortality and/or reproductive rates or other natural regulatory mechanisms. Furthermore, attempts to unravel these mechanisms from the more or less haphazard historical changes during periods of intensive whaling—whether "managed" or not—have not been particularly successful.

One approach to management in a situation of great uncertainty has been the proposal for the cessation of whaling for a period during which more research would be done and, specifically, changes in stocks would be monitored. Such an approach would theoretically give important data on the recovery rate and hence on the population dynamics of species and stocks which had been substantially reduced by whaling. It would not tell us anything about stocks which might *not* have been much affected by whaling, except perhaps the numbers of animals in them. In general, a total moratorium, whatever its pros and cons on other grounds, is not the most efficient way to better understanding of whale population dynamics, even if it were accompanied by a new research and monitoring program of unprecedented scale and scope. In any case, such an approach has not been found to be politically acceptable at this time.

An alternative approach, and the one here recommended, is to set quotas for different stocks of the species on the basis of a wide range of estimates of the state of the stocks. Thus, for one stock or group of stocks, an "optimistic" estimate might be applied, creating a definite risk that it would be overexploited, whereas another, not considered necessarily to be in a state essentially different from the former, would be regulated in accordance with the more pessimistic view; this would mean a very small quota or, in all likelihood, a zero quota. Such a regulatory regime would have to be maintained for several years in order to give a chance of detecting stock changes and making comparisons—but this is true also of the simplistic approach to stock classification. A third stock or group might be regulated according to a central value, but this would not, it seems to me, give us much more information.

This is not, of course, the first time that experimental regulation of the exploitation of a marine resource has been suggested—Burkenroad and others did so in the 1940s and after, and there have been attempts to put this idea in practice. However, this approach is normally to a single stock and does not include a "control" situation. The problem is different with most fish, in any case, because having generally shorter life-spans, their responses to changes in exploitation are more rapid. Ideally, I suppose one would devise an experimental design to give comparisons both over time and as between different stocks so that the conclusions would not depend too much on an assumption either that environmental conditions did not change over time, or that all of a group of stocks could, over the experimental period, be described by the same model with the same parameter values.

Put another way, this alternative approach would mean that very different rates of exploitation were permitted on stocks which there was no reason to believe were in different categories.

In the long run, taking a calculated risk of depleting one or more stocks need not be a departure from the principle of management for sustained yields, provided that the experimental regime is applied as a whole to a group stocks which are not *completely* separated. Such a group might be the sperm whale stocks in the Southern Hemisphere. So, if one or more of them became depleted, and eventually had to be protected, the danger of irreversible depletion would be lessened by the possibility of migration into them of individuals or groups of whales from another stock which had been subjected to a much more conservative regime. In any case, neither the species as a whole nor any group of stocks of that species would become endangered, although there is a risk that genetic diversity in the species might be somewhat reduced.

There are, of course, some practical problems. Different stocks are of interest to different countries, especially to those having land station operations, and their interests would have to be protected. The experimental regime would be applied mainly to pelagic operations, which are, in any case, the ones in which faulty regulation is most likely to endanger the resource. A diversified, experimental regime could, in fact, be a net advantage to pelagic operations, since they might be permitted a total quota not much different from the one they would secure under the present management rules, but they would be able to take the whole quota or most of it from a smaller area. The table below shows, purely for illustration purposes, how the Southern Hemisphere sperm whale quotas were, in fact, allocated in 1975–1976 and how they might have been allocated on an experimental basis.

	As allocated		**Experimental approach**	
Division	**Males**	**Females**	**Males**	**Females**
3, 4	1,420	1,140	0	0
5	900	630	0	0
6	0	270	0	0
7	450	360	0	0
3–7	2,770	2,400	0	0
8	1,260	810	2,380	1,600
9, 1, 2	1,840	1,660	3,490	3,270
1, 2, 8, 9	3,100	2,470	5,870	4,870
All	5,870	4,870	5,870	4,870

These higher quotas in Divisions 1, 2, 8, and 9 may be greater than would be compatible with even the most "optimistic" estimates, in which case they would have to be lowered somewhat; whether such a reduction should be balanced by small quotas in Divisions 3 to 7 depends on the economic and enforcement considerations mentioned below, and on the location of the central estimates originally used with the range of estimates.

Of course, other, perhaps better, formulae could be developed than the calculation of a total quota from central estimates and then reallocating this by divisions. One consideration in this connection must be whether it is true, as has been said, that from the point of view of scientific knowledge, even a small quota is much better than zero. It seems to me that the difficulties that have been met in interpreting data from very small samples, and small effort by few whaling units, are such that any such theoretical advantages would be outweighed by the ease of implementing—and generating confidence about the implementation of—zero quotas in some areas, as well as by the economic advantage of taking all the permitted catch in one region.

The new management policy has been referred to as a compromise between those seeking a total moratorium on commercial whaling for a certain period and those seeking the continuation of relatively unrestrained whaling: that is, whaling until it can be proven that a stock has already been reduced below some "optimal" level. However, other considerations than simply pessimism or optimism in the assessment of stocks, or even of the "ethics" of whaling, have arisen. Thus, in some places, "whale watching" is now an activity of great economic value. This means that the economic interests of different social groups (the balance being reflected in the policies of nations) can be qualitatively different. This is not to say that all the different interests are necessarily incompatible—whale-watchers and whalers both need an abundance of whales—but they may be willing to accept different levels of risk of depletion. The "blunt knife-edge" classification system is not really a compromise between such different interests—a stock is now only protected when it seems practically certain that it is depleted; that is, reduced to well below its MSY level. It is significant that the whale stock for which "whale watching" is of the greatest economic value at present—the California gray whale *Eschrichtius robustus*—is not depleted—but is still protected only because it, with other previously protected species, was arbitrarily exempted for the time being from the new management policy. Experimental application of a diversified regime offers the possibility of compromise between different kinds of "valuation" of whales, as well as between the optimists and the pessimists.

My last point concerns specifically the sperm whale. Its very wide distribution, and the fact that it does not yet appear to have been depleted as severely as most of the baleen whales, make it particularly suitable for application of a diversified management regime. In addition, however, it is used essentially to provide industrial products rather than human food. Vegetable substitutes for these are already available from several sources on a pilot scale. At the same time, although I am not aware of any current research on this matter, there seems to be a possibility of making sperm whale meat more acceptable for general human consumption than its normal taste, and possibly its contaminant content, presently allow. By the time an experimental regime has shown results—a minimum of 5–10 years, I think—the technical and economic possibilities of substitution of the sperm oil and processing of the protein will have been clarified and the management regime could be reviewed in the light of this, and of the further knowledge we should have by then about the ecology and behavior of what some people suspect is a species of exceptional intelligence.

Appendix

From: *Report Supplement 1*

8. Management of Whale Stocks

The Group agreed that a thorough, wide-ranging and, hopefully, creative discussion should take place at Bergen on the scientific aspects of management problems (and, necessarily, of the relation of these to the definition of objectives and the application of procedures) as provided under Item 5 of the Provisional Agenda for the Consultation, in the proposed Working Groups 16, 17, and 18. The experience and current problems of whale management obviously provide an important—perhaps the most important—input to that discussion.

The history of whale research shows that, being constrained to learn mainly from casual observations during, and as a result of, whaling operations, many gaps are left in the knowledge required for the application of a scientifically based policy for sustained use of the resource, or even for developing an appropriate policy. Thus, the application by the IWC of its new management policy requires very detailed information such as precise and calibrated effort measures which can give satisfactory abundance estimates about species selection in whaling, etc. It is, therefore, quite essential that the industry be required to provide such information as the scientific advisers may specify—promptly, fully, and accurately. However, while such information and improved research along traditional lines—including the intensive development and use of methods of census and population study which do not depend on whaling—are necessary components of the total management activity, they will still not be sufficient to provide the reliable assessments that are required satisfactorily to implement even present policies. Such knowledge as we have of density-dependent changes in vital rates—and that, it must be admitted, is very little—comes almost entirely from observations of stocks which have been changed drastically by whaling, yet it is such changes that determine the existence, and levels, of sustainable yields. This leads to a dilemma when an attempt is made to stabilize exploitation stocks at "optimal" levels.

IWC scientists, and others, fully recognize the inadequacy of present assessments, but naturally feel bound to work for consensus on estimates in a central range, to hedge these with conservative margins of "safety," and to propose management decisions on the basis of them. Such a procedure may be acceptable if "central" estimates are judged to be not only the "best," but significantly better than other estimates within the possible range. This may be so in some cases, but is evidently not so in others, where there may be a wide range of equally unreliable estimates, or only a few or even a single—and not very good—one. In these situations, scientists can explore the sensitivity of assessment models, but each management decision, by present practice, must finally be based on a single value. Not only is this value then applied to one stock, but it is usually being applied, in effect, to a number of stocks of the same species, since separate sets of parameter values for the various stocks are practically never available. This means that the management decision is as likely as not to lead either to unintentional underexploitation or to overexploitation of all or several stocks. If a "best" estimate turns out to be, even by chance, close to a "correct" one, application of it will presumably lead to stabilization of the stock, but by the very nature of this process, knowledge of the dynamics will hardly be improved. In particular, success in holding a stock at a particular sustainable level tells us little or nothing about other, possibly more productive, sustainable levels.

The Group suggests that some alternative procedures should be considered. For example, one set of assessment results might be applied to one stock, and another set to another stock, and the consequences observed. Management in such a way would be an approach to a controlled experiment, and intelligently done could lead to a quantum jump in knowledge of stock responses. A whole range of alternatives are possible—one stock could be protected and another, not judged to be in a very different state, could be exploited, possibly quite intensively. The set of decisions might be chosen, for example, in such a way that the total quota was not much different from that which would be determined from a single "best" estimate. This approach might also have other advantages. It would be far less likely that all stocks of a species in an ocean region would be unintentionally overexploited. The permitted catches would be taken in a more limited area and, therefore, possibly at less cost. Even different objectives of management upheld by different groups of people might be accommodated and conflict between them correspondingly reduced. On the other hand, the strategy adopted must give due weight to the fact that a whale stock can be reduced very much faster than it can grow. These are, however, to us, secondary questions; the main aim is to manage so that knowledge is sure to be improved at the same time as the resource is used. Other possible patterns of management than that suggested above should be looked into, such as exploiting a series of stocks in turn. This, and others, would require a decision-making process longer in term than the present one, and the implications of this also need consideration.

The Group agreed that these suggestions should be put to the forthcoming meeting of the IWC as well as to the Bergen Consultation, and that, to this end, the present report should be transmitted, followed by a more extended explanation which S. Holt was invited to write.

From: *General Conclusions*

3. Management

Whaling has been carried on throughout recorded history. As technology developed, the intensity and range of catching operations increased, so that whaling has extended from the slow-swimming species near shore to the fast rorquals in distant polar seas. Various major episodes can be identified, such as the Greenland or northern fishery for right whales *Eubalaena glacialis*, the southern fishery for sperm whales, and the Antarctic pelagic fishery of modern times. Each of these fisheries has shown similar patterns of development, expansion, and eventual collapse, in nearly all cases as a result of overexploitation. In the case of the rorquals, this sequence has been associated with successive shifts of primary interest to smaller and smaller species, a process which, it seems, cannot go any further unless there is an unexploited and, as yet, unrecognized "resource" of an even smaller species of whale. Apart from partial protection afforded to right whales and the gray whales, they were all essentially unregulated, other than by economic factors, until the International Whaling Commission was set up by the convention signed in 1946. Even then, the IWC exerted little influence on the national and commercial interests until the stocks of the major baleen whale species had been very seriously depleted. The regulations promulgated by the IWC apply, of course, only to its Member States, which states now take about 90% of the world catch of whales by weight. The IWC set up the Committee of Three Scientists to provide independent assessment and advice in the early 1960s, and procedures relating to the concept of maximum sustainable yield (MSY) developed then, and subsequently within the Scientific Committee, have formed the basis of an increasingly restrictive policy designed to manage more effectively the exploitation of whale stocks, whatever their location or state. This has now reached the point where complete protection is automatically afforded to much-reduced populations; catches not exceeding 90% of the estimated maximum sustainable yield are permitted by the Commission, on the advice of the Scientific Committee, from populations judged to be at or above levels appropriate for MSY; and rules are being developed for restraining the exploitation of currently unexploited stocks while scientific information is being obtained about them. Application of such a policy places heavy responsibilities on scientists, industry, and governments with regard to the collection, analysis, and interpretation of statistical and scientific information.

In fact, not all the necessary catch and effort data relating to the whaling operations carried out even by Members of the IWC are available for such analysis, and the biological responses of populations to exploitation are insufficiently understood to produce knowledge adequate to ensure the intent of the new management policy is, in fact, fulfilled. Thus, there are some doubts about the validity of the assessments and of the models, and the implementation of the conclusions. Nevertheless, this policy is a considerable advance on the position taken in earlier years. It clearly requires a considerably increased and improved scientific input based upon expanded research and better data from industry. It is also necessary that all whaling activity—including the 10% not conducted by IWC Members—should be bound by the same set of decisions so that management is fully effective for all whale stocks throughout the world, no matter by whom these stocks are used. This set of decisions could, the Group believes, with advantage be far more flexible than is envisaged by the new management procedure of the IWC, and specifically should be such as to lead to substantially better understanding of the dynamics of whale populations. This might be done by viewing every regulation essentially as an element of a large-scale controlled experiment. We suggest that, properly planned, this approach could be acceptable both to whaling industries and to those scientists and others who believe that in our present ignorance, we could well be endangering the resource while acting on "best"—which might be more appropriately labeled "least bad"—estimates of stock sizes and of sustainable yields. The Group recommends that the virtues of, and the difficulties associated with, such an approach should be explored more thoroughly than has been done so far (see more detailed consideration of this matter in Section 8 of ACMRR/MM/SC/2 Supplement 1).

The Working Party may be expecting the Group to make explicit judgment of the assessments that have been made by the Scientific Committee of the IWC and on which the Commission is currently acting. The Group does not feel able to make any general statement of this kind because the various assessments are of greatly varying quality and are continually being corrected, refined, and reanalyzed. Several valid general criticisms of the present situation can be made, and they are reflected in our summary and in the full report: the inadequacy of some statistics, the insufficiency of the effort for research and analysis in relation to the demands made for scientific advice.

With respect to particular assessments, it cannot be said that they are, in general, too optimistic or too pessimistic; if we make a criticism, it is that the Scientific Committee, whether or not by intent, but certainly by the manner in which it has, under pressure, communicated its conclusions, tended to obscure the uncertainties of its deductions and calculations. This has an understandable origin—too often in the past, honest uncertainty has been taken as a sign of weakness, and management decisions or non-decisions have been inappropriate to the situation. We believe that ignorance must be honestly faced and appropriate decision procedures developed which take account of it. It is better than nothing, but still not sufficient, merely to take "central" values for assessments—usually arrived at by a process which is less than scientific—and then to adjust them by some unmeasured—and probably unmeasurable—margin of error.

The problem is not only in estimating the parameters of models; it is equally in the construction of realistic models. It may well be that some attitudes to management criteria are demanding more than science can provide at the present stage. This too must be taken into account when management of whaling operations is discussed.

Chapter 13

Does the Bottlenose Whale Necessarily Have a Sustainable Yield, and If So Is It Worth Taking?[1]

Assume that the stock in, say, the 1880s had been essentially unaffected by whaling and was in a steady-state. It recruited at 2 years of age, the females matured at 11, and there was a constant natural mortality, not age-dependent. The sex ratio was 0.5 and mature females gave birth every second year. Then, $r = M = e^{-2M} \times 0.5 \times 0.5e^{-M(11-2)}$ from which $M = 0.091$ (round to 0.09) and the percentage mature females of total recruited females is 44. Such a value is not inconsistent with the numbers in the Labrador Sea in 1971 (given by Christensen 1973, Figure 7).

Now, suppose whaling takes place for a few years and it takes a significant proportion of that stock. The age composition changes such that if the age at maturity remains the same, the percentage mature would be reduced, and the number of births per recruited female correspondingly. If whaling were now to stop, this stock would continue to decline, but not to extinction; the age composition readjusts and the stock settles at some new level. Such a model shows unstable equilibrium.

However, we would expect *some* density-dependent change in the birthrate, otherwise the stock would have behaved very erratically in the past and not be likely to have survived for long. In one reasonable model, applied to other whales, it would be assumed that female whales grow faster and mature earlier in a reduced stock. We can calculate how strong such a response would need to be to maintain $r = M = 0.9$. We can write in this case: $e^{-0.09(11-2)} = e^{-(F+0.09)(A_m-2)}$ where A_m is the new age at sexual maturity. We can find, for example:

$F =$	For $A_m =$
0.011	10
0.026	9
0.045	8
0.072	7
-	-
0.720	3

If, with the stock reduced by an F of, say, 0.026, the age at maturity became less than 9 years, the stock would begin to increase when whaling stopped, i.e., there would be a sustainable yield at that level of stock. If the age at maturity went down, but only, say, to 10 years, there would be no sustainable yield and the stock would continue to decline for a time after whaling stopped.

[1]Editors' note: written by Sidney J. Holt, FAO Rome, as Document SC/28/Doc 32 for the 1976 meeting of the IWC Scientific Committee and published in Reports of the International Whaling Commission 27:206–208, 1977.

It is not necessary to maintain the assumption of a one live birth every two years per mature female. Suppose in the unexploited stock, live births are only 80% of the maximum theoretically possible (observed pregnancy rate of 0.4). Such a stock would attain equilibrium with $M = 0.081$ (round to 0.08).

Such a value is not inconsistent with the age distributions for Labrador in 1971 given by Christensen (1973, Figure 7). There, in a small sample of 128 whales, the percentage mature was 48% in each sex; accepting the ages at maturity cited by the author, M is about 0.065. With $M = 0.08$, one expects a ratio of 41%; such discrepancies are certainly not significant. If M had been 0.065 in the eastern North Atlantic, then balance would be achieved with a birth-rate 53% of the maximum possible (pregnancy rate = 27%).

Now, if such a stock was reduced by whaling, there could be a density-dependent response by an increase in pregnancy rate—but a limited increase, obviously—as well as a shift in the age at maturity, as noted in baleen whales. Depending on the rate and balance of these responses, we can generate sustainable yields, "negative sustainable yields," or none. Nevertheless, over a wide range of possible values, the stock would not proceed to extinction after whaling stopped; it could rise somewhat, continue to decline for a time, or stay as it was at the time whaling stopped.

Is such a model compatible with the observed course of bottlenose *Hyperoodon ampullatus* whaling in the Northern Atlantic? I think it is. From Christensen (1976, Figure 5), by drawing a straight line between the points for 1885–1897, we can see, very roughly, that a cumulative catch of about 25,000 caused the stock index to decline from 49 to 40, i.e., by just under 20%. This suggests an "initial" stock size of about 130,000 whales, if the natural deaths and births continued to balance each other during the period. Actually, because of the two-year recruitment time-lag, with M constant, the number of natural deaths could have been a little less than the recruits—approximately 2,000 whales less over the period. Without this small "supplement," the stock would have been reduced slightly more than was observed; we should, therefore, adjust the calculated initial stocks downwards, but not much. The number left in 1896 would be about 100,000; F over the decade averaged 0.022.

Now, in the next decade, intensive whaling continued. A cumulative catch of about 19,000 whales "caused" a decline in catch per vessel from 40 to 27, that is, by 32%. This would suggest a 1896 stock level of 60,000 whales declining to 40,000 by 1906, whereas on the evidence of the first decade, we should have expected 100,000 to decline to 81,000. From a "simple" population model, the decline would have been less than this because of a density-dependent response of r. However, by now, the number of recruits can be declining faster than the number of natural deaths because the change in age composition has begun. With no other density-dependent reactions, the percentage mature could have declined from 44 to 36%, that is, by a factor of 0.83.

If we take 100,000 as the 1896 value, then by 1906, it was reduced, by 32%, to 68,000, i.e., by 32,000 whales of which only 19,000 are accounted for by capture. Is the other 13,000 a "recruitment gap?"

The average annual number of recruits to the stock in the first decade is $(130{,}000 + 100{,}000)/2 \times 0.09 = 10{,}350$, and 10,350 and 17% [i.e., $(1 - 0.83)100$] of this is 1,760. In the next decade, the recruitment gap could, therefore, have been nearly 18,000 whales—enough to account for the stock decline being steeper than expected from the catch rate.

Actually, it is only after 1913 that the catch-per-vessel value begins to behave very erratically. So, we can look at the longer period 1896–1913 throughout which whaling was intense, although it began to fall off in the last years of the period. In this period, catch per vessel declined from 40 to about 25; the cumulative catch was 25,000; this gives, crudely, a 1896 stock of 67,000 declining to 42,000. If the 1896 stock was really 100,000, the 1913 stock would have been 63,000, a decrease of 37,000 whales, of which 25,000 is accountable to catch. A "recruitment gap" can easily account for the remaining 12,000 (1,760 × 17 years = 30,000) even if there had been a partial compensatory change in age at maturity, in pregnancy rate, or in both.

Between 1914 and 1966 (a cut-off date taken to exclude the Labrador catches), the total catch was about 7,000 whales averaging about 135 whales a year. The average catch per bottlenose whaler in successive periods was:

(1913)	25
1914–1927	23
1930–1939	14
1946–1949	12
1953–1961	16
1962–1966	40

These figures do not tell us much. The jump up in the 1960s is too great and sudden to attribute to "stock recovery." The level did seem to decline in the four or five decades after the end of World War I; it *might* have begun rising in the 1950s.

In the 1914–1927 period, the average annual catch was 190 whales. This could have exceeded any sustainable yield at the time, accounting for a lower catch per unit effort over the next three decades during which the average annual catch was only 51 whales. If 51 was below the sustainable yield of the surviving stock, we might expect some subsequent increase in catch per unit effort, but not so suddenly as it occurred. Let us suppose there had been a density-dependent reaction as a change in age at maturity, say, from 9 to 6 years. This leads to an $r - M = 0.11 - 0.07 = 0.04$. The sustainable yield of a stock of 63,000 would have been 126 whales. If only 52 were taken, the stock would have increased, over three decades, by 30–35%. This does not seem to have happened. We can suppose stronger responses of $r - M$, but then the expected recovery would be so rapid that surely we would see it reflected in the catch per vessel.

Between 1913, when the stock might have been 63,000, and when whaling stopped in 1970, 8,000 whales were taken (excluding Labrador). Is the stock now about 63,000 – 8,000 = 55,000, less than that number, or more and, if either, by about how much?

"Logistic" theory predicts that if the SY of the 1914 stocks was greater than 100–150, the stocks would have started to recover; if less, to decline further. The catches per vessel in later years, taken at their face value, do not suggest a very full recovery, if any; such a view is supported by the fact that effort was put into whaling further west. If r had risen by an amount only one-quarter of the natural mortality rate, with the intervening catches at less than half of the SY, one could look for an increase in stock size by nearly 5% over the period; there is no indication of such a trend. In fact, the catch rates in the intervening years (1914–1939) suggest that the stock was further reduced, by continued whaling at a lower level (a decline perhaps supplemented in the early few years by a recruitment lag), to around 38,000 whales. This would put the present number at no more than 40,000 whales. Whether the 51 whales taken annually from 1940 to 1966 were much less than MSY, we have no means of knowing. By the criteria established under the new management policy, which relate not to SY levels, but to relative *stock* levels, it seems the bottlenose would be placed in the Protection Stock category, at least in the eastern North Atlantic.

The values of stock sizes quoted in this note, and of SYs, would all be decreased if there was any increasing trend in "whaling power per unit effort" over the period, especially between 1885 and 1906.

The Effect of Unbalanced Sex Ratios

Benjaminsen (1972) gives, in his Tables I and III, data for the sizes and numbers of mature and immature whales by sex in a 1967 sample of 51 whales from northeast Iceland, which are summarized below:

	Numbers			Mean lengths (ft)		
Stage	♂	♀	♂+♀	♂	♀	♂+♀
Mature	17	12	29	26.8	24.2	25.7
Immature	15	7	22	21.4	21.3	21.4
All	32	19	51	24.3	23.1	23.9

53% of the males and 63% of the females were mature. The numbers are, of course, very small, and we do not know if it was a random sample of the catch. I simply note that the ratio of 63% mature females would be consistent for $M = 0.09$ with a reduction in age at maturity from 11 to 7 years. Males were bigger than females (by about 16% by weight if weight is roughly proportional to the cube of the length) and more numerous in catches.

Benjaminsen's Figures 4 and 5 show that over the period 1938–1969, very roughly 20% more males than females were taken at Svalbard, and at Iceland, the males exceeded females by 2:1. Figure 6 shows that the mean length of males exceeded that of females in catches in all areas. The difference was clearest at Iceland and Svalbard, but was about 2 ft in all of them. The mean length of females in all areas except Labrador was about 23–24 ft; of males 25–26 ft. In Labrador, the length might have been higher; this would be consistent with the fact that exploitation at Labrador began much later, and that half of growth in length (95% by weight) occurs after recruitment at 2 years.

So, in the eastern North Atlantic, the mean lengths in catches of females exceeded those of males by about 10%. However, the size of females at a given age exceeds that of males by 15–20% (growth curves given by Benjaminsen, Figure 10), which is more than enough to account for the mean size differences in catches, without postulating major differences in the age composition of the sexes. Considering the wide overlap of the size distributions, relative to the differences between their means, it does not to me seem significant, except insofar as the sexes are segregated in May and June in Iceland.

Without information on the *changes* in sex ratio over a long period—including the early years of whaling, further interpretation is difficult. If males were selectively hunted over a period, we should expect the sex ratio to shift away from the unexploited ratio in favor of females. There is no evidence of this having happened. It seems quite likely, however, that the proportion of females in the stock in the post-war period is less than 50% and, considering that 1711 were taken at Svalbard and 2125 at Iceland, might be as low as 38%. This means that values of r quoted would need to be reduced by a factor of 0.76 and $r - M$ values accordingly. To maintain stability, and give sustainable yields at a given level, the density-dependent responses, by changes in pregnancy rates and/or ages at maturity, would have to be stronger than I have postulated.

References

Benjaminsen, T. 1972. On the biology of the bottlenose whale, *Hyperoodon ampullatus* (Forester). Norwegian Journal of Zoology 20(3):233–241.

Christensen, I. 1973. Age determination, age distribution and growth of bottlenose whales, *Hyperoodon ampullatus* (Forester) in the Labrador Sea. Norwegian Journal of Zoology 21:331–340.

Christensen, I. 1976. The history of exploitation and the initial states of the northern Atlantic bottlenose whale, *Hyperoodon ampullatus*. Paper NA 76/1 presented to meeting of IWC Working Group of North Atlantic Whales, University of Oslo, 5–13 April 1976. Unpublished. [Available from the IWC Secretariat].

Chapter 14

Assessment of Southern Hemisphere Minke Whales[1]

The Scientific Committee, at its 1975 meeting, calculated the relation between initial and current stock size in each area assuming that incoming recruits each year were a constant fraction of the number in the initial *stock*, i.e., a constant number. The same assumption is contained in the calculations made during the 1976 meeting with a modified formula. However, the stocks have now begun to decline as a result of whaling and it is more "conservative" and, I think, more reasonable to assume that recruitment is declining with stock size. A simple and perhaps extreme assumption is that the number of recruits is proportional to the stock size A_m years earlier. If there had, in fact, been a significant and instantaneous density-dependent response of the rate of reproduction to the decline in stock size, this procedure would slightly underestimate the recruitment. Although we cannot say where the true value would lie, it seems likely to be nearer that given by an assumption that recruitment is essentially independent of stock size. It would be prudent to assume proportionality until such time as evidence is available (e.g., changes in pregnancy rate and/or age of maturity) that there has, in fact, been a density-dependent change.

I have calculated the consequences for Area II where whaling has been going on for the longest time, applying the expression: $N_i = (N_{i-1})\, e^{-M} + (1 - e^{-M})N_{i-A_m}$ and taking $M = 0.125$ and $A_m = 6$.

The calculation is not sensitive to the value of A_m if the stock is not declining very quickly. The results are shown in Table 1, column B, including an extrapolation to future years.

The estimation procedure used by the Scientific Committee is such that it is appropriate, for this area, to adopt the current stock size also in applying my model. This leads to a slightly *higher* estimate of initial stock number (and hence higher quota as calculated from the rules for Initial Management Stocks) and hence a lower ratio current/initial stock size. The difference is not yet very much, but it will be seen that in the coming years, the two series diverge rather rapidly.

It would be more reasonable, in fact, to assume that the recruitment is proportional to the number of *females* rather than to the total number. This only makes a difference if the rates of depletion of females and males are different, but this seems to have been the case in general, and especially in Area II, where the proportion of females in catches differs substantially from 50% and also has changed with the beginning of pelagic operations in the area (Table 2). I have calculated the consequences (column D) assuming that the proportion of females at birth is 0.50 and the natural mortality rates of females and males are equal; over the range of stock sizes considered, the calculation is not sensitive to these assumptions. Again, the effect is not yet great, but can soon become significant.

---

[1]Editors' note: written by Sidney J. Holt as Document SC/28/Doc 35 for the 1976 meeting of the IWC Scientific Committee and published in Reports of the International Whaling Commission 27:214–215, 1977

Table 1. Southern Hemisphere minke whales *Balaenoptera bonaerensis*—Area II.

Years	Stock sizes in thousands			
	A	B	C	D (% females)
1965–66	37.1	37.7	37.7	37.8 (50)
1976–77	31.5	31.5	31.5	31.5 (47)
1977–78	30.5	30.2	30.3	30.3 (46)
1978–79	29.6	29.1	29.2	29.1 (45)
1979–80	28.9	28.0	28.2	28.0[a](45)
1980–81	28.2	27.0[a]	27.1[a]	26.9 (44)
1981–82	27.6	25.8	26.0	
1982–83	27.1	24.8	25.0	
1983–84	26.6[a]	23.7	24.0	
1984–85	26.2	22.5	23.0	
1985–86	25.9	21.4	22.0	
MSY level (60%)	22.3	22.6	22.6	

Notes to Table 1

[a] Year in which stock enters "Sustained Management" category under IWC rules (20% more than MSY level of 60% of initial).

A – Recruitment constant; annual catch from 1976–1977 on = 37.7 × 0.05 = 1,885.

B – Recruitment proportional to preceding stock number (males plus females); catch from 1976–1977 on = 37.7 × 0.05 = 1,885.

C – As B, but with a quota of 1,855 as in A.

D – Recruitment proportional to preceding number of females only, and assuming no change in sex selectivity of whaling, so percentage females in pelagic plus land station catches = 66. Future catch = 37.8 × 0.05 = 1,890.

Table 2. Percent females in Southern Hemisphere catches.

Season	Antarctic pelagic		Pelagic plus land stations
	All areas	Area II	Area II
1969–1970	-	-	68.1
1970–1971	-	-	68.0
1971–1972	46	-	68.0
1972–1973	50	-	67.9
1973–1974	62	60.2	66.3
1974–1975	61.0	60.0	63.8
1975–1976	57.3	67.9	67.9

Simulation of future changes under the IWC rules is, in this case, not so straightforward, but can be done under the simple assumptions that the differences between the sex ratio in catches in early years, by land stations and by pelagic operations, and the ratio 0.5 effectively measure the catching selectivity whatever the cause of it (presumably segregation of sexes or their differing vulnerability). On this assumption, the sex ratio in future catches can be calculated; it tends to move towards the ratio in the unexploited population, and you can pass it, which is what has been noted in the history of whaling for other baleen species. In fact, in Area II, the land station catch has every year contained 67–68% females. The pelagic catch ratio in the Antarctic as a whole has been more variable, averaging 56%.

To predict future stock sizes in the short term, I have taken the sex ratio as 66% in consideration of the increased proportion of total catch taken by pelagic operations and of the necessarily changing sex ratio in the stock (see last column of table).

It will be noted that these more "conservative" calculation lead to *higher* catch quotas. This is partly because the modified DeLury model was applied to Area IV (a procedure which is not objectionable because whaling has been conducted for only five years in that area) and the current stock numbers in the other areas calculated from that; if the DeLury method is applied to a longer series, then it underestimates the initial, and overestimates the current, stock sizes.

In view of the conclusion that at the present level of whaling the Area II minke whale *Balaenoptera bonaerensis* stock will soon need to be reclassified as a Sustained Management Stock, it may be timely to calculate the highest optimistic, but reasonable, MSY that it might be possible to take from it. Following the arguments in Holt (1977), we find, for the 60% level: MSY = $0.177MP$ = 840.

This estimate depends, of course, critically on the value of M, and one might ask how confident we feel that in the Southern Hemisphere minke whale, M is as high as 0.125, that is, double the value used for sei whale *B. borealis* assessments (0.06) and threefold the fin whale *B. physalus* estimate (0.04). The effects of assuming a somewhat lower value $M = 0.08$ are that an initial stock of 38,400 whales, declining to 31,500 by 1976–1977, enters the Sustained Management category by 1979–1980 (the same year as indicated in column D). In this case, however, the "reasonable optimistic" MSY is only 540 whales.

A further factor for future consideration is the nature of the expected density-dependent change in the recruitment rate as the stock is reduced. If this was mainly caused by a decrease in the age at maturity, the effect is to change the lag of six years assumed for all the above calculations. At first, we would expect the annual recruitment to be boosted by additional numbers as the generations are caused to overlap; that is, in a particular year soon after exploitation starts, the normal number of mature animals 6 years of age will be augmented by some born a year later and maturing at 5 years. This effect is soon dissipated, however, and the stock declines faster than would be expected if age at maturity were constant.

References

Holt, S. J. 1977. A possible basis for determining limits to the allowable catch when the degree of density dependence of the new recruitment rate is not known. Reports of the International Whaling Commission 27:230–233.

Chapter 15

A Possible Basis for Determining Limits to the Allowable Catch When the Degree of Density Dependence of the Net Recruitment Rate is Not Known[1]

This note is concerned mainly with whales classified as Initial Management Stocks and with steady-state situations. The management of recovering or depleted stocks could be analyzed by corresponding methods. Consideration of models with time-lags between birth and recruitment involves no basically different principles, only more complicated arithmetic. The natural mortality rate M is assumed to be constant.

An assumption that annual recruitment in a reduced stock remains at the same level as in an unexploited stock in steady-state implies that the birth rate and the sustainable yield increase steadily as the stock level declines. This situation is illustrated in line A of Figure 1. The figure is drawn such that the properties of the various models can be seen irrespective of the values of M and of the initial stock size P. A feature of the model illustrated by line A is

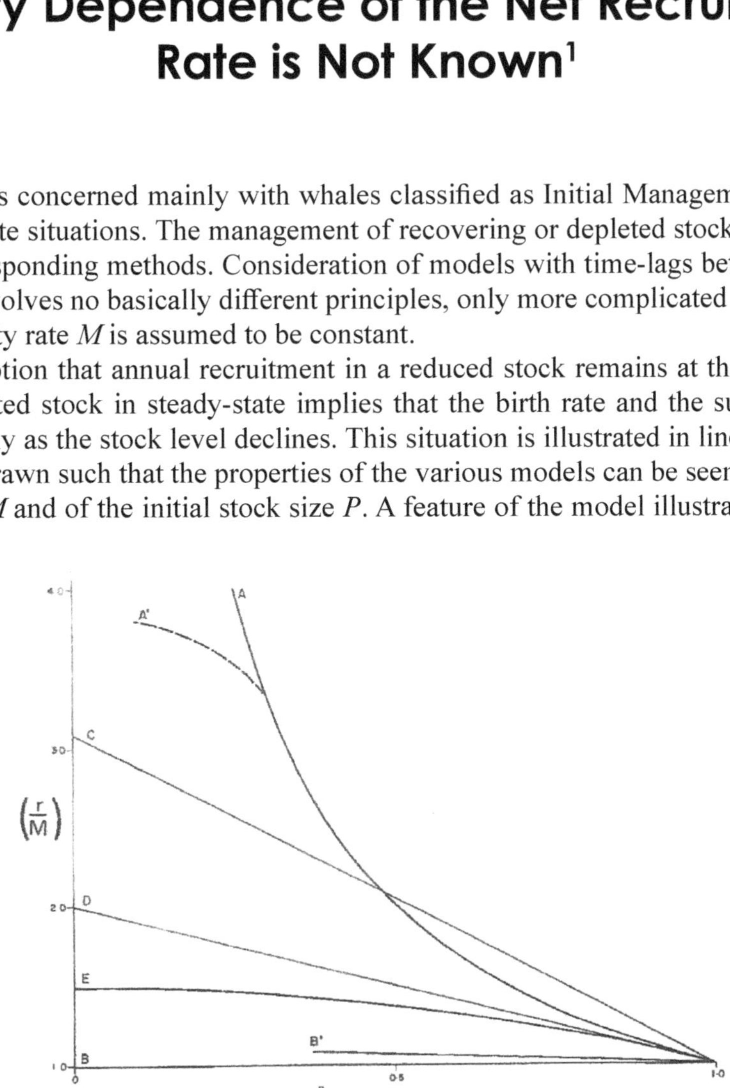

Figure 1. Graphical representation of various models relating the ratio of the recruitment and natural mortality rates to the size of stock, as a fraction p of the initial stock P.

[1]Editors' note: written by Sidney J. Holt as Document SC/28/Doc 43 for the 1976 meeting of the IWC Scientific Committee and published in Reports of the International Whaling Commission 27:230–233, 1977.

that for any given decrement in stock size, the rate of recruitment increases more than proportionally. This could, I suggest, only be expected in a situation where there were, initially, severe overcrowding with consequent stunted growth and other biological stresses. This is the model implied by the simple DeLury procedure for estimating stock size, which is generally satisfactory only if data are used for period shorter, or not very much longer, than the time-lag between birth or recruitment and maturity. It leads to highly "optimistic" predictions when applied to whale populations.

Model A could be made slightly more plausible by modifying it as in A' if, as for whales, the highest possible value of r can be estimated from biological data. This theoretical possibility is, however, not relevant to the discussion of management under the Commission's present procedure. Model B is a "pessimistic" one. It implies there is no density-dependent response to stock reductions caused by whaling so that r is the same at all stock levels. Such a stock would be unstable, having no sustainable yield at *any* level. Stability is introduced, however, giving sustainable yields at all stock levels, if there is only a very small increase in reproductive rate with declining stocks; an example is shown as B'. In this case, however, responses to minute changes in exploitation rate or natural parameters would show extremely slow returns to equilibrium. B' represents, of course, only one of the innumerable possibilities of the simple logistic model. Line C represents one of the more usual versions of the simple logistic. It has, as always, MSY at $0.5P$, but is peculiar for stock sizes between 0.475 and $1.0P$ in that the recruitment is higher *in absolute* numbers in that range of stock sizes, than in the unexploited stock. I suggest that in the absence of definite evidence or plausible arguments to the contrary, a model having such a property should not be applied to whales in order to calculate quotas.

Line D is drawn with a slope at P equal to that of the hyperbola, line A, implying that the maximum population growth rate in residual stocks would be equal to MP. This model could still be regarded as representing the upper limit of "reasonable optimism." Any other line (not necessarily straight) lying between D and B would then be "reasonable"[2] unless other biological constraints were seen. In choosing between the various possibilities, it might be presumed that, when initial management stocks are under consideration, an attitude of watchful optimism would be acceptable, and D might be taken as the applicable model, at least for management guidance. The properties of the various models outlined are summarized in Table 1.

---

[2]Allen (1977) pointed out that this suggestion to consider as "reasonable" only models which do not predict increases in recruitment with decreasing stock size, is essentially to impose an additional constraint on the "more flexible model" described in Allen (1976). In that model, the following relations hold (using my notation): $(r/M) = (r/M)_0 - \{(r/M)_0 - 1\}p^{a}$, $p_{msy} = [1/(\alpha+1)]^{1/\alpha}$, and $s = \alpha\,[1 - (r/M)_0]$ where α is a density-dependent exponent in the model; α can take values from –1 to +∞. For the simple, symmetrical logistic with $p_{msy} = 0.05$, $\alpha = 1$. For the special case $s = 1$, we find: $(r/M)_0 = (\alpha+1)/\alpha$, $(r/M)_{msy} = (\alpha+2)/(\alpha+1)$, and $MSY/PM = [1/(\alpha+1)]^{(\alpha+1)/\alpha}$. The table shows selected values for these quantities. With the constraint $s = 1$, the model for $\alpha = -1$ becomes the hyperbola (A) of Figure 1.

α	p_{msy}	$(r/M)_0$	$(r/M)_{msy}$	MSY / PM
5.0	0.70	1.20	1.17	0.116
2.4	0.60	1.42	1.29	0.177
2.0	0.58	1.50	1.33	0.192
1.0	0.50	2.00	1.50	0.250
0.2	0.40	6.00	1.83	0.335
0.0	0.37	∞	2.00	0.368
–0.5	0.25	∞	3.00	0.500

Table 1.

A (constant recruitment)	
SY = $PM(1 - p)$	MSY = PM at $p = 0$
A' stabilizes at low levels, with MSY at a level below $0.3P$	
B (recruits proportional to number of parents)	
SY = 0	MSY = indeterminate
B' and C (general logistic)	
SY = $PMsp(1 - p)$	MSY = $0.25PMs$ at $p = 0.5$
where $s = 1 - (r/M)_0$ i.e., the slope of the line with changed sign	
D (particular case of logistic, as C, but with $s = 1$)	
SY = $PMp(1 - p)$	MSY = $0.25PM$ at $p = 0.5$
E (example of non-linear logistic family; particular case with slope = −1 at P)	
SY = $PMp(0.5 - p^2/2)$	MSY = $0.192PM$ at $p = 0.58$

Suppose, for example, we conclude that a stock has been brought down by a few years of whaling to a level 70% of its initial size of 100,000 and that from early age compositions, we have an estimate of M of 0.04. Then, we find that the most optimistic reasonable estimate of sustainable yield from a stock of that size is from Model D: 100,000 × 0.04 × 0.70 × 0.30 = 840 (i.e. about 0.8% of initial stock size; 1.2% of current stock size). The simple logistic model would predict for this stock an MSY of 1,000 at 0.5 of initial size. However, if the line were really a curve (Model E), with the same slope at $p = 1$ as has D, the estimate of sustainable yield at 70% of initial size would be 710 (0.7% of initial stock size; 1.0% of current stock size). The MSY, now at 0.58 of initial level, would be 770 (1.3% of MSY stock size, 0.8% of initial size). At 0.58 of initial stock, Model D would be, again, more optimistic, predicting an SY of 980 (1.0% of initial stock size, 1.7% of MSY stock size).

Since models for the more fully investigated baleen stocks give MSY at levels above 50% of initial stock, we should, when in ignorance of density-dependent responses, be rather more cautious than Model D suggests. However, this model does set an upper limit to optimism, and in applying it as a working basis for calculating maximum allowable catches from initial management stocks, we should not necessarily be accepting the implication that numerical MSY can be taken at a stock level as low as $0.5P$.

It is not difficult to extend this approach to include allowance for time-lags, provided the ages at recruitment and at maturity in the unexploited—or lightly exploited—stock can be estimated, even if only roughly. Obviously, the theory is a poor substitute for observations of changes in recruitment rate. However, not only are such changes difficult to measure, especially in Initial Management Stocks, but they can never be identified with certainty as having been caused by the changing density of that stock.

The Choice Between 60% and 50% Stock Levels for MSY

The linear logistic (50%), reasonably optimistic, model gives $MSY_{50} = 0.25MP$. A corresponding non-linear logistic (MSY at 60% of initial level) gives $MSY_{60} = 0.177MP$. Therefore, if we had no strong basis, in observations or by analogy with other species, for choice, and we *correctly* chose the 60% model, then we must expect, eventually, $0.073MP$ less yield (29% less) than if we *correctly* chose the 50% model. However, if we chose the 60% model and it turned out that the stock, in fact, behaved according to the 50% model, we shall have sought to stabilize at a value $SY/MP = 0.177$ instead of the correct value 0.240, and will thus take $0.063MP$ less (26% less) than the sustainable yield that could actually be available. If the stock had previously been reduced by whaling to about 60% of initial level, it would, in this case, increase slowly, at the rate of about $0.4M$% annually, towards a new equilibrium level of $0.77P$. If the stock had been only lightly exploited, if at all, it would be brought slowly down to the stable 0.77 level, at a rate depending on the size and time-pattern of early catches. If, on the other hand, we chose 50% and it turned out that the stock behaved according to the 60% model, we shall have sought to stabilize at $SY/MP = 0.250$ instead of the correct value of 0.173, thus taking $0.077MP$ *more* (45% more) than the sustainable yield actually available; the stock would decrease rapidly, at first at the rate of $7.7M$% annually, then increasing. In this case, if we persist in our mistaken choice, the stock would move not towards equilibrium at a lower level, but to extinction.

Clearly, the choice made between 50% and 60% models has important consequences, especially if a mistaken choice is not quickly discovered. A major difference is in the time required to re-establish the correct situation after discovery of the mistake. In the first case, three years of whaling at an annual catch rate of 5% of initial stock size, or proportionately more or fewer years at a lower or high catch rate, would serve to bring the stock from the 50% to the 60% level. In the second case, if whaling were to cease completely until the stock had recovered from the 50% to the 60% level at an average rate of about $0.36M$ per year, recovery would take 7–8 years with $M = 0.04$. No recovery at all would be expected unless whaling were reduced to less than two-thirds of its former level of effort. If whaling effort were only halved, recovery would take more than 16 years. It seems to me that the more cautious choice of the 60% model is overwhelmingly favorable from the point of view of establishing a sustained whaling industry.

The above analysis has been made with the "reasonably optimistic" 60% and 50% models, that is, for $s = 1$. The parameter s can be regarded as an index of "optimism," thus $s \geq 1$ is unreasonably optimistic, $s \leq 0$ is unreasonably pessimistic; s below, but near to, 1 is optimistic, but reasonable; s above 0, but small is pessimistic, but reasonable; $s = 0.5$ is neutral. This last might be appropriate for Southern Hemisphere sei whales *Balaenoptera borealis* (see Table 2), which are already quite seriously depleted, about which there is a fair amount of information, but for which the strength of the density-dependent response has not yet been measured. The consequences of adopting $s = 0.5$ instead of $s = 1$ can readily be seen by inspection of Figure 1: recovery times from an erroneous 50% to a correct 60% level are doubled; the sustainable yields are halved, and so on.

Table 2. Southern Hemisphere sei whales. Estimated ratio current/initial total stock size (%).

Ratio	Model	Recruits dependent on total stock	Recruits dependent on female number		
			M♂ = M♀	M♂ ≠ M♀ No LS (1)	M♂ ≠ M♀ LS incl. (2)
Reasonable, but optimistic, assessment ($s = 1$)					
1974–1975/	MSY at 50%	53	51	48	45
1961–1962	MSY at 60%	51	50	47	44
1976–1977/	MSY at 50%	50	48	45	41
1949–1950	MSY at 60%	48	46	43	39
Neutral optimism - pessimism ($s = 0.5$)					
1974–1975/	MSY at 50%	50	49	45	41
1961–1962	MSY at 60%	50	48	43	37
1976–1977/	MSY at 50%	46	45	41	37
1949–1950	MSY at 60%	45	44	40	36

Note: Effects of land station catches *outside the Antarctic* (1) not taken and (2) taken into account.

From a set of stocks with present total numbers at 40% of an initial number of 162,000, with $M = 0.06$, and taking the "optimistic, but reasonable" view ($s = 1$), the sustainable yields and MSYs are as follows:

Model	Present SY	MSY
MSY at 50%	2,330	2,430
MSY at 60%	1,240	2,300

If the "neutral" view is taken (s = 0.5), as seems preferable considering the stocks as a whole are well below total MSY level, the above values would all be approximately halved. In any case, the present SY values thus obtained from the 60% model are far below the present quota levels for sei whales (set at nearly 1900 whales for 1976–77), excluding the SY of the protected stock in Area III).

The Scientific Committee has concluded that the Southern Hemisphere minke whales *B. bonaerensis* are now at about 80% of their initial numbers, and has taken, for the purposes of giving advice on quotas, a value of $r - M = 0.07$ and $M = 0.125$. (See also Holt 1977). This implies $r/M = 1.56$) at $p = 0.80$; the situation is illustrated in Figure 2. The straight line is the corresponding simple logistic, and the line convex upwards is the logistic form with $p_{msy} = 0.60$. It will be seen that the Committee's calculations imply, in both cases, that the number of recruits would increase continuously as the stock is reduced from its initial numbers all the way down to presumed MSY levels. This assumption seems unreasonable in the absence of any evidence to justify it.

Figure 3 shows how the stock level for MSY could be simply determined graphically for any logistic-type model including, especially, non-linear versions. In general, the locus $[r/M_{msy}, p_{msy}]$ is that at which the slope of the graph of r/M against p is equal to, but opposite in sign from, the diagonal between that locus and the point $[r/M = 1, p = 0]$. For

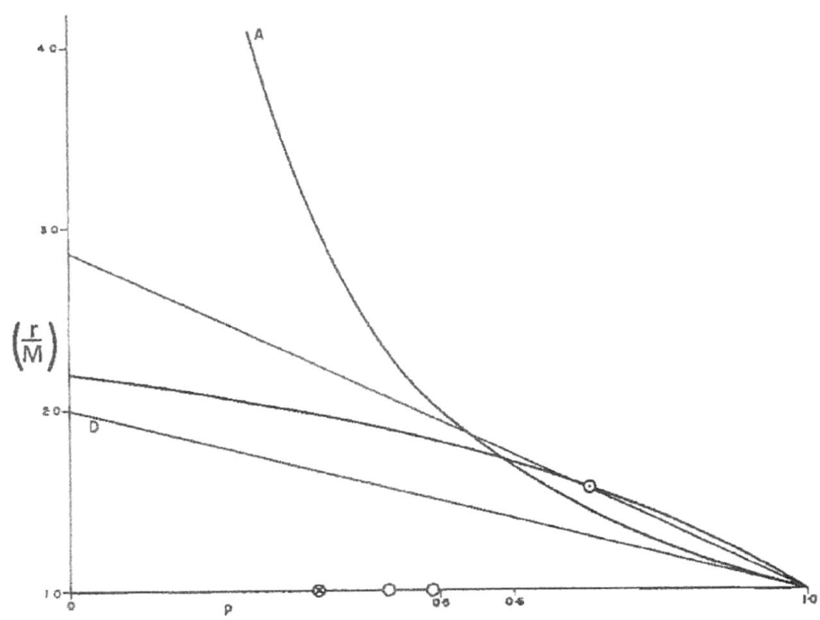

Figure 2. Southern Hemisphere minke whales.

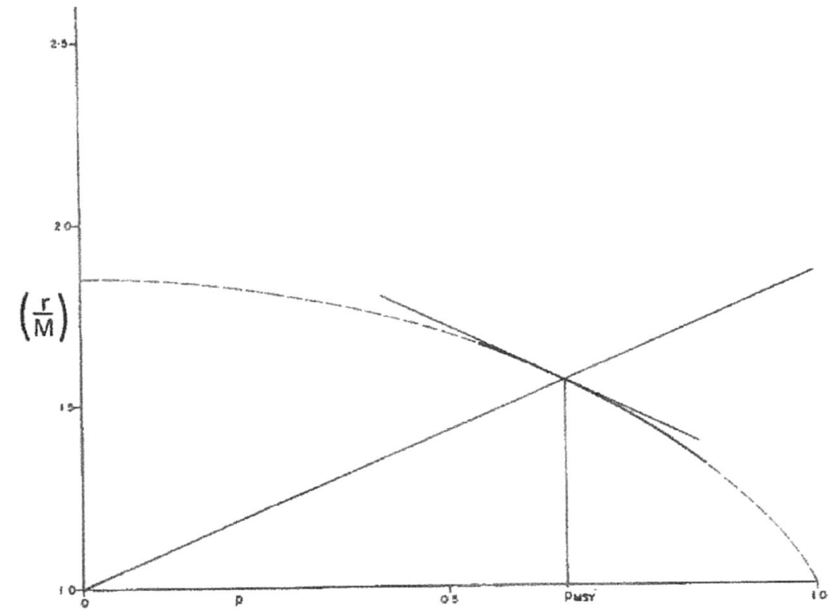

Figure 3. Graphical determination of p_{msy} and MSY.

logistic models, this diagonal passes through the point $[(r/M_0, p = 1]$. The following steps are taken:

1) Read off the value of $(r/M)_0$, that is, of r/M for $p = 0$.
2) Draw a straight diagonal between the coordinates $(r/M)_0$, $p = 1$ and (r/M), $p = 0$.
3) Drop a vertical line to the p axis from the intersection of the diagonal with the model line and read off $p = p_{msy}$.
4) Multiply p_{msy} by P to obtain P_{msy} if desired.

Then, to calculate MSY, we proceed as follows:

1) Read off $(r/M)_{msy}$ for P_{msy}.
2) Multiply by M, and again by P.

This procedure immediately reveals certain comparative features of various models, and in particular shows, for given values of M and P, under what conditions a model defined by a line which is convex upwards (giving $p_{msy} > 0.05$) has a higher or a lower MSY value than the simple logistic. It also enables us to compare at once the differences in expected sustainable yields at stock levels other than MSY from various models which predict the same MSY. For example, the 60% model predicts for a stock at 80% of initial level (i.e., the presumed present overall state of Antarctic minke whales) an SY 6% higher than does the 50% model which has the same values of M and P and gives the same MSY.

As R. M. May has pointed out in a series of papers (e.g., May 1972, 1974, 1976), if time-lags in density-dependent response of reproductive rate to change in population abundance are introduced into logistic models, the models may exhibit, under certain conditions, stable cycles and chaotic fluctuations as well as stable equilibrium points. In the notation used in this paper, the behavior of the model depends on the magnitudes of s, M, the age at maturity A_m, and the shape of the curve of r/M against P. Defining $C = -sMA_m$, then there are stable points for $2 > C > 0$, stable cycles for $2.7 > C > 2$, and chaotic behavior for $C >$ c2.7.

For a population described by the modified logistic with MSY level at 60% of initial stock, and with the same number of recruits at the MSY level as at the unexploited level, and with $M = 0.06$ and $A_m = 8$, we find $C = 2.31 \times 0.06 \times 8 = 1.11$. As the population is reduced by whaling, the likelihood of oscillation resulting from excessive delayed feedback diminishes. Thus, it does not seem that we need to worry about oscillation in the sei whale even if, contrary to the arguments presented in this paper, the recruitment is maintained at a relatively high level even in a reduced stock. However, if M is substantially higher—as it has been taken to be for minke whale assessments—and/or the age at maturity much higher and thus the recruitment response delay longer, as in sperm whales *Physeter macrocephalus*, we need to watch for the appearance, theoretically, of stable cycles or even chaotic behavior, especially in unexploited stocks; this then raises the question of what is meant by "initial population" even when only single-species models are being considered.

References

Allen, K. R. 1976. A more flexible model for baleen whale populations. Reports of the International Whaling Commission (Scientific Report) 26:247–263.

Allen K. R. 1977. Comments on a stock recruitment relationship proposed in SC/28/Doc 43. Reports of the International Whaling Commission 27:234.

Holt, S. J. 1977. Assessment of Southern Hemisphere minke whales. Reports of the International Whaling Commission 27: 214–215.

May, R. M. 1972. Will a large complex system be stable? Nature 238:413–414.

May, R. M. 1974. Biological populations with nonoverlapping generations: stable points, stable cycles, and chaos. Science 186:645–647.

May, R. M. 1976. Simple mathematical models with very complicated dynamics. Nature 261:459–467.

Chapter 16

Some Implications of Maximum Sustainable Net Yield as a Management Objective for Whaling[1]

Provision for MSY is an economically desirable objective of management of fishing and whaling only in very special cases, if ever.

If the rate of discount is high enough in relation to the rate of natural increase of the stock, it may pay an exploiter, in the long run as well as in the short run, to conduct whaling more intensively, and not necessarily sustainably, than would be called for by an MSY policy (Clark 1976). However, if the capital costs of whaling vessels have already been recovered by previous highly profitable whaling, during which initial stocks were depleted, it may be useful to consider the case of zero discount rate and simply compare the values of catches with the costs of operation and maintenance of vessels, as was done by Beverton and Holt (1957) and by S. Gordon and A. Scott (Anderson 1975) in earlier fisheries literature on bioeconomic modelling.

The simplest model to consider is a symmetrical logistic, giving MSY at a stock level half the initial size. Let us suppose that at that level, the value of the catch is equal to the cost of taking it, i.e., net yield is zero. It is easily shown that the maximum *net* yield (MEY) in such a case would be obtained from a stock held half way between the "initial" and "MSY" levels, i.e., 50% higher than the MSY stock level, and that, in those circumstances, the SY would be 75% of the MSY.

Now, suppose that the net yield at MSY level will normally be positive; if it were not, there would be no economic incentive to conduct whaling at that level and the Commission's new management policy would be irrational. If the net yield at MSY level is positive, then the MEY occurs at stock levels lower than 75% of initial stocks, but always, of course, higher than 50%. In general, if we write v = value/cost at MSY level, then the stock level for MEY, expressed as a fraction of initial stock, is given by $p_{\mathrm{MEY}} = (2v + 1)/4v$. These results are based on the assumptions that (i) the total cost of whaling is proportional to whaling effort exerted and (ii) the price per whale (or of whale products if a weight criterion for management were adopted) is constant, at least over the range of stock levels from MSY to MEY. To the extent that the prices of whale products vary inversely with the level of catch, then the MEY stock levels would become relatively higher. Offsetting this, however, is the fact that the costs of whaling might vary less than proportionately with whaling effort because the cost of moving an expedition between its home base and the whaling area does not depend on the effort exerted in the whaling area.

---

[1]Editors' note: written by Sidney J. Holt, FAO/UNEP Marine Mammal Project, as Document SC/29/Doc 17 for the 1977 meeting of the IWC Scientific Committee and published in Reports of the International Whaling Commission 28:191–193, 1978.

Of course, it is possible that the costs of catching whales at MSY level would exceed their market value and that to be viable on that sustained basis, the whaling industry would have to be subsidized. It is, therefore, worth looking briefly at the case where net economic yield at MSY level is negative, i.e., where $v < 1$. Then, $p_{MEY} > 0.75$, but MEY is not necessarily positive. If $v < 0.5$, EY is not positive at any real value of p. At intermediate values of v, the EY becomes positive for values of $p > 1/2v$. For example, if $v = 0.67$, i.e., the value of the MSY is only two-thirds of the cost of whaling effort, then the industry would break even if the effort were halved and the stock maintained at 75% of initial level.

Very similar results are obtained from an asymmetrical logistic model having an MSY at, for example, 60% of initial stock level. So, to get about 75% of MSY, one would use one-half the whaling effort required to take MSY, and would roughly maximize the net yield. The new *rate* of yield, [(value/cost) – 1] × 100%, would be considerably higher than at MSY level, depending on the value of v. For example, if $v = 1.2$ (i.e., 20% "profit" rate at MSY), then MEY/MSY = 0.82 at $p_{MEY} = 0.71$, and the ratio value/cost = 1.72 (i.e., 72% "profit" rate). This would be taken with a whaling effort less than 60% of that needed for MSY, and the net yield would be nearly twice that obtainable by whaling at MSY level (see Figure 1).[2]

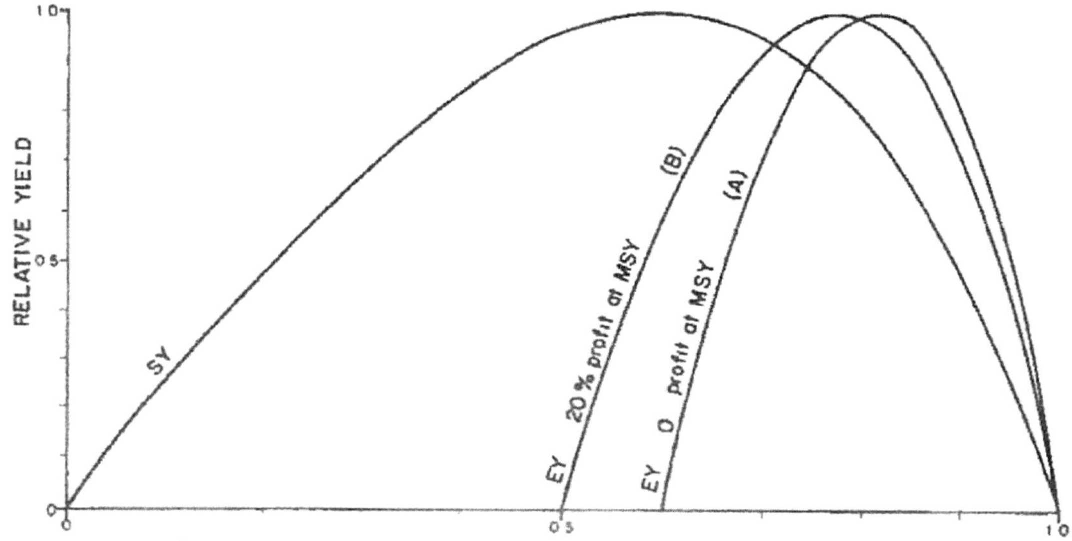

Figure 1. Sustainable yield and net yield as a function of stock size. The curves shown are derived from an asymmetric yield curve having MSY at 60% of initial stock size, $SY \propto p(1 - p^{1.4})$. Curve (A) is based on an assumption that at MSY level, the value of the sustainable yield is equal to the cost of taking it. In curve (B), a 20% excess of value over cost at MSY level is assumed. For curve (A), a 10% safety factor in a catch quota may lead to the stock stabilizing at $p = 0.74$, with a whaling effort 73% of that required for MSY. Corresponding values for 20% and 30% safety factors are $p = 0.79$, 61% effort; and $p = 0.83$, 51% effort, respectively. Relative to the yield and effort at MSY, we find, at MEY: (A) 73% yield for 54% effort, and (B) 84% yield for 65% effort.

[2]The curve of net yield is always more nearly symmetrical than the curve of yield against stock size; in fact, over a wide range of values of α, its degree of asymmetry is about half that of the yield curve. That is, p_{MEY} is closely approximated by the expression: $p_{MEY} = p_0 + (1 - p_0)(0.5 + p_{MSY})/2$.

Another way of locating a curve of net yield against stock size is to identify a relative stock level (p_0) at which net yield is zero. For the general logistic model: $EY = [(p / p_0) - 1] (1 - p^{\alpha})$ from which $p_0 = [(\alpha + 1) p_{MEY} \alpha - 1] / \alpha p_{MEY}^{\alpha - 1}$.

In Figure 2 are shown sample curves of the relation between p_0 and p_{MEY}.

In fishery management practice, it is usual to express sustainable yields and net yields as functions of fishing mortality rates, and thus of effort, rather than as functions of stock size, as is the custom of the IWC Scientific Committee. The curves of Figure 1 are shown in Figure 3 plotted on a scale of whaling mortality rate *F* adjusted relative to the mortality rate corresponding with MSY. The degree to which the stock levels for MEY differs from that for MSY does not depend only on the break-even level p_0, but also on the actual shape of the curve of sustained yield between the MSY stock level and the initial level. Simple models, other than the generalized logistic, which give flatter or more peaked yield curves, lead to MEY levels further from or nearer to the MSY level. Thus, the calculations and percentages given here are only illustrative. However, to the extent that a particular model is adopted, such as an asymmetric logistic, when formulating advice on quotas based on MSY estimates, then it would be consistent to inform the Commission of the beneficial consequences, with respect to whaling effort and net yield, of departing somewhat from the MSY criterion.

Apart from the obvious saving of natural resources, such as fuel used in excessive whaling, by adopting an MEY policy, there is an important consequence of the above argument in relation to the 10% safety factor at present used in applying the "new management policy" (NMP) of IWC. If assessments of a sustained management stock are correct and the quota is set at 90% of MSY, then the stock would increase somewhat. In the case of the asymmetric logistic, it would theoretically come to equilibrium at a level 23% above the MSY level. This would be *below* the MEY level unless the "profit" rate at MSY level exceeded 67%. Tak-

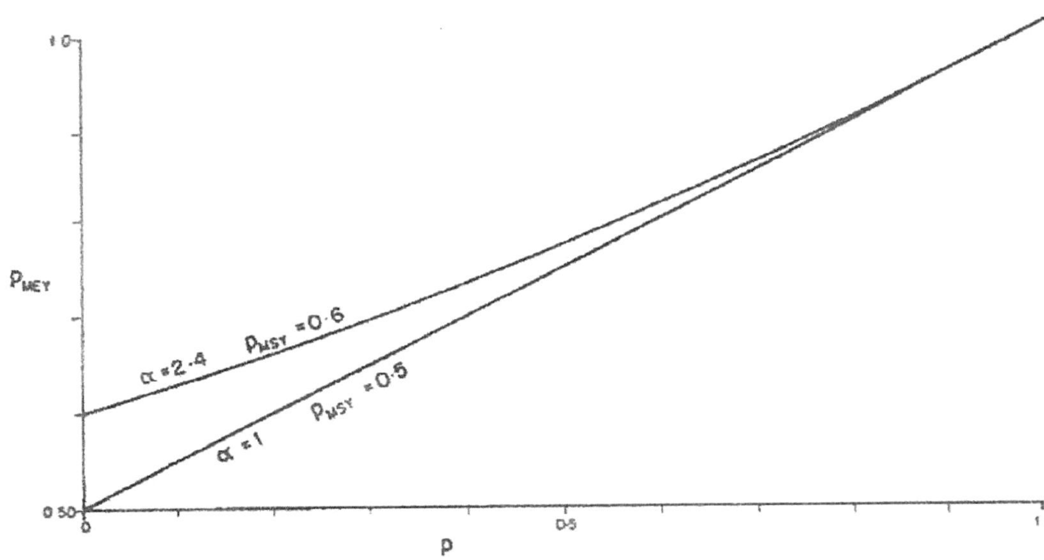

Figure 2. Relation between the break-even stock level p_0 and the level for maximum net yield P_{MEY}. In the case of the curve for $P_{MSY} = 0.6$, the break-even point is at 0.44 of initial stock, the rate of profit at MSY level is 36%, and at MEY level double that, i.e., 73%, there being only a 12% loss of catch as between MEY and MSY.

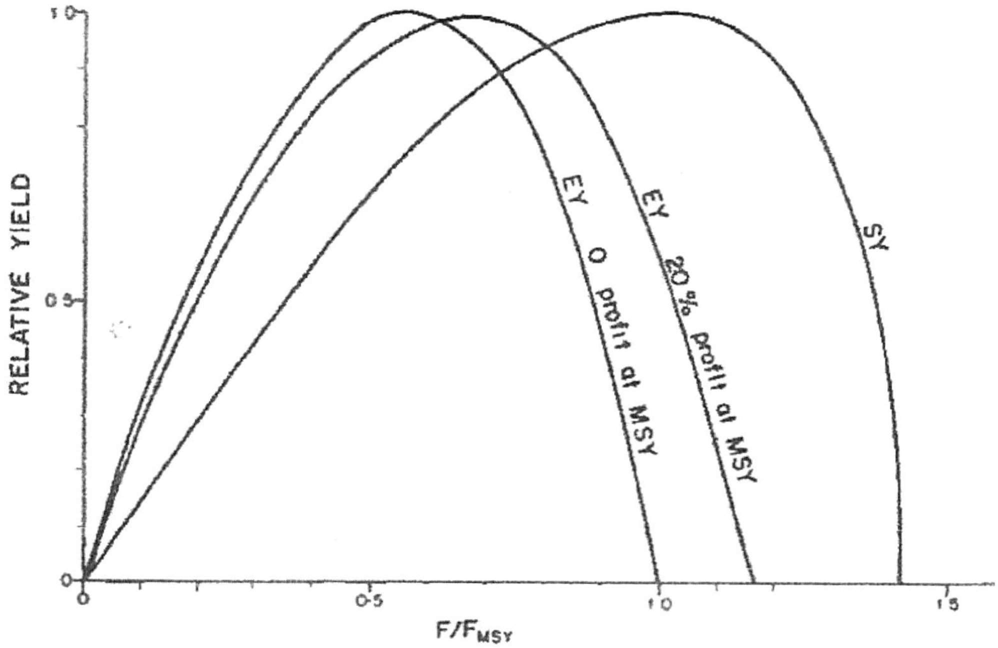

Figure 3. The functions illustrated in Figure 1 are here shown plotted against a linear scale of sustained whaling effort expressed in terms relative to the effort required to give MSY.

ing a 10% safety margin is not, therefore, likely to be an economically harmful procedure. Moreover, if the quota were set at around 80% of estimated MSY, it would be economically advantageous. This is quite apart from discussion of whether the present 10% "safety" factor is sufficient in view of the very great uncertainties in estimation of sustainable yields. If these uncertainties were also taken into account, it seems that a safety factor of the order of 30% rather than 10 or 20% (i.e., set quota at 70% of estimated MSY) would be reasonable.

It is not strictly logical to treat each stock of whale separately when several species are, in fact, being caught on the same expeditions. However, if the several stocks are all being exploited at approximately their MSY levels, as is intended by the NMP, and if, as is generally assumed, the baleen whales all have approximately the same shaped yield curves, the above arguments apply equally to the sums of the sustainable yields and to the total net yields.

Whaling is not the only industry based on marine mammals which might benefit from a move away from MSY as the central objective of management towards a more economically valid objective. Thus, most of the effort to resolve the "tuna–porpoise problem" in the eastern tropical Pacific has been directed to studies of the biology and dynamics of the dolphin species (family Delphinidae) involved and of techniques to avoid their capture in purse-seines. However, the international Yellowfin Tuna *Thunnus albacares* fishery is, like whaling, at present regulated by the MSY criterion. If, instead, there were an attempt to establish MEY as an objective of management, not only would the tuna industry itself benefit (with little loss of total yield), but because of the substantially reduced purse-seine fishing effort, the incidental catch of dolphins would be reduced correspondingly. In addition, of course, there would be less likelihood of inadvertently overfishing tuna because of assessment errors.

In the case of both the whales and the tuna, the roughly 50% larger stock for MEY than for MSY could contain more age classes at any given time and hence would be more stable. Beddington (1978) has shown that the expected increase in stability could be considerable. Add to this, consideration of the fact that the average size of whales in the population and in the catch would be somewhat greater when whaling was less intense (even for baleen whales), and it seems that a policy of setting quotas at, say, 80% or even 70% of the estimated MSY, instead of at 90%, as at present, would have several advantages.

References

Anderson, L. G. 1975. Criteria for maximum economic yield of an internationally exploited fishery. Pages 159–182 *in* H. G. Knight, editor. The future of international fisheries management. West Publishing Company, St. Paul, Minnesota.

Beddington, J. R. 1978. On the risks associated with different harvesting strategies. Reports of the International Whaling Commission 28:165–168.

Beverton, R. J. H., and S. J. Holt. 1957. On the dynamics of exploited fish populations. Her Majesty's Stationery Office, UK.

Clark, C. W. 1976. Mathematical bioeconomics: the optimal management of renewable resources. John Wiley & Sons, New York.

Chapter 17

A Simple Model of Pelagic Whaling[1]

In his account of sperm whaling operations, Ohsumi (1980) provided information which permits a reappraisal to be begun of the validity of the "catch per catcher's day's work" as an index of the abundance of whales. The basic assumption in the application of this index is that in a certain area, over a short time, and with a particular kind of catching unit, the catch per unit time is proportional to searching success and hence to population density. There are, of course, numerous ways in which this assumption departs from reality. It is widely accepted that, for both whaling and some pelagic fishing operations, a more suitable measure of time for this purpose than the total time of operation would be the time spent searching. Such a measure is not yet available for whaling on a historical basis (although information could presumably be extracted from logbooks of catcher boats), and the effective searching time must be inferred from statistics for total day's work. This paper indicates the nature and order of magnitude of important biases in the application of such statistics by methods used to date and suggests how better abundance indices might be obtained from historical data.

The catch per unit searching effort in whaling can be regarded as the product of the sightings per unit search time, the probability that a whale sighted will be chased, and the probability that a chased whale will be killed. It is said that individual whales and pods are, despite the availability of electronic equipment, still found visually, although in recent years, the probability of finding them may have been increased by the application of correlations with oceanographic features and by the accumulation, through whaling experience, of knowledge about their distribution and movements. Such improvements in searching and finding will cause declines in whale abundance as deduced from simple CPUE statistics to be underestimates of the true degree of depletion, but we have no obvious method of estimating this bias.

The time spent searching for whales is essentially the hours of daylight less the time spent chasing, killing, and handling the catch and any daylight time spent towing or resting. Most resting time and some towing time is, however, expended at night, except when catch rates were very high in the early years of pelagic whaling and "stop catch" was declared for a period for an expedition to permit processing of a backlog. In the early effort data provided by the BIWS, "stop-catch days" were identified and could be subtracted from the total day's work to improve abundance estimates; we are not here concerned with this.

---

[1]Editors' note: written by Sidney J. Holt, FAO, Rome, Italy, as Document SC/30/Doc 5 for the 1978 meeting of the IWC Scientific Committee and published in Reports of the International Whaling Commission 29:155–157, 1979. In a footnote to the title, Holt noted, "Since the work reported here was begun, it has come to my attention that John Beddington, Charles Fowler and perhaps others were pursuing similar lines of analysis. I have consulted the two scientists mentioned, and others, and in order to simplify consideration of this matter by the Scientific Committee I have adopted Beddington's notation where that is applicable (see Beddington 1979 and Fowler 1978)."

If the total time spent chasing, killing, and/or handling whales occupies a significant part of daylight hours, then changes in the abundance of whales, and hence in the catch rate, will affect the fraction of time available for searching and hence the frequency of sightings. In the simplest model of this phenomenon, the time spent each day chasing, killing, and handling might be considered to be proportional to the number of whales caught that day. This leads to a non-linear relation between catch per catcher's day's work and abundance such that, as a stock declines, the relative decline from an initial abundance will be underestimated. Further, if the absolute stock size or density is estimated by the DeLury or modified DeLury method, it will be underestimated. This means that unless appropriate corrections are made, there could be a tendency to classify some stocks that should be Initial Management Stocks as Sustained Management Stocks, and some as Sustained Management Stocks that should be Protected Stocks. In addition, the quotas set for Initial and Sustained Management Stocks, as some given percentage of the estimated initial, present, or MSY stock size, could be too high.

We define T_d as the total daylight time, T_s as the time available for searching, T_c as the time spent chasing, T_h as the daylight time spent handling and towing, w as the time spent chasing or handling a whale, C as the catch, and D as the fraction of a 24-hour day in which it is light enough to search and chase, so that $T_d = T_s + T_c + T_h$ and $(T_c + T_h) = wC$.

Further defining, where $C = pAT_sq$, A = abundance of whales, C = catch, p = probability of successful chase, and q = efficiency of search, then the catch per catcher's day's work (CPUE) is $CD/T_d = DpqA \,/\, (1 + wpqA)$. If D, pq, and w are constant, this reveals a non-linear relationship between CPUE and abundance, as in Figure 1.

If we further define A as the abundance relative to an initial abundance, and $CPUE'$ as the corresponding catch per catcher's day's work, then we have $(1 + wpq) = Dpg$ and $CPUE' = (1 + wpq)A'/(1 + wpqA')$. It may be readily shown that the bias $(CPUE' - A')$ is maximal for $A' \leq 0.5$. In fact, A'_m, the value of A' at which bias is maximal, is given by $A'_m = [\sqrt{(1 + wpq)} - 1]/wpq$.

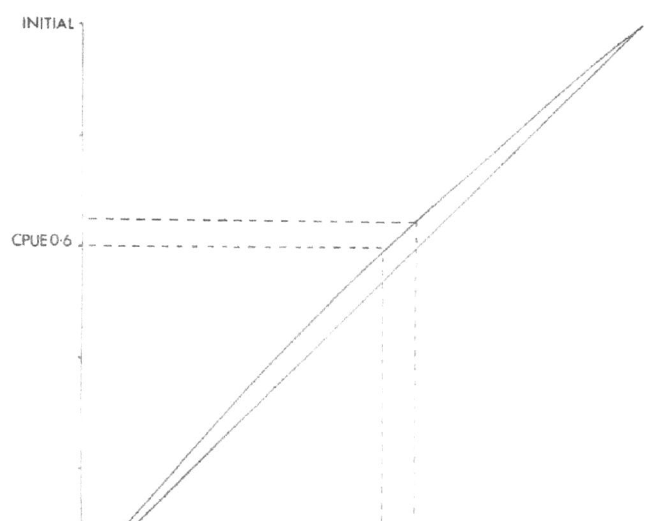

Figure 1. Relation of catch per unit effort, as % of initial, with abundance as % of initial, for case when, at initial abundance, about 14% of total daylight time is occupied with chasing and handling.

Thus, even for small values of *wpq*, e.g., when *w* is a short time, the greatest bias is in the mid-level of A' and somewhat below the value usually taken as that giving maximum sustained yield, so that as a stock declines toward MSY level, the bias increases continuously.

Ohsumi's data show that the fraction of the hours of daylight spent chasing and handling each sperm whale *Physeter microcephalus* in the North Pacific in 1977, when an average of two were caught per day, was 0.07. In the early years, the catch rate was about twice that in recent years and from such information, the specific curve of the form of Figure 1 can be calculated. It shows that when the CPUE has fallen to, say, 60% of the initial value, the abundance will have fallen to about[2] 54% or less of its initial value. When the CPUE has fallen to 44% of initial (MSY level for male sperm whales predicted by the Allen model with parameter values as adopted by the IWC Scientific Committee), the abundance will actually be about 35% or less of the initial.

Conversely, if we seek to reduce a stock to not less than 44% of its initial level, it would be necessary to hold the CPUE at not less than 55% of the initial value. In the first example given, the current stock and hence the quota would have been determined 15% too high; in the second example, of course, the stock would have been classified as just in the Sustained Management category, when it should have been protected, and by a substantial margin.

A further factor is revealed by Ohsumi's data. His Figure 2 shows that the chasing time per whale (which is the main element in daylight hours not spent searching) increases in direct proportion with the length of the whale, as might be expected from the relation of swimming speeds and diving duration with size. Over the period during which pelagic sperm whaling was conducted, the average size of male sperm whales caught declined and also, with increasing catches of females, the average size of all sperm whales caught declined even further. This would introduce an additional bias, in the same direction, in estimates of changes in abundance from CPUE data treated in the usual manner.

The analysis is complicated in the case of baleen whaling by the fact that chasing and handling an individual of one species of whale takes time that might otherwise be spent searching for that and other species simultaneously. The catch per catcher's day of the four large baleen species in the Antarctic has ranged from as much as 4 in the mid-1950s to less than 1 in the 1970s. The change has not, however, been smooth as the concentration on different species (essentially pairs of species) has changed over the post-Second World War period. During this time, the average size of all baleen whales caught has steadily declined from just over 70 ft to 50 ft or less. If this or other interspecific changes are associated with substantially decreased chasing and handling times per whale, the two factors of declining catch rate and size would introduce substantial bias in the estimates of stock abundance changes based on catcher's day's work. If it took, say, 20–30 minutes on average to catch and kill a baleen whale (before the introduction of Asdic), it could mean that fin *Balaenoptera physalus* and sei whale *B. borealis* estimates on which the Scientific Committee has based its classification and quota recommendations could be biassed on the optimistic side by about 10–20%.

---

[2]The approximations expressed here arise from uncertainties in how to deal with towing time—some of which was not recorded or took place in the dark when searching and chasing were not possible. Furthermore, although total towing time might increase with numbers to be towed, it presumably would not do so proportionally; also when whales are scarce, distance and hence time spent towing might be increased as catchers might be successful on average further from the mother ship.

The above analysis suggests how one might better approach the problem of "effort corrections" for factors such as tonnage or engine power, weather, and Asdic. In each case, the factor in question may affect searching efficiency, chasing time, or both as well as the probability of successful chase. The effect on catches per catcher's day and hence the biasses in estimation of abundance will be linear and/or non-linear. Thus, higher operational (cruising) speeds should increase search efficiency proportionally, and increased maximum speeds would reduce average chasing time, possibly in approximately inverse proportion. Asdic, which apparently is not used in searching, would both decrease the chasing time and increase the probability of successful chase—the former especially in the case of baleen whales and the latter in the case of sperm whales. Changes in the weather conditions presumably affect both searching efficiency and chasing time and success.

When a terrestrial mammal population is depleted, there has often been noted a tendency for the area inhabited by the population to be diminished. As whales are social animals in communication with each other, possibly over considerable distances, it may be reasonable to suppose that as stocks have been reduced, the feeding range has been reduced. If, as expected, the whaling expeditions are, at least to some extent, successful in "homing" in on the highest concentrations of whales, the catch per unit effort might be an index of relative density, but not of total abundance relative to that at some other date. The direction of bias would, yet again, be in the same direction—leading to overestimates of stock size and underestimates of the degree to which initial stocks had been reduced. It has not been possible to make an analysis of this effect in time for the present meeting, but such should be possible from detailed catch and effort data. In later years, when the number of pelagic expeditions operating in the Antarctic was substantially reduced, the area in which whaling effort was expended has been, of course, in any case reduced and its distribution would not be an indication of the contraction of whale stocks. However, a few years earlier, in the period when the fin whales of the Southern Hemisphere were being reduced (1948/49–1964/65), the number of catchers and the total effort did not change greatly from one year to the next; in such a period, it might be possible to determine the degree of stock contraction which occurred, if any.

Table 1. Antarctic pelagic baleen catches. i = indicator of daily chasing/handling time = catch per day × mean length of whales caught = average IHP of catchers.

	Catch per catcher day				Catch per catcher day		
Season	**Blue+fin+sei**	**+minke**	**i**	**Season**	**Blue+fin+sei**	**+minke**	**i**
1946/47	1.51		0.88	1962/63	1.12		0.24
1947/48	1.46		0.79	1963/64	1.09		0.22
1948/49	1.30		0.62	1964/65	1.56		0.31
1949/50	1.34		0.58	1965/66	1.52		0.25
1950/51	1.31		0.52	1966/67	1.30		0.20
1951/52	1.52		0.58	1967/68	1.28		0.20
1952/53	1.47		0.53	1968/69	1.06		0.17
1953/54	1.79		0.64	1969/70	1.12		0.18
1954/55	1.70		0.59	1970/71	1.13		0.18
1955/56	1.82		0.61	1971/72	0.97	1.35	
1956/57	1.81		0.58	1972/73	0.73	1.47	
1957/58	1.79		0.53	1973/74	0.85	2.01	
1958/59	1.75		0.47	1974/75	0.62	1.53	
1959/60	1.45		0.36	1975/76	0.36	1.45	
1960/61	1.39		0.34	1976/77	0.50	2.64	
1961/62	1.08		0.25				

Certain other factors could also be taken into account from existing data which may be having an effect on observed changes in CPUE, although in directions which cannot immediately be foreseen. One of these is the change in the duration of daylight with changes in mean latitude of operation which have occurred. The other is the mean length of day as a function of the dates of beginning and end of season. It does not seem likely that these will have affected greatly the estimates of abundance of any particular baleen species; it may, however, have affected sperm whale estimates because the time and place of catch of this species have changed substantially over the period of pelagic whaling in the Southern Hemisphere. It is evident that other factors must be brought into a fully realistic model of whaling operations, including particularly the consequences of cooperation among catchers and of the distribution of whales in pods rather than singly. Technological advances tend to reduce the time spent catching each whale and to increase the efficiency of utilizing the remaining searching time. It is not unreasonable to assume that even when effort data have been "corrected" for some such factors, there will be other, as yet unknown, factors not taken into account and which are more likely than not to lead to further excessive optimism in assessments. In this situation, even if all other factors are not allowed for, estimates would always be improved if the non-linear relation between CPUE and abundance, arising from the fact that time available for searching increases with declining catch rates, is applied using the best available estimates of chasing/handling time per whale in the past, no matter how approximate these are. Furthermore, if we wish to be reasonably conservative in our estimates, it would be better to use values of chasing time per whale from earlier rather than from later years, i.e., from before the introduction of the present types of powerful catchers equipped with electronic devices.

References

Beddington, J. R. 1979. On some problems of estimating population abundance from catch data. Reports of the International Whaling Commission 29:149–154.

Fowler, C. W. 1978. A rationale for modifying effort by catch, using the sperm whale of the North Pacific as an example. Paper SC/30/Doc 52 presented to the IWC Scientific Committee, June 1978. Unpublished. [Available from the IWC Secretariat]

Ohsumi, S. 1980. Criticism on Japanese fishing effort for sperm whales in the North Pacific. Pages 19–30 *in* G. Donovan, editor. Sperm Whales: Special Issue, Reports of the International Whaling Commission (Special Issue 2).

Chapter 18

Reclassification of Protected Baleen Whale Stocks[1]

The recovery of stocks which have been classed as "protected" under the New Management Policy can be simulated by back-calculating from the population models and parameter values on which the original decision to protect was based. Such simulation leads to the conclusion that certain, at present, protected Southern Hemisphere fin whale *Balaenoptera physalus* stocks and possibly one or more of the sei whale *B. borealis* stocks might be classified as Sustained Management Stocks in the 1977/1978 season or soon thereafter. There is, we think, no question that any of the stocks of humpback *Megaptera novaeangliae* or blue whales *B. musculus* could have recovered sufficiently for reclassification to be seriously considered in the near future. The case of the eastern Pacific gray whale *Eschrichtius robustus*, which has increased in numbers substantially under protection, calls for special consideration for several reasons, including (a) its dependence for continued existence on the safeguarding of critical breeding habitat from excessive human interference, and on the abstention of the North American countries from hunting it within their coastal waters, and (b) the fact that it is, so far, uniquely the source of high non-consumptive values which are realized through a recognizable industry. This note is, therefore, concerned only with the problem of the fin and sei whales, some Protected Stocks of which have judged to be close to the boundary stock size between Protection and Sustained Management.

The shortness of the time elapsed since stocks of these species have been protected means there is no possibility of estimating the degree of change in stock size from sightings by commercial expeditions—the only source of direct observations. Reclassification can, at present, be discussed only in terms of population models. In that short time, however, there have been considerable developments in such models and their application, and important questions have been raised as to their viability and adequacy for prediction. Here, we briefly examine some of these matters with reference to the reclassification problem.

a) The presumed linear relationship between stock density and catch per unit effort as measured by catch per catcher's day's work adjusted by various multipliers, mainly for size of catcher and average weather.

Until now, very few data have been made available to the Commission concerning the break-down of total time of pelagic operations into its components such as searching, chasing,

[1]Editors' note: written by Sidney J. Holt, Department of Fisheries, FAO, Rome, Italy, and J. Gordon Clark, Friends of the Earth, as Document SC/30/Doc 6 for the 1978 meeting of the IWC Scientific Committee and published in Reports of the International Whaling Commission 29:159–162, 1979.

handling, towing, resting. At the IWC Special Meeting in Cronulla, November 1977, important data were submitted by Ohsumi (1979) concerning this breakdown with respect to sperm whaling *Physeter macrocephalus*. From these, we can deduce that the relationship is definitely not linear and that the direction of bias introduced by the linear assumption is such that the degree of depletion of stocks has almost certainly been significantly underestimated. The density of whales is expected to be essentially proportional to sightings per unit searching time. The amount of searching time available in the daylight hours is increased as the stock density—and hence the catch per day—declines because the time spent chasing and handling, and possibly towing, increases roughly in proportion to the number of whales caught. This effect might be offset, but surely only partially, and in a limited way, by the decrease in the length of the working day as a result of changed labour conditions in the Japanese fleet, as referred to by Ohsumi.

Clearly, specific data on baleen whaling operations are needed, but if we assume that the average time taken to chase, handle, and tow a baleen whale is about the same as for a sperm whale in the North Pacific, we can deduce that the densities of stocks which might have been reduced by whaling have been significantly overestimated with respect to the "initial" densities. The overestimation would, in some cases, be of the order of 20–30% (Beddington 1979; Holt 1979; Fowler 1980). That is to say, stocks which, in recent years, were classified as Protected because they were thought to be at levels a little less than 60% of initial levels are likely, in reality, to have been, at the time of such classification, at substantially lower levels. This effect is enhanced by the fact that it takes a shorter time to chase and catch a smaller whale than a larger one (in sperm whales, the time is roughly proportional to length, other things being equal), and the average size of whales taken has declined because of the decline in average size of each species and an increase in the proportions of the smaller species in the baleen catch.

Further, if, as suggested by Holt (1978), the change of mean density in the area where whaling is carried out is an underestimate of the decline in absolute abundance of whales, then the stock size at the date of protection would be less than was thought.

b) The effects of the introduction of Asdic in pelagic whaling in increasing the effectiveness of a catcher's day's work have not been taken into account in assessments made so far.

From the information that has been made available to the Scientific Committee about sperm whaling operations, we might presume that Asdic is also not of any great significance in increasing searching efficiency in baleen whaling, although it would be desirable to have direct evidence on this point. We know, however, that Asdic is used in the chasing operation and, in the absence of further information, it must be presumed that it has increased overall efficiency considerably; if it had not done so, it would hardly be likely that the expense and trouble of installation and of training crews in its use would have been worthwhile to the whaling enterprises since it has been argued that it is of relatively little importance in sperm whaling. This means that the chasing and killing efficiency itself must have been greatly increased either by reducing several-fold the average time taken to chase a baleen whale (which seems the more likely effect, considering that it has been said that Asdic is used primarily to induce and maintain a fast shallow flight reaction, with consequent rapid exhaustion of the whale) or by increasing the likelihood that a sighted whale will, in fact, be captured, or both. Operational data are needed to substantiate or correct these hypotheses and could, no doubt, be provided for consideration at future meetings. In the absence of such data, it must be presumed that at the time of protection, which occurred later than the general introduction of Asdic (1958–1962 in Japanese pelagic operations, 1960–1970 in coastal operations), the relative abundances of stocks under consideration were overestimated.

c) Beddington (1979) has suggested that efficiency of catcher's day's work changes with changes in the numbers of catchers per expedition.

This factor, too, will have introduced biases into the abundance estimates on which protection decisions were based.

d) Since the various fin and humpback whale stocks were protected, it has proven necessary, especially in considering the status of sei whale stocks, to take into account the interactions between the baleen whale species.

These interactions are presumed to derive mainly from the strong dependencies on krill *Euphausia superba*, their main food resource. In particular, unprotected stocks of whales, even if as in the case of the minke whale *B. bonaerensis* they are being reduced, may be benefiting from the relief of predatory pressure on the krill by feeding more intensively themselves and either increasing in number or biomass, or by not decreasing under whaling as fast as might have been expected. We must also note that over the period of intensive whaling for fin and humpback whales, other predators on krill in the Southern Hemisphere, particularly crabeater seals *Lobodon carcinophagus*, have been increasing. Thus, the single-species models used as the basis for classification of Protection Stocks cannot simply be "put into reverse gear." It seems unlikely that the Protected Stocks will be increasing as rapidly as would be expected from the original assessments.

e) Whaling, albeit on a small scale, has continued on some Protected Stocks.

We know that individuals from Protected Stocks continue to be taken by commercial operations not bound by decisions of the IWC and not subject to the Commission's Observer Scheme; for example, humpback whales by Spain and possibly humpback and blue whales by Chile (Casinos et al. 1977; Ottaway 1978). In addition, *any* catches, for example, by "aboriginal" whaling for local consumption could affect adversely the recovery of very small residual stocks, especially as such whaling operations usually involve the loss and likely death of numerous whales which are struck, but not caught. Such seems to be the case in the Kingdom of Tonga, for example, with respect to a South Pacific breeding stock of humpback whales, where the numbers caught or struck are likely to have reduced the recovery rate, or even completely impeded recovery.

It is particularly serious that subsistence catches, taken especially by unmechanized means, include high proportions of calves and pregnant and lactating females. Humpbacks, for example, are taken by native fisheries in Bequia (Caribbean Islands) and in Greenland. The Bequia fishery concentrates on mothers accompanied by calves or yearlings and may be preventing recovery of the stock (Winn 1976); the Greenland fishery taking 5–10 a year may be having a similar effect (IWC 1978).

While these examples do not involve fin and sei whales, other operations of which we are not aware may do so, and these latter species are at risk from unmonitored whaling operations such as are conducted by MV "Sierra" and from the possible increase in whaling by non-Member countries of the IWC. Exact prediction of the consequences of such activities depends, in part, on which catch data were taken into account in the original assessments, but we clearly should not necessarily assume that all stocks are recovering at the originally predicted rate.

Since such information as we have strongly suggests that both the abundances of Protected Stocks and their rates of recovery may be less than previously assumed, it would seem inadvisable to reclassify any of the stocks which are now protected, solely on the basis of extensions of the original calculations. To permit classification to be reconsidered, the following work is needed at least:

a) A full re-analysis of existing catch, effort, and biological data.

b) Submission and analysis of new data including, especially, detailed operational information.

c) A re-examination of the models used to see to what extent conclusions will be modified if, as seems necessary, a multispecies approach to baleen whaling management is adopted.

d) The preparation of a program of sightings for scientific purposes. This could be supplemented by marking and by planned scientific sampling on a minimal scale. However, such sampling would be unlikely to provide the information required with minimal impact on the Protected Stocks unless it were more precisely planned and more strongly coordinated than results from the application of the IWC provision for issuance of scientific permits. It would be timely, therefore, for a start to be made in designing an appropriate international program of research on Protected Stocks.

e) Acquisition of more data pertaining to continuing catches by non-Members of IWC, and to all catches for subsistence purposes. This the authors have been doing for Tongan humpback catches, with the collaboration of the Government fishery officer there, Mr. W. Wilkinson, and also with the FAO Fishery Expert, Mr. Davidson Thomas, who, with his colleagues, has logged sightings in the area by local fishery research vessels operating an FAO/UNDP development project. The Scientific Committee may also need to assure itself that the present arrangements under the Commission for "observing" whaling operations are, in fact, ensuring the full availability of information from the fleets about sightings and possible occasional catches of protected species.

If, following careful analysis, planning, and research, taking into account the factors indicated above, it appears that some, at present, Protected Stocks should be reclassified in accordance with New Management Policy, consideration should be given to an experimental approach to resumed whaling, along the lines suggested by ACMRR in the light of the Bergen Consultation (FAO/ACMRR 1977). Thus, even if improved estimates are made of the current sizes of Protected Stocks, both in absolute terms and in relation to "initial" sizes, considerable uncertainty will remain as to the level of yield they can sustain. In this case, if the "best" estimates of two stocks of, for example, Southern Ocean fin whales, indicate that they both have increased to levels just above presumed MSY levels, it could be desirable to open only one of them to whaling, while keeping the other protected for a few years, even if the quota for the first stock were to be set somewhat higher than would otherwise have been done. It is suggested that such possibilities should be critically evaluated by the Scientific Committee in the near future.

References

Beddington, J. R. 1979. On some problems of estimating population abundance from catch data. Reports of the International Whaling Commission 29:149–154.

Casinos, A., S. Filella, S. Pelegri, and J. Pelegri. 1977. Notos sobre los cetáceos de las aguas ibéricas. I. Sobre un ejemplar de *Megaptera novaeangliae* (Borowski, 1781) (*Cetacea, Balaenopteridae*). Miscelanea Zoologica 4(1):299–303.

FAO/ACMRR. 1977. Report of the Advisory Committee on Marine Resources Research Working Party on Marine Mammals. FAO Fisheries Report 194:26–27.

Fowler, C. W. 1980. A rationale for modifying effort by catch, using the sperm whale of the North Pacific as an example. Reports of the International Whaling Commission, Special Issue 2:99–102.

Holt, S. J. 1979. A simple model of pelagic whaling. Reports of the International Whaling Commission 29:155–157.

International Whaling Commission. 1978. Report of Scientific Committee. Reports of the International Whaling Commission 28:38–92.

Ohsumi, S. 1980. Criticism on Japanese fishing effort for sperm whales in the North Pacific. Reports of the International Whaling Commission, Special Issue 2:19–30.

Ottaway, M. 1978. How whale killers sail round the law. Sunday Times, London, 19 March 1978.

Winn, H. 1976. FAE/ACMRR/SC/WG.1.2 Document of Scientific Consultation on Conservation and Management of Marine Mammals and Their Environment, Bergen, Norway, 31 August–9 September 1976.

Chapter 19

Proposal for a Modified Management Policy[1]

A. Comment on "Towards an Improved Whale Management Procedure" by K. R. Allen (1979)

This attempt to find solutions to some of the problems arising from application of the New Management Procedure (NMP) is to be welcomed. It is understandable that, given the present circumstances, a spreading of the workload of the Scientific Committee over time would seem desirable. However, the present difficulties arise essentially from the fact that the research and evaluation efforts made by countries are not commensurate with the demands made by the Commission; if more than a handful of scientists, mostly working part-time only, were involved in whale assessments, there would be little difficulty in getting through the full range of business in an annual meeting, since there would have been far more intensive preparatory work. A second weakness, at present, is the state of the database and accessibility to it. The Committee works inefficiently because detailed data on effort, age distribution, and so on are not made available fully and promptly by countries involved nor is there yet a well organized database in the Secretariat of the Commission.

Dr. Allen's proposal to spread assessments and reviews over a five-year cycle seems to me to accept the inevitability of insufficient scientific effort and poor data submission and availability, and diverts attention from the primary need to improve these.

The proposal also continues an independent single-species stock approach, which is increasingly seen as a weakness of the NMP. It is true that in focusing on observed stock changes, a degree of empiricism is accepted which would take into automatic account any observed consequences of some interspecific relations among the whale species. However, remnants of possibly fallacious assumptions remain in the paper—for example "Stocks which have been previously unexploited must decrease if any exploitation is to be allowed" (p. 144). This is not necessarily true if the previously unexploited stock had been increasing as a result of changes in its available food supply.

Another difficulty in revising assessments in rotation lies not in the biological interactions, but in the use of common effort data in multispecies whaling. Practically all estimates of (a) "initial" stock size and (b) relation of present level to some earlier level, depend on the relation of catches, through a time-series, to catches per unit effort (CPUE). Changes in our appreciation of CPUE as an index of abundance—for example, effects of technological improvements, or of changes in effective searching time and efficiency—affect *all* assessments. If we find that there is a substantial bias in all previous estimates, should we really wait up to

[1]Editors' note: written by Sidney J. Holt, FAO and UNEP, presented as Document SC/30/Doc 51 for the 1978 meeting of the IWC Scientific Committee and published in Reports of the International Whaling Commission 29:327–333, 1979.

five years before we correct this bias for some quota determinations? We have experienced other changes in research results which affect many assessments—for example, the periodicity of ear plug laminations.

Some of the above objections might not matter much, in practice, if we really felt we could observe the direction of stock change over a short period of years and hence estimate the replacement yield. However, I have great doubts regarding the reliability of present CPUE indices, except as gross indicators (and, usually biased indicators) of extreme depletion. This, combined with time-lags even in the single-species system, makes unexpected population "crashes" seem quite likely if the new procedure were to be applied. It is true that large and sudden decreases in quotas create disturbances in whaling industries, but I suggest that human institutions can respond more quickly, more flexibly, and over a wider range than can whale populations.

If the maximum sustainable yield level (MSYL) is not to be taken as the optimum level (p. 143), then the entire new procedure still rests on a definition of that level, and an estimation of it (p. 143). The quotas that emerge from it depend substantially on the chosen level and the absolute value of the estimate. Although the MSYL has been rather arbitrarily chosen, there was some logic in choosing a yield model giving a curve skewed to the right and not merely the selection of one because it is more "conservative" than the symmetrical logistic (e.g., Ohsumi's fin whale *Balaenoptera physalus* model; consideration of the biological features of mammalian population dynamics). Also, the "60% model" is not, in all respects, more conservative than the 50% model; it provides for earlier protection, but at the same time can overestimate the yielding capacity at high stock levels.

The indices or categories of "reliability" proposed are entirely subjective. If the quota is to be a variable portion of replacement or sustainable yield, it would be better to determine that fraction by criteria of data availability rather than presumed reliability of estimates; this would, at least, give a direct incentive for improvement. Thus, one could envisage a points system based on length of time-series, fineness of effort data, number of age determinations reported, and so on.

B. Alternative Proposal

The following proposal is still open to some of the criticisms made in section A above regarding Allen's proposals (1979) such as derivation basically from a single-species case. However, it does, I think, simplify classification; have some of the advantages of the "step procedures" that have been suggested, without their disadvantages; provide a means of quota decision when little information is available; provide automatic, and variable, safety factors; and not depend critically on the assumed or estimated levels of MSY. Criticism of the NMP within the IWC has focused on the fact that changes in scientific assessments can lead to sharp changes in quotas, and also on whether the "90% rule" is an adequate safety measure. Externally, it has been observed that devotion to the attainment of MSY is not economically justified in the long run, and it attaches no value to the enhanced stability that would be expected in stocks held somewhat above MSY levels. In addition, I have observed (Holt 1977) that for some stocks, it would also be reasonable to ensure maximum sustained product yields by holding stocks above MSY levels by number.

The response within the IWC has been to examine various "step procedures" and different safety factors. As to the criticisms of MSY, it is perhaps unrealistic to suppose that there could be, at this time, and considering the state of most of the whale stocks, any international agreement to seek an economic optimum state; such an approach would also have to take account of Colin Clark's bioeconomic models incorporating discount rates that might be higher than the maximum net increase rates of whales (Clark 1973). In addition, of course, we have to develop a procedure that can be applied where interspecific interactions via predation on a common food basis and/or environmental trends make necessary the application of multispecies models and consideration of continuously changing states.

My purpose here is to suggest, for the single-species model, that it is possible to use, as the basis for a procedure, a biologically determined characteristic of yield models, but one other than the MSY. This characteristic is the *biological production* of the stock, defined as the sustainable yield plus the total natural deaths, which is equal to the number (or weight) of annual recruits in a steady-state.

For illustration, I take the modified logistic giving an MSYL/initial level of 60%, which has been applied both to baleen whales and to female sperm whales *Physeter macrocephalus*. This model, giving the familiar right-hand skewed curve of sustained yield against stock size, is not discordant with such data as exist for whales and also for some other large mammals, and it has been said that its adoption by the IWC instead of the symmetrical logistic is a conservative procedure. This is true with respect to the level at which a stock becomes protected, but the reverse can be true with respect to quota level for Sustained and Initial Management Stocks. This is more evident when the recruitment–stock size function is an interpolation between an estimate of r in the unexploited or lightly exploited stock r_1 and the estimate from biological considerations of the highest value that r might reach in the extremely depleted stock r_0. In this case, then, for the same values of r_0, r_1, M, and initial stock size, the 60% model predicts much higher sustainable yields, and hence higher quotas under the NMP for all stock sizes higher than about 55% of initial level.

The actual quotas calculated on both models are critically dependent on the assumed or estimated value of r_0. I have observed (Holt 1977) that values used by the IWC Scientific Committee to establish sei whale *B. borealis* quotas are such that, as the initial stock is reduced, not only does r increase, but also the recruitment $R = rP$ increases before decreasing. I suggested that, in the absence of information to support such a value, it would be safe to adopt an upper limit of r_0 such that R would never be expected to exceed the recruitment in the unexploited stock R_1.

Suppose, however, we act on the hypothesis that R can increase up to a point. This means that the curve of gross production, $R = P(F + M)$ against P, has a maximum R_{max} = MSP [*Editors' note: maximum sustained production*]. This occurs at a stock level somewhere between P_{MSY} and P_1.

Figure 1 shows the curves of sustainable yield Y and of R for a minke whale *B. bonaerensis* stock, if it was in a steady-state, with $r_1 = M = 0.095$ and $r_0 = 0.155$. These values would correspond with the situation in which the stock was originally (that is, before pelagic whaling reduced its competitors) in a steady-state with r_1 equal to the present—and unchanged—M_1 and that the present high r, determined from catch curves, for example, is nearly as high as is biologically possible, given the new level of food supply, and that r would now decline toward the original value r_0 as the stock grows toward a new asymptote, unimpeded by whaling.

The maximum sustained production occurs in this case at 79% of the (eventual) unexploited level. Whaling to hold the stock at MSP level would give a value: $Y_{opt} / Y_{max} = 0.82$ compared with a value of 0.90 achieved by the NMP. This ratio varies greatly, however, with

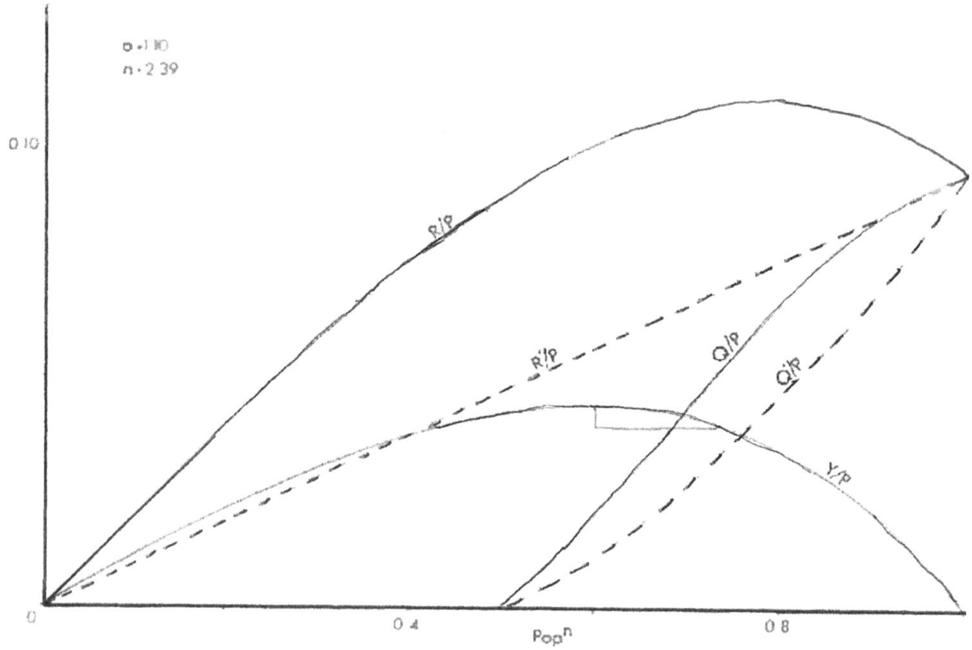

Figure 1. The curve of sustainable yield, biological production, and quota as a fraction of population *P*.

parameter values and, for low enough reproduction coefficients, reaches zero—i.e., MSP is at the unexploited level.

The above difficulty is eliminated by a policy seeking to take some proportion of the sustained production. The proportion would need to decrease with decreasing stock size.

There are several possible rules to govern this decrease. A desirable rule might make an automatic compromise between two different objectives; maintenance of high yields, and maintenance of the health of the whale population within an ecosystem and of its functional role in that system. The former objectives dictates MSY as an appropriate criterion if we are limited to biologically determined criteria, modulated essentially by safety considerations such as the avoidance of overshoot, the deleterious effects of errors in estimation of MSY and/or of MSYL, and the minimization of large changes in quotas resulting from trivial changes in assessments. The latter objective could be met by MSP; surely a population kept at that level would be "healthy," fulfill its role, and be reasonably stable. A rule that ensures that quotas are set to stabilize a stock somewhere between these two levels could be to set a quota as a fraction of the biological production ranging linearly from unity at MSP level down to zero at MSY level. However, we are not sure of either of these levels. It is convenient to take the highest possible level for MSP, which is unity (the unexploited population), and the lowest possible level for MSY. It is recognized that whale population curves are, if anything, skewed to the right, so that the lowest MSYL is 0.5 of that initial.

Appendix 1 shows the results of applying such a rule in a range of situations and demonstrates its advantages. One result is the merging of the Sustained Management and Initial Management categories. Another is, of course, that a stock enters the protected category when it is at or below 50% of its initial level, as compared with 45 and 54% by the NMP for the 50% and 60% MSY assumptions, respectively. However, the grading of quotas is spread over a much wider range of stock sizes, which guards against large abrupt quota changes in either direction (except there being a radical reassessment of the state of the stock). The rule gives quotas for what are, at present, defined as Initial Management Stocks which diminish gradually as the stock is reduced, and leads to stabilization at levels fairly close to those attained under the NMP, but with an effective variable safety factor which derives directly from the application of the simple rule, and varies, for a broad range of parameter values and assumptions, from about 9 to about 25%.

It may be worth noting that the stabilized stock and yield levels which are likely to be obtained by application of the quota rule suggested here are not far from the likely levels for maximum net economic yield with zero discount rate (Holt 1978). It would be interesting to examine how they relate also to population stability expressed as a function of p [*Editors' note: population size relative to maximum natural size—see Appendix 1*].

Appendix 2 develops the mathematical argument of Appendix 1 for the more general case in which the natural mortality M, as well as the recruitment rate r, is density-dependent.

References

Allen, K. R. 1979. Towards an improved whale management procedure. Reports of the International Whaling Commission 29:143–146.

Clark, C. W. 1973. The economics of overexploitation. Science 181:630–634.

Holt, S. J. 1977. A possible basis for determining limits to the allowable catch when the degree of density dependence of the net recruitment rate is not known. Reports of the International Whaling Commission 27:230–233.

Holt, S. J. 1978. Some implications of maximum sustainable net yield as a management objective for whaling. Reports of the International Whaling Commission 28:191–193.

Appendix 1

For the general class of logistic population growth–yield models, we can write:

$$r_p = (r_0 - r_1)(1 - p^n) + r^1 \qquad (1)$$

where r = rate of recruitment, p = population size relative to maximum natural size ("initial"), subscript 1 = value in "initial" stock, subscript 0 = value when stock is vanishingly small, i.e., the highest possible value of r_0.

If the regulation of the population is through the density-dependence of recruitment, only then $r_1 = M$ constant. There is a maximum in the number of recruits, for $p < 1$, if $r_1 < -(dr/pdp)$. For the "60% logistic model", this requires that $(r_0/r_1) > 3/2$, and for the 50% logistic model $(r_0/r_1) > 2/1$.

For the yield curve with MSY at $p = 0.60$, $n = 2.39$, while for the symmetrical logistic, $n = 1$. The biological production of the stock, when balanced, is the combined annual numbers of deaths by whaling and natural causes, and is equal to $R = rpP$ where P = initial population number, so

$$R_p = P_p\,[r_0 - (r_0 - r_1)\,p^n] = P_pM\,[1 + a\,(1 - p^n)] \qquad (2)$$

where $a = (r_0 - r_1) / r_1$ and is positive. R has a maximum value R_{max} at

$$p_{opt} = [(a + 1)/a(n + 1)]^{1/n} \tag{3}$$

so that

$$R_{max} = P \times M[(a + 1)^{n+1/n} /(n + 1)] \times (n/a^{1/n}). \tag{4}$$

From equation (3), note that $p_{opt} \geq 1$ when $a \leq 1/n$.

That is, the maximum production is, in such cases, given by the unexploited population, although it could be increased further if the natural mortality were reduced or if the carrying capacity of the environment for the population increased.

For the same population model

$$p_{msy} = [1/(n + 1)^{1/n} \tag{5}$$

so that

$$p_{opt} = p_{msy} \times (a + 1/a)^{1/n}. \tag{6}$$

Thus, $p_{msy} \leq p_{opt} \leq 1$.

Note that the sustainable yield

$$Y = R - MPp \tag{7}$$

so that

$$Y_{max} = PM[an/(n + 1)][(1/(n + 1)]^{1/n} \tag{8}$$

and

$$Y_{max}/R_{msy} = an/[an + (n + 1)]. \tag{9}$$

The fishing mortality rate is given by:

$$F = Y/Pp \tag{10}$$

and, in general, $Y/R = F/(F + M) = E$ for a given value of p.

Although assessments for whales are usually considered in terms of population levels, it may be useful also to follow the practice more common in fisheries management and examine relationships of yield and biological production with the fishing mortality. For this purpose, we establish, from equations (2) and (10), that $F = aM(1 - p^n)$ from which the level of fishing mortality which, if maintained, leads to population extinction, is $F_{ext} = aM$ and from equations (3) and (5), $F_{msy} = aM[n/(n + 1)]$ and $F_{opt} = aM[(n - 1)/(n + 1)]$. Thus, the relation of F_{msy} with F_{opt}, and of both of these with the extinction mortality, are independent of a and M, and given by $F_{opt}/F_{msy} = (n - 1)/n = 0.58$ for $n = 2.39$, $F_{msy}/F_{ext} = n/(n + 1) = 0.7$ for $n = 2.39$, and $F_{opt}/F_{ext} = (n - 1)/(n + 1) = 0.4$ for $n = 2.39$.

This brings out a point which is commonly not appreciated—that the more the curve of sustainable yield against population size is skewed to the right, as it appears to be for whales (and, indeed for large mammals generally), then the closer the fishing mortality to give MSY is to that which would, if sustained, exterminate the population.

Recent analyses of data for the minke whale of the Southern Hemisphere provide an illustration of the relationships conveyed by the above equations.

$M = r_1 = 0.095$. Assume $n = 2.39$.

The recruitment rate has, with an improvement in environmental carrying capacity, increased to about 0.15, so that $r_1 \geq 0.15$ and, therefore, $a \geq 0.58$. Now, $1/n = 0.42$ (corresponding, in this case, with $r_0 = 0.135$) so that these stocks would have had maximum production at sizes somewhat less than their unexploited levels under the original environmental conditions, with $p_{opt} \leq 0.91$. From biological considerations, r_0 could be as high as 0.20. This value would give $a = 1.10$, p_{opt}=0.79. Similarly, if $r_0 = 0.25$, $a = 1.63$, $p_{opt} = 0.73$. The yield and production curves for these parameters with $r_0 = 0.20$ are given in Figure 1.

It may be useful to determine what increase in the carrying capacity for minke whales in the Southern Ocean is represented by the increase from $r_1 = 0.095$ to $r_2 = 0.15$. If we define the carrying capacity (i.e., the asymptotic level of growth curves of unexploited populations) as J, and designate J_1 = original carrying capacity, J_2 = new carrying capacity, and $j = J_2/J_1$, then it is readily shown that if r_0 = constant, then $j = [(r_0 - M)/(r_0 - r_2)]^{1/n}$. If $r_0 = 0.20$, $M = 0.095$, $n = 2.39$, then $j = 1.36$, and if $r_0 = 0.25$, $j = 1.61$.

It is interesting to note that if $r_0 = 0.20$, the stock was, at that time, very close to the new MSY level: $1/1.61 = 0.62$.

Returning now to the proposed modified management policy, based on R_{max}, we take the minke whale parameters, with $r_0 = 0.20$, for the purposes of demonstrating its application to a single-stock, steady-state situation. This is illustrated in Figure 1, which shows the curve of sustainable yield, of biological production, and of quota, all as a fraction of population P. The quota is calculated as $Q = 2R(p - 0.5)$. It will be seen that if the assessment was correct, the stock would stabilize at $p = 0.70$, with $Y = 0.95Y_{max}$. Under the present procedure, stabilization would be achieved at $Y = 0.90Y_{max}$, with $p = 0.74$, $Q = 0.042p$. A stock would be protected when $p \leq 0.54$, but under the proposed new procedure, only when $p \leq 0.50$. However, under this proposed scheme, quotas would change more gradually over the whole range of exploited stock sizes.

Still, with the 60% MSYL model, but with the higher a, we find for $a = 1.63$, $Q = 0.060$ stabilizing at 91% of MSY, $p = 0.73$, and for $a = 2.00$, $Q = 0.071$ stabilizing at 89% of MSY, $p = 0.75$. In this last case, the quota would actually increase somewhat as the stock was reduced from $p = 1$ to $p = 0.95$, but thereafter would decrease as in the other examples.

If we now assume a 50% MSYL model which, if 90% of MSY is taken, stabilizes at $p = 0.64$, we find that for $a = 1.1$, proposed procedure $Q = 0.025$P stabilizing at 90% MSY, $p = 0.63$; for $a = 1.63$, proposed procedure $Q = 0.034P$ stabilizing at 87% MSY, $p = 0.67$; and for $a = 2.63$, proposed procedure $Q = 0.051P$ stabilizing at 82% MSY, $p = 0.71$. Although the results do not much differ from the present procedure, it will be seen that higher safety limits are automatically achieved when quotas are high because a is assumed or found to be relatively large, and especially when a 50% MSYL is adopted rather than a 60% MSYL.

An advantage of the new proposal is seen if we begin with an unexploited or very lightly exploited stock about which we know little; we must suppose that we have estimates of M and of the initial stock size. Then, the initial quota is MP, but as the stock is reduced, we do not know whether the biological production will first increase, decrease, or remain little changed. Now all production curves lie above the diagonal line: $R' = pM$ (shown dashed on Figure 1). If we have no other information, we can establish a conservative quota by applying the proposed rule to this line, so $Q = PMp(2p - 1)$. If the new stock followed a 60% MSYL model, this would stabilize at $p = 0.76$, 80% of MSY, $Q = 0.037P$. Similarly, for the 50% MSYL model, stabilization is at $p = 0.68$, 85% of MSY, $Q = 0.023P$. If one were prepared to assume that 60% MSYL or, indeed, any other model applied, then a minor upward adjustment of Q could be made, corresponding not with the straight diagonal, but with an R curve that has a slope at $p \to 0$, the same as that of the Y curve, so that the former would not intersect with the latter.

Suppose that, subsequently, a measure of increase in r was obtained as stock is reduced giving an estimate of the slope of the right-hand side of the curve of r against p, as $p \to$ unity. If we are prepared to assume that whale yield curves are, if anything, skewed to the right, then this slope gives an estimate of the upper bound of r_0, by linear extrapolation back to $p = 0$, corresponding with a 50% MSYL curve. A revised, lower, estimate of r_0 could be obtained if, from biological considerations, this were considered to be unrealistically high, or if one were prepared to assume that the MSYL was not less than some given value greater than $p = 0.5$.

It is easily shown that for any given slope of the r-curve at $p \to$ unity, the following relation holds between a and r for any pair of models indicated by subscripts u and v: $a_u n_u \approx a_v n_v$ $\approx$ ... Thus, referring to Table 1, the 50% MSYL model corresponding with the 60% MSYL model for $a_{50} = 1.10$, with the same slope of r at $p \to$ unity, has $a_{60} = 2.39 \times 1.10 = 2.63$, shown in the right-hand columns of R/P and Y/P. Once the stock has been reduced to 80% of its "initial" level, the 60% model is noticeably more "conservative" (5+% difference) for the same set of parameters.

The situation is otherwise if, instead of having estimates of the "initial slope" of the r-curve, we were to adopt an estimate of r_0 from other considerations, such as interspecific comparison. Then, in each case, the 60% MSYL curve is always much less conservative, giving e.g., 20+% higher quota at 80% of initial stock or less. The same argument applies, of course, to the application of the present management procedure.

Table 1. Exemplary calculations of sustainable production and yield (M = 0,095 and maxima are underlined).

		R/P						*Y/P*					
	a	**1.10**	**1.63**	**2.00**	**1.10**	**1.63**	**2.63**	**1.10**	**1.63**	**2.00**	**1.10**	**1.63**	**2.63**
p	*n*	**2.39**	**2.39**	**2.39**	**1.00**	**1.00**	**1.00**	**2.39**	**2.39**	**2.39**	**1.00**	**1.00**	**1.00**
1.00		0.095	0.095	0.095	0.095	0.095	0.095	0	0	0	0	0	0
0.95					0.095						0.005		
0.90		0.106	0.117	0.124	0.095	0.099.	0.108	0.021	0.031	0.038	0.009	0.014	0.023
0.81						0.101						0.024	
0.80		0.111	0.127	0.139	0.093	0.101	0.116	0.035	0.051	0.063	0.017	0.025	0.040
0.79		0.111						0.036.					
0.73			0.129						0.060				
0.71				0.143						0.075			
0.70		0.109	0.129	0.143	0.088	0.099	0.119	0.042	0.062	0.076	0.022	0.033	0.052
0.69							0.119						0.053
0.60		0.101	0.123	0.137	0.082	0.094	0.117	0.044	0.066	0.080	0.025	0.037	0.060
0.50		0.090	0.110	0.124	0.074	0.086	0.110	0.042	0.063	0.077	0.027	0.039	0.062
0.40		0.075	0.093	0.106	0.063	0.075	0.098	0.037	0.055	0.068	0.025	0.037	0.060
0.30		0.058	0.072	0.082	0.050	0.061	0.081	0.030	0.044	0.054	0.022	0.033	0.052
0.20		0.040	0.049	0.056	0.036	0.044	0.059	0.021	0.030	0.037	0.017	0.025	0.040
0.10		0.020	0.025	0.028	0.019	0.023	0.032	0.011	0.015	0.019	0.009	0.014	0.023

In Table 2 are given the values of F corresponding with the entries in Table 1; the right-hand columns show these as % of the highest sustainable value, F_{ext}. Figure 2 illustrates one of the pairs of curves of production and yield against fishing mortality for $a = 1.10$, $n = 2.39$.

Table 2. Values of F corresponding with the entries in Table 1 (values for Y_{max} underlined; values for Y_{opt} double underlined).

		F						F/F_{ext}%	
	a	1.10	1.63	2.00	1.10	1.63	2.63		
p	n	2.39	2.39	2.39	1.00	1.00	1.00	2.39	1.10
1.00		0	0	0	0	0	0	0	0
0.95					0.010				
0.90		0.023	0.028	0.042	0.011	0.016	0.025.	0.22	0.10
0.81						0.029			
0.80		0.044	0.064	0.078	0.022	0.031	0.050	0.41	0.20
0.79		0.046							
0.73			0.082						
0.71				0.106					
0.70		0.060	0.089	0.109	0.033	0.047	0.075	0.57	0.30
0.69							0.078		
0.60		0.073	0.110	0.133	0.044	0.062	0.100	0.70	0.40
0.50		0.084	0.126	0.154	0.055	0.078	0.125	0.81	0.50
0.40		0.093	0.138	0.170	0.066	0.093	0.150	0.89	0.60
0.30		0.100	0.147	0.180	0.077	0.109	0.175	0.95	0.70
0.20		0.105	0.150	0.185	0.088	0.124	0.200	0.97	0.80
0.10		0.110	0.150	0.190	0.099	0.140	0.225	0.99	0.90
0.00		0.110	0.155	0.190	0.110	0.155	0.250	1.00	1.00

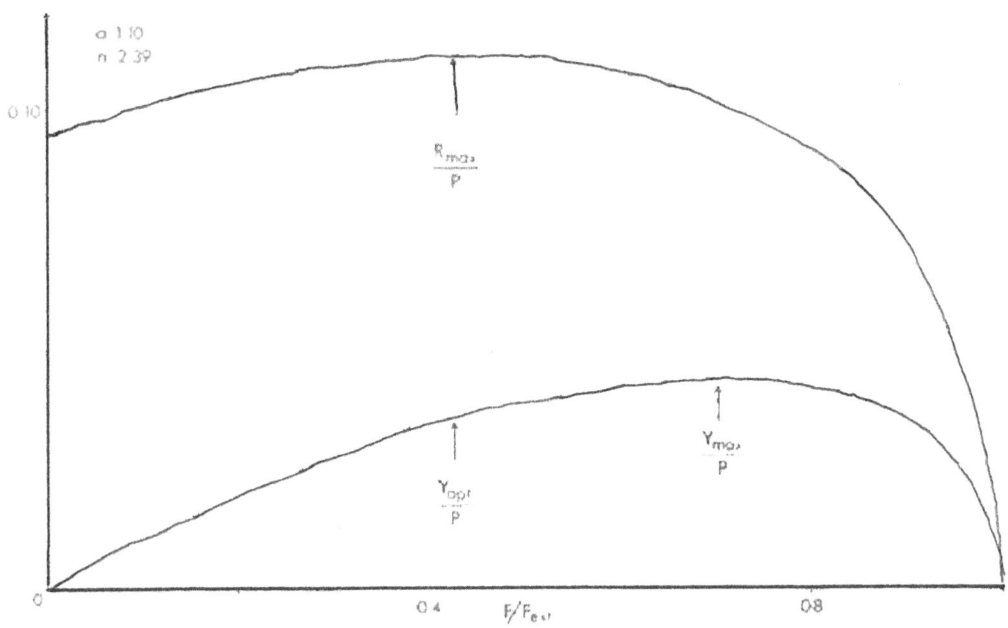

Figure 2. One of the pairs of production and yield against fishing mortality.

Appendix 2

In the text of this paper, I refer to the generalized logistic model in which the natural mortality M is constant. Here, I consider this model with both r and M density-dependent.

In this general case, we write $M = M_1 + (M_0 - M_1)(1 - p^m)$. Writing $b = (M_0 - M_1/M)$ and remembering that $M_1 = r_1$, we have $r = r_1(a + 1 - ap^n)$ and $M = r_1(b + 1 - bp^m)$.

The expression for gross production R is the same as the case for M constant, but the expression for yield Y is now different, thus $Y = pP(r - M) = pPr_1(a - b - ap^n + bp^m) = r_1pP[(a - b)(1 - p^n) - b(p^n - p^m)]$. First consider the special case when $n = m$, that is, when the exponent of density-dependence is the same for births (or recruitment) and for natural deaths. Then, the expression becomes essentially of the same form as that for M = constant: $Y = r_1pP(a - b)(1 - p^n)$.

Since $p_{max} = [1/(n + 1)])^{1/n}$, $Y_{max} = r_1P(a - b)[1/(n + 1)]^{1/n} \times n/(n + 1)$.

In the more general case, where $n \neq m$, there is no exact analytical solution. However, it can be said that the yield curve lies between, and has a maximum at a population level between, that for a single exponent of value n and that for an exponent of value m, and does not depart far—especially near the maximum—from the curve with an exponent $(n + m/2)$.

Chapter 20

History of the Regulation of Southern Minke Whaling Scientific Advice and a Response of Industry[1]

At its 24th meeting in 1972, the Commission abolished the Blue Whale Unit and reported that "it had become apparent that *a more effective way of ensuring the decline in stocks* was to set limits for the different species of whales" (*Holt's emphasis*). This unfortunate slip foreshadows my contention, in Document SC/32/Mi2, that inappropriate allocation of quotas to Areas could encourage excessive effort to be directed to some stocks both within the Indian Ocean and outside it. In view of the fact that the Seychelles has proposed (Document IWC/32/18) a regulatory action which would stimulate a substantial response from the whaling industry [*Editors' note: change to the boundary of the Indian Ocean Sanctuary*], it may be useful to recapitulate the story so far and, in doing so, try to discern patterns of response.

The relevant information is summarized in Tables 1 and 2. A pelagic catch quota was first set at the 1972 meeting, for the 1972/73 season—a total of 5,000 whales. Land station operations in Area II (Brazil) and Area III (South Africa) were unregulated.

The Commission's quota for 1973/74 was the same, but was objected to by the pelagic whaling nations which latter announced a much higher bilaterally agreed number (8,000) within which they would constrain their catches.

The 1974/75 quota of 7,000 was a compromise somewhat closer to what the whaling nations wished to have. For the first time, it was geographically allocated by pairs of Areas: I + II, III + IV, and V + VI. It applied only to pelagic operations.

The quota of 6,810 which was agreed in 1975 included the land station catch in Area II. 1975 was the last year of land station operations in Area III. The overall quota for 1975/76–1976 was allocated by Area, as it has been each year since. These allocations have, however, varied substantially from year to year.

From the time of the first geographical allocation, a flexibility was introduced which later became known as an "allowance for Area" because that was the term by which Japan requested this concession in 1976. This is an amount by which the sum of the maximum permitted catches in each region may exceed the total quota. In 1974/75, it amounted to 21.4%, and in recent years, it has been stabilized at 10%. Examination of the way this flexibility has been used gives some insight to operational decisions, the consequences of which may affect the future regulatory regime.

---

[1]Editors' note: written by Sidney Holt for the 1980 meeting of the IWC Scientific Committee as Document SC/32/Mi15 and published in Reports of the International Whaling Commission 31:319–321, 1981.

Table 1. Southern Hemisphere minke whales *Balaenoptera bonaerensis*: quotas and catches.

	Quota (maximum permitted catches – mpc)							Catches						
		Agreed		IO%		Other regions					%IO of		Other regions	
Season	Total	pelagic	IO	A	B	Total	Pelagic	Total	Pelagic	IO	Total	Pelagic	Total	Pelagic
1971/72								3,856	3,019	3,014	81.7	99.8	707	5
1972/73	None	5,000	Not allocated by region					6,567	5,744	5,744	90.1	100.0	823	0
1973/74	None	5,000[a]	Not allocated by region					8,595	7,713	6,384	74.3	81.3	2,211	1,329
1974/75	None	7,000	4,140	59.1	48.7		4,360	8,149	7,000	3,700	45.4	51.2	4,449	3,300
1975/76	6,810	6,034[b]	3,291[b]	48.3	40.7	4,800	4,024	6,810	6,034	3,035	44.6	50.3	3,775	2,999
1976/77	8,900	7,900	4,603	51.7	47.8	5,029	4,029	8,900	7,900	4,476	50.3	56.7	4,424	3,424
1977/78	5,690	5,000	2,789	49.0	44.5	3,472	2,782	5,690	5,000	2,764	48.6	55.3	2,926	2,236
1978/79	6,221	5,458	3,899	62.5	57.0	2,944	2,181	6,204	5,465	3,866	62.3	70.7	2,338	1,599
1979/80	8,102	7,280?	4,761	58.8	53.4	4,251	3,329	?	7,058	4,569	≤56.4	63.8	2,589+	2,589
Estimates of 1979/80 RY	No allowing for unbalanced sex ratio							Allowing for unbalanced sex ratio						
(1)	11,110		5,940			5,170		8,102		4,328			3,774	
(2)	7,040		4,840			2,200		6,317		3,398			2,919	
(3a)	5,240		3,480			1,760		3,807		2,529			1,278	
(3b)	7,040		4,840			2,200		4,942		3,398			1,544	

Notes: IO = Indian Ocean, mpc = maximum permitted catch, IO% = mpc in IO as % of: A—total quota, B—sum of mpcs.

[a] USSR and Japan objected to quota and announced bilateral agreement on 8,000 pelagic.

[b] Brazilian allocation of 642 out of 2,160 in Area II not agreed, but majority decision of whaling countries concerned was followed. No dispute over South Africa 134 in Area III, but that allocation was not taken.

Estimates of 1979/80 replacement yield (RY): (1) majority view of 1979 Scientific Committee; (2) minority view in 1979 Scientific Committee; (3) from Table 18, IWC/32/Mi2—last two rows "revised estimates."

Table 2. Southern Hemisphere minke whales: advice and notes on management decisions.

Season	Scientific Committee advice		Notes on regulatory decisions
1971/72	Preliminary MSY estimate of 5,000, being about 3% of an unexploited stock estimate. Again strongly urged abolition of BWU.		Catches unregulated. USA asked for discussion and suggested a provisional quota. Commission agreed "to keep a careful watch on level of catches."
1972/73	Reiterated 5,000 MSY, but believed this and the 1971 stock estimate too low (based on paper by Best and Ohsumi).		Pelagic quota. Land stations unregulated. No geographic allocation.
1973/74	Could not choose between MSY estimates of 12,230 (Ohsumi and Misaki) and 5,000 (Chapman). Advocated cautious management of undoubted surplus.		USSR and Japan objected to 5,000 quota decision and bilaterally set pelagic maximum catch at 3,000 which they failed to take, by 4%.
1974/75	No advice possible on stock size and MSY. Advises caution. Suggests 5% of exploitable population, being 4.6% of 1973 total estimate, giving *maximum safe* quota of 7,000, of which 52% in IO.		Land stations still unregulated. Increased pelagic quota allocated by pairs of Areas, with 21.4% flexibility.
1975/76	New stock estimates: total initial = 139,700, of which 45% in IO, current = 122.0, of which 41.1% in IO. No MSY estimates, so advised 5% initial stock except in Area IV where 7% of MSYL less 10%. This implies 6,710, of which 2,810 (41.9%) from IO. Later changed, by adding 100 to Area II, to 6,810, of which 41.3% IO.		NMP being implemented; quota allocated by Areas and applies also to the remaining land station in Area II. 18.8% flexibility. Area IV stock SMS; all others IMS.
1976/77	Two alternatives; 9,130, of which 4,500 (49.9%) IO and 8,686, of which 4,116 (47.7%) IO. These based on 5% of initial stocks, except more conservative 90% of 7% of MSYL in Area IV. Recommended caution in setting "allowances."		Flexibility now 8.2% based on "10% allowance" all Areas except IV which reclassified as IMS. However, because by agreement 7,900 was reserved for pelagic operations, the overall flexibility they exercised was 9.3%; in Indian Ocean 7.0%.
1977/78	Majority 5,690, of which 2,535 (44.6%) IO. Two minorities: 7,729, of which 4,213 (54.5%) IO and 8,585, of which 3,950 (46.0%) IO.		10% allowance in all Areas. However, 5,000 being reserved for pelagic operations, their actual flexibility was 11.4%, and 10% could be taken up in the Indian Ocean. Unclassified.
1978/79	Three alternatives: (1) 5,655, of which 3,081 (54.5%) IO, (2) 6,221, of which 3,545 (57.0%) IO, (3) 6,330, of which 3,676 (51.1%) IO. These all aimed to restrain excessive catch of females. Based on RY of increasing stocks.		10% allowance all Areas. Land stations' allocation not yet published, but on basis of it being as in previous year, 60% of Area II maximum permitted catch (=763), this again gave the pelagic fleets 11.3% effective flexibility. Brazil catch fell 24 whales short of this allocation, and pelagic fleets fell 100 whales short of the remainder. Unclassified.
1979/80	Majority RY of 11,110, of which 5,933 (53.4%) IO. Minority 8,998, of which 4,838 (53.8%) IO. Majority catch limits, reasonably to ensure female replacement yield not exceeded: 8,102, of which 4,328 (53.4%) IO.		10% allowance all Areas. Land stations' allocation not known at time of writing. If on same basis as in 1977/78 and presumed 1978/79, it would be 822 out of 1,370 mpc in Area II; 11.3% flexibility again. Unclassified.

In the first two years for which pelagic quotas were set by the Commission, they were substantially exceeded, and nearly all the pelagic catch was taken in the Indian Ocean. The 1974/75 quota was honored, and just over half the catch was taken in the Indian Ocean. More than half the annual pelagic catch has been taken in the Indian Ocean since then; the proportion has been variable, partly in accordance with estimates of relative abundance of whales, and reached 71% in 1978/79.

Since 1975/76, the quota and Area allocations have been set for all operations combined. Subsequently, agreements were negotiated by participating countries to share these quotas. This essentially involved land stations in Brazil and, for one year, in South Africa, being each allocated a share and the two pelagic nations sharing the remainder between them. Thus, one can calculate an effective pelagic quota and an Area breakdown of this and examine how the "Area allowances" were used.

This is brought out in Table 3 which shows that, in the Indian Ocean, a much higher proportion of the maximum permitted catch (mpc)—in fact, nearly 100% in later years—was taken in that region than elsewhere. That is, whalers used the allowance system systematically to take relatively more from their preferred region—the Indian Ocean, and, therefore, in all years—to exceed the recommended catches there, regardless of the basis of those recommendations. This means that for this region, about which somewhat more is known concerning stock sizes and appropriate yields than elsewhere, catches have, for six years, exceeded recommendations. The average annual excess is approximately 14%.

Pelagic minke whaling *Balaenoptera bonaerensis* began to spread to other regions than the Indian Ocean in 1973/74. The regional pelagic allocation in 1974/75 encouraged this. Unselective operations within these allocations would have given the Indian Ocean 48.7% of the catch, but the flexible system permitted 59.1%. The minimum the whalers could have taken there and still reached the overall quota was 2640 or 38%. They actually took 51.3%, showing a clear, but not overwhelming, preference for the Indian Ocean.

One could construct an index of preference from this type of information and examine changes from year to year. This is hardly necessary, however, because it is evident from Table 3 that in all subsequent years, the overwhelming preference was for the Indian Ocean. That is, the pelagic catches there were always practically the maximum permissible under the quota system, at the expense of lower-than-recommended and allowable catches elsewhere. This preference has even extended to the 1979/80 season where the pelagic allocation of the quota appears not to have been all taken, yet neither country took any whales out of Area VI.

Table 3. Ratios of catches to maximum permitted catches, taking account of land station/pelagic agreements.

	Land and pelagic		Pelagic		All areas	
Season	Indian Ocean	Other regions	Indian Ocean	Other regions	Land and pelagic	Pelagic
1972/73						1.15
1972/74 (5,000)						1.54
(8,000)						0.96
1974/75	0.89		0.89	0.76		
1975/76	0.92	0.79	0.96	0.75	1.00	1.00
1976/77	0.97	0.88	0.97	0.85	1.00	1.00
1977/78	0.99	0.84	0.99	0.80	1.00	1.00
1978/79	0.99	0.79	0.99	0.73	0.99	1.00
1979/80	0.96	0.62	0.96	0.78	0.88+?	0.98?

The proportion of total and of pelagic catches taken from the Indian Ocean reached a minimum in 1975/76 and, with fluctuations, has since risen again. This has been permitted by the varying quota allocation, but the preference to stay in the Indian Ocean as much as possible is consistent.

If Scientific Committee calculations were correct, we would need to worry only about the continuing 10% excess. But, in fact, there is good evidence that they were biased by an overestimation of the Area III stock and hence of the whole Indian Ocean; our concern, therefore, has to be about an excess of about 25%—at least.

If there is any validity to my calculations of replacement yield (Document SC/32/Mi2), then the excessive catch could be much greater.

What of the situation elsewhere? Despite their reluctance to spend more time in other regions than was required by the quota system, the whalers have been constrained to exert a substantial effort in them. If the quota allocations to them had been reasonable ones, those regions would not, in all, have been overexploited. However, in the first years of regulation, the allocation to other regions was higher than now, both relatively and in absolute numbers. And there is reason to believe that the numbers of whales there, and their replacement yields, have been exaggerated even more than they have in the Indian Ocean. Thus, catches have probably been, for several years, much higher than replacement yields—possible two- to threefold higher. So, subdivision of management units and species allocations to them on a poor scientific basis, combined with the exploitation of a flexibility in regulations which was introduced for other reasons, can lead to excessive whaling effort on, and eventual depletion of, all the stocks.

To the above concerns must be added the fact that from the Indian Ocean, 1.52 females have been taken for every male, whereas elsewhere, this ratio has been 1.40:1. This difference is not great, but it contributes to the probable relative reduction in the reproductive potential of the Indian Ocean population(s) (Table 4).

Table 4. Sex ratios in pelagic catches, 1972/73–1979/80.

Region	% females	No. caught
Indian Ocean (Areas III and IV)	60.4	51,937
Other regions	58.2	17,606
Area II	76.7	4,071
Areas I, V, VI	52.6	13,535

Chapter 21

Maximum Sustainable Yield and Its Application to Whaling[1]

Scientific analyses should include an assessment of their own limitations. It is incumbent upon the practitioners of science to develop procedures which can express the information content of their prediction in a readily understandable manner. The effort may require as much thought and ingenuity as the development of the original prediction (Kraus 1974).

The International Convention for the Regulation of Whaling, 1946 provides, in its Article V(2)(b), that regulations with respect to the conservation and utilization of whale resources "shall be based on scientific findings." The Convention established the International Whaling Commission (IWC) with power to set up committees. One such committee is the Scientific Committee which deals with the above provision. The other standing committee, to which all Member States may nominate members, is the Technical Committee; it is usually composed of the Commissioners themselves; it acts, in effect, as a committee of the whole Commission and gives most attention to another requirement of Article V, namely that regulations "shall take into consideration the interests of the consumers of whale products and the whaling industry," especially, in practice, the latter.

From early days, the Scientific Committee has usually been well aware when whale species and stocks were being overexploited and has issued warnings, pleas, and suggestions as to the level of permissible catch. It was not until the early 1960s, however, that numerical and algebraic models began to be applied systematically to predict the quantitative consequences of proposed regulations. Such applications to the blue whales *Balaenoptera musculus* and humpback whales *Megaptera novaeangliae* in the Antarctic, which were already greatly depleted, and to the fin whale *B. physalus*, then in process of rapid depletion, seemed convincing both to scientists and Commissioners. This temporarily silenced the disagreement among scientists who had previously contributed to delays in taking remedial action. Agreement by consensus on "best estimates" for whale stocks became a major objective of the Scientific

---

[1]Editors' note: this paper, written by Sidney Holt, was presented at the Airlie House Workshops on the Concepts of Management and Conservation of Wild Living Resources held in Virginia (USA) in 1975 as Document FAO/ACMRR/MM/IV/4, revised for the Bergen Scientific Consultation in 1976, and eventually published in 1981 in *Mammals in the Seas, Volume III, General Papers and Large Cetaceans* (Food and Agriculture Organization of the United Nations, Rome). At the time of this writing, Holt was employed by the Food and Agriculture Organization of the United Nations in its Fisheries Biology Branch. Permission to reproduce this article has been given by FAO. The views expressed in this publication are those of the author and do not necessarily reflect the views or policies of the Food and Agriculture Organization of the United Nations. The version presented here has undergone very minimal editing from the published version.

Committee. This has, I suggest, over the years, led the scientists concerned to underemphasize the limitations of the methods and data used. It has not been easy to examine these limitations critically because few scientists are occupied full-time (or even most of their time) with this matter; research funds and facilities are chronically inadequate, and the rules of the game include unwritten rules of international courtesy.

While understandable, this has regrettable consequences, particularly so as scientific attention moves from the obviously depleted stocks to those which have only been lightly or moderately exploited. It was perfectly obvious, by the time the blue whale had been reduced so much that it was rarely seen, that it could not support continued hunting. If appropriate action had been taken at that time with respect to the fin whale, we should not now be arguing about whether it, too, needs complete protection. According to Gambell (1974), "the future for the whales looks bright" because "there is a prospect of management being grounded on rational policies," yet at this time, gambles are still being taken with the most important stocks which remain—sperm whales *Physeter macrocephalus*—as well as with the smallest baleen whale, the minke *B. acutorostrata* and *B. bonaerensis* (Holt 1974a). I believe this results from overconfidence in the methods of manipulation of data compounded by unrealistic demands for precise scientific advice.

The 1946 Convention does not refer to maximum sustainable yield (MSY) nor to the concept of sustainable yield. The Preamble to the Convention does, however, refer to achieving optimum levels of whale stocks as rapidly as possible, and Article V, paragraph 2 lays down that amendments to the Schedule should be such as are necessary, *inter alia*, "to provide for ... *optimum utilization* of the whale resources," (*Holt's emphasis*) although no definitions are given. The first explicit reference to the sustainable yield criterion for management is in the Resolution adopted by the IWC at its 12th meeting in 1960. This established the Special Committee of Three Scientists to study and report "on the level of sustainable yield that can be supported by the Antarctic whale stocks" and "on conservation measures that would increase this sustainable yield so that it may be brought to an optimum." In its First Interim Report submitted to the Commission in June 1961, the Special Committee still does not refer to MSY, although it mentions the methods of analysis "to determine optimum yield" (Chapman et al. 1962). In its Final Report (Chapman et al. 1964), the Special Committee used the term "optimum level" to mean the stock level corresponding with MSY. The Committee presented assessments of current sustainable yields in numbers of fin, blue, humpback, and sei whales *B. borealis* in the Antarctic, but remarked that "consideration must be given to assessing stock levels in units more clearly related to economic needs, taking into account both the yield of economically valuable products and the relative cost of production ... related to the catch per unit effort." An "unofficial" group of whale specialists in a meeting in 1970 (Holt et al. 1974) continued to use the term "optimal stock size" as the size giving MSY (implicitly, in numbers) (Schevill 1974).

Meanwhile, in 1966, the 1958 Geneva Convention on the Conservation of the Living Resources of the High Seas had come into force. Its Article 2 contained the following definition of MSY drafted in 1955 by the UN International Technical Conference on the Conservation of the Living Resources of the Sea: "Conservation means the aggregate of measures rendering possible the optimum sustainable yield ... so as to secure a maximum supply of food and other marine products." This definition of conservation has been substantially included in practically all draft articles on fisheries submitted to the 1974–1975 UN Conference on the Law of the Sea.

At its Annual Meeting in June 1974, the IWC decided to categorize whale stocks for management purposes. The operative clauses of the decision, which identify three categories of stocks for management purposes refer to, but do not define, MSY in the following terms:

"*Initial management stocks* ... may be reduced in a controlled manner to achieve MSY levels or optimum levels; *sustained management stocks* ... should be maintained at or near MSY levels and then at optimum levels; *protection stocks* ... which are below the level of sustained management stocks and should be fully protected."

The full text of this "decision" is appended (Appendix 2). The advisory role of the Scientific Committee is defined therein and should be noted; in particular, that the measures to be applied to Initial Management Stocks be such as are "necessary to bring the stocks to the MSY level and then optimum level *in an efficient manner and without risk* of reducing them below this level" (*Holt's emphasis*). It should be noted also that the term "optimum yield" is used—but not defined—in the decision, and apparently implies that the corresponding "optimum level" of stock would be somewhat higher than that giving the MSY.

At a special meeting in December 1974, the Scientific Committee attempted to respond to the request from the Commission for "advice on the criteria which should be used in defining (the three) categories of whale stocks." Since it was made clear that species which were already fully protected ("Protected Species": blue, pygmy blue *B. m. brevicauda*, humpback, right whales *Eubalaena* spp., and gray whales *Eschrichtius robustus*) were automatically to be designated as Protection Stocks, the Committee discussed the "criteria" having in mind certain "difficult" cases of deciding whether particular rather lightly (or recently) exploited stocks or species would become categorized as under "initial" or "sustained" management, and whether some other stocks which have been more heavily exploited for a longer time would be under "sustained management" or "protection." In the latter case, the difficulty proved to be to interpret the term "near" (MSY) levels; in the former case, "near" is also a problem, but the difficulties come mainly from the very feeble knowledge about species (particularly the minke whale) which have only recently become the object of intense pelagic whaling. "Near" could conceivably be defined scientifically in terms of statistical errors of estimates of MSY (though such a definition would not necessarily be acceptable—or even useful—to the Commission itself), but these errors are unmeasured, and are perhaps unmeasurable. Sensing this, the Scientific Committee has sought ways of giving, in effect, blurred definitions of some stock categories, and devises a proposal for a graduated range of whaling effort on stocks below the "best" estimates of the MSY level.

Another general problem involves taxonomy. The IWC and the national (Norwegian) agency which provides statistics of catch and whaling effort to the IWC—the Bureau of International Whaling Statistics (BIWS) —have a long history of lumping different taxa. Important cases have been the lumping of the blue whale and pygmy blue whale subspecies, and of the sei whale and Bryde's whale *B. edeni*. However, although the question of the validity of the pygmy blue whale as a distinct subspecies impeded for a year or two the protection of blue whales in the late 1960s, much more serious questions are the identification within each species of distinct breeding stocks, as biologically defined, and of stocks as "defined" for purposes of the IWC decision, as those "units which can be most effectively managed individually."

The purpose of the 1946 Convention was "to provide for the proper conservation of whale stocks and thus make possible the orderly development of the whaling industry." To this end, the IWC reviews and amends each year a "Schedule" of regulations. It is constrained to base its actions on scientific findings. The possible actions are of several kinds and include closing certain areas ("sanctuaries") to whaling, establishment of minimum size limits, restricting capture of lactating females, opening and closing whaling seasons, applying different rules to land stations and pelagic expeditions, and special rules to aboriginal peoples catching whales on a subsistence basis for local consumption as human food. The actions the Commission considers most important are, however, the setting of overall catch quotas for each season. These are now defined as the total numbers of each species of large whale that may be taken in various areas; in the case of sperm whales, the quotas are also established for each sex. A species may be totally "protected" throughout all or part of its range; this is effectively the same as setting zero quotas, although there are legalistic differences between the two actions. It is well known that until recent years, the Commission set quotas for the Antarctic in terms of "Blue Whale Units" (BWU) —a weighted average of numbers of four types (six taxa) of baleen whales.[2]

Since the 1960s, the quota (whether in BWU or by species) has been divided between those IWC nations which conduct pelagic whaling by agreements concluded concurrent with, but formally outside, IWC meetings—and there is mutual interaction between these debates and Commission debates on quotas. Each nation—and whaling company—will, among other things, have made prior calculations of what catches are needed in the coming season to at least pay the costs of, and yield a modicum of profit to, the expeditions (a factory ship plus a number of catcher boats) that it wishes to send out. As the total number of expeditions has decreased over the years, it becomes more difficult to adjust the whaling effort to an exact quota—one expedition less, or more, can mean a quota allocation not reached, or reached too easily and shared too thinly. Furthermore, although the BWU is not now used, each whaling enterprise continues to make its operational decisions on the basis of the expectation of catches of a number of species of baleen whales differing greatly in size, as well as of sperm whales. It is evident, therefore, that at governmental level, the debate on a quota for one species—say, fin whale—is influenced considerably by expectations of what might be acceptable for another, such as minke.

Lastly, the limited applicability of IWC decisions should be noted. They apply, of course, only to the vessels operating under the flags of Member states of the Commission; Schedule amendments require a three-quarters majority decision and are not binding on a state which has invoked the "90-day objection clause" in the Convention; and there has long been controversy over the application of quota decisions equally to land stations as to pelagic expeditions. A considerable amount of whaling is carried out by non-Member states, and certain Members participate in joint whaling enterprises with non-Members, the vessels of which fly the flags of the latter.[3]

The term "MSY" describes a property not of an animal population, but of a mathematical model which is applied to data from a population. The model may or may not be based on, or derived from, those data. In this simple form—and that is the form which has been, so far, applied to whales—the model assumes that for a given value of population size, there is a catch (the sustainable yield – SY) which can be taken regularly for an indefinite period. Thus, a curve of SY against population size can be computed; SY is zero for an unexploited population and also for an extinct population; somewhere in between is the greatest SY—the MSY.

In whales, both population size and catch have, for this purpose, been defined as numbers of animals. The population size is variously defined as total number, or the number of mature animals, or the recruited number—that is, the number of animals having a size, distribution, and behavior such that they are liable to capture by the whaling effort, as it is deployed and

---

[2]BWU = 1 blue whale = 2 fin = 6 sei = 2½ humpbacks – based on the oil yield from the different species.
[3]For a detailed historical account of whaling management, see Vamplen (1972).

regulated. Evidently, if "recruitment" occurs several years after birth, the number of "recruits" (which is the most important factor in sustaining the catch at a certain level) will depend on the size of the mature population several years before, rather than during, the year of catch; this time-lag of dependence of SY on population size can readily be—as has been—incorporated into the simple models applied to whales. Consideration of it led Gulland and Boerema (1973) to define another quantity, the *replacement yield* (RY), as "that catch which, if taken (in a given year) will leave the abundance of the exploitable (i.e., recruited) part of the population at the end of the year the same as at the beginning." This led, further, to a definition of (a) *equivalent sustainable yield* (ESY) from a population which has been changing, "as the SY from a population of the same abundance (or with the same abundance of the exploited phase) which has remained at this level of abundance for a long time"; and (b) *maintainable yield* (MY) as "the largest catch that can be maintained from the population, at whatever level of stock size, over an indefinite period." MY is equal to the corresponding SY for populations at levels below that for MSY, and to MSY for populations at or above this level. The following observations about MSY apply fully to these other "properties" of the simple model, few of them are new. Serious criticisms of the concept have been made more often by scientists than by whaling and fisheries administrators despite the fact that MSY is often referred to as a scientifically based management objective (see, for example, Beverton and Holt 1957).

The MSY and the population which will sustain it may be, and variously are, defined in absolute or in relative terms. Thus, MSY can be expressed as a multiple (or fraction) of the current yield, or of the SY from a population of the current size and vice versa. The MSY stock level may, likewise, be expressed as a number or as a fraction (or multiple) of the current stock level, or as a fraction of the stock level before exploitation started. This latter statistic is of crucial importance in discussion of whaling regulations because series of data for catch per unit whaling effort are often available, and these are assumed to indicate the trend of stock level from soon after the beginning of exploitation. The simplest mathematical model based on the logistic curve of population growth gives a symmetrical parabolic SY to population curve, so that the MSY stock level is half the initial level. Modified models give skewed curves, usually such that this ratio is greater than 0.5, commonly 0.7–0.8.[4] Others give curves, skewed or symmetrical, which are more flat-topped than the parabola (see Appendix 1). The existence of this range of "plausible" simple models would alone make the level of MSY difficult to identify, either in absolute or relative terms, but there are other difficulties.

The 1974 IWC decision recognized that "management of whale stocks should also include such considerations as total weight of whales ..." This recognition followed presentation of evidence (Holt 1974b) that, notwithstanding the application of size limits, the average size of each species and sex of whales in catches has declined over the period of intensive exploitation. One expects to observe such a trend especially in situations where the exploited animals are rather long-lived and continue to grow substantially after "recruitment," as is the case for all large whales, but especially for the male sperm whale. The effect has been obscured because whales are not easily and routinely weighed; their lengths are measured and mean lengths change in the same direction as mean weight —but only in proportion to the cube root of the weight, and hence less noticeably. The changes are, in any case, evident in the statistics for oil yield per whale, but these statistics are affected also by changes in the relative

[4] Fish stocks, on the other hand, are said more often to be described by curves skewed the other way, with a ratio <0.5.

utilization of oil and meat. Although there is some resistance, both among whale biologists and administrators[5], to the regulation of whaling to maximize weights rather than numbers, the whaling industry has always based its activities and policies on consideration of quantities (and hence total values) of products. The BWU was based on relative oil yields of different species and, although the interspecific size range is wider than the intraspecific range, the latter is not negligible. In the special case of sperm whales, the males of which are much larger than the females and are distinguishable from them before capture, the calculation of total quotas includes a "weighting factor" to allow for the size difference.[6]

In all models studied, the curve of SY (weight) against population size, whether the latter is expressed in number or weight, is skewed to the right relative to the corresponding curve of SY (number); MSY (weight) corresponds with a relatively higher stock size than that which gives MSY (number). The SY (weight) curve is also generally a little more flat-topped than the curve of SY (number). These differences are not very important for the baleen whales as they are, at present, exploited,[7] but they are significant—a shift in population size for MSY in sperm whales of from 7 to 18%. This is essentially because in sperm whales, a greater proportion of the growth in weight occurs after "recruitment"; this proportion is regulated by the setting of minimum size limits. So, if the catching of whales is to continue to be regulated on the basis of MSY calculations, it would be better from *all* points of view to aim at MSY (weight) rather than MSY (number). This would result in higher value from the resources, for less whaling effort, and higher stock levels (see Appendix 3).

MSY has commonly been referred to in international discussions as a "biologically determined criterion" for sound resource management. This is misleading. MSY is a function at least of the sizes of animals which first become liable to capture, as well as of the population size as determined by the overall rate of exploitation. Models which have been applied indicate that the yield in weight-per-recruit has a maximum for a finite fishing mortality rate, i.e., exploitation rate or $E = F/(F + M) < 1$, only if the ratio of size (weight) at first capture to final size is less than a critical value $= [3/(3 + M/K)]^3$ (Beverton and Holt 1966). The catch, if *all* surviving animals are caught as soon as they reach that critical size, is theoretically equal to the maximum biomass reached naturally by a year class during its lifetime; this quantity may have some biological significance, perhaps because the critical size may bear some relationship with—may even be close to—the mean size (and age) at sexual maturity (Holt 1962). But where—as in whales—the number of recruits is closely related to the numbers of parents, such a simple interpretation of yield-per-recruit models cannot be applied. There could, perhaps, be calculated a combination of size at first capture and exploitation rate (less than 1) in whales which would predict a unique MSY for each stock; unfortunately, models in which a simple yield-per-recruit equation is combined with a linearly density-dependent reproduction rate, have not yet been used to explore this possibility. In any case, it is sure that:

a) the size limits for whales have, so far, been set by the IWC without any reference to such models, and
b) calculations of curves of SY against whale stock size have always "taken as given" the size at recruitment equal to the size limit in force, if any.

---

[5]One reader of the first draft of this paper commented "some people will not agree to thinking of whales as meatballs." Another said, however, "we should not think of whales and other wild animals merely as statistics."

[6]Not now (i.e., since 1976).

[7]Except, perhaps, for the minke whale, for which no minimum size limit has been established by the IWC.

Another reason why MSY should not be regarded as a "biological" criterion is that since the logistic curve of population growth was first applied in the way outlined above, the argument for managing a fishery so as to hold the stock at the MSY level has been made, implicitly or explicitly, in terms of the *efficiency* of the catching operation. Thus, if the stock is reduced and held, by fishing, below that level, the fishing effort is being "wasted" because the *same* catch could be sustained with less effort by allowing the stock to recover. So, MSY stock level is not so much a level at which to aim as a level below which we should avoid going. Furthermore, once the idea of *sustaining maximum efficiency* is introduced, there is no reason to prefer the MSY level to any other higher level. Graham (1935, 1939) argued for restoring overfishing stocks to the MSY level or above, not to increase net profits, but to allow the same fishermen to take the same catch while working much less hard.[8] Such humanistic views have been unfashionable for some decades, but the time may be ripe to reconsider and develop them. "Overfishing" is now seen to waste not only the working lives of fishermen, but also the limited supply of non-renewable resources of fuel and materials which are expended in seeking and capturing fish—and whales. A purely economic approach to this problem has not been fruitful in international affairs, but provided we are wary of the consequences of "artificial" price structures and of high discount rates, such an approach gives some insight into the behavior of the resource-exploitation system with respect to conservation. Meanwhile, our economic system, having led us to be profligate with non-renewable resources, is now leading us to "mine the biota" too (Woodwell 1974).

Curves of *net value* of sustained catch (i.e., value minus costs of catching) against stock size have maxima at higher stock levels than the MSY level (Figures 1 and 2 in Appendix 1). Thus, in steady-states, there is an overall economic advantage in holding the stock rather higher than the MSY level; this would be advantageous also to each operating unit which thereby obtains a higher sustained catch per unit of effort (Chapman et al. 1965; Gulland 1968) (see also Appendix 1). This idea promised to offer an alternative to the MSY concept as an objective of fishery management when it was put forward by Gordon (1953), Beverton and Holt (1956), and others. Unfortunately:

a) there seems no way to get international agreement on the economic criteria, particularly before there is agreement on allocation of catches among participants; and
b) this simple economic model takes no account of *discounted* present values of future catches.

It was the recognized difficulty—if not impossibility—of getting agreement on economic criteria that led negotiations before the 1958 Geneva Conference to concentrate on physical measures of desirable gross yield, the argument being that to aim at this would favour the more efficient operators. Clark (1973a) and others have shown that for resources having net growth rates of the order found in whales, and with current levels of discount rates, attainment of an economic "optimum" may not require *sustained* harvesting at all (Appendix 5). This must lead us to examine the roots of our beliefs concerning the "proper" harvesting of natural renewable resources.

---

[8]"The benefit of efficient exploitation lies more in economy of effort than in increase in yield, or preservation of future stocks, though both of these purposes may also be served."

A beginning was made, as far as marine fisheries are concerned, by Rothschild (1971) who, however, quotes the views of R. Dorfman and a Panel of the U.S. Commission on Marine Science, Engineering and Resources, to the effect that a suitable economic goal might be the attainment of "efficiency involving distribution of products to the people that need them, full utilization of resources and high consideration of the needs of the consumer." Rothschild's paper is concerned more with procedures than with goals, but if we can agree on goals, his approach will be useful in attaining them. He states that "conservation of resources is essentially an allocative process and the calculus that is conventionally used in fisheries is not a powerful technique for allocation." Unfortunately, he does not—at least in the paper quoted—pursue this idea. In particular, he does not identify the most important aspect of the "allocative process" involved in management for sustainable yields—that is, the allocation of benefits as between the present and future generations of producers and consumers. He does suggest that in necessarily "taking a broader view temporally as well as spatially of fishery management," we should seriously engage in "technological forecasting." No doubt this is desirable, and good information flow within the subsystem "whales and whaling" and from outside it, combined with intelligent analysis aimed at predicting as well as evaluating technological changes, could improve management.[9] The present need may, however, be for forecasting at another level—what values will our descendents place on whales, what properties will they value, how will they view matters of resource ownership, how will they value the ecosystem of which the whales are a part, what needs will they feel for the unique products, if any, obtained from whales? Just as, in relation to technological forecasting, Rothschild, quoting J. R. Bright, observed that "a radical new technological advance is more visible to society first in written words, then in ... more effective material forms, long before it achieves widespread usage," so we may now perhaps perceive the seeds of change on which social forecasting could be based. The support, in several countries and from many quarters over recent years for the movement for a moratorium on commercial whaling, notwithstanding the opposition of most of the scientists concerned,[10] indicates the need to look, as ecologists, at the whole in order to comprehend the parts, and reminds us that development of resource management criteria is an ideological and general problem, rather than merely a specialized scientific and technical one. Nevertheless, the criteria must be capable of practical application, and it is difficult to visualize any for which world application would not involve scientific analysis.

The above comments on the existence or otherwise of the "M" in MSY and on the appropriate measure of the "Y" are maybe valid, but they are not new (see, for example, among many other critical publications, Gulland 1968a). They are also relatively trivial; the concept of MSY can be modified to accommodate them—at least in theory—although the history of the IWC

---

[9]Scientific advice regarding a proposed regulation of fishing has not always been inadequate in that, although the resource is treated as a dynamic system, the fishery is often treated as if it were static. For example, assessments of the benefits of introducing a higher size limit or large size of trawl mesh may assume that the fishing effort would remain unchanged. Yet, to the extent that the proposed measure is effective in increasing the stock size so more effort would be attracted into that fishery; it is not difficult to predict approximately how much. In general, I believe one should assume that when one aspect of the industry is regulated, the industry will always, if the regulation has a significant effect, tend to change in such a way as partially to negate the effect. This has happened several times in pelagic whaling—reduced quotas for certain species as they became depleted encourage whaling on regulated species so that the total value of catch per unit of effort is maintained at profitable levels.

[10]Supporters of a moratorium have been accused of "being emotional" about whales. Apart from the question of whether emotion is sinful, one observes that scientists are as liable as anyone to make ideological statements when they think they are giving "objective" scientific advice.

(and of some other international fishery bodies) shows how difficult it is and how long it takes to get acceptance of such innovations. The validity of the "S" in MSY is, however, more difficult to appraise. For any level of yield to be sustainable, it is necessary for the net rate of increase of stock (recruitment or reproduction plus growth minus natural mortality, all expressed as a ratio to population size) to increase with decreasing population size, and vice versa, at least over some range of population size. The practice in making whale stock assessments has been to assume that natural mortality is not age-specific and is density-independent. Thus, the density-dependence is sought in the reproduction (or recruitment) rate. In some whale species, the reproduction rates are rather better known than for most exploited wild animal stocks, although there are some technical problems and some controversy over interpretations.[11] Reliable information about the changes in reproduction rate with population size or density is, however, scant. Such changes can be inferred from the changes in population structure as exploitation proceeds; these inferences, taken with estimates of maximum feasible reproduction rates (percentage pregnancies, number of calves, etc.) have led to fairly firm statements about whether a certain catch from a much reduced stock will exceed the replacement yield, i.e., whether it should lead to a decline or increase in stock in the short term. Tests of such predictions have not been made systematically. In theory, such tests should be easy to make because the catches are recorded and good effort data provide measures of stock abundance (catch per unit effort). In practice, given the likely rate of change in a whale stock—or even the maximum possible rate when a depleted stock is protected—the abundance estimates are too variable. They are affected by the fact that the whaling effort is expended, in the same area and period, variably on several species; its effectiveness depends on the weather, and its efficiency changes for a number of technological reasons, few of which have been taken into account. The abundance estimates could and should be much improved, but we cannot expect too much from them in attempting to measure the density-dependence of reproduction or in testing the assumption of constant natural mortality.

In recent years, it has been suggested, mainly on the basis of work by R. Gambell and C. Lockyer on fin whales, that the mechanism of density-dependence is in the change of growth rate in young animals (resulting from density-caused changes in available food) which leads to changes in the age of sexual maturity. This hypothesis—with supporting data—is now being examined critically by the FAO/ACMRR Working Party on Marine Mammals. My concern here, however, is to see how close a determination of the reproductive relationships brings us to an understanding of the MSY problem. This is examined in Appendix 4. My conclusion is that even with far better data than are ever available for wild animal populations, the level of stock corresponding with MSY could not be determined with any useful accuracy, although a value for MSY itself might be inferred. For example, Doi (1973), having discussed the Antarctic sei whale, "emphasizes that it is possible to estimate the maximum sustainable yield, to diagnose the population status and to determine the management course judging from a reproduction curve." Unfortunately, this is not so.

---

[11]The reader is referred to Cushing (1974) for a succinct description of the problem in marine fisheries. He identified the stock–recruitment relationship as the central puzzle of fisheries research and states that great collapses of "managed" fisheries "are attributable to the unstated concept that fishing could continue until recruitment was seen to fail. Then, failure can be attributed either to natural causes or to fishing, and stocks collapsed while scientists disputed the two possibilities." He adds that failure of "management" has been no less than that of science. As in many other situations, the devil is in the "unstated concepts" as well as in some of the stated ones, such as "the best regulation of fishing is to leave it alone until there is evidence of adverse effects." See Gulland (1971) for a general discussion of the consequences of acting on such null hypotheses.

According to the simple idea of SY, the net rate of increase should be zero in the initial stock, just before exploitation starts. But this has never been tested. It seems to me quite possible that the abundances of whale stocks were changing in the period before intense exploitation, and that their population structures at the time were, therefore, not characteristic of steady-states. Such changes could be cyclic or long-term trends, and related to climatic trends and cycles or other "external" factors, or to man's previous exploitation of other species. Furthermore, the factors causing such changes could usually be expected to continue their effects after exploitation started. The existence of steady-states is something to be tested, not assumed; as Bradley (1972) has written: "In a world of dynamic change, it is constancy, not stability, that requires explanation." It is also necessary, for example, by simulation, to examine the likely errors of applying steady-state models to real situations in which the stocks might have been changing naturally.

While numerical calculations of whale stock changes have, as mentioned, included time-lags between birth and recruitment, little account has been taken of the fundamental properties of models in which such time-lags have been incorporated. Thus, Cook (1965), May (1973), and May et al. (1974) have shown that such models, even variants of the simple logistic, display steady-states under certain conditions, but stable-limit cycles, damped oscillations, or even wild behavior under other conditions. Marchesseault et al. (1976) have recently formulated and examined some of the properties of a simple model incorporating a constant recruitment delay and applied it to a lobster *Homarus americanus* population. Their model is very close to the one which is implicit in some of the numerical calculations for whale stocks. The model predicts, in effect, that if the age of recruitment (or maturity) is *under*estimated (in the lobster application, the cases compared are actually recruitment at 5 years and at 0 years, i.e., the "delay" model is compared with the simple "instantaneous" logistic), the amount of fishing effort required to obtain MSY is overestimated, and so is the MSY itself. Thus, if the age of recruitment is underestimated, scientific advice based on a delay-logistic model would encourage quite serious overexploitation, whether regulation is by effort control or catch quota. The MSY stock level is not affected by the time delay, although it must always be borne is mind that the delay model predicts the relationship between sustainable yield and *recruited* or *mature* stock, not the total stock. The relevance of this to a critique of whale assessments is that the age determination, and hence estimation of the ages at recruitment and at sexual maturity and full reproductive activity, has been somewhat controversial. While there have, in recent years, been important amendments to the age determinations from ear plug rings, age determination for some toothed and baleen whales are not yet sure, and for the blue whales, there are practically no data, since they were greatly reduced in numbers before the ear plug method was invented.

All applications of the MSY concept have assumed reversibility so that if intensified whaling "overshoots" the MSY level, it is only necessary to reduce the intensity sufficiently and for long enough for the stock to recover. It would not grow back along the sustained yield curve, but would approach it as the population structure tended to the steady-state one, at an overall rate given by the assumed linear dependence of net increase rate on total stock abundance or density. I find several difficulties with this assumption. Firstly, it has been said that greatly depleted whale stocks have not always recovered as the model would predict (Talbot 1974)[12] after they have been protected. This could be for many reasons: because there

[12]It is difficult to pin down the facts about this. Gray whales have recovered; humpbacks seem to have increased in some areas; right whales *E. australis* seem at least to be holding their own in the Southern Hemisphere. We do not know what is happening in most cases because such data as there are on sightings are highly variable.

have been small permitted catches of the same order of magnitude as the sustainable yield of the depleted stocks; because the models may be erroneous in ignoring a number of properties of the populations, apart from their age structure—one such property is social structure, disrupted in the course of depletion; because the area of distribution may contract as the total population size decreases, thus tending to maintain *higher* densities—and lower net increase rates—than the simple model would predict.[13] Considering how little is yet known about the social behavior of cetaceans and their physical capabilities, it seems possible that the simple stock models could lead us astray in our quest for the "real" MSY. Perhaps the most likely reason for non-recovery, or recovery at other than the predicted rate, is, however, the occupancy of an ecological niche by a competing species—perhaps another whale species. This has been suggested to occur between the blue and minke whales and the fin and sei whales.

The "S" in MSY stands not for *sustained*, but for *sustainable*, that is, it specifies a future option: if we do certain things now, someone else can, if he wishes, take certain yields in the future. If he does not—the model implies—the "quality" of the resource will deteriorate (the stock will grow toward its asymptotic size, and in doing so lose some of its "productivity"). This leads to the idea that is "beneficial" for a stock to be reduced, by harvesting, to some fraction of its initial size, and that it is wrong to "waste" a renewable resource by leaving it unharvested.[14] It is not, however, reasonable simply to assume that future generations will place the same values on a resource as we do. This cuts both ways: one IWC delegate responding to an argument that future generations may "need" food, including whale meat, more than we do, said that *they* will decide what they need and how to get it; therefore, why should we give priority to leaving whales in the sea? Nevertheless, the implications of the assumption need to be explored with as much technological, economic, and social forecasting as practicable.[15] For whaling, this at least includes looking at the likely availability of similar products from other sources, the relative costs of fuel (pelagic whaling, like distant-water fishing, involves high consumption of non-renewable energy sources) and the evolution of human dietary habits. But should we not go further and try to anticipate qualitatively new "values" for natural "resources" whether as tourist attractions, unique objects of scientific research, companions of man, symbols of our atonement for past sins against nature or other (McIntyre 1974). Since we can glimpse such values, but not evaluate them, our present actions might be judged in terms of the number of future options they leave open. Intuitively, to refrain from interference now is likely to leave more options than would drastic interference. Our intuition may be wrong; natural processes we do not yet understand may remove some of these options tomorrow; nevertheless, failing other guidance, we live each day as if tomorrow will come.

---

[13]In this connection, I do not agree with Talbot's (1974) general criticism that "what is 'known' of the population dynamics of whales is inferred from fish and other animals." It is inevitable that models and methods developed in one field will be applied in another; in fact, however, the earliest application of the "logistic" or "sigmoid" model was to whales (Hjort et al. 1933). One way in which the applications of "surplus stock" models to whaling and to fishing have differed is that, whereas in the former, yield is expressed as a function of stock size, in the latter, it is treated as a function of the fishing mortality rate. This is really only a matter of style, but it does set the mind on a different track when considering alternative management actions.

[14]This latter theory has often been invoked in fishery discussions by those who wish to be free to exploit the resources off the coasts of other countries which do not have the technical or economic capacity to exploit them.

[15]We should not overlook the fact that values have changed drastically even in the recent history of whaling. It is not many years since European nations valued baleen whales practically only for their oil, or for their oil and baleen plates.

So, the MSY criteria could be judged in terms of its success in restraining interference and thus, hopefully, balancing a variety of future options against some products now. In this respect, it scores fairly well, although maximizing weights rather than numbers, and maximizing sustainable net economic yield would be better. Suggested alternative criteria would need to be judged against these virtues of MSY. If we believe that man has a long future before him and will behave no less wisely in future than now, we should give more weight to future options than to present ones—provided that those uses are not necessary for human survival which, in the case of whales, they clearly are not.

One criticism of the simple stock models is that they are deterministic rather than stochastic. We are concerned always with the evaluation of risks, and risks cannot be expressed deterministically. This is important in the application of models to fish stocks which fluctuate greatly, mainly because of variability in the survival of young, but not so much to animals which bear one or at most two young at a time and care for them, and in which the natural mortality is not mainly caused by predation.[16]

The 1974 IWC "decision" concerns stocks, not species of whales. Although it seems to impose an operational definition of stocks for management purposes, that definition is apparently tautological and the discussion by the Scientific Committee continues to be concerned with the identification of biologically distinct populations and with the degree, time, and place of any mixing between them. Knowledge of these identities and processes is still very scant, and as far as the measurement of SY and the location of MSY are concerned, the practical questions are:

a) What may be wrong with regulating the catch as if from a single stock when, in reality, there are several more or less separate ones?
b) What may be wrong with regulating catches as if from several separate stocks when there is really only one (or there is substantial mixing between several)?

In case a), we can be sure that the yield will only by chance be the sum of the SYs or MSYs of the separate stocks; usually some stocks will be unwittingly "overexploited," others "underexploited." However, at least one compensating mechanism exists in the whaling operations. If, for example, the stocks inhabit different (even if overlapping) areas—as is usually assumed—then the underexploited ones will increase relative to the overexploited ones, presumably become more dense, and thus attract relatively more attention from whalers. Such compensation is, however, limited by the fact that the attention of whalers is not determined solely—or even mainly—by those densities; in particular, they are simultaneously seeking other species of whales. One way of reducing the risks of excessive whaling effort on (unrecognized) stocks is to allocate the total quota by areas; such action has been proposed by the IWC Scientific Committee.

It may be noted that essentially the same problem existed with the BWU in the Antarctic, and was the strongest reason for its eventual abolition. One might ask why the "compensating mechanism" did not operate to protect the blue whale? It probably did, to a certain extent.[17]

---

[16]Few stochastic models have been formulated. Except for the simple case of exponential growth (Mann 1973), they are mathematically complex.

[17]Gulland (1968b) pointed out that, paradoxically, the BWU—at the very time it was most under attack as a conservation measure—did operate to "protect" the fin whale relative to the sei whale because the meat from 3 sei whales was more valuable than the meat from 1 fin whale (BWU = 1 blue = 2 fin = 6 sei = 2½ humpbacks).

However, given a certain biomass density of blue whales in the sea and an equal density of fin and other baleen whales, the whalers who are permitted to take a certain catch (in BWU or total weight) prefer to chase and capture fewer of the large species than more of the smaller ones.[18]

A policy of dividing up a total quota by areas in case these contain different stocks can lead to a different classification of a species, according to the IWC categories. This is true also where there is evidence of separate stocks, but if—as is usually the case—there are no separate measurements of their parameters, so that to all of them are applied the same overall parameter values, then the same species might be classified in one area as "sustained management" and in another area as "protected." But it is likely that in the former area, the species had not been reduced by whaling so far below its initial abundance because it has a lower initial abundance in that area and thus attracted less effort; nevertheless, its small residual number may be less able to sustain a continuing catch than in the other area, and its very existence might even be threatened by continued whaling. This and related problems have led E. Mitchell and others to suggest a rider to the 1974 decision, to the effect that no stock should be reduced to a level at which its existence is threatened or from which it is likely not to be able to recover even if it were no longer exploited. A term suggested for such a level is "minimum allowable stock," defined in absolute, not relative, terms. Unfortunately, no one has a clear idea how such a number could be estimated; it would have to be defined probabilistically and in relation to transient rather than steady-states and, therefore, cannot be regarded as another point on the SY–stock curve. Furthermore, probabilities of extinction or non-recovery surely also depend on population properties other than number. Nevertheless, as an additional criterion—to ensure that management does not restrain the overall impact of whaling while permitting qualitative changes in the ecosystems—it might be worth consideration.[19]

A common approach to empirical determination of the shape of a single-species SY curve is to plot estimates of net increase rate against population, fitting a straight or curved line as the data indicate, or some variant of this procedure. The method depends critically on measures of abundance (or density) derived from catch-per-unit-effort data, except when, rarely, direct measures are available, e.g., from sightings or echosurvey. In handling such data, the attempt is usually made to define "effort" such that it is strictly proportional to the instantaneous fishing mortality coefficient. In particular, trends in efficiency (usually, it is assumed, increasing) of fishing units must be allowed for. These require analyses of technological factors such as vessel size, power, and search and catching equipment, and this has been done, partially, in connection with pelagic whaling. Physical factors such as hours of daylight and weather are also important, at least with respect to the variability of the effective effort, but also, in certain circumstances, with respect to trends. Thus, changes in the latitudinal distribution of Antarctic pelagic whaling and the dates of the season, which are related to changes in hours of daylight and prevailing summer weather, have not yet been adequately taken into account in whale assessments. However, a bigger problem, which has certainly *not* so far been resolved, is how to apply the same total effort data to a number of species caught concurrently, when the abundances and values of individuals of those species show differing trends over the period of

[18]Except when, as in the case of sei and fin whales in the BWU, the weighting factor, having been based originally on relative oil yields, is quite inappropriate when other properties become more highly valued—such as meat quantity and quality. Decline in mean size of intensely exploited species also unbalanced the weighting factors.

[19]See the arguments for protection of blue and humpback whales in Chapman et al. (1964)—Supplementary Report.

study. To the extent that the geographical distributions of the various species differ, the problem is partially resolved by calculating catch per unit effort area by area. Much more could be done if whaling effort figures were split into "searching time" and "capture time," but such information is not available. Even if it were available, the decisions of whalers, which are manifest in performance data, have, over the past 40 years, been modulated in a complex way by changing values of different products and by the existence of IWC regulations (such as protection of certain species) as well as by changes in the relative abundances and sizes of the various species of baleen whales and of the sperm whale. Thus, the calculated trends in abundance of each species, even when all practicable adjustments have been made, must be rather uncertain. This remains true even where calculations take account of whale sightings, especially when, as is usually the case, the sightings used are those reported by whaling expeditions.[20] It may well be that the total catch (in weight) of all species, per unit adjusted effort, is a better indication of whale biomass than the catches per unit effort by species are of species abundances; indeed, Beverton and Holt (1957) suggested that "paradoxically (the 'sigmoid curve method') is perhaps most satisfactory when applied to the behavior of several species at once, or even of an entire community as done by Graham (1935, 1939) and Baerends (1947)." Rough application to the data of total Antarctic baleen catch in weight per unit effort indicates, as might be expected, that these stocks are, as a group, far below the level of MSY (Appendix 5).

The question posed above of managing one stock, believing it to be several, is a special case of a more general one, which is now assuming importance in the IWC: that is, how are catches to be regulated if different species, simultaneously hunted, are biologically interdependent? It is tempting, if one has become used to thinking in terms of steady-states, to construct models for interdependent species or stocks, to give an SY surface of $n + 1$ dimensions, which may have one or more peaks on it. This could readily be done, leaving us only (?) with the problem of estimating the set of perhaps $n(n - 1)$ parameters representing the interdependences. Although it would be interesting to see the results of such a theoretical exercise, I think it unlikely that such a multispecies model would resemble reality. Most interactive models predict cyclic population behavior, as in the simple predator–prey model of Volterra.[21] This is true also if the common prey of two competing predators is treated dynamically, and a three-element model may lead—at best—to predictions of cycles and trends to which the MSY concept could not be applied. This will not help us decide whether to take krill out of the Antarctic or baleen whales, or both and in which proportions, or neither (Bondar and Bobey 1974). Nor will it help us approach sperm whaling rationally in relation to possible intensive exploitation of ocean squids in the future. Nor can such modeling be an adequate response to those who have suggested that perhaps by reducing drastically the number of minke whales in the Antarctic we may facilitate the more rapid recovery, under protection, of one of its competitors, the blue whale. Present knowledge of the structure and behavior of marine ecosystems is so limited that the analytical methods of population dynamics at present give poor guidance for ecosystem management. Recent work on some ecosystem properties such as

[20]A very recent analysis both of sightings data and the division of effort between searching and catching is given by Schweder (1974).

[21]Renewed interest in the problem of estimating the parameters in interactive models is indicated by the publication, as this paper was being revised, by Schwartz and Bremmermann (1975). The authors discuss the best estimates, and their errors, for parameters of a Lotka-Volterra system and logistic growth of four competing species. See also May et al. (1974).

"resilience" and "stability" sustains some hope that such properties will be predictable from studies of the major components or the relations between these (Holling 1973), but we are far from being able to apply such understanding to marine resource management.

Another approach to management of marine populations and ecosystems might be to consider the flows of materials and energy through them. Could we specify a safe limit to the proportion of the total biological production or biomass that may be removed (in terms of current knowledge), with the proviso that the pattern of removal be such as not to change the ecosystem qualitatively? Or can studies of metabolic efficiency help define management objectives? For example, it has been suggested that whales, being warm-blooded, large, and long-lived, may be biologically efficient in the sense of maintaining stability at the expense of great energy consumption for maintenance, but for the same reasons highly inefficient as intermediaries for man's use of the biological productivity of Antarctic waters (Steele 1974). We would need to take into account, however, the energy required to gather that production, which, in the case of whales, comes from within the system and, in our case, at present, from outside it. The provisions and definitions of the US Marine Mammal Protection Act of 1972 (see summary in Appendix 8) have, as yet, no explicit counterpart in international agreements. Some of the provisions do not seem to be incompatible with possible interpretations of some existing conventions, nor with the above suggestions. I suggest the following interim strategy for the formulation of improved goals in the management of wild stocks:

a) accept the concept of MSY stock level, not as a sole objective, but as a lower bound to permissible exploitation, and as part of a multifaceted definition of "conservation"; in setting this bound, take account of the great uncertainty of estimation of any MSY stock level including the various biases noted in the preceding sections;
b) modify the yield definitions in "conservative" directions—e.g., by weight rather than numbers, net yield rather than gross yield;
c) calculate MSY of stocks and species separately and in combination, and select that application which would be most "conservative." (This might lead to the classification of all the Antarctic baleen whales as "a protection stock" in terms of their present combined biomass as a fraction of their "initial" biomass);
d) test the assumption of initial steady-states and apply an MSY concept appropriately modified in accordance with indications of natural trends and cycles;
e) examine critically the application of MSY in each case with respect to possible or likely options of future uses and values of the resource; predict, as well as present data and methods permit, relevant technical, economic, and social trends;
f) examine each application with respect to the problems of providing adequate scientific advice according to the guidelines given by ACMRR (FAO 1974);
g) in connection with f) above, link management for high sustainable yields with the adequate financing of research, not only on the populations being exploited, but on the ecosystems of which they are part, as well as on the properties of whales, other than those for which they are, at present, most valued. This would ensure that present beneficiaries are obliged to put back some of their benefits not only to improve present use patterns, but also to give scientific information relevant to future options.

It may be difficult, if not impossible, to get general agreement on some of the modifications of the MSY concept outlined above, particularly the criterion of net yield rather than gross yield. Operators and states have been unwilling to make public, for international use, any economic data such as statistics of costs pertinent to this matter. This problem is tied up with the question of resource "ownership"; if there were a single corporate "owner" (custodian or trustee) of the whale resources, some such difficulties could be resolved (Gulland 1968c). Furthermore, nations and peoples who do not, at present, exploit the resources and those that do not now consume the products, will presumably give a relatively higher value to future options than do the present exploiters. Their interest in the resource should, therefore, be supported since, in a sense, they might act as agents, in the present, on behalf of the future generations both of man and whales.[22] This is one reason why the present UN Conference on the Law of the Sea, in which practically all nations are participating, should be a valid forum for the reconsideration of the objectives of conservation and management. So far, in the preparatory work, and in the first session, attention has been focused almost entirely on the question of who gets the benefits, and this is seen as a matter of jurisdiction. Equitable allocation is certainly a necessary, though not sufficient, conditions for realizing long-term benefits from living marine resources.[23] Jurisdiction, research, international advice, and apparently reasonable conservation measures did not save the Peruvian anchoveta fishery from collapse. Antarctic whaling is regulated, national allocation of catches are made, and competition is regulated within participating nations as well as between them. The resources have been "appropriated"—if only temporarily by present participants—that is, in effect, if not formally by the elimination of other competitors, by possession of the "veto" in the IWC, and by the reduction of stocks to levels so low that it is in the interest of no one else to start whaling. All this has not assured a future for that industry.

Perhaps the problem of whale management—even of marine fishery management generally—is one of the subset of "insoluble" problems identified by Crowe (1969). There are problems relating to "the commons" concerning which natural scientists "pass technically insoluble problems over to the political and social realm for solution" and conversely, social scientists, seeing no political solutions, defer to future technical advances. Nevertheless, in attempting to reformulate international conservation objectives, fishery scientists are acting in the hope that the political problems will not, in the long run, prove completely insoluble. I doubt, however, if they can construct viable alternative objectives without incorporating some of the findings of other disciplines—some such as general ecology, close to their own; others far removed.

---

[22]Cetologists also are potentially a powerful lobby for the future, at least when the continuation of their research does not depend indirectly or directly on the present whaling industry for funding or materials, or both. Research now needs to be funded with future values in mind; it is difficult to see how to ensure that without substantial funding from international sources (as well as from those who have profited from past and present exploitation) both for an expanded program and for continuing some current activities.

[23]"There are several identifiable valid objectives or goals of international fisheries management, including optimization of biological, economic, and social yield. None of these goals can be achieved, however, unless there exists a jurisdictional base for management permitting authorities to regulate the disposal of resources." So "the overall goal should be to maximize opportunities of individual states ... to achieve benefits from living marine resources with respect to which they have exclusive or shared management rights" (ASIL 1974). No hint of a "common heritage" here, but I suppose such "management rights" as are shared by all states would, collectively, form that agreed principle. Unfortunately, the valuable study quoted, drawn up by a group composed, it seems, exclusively of lawyers and economists, reveals little awareness of the subtleties of marine ecology and the uncertainties and difficulties of stock predictions. It recognizes, however, that present research and scientific training efforts are quite inadequate to the demands arising from modern fishing activities.

It is perhaps appropriate to close this paper with two quotations. Bradley (1973) attempted to "evolve from relevant insights in behavioral theory, economics, and political studies a regulatory system similar to the systematic interrelations that exist in the ecosystems of the natural environment." He observes that: "In the past, the belief persisted that an increased technological capacity to alter the environment brought increased control. This belief, far from dead, is a manifest delusion. First, the difficulty in devising any physical control system lies not only with generating enough power but also with generating enough of the proper kinds of information. Since the environmental resource system is a system, any change in a given parameter will have numerous, unpredictable repercussions throughout the system; so even if the effect of the intervention is to bring under control the variable directly affected, the total system is likely to be less predictable than before, while all the learned skills based on the former given are depreciated. Further, interventions and the further interventions to which their unexpected results tend to lead, are likely to be self-multiplying. The rate of change increases at an accelerated speed, without a corresponding acceleration in the rate at which further responses can be made; and this bring ever nearer the threshold beyond which control is lost."

David Bella has expressed more briefly a related idea (Bella and Overton 1972; Bella 1974): "We are faced with an *environmental predicament*; our ability to modify the environment increases faster than our ability to foresee the effects of our activities," and further, "the most significant aspect of this predicament may be our inability to foresee our own reactions to the actions that we are capable of taking."

Acknowledgements

Brian Rothschild, John Gulland, Geoffrey Kesteven, Saul Saila, Ray Dasmann, Basil Parrish, and Victor Scheffer, among others, kindly commented on earlier drafts of this paper. The views expressed here are, of course, my own.

References

ACMRR (FAO). 1974. The scientific advisory function in international fishery management and development bodies. Supplement I to the Report of the Seventh Session of the Advisory Committee on Marine Resources Research. FAO Fisheries Report, 142(Supplement I).

Allen, K. R. 1974. Some properties of the Schaefer model for baleen whale populations. Paper presented to the IWC Scientific Committee, December 1974. Working Paper 2.

American Society of International Law (ASIL). 1974. Principles for a global fisheries management regime. Studies in Transnational Legal Policy 4.

Baerends, G. P. 1947. De rationeele exploitatie van den zeevischstand, in het bijzonder van den vischstand van de Noordzee. Visscherij en Voldselvoorz Verslagen en Mededeelingen van de Afdieling Visscherijen No. 36.

Bell, F. W., D. A. Nash, E. W. Carlson, F. V. Waugh, R. K. Kinoshita, and R. F. Fullenbaum. 1975. A world model of living marine resources. Pages 291–323 *in* W. C. Labys, editor. Quantitative models of commodity markets. Ballinger Publishing Company, Cambridge, Massachusetts.

Bella, D. A. 1974. Fundamentals of comprehensive environmental planning. Engineering Issues: Journal of Professional Activities 100(E.11):17–35.

Bella, D. A., and W. S. Overton. 1972. Environmental planning and ecological possibilities. Journal of Sanitary Engineering Division ASCE 98(SA3):579–592.

Beverton, R. J. H., and S. J. Holt. 1956. The theory of fishing. Pages 372–441 *in* M. Graham, editor. Sea fisheries: their investigation in the United Kingdom. Edward Arnold, London.

Beverton, R. J. H., and S. J. Holt. 1957. On the dynamics of exploited fish populations. Her Majesty's Stationery Office, UK.

Beverton, R. J. H., and S. J. Holt. 1966. Manual of methods for fish stock assessment. Part II. Tables of yield functions. FAO Fisheries Technical Paper 38 (Revision 1).

Bondar, B., and P. J. Bobey. 1974. Should we eat krill? Ecologist 4(7):265–266.

Bradley, M. D. 1972. Analyzing human and general ecosystems. Environmental Affairs 2(2):303–313.

Bradley, M. D. 1973. Decision-making for environmental resources management. Journal of Environmental Management 1:289–302.

Chapman, D. G., K. R. Allen, and S. J. Holt. 1962. First Interim Report of the Special Committee of Three. Reports of the International Whaling Commission 13:36–38.

Chapman, D. G., K. R. Allen, and S. J. Holt. 1964. Special Committee of Three Scientists Final Report. Reports of the International Whaling Commission 14:40–106.

Chapman, D. G., K. R. Allen, S. J. Holt, and J. A. Gulland. 1965. Report of the Committee of Four Scientists. Reports of the International Whaling Commission 15:47–50 (see especially pp. 49–50).

Clark, C. W. 1973a. The economics of overexploitation. Science 181:630–634.

Clark, C. W. 1973b. When should whaling resume? MS. 23 pp.

Clark, C. W. 1974. Fisheries management: maximum sustained yield vs. optimal sustained yield. MS. 28 pp.

Cook, L. M. 1965. Oscillation in the simple logistic growth model. Nature 207:316.

Crowe, B. L. 1969. The tragedy of the commons revisited. Science 166(909):1103–1107.

Cushing, D. H. 1974. A link between science and management in fisheries. Fishery Bulletin US 72(4):859–864.

Doi, T. 1973. The theoretical treatment of the reproductive relationship between recruitment and adult stock. Rapports et Procès-Verbaux des Réunions du Conseil Permanent International pour l'Exploration de la Mer 164:341–349.

Fox, W. W. 1970. An exponential yield model for optimizing exploited fish populations. Transactions of the American Fisheries Society 99:80–88.

Gambell, R. 1974a. The unendangered whale. Nature 250:454–455.

Gambell, R. 1974b. A review of reproduction parameters and their density-dependent relationship in southern hemisphere sei whales. Paper presented to the Special Meeting of IWC Scientific Committee, December 1974, Working Paper 9.

Gordon, H. S. 1953. An economic approach to the optimum utilization of fishery resources. Journal of the Fisheries Research Board of Canada 10(7):442–457.

Graham, M. 1935. The modern theory of exploiting a fishery and application to North Sea trawling. Journal du Conseil Permanent International pour l'Exploration de la Mer 10:264–274.

Graham, M. 1939. The sigmoid curve and the over-fishing problem. Rapports et Procès-Verbaux des Réunions du Conseil Permanent International pour l'Exploration de la Mer 108(1):57–66.

Gulland, J. A. 1968a. The concept of the maximum sustainable yield and fishery management. FAO Fisheries Technical Paper 70.

Gulland, J. A. 1968b. The concept of marginal yield from exploited fish stocks. Journal du Conseil Permanent International pour l'Exploration de la Mer 32(2):256–261.

Gulland, J. A. 1968c. The management of Antarctic whaling resources. Journal du Conseil Permanent International pour l'Exploration de la Mer 31(3):330–341.

Gulland, J. A. 1971. Science and fishery management. Journal du Conseil International pour l'Exploration de la Mer 33(3):471–477.

Gulland, J. A., and L. K. Boerema. 1973. Scientific advice on catch levels. Fishery Bulletin US 71(2):325–335.

Hjort, J., G. Jahn, and P. Ottestad. 1933. The optimum catch. Hvalrådets Skrifter 7:92–127.
Holling, C. S. 1973. Resilience and stability of ecological systems. Annual Review of Ecology, Evolution, and Systematics 4:1–23.
Holt, S. J. 1962. The application of comparative population studies to fishery biology. Pages 51–71 *in* E. D. Le Cren, and M. W. Holdgate, editors. The exploitation of natural animal populations; a symposium of the British Ecological Society, Durham, 28th–31st March 1960. Blackwell Scientific Publications, Oxford.
Holt, S. J. 1974a. Whales: conserving a resource. Nature 251:366–367.
Holt, S. J. 1974b. The maximum sustainable yield in weight and related matters. Paper presented to the Special Meeting of IWC Scientific Committee, December 1974, Working Paper 12.
Holt, S. J., et al. 1974. Report of the Working Group on Biometrics and Management. Pages 14–20 *in* W. E. Schevill, editor. The whale problem: a status report. Harvard University Press, Cambridge, MA.
Kraus, E. 1974. The unpredictable environment. New Scientist 63.
Mann, S. H. 1973. On the optimal size for exploited natural animal populations. Operations Research 21(3):672–676.
Marchesseault, G. D., S. Saila, and W. J. Palm. 1976. Delayed recruitment models and their application to the American lobster fishery. Journal of the Fisheries Research Board of Canada 33(8):1779–1787.
May, R. M. 1973. Stability and complexity in model ecosystems. Princeton University Press, Princeton, New Jersey.
May, R. M., G. R. Conway, M. P. Hassell, and T. R. E. Southwood. 1974. Time delays, density-dependence and single-species oscillations. Journal of Animal Ecology 43:747–770.
McIntyre, J. M. Editor. 1974. Mind in the waters. Charles Scribner's Sons, New York.
Ohsumi, S. 1974. Conversion of sustainable yield curve by numbers into that by weight in the fin and sperm whale populations. Paper presented to the Special Meeting of IWC Scientific Committee, December 1974, Working Paper 24.
Pella, J. J., and P. K. Tomlinson. 1969. A generalized stock production model. Bulletin of the Inter-American Tropical Tuna Commission 14:421–496.
Rothschild, B. J. 1971. A system review of fishery management with some notes on the tuna fisheries. FAO Fisheries Technical Paper 106.
Schevill, W. E. Editor. 1974 The whale problem: a status report. Harvard University Press, Cambridge, Massachusetts.
Schweder, T. 1974. Transformation of point processes: applications to animal sighting and catch problems with special emphasis on whales. Doctoral dissertation, University of California, Berkeley, California.
Steele, J. H. 1974. The structure of marine ecosystems. Blackwell Scientific Publications, Oxford.
Swartz, J., and H. Bremmermann. 1975. Discussion of parameter estimation in biological modelling: algorithms for estimation and evaluation of the estimates. Journal of Mathematical Biology 1:241–257.
Talbot, L. 1974. The great whales and the International Whaling Commission. Pages 232–236 *in* J. M. McIntyre, editor. Mind in the waters. Charles Scribner's Sons, New York.
Tillman, M. F. 1974. Re-examination of population estimates for the North Pacific sei whale. Document 29 to IWC Scientific Committee Meeting, La Jolla, December 1974.
Vamplen, W. 1972. The evolution of international whaling controls. Maritime History 2(2):123–139.
Wada, S. 1975. Report of the Scientific Committee, Annex L: indices of abundance of large-sized whales in the North Pacific in the 1973 whaling season. Reports of the International Whaling Commission 25, 129–165.
Woodwell, G. M. 1974. Biotic energy flows. Science 183:367.

Afterthought

In this paper, I have criticized the theory and practice of fishery science as applied to conservation of living marine resources, especially whales. My intention was to stimulate discussion among colleagues, especially at the "MSY Workshops" at Airlie, Virginia, for which the paper was primarily written. I felt justified in taking a critical position because I have been involved in, and a party to, the formulation of scientific advice which seems, in retrospect, not always as good as it might have been.

However, it must be said that "scientific advisers," especially at the international level, labor under considerable disadvantages. They act as detectives, not only of the properties of natural systems—if that were not difficult enough—but also of the behavior and intentions of the industry they are trying to assist.

Information about the economic and other considerations which determine the tactical deployment of pelagic whaling effort, for example, would reveal the meanings of abundance indices which are now obscure. Many of the mathematical contortions in which we engage in efforts to understand the dynamics of natural populations and of fisheries are necessary, not so much because of nature's secrecy, but because of man's. Scientific advisers are too often expected, collectively, to offer economically and socially meaningful advice, while denied access to the kinds of information on which operational decisions are based.

Twelve years after the desirability of such information was formally recognized by the IWC, its Scientific Committee could only report in 1975 lamely, "we have no information on this aspect at present," referring to economic criteria which might be applied to determine an "optimum" management policy.

Uncertainty lies in this area as much as in our understanding of ecosystems, yet it need not. Now, however, the rules of the game do not favor the future.

In revising this paper for publication, I have taken into account some of the points raised in the Airlie Workshops.

Appendix 1

Relations between the stock levels and fishing (whaling) effort for MSY and those for maximum net yield, "commercial" extinction, and "biological" extinction

In the simplest model, the yield curve is a symmetrical parabola: $Y = 4P(1 - P)$ (Figure 1), where P is stock size as a fraction of initial size ($P_{MSY} = 0.5$) and Y is sustainable yield as a fraction of MSY.

Define cost of whaling to maintain a given P as proportional to F, the fishing mortality coefficient, which is, in turn, proportional to $(1 - P)$. Write net yield as $y = Y - a(1 - P)$, where a is the ratio of the cost of exerting a unit of effort to the value of a unit of catch. The curve of net yield against stock size has a maximum at $P_1 = 0.5 + a/8$.

Define the stock level for zero net yield to which the unregulated industry will "naturally" tend ("commercial extinction") so that the yield at the point $Y_0 = a(1 - P_0)$, but $Y_0 = 4P_0(1 - P_0)$ so that $a = 4P_0$. Hence, the curve of net yield is a squeezed and displaced symmetrical parabola (Figure 1) with maximum at $P_1 = (1 + P_0)/2$ and a net yield at that level of $y_{max} = 1 - P_0$. The ratio $y_{max}/y_{MSY} = (1 - P_0)^2/(1 - 2P_0)$. This ratio is 1.067 for $P_0 = 0.2$, for example; that is, the maximum net yield would be nearly 7% greater than the net yield at MSY. The yield at the level of y_{max} is $Y = 1 - P_0^2 = 0.96$ in the example, so, in this case, if one were prepared to accept 4% less than the MSY, one could gain 7% in net yield. This would be obtained with only $(1 - P_0)$ of the effort required for MSY, i.e., 20% less if $P_0 = 0.2$ (see also below).

If the catch per unit effort is proportional to P, then we have the following relation: (CPUE at y_{max})/(CPUE at *MSY*) = $1 + P_0$ and (CPUE at y_{max})/(CPUE at y_0) = $[1 + (1/(P_0)]$. In the example of Figure 1, the gain in CPUE by moving from the MSY point to the y_{max} point is 20%.

Similar relations are found for models in which the curve of sustainable yield against stock size is asymmetrical (see below). In general, $P_{MSY} \leq P_1 \leq (1 - P_0)P_{MSY} + P_0$. That is, P_1 is always greater than P_{MSY} by an amount somewhat less than $100 \times P_0(1/P_{MSY} - 1)$%. Figure 2 illustrates a situation in which P_{MSY} is displaced 16% to the right.

If we treat, for example, all Antarctic baleen whales as a single "stock" for which there is a unique sustainable yield curve, and we know that whaling continues to be profitable at a certain stock level, we can readily estimate an upper limit for the amount by which the stock for maximum net yield would be above that for MSY (see Appendix 5).

We may further deduce what stock level, $P_2 > P_1$, will give an SY such that the *net* yield is equal to that which would be obtained at MSY level. For the symmetrical parabola, this is $0.5 + P_0$. In our example, $0.5 + P_0 = 0.7$, i.e., 40% higher than at MSY. The SY at this stock level, Y_2, is $(1 - 2P_0)(1 + 2P_0)$, in the example 0.84, i.e., 16% less than MSY, but the catch per unit effort is 40% higher.

In the skew yield curve of Figure 2, P_1 is 15% higher than P_{MSY} and gives a net yield 2.5% higher than at MSY. The stock that will give the same net yield as at MSY is 23% higher than P_{MSY}; the gross yield at that level is 0.91 of MSY, i.e., 9% less.

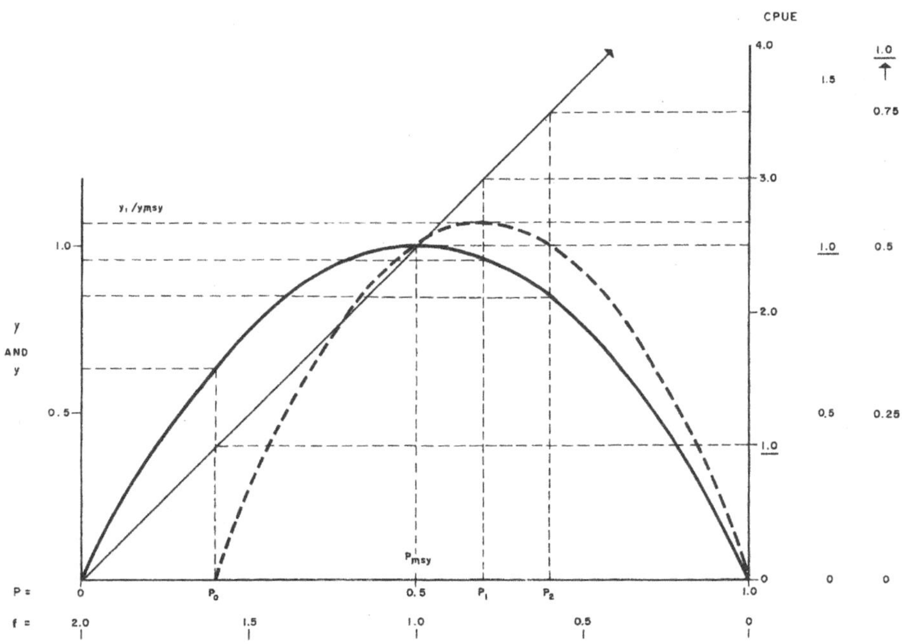

Figure 1.

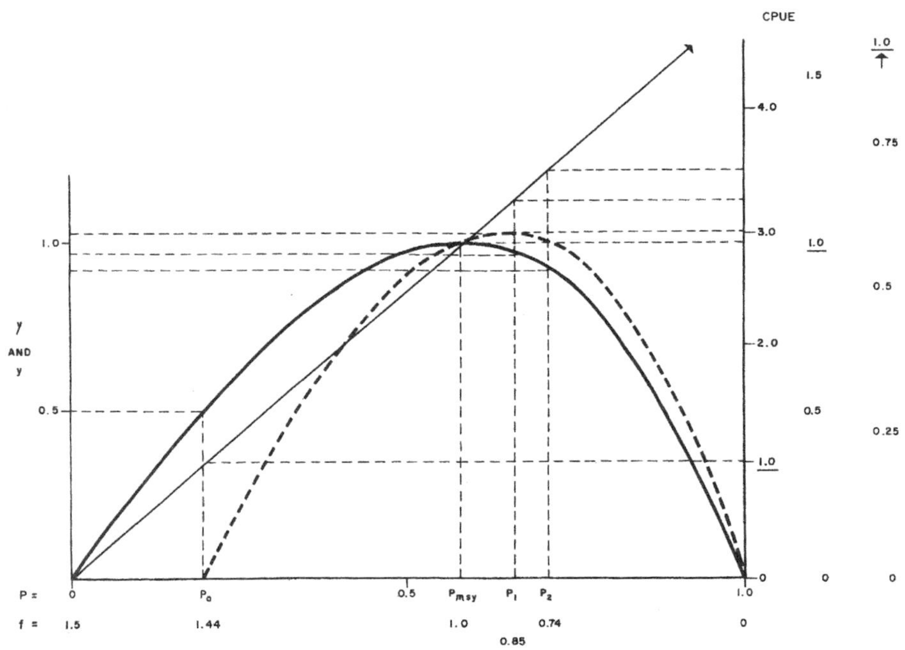

Figure 2.

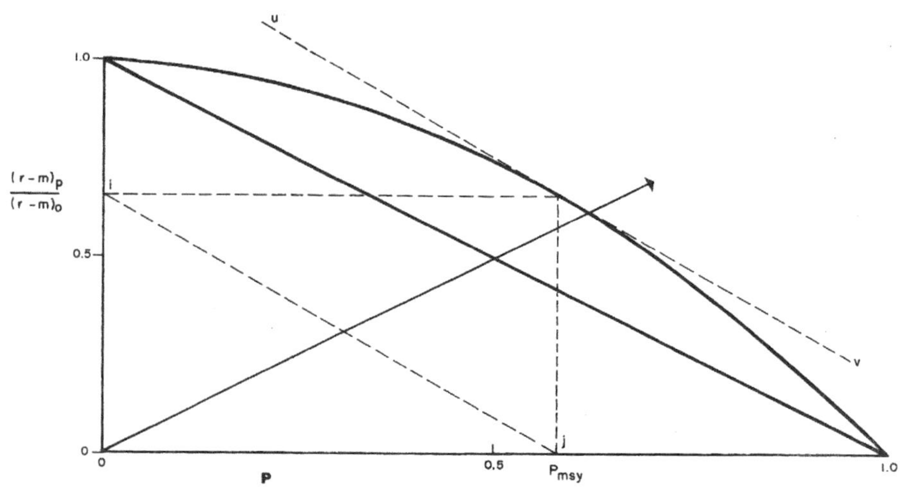

Figure 2a.

Lastly, note that for the symmetrical parabola, the SY level at the threshold of "commercial extinction" is $4P_0(1 - P_0)$ of MSY; in the example, 64% of MSY. For the skew curve, it is less (50% of MSY).

Assume the curve of SY against P can be represented by the expression: $Y = P(1 - P^{n-1})/b$ (e.g., Pella and Tomlinson 1969). The effect of skewness on deductions concerning the relation of y_{max} to MSY can then be examined. In the above equation, as n increases above 2, the SY curve becomes skewed increasingly to the right. This curve has a maximum at $P_{MSY} = (1/n)^{(1/n-1)}$ and putting, as before $Y_{MSY} = 1$, we find $b = P_{MSY} - P_{MSY} = P_{MSY}[(n-1)/n] = (n-1)/n^{n/(n-1)}$.

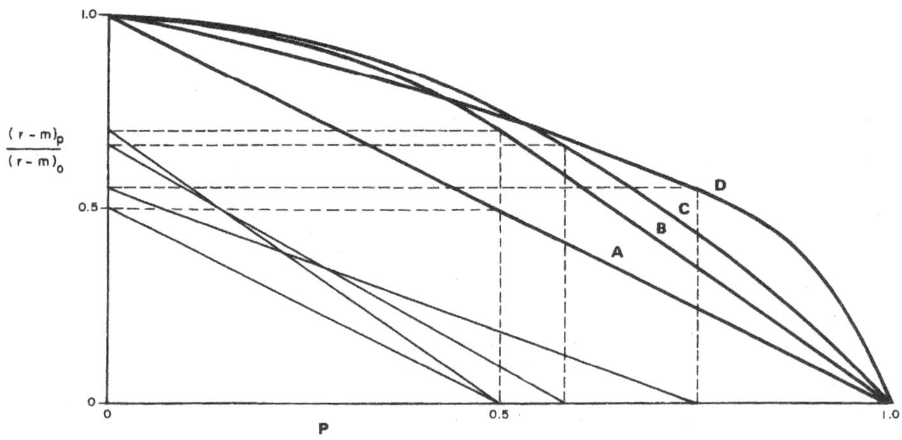

Figure 2b.

Now, writing $y = Y - P_0F$ gives $y = (P - P_0)(1 - P^{n-1}) / b$. Define $\mathring{y} = y / y_{MSY} = (P - P_0)(1 - P^{n-1})/(P_{MSY} - P_0)(1 - P_{MSY}^{\,n-1})$. $\mathring{y}$ has a maximum at P_1, given by the expression: $nP_1^{\,n-1} = P_0(n-1)P_1^{\,n-2} + 1$. For the case $n = 3$, we find $P_1 = (P_0 + \sqrt{P_0^2 + 3})/3$ from which $P_1/P_{MSY} = (P_0 + \sqrt{P_0^2 + 3}) / \sqrt{3} \cong (1 + P_0)/\sqrt{3}$ and $\mathring{y}_{max} = [3(P_1 - P_0)(1 - P_1^2)/2(1/\sqrt{3}P_0)$.

The following table summarizes the properties of the model for $n = 3$ ($A = +0.15$) for various values of P_0.

P_0	P_{MSY}	P_1	P_2	Y_1	Y_2	$\mathring{y}_{max}$
0.1	0.577	0.612	0.645	0.994	0.979	1.006
0.2	0.577	0.648	0.716	0.977	0.907	1.033
0.3	0.577	0.686	0.788	0.944	0.776	1.105
0.4	0.577	0.726	0.863	0.892	0.572	1.304
0.5	0.577	0.768	0.939	0.818	0.288	2.132

Numerical solutions for a range of other values of n and P_0 are shown graphically in Figure 3. A linear scale is used for an index of skewness: $A = 2P_{MSY} - 1$. This index is zero for $n = 2$ and has a range $0 \rightarrow 1$ for $n = 2 \rightarrow \infty$ over the calculated range $P_1 \cong P_{MSY} + P_0(1 - P_{MSY})$. Suppose, for example, MSY is thought to be obtainable from 0.7 of the initial stock; the curve for $n = 6$ would be appropriate. Then, for $P_0 = 0.2$, we find $P_1 = 0.74$, that is, 5% higher than P_{MSY}. At that stock level, the sustainable yield is 0.990, i.e., 1% less than MSY, and $\mathring{y}_{max} = 1.01$. Lastly, for $\mathring{y} = 1$, we find $P_1 = 0.77$ (10% higher than P_{MSY}) and $Y = 0.964$.

We now consider the changes in whaling *effort*, corresponding with movements along the *SY* and *Y* curves. If $Y = FP$, then $F = \frac{1 - P^{n-1}}{b}$ and specifically $F_{MSY} = \frac{1 - P_{MSY}^{\,n-1}}{b}$. It is convenient to define effort f as the fraction or multiple of the effort which is required to sustain MSY. So, $f = F/F_{MSY}$. In the context of this paper, three critical values of f may be calculated: $f_e \neq 0$ leading to $Y = 0$, i.e., extinction of the stock; f_0 ($\leq f_e$) leading to $y = 0$, i.e., "commercial extinction"; and f_1 leading to y_{max}. From these definitions, we have: $f_e = 1/(1 - P_{MSY}^{\,n-1}) = n/(n-1)$ from which $P_{MSY} = (f_e - 1/f_e)^{f_e - 1}$, $f_0 = f_e(1 - P_0^{\,n-1})$, and $f_1 = f_e(1 - P_1^{\,n-1}) = n/(n-1)(1 - P_1^{\,n-1})$. Values of these are plotted in Figure 4. It will be seen that, theoretically, stock extinction corresponds with an effort double that giving MSY or less. For the moderately

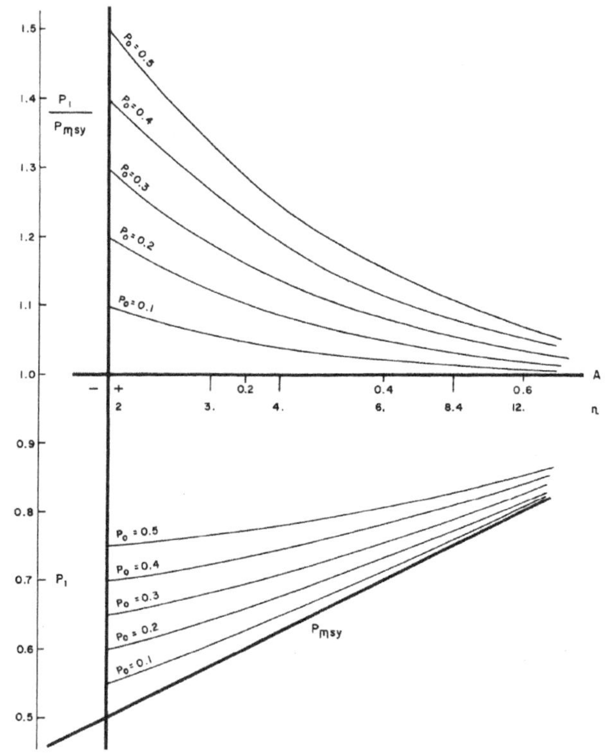

Figure 3.

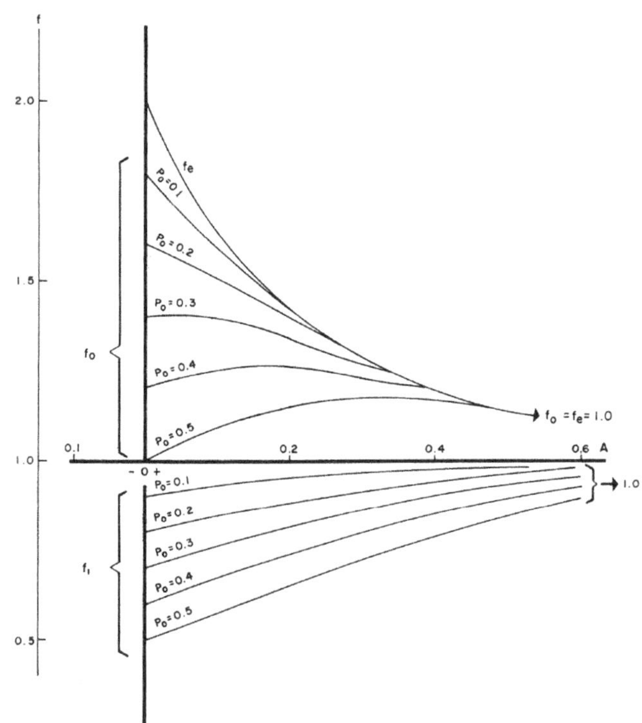

Figure 4.

skewed SY curve ($n = 3$), extinction occurs with an effort 50% greater than for MSY.[24] This is not to suggest that we may place any reliability on this deterministic model for predicting what is essentially a probability; the examples are given to show that if the curve of yield against stock size is skewed to the right, then with relatively little increase in effort beyond what is needed to obtain MSY, the stock will decline rather precipitously.

"Commercial extinction" occurs, of course, at lower f values; how much lower is shown for four values of P_0. Figure 4 also shows for these P_0 values the relative effort f_1 required to maximize net yield.

For $n < 2$, the yield curves are negatively skewed, as appears to be the case for some fish stocks. For $n = 1$, the expression becomes the derivative of the well-known Gompertz growth curve (Fox 1970) and has a maximum at $P = 1 / e = 0.368$.

I will not attempt to examine here the implications of departures from the simple assumptions that value and costs are proportional, respectively, to catch and effort. One direction of bias is, however, worth noting, as it may apply to any practically *unique* product from whales, e.g., sperm oil. In such a case, we expect the price to vary inversely with the production. Then we find (a) y_{max} is given by higher stock levels than would be deduced from the constant price model; and (b) "commercial extinction" is delayed as effort increases and $f_e \rightarrow f_o$.

Appendix 2

Decision of the 1974 annual meeting of IWC

"The International Whaling Commission

Noting that whale stocks are a common concern to mankind

Concerned that some species of great whales are at present considerably depleted below their optimum population levels

Recalling that the historic decline in whale populations occurred not only because of excessive exploitation, but also because knowledge was inadequate to protect the species

Motivated by the need to preserve and enhance whale stocks as a resource for future use when food needs of the world will be greater because of increased human population and by the need to maintain marine ecosystems in a well-balanced condition capable of high productivity

Taking into consideration the long-range interests of the consumers of whale products and of the whaling industry as cited in Article V.2 of the Convention, and

Recognizing that the management of whale stocks should be based not only on the concepts of maximum sustainable yield in numbers by species, but should also include such considerations as total weight of whales and interactions between species in the marine ecosystem,

---

[24]But a model which gives a finite SY for $F \rightarrow \infty$ is unrealistic, e.g., Bell et al. (1975).

Decides that

1. It shall classify all stocks of whales into one of three categories according to the advice of the Scientific Committee.
    a) *Initial management stocks* which may be reduced in a controlled manner to achieve MSY levels or optimum levels as these are determined;
    b) *Sustained management stocks* which should be maintained at or near MSY levels and then at optimum levels as these are determined;
    c) *Protection stocks*, which are below the level of 'sustained management stocks' as described in b), which should be fully protected.
2. The Committee should define stocks for this purpose, as the units which can be most effectively managed individually.
    a) Commercial whaling shall be permitted on 'Initial Management Stocks' subject to the advice of the Scientific Committee as to measures necessary to bring the stocks to the MSY level and then optimum level in an efficient manner and without risk of reducing them below this level.
    b) Commercial whaling shall be permitted on 'Sustained Management Stocks' subject to the advice of the Scientific Committee.
    c) There shall be no commercial whaling on species or stocks classified as 'Protection Stocks' including those species listed for full protection in the current schedule.
3. *Also decides* to implement this resolution by
    a) Requesting the Scientific Committee to provide advice on the criteria which should be used in defining categories of whale stocks which should be treated as in Section 1 above, this advice to be provided as soon as possible with a view to its incorporation in the Schedule.
    b) Directing the Scientific Committee to arrange to provide the Commission with annually updated advice on these criteria and on the allocation of stocks to the categories.
    c) Making all necessary amendments to the Schedule not later than the 27th meeting of the Commission."

Appendix 3

(From background paper to Airlie Workshop)

Number or weights?

From the tables of yield functions of Beverton and Holt (1966) may be derived the mean weights of individuals in the catch, as a function of the size at recruitment (relative to final size) of the exploitation rate E and of the ratio of the natural mortality coefficient to the growth parameter K. These mean weights can be expressed relative to, for example, the initial mean weight (i.e., for $E \to 0$) or the mean weight for a value of E corresponding with the effort required to give MSY in accordance with the logistic or other model. In general, the former are more useful, since the relative decline in mean size can readily be observed as exploitation intensifies; for our present purpose, the latter is, however, preferable.

Taking the Antarctic fin whale as an example, parameter values used by Allen (1974) are $M = 0.04$, $K = 0.26$ from which $M/K = 0.15$, and C (ratio of length at recruitment to final length) = 0.8. The required values of mean weights may be obtained by interpolating in the tables for $M/K = 0.25$ and $M/K = 0$ and are summarized below:

P	Y	$\overline{W}$	$Y \times \overline{W}$	f
0.05	1.0000	1.0000	1.0000	1.00
0.51	0.9996	1.0022	1.0018	0.98
0.52	0.9984	1.0045	1.0029	0.95
0.53	0.9964	1.0067	1.0031	0.94
0.54	0.9936	1.0089	1.0025	0.92
0.55	0.9900	1.0112	1.0011	0.90
0.56	0.9856	1.0134	0.9988	0.88
0.58	0.9744	1.0179	0.9918	0.84
0.58	0.8704	1.0303	0.8968	0.64

So, although there would not be much gain in sustainable yield of products by maximizing weight rather than numbers caught (less than 1%), the weight maximum could be obtained with 6% less effort than MSY (number). Further, the same total weight as that obtained from MSY (number) could be sustained with 12% less effort. Calculation shows that a sustainable catch of only 1% less than MSY (weight) could be taken with 16% less effort than needed to take MSY (number); and 10% less with 36% less effort. With further decreasing effort, the catch falls rapidly. So, if fin whales were considered in isolation, the net economic yield (discount rate zero) would probably be obtained in the region of $P = 0.7, f = 0.6$.

The above calculations have been made assuming, as Allen (1974) did, that the curve of sustained yield in numbers is symmetrical. If $P_{MSY} = 0.6$ as Ohsumi (1974) assumed ($n = 3.6$), the same calculations can be made and the estimates of relative losses and benefits are of the same order. The SY (weight) curve is not only skewed slightly to the right of the SY (number) curve, but it is slightly flatter in the region of its maximum, so that one could afford to keep the stock relatively rather large without risk of forfeiting catch.

This suggests that apart from permitting depleted stocks to recover more than the MSY (number) policy would allow, it would be wise, from both biological and economic points of view, if any "controlled reduction" of "initial management stocks" of, e.g., minke whales, were limited to, say, 75% of the initial stock size. But even the MSY (weight) of minke whales may not be enough to pay the costs of catching them, if they alone are caught. Thus, it would be better not to reduce these stocks, at least until the stocks of depleted species have had a chance to recover. Otherwise, there is bound to be a virtually overwhelming economic incentive to take catches increasingly exceeding the sustainable yield as the stock density declines, with serious danger of "overshooting" even the MSY (number) level.

Any steady-state mean weight will, if exploitation is unselective above the minimum legal size, be reached only after a transient period. In whaling, however, it appears there is, at least initially, some selection in favour of larger individuals. These two factors work in opposite directions, but do not necessarily "cancel out." Bearing in mind these uncertainties, the changes in mean weight, which the yield tables predict in relation to the rate of exploitation, are of the same order of magnitude as those observed in the Antarctic baleen whales. The male sperm whales of the Southern Hemisphere seem, however, to have declined in mean size more than would be expected, but no more than could be accounted for by selective capture of larger individuals.

Appendix 4

In the process of assessing whale stocks, as with fish, one examines "reproductive relationships." By this is usually meant the determination of a curve relating the total numbers of animals, or of the mature animals, to the numbers of recruits derived from them; this is referred to in the literature of fishery science as the "stock–recruitment problem." Since the assessment of whale stocks is mainly a matter of estimating and predicting numbers, a knowledge of the reproductive relationship in each species and stock should permit calculation of limiting values of sustainable yield and even the order of magnitude of MSY, provided we are content to assume that the natural mortality is not also density-dependent. What is not generally realized is how precise this knowledge would need to be for predicting the MSY stock level with useful accuracy. Thus, very closely similar stock–recruit curves can lead to widely differing conclusions as to the level of stock, relative to "initial" stock size, which will give MSY. This is illustrated in Figure 5a–b.

In Figure 5a are shown two stock–recruit curves. The scales and parameters of these are adjusted so that the number of recruits each predicts for the unexploited stock ($P = 1$) and a stock half that level ($P = 0.5$) are the same; recruitment from $P = 0.5$ is assumed to be three-quarters of that from $P = 1$. Curve A represents the expression $1/R = a + b/P$ which has been commonly discussed in the literature (e.g., the Report of the IWC Committee of Four Scientists). It is asymptotic to a relative value $R = 1.5$.

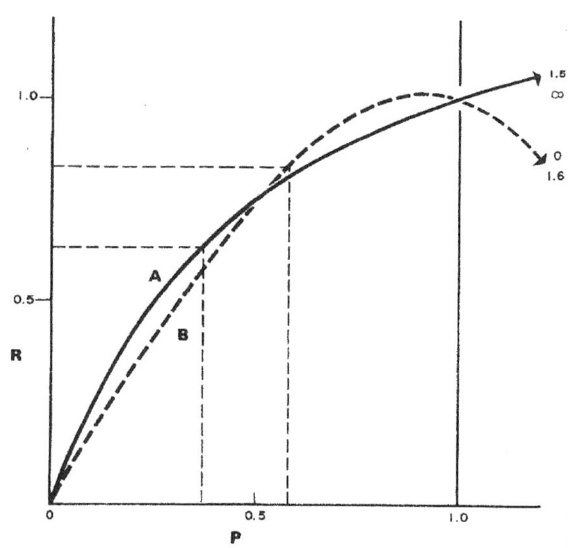

Figure 5a.

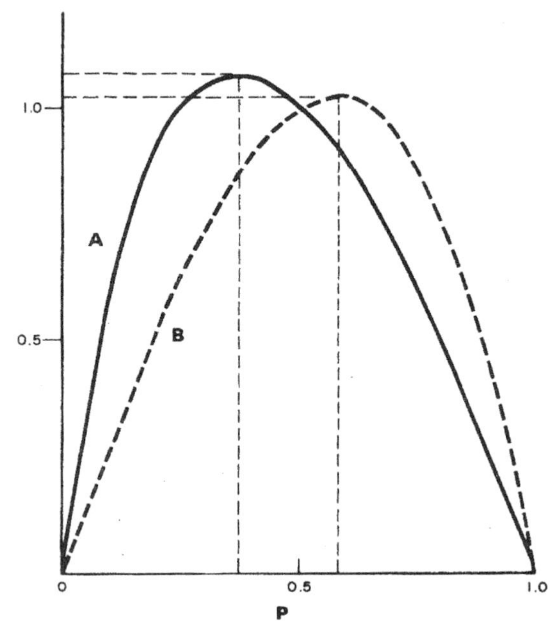

Figure 5b.

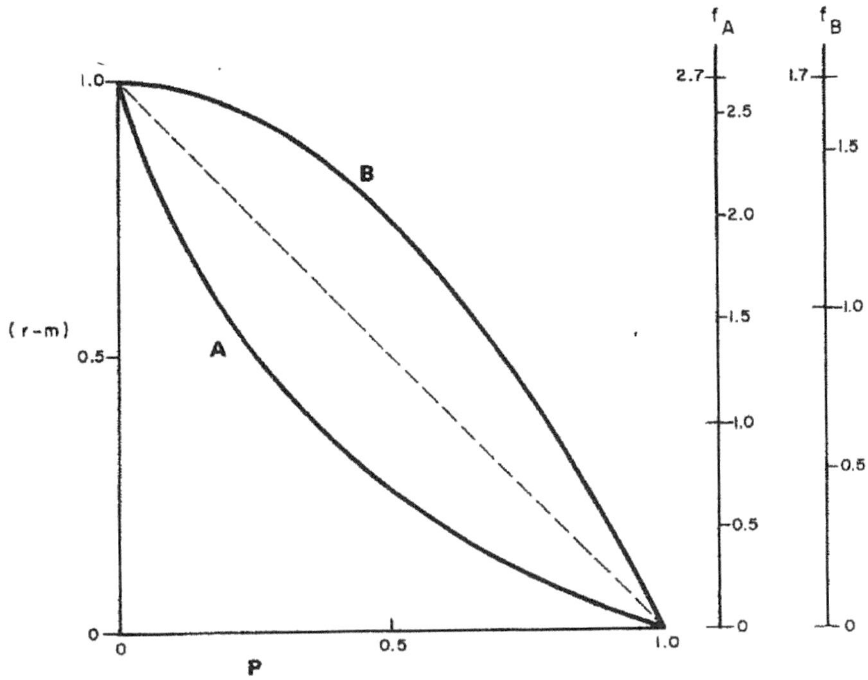

Figure 5c.

Curve B represents the expression $R = P(1.66 - 0.66P^2)$. This is the function which may be deduced from the yield model illustrated in Figure 2 (i.e., $n = 3$) if the natural mortality coefficient is assumed to be constant, independent of density. This function happens to have a maximum at $P = 0.91$. Thereafter, recruitment is predicted to decline, reaching zero at a *theoretical* stock level 58% bigger than the "real" range $0 \leq P \leq 1$.

The curve for the function corresponding with the symmetrical logistic with constant M ($R = 2P - P^2$) is not shown; it lies between curves A and B, has a maximum at $P = 1$, and reaches $R = 0$ at $P = 2$. I do not know of data for any marine animals which would allow one to discriminate between curves A and B.

Figure 5b shows the corresponding yield curves. Curve B is the same as in Figure 2 except that the vertical scale is adjusted so that SY = 1 for $P = 0.5$. The vertical scale of curve A is similarly adjusted. The sustained yield is, from the definition of the stock–recruitment curves, the same in all cases—including for the symmetrical logistic—at $P = 0.5$.

It is evident that the MSY predicted from A is 7.2% higher than that predicted from the logistic; MSY for B is 2.6% higher. Given similar stock–recruitment curves, the MSY cannot alter much. MSY is obtained, however, from very different stock levels—0.37 for case A and, of course, 0.58 for case B. The initial stock sizes and the SY at half initial stock are the same in all cases, yet the MSY stock levels differ over a range ±25%. Real data would not permit even that poor degree of discrimination. These differences would be foreseen if it were possible to calculate $r = R/P$ or better $(r - M)$, and plot it, rather than R, against P. The graph for case A is strongly concave upward, having, at $P = 0.5$, a value of $(r - M) = 0.25$ of the maximum $(r - M)$ for $P = 0$ (Figure 5c). The range of change of r between $P = 1$ and $P = 0$ is also different: threefold in case A, 1.66-fold in case B, and twofold in the case of logistic.

Yet, over the critical range which might be observed, say, $0.3 < P < 0.7$, the r values differ only by, at most, about 5%.

From this, I think one must conclude that a calculated value of stock level for MSY depends practically entirely on the particular function used to "smooth" the data. Furthermore, when data are highly variable, one is tempted to decide that the only reasonable line to draw through them is a straight one. This procedure is sufficient to lead to a particular MSY level; to $P_{MSY} = 0.5$ if the data analysis involves, implicitly or explicitly, a regression of estimates of $(r - M)$ on estimates of P. In fact, if an exploited whale stock became stabilized at about half its initial size, it would be practically impossible to predict whether the stock should be allowed to increase or reduced still further in order to obtain MSY. Only an experimental approach to the problem would reveal this, but either way, the gain in sustained yield would be difficult to detect. Since there are other advantages in taking a given sustained yield from a larger rather than from a smaller stock, it is clearly better to graduate recruit–stock data, using arbitrary functions, with functions more like B than A.

Data for some fish stocks suggest a concave $(r - M)$ curve, and for some mammals, including the Southern Hemisphere fin whale, a convex curve. I do not know of data for any other large whales, but it seems reasonable to assume, pending evidence to the contrary, that for these animals, an MSY stock level would be greater than half the initial numbers. There are two—at least—important questions concerning the approach to managing these stocks to obtain MSY. The first is "how much greater than half?" The second is "how can one determine this is a lightly exploited stock?" To examine these questions, focus attention on the bottom right-hand corner of Figure 5c. Bear in mind that the scale of $(r - M)$ is the same as the scale of the *whaling effort* which is required to stabilize the stock at any given level. This is illustrated on the right-hand side of Figure 5c showing the effort f on scales such that $f = 1$ for the level of MSY. At the top of these scales are shown the levels of f, relative to that needed to obtain MSY, which, *if maintained*, lead to extinction of the population. Of course, in practice, such levels would not, for economic reasons, be maintained, except possibly if a depleted species or stock were left unprotected while whaling continued on more abundant species or stocks in the same region.

The slope of the $(r - M)$ curve near the initial state determines the possible range of the MSY stock level. In the example, that slope for curve B is –2. Whatever the exact shape of such a curve (provided its slope changes monotonically), the MSY level is between 0.5 and 0.75.

If the slope at the bottom right-hand corner were, say, –3, we could establish that MSY level is in the range 0.67 – 0.83, and so on. In general, the upper limit of MSY stock level, for convex curves, is 2 + 0.5/Slope. The lower limit is either 0.5 or 1 + 1/Slope, whichever is higher.

Two measures are sufficient to locate the stock status on the diagram. That is to say, if one had an index of initial stock abundance, and the ratio between the whaling efforts which stabilize the stock at two levels somewhat below the initial level, a first estimate of the range of the MSY level is possible, at least in theory.

At first glance, it might seem worth exploring the application of such an approach. In practice, different kinds of information might be combined and one should be able to make analyses which do not require waiting for stabilization, provided that the age compositions of the catches are known. The basic requirement would be good and prompt measures of the changing stock size from sightings and effort. However, the discrepancies noted in estimates by different methods used so far suggest that methods are not yet adequate for this purpose.

Appendix 5

The consequences of discounting present values of future catches

In the model given in Appendix 1, paragraphs 2 and 3, the discount rate d is zero. Clark (1973b) has used the same basic model (a symmetrical, parabolic yield curve, and constant price, and costs per unit effort) to study the consequence of $d \neq 0$.

When $d \rightarrow \infty$, then $P_1 \rightarrow P_0$. The economically optimal stock level P_d, for any given discount rate is, from Clark (1974), $P_d = \frac{1}{4}[1 + P_0 - db + \sqrt{(1 + P_0 - db)^2 + 8P_0 db}]$.

This shows that the "economic optimum" lies somewhere between the rent-dissipating common property equilibrium level of stock P_0, and the rent-maximizing level P_1. Furthermore, it depends on two ratios: the cost/price ratio which is equal to P_0 and, since b is the reciprocal of the intrinsic rate of increase of the stock at $P \rightarrow 0$, the ratio of the discount rate to the stock increase rate $r = 1 / b$.

From the above equation, it is interesting to calculate, for each value of P_0, what discount rate would be required for $P_d = P_{MSY} = 0.5$. The values are:

P_0	db
0.1	0.125
0.2	0.333
0.3	0.750
0.4	2.000

To take a concrete example, if, for the Antarctic fin whale, $r = (R_0 - M) = (0.12 - 0.04) = 0.08$, then we find that the economic optimal population P_d will be less than P_{MSY} when the discount rate d exceeds 1% for $P_0 = 0.1$, or 2.7% for $P_0 = 0.2$, or 6% for $P_0 = 0.3$, or 16% for $P_0 = 0.4$.

Clark (1974) suggests that, in this case, $P_0 \cong 0.1$, but even if it were considerably higher, it is clear that the economic optimum, as here defined, would be at a smaller stock level than P_{MSY}.

Now, this is an oversimplified treatment of the fin whale case. At least two other factors come into play. One is the extent to which the costs of whaling are not proportional to the *fin whale* stock level, because other species are also being caught. Then, $P_d = 0$ whenever $d > r$, so naturally the fin whale would, if not protected, be hunted practically to extinction. Another factor is the fixed cost of exerting *any* whaling effort—for example, the cost merely of moving ships from the Northern Hemisphere to the Antarctic whaling grounds.

This affects the price–cost ratio and, hence, tends to increase P_0 to P_0^1. It does not affect the position of the economic optimum stock if $d = 0$ in the proportion $(P_1 - P_0)/(P_1 - P_0^1)$.

Appendix 6

Baleen whales in the Southern Hemisphere treated as a single resource

Gambell (1974b) has tabulated the calculated biomass of Southern Hemisphere stocks of baleen whales, summarized below, with the catches taken in the same period. It has been suggested that the blue whale and the minke whale overlap substantially in their diets and are competitors for a limited food supply. (That food is a limiting factor in the Antarctic is presumed from the fact that the early growth rate of fin whales has increased as they have been depleted, so that they now become sexually mature much earlier than before.)

Here, I make the very extreme assumption that the minke has attained its recent abundance, starting from an initial very small population, and benefiting from the "surplus" of krill resulting from the decline of the blue whale; furthermore, that other competitors, not necessarily whales, have not so benefited. Thus, were it not for the intensive catching in the 1970s, the minke might now be in a period of increase, although according to the SY model, the blue whale should be, relatively, in an even more active period since it has been protected.

The new recruitment over a decade is roughly the consequence of reproduction of the stock at the level existing at the beginning of the decade or a few years earlier. Thus, these figures suggest a combined MSY of a little over 7 million tons per decade from a baleen whale biomass of the order of half the initial level.

The present level is less than 40% of MSY level ($P < 0.2$). P_0 is presumably somewhat higher than 0.2 because the profitability of Antarctic whaling is maintained by catches of sperm whales, which, in weight, in the decade 1961–1970, totaled 1.3×10^6 tons. However, the 1970 stock available legally to whalers is not 8.1, but 7.4×10^6 tons because blue and humpback whales are protected. Furthermore, since 1970, the minke stock has been intensively exploited—it is perhaps already less than 1 million tons; whaling, as now regulated, appears still to be profitable, so that, even taking account of the sperm whale catch (the significance of which varies greatly among countries whaling in the Southern Hemisphere), P_0 is probably, under present price cost conditions, in the range 0.15 to 0.2.

Table 1. Whale stock (tons $\times 10^6$).

Item	1920	1930	1940	1950	1960	1970
Blue	17.1	14.3	3.6	1.6	0.4	0.5
Fin	22.3	21.2	15.2	13.7	6.3	4.4
Sei	2.7	2.7	2.7	2.7	2.7	1.5
Humpback	1.5	1.0	0.6	0.6	0.2	0.2
Minke[a]	0.5	0.6	0.8	1.0	1.2	1.5
Total	44.1 (1.00)[c]	40.0 (0.91)	22.9 (0.52)	19.6 (0.44)	10.8 (0.24)	8.1 (0.18)
Catch	9.0	21.7	8.7	15.9	7.5	
R[b]	4.9	4.6	5.4	7.1	4.8	

[a] Assume exponential threefold increase from small initial stock to Gambell's figure for 1970.
[b] R = difference between catch and decline in stock.
[c] Proportion of 1920s level.

Appendix 7

Ineffective catch quotas

That the establishment of catch quotas does not necessarily place an effective restraint on whaling is shown by the recent history of such quotas for pelagic whaling in the North Pacific (Table 1).

Table 1. Quotas and catches for North Pacific fin, sei, and Bryde's whales, 1967–1975.

	Fin		Sei and Bryde's		
Year	Quota	Catch	Quota	Catch	
1967		2,272		6,116	(63)
1968		1,882		5,910	(171)
1969	1,600	1,276	(6,116)	5,247	(89)
1970	1,453	1,012	5,541	4,643	(39)
1971	1,308	802	4,710	3,912	(919)
1972	1,046	758	3,768	2,528	(201)
1973	650	460	3,000	2,580	(724)
1974	350	413	3,000	2,641	(1,362)
1975	300		2,000		

First note that the catches of both "species" (sei and Bryde's whales are different species, but because they were not readily distinguishable by gunners, were treated as one in whaling regulations) decline continuously and at the same rate both before and after the beginning of the quota system in 1969. In the last column are given, in parentheses, the catches of Bryde's whale, as identified and reported after capture, and which are included in the total "sei" catches; so, the real catch of sei whales declined even in the last two years.

It is evident that the quota was not reached, for either species, in any year of regulation, except when the fin quota was exceeded in 1974 by 18%. (The regulations permit an excess of up to 10% on one species provided there is "an appropriate reduction" in the catch of the other species.) Furthermore, the catch was, on no occasion other than the exception mentioned, more than 88% of the quota, and it averaged 77%.

Now, for a catch occasionally not to reach a catch limit is not in itself wrong or unexpected. When a quota is determined, the national industries have to make appropriate operational decisions, including how many units—expeditions and catchers—to deploy. But, to fail to reach the quotas so consistently and by so much is a strong indication that the quotas were being set far above the real availability of whales.

Whaling quotas are determined in June each year for the following season, that is, while the North Pacific whaling season is in progress and thus before current performance can be assessed. There is, therefore, an additional uncertainty regarding the state of the stocks which does not exist for Southern Hemisphere whaling. This might be offered as explanation of the excessive quotas were it not for the fact that the "regulated" catch in any one year was, more often than not, less than the quota for the *following* year, although the quotas themselves were continuously declining. In fact, next year's quota was set each June at a level lower than the previous season's catch, but higher than the catch eventually taken in the year of decision.

It is difficult to conclude other than that the quotas were having no managing or conserving effect except perhaps to facilitate the division of the catch between participating countries by restraining the competition between them. An outside observer might be forgiven for wondering whether the entire "regulation" was not a facade simply for systematic "phasing down" of North Pacific whaling.

This, and similar experience elsewhere, suggests a novel approach to management—that is, to penalize countries or enterprises in some way (e.g., by proportionately reducing the quota for the following year) to the extent that they do not succeed in taking the quotas they themselves have negotiated.

It would be interesting to know what was happening to the stock while this was going on. Unfortunately, the different indices of stock size are highly variable and not entirely consistent with each other. They are, however, summarized in Tables 2 and 3. It seems that both "species" continue to decline, but at what rate is not at all clear; if the quotas had been reached, the declines would have been significantly faster.

Table 2. North Pacific indices of abundance relative to 1966 (from Wada 1975).

	Fin			Sei (and Bryde's)		
Year	CPUE	CPUE "adjusted"	Sightings	CPUE	CPUE "adjusted"	Sightings
1967	0.69	0.58	0.80	1.62	0.99	1.25
1968	0.56	0.78	1.00	1.67	1.45	1.41
1969	0.27	0.59	0.78	0.99	0.92	0.58
1970	0.25	0.36	0.86	0.86	0.66	0.88
1971	0.27	0.48	0.60	0.69	0.59	0.69
1972	0.24	0.50	0.65	0.65	0.52	0.52
1973	0.21	0.53	0.69	0.66	0.59	0.58

Table 3. North Pacific sei whale abundance relative to 1968 (from Tillman 1974).

Year	Sightings	Modified DeLury method (0.06 M 0.08)
1963		2.0
1967		1.5
1968	1.00	1.0
1969	0.85	
1970	0.64	
1971	0.50	
1972	0.38	
1973	0.41	
1974		0.9

Appendix 8

Extracted text from the US Marine Mammal Protection Act

Species and population stocks which are, or may be, in danger of depletion or extinction as a result of man's activities should not be permitted to diminish beyond the point at which they cease to be a significant functioning element in the ecosystem of which they are a part ... or to diminish below their optimum sustainable population (OSP). Marine mammals, being resources of great international significance, aesthetic and recreational as well as economic, should be protected and managed with the primary objective of maintaining the health and stability of the marine ecosystem. Where consistent with this primary objective, the goal should be to obtain an optimum sustainable population, keeping in mind the optimum carrying capacity (OCC). "Depletion" is defined as a decline of the number of individuals within a species or population stock to a significant degree over a period of years, or a decline which, if it continues or is likely to resume, will lead to the species being classed as "endangered" (and, therefore, protected under the 1969 Act) or is the state in which the stock is below the OCC of its environment. "Conservation" and "management" mean the collection and application of biological information for the purposes of increasing and maintaining numbers at the OCC, including the entire scope of activities that constitute a modern scientific resources program, including, but not limited to, research, census, law enforcement, and habitat improvement, as well as, where appropriate, periodic or total protection or regulated taking of species or populations. "Taking" means to harass, hunt, capture, or kill, or to attempt to do so. The OCC is the ability of a given habitat to support the OSP in a healthy state without diminishing the ability of the habitat to continue that function. The OSP is the number of animals which will result in the maximum productivity of the population or species. "Stock" means a group of the same species or smaller taxa in a common spatial arrangement that inter-breed when mature.

Chapter 22

Aspects of Determining the Stock Level for Maximum Sustainable Yield[1]

General Theoretical Considerations

In the literature of fishery science, it has frequently been asserted or implied that if the curve of recruitment against stock size is known, then the state of a stock can be assessed with respect to management for maximum sustainable yield (MSY). This is true, in theory, for numerical assessments if the coefficient of adult natural mortality is not considered to be density-dependent.

Thus, Doi (1973) emphasizes "that it is possible to estimate the maximum sustainable yield, to diagnose the population status and to determine a management course, judging from a reproduction curve." It is not commonly realized, however, just how precise knowledge of the reproduction curve has to be for useful estimation of MSY stock level from data. My purpose here is to demonstrate this graphically. To this end, I have examined a number of commonly adopted algebraic functions which have been suggested or used to graduate data for stock and recruitment, and calculated the curves of sustainable yield against stock size that they imply. I have also calculated the stock–recruitment functions implied by a number of commonly adopted forms of sustainable yield curve.

The resulting reproduction curves are shown in Figure 1. For illustration, all functions have been reduced to a one-parameter form. Stock size P is expressed as a fraction of "initial" (unexploited) equilibrium stock. Recruitment R is likewise expressed as a fraction of the level of recruitment to the unexploited stock. The value of the single parameter is taken, in each case, such that the recruitment at half the initial stock level is three-quarters of the recruitment at initial level. This is an arbitrary value, but the conclusions are essentially the same if other values are assumed. This value happens, however, to correspond roughly with that implied by a number of whale stock assessments. The natural mortality coefficient is assumed constant.

---

[1]Editors' note: written by Sidney J. Holt, Threshold Foundation, 7 Regency Terrace, Elm Place, London SW7, England initially for the meeting in 1975 of the FAO Advisory Committee on Marine Resources Research (ACMRR) and eventually published in 1981 in *Mammals in the Seas, Volume III, General Papers and Large Cetaceans* (Food and Agriculture Organization of the United Nations, Rome). This paper is an elaboration of Appendix 4 in Holt (1975a), which is also Chapter 21 in this book. The lengthy caption to Figure1 comes from Holt (1975b), a document submitted to the 1975 meeting of ACMRR mentioned above. At the time of the 1975 writing, Holt was employed by the Food and Agriculture Organization of the United Nations in its Fisheries Biology Branch. Permission to reproduce this article has been given by FAO. The views expressed in this publication are those of the author and do not necessarily reflect the views or policies of the Food and Agriculture Organization of the United Nations. The version presented here has undergone very minimal editing from the published version.

Thus, all the reproduction curves pass through the points:

$P = 0$	$R = 0$
$P = 0.5$	$R = 0.75$
$P = 1.0$	$R = 1.0$

and, of course, predict the same sustainable yield (SY) for $P = 0.5$.

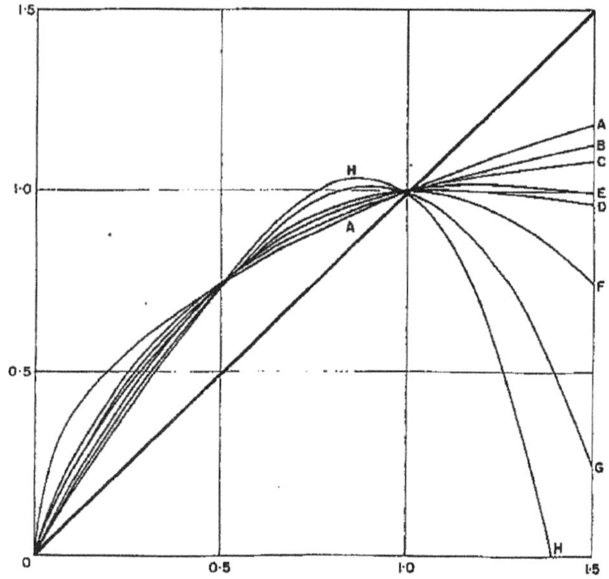

Figure 1. Reproduction curves A–H. Note that because the scale is such that $R = 1$ when $P = 1$, then SY = $R - P$. Description of curves are as follows:

A: $R = P^b$; $P_{msy} = 1/b^{1/(b-1)}$; $b = 0.415$; this form was applied by Cushing (see Paulik 1973); not asymptotic.

B: $R = P/[P + b(1 - P)]$; $b = 0.33$; form derived by Beverton and Holt (see Ricker 1973); asymptotic.

C: $R = b(1 - e^{-aP})$; $b = 9/8$; $e^{-a} = 1/9$; see Chapman (1973); asymptotic.

D: $R = 7.91(0.6 - 0.1P)Pe^{-(0.7 + 0.675P)}$; this equation was derived by Doi (1973) for the Antarctic sei whale *Balaenoptera borealis* stock; his choice of age at maturity (he does not specify this in the paper, but he, in fact, assumed 10 years) is such that at $P = 0.5$, $R = 0.77$, not 0.75; this makes no difference here; Doi's assumptions were (i) age at recruitment constant, (ii) pregnancy rate and pre-recruitment mortality both vary linearly with mature stock number, and (iii) mature mortality rate constant; R has a maximum at $P = 1.14$, $R = 0$ for $P = 6.0$.

E: $R = aPe^{-bP}$; $b = 0.81093$, $a = e^b = 2.25$; derived by Ricker (see Ricker 1973 and other references); this expression has a maximum R for finite P; in the example used here, the maximum occurs at $P = 1.23$.

F: $R = P(1 - P) + P$; this is the reproduction curve implied by the logistic growth curve and the symmetrical parabolic SY curve if M is constant; it is peaked and symmetrical with R maximum at $P = 1.0$, and $R = 0$ when $P = 2$.

G: $R = bP(1 - P^2) + P$; $b = 0.66$; the Pella and Tomlinson yield curve (see Holt 1975a) with $n = 3$ implies this form of stock–recruitment curve if M is constant; it is peaked, but symmetrical with, in this case, maximum R at $P = 0.91$ (i.e., within the "natural" range of P) and $R = 0$ at $P = 1.58$.

H: $R = bP(1 - P^3) + P$; $b = 0.57143$; Pella and Tomlinson with $n = 4$; maximum R at $P = 0.88$ and $R = 0$ at $P = 1.40$.

Except perhaps for the extremes of functions from Cushing (1973) and from Pella and Tomlinson (1969) ($n = 4$; see legend to Figure 1), these curves are close to each other in their middle range. It is doubtful if, being fitted to data, discrimination among them would be possible.

The MSY stock level is the value of P for which the tangent to the reproduction curve has a slope of 45°, i.e., is parallel to the diagonal line. In Table 1, the calculated values of P_{msy} are listed, with the corresponding MSY values expressed as a multiple of the MSY for the logistic model, i.e., the SY for $P = 0.5$.

The possible range of P_{msy} under the particular restraints here applied is zero to 0.5/0.75 = 0.67 (see Chapman 1973 and Larkin 1973 for discussion of non-parametric models). It

Table 1. Calculated values of P_{msy} with corresponding MSY multiples for the logistic model.

Reproduction function	P_{msy}	MSY (relative to logistic)
A Cushing (1973)	0.22	1.340
B Beverton and Holt (1957)	0.37	1.072
C Chapman (1973	0.41	1.032
D Doi (1973)	0.44	1.011
E Ricker (1973)	0.45	1.012
F from logistic	0.50	1.000
G from Pella and Tomlinson (1969), $n = 3$	0.58	1.026
H from Pella and Tomlinson (1969), $n = 4$	0.63	1.080
	0	2.000
Extremes	0.66	1.330

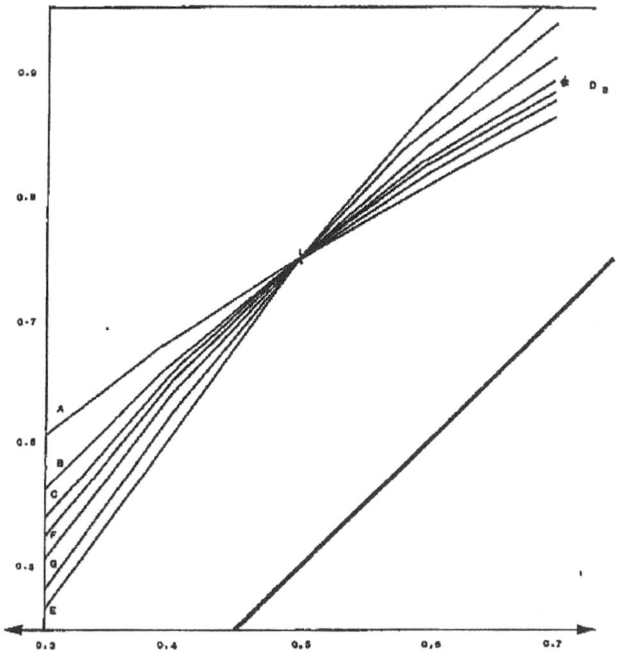

Figure 2. Showing the same curves as Figure 1, with scales extended and mid-section only. Vertical marks are the levels of P_{msy}.

seems unlikely that the MSY level could be deduced from excellent data for a reproduction curve with less than 20% error and, I suspect, not even within 30% error in practice. It should be noted that the value of MSY does not (with the exception of the Cushing function) vary more than a few percent; it is always somewhat higher than predicted by the symmetrical logistic, the divergence increasing as P_{msy} move away from 0.5.

There are, of course, more discriminating methods of analyzing stock–recruitment data—by, for example, the common method of plotting $r - M$ against P, where $r = R/P$. Nevertheless, if any mathematical expression, linear or other, is fitted to such data, the level of MSY predicted depends essentially on the properties of that expression. If the data are graduated by eye, then the location of the tangent can only be judged very approximately.

Cushing (1973) has examined the variability of a sustainable yield curve derived from the Ricker (1973) reproduction function, and using data for Northeast Atlantic Cod *Gadus morhua*. If he had not assumed this particular form of the curve of R against P, but fitted others which gave no worse fits to his data (e.g., the logistic), he would have deduced a considerably greater uncertainty in locating the sustainable yield curve, and different values for the sustained fishing mortalities required for MSY or for stock extinction. Cushing plots his sustainable yield against the intensity of fishing; in Table 2 are listed critical values of relative fishing effort for the models here examined.

This table lists the values of fishing effort which would, if sustained, give MSY and those which, theoretically, lead to extinction of the stock (f_{ext}) both expressed as ratios of the effort which would hold the stock at half its initial size ($f_{0.5}$). The restraints are the same as in Table 1.

Compare the last column in Table 2 with P_{msy} in Table 1.

Some Consequences for Whale Assessments

Data for whale recruitment present further difficulties. As recently as 1970, Allen (1973) showed (his Figure 106) that for the fin whale *B. physalus* in the Antarctic, there was *no* density dependence of the rate of recruitment; that is, recruits and parent stock were, as near as the data could detect, proportional to each other over a range down to 20% of the initial stock size. If this is so, P_{msy} is strictly indeterminable, and the SY at any size negligible. Allen concludes that the proportionality is an artifact due to the use of biased age–length keys to estimate age compositions. One wonders if, in a situation where all values of P_{msy} are equally probable over a wide range, it is proper or useful to express the "best estimate" as being in the middle of that range.

Table 2. Implications of reproduction function for scales of fishing mortality and effort, and critical values.

Function	$f_{ext}/f_{0.5}$	$f_{msy}/f_{0.5}$	f_{ext}/f_{msy}	$f_{0.5/2}/f_{msy}$
A	∞	2.85	∞	0.18
B	4.00	1.45	2.78	0.34
C	2.40	1.26	1.90	0.40
D	2.72	1.50	1.81	0.33
E	2.50	1.12	2.22	0.45
F	2.00	2.00	2.00	0.50
G	1.32	0.88	1.51	0.57
H	1.14	0.86	1.33	0.58

References

Allen, K. R. 1973. Analysis of the stock–recruitment relation in Antarctic fin whales (*Balaenoptera physalus*). Rapports et Procès-Verbaux des Réunions du Conseil Permanent International pour l'Exploration de la Mer 164:132–137.

Beverton, R. J. H., and S. J. Holt. 1957. On the dynamics of exploited fish populations. Her Majesty's Stationery Office, UK.

Chapman, D. G. 1973. Spawner–recruit models and estimation of the level of maximum sustainable catch. Rapports et Procès-Verbaux des Réunions du Conseil Permanent International pour l'Exploration de la Mer 164:325–332.

Cushing, D. H. 1973. The variability of a yield curve calculated from a stock and recruitment relationship. ICES Document CM 1973/F:29.

Doi, T. 1973. A theoretical treatment of the reproductive relationship between recruitment and adult stock. Rapports et Procès-Verbaux des Réunions du Conseil Permanent International pour l'Exploration de la Mer 164:341–349.

Holt, S. J. 1975a. The concept of maximum sustainable yield (MSY) and its application to whaling. Document FAO/ACMRR/MM/IV/4, Airlie House Workshop, Virginia.

Holt, S. J. 1975b. Aspects of determining the stock level for maximum sustainable yield. Document ACMRR/MM/EC/29. Advisory Committee on Marine Resources Research, Marine Mammal Symposium, Bergen, Norway.

Larkin, P. A. 1973. Some observations on models of stock and recruitment relationships for fishes. Rapports et Procès-Verbaux des Réunions du Conseil Permanent International pour l'Exploration de la Mer 164:316–324.

Paulik, G. J. 1973. Studies of the possible form of the stock–recruitment curve. Rapports et Procès-Verbaux des Réunions du Conseil Permanent International pour l'Exploration de la Mer 164:302–315.

Pella, J. J., and P. K. Tomlinson. 1969. A generalized stock production model. Bulletin of the Inter-American Tropical Tuna Commission 14:421–496.

Ricker, W. E. 1973. Critical statistics from two reproduction curves. Rapports et Procès-Verbaux des Réunions du Conseil Permanent International pour l'Exploration de la Mer 164:333–340.

Chapter 23

The Classification of Whale Stocks and the Determination of Catch Limits, under the Current IWC Management Procedure, with Limited Information[1]

Paragraphs 10(a), (b), and (c) of the current Schedule of the International Whaling Commission (IWC 1983) establishes, for commercial whaling, a set of rules whose application depends formally on the possession of scientific information which is not, in general, available to the IWC Scientific Committee. In lieu of this, the Committee has often presented its advice on the basis of its appreciation of broad features of the biology of whales and of the dynamics of whale populations. Its views have been developed over several years of discussion and rest substantially on a few rudimentary comparisons between species, particularly balaenopterid species. It has been recognized that these views have not always been interpreted in a consistent manner in the assessments of particular baleen whale stocks (see IWC 1984a, pp. 54–56).

One basic principle in Paragraphs 10 and 13 of the Schedule is that if a stock is found to be depleted to below a certain reference level, whaling shall cease until it has demonstrably recovered. In the case of commercial whaling, the reference level is 90% of the level giving maximum sustainable yield (MSYL), while in the case of whaling to satisfy aboriginal subsistence needs, it is "a certain minimum level," which, from the context of the Commission's discussions which led to the establishment of this provision, is related closely to the notion of a threshold level below which it is presumed that extinction of the stock is a serious risk.

A second principle is that if a stock is found to be above the pertinent reference level, it may be exploited in such a way that it will move towards or remain near to the MSYL. To achieve this, the catch limits must be less than MSY, with two provisos. The first is that for stocks regulated in accordance with Paragraph 13(a)(2), the catches each year must be less than the current replacement yield. The second is that catch limits for commercial whaling on a stock which is more than 20% larger than the MSYL shall not exceed 5% of the estimated initial size of the exploitable stock, unless there is positive evidence that a continuing higher percentage will not reduce the stock to below MSYL [Paragraph 10(b)].

A third principle is that some account must be taken of uncertainty in the estimation of MSY, SY (sustainable yield), and RY (replacement yield). This is not made explicit, but it is, nevertheless, the intention of the provisions that the catch limits for commercially exploited

[1]Editors' note: written by Sidney Holt, 2 Meryon Court, Rye, East Sussex, UK, as Document SC/36/O 12 for the 1984 meeting of the IWC Scientific Committee and published in Reports of the International Whaling Commission 35:487–494, 1985.

Sustained Management Stocks (SMS) and Initial Management Stocks (IMS) shall not exceed 90% of MSY, and shall be substantially less if the stock is considered to be between MSYL and the threshold. It is important to note that the Schedule does not preclude the Committee from proposing catch limits less than 90% of the estimated MSY if it judges that the uncertainty of the estimate justifies a larger safety factor. With respect to aboriginal subsistence whaling, the Committee is enjoined to advise on a range of rates of increases towards MSYL for stocks which are considered to be above the threshold, but below MSYL. This implies the calculation of catch limits corresponding to various notional rates of increase. Safety factors are not explicitly mandated, but it must be presumed that, since they must be set so that they "*will* allow whales stocks to move to MSYL" [*Holt's emphasis*], the Committee should itself determine a safety factor (on the basis of its appraisal of the error range in its estimates of SY or RY) such that the stock would, in all likelihood, move in the desired direction.

The pragmatic approach to these provisions has included the following elements:

1) an assumption that the MSYL for baleen whales is not less than 60% of the initial stock size;
2) a presumption that the stock was stable before substantial whaling started, unless there is clear evidence to the contrary;
3) a recent estimate of the stock size;
4) an estimate of, or reasonable guess at, the natural mortality rate which is presumed to be constant in the recruited stock, both with respect to age and density.

In addition, for purposes of both classifying stocks and estimating sustainable and/or replacement yields, it is necessary to simulate population trends using catch data and values for the age at sexual maturity and at recruitment. Such simulation requires information about the density-dependence of the recruitment rate, which may be manifest through density-dependence of the reproductive rate, the prerecruit mortality rate, or both. In practice, it has been assumed that these last dependencies can be expressed by a simple two-parameter function relating number of recruits to the size of the parent stock.

In this paper, I refer to r, the recruitment rate, as the ratio of the number of recruits to the number of parents, to the net recruitment rate as the difference between that ratio and M, the natural mortality rate (expressed as instantaneous rates), and to recruitment as the annual number of recruits.[2]

At a special meeting in 1978 (IWC 1979), the IWC Scientific Committee adopted a value of M for minke whales *Balaenoptera bonaerensis* based on an apparent relationship between M and maximum body length derived in Ohsumi (1979) using published M values for both baleen whales and odontocetes.[3] It was later realized that there are several problems associated with that analysis (e.g., see Chapman 1983; IWC 1984a) including determination of which species it is appropriate to use in the regression, comparability and appropriateness of the various published M values, choice of independent variable, and predictive value of such analyses. These are discussed more fully in de la Mare (1985) and Mizroch (1985).

---

[2]Rates of recruitment are expressed in two ways (IWC 1978): r_{I} the proportion of the total parent population, as used here; and r_{II} the proportion of the current recruited population.

[3]At the 1977 Scientific Committee meeting, I had suggested that there might be a linear relationship of M with either maximum length or body weight in balaenopterids, as an approach to obtaining an M value for catch composition in the early years of exploitation.

Such difficulties do not necessarily mean that interspecific comparisons cannot be useful; simply that they must be examined and applied carefully, and with consideration to the implications of the particular application. For example, the value of M is often not critical in the estimation of the degree of depletion and hence in classification, but it is all important in the estimation of sustainable yields. In this connection, an aspect of the assumption that M is independent of age (above the age at recruitment) is worth noting. In mammals, and in many fish, adult mortality increases with age; cetaceans are no exception (e.g., see Ralls et al. 1980; de la Mare 1985). If this is to be approximated by a constant "average" mortality, it becomes important how that average is calculated. Heinke's (1913) procedure gives an unbiased estimate of the average mortality rate, whereas fitting a straight line to an age distribution does not. If comparative studies are to be used for assessments, then it is essential that the values on which they are based have been obtained by appropriate methods, if not always by the same methods.

In no case is the function relating the gross or the net recruitment rate to the parent stock size known empirically. So, assumptions must be made about its shape and its scales if population trajectories since the onset of whaling are to be calculated for the purpose of classification and the setting of catch limits. In each of several algebraic expressions used for this purpose, the shape is determined by an assumption about MSYL as a proportion of the initial stock size.

Scale is fixed by reference to two or more points which may be determined empirically or inferred from interspecific comparisons, or may be inferred from known or presumed characteristics of the species. One of these points is always $r_1 + M$, for the natural steady-state. One other may be the net recruitment rate at MSYL, i.e., $r_x - M$ at $p = p_x$. This has been commonly taken to be 0.04 because it was thought that this value had been obtained for fin whales *B. physalus* in the Southern Hemisphere. In principle, any estimated or presumed value of $r - M$ for a determined value of $p \neq 1$ can be used to fix the scale of the stock–recruitment curve. A particular case is $p = 0$, for which an upper limit to r_0 can be deduced in certain circumstances. This is because, in cetaceans, the pregnancy rate cannot exceed unity, the prerecruit mortality rate is considered unlikely ever to become lower than the immediate post-recruit mortality, and there are biological limits to the length of the reproductive cycle and to any possible reduction in the age at female maturity.

Another possible limit to scale is given by consideration of the feasible strength of the density-dependent response of the recruitment rate as the stock is initially reduced from its unexploited size. Holt (1977) suggested that this response, measured as the slope of the stock–recruitment curve at $p = 1$, would not be steeper than the slope of the hyperbola which describes the relation $r \times p =$ constant. If it is steeper, then the recruitment will initially increase as the population decreases. Such a phenomenon—super-compensation—has been observed in some fishes. It is usually understood, however, that it is found only when there is some special biological process, of which the best known is cannibalism, involving consumption of prerecruits by post-recruits or within the prerecruit phase of life (see, for example, the recent review by Pitcher and Hart 1982).

It is difficult to envisage a biological mechanism which could lead to a whale stock having, in the initial stages of depletion, such a vigorous density-dependent response that the numbers of recruits will actually increase. In fact, when the Scientific Committee has made explicit decisions that population simulations should be performed, it has usually been assumed that as the stock declines, the number of recruits will decline, even if slowly at first. It has not, however, always been realized that, in many assessments, there has been an implicit assumption that the number of recruits increases considerably, over a wide range of stock sizes, as the stock is reduced. This presumption generates high estimates of sustainable yield and, at the same time, severely masks likely declines in stock.

Two "rules" have been suggested for, essentially, the strength of density-dependence of the recruitment rate and its relation with M. It has been asserted that both are based, though loosely, on interspecific comparisons. I shall show that both imply that the density-dependent response of the recruitment rate is extremely strong. It is, therefore, necessary to examine these since, if they are valid, they contradict my contention that there is no reason to suppose that the recruitment would increase with declining stock size.

If the shape and scale of a stock–recruit function has been established by the procedures outlined above, the resulting predictions will depend, to some extent, on the precise form of the function used. For example, the Pella–Tomlinson (1969) function incorporated in the Allen (1976) "flexible model," which has been much applied in whale stock assessments, will, in certain circumstances, give different predictions from those that would have been obtained if an equally flexible, but formally different, function had been used, such as that defined by Shepherd (1982). The practical significance of any such difference depends on whether the function is forced through a value or $r - M$ at MSYL, through a chosen value of $r_0 - M$, or is constrained in some other way, as by having the slope at $p = 1$ no steeper than that of the hyperbola. I have examined the difference for this last case, for two alternative functions, since it has a bearing on decisions concerning the maximum plausible value of $r_x - M$.

The Gulland "Rule"

Gulland (1971) used the expression

$$Y_x = M \times P/2 \tag{1}$$

to estimate the MSY of fish stocks about which little is known other than a population size from surveys, and a natural mortality from, say, the age composition of the unexploited stock. This was derived from an assumption that $Y_x = M$. Gulland thought that such a relationship might hold in fish, but he cited no empirical evidence for this and, so far as I am aware, none has been presented since.

Equation (1) has been used by the Food and Agriculture Organization of the United Nations (FAO) to obtain an idea of the likely yields from stocks that have not yet been exploited or have been only lightly exploited. Beddington and Cooke (1983) have recently examined Gulland's rule with respect to an age-structured population model and shown that, in many situations, it would overestimate MSY, sometimes very greatly so. Shepherd (1982) examined the behavior of the ratio f_x/M, where F_x is the coefficient of fishing mortality for MSY as a function of the density-dependence in his stock–recruitment function. He showed that the ratio can take a very wide range of values, so that the rule could be misleading. The same is true for the Pella–Tomlinson stock–recruitment mode, of which the simple logistic, from which Gulland deduced his rule, is a special case.

Ohsumi (1977) proposed that Gulland's rule be applied to the western North Pacific stock of Bryde's whales *B. edeni*, and that catch limits be based on it, noting that "The rate of exploitation for MSY [$= F_x = r_x - M$] is considered in general to be near the value of M at the MSY population level". Although the rule has not been applied by the Committee, it does have, in this context, a formal relation with the other general rule which has been applied to a number of baleen whale stocks, so that it is worth clarifying its properties and implications.

Gulland's rule was based on an examination of the properties of the symmetrical logistic form of the Pella–Tomlinson (P–T) model, with $n = 1$. Gulland argued that the rule should hold approximately also for skewed yield curves. He evidently had in mind such skewed curves as have been postulated for fish, in which $p_x < 0.5$, but the arguments he made for this would, if valid, be equally applicable to skewed curves such as have been applied to baleen whales, where $p_x > 0.5$. One of the skewed curves he considered came from the simple form of the Beverton and Holt model with constant recruitment (Beverton and Holt 1957).

There is no obvious theoretical or biological justification of the rule nor, apparently, empirical support for it. Figure 1 illustrates the *unique* case where the rule is strictly valid for a simple density-dependent response of the recruitment rate, such that $A = 2$, where A is defined as the ratio $(r_0 - M)/M$. Over this recruitment rate curve is shown the hyperbola for

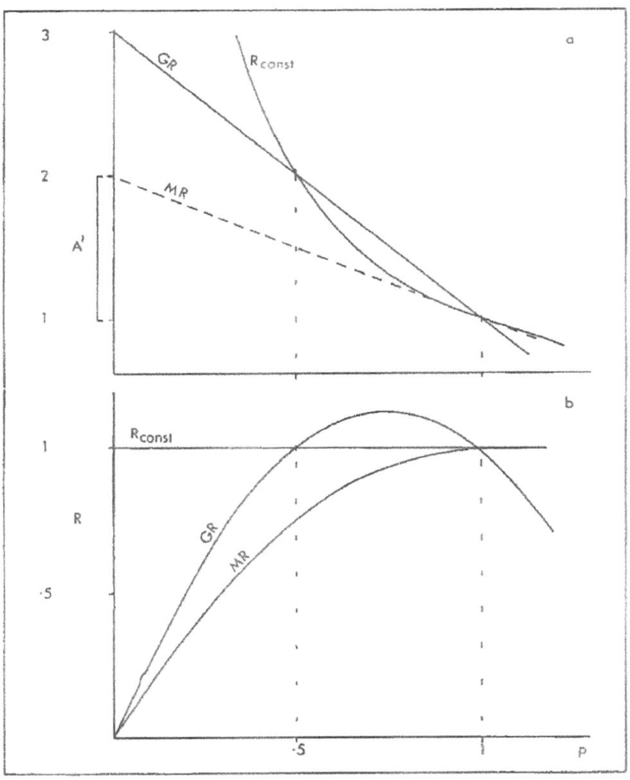

Figure 1. Recruitment rate (a) and stock–recruitment (b) curves for the case $p_x = 0.5$. The "Gulland rule" (GR) is that $F = (r_x - M) = M$ in this case. In (a), the intersection of the GR line with the hyperbola defining constant recruitment shows that for all population sizes greater than MSYL, recruitment would exceed that in the unexploited stock. (b) shows that recruitment reaches a maximum of 10% greater than in the unexploited stock at exactly $p = 0.75$. The line for the "Modified rule" (MR) has the slope of the hyperbola at $p = 1$. If an assumption that recruitment increases with declining stock is not considered to be tenable, then, for this case, the GR overestimates the possible MSY by at least 100%.

constant recruitment. Clearly, over the entire range $p_x = 0.5$ to $p = 1$, the number of recruits exceeds the number in the unexploited stock. Recruitment is at a maximum at $p = 0.75$, at which point $r/r_1 = 1.125$. For those fish—or whales—for which there is no expectation of such "super-compensation," the linear relation of r with p must have a slope no steeper than that of the hyperbola at $p = 1$, which is $-M$. This is shown as a broken line in Figure 1; it corresponds with $F_x/M = 0.5$. The rule would overestimate MSY by at least 100% in such a simple case.

Francis (1974) showed analytically that the Gulland rule can hold true "if and only if the level of recruitment which occurs at the unexploited equilibrium level is equal to that which occurs at the population level which produces the maximum sustainable yield." This, in fact, is why Gulland found that his rule was fairly consistent with the Beverton–Holt constant recruitment model for a wide range of parameter values. Francis' paper does not, however, indicate clearly how misleading the rule might be if the formal condition for its validity is not met.

The rule can be further explored for skewed curves using the general P–T model. We can write this model in the form

$$r = M[1 + A(1 - p^n)]. \tag{2}$$

Following Shepherd (1982) and others, I refer to A as expressing the *resilience* of the stock, and n, which determines the MSYL, as defining the degree of *compensation*.

From equation (2), we can write

$$Y = A \times M \times p \times P(1 - p^n) \tag{3}$$

where Y is the MSY and P the initial population size.

In the P–T model, $p_x = 1/(n + 1)^{1/n}$ so that

$$Y_x = A \times M \times P \times n \,/\, [(n + 1)^{(n+1)/n}] \tag{4}$$

But, if $Y_x = M \times P/2$, then

$$A = [(n + 1)^{(n+1)/n}] \,/\, 2n \tag{5}$$

Figure 2 shows this recruitment rate function for $n = 2.39$ (60% MSYL). It implies recruitment at a maximum for $p = 0.7501$ (i.e., very close to that for the symmetrical logistic), but with $R/R_1 = 1.19$.

In general, for the P–T model, maximum recruitment, R_h, occurs at

$$p_h = [(2n + 1)/(n + 1)^2]^{1/n} \tag{6}$$

and

$$R_h/R_1 = p_h\,[1 + A(1 - p_h^n)] \tag{7}$$

for values of A given by equation (4).

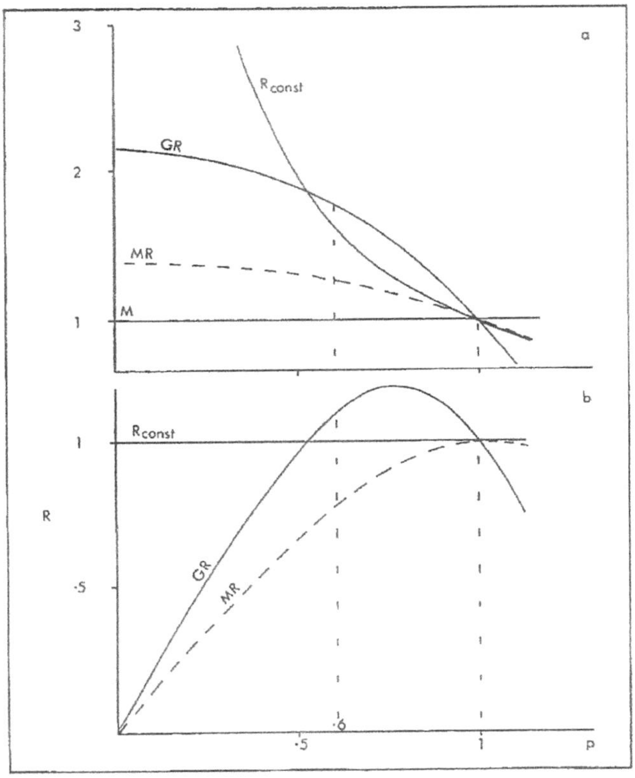

Figure 2. As for Figure 1, but $p_x = 0.6$. The Gulland rule implies a recruitment exceeding that in the unexploited stock over virtually the same stock range – $p = 0.5$ to 1.0 – and 15% higher at the maximum, which is very close to $p = 0.75$. Even at MSY, a recruitment is implied 8% higher than in the unexploited stock. The "Modified rule" implies little change in recruitment until the stock has been reduced by about 10%. If the assumption that recruitment increases with declining stock is not considered to be tenable, then, for this case, the Gulland rule overestimates the possible MSY by at least 175%.

If we accept the constraint that R cannot exceed R_1 in the relative population range $p = 0$ to 1, A in the P–T model must be adjusted so that the slope of the recruitment curve does not exceed $-M$ at $p = 1$. This slope is, from equation (2), $-A \times M \times n$, from which we define the maximum acceptable value of A as $A' = 1/n$. The recruitment rate curve for $A = A'$ and $n = 2.39$ is shown as a broken line in Figure 2. The Gulland rule, for the 60% MSYL yield curve, and no increase in recruitment with declining stock, is $Y_x = 0.177M \times P$, so the application of the original rule to baleen whales would lead, in such a case, to a more than threefold overestimation of the likely MSY. If there is no upper constraint on R, then the overestimation can be even greater.

The 4% Rule

Ever since the IWC management procedure has been applied by the Commission, advice has been given by the Scientific Committee that a reasonable estimate of MSY is obtained by the expression

$$Y_x = 0.04 p_x P. \tag{8}$$

This rule has, for example, been used to determine catch limits for the Northwest Pacific Bryde's whale stock since catches from it were regulated. It is supposedly based on the finding that, for the fin whales of the Southern Hemisphere, $(r_x - M) = 0.04$. From examination of the records of the Scientific Committee back to 1968, I have come to the conclusion that, in fact, this value for the net recruitment rate is, if valid at all, referable not to an MSYL of 60% or even of 50%, but to a stock level $p < 0.2$.

Furthermore, the value 0.04 was itself a compromise between a lower value obtained directly by K. R. Allen (1970, 1971, 1972) from the fin whale data, and a higher value given by Japanese scientists (IWC 1971) on the basis of a more complex model involving an array of untested—indeed untestable—assumptions. Notwithstanding this difference, the Japanese model also gave values for $(r - M)$ at what is now the accepted MSYL, much lower than 0.04.

If $(r_{<0.2} - M) \leq 0.04$, then, from the P–T model, $A \times M \leq 0.041$ and $(r_x - M)$, 0.029.

For fin whales, $M = 0.04$. This means that for $A \leq 1.02$, we are still considering a density-dependence strength considerably higher than permissible with $A' = 0.42$ ($1/n$ for $n = 2.39$). It should be noted, however, that Allen (1973) subsequently published an analysis of the same fin whale data in which a gross recruitment rate of about 0.05 was found for the period 1955–1962, implying a net rate as low as 0.01 when the stock was at about $p = 0.2$ or less. He found a higher value, $r = 0.06$, for earlier years when the stock was somewhere between $p = 0.2$ and 0.7. This was recognized to have been possibly an artifact arising from the use of a single recent age-length key, but even if it were not—which would imply something wrong with the estimate for later years—it could not possibly justify the use of a value of $(r_x - M)$ greater than 0.02.[4]

Allen's estimates were of r_{II} rates. In considering sustainable yields, we need to use valid r_I rates which, in a declining stock size, are lower than r_{II} rates. So $(r_x - M) = 0.2$ is, for this reason, also an overestimation.

The application of a "2%" rather than a "4%" rule would imply for the fin whale, $A = 0.71$ (Figure 3), while for a "1%" rule, $A = 0.35$. For $A = A' = 0.42$ and $M = 0.04$, the MSY is 1.2% of the MSYL stock. Thus, the fin whale data do not, on closer inspection, contradict a hypothesis that the recruitment does not increase with decreasing stock size.

In fact, any incompatibility between my hypothesis and a somewhat less conservative interpretation of fin whale data is more apparent than real. Suppose, for example, we accept one of Allen's estimates, that $(r_x - M) = 0.0216$. In this case, for $M = 0.04$, recruitment exceeds R_1 in the limited range $p = 0.7$ to 0.99, and it does so by no more than 5%.

Interspecific Applications

In the light of the above discussion, there are two possible approaches to the application of the very meager information available to us from the fin whale study, assuming that the results quoted in the late 1960s and early 1970s were to be revised using more appropriate methods than were then available. One of these is an amended equation (8), the constant 0.04

[4]In the original paper considered by the Scientific Committee at its 1984 meeting, the full history of the assumption and use of the 4% rule was detailed in an appendix, which is not here published. The text of the appendix is available from the Secretariat and the author.

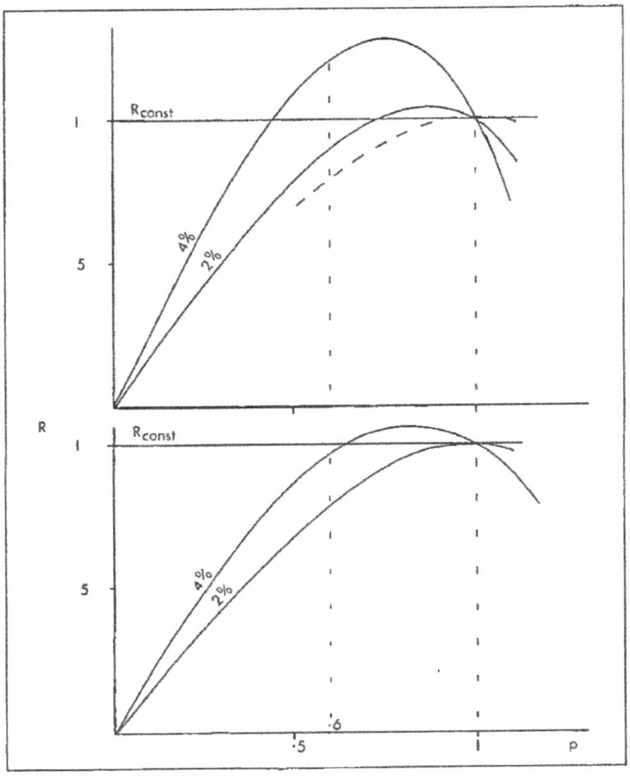

Figure 3. (a) and (b) show stock–recruitment curves for which the 4% rule would be valid for *M* values characteristic of the fin whale *Balaenoptera physalus* ($M = 0.04$, a) and the sei whale *B. borealis* ($M = 0.06$, b) for the case in which $p_x = 0.6$. If the reinterpretation of the southern fin whale data proposed in this paper were accepted, then $(r_x - M)$ would not exceed 0.02. The implications of this are illustrated as the line labeled "2% rule" in (a). Recruitment exceeds that in the unexploited stock through the stock range $p = 0.74$ to 1.0, but never by more than about 4–5%. This reinterpretation also implies that in the fin whale, the recruitment has been reduced by about 20% when the stock had been reduced to MSYL. Part of the stock–recruitment curve for the "Modified rule" is also shown. (b) shows that for a stock with $M = 0.06$, the 2% rule would not imply any increase in *R* with declining stock size; it is, in fact, very close to the stock–recruitment curve given by the "Modified rule" in this case, which is not shown. For stock with *M* higher than about 0.06, the 2% rule is more conservative than the "Modified rule." It must be borne in mind, however, that the latter expresses only a maximum acceptable net recruitment rate, not an estimate of that rate.

being substituted by another value, perhaps of the order of 0.02. The other approach is to apply a modification of the Gulland rule, appropriate to species in which MSYL = 0.6*P*, and with the necessary constraints on the vigor of the density-dependent response of the recruitment rate. One of these alternatives will be more conservative than the other, depending on the value of *M*.

If possible, the decision of which to adopt—if either—should be based on biological considerations. The question to be formally answered is: is the value of *A* or the value of the product $A \times M$ the more likely to be constant across species, at least of the genus *Balaenoptera*?

If, in fact, the smaller species have higher natural mortality rates than the larger ones (which, though certainly plausible and in keeping with what is known about lifespan in other animals, cannot yet be taken as proven, given the difficulty of obtaining any unbiased estimates of M in whales), then to maintain natural equilibria, they must have shorter breeding cycles, lower prerecruit mortality rates and/or earlier sexual maturity. They are all constrained by a birth rate of one calf per cycle and by the need for gestation and lactation periods to fit in with the annual patterns of migration and breeding. It seems, therefore, that having a high r_1 does not necessarily mean that a proportionately high A is attainable. This would suggest that it may be more appropriate to consider the product $A \times M$ as constant across species, i.e., that the *difference* between the maximum and the minimum recruitment rates, rather than their ratio, is an interspecific constant. This is what has, in effect, been assumed in the practice of applying across species a certain value of the net recruitment rate at MSYL.

This choice would, however, imply that stocks with high M would be less resilient than those with low M, and it is not obvious why this should be so. Assuming equal resilience means taking A as interspecifically constant. $A' = 1/n$ gives us an upper limit for A if the stock–recruitment curve is to be asymptotic. This value implies that the gross recruitment rate would not increase by more than 42% as the stock declined to very low levels. This would not be unreasonable if the density-dependent response was more likely to be a function of the post-natal and juvenile mortality rate than of changes in age at maturity and/or pregnancy rate. It is worth recalling that the application of the existing 4% rule, in the case of the fin whale—which has been done, for example, with respect to the Spanish–British Isles stock—implies that the recruitment rate is so resilient that it could increase by a factor of 2.4; this factor cannot be obtained by decreasing the prerecruit mortality from an initial value commensurate with the balance equation to a value in a depleted stock no less than the adult mortality rate.

Application of this "modified Gulland rule" to a stock with $M = 0.06$ would imply that the MSY would be 1.77% of the stock size at MSYL; the corresponding percentage for a stock with $M = 0.08$ would be 2.36%.

Alternative Stock–Recruitment Functions

The P–T function is only one of several that might be used to establish a theoretical stock–recruitment curve, given two or at most three points on it plus an assumption that $p_x =$ 0.6 or any other given value. One well known family of functions is due to Ricker (1954) and another to Beverton and Holt (1957). More recently, Shepherd (1982) has explored a family of curves of which the Beverton–Holt model is a special case, just as the logistic is a special case of the P–T model. Shepherd's general production model for fish stocks provides one basis for examining a legitimate question about the P–T model: the degree to which, in using it, the sustainable yield in general and Y_x in particular, may depend not only on the value of the resilience A, but also on the functional form of that particular model.

Shepherd only explored the properties of his model for sets of parameter values that led to $p_x < 0.05$, and with high resiliencies more appropriate to fish than cetaceans. Because of these restrictions, he was led to conclusions which are, in fact, not so general as he apparently thought. In particular, his conclusion that "in the absence of other information it would ... be reasonable to manage the fishery so that the yield/(current) biomass is of the order of the natural mortality" and, further, that "management at this level of exploitation should safeguard against collapse (of the stock and the fishery)" is not valid for parameter sets outside the range of his exploration. Shepherd realized that this must be so when he qualified the latter conclusion with "... except in extreme cases where $A < 1$ [*Holt's notation*]." Of course, what is an extreme case for a highly fecund fish may be normal for a whale.

Shepherd's stock–recruitment relationship may be written in the following form, using the notation of this paper:

$$r + M[(1 + A)/(1 + Ap^{n'})] \tag{9}$$

where n' is, like n, an exponent expressing degree of compensation. The two are not, however, identical in value for yield curves having the same MSYL; for a given P_x, $n' > n$.

The MSYL and p_x are not, in this model—unlike the P–T model—functions only of the exponent of compensation; they depend also on A. For any chosen value of p_x and a given value of A, a corresponding value of n' must be calculated. Straightforward algebra shows that in the Shepherd model, the three variables are related by the expression:

$$A = [1 - (n' + 1)\, p_x n'] / [P_x n'(n' - 1 + p_x n')] \tag{10}$$

and that

$$Y_x = M \times P \times A[p_x(1 - p_x^{n'})/(1 + A_x p_x^{n'})] \tag{11}$$

For given values of M, P, p_x, and A, the MSY from the Shepherd (S) model is always greater than from the P–T model. The discrepancy become greater for lower p_x and higher A; since SY is proportional to M and to P, it is not affected by their values. A full comparison of the properties of the two models will be published elsewhere. One result is, however, directly relevant to the questions addressed in this paper. If we fix, not A, but the initial slope of the stock–recruitment curve, then, with other parameter values being equal, A(S) > A(P–T model). For $A = A'$, the difference is 23% in both cases. That is, by applying the same parameter values and the same constraints to the resilience, with $p_x = 0.6$, the Shepherd model will predict an MSY 23% greater than that from the P–T model (Figures 4 and 5). Thus, the choice of stock–recruitment function is not a trivial one, although even this discrepancy is small in relation to our uncertainty about the general level of MSY.

If the scale of the stock–recruitment curve is normalized for very low p rather than $p = 1$, the discrepancy is much less. For example, by scaling to an estimated net recruitment rate in a depleted population, as in the fin whale case, we are, for practical purposes, fixing A directly. For the values of A in the range we have been considering for baleen whales, application of the S model will give an estimate of MSY less than 10% greater than that obtained from the P–T model.

Thus, the P–T model is the more "conservative" of the two. However, other simple stock–recruitment functions could be even more conservative; we cannot draw any general conclusions from this particular comparison. We merely note that predictions of MSY, and indeed all stock simulations, depend on the functional form of the stock–recruitment curve assumed, as well as on its scales, and that there seems to be no chance that this form could actually be observed in whales so that a choice could be made on the basis of empirical evidence.

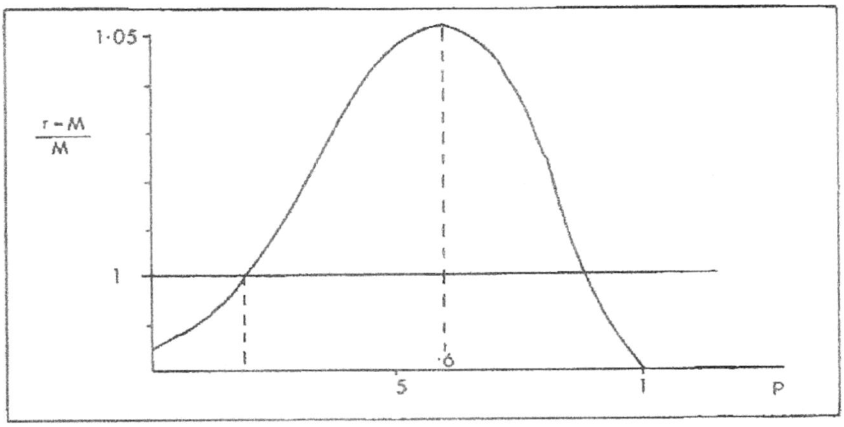

Figure 4. This illustrates the difference between the Shepherd and the Pella–Tomlinson stock-recruitment models when both are scaled to the same net recruitment rate ($r-M$), at stock size 0.2*P*, and exponents to meet the requirement that $p_x = 0.6$. The y-axis is the ratio of the net recruitment rate from the (S) model to that for the (P–T) model relative to *M*. Thus, for example, if the net recruitment rate for the fin whale had been 0.040 at $p = 0.2$, then the (P–T) model would predict that $(r_x - M) = 0.0288$. The (S) model would predict that $(r_x - M) = 0.0288 \times 1.052 = 0.0303$. If these models are scaled to each other at a low value of *p*, the difference between them is negligibly small, considering the other sources of error and variation in whale assessments.

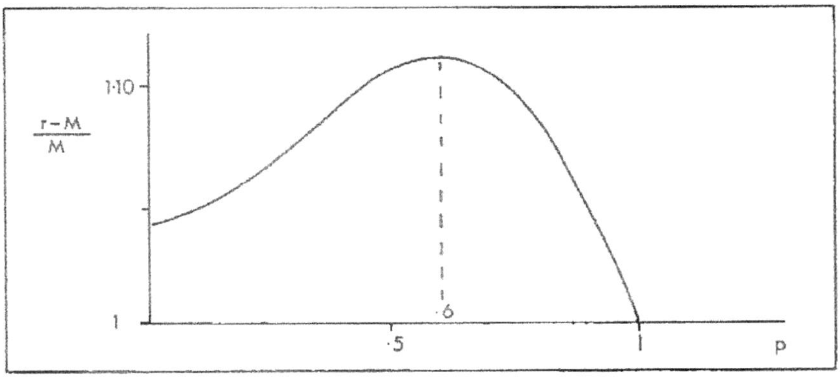

Figure 5. Here, the difference between the two models is illustrated when the recruitment rate curves (and hence also the net recruitment rate and the stock–recruitment curves) have the same slope at $p = 1$. Note that the y-axis scale is half that in Figure 4. As in Figure 4, the calculations on which the curve is based are for $p_x = 0.6$. If the stock–recruitment curves are scaled so that the density-dependent responses are the same for *p* close to 1, of which the assumption that there is no increase in recruitment with declining stock is one species case, the (S) model predicts an 11% higher value of the net recruitment at MSYL than does the (P–T) model. Note that the discrepancy between models is, as in Figure 4, greatest at the MSYL; this is a general feature of the relation between the two models.

Conclusions

Two "rules" have been suggested to, and one of them used by, the Scientific Committee as a basis for scientific advice on baleen whale catch limits. Neither, in their original form, are derived from empirical observation.

The scarce data permit rough reformulation of one of the rules (MSY = 4% of 60% of initial stock), although a thorough review is needed of the data on which it was originally based—from the Antarctic fin whale. Meanwhile, use of a modification of the other "rule"—that proposed by Gulland (1971) for application to fish stocks—could be an instrument to reduce the likelihood that more whale stocks will become depleted as a result of setting catch limits much higher than the density-dependent response of the recruitment—which has still not been measured in any whale stock—is likely to sustain. This procedure would also take into account the likelihood that species with relatively higher adult natural mortality rates have at least as vigor-us a density-dependent response as those with lower rates.

If the Scientific Committee is to continue to make assumptions as to the probable values of the main parameters in the models it adopts, in the absence of empirical evidence, whether or not this is done with interspecific comparisons in mind, an examination of the evolutionary implication of its assumptions is necessary. So also is a rigorous examination of the basis for interspecific extrapolation. A fundamental implicit assumption in many assessments made since the NMP was adopted is that the density-dependent response to stock reduction, on which all estimates of sustainable yields rest, is so strong that *reductions* from initial size lead to large *increases* in the numbers of recruits, and that the gross recruitment rate in a greatly reduced stock is several times that in the unexploited stock. This assumption not only leads to conclusions that stocks are less depleted than they are, and that sustained yields are much higher than we have any reason to believe they could be, but also to exaggeration of the likely rates of recovery from depletion.

A starting point for reconsideration of the implications of this assumption might be the arguments by G. Bateson in 1963 and L. B. Slobodkin in 1968, reformulated by the latter author in Slobodkin (1980). These are, firstly, that "successful evolution requires the maintenance of flexibility in response to environmental perturbation" and, secondly, that "this response must be maintained in the most parsimonious way." From this, Slobodkin passes to "the parsimony argument is that organisms must not make an excessive or unnecessary commitment in responding to perturbation" and pleads for "a re-emphasis in ecological theory on the actual process of responding to perturbation ... including specification of the physiological, behavioral and anatomical machinery involved in developing a solution."

Even fish, with extremely variable annual recruitment, and extraordinarily high juvenile mortality rates critically dependent on an unpredictable local environment, appear only in a few special cases to have evolved mechanisms of population regulation which lead to supercompensation, and hence very rapid recovery from environmental perturbations. The idea that this could occur in the baleen whales is contradictory with our presumption that before exploitation by humans began, these animals had relatively stable populations, having solved the problem of survival in a hazardous and variable environment by increase in size and mobility, and by ensuring a relatively high infant survival rate from a very low birth rate by efficient nurture of few offspring.

There is little, if any, evidence that baleen whale populations are food-limited. There have been no studies, as far as I am aware, comparing fat reserves in animals from unexploited and exploited populations to either support or refute the theory that the increased food available to females in exploited populations might lead to improved infant survival rates. It has been proposed that indirect evidence concerning reduced age at sexual maturity in reduced populations was largely an artifact of the methods of interpreting data (e.g., see IWC 1984b; Cooke 1985). Reanalysis of pregnancy rate data for fin whales reveals no trend in pregnancy rates over the period 1931–1975 (Mizroch and York 1984). In short, there is no biological or direct observational evidence of a significant density-dependent response when a baleen whale population is greatly reduced from its unexploited state. Thus, application even of the more conservative rules for catch limits here proposed should be with the greatest caution, pending the acquisition of substantial new knowledge about these processes.

Acknowledgments

This work was supported by the International Fund for Animal Welfare. Thanks for helpful comment on the draft by J. Cooke, Institute of Animal Resource Ecology, 2204 Main Mall, Hut B8, University of British Columbia, Vancouver, BC, Canada V6T 1WS, W. de la Mare, 2c Stanwick Road, London W14, and J. Breiwick, National Marine Mammal Laboratory, 7600 Sand Point Way, NE, BIN C15700, Seattle, WA 98115, USA.

References

Allen, K. R. 1970. A re-examination of recruitment in Antarctic fin whale stocks. Reports of the International Whaling Commission 20:48–53.

Allen, K. R. 1971. Notes on the assessment of Antarctic fin whale stocks. Reports of the International Whaling Commission 21:58–33.

Allen, K. R. 1972. Further notes on the assessment of Antarctic fin whale stocks. Reports of the International Whaling Commission 22:247–253.

Allen, K. R. 1973. Revised estimates of Antarctic fin whale populations. Reports of the International Whaling Commission 23:99–114.

Allen, K. R. 1976. A more flexible model for baleen whale populations. Reports of the International Whaling Commission 26:247–263.

Beddington, J. R., and J. G. Cooke. 1983. The potential yield of fish stocks. FAO Fisheries Technical Paper 242.

Beverton, R. J. H., and S. J. Holt. 1957. On the dynamics of exploited fish populations. Her Majesty's Stationery Office, UK.

Chapman, D. G. 1983. Some considerations on the status of stocks of southern hemisphere minke whales. Reports of the International Whaling Commission 33:311–314.

Cooke, J. G. 1985. Has the age at sexual maturity of Southern Hemisphere Minke Whales declined? Reports of the International Whaling Commission 35:335–340.

de la Mare, W. K. 1985. On the estimation of mortality rates from whale age data, with particular reference to minke whales (*Balaenoptera acutorostrata*) in the Southern Hemisphere. Reports of the International Whaling Commission 35:239–250.

Francis, R. C. 1974. Relationship of fishing mortality to natural mortality at the level of maximum sustainable yield under the logistic stock production model. Journal of the Fisheries Research Board of Canada 31:1539–1542.

Gulland, J. G. 1971. The fish resources of the ocean. Fishing News Books, Surrey, UK.

Heincke, F. 1913. Investigations on the plaice. General report 1: plaice fishery and protective regulations, part 1. Rapports et Procès-Verbeaux des Réunions du Conseil Permanent International pour L'Exploration de la Mer 17A:1–153.

Holt, S. J. 1977. A possible basis for determining limits to the allowable catch when the density dependence of the net recruitment rate is not known. Reports of the International Whaling Commission 27:230–233.

International Whaling Commission. 1971. Report of the Scientific Committee. Reports of the International Whaling Commission 21:24–43.

International Whaling Commission. 1978. Report of the Special Meeting on Southern Hemisphere sei whales. Reports of the International Whaling Commission 28:335–343.

International Whaling Commission. 1979. Report of the Special Meeting on Southern Hemisphere minke whales, Seattle, May 1978. Reports of the International Whaling Commission 29:349–358.

International Whaling Commission. 1983. International Convention for the Regulation of Whaling, 1946. Schedule. [Dated November 1983]. International Whaling Commission, Cambridge, UK.

International Whaling Commission. 1984a. Report of the Scientific Committee. Reports of the International Whaling Commission 34:35–181.

International Whaling Commission. 1984b. Report of the minke whale ageing workshop. Reports of the International Whaling Commission 34:675–699.

Mizroch, S. A. 1985. On the relationship between mortality rate and length in baleen whales. Reports of the International Whaling Commission 35:505–510.

Mizroch, S. A., and A. E. York. 1984. Have pregnancy rates of Southern Hemisphere fin whales, *Balaenoptera physalus*, increased? Reports of the International Whaling Commission, Special Issue 6:401–410.

Ohsumi, S. 1977. Bryde's whales in the pelagic whaling ground of the North Pacific. Reports of the International Whaling Commission, Special Issue 1:140–150.

Ohsumi, S. 1979. Feeding habits of the minke whales in Antarctic. Reports of the International Whaling Commission 29:473–476.

Pella, J. J., and P. K. Tomlinson. 1969. A generalized stock production model. Bulletin of the Inter-American Tropical Tuna Commission Bulletin 13:419–496.

Pitcher, T. I., and P. J. B. Hart. 1982. Fisheries ecology. Croom Helm, London, UK.

Ralls, K., R. L. J. Brownell, and J. Ballou. 1980. Differential mortality by sex and age in mammals, with specific reference to the sperm whale. Reports of the International Whaling Commission Special Issue 2:233–243.

Ricker, W. E. 1954. Stock and recruitment. Journal of the Fisheries Research Board of Canada 11(5):559–623.

Shepherd, J. G. 1982. A family of general production curves for exploited populations. Mathematical Biosciences 59:77–93.

Slobodkin, L. B. 1980. Growth and regulation of animal populations. Dover Publications, New York.

Chapter 24

Some Considerations Regarding Second-Stage Testing of Management Procedures[1]

Some implications of features of different stock–recruitment functions are identified in Document SC/41/O 8 and especially in its Appendix 3. In particular, it is shown that the use of the P–T (Pella and Tomlinson 1969) model with MSYL < 0.5 and high MSY% is unsatisfactory as it implies unreasonably high stability at very low population levels. This exaggerates the ability of a management procedure that does not include a "protection level" to avoid inadvertent extinction of the stock.

Changing Carrying Capacity

If population models are to be used for second-stage screening involving variation or trends in carrying capacity, the stock–recruitment function adopted needs to be well behaved at high population levels, including especially $p > 1$. Neither the P–T nor the May (1980) function are good in this respect, particularly with parameters that give high MSYL or MSY%, since they can lead to negative recruitment rates.

The Shepherd (1982) model does not have this feature; for all values of A and n, as $p \rightarrow \infty$, then $r \rightarrow 0$. Consideration should, therefore, perhaps be given to using this model for future testing. More flexibly, the family of functions defined in document SC/41/O 8 Addendum might be adopted. In either case, it would be necessary to specify one or more values of the resilience A. One such value might be that which is consistent with a biologically feasible upper limit to r_0, while another might be the value of A' corresponding with the set of other parameter values being used for a given test.

Inequality of Age at Maturity and Age at Recruitment

The age at recruitment is determined by various factors, including a) the availability of migrating and/or segregated whales of different ages, sizes, sex, sexual condition to whaling operations of restricted range and/or season; b) deliberate selection of the largest whales by whalers seeking to maximize commodity yield from a given numerical quota: c) the existence of minimum size limits; and d) rules affecting the catch of calves and lactating females.

Both numerical MSY and MSYL are functions of the age at recruitment. It would be in accord with the aims of the 1946 Convention to establish a management procedure which

[1]Editors' note: written by Sidney Holt in 1989 for the IWC Scientific Committee as Document SC/41/09, but not published in *Reports of the International Whaling Commission.*

would seek the global MSY (GMSY); this would involve the setting of appropriate minimum size limits or other regulations supplementing numerical quotas.

MSY Defined by Weight

The 1974 resolution under which the New Management Procedure was adopted envisaged that management would progress to a point where maximum sustainable yield in weight was the objective, this quantity, regarded as "optimal," being a proxy for commodity production or overall value.

This criterion is obviously important in the exploitation of species such as the sperm whales *Physeter macrocephalus*, where one sex is much bigger than the other. In other species, the degree of difference between the MSY number and weight criteria depends on what fraction of the life-long total growth in size occurs during the exploitable phase of life, which is itself determined by the factors enumerated above (Holt 1981). Historically, this fraction has been higher in the sperm whales and bottlenose whales *Hyperoodon ampullatus* than in most of the baleen whales. The setting of minimum size limits for the larger baleen whales may have contributed to this distinction. The absence of minimum size limits for minke whales *Balaenoptera acutorostrata* has certainly meant that, in the North Atlantic, the age at recruitment has been considerably lower than that which would maximize weight/product yield. The same is possibly true in the Southern Hemisphere.

Although the IWC has, since 1974, been formally committed to the principle of optimizing sustainable yield, rather than merely maximizing numerical yield, it proved not to be feasible to pursue that objective while commercial whaling continued on stocks that were depleted below optimal levels. The situation is now different, and it is timely for the Scientific Committee again to examine management procedures which would guide the stocks toward optimal states. Examination of this requirement, for baleen whales, should be an element in the second-stage screening now being planned.

References

Holt, S. J. 1981. Objectives of management, with particular reference to whales. Pages 349–357 *in* Mammals in the seas, Volume III, General papers and large cetaceans. FAO, Rome.

May, R. M. 1980. Mathematical models in whaling and fisheries management. Pages 1–64 *in* G. F. Oster, editor. Some mathematical questions in biology, Volume 13. American Mathematical Society, Providence, Rhode Island.

Pella, J. J., and P. K. Tomlinson. 1969. A generalized stock production model. Bulletin of the Inter-American Tropical Tuna Commission 14:421–496.

Shepherd, J. G. 1982. A versatile new stock-recruitment relationship for fisheries, and the construction of sustainable yield curves. Journal du Conseil International pour l'Exploration de la Mer 40(1):67–75.

Chapter 25

Are Measurements of Increase Rates of Depleted Stocks Useful in Estimating MSY Rate?[1]

Introduction

The original of this paper was a contribution to the Scientific Committee's Working Group on Biological Parameters and MSY Rates. It was revised for the 1989 meeting of the Committee and subsequently again in the light of discussions at that meeting and other papers contributed to it.

What Butterworth (1989) referred to as "existing estimates (or opinions)" [see Appendix 1] concerning MSY rates (hereafter MSYR[2]) derive essentially from four procedures: (a) estimation of net recruitment rate from age compositions of catches and abundance data; (b) fitting a population model to a time-series of abundance estimates and corresponding catch data, preferably spanning the presumed MSY level (hereafter MSYL) as the stock changes in abundance as a consequence of exploitation; (c) observations of rates of increase of stocks from which zero or small catches are being taken and which usually have been greatly depleted by past whaling; and (d) attempting to judge the magnitude of demographic responses to changes in population abundance, whether or not caused by human actions, and hence sustainable yield rates. Apparent changes in age at sexual maturity and pregnancy rate have been studied for this purpose, but useful conclusions have been few because such study is bedeviled by sampling biases and by the virtually complete absence of information about juvenile mortality rates. This paper is focused on the theoretical basis for opinions derived from procedure (c), bearing in mind that it is really a special case of (b). Some of the opinions based on procedure (a) were examined by Holt (1985a) and shown not to be well founded. Appendix 1 of this paper makes some comments on (b). Chapman (1987) made cogent comments which are relevant to this discussion and parallel to some of the arguments made here.

Steady-State Considerations

First consider the properties of the population model adopted for current first-stage screening of proposed revised management procedures (IWC 1988). This incorporates a Pella–Tomlinson (1969) stock–recruitment relationship (P–T) and assumes a constant post-recruit natural mortality rate M. In the simple form of the model, it is assumed that recruitment coincides with a fixed age at first parturition.

---

[1]Editors' note: written by Sidney Holt, Podere Il Falco, Loc. Acquaioli, 06062 Citta della Pieve (PG), Italy, as Document Sc/41/O 8 for the 1989 meeting of the IWC Scientific Committee and published in Reports of the International Whaling Commission 40:517–524, 1990.

[2]MSYR has been used for this quantity when it is expressed in fractional or exponential terms.

Define p as the size of a population relative to its initial abundance; p_x as the value of p for maximum sustainable yield, otherwise referred to as MSYL; r as the gross recruitment rate (ratio of the number of recruits to the size of the recruited—mature—stock), so that r_x is the gross recruitment rate at MSYL and r_o the maximum value that r tends to as p tends to zero.

Then $(r_x - M) \times 100$ = MSY%, and $(r_o - M) \times 100$ = the intrinsic rate of natural increase J%, which may also be expressed as a fraction J. If, in the first instance, it is assumed that a depleted stock is very small relative to its pre-exploitation abundance, and that it has been in that reduced state for some years, then the quantity $(r_x - M) / (r_o - M)$ is of interest; I define this ratio as f_x, the fraction that the MSY rate is of the maximum rate of increase.

If we specify the P–T model as $r = A \times M \times p(1 - p^z)$, then the following relations hold: $f_x = z / (z + 1)$ where z is the exponent of the P–T function, which determines MSYL; $p_x = [(z + 1)^{1/z}]^{-1}$; $(r_o - M) / M = A$ (definition of resilience) so that $J\% = A \times M \times 100$).

The relationship between f_x and p_x is illustrated in Table 1. Note that for $p_x > 0.5, f_x > p_x$, and vice versa.

It is obvious that if we had a good estimate of J from observations of the rate of increase of a very depleted stock, believed that the P–T function faithfully reproduced the relevant characteristics of whales, and were confident from other available information that MSYL was about 0.6 or higher, then a useful working value of f_x might be derived. If, however, we thought MSYL was 0.5 or lower, then the sensitivity of such a calculation to the value of MSYL is so great that there would be a danger of adopting an erroneous working value of f_x. Furthermore, if we presumed MSYL was high and were wrong about that, we could find ourselves using a wildly wrong value for MSY%.

Clearly, if this approach is to be pursued, the degree of dependence of these conclusions on the specific properties of the P–T model must be examined. Another yield model with the same flexibility has been proposed by May (1980). The curves of sustainable yield against normalized population size given by this model are, when graphed, visually similar to corresponding curves from the P–T model. Using my notation, May's recruitment rate function:

$r = M[A(1-p)/(1 + a \times p) + 1]$ gives $f_x = (1 - p_x)/(1 + ap_x)$ where a is a parameter, corresponding with z, which determines the MSYL: $p_x = [(1 + a)^{0.5} - 1]/a$ from which $f_x = p_x$. Some solutions to the above expression are given in Table 2 for reference.

p_x	z	f_x
0.80	11.214	0.918
0.70	5.043	0.835
0.60	2.390	0.705
0.50	1.000	0.500
0.45	0.540	0.351
0.40	0.189	0.159
0.37	0	0

Table 1. P–T model.

p_x **and** f_x	a	q
0.8	–0.937	3.984
0.6	–0.556	1.501
0.5	0	1.000
0.4	1.250	0.667

Table 2. May model.

Unlike the P–T model, the May model has no trouble dealing with low MSYL, but for $p_x < 0.5$, it would predict, from estimates of $J\%$, an MSYR much higher. For higher MSYL, the May model predicts lower MSYR than the P–T model. The relation between the two is easily seen if, for the May model, we make a simple transformation of the MSYL-determining parameter a thus: $q = 1/(1 + a)^{-0.5}$ so that $f_x = p_x = q/(1 + q)$. [See Table 2 and cf. preceding corresponding expression for the P–T model.]

The P–T and the May model are each particular cases within families of models in which the graph of net recruitment rate decreases monotonically from a maximum, when p is zero, to zero when p is unity. This graph is a straight line for $p_x = 0.5$, when both are identical. For $p_x > 0.5$, the graphs are convex, as viewed from above, and concave for $p_x < 0.5$. The graphs for the May model "bulge" less than do those for the P–T model, for the same value of p_x. I shall refer to the amount of "bulge" as the degree of convexity or concavity. Figure 1 shows the shapes of the two net recruitment rate curves for $p_x = 0.6$ and Figures 2a and 2b the corresponding sustainable yield curves.

For applying the New Management Procedure, the IWC Scientific Committee has presumed that the MSYL of baleen whale stocks is higher than 50%, and specifically that $p_x = 0.6$. This is based on the idea that density-dependent response of whales to population change occurs more when the population is near carrying capacity than when it is depleted. In the terminology of this paper, the strength of response may be indicated by the slope of the graph of f against p (with sign reversed). Fowler (1984) and others have given their persuasive reasons for this presumption, which implies that the graph is convex. There is no reason to believe, however, that the value of 0.6 is approximately correct. It expressed a consensus since 1976 for management purposes; this value was close to that implied by certain population models current at the time and since discarded. Higher values, up to 0.8, have since been suggested from time to time, but we have as yet no solid basis for pinpointing the MSYL for any baleen whale stock more precisely than this general range.

Butterworth and Best (1990) have shown that if $p_x > 0.5$, and the curve of f against p is convex at all values of p from 0 to 1, then $f_x > 0.5$ regardless of the particular stock–recruitment function. Their definition of convexity, which in my terms is that the second deriva-

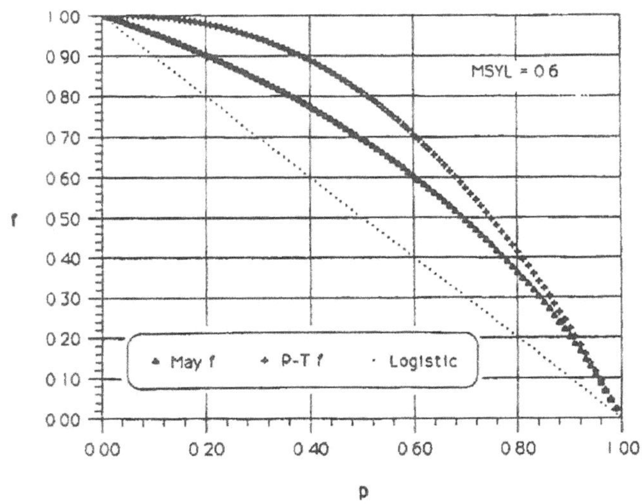

Figure 1. P–T and May functions.

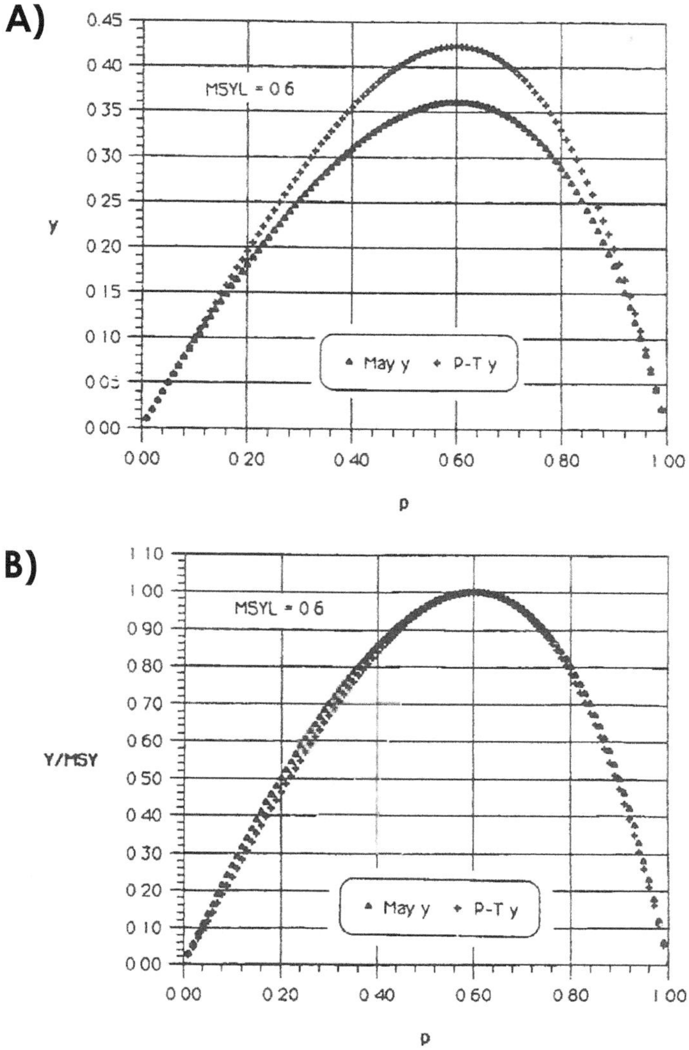

Figure 2. A) $y = p \times f$ for P–T and May functions. B) Y/MSY for P–T and May functions.

tive f' (p) is negative over the range of $0 \leq p \leq 1$, is, however, unnecessarily restrictive in the present context. In particular, the graph of f against p may be inflected yet still be generally convex in the sense that $f \geq (1 - p)$ for $0 \leq p \leq 1$. Figure 3 shows an example for $p_x = 0.6$. This has been drawn using a suitable simple expression: $f = (1 - w \times p^u)(1 - p^v)$ where $w = 0.3$, $u = 0.8$, and $v = 0.55$. In this example, $f_x = 0.48$, i.e., < 0.5.

Any function which produces an inflected curve in which, for small values of p, f decreases close to, but always above, the diagonal line representing the symmetrical logistic ($p_x = 0.5$) and then, after the inflection, turns down more sharply as p approaches 1, can show this phenomenon. It is easy to see graphically, bearing in mind the procedure for determining the MSYL geometrically,[3] that if $p_x \leq 0.5$, then $f_x \leq (1 - p_x)$ for all such functions.

---

[3]Holt (1977, pp. 232–234 and Figure 3). At $p = p_x$, the slope of the graph of f against p is equal to that of the diagonal [0,0]; $[f_x, p_x]$, but with opposite sign.

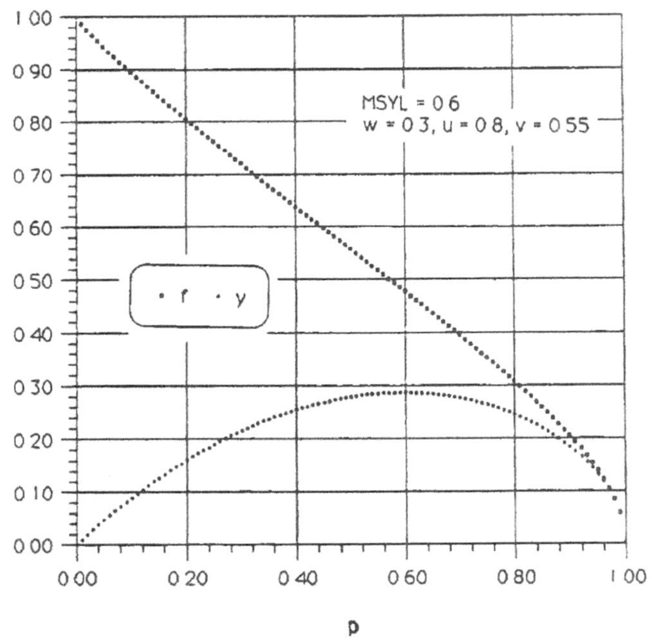

Figure 3. *f* and *y* for Holt function (this paper).

A published function which incorporates a monotonically decreasing recruitment rate with increasing population size, and which gives an inflected curve, is that of Shepherd (1982). Some of its properties were explored by Holt (1985a). The stock–recruitment function in this model is $r = M(1 + A)/(1 + Ap^n)$ where n is an exponent playing a similar role to that of z in the P–T model except that p_x is generally a function of both n and A. With $n = 1$, this becomes the "first form" of the function proposed by Beverton and Holt (1957), which with $n > 1$ and $n < 1$ functions proposed by Ricker (1954) and Cushing (1971), respectively, are mimicked. Shepherd proposed this model to eliminate certain undesirable properties of the P–T model. [From his point of view, this was specifically the fact that the latter can show negative recruitment above certain population sizes that are (temporarily) higher than carrying capacity. In a later paper (Shepherd 1987), he noted that a related advantage of his model was that it avoids the need in the P–T model for the implicit assumption that natural mortality is density-dependent at these high population levels.][4] The resulting yield curves are visually similar to those from the P–T and May models.

The Shepherd model gives $f_x = (1 - p_x^{\,n})/(1+Ap_x^{\,n})$. It is important to note here that, unlike the P–T and May models, f_x is not independent of A and hence of M, and thus since A (or M) is not specified, there is an infinite number of possible solutions for each MSYL.

In particular, taking the familiar $p_x = 0.6$ as an example, a range of f_x from 0.5 to 1.0 can be obtained by choosing a range of values of A and n (see Table 3; Figure 4).

In principle, the locus of f_x in the Shepherd model, for a given value of p_x, could be fixed if a value of M, hence of A, hence of n, were known or assumed. However, M would have to be known accurately to give results, even if the value of J had been precisely estimated and the MSYL was also exactly known. Suppose, for example, that J had been estimated at 10%,

[4]Editors' note: comment by Holt.

Table 3. Generalized Shepherd model. [Some parameter sets for p_x = 0.6.]

b	n	f_x
1.00	4.19	0.80
0.31	3.00	0.74
0	2.40	0.71
–0.90	0.20	0.52
–0.95	0.10	0.51
–1.50	–0.75	0.39
–2.00	–1.15	0.31
–5.00	–1.90	0.13

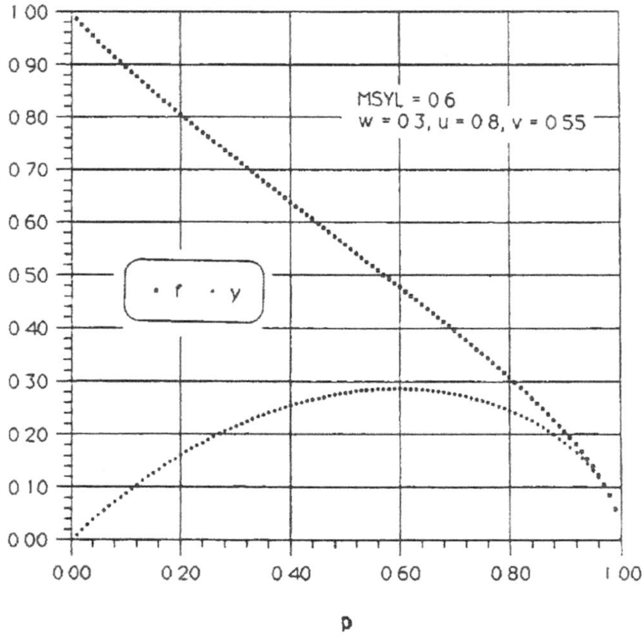

Figure 4. *f* and *y* for Shepherd function.

and M was assumed to be 0.1, p_x = 0.6; then f_x would be estimated as about 0.5. If, however, M was really 0.15, then f_x would be estimated as about 0.8.

The Shepherd model is, in fact, a special case within a family described by $f_x = (1 - p_x^{\,n})/(1 + bp_x^{\,n})$ where b is defined as A/B, B being a scaling constant that may be positive or negative. This becomes the Shepherd model when $B = 1$. When $A \ll B$, we have the P–T model. When $B = 1$, $n = 1$, we have the May model.

A feature of the generalized Shepherd model is that it can give, for $p_x > 0.5$, $f_x < 1 - p_x$. This occurs when the inflected graph of f against p is generally concave. For certain intermediate parameter values, the graph is concave-convex in the sense that it lies below the logistic straight line for low p and above it for high p.

Implications for estimating MSYR from J

This review of four stock–net recruitment curves demonstrates that a wide range of characteristic properties can be exhibited by simple functions, none having more than three parameters. Other simple functions could be formulated with similar general properties—and hence, in principle, equally suitable as any of these four for use in whale population modeling—but with different specific properties which are relevant to the problem of predicting MSYR from J; I have given one example. Butterworth (1989) wrote that "Fowler's (1981, 1984, 1988) hypothesis of convexity in net growth rate as a function of stock size implies that MSY% must be minimally 50% of [estimates of about 7% as the annual growth rate of depleted right whales *Eubalaena australis* in the South Atlantic][5] (assuming transient age structure effect is not substantial)." I can find nothing in the papers by Fowler there cited which would justify attributing to his hypothesis the restricted definition of convexity adopted by Butterworth and Best (1990) rather than the more general one which I use here. Thus, for generally convex net recruitment rate curves, f_x lies in the range 1 to approximately $1 - p_x$, and thus depends on both the value of p_x and the chosen function.

In practice, estimates of stock recovery rates are made by observing increases in depleted stocks over a considerable period, and in situations such that the exact degree of depletion is not known. Estimates of degree of depletion come from the application of population models to historical series of catches and estimates of abundance. There is an inescapable circularity in basing estimates of MSYR on any model which gives the relation between those rates and the rate of increase at a depleted stock level if that level has itself been estimated from an application of the same or a similar model to the history of exploitation. But it may be that a rough idea of the degree of depletion is available from other kinds of information. We might, for example, reach the opinion that the number of whales in a residual stock was approximately a fraction d of the original number. The consequences of such a judgment can easily be examined using a chosen model. The results will be highly model-dependent in directions which are obvious from the analyses given above. For any given model, the implications of misjudgments of the degree of depletion, perhaps in the range $0.05 < d < 0.15$, can also be examined. For example, using the P–T model, we define ${}_df_x = (1 - p_x^z)/(1 - p_d^z)$.

Then, for $p_x = 0.6$, $d = 0.05$, we would estimate ${}_df_x = 0.706$, while for $d = 0.15$, ${}_df_x = 0.713$ ($f_x = 0.705$); clearly such differences are not worth worrying about, considering the other problems we face. If, however, we think the MSYL is lower, then the matter of judging how depleted the stock is becomes more critical. For example, if $p_x = 0.4$, then ${}_df_x = 0.37$ and 0.53, respectively ($f_x = 0.16$). These substantial differences derive from the fact that although the MSYL is closer to the observed level in this case, the—in this case—concave P–T recruitment rate function becomes rather steep as small stock sizes are approached.

The difficulties of using population models to predict MSYR from J or some approximation to it do not end here. There has been considerable discussion of the possibility of depensation in greatly depleted populations, so that the rate of increase would, for many years, be less than that expected from the simple models. But data on such situations are sparse and those that exist are open to many interpretations. In his review of right whale recovery in the South Atlantic, Best (1988) listed, with references, many of the "explanations" that have been offered by a variety of authors for an apparent failure of some depleted whale "stocks" to recover under long protection. One of these could be depensation, but others such as interspecific competition, incomplete protection, environmental degradation (reduction in carrying capacity), and inadequate monitoring have been cited. Another difficulty can arise if the group of animals being monitored is not, in fact, a biologically self-contained population. Best suggested that in the situations with

[5]Editors' note: Holt's insertion as his explanation.

which he was familiar "The apparent failure of other right whale stocks [than those at Tristan da Cunha] to recover under protection may have been the result of extreme depletion in 1935, so that any subsequent recovery was initially undetectable." He gave his opinion that this undetectability is "the most parsimonious explanation" of his data.

I find Best's argument incomplete. Difficulty in detecting a certain rate of change is a function of the abundance of the animals being observed because typically the smaller the population the larger the sampling variance for an abundance estimate and hence the greater the standard error of a rate of change estimate derived from a time-series of abundance estimates. But, while admitting that change in small populations might also be undetected (although not, in principle, be undetectable) because for a variety of reasons they might be inadequately monitored, it must be said that these factors could be compensated by the fact that the more depleted populations could be expected to have higher actual increase rates than those less depleted. This last would, in principle, enhance detectability, and especially if the true stock–recruitment rate function had a form implying a relatively low value of f_x. Further, Best provides no data which permit distinction between two very different kinds of undetectability: one being that due to the rarity of the animal being monitored; the other to a very slow real rate of change. Neither of these "explanations" of a failure to detect change is *a priori* more parsimonious than the other.

Transient Phenomena

In the simple partially age-structured model, with a time-lag equal to the age at first parturition, which the Scientific Committee has been using, the rate of increase of a depleted stock (relative to the theoretical SYR) (sustainable yield rate) soon after the cessation of whaling will depend, *inter alia*, on the rate of decline of the stock in the final stages of exploitation. If the rate of decline had been high, then the incoming recruitment immediately after cessation will be substantially higher than the steady-state recruitment would be, so the rate of increase will be higher than *J*. But with each passing year, the recruitment will decline, so the rate of increase will slow, approaching the SYR. However, as the growing population passes the level from which its current recruitment is coming, the rate of increase will continue to slow to less than the SYR. At the next stage, as the recruitment begins to come from an increasing number of parents, the increase rate will begin to converge on the SYR, but such convergence can be quite slow. Thus, it should not be assumed that the stock which has regained its MSYL and remains unexploited will, by then, be growing at MSYR.

The pattern of recovery thus depends on the prior history of exploitation and on the particular values of age at recruitment and first parturition, and of the parameters which determine the MSYL. These effects will, however, be small compared with the general model problems identified earlier. The fact that any estimates of increase rate are not made instantaneously, but must be derived from many years of observation, further compounds the problem of interpretation.

Estimation of *J*

Estimates of rates of change of animal populations from censuses over time usually have wide confidence limits associated with them; whales are no exception. Best (1988) cites a value of 4.5% for southern right whales at Tristan da Cunha, but with a standard error of 2.7%;

thus the confidence limits embrace zero. Best's analysis was based on incidental sightings rather than scientifically designed censuses, so would be expected to have wide confidence limits, but other scientifically based estimates also display considerable uncertainty. Thus, an increase rate of 6.8% was reported by Whitehead and Payne (1986) for the right whale off Peninsula Valdez, Argentina, but their 95% confidence limits were 0.0–13.6%. A more recent study narrows these somewhat to 4.2–11.0% (Payne et al. 1990). Nevertheless, the combination of such uncertainty with the demonstrated model and p_x dependence of f_x makes estimation of MSYR from such measures of recovery unrealistic.

Reilly (1987) gave estimates of the rate of increase of gray whales *Eschrichtius robustus*, using two regression methods: linear, weighted by the reciprocal of the variances, and weighted exponential. These were, respectively, 2.29% (95% CL 0.33–4.24) and 2.51% (CL 0.5–4.43). In an earlier paper, Reilly (1984) used the ratio of the annual catch by USSR on behalf of aboriginals to his estimate of absolute population size to calculate an exploitation rate of 1.2%, from which "best estimates" of SY% would be roughly in the range of 3.5–3.7%, although still with wide confidence limits. However, it is not appropriate simply to add 1.2% to the limits of the increase rate because there is also uncertainty about the total population estimates. Reilly also attempted to calculate, using biological information, an estimate of *J*%; he arrived at a "best" value of 8.8%, but acknowledged considerable uncertainties in this.

There appears to be a consensus that the gray whale "stock" is indeed a biological population and that it has recovered substantially from earlier depletion. It is tempting, therefore, to try to assemble a coherent picture of the history of this stock and from this arrive at a useful estimate of MSYR. (I do not attempt to deal here with the important theoretical question of interspecific analogy whereby one might, from such an estimate, reach an opinion about the magnitude of MSYR in baleen whales generally.) However, apart from the apparent circularity of fixing the current value of p by applying a population model to catch and abundance data—which itself requires some knowledge of MSYR or a proxy of that quantity—we run into a difficulty, pointed out by Lankester and Beddington (1986) and also by Holt (1985b), that, given the catch history as far as available records permit, and the population size and its recent rate of increase noted by Reilly, the stock trajectory cannot be simulated with a population model of the kind so far applied in the IWC. Until this anomaly is resolved, it does not appear to be practicable validly to estimate MSYR for the gray whale from the observed rate of recent increase, regardless of the problems of confidence limits and model dependence. It seems unlikely that the anomaly can be resolved unless new kinds of information become available which challenge current ideas about history of gray whale exploitation and/or the biology of the species.

Conclusion

Several factors make hazardous the estimation of MSYR from proxies of J or of SYR at low levels of abundance: uncertainties in observing rates of increase, the circularity of deductions about degrees of depletion, and the complex relation of equilibrium rates with non-equilibrium rates in an age-structured population. In addition, the foregoing discussion implied that population models are being applied to stocks that are, in fact, biologically coherent populations. If, as is generally the case, population identity is highly uncertain, then another hazard is introduced, which deserves examination, but is not attempted here.

In principle, methodological improvements could reduce some of these difficulties. Discussion of criteria to be applied in seeking to transfer predictions from one species and situation to another might lead to a measure of consensus about the limits of plausibility of such transfer. But, the fundamental dependence of any such estimation on the form of the stock–recruitment rate function adopted, and the wide variety of results that can come even from simple, and in most other respects similar, models must lead to the conclusion that such efforts are destined to failure.

The best that could be hoped is that reviews of monitored rates of "recovery" will suggest approximate upper bounds to the MSYR. While this may be of general biological interest, it is of little use for the management of renewed commercial whaling, which would call rather for the determination of lower bounds if useful working estimates are not available. Although lower bounds might, in certain cases, be postulated on the basis of assumptions about the general form of published theoretical stock–recruitment rate curves, these are of limited value in the absence of independent information about the true MSYL. Furthermore, we must take into account the likelihood that any particular simple function would not necessarily hold at the extremes of population ranges even if it was a useful descriptor of the dynamics of the population over its middle range.

The circumstances discussed in this paper assume monotonic gross recruitment rate curves and a constant natural mortality rate in the exploited phase. The conclusions regarding poor predictability of MSYR from J apply *a fortiori* to the case where natural mortality is density-dependent. It should also be noted that, in that case, MSYR can be relatively high in all models, even in $A < A'$, where A' is the limiting value for super-compensation.

The claim by Butterworth and Best (1990) that the "rapid" increase rate of the right whales in the South Atlantic is at variance with my hypothesis that super-compensation would not be expected in baleen whales depends on their consideration only of the P–T model. They do recognize that density-dependence of natural mortality could affect the validity of their claim, but not necessarily only those which permit p_x to be less than 0.5 can lead to a different conclusion. This should be obvious from the fact that existence or absence of super-compensation is determined by the slope of the right-hand limb of the stock–recruitment rate curve and the natural mortality rate. The P–T function for $p_x > 0.5$ gives a steeper right-hand limb than that of May, so that other things being equal, a value of J which implies super-compensation in the P–T model would not necessarily do so in the May model applied to the same data. [For reference, $A'(\text{P–T}) = 1/z$, and $A'(\text{May}) = 2 - a = (3q^2 - 1) / q^2$]. The Shepherd model, and especially the generalized Shepherd model, can have considerably less steep slopes in the right-hand limb than either the P–T or May models, and, of course, these slopes vary with the assumed value of p_x. I believe there is little point in attempting to "test" the hypothesis of "no super-compensation" by consideration of rates of population recovery from levels below MSYL.

Acknowledgments

The study reported here was supported by the International Fund for Animal Welfare. In approaching this problem, I benefited from discussions with William de la Mare and Justin Cooke. I am particularly grateful also to Douglas Butterworth whose careful review of a draft of this paper and suggestions for corrections and improvements led to its substantial revision.

References

Best, P. B. 1985. The recovery of right whales off South Africa. Paper SC/37/PS28 presented to the IWC Scientific Committee, June 1985. Unpublished. [Available from the IWC Secretariat]

Best, P. B. 1988. Right whales *Eubalaena australis* at Tristan da Cunha—A clue to the "non-recovery" of depleted stocks. Biological Conservation 46:23–51.

Beverton, R. J. H., and S. J. Holt. 1957. On the dynamics of exploited fish Populations. Her Majesty's Stationery Office, UK.

Butterworth, D. S. 1989. A note on the consideration of the MSY% in the Comprehensive Assessment. Reports of the International Whaling Commission 39:144–145.

Butterworth, D. S., and P. B. Best. 1990. Implications of the recovery rate of the South African right whale population for baleen whale population dynamics. Reports of the International Whaling Commission 40:438–448.

Chapman, D. G. 1987. Southern Hemisphere minke whale recruitment rate. Reports of the International Whaling Commission 37:79.

Cushing, D. H. 1971. The dependence of recruitment on parent stock of different groups of fishes. Journal du Conseil Permanent International pour l'Exploration de la Mer 33(3):340–362.

de la Mare, W. K. 1987. Fitting population models to combinations of CPUE series and absolute abundance estimates. Reports of the International Whaling Commission 37:379–381.

Fowler, C. W. 1981. Density dependence as related to life history strategies. Ecology 62(3):602–610.

Fowler, C. W. 1984. Density dependence in cetacean populations. Reproduction of Whales, Dolphins and Porpoises. Reports of the International Whaling Commission, Special Issue 6:373–379.

Fowler, C. W. 1988. Population dynamics as related to rate of increase per generation. Evolutionary Ecology 2:197–204.

Holt, S. J. 1977. A possible basis for determining limits to the allowable catch when the degree of density dependence of the net recruitment rate is not known. Reports of the International Whaling Commission 27:230–233.

Holt, S. J. 1985a. The classification of whale stocks and the determination of catch limits under the current IWC management procedure, with limited information. Reports of the International Whaling Commission 35:487–494.

Holt, S. J. 1985b. The gray whale revisited. Paper SC/37/PS20 presented to the IWC Scientific Committee, June 1985. Unpublished. [Available from the IWC Secretariat]

Holt, S. J. 1985c. Classification of North Atlantic minke stocks. Paper SC/37/Mi4 presented to the IWC Scientific Committee, June 1985. Unpublished. [Available from the IWC Secretariat]

Holt, S. J. 1986. Northeast Atlantic minke stock: Reassessment. Paper SC/38/Mi3 presented to the IWC Scientific Committee, June 1986. Unpublished. [Available from the IWC Secretariat]

Holt, S. J. 1989. 'The state of the Northeast Atlantic minke whale stock'—a critique. Reports of the International Whaling Commission 39:213–218.

International Whaling Commission. 1987. Report of the Sub-Committee on Northern Hemisphere Minke and Other Baleen Whales. Reports of the International Whaling Commission 37:89–112.

International Whaling Commission. 1988. Comprehensive Assessment Workshop on Management. Reports of the International Whaling Commission 38:163–168.

Lankester, K., and Beddington, J. R. 1986. An age structured population model applied to the gray whale (*Eschrichtius robustus*). Reports of the International Whaling Commission 36:353–316.

May, R. M. 1980. Mathematical models in whaling and fisheries management. Pages 1–64 *in* G. F. Oster, editor. Some mathematical questions in biology, Volume 13. American Mathematical Society, Providence, Rhode Island.

Payne, R., V. Rowntree, J. S. Perkins, J. G. Cooke, and K. Lankester. 1990. Population size, trends and reproductive parameters of right whales (*Eubalaena australis*) off Peninsula Valdes, Argentina.

Individual Recognition of Cetaceans: Use of Photo-Identification and Other Techniques to Estimate Population Parameters. Reports of the International Whal-ing Commission, Special Issue 12:271–278.

Pella, J. J., and P. K. Tomlinson. 1969. A generalized stock production model. Bulletin of the Inter-American Tropical Tuna Commission 14:421–496.

Reilly, S. B. 1984. Observed and maximum rates of increase in gray whales, *Eschrichtius robustus*. Reproduction of Whales, Dolphins and Porpoises. Reports of the International Whaling Commission, Special Issue 6:389–399.

Reilly, S. B. 1987. Reanalysis of rate change in the California–Chukotka gray whale stock, 1967/68–1970/80. Reports of the International Whaling Commission 37:347–349.

Ricker, W. E. 1954. Stock and recruitment. Journal of the Fisheries Research Board of Canada 11:559–623.

Shepherd, J. G. 1982. A versatile new stock-recruitment relationship for fisheries, and the construction of sustainable yield curves. Journal du Conseil International pour l'Exploration de la Mer 40(1):67–75.

Shepherd, J. G. 1987. Towards improving stock-production models. *In* Report of the Working Group on Methods of Fish Stock Assessments (WGMG), 9–16. June 1987, Copenhagen, Denmark. ICES CM Document 1987/Assess:24.

Walløe, L., R. M. Anderson, R. J. H. Beverton, and A. Semb-Johansson. 1987. The state of the Northeast Atlantic minke whale stock: Report of the group of scientists appointed by the Norwegian Government to review the basis for Norway's harvesting of minke whales. Økoforsk, As, Norway.

Whitehead, H., R. Payne, and H. Payne. 1986. Population estimates for the right whales off Peninsula Valdez, Argentina, 1971–1976. Right Whales: Past and Present Status. Reports of the International Whaling Commission, Special Issue 10:169–171.

Appendix 1: An Opinion about MSY%

Butterworth (1989) suggested that it would be helpful if Scientific Committee members were to indicate sources of estimates and opinions concerned the value of MSYR. This Appendix represents my attempt to do so insofar as the minke whale *Balaenoptera bonaerensis* and *B. acutorostrata* is concerned. I described in detail in the original of Holt (1985a) my opinion concerning the larger species to which a value of 4% had customarily been assigned for purposes of applying the New Management Procedure (NMP). This description was focused on several years of discussion in the Committee about the fin whale *B. physalus*. In fact, the crucial debate was about the sei whale *B. borealis* in the Southern Hemisphere (de la Mare, personal communication); this has not yet been documented and critically reviewed in the context of the Comprehensive Assessment.

Butterworth (1989) wrote that "The wide divergence of opinions ... about appropriate values for MSY% for baleen whales seems at present to be conventionally encompassed by the range 1%–4%." I do not agree, for reasons that will become clear. In fact, the values 1%–4% now being used for screening proposed revised management procedures were chosen to give alternative "high" and "low" values; they do not represent the upper and lower limits of documented estimates.

The range 1–4% stems from a discussion in the sub-committee on Southern Hemisphere minke whales *B. bonaerensis* in 1984. There, D. G. Chapman and I, "from results from SC/36/O 12 [subsequently published as Holt (1985a) and from analyses from Northeast Atlantic minke whales *B. acutorostrata* (Holt, 1985c)] concluded that recruitment rates were likely to be less than 2%." In that same discussion, P. B. Best (1985) expressed his opinion "that the available data for baleen whales indicated rates in the range 1% to 4%," and he felt that, from its life history strategy, the minke whale might be at the upper end of this range. These opinions were subsequently expressed in almost identical language in the full Committee. Some members of the Committee then used an assumed 2% rate to derive "replacement yields" which provided their basis for proposals for catch limits for the 1984/85 season.

Best did not fully explain the logic of his opinion, although I understand that he thought that as—in his view—the minke whale has a yearly calving interval, in apparent contrast with species having two-yearly intervals, this gave a potential for greater density-dependent response of reproductive rate to population change. While this may be theoretically true (but not complete, because the recruitment rate depends also on the natal and juvenile mortality rates and their possible density-dependence), it does not, I think, help us evaluate the response that actually occurs.

My own opinion was based on my experience of using the published Norwegian Barents Sea CPUE series, the marking estimates of the Northeast Atlantic minke whale, and the associated historic catch series—all of which had previously been accepted by the Scientific Committee and used to estimate the level to which this stock had been depleted. Taking $p_x = 0.6$ and $M = 1$, as was then customary, I obtained "best" values of $A = 0.28$ and 0.07 ($A' - 0.42$) corresponding with MSYR = 1.97% and 0.49%, respectively. I considered the larger values to be the better estimates because they came from using a lower population target, which came from a marking estimate in which mark loss had been taken into account. On the basis primarily of that assessment, the Committee classified this stock as a Protection Stock.

My assessment was challenged by Norwegian scientists at the 1986 meeting on two grounds: first, that the catch statistics had been falsified by progressively increasing non-reporting of catches and that this reporting had exaggerated the CPUE decline; second, I had improperly excluded early CPUE data during which substantial numbers of bottlenose whales *Hyperoodon ampullatus* had been caught as well as minke. Subsequently, an investigation conducted under the leadership of L. Walløe (Walløe et al. 1987; see also Holt 1989) could not confirm the alleged trend in underreporting. Detailed catch data were not, at the time of writing, publicly available to test the second criticism.

Meanwhile, two other facts had emerged. One was that the age at recruitment had changed greatly during the history of this industry. It had been initially lower than the seven years assumed on the basis of Barents Sea catches because originally the main operations had been coastal ones form Vesterfjord, where a high percentage of calves and juveniles were taken. Although this practice continued, the offshore catching later became more important. Examination of the consequences of this change calls for new assessments that have not yet been made. The other emergent fact was the demonstration that CPUE in Vesterfjord has declined continuously; this series of data was not confounded by possible multispecies operations nor by the large changes in vessels and locations of operations that had occurred in the Barents Sea.

At the 1986 meeting, I presented a reassessment, using the Vesterfjord data as well as the Barents Sea data, and an improved method of fitting the trajectory (Holt 1986). This showed that the two CPUE series used were not incompatible, and tended to confirm the previous low value of MSYR. The "best" estimate for A was 0.079, corresponding with MSYR = 0.6%.

Meanwhile, de la Mare (1987) produced a statistically more appropriate procedure—the so-called "Hitter–Fitter" program—for making assessments of the kind I had done by crude methods. Application of this procedure to the Northern Hemisphere minke stocks during the 1986 meeting (IWC 1987) showed, as de la Mare had warned, that while it could reliably estimate degree of depletion (provided, of course, that relative and/or absolute abundance data were reliable), it has little statistical power for these fisheries for the estimation of A and hence of MSYR.

Thus, I would not now put much weight on estimates of resilience by any such procedure. Nevertheless, there appears to be no intrinsic reason why estimates so obtained would be biased downward. So, a value of 1%, while it may be convenient for screening proposed management models, does not, in my opinion, represent a scientifically justifiable lower bound; the true value could as well be lower as higher than that.

Appendix 2: Implications of Different Stock–Recruitment Functions for Second-Stage Screening Protocols of Proposed Management Procedures

The different specific shapes of stock–recruitment curves at low p have important implications for modeling the recovery of depleted stocks. This is particularly so if $p_x < 0.5$ and if the P–T model is used. This is because that model gives a very steep curve near the y-axis.

IWC (1988) proposed a test of the consequences of the population model being incorrectly specified, involving left- and right-peaked yield curves, using the P–T model with MSYL 0.4 and 0.8 and maintaining an MSYR range of 1 to 4%.

For MSYR = 0.4% and MSY% = 4, J% is 25 from the P–T model. A stock reduced to 5% of its initial level will have a sustainable yield rate of 16%; if reduced to 10% of initial level, it will have a rate of increase of 15%. Thus, a management procedure which left open the possibility of occasional depletion of the stock (for example, a procedure in which no "protection level" close to MSYL is provided) may appear to work far better than it would it practice simply because the screening model allows an unrealistically vigorous density-dependent response.

For a generalized baleen whale model to serve as a test bed for a new management procedure, it would seem reasonable to look at a simple case in which the sex ratio is 50:50 and mature females conceive every second year. This gives a conception rate of 0.25 in the mature/recruited population. Then consider the (wildly optimistic) case in which all fetuses are carried to full-term and all births are live. Then assume the annual survival rate from birth to recruitment is no less than that among recruited animals—again, an excessively optimistic assumption. With M = 0.05 and age at first parturition = 8 years, the gross recruitment rate could not exceed 16.5%, so J% could not exceed 11.5%. So, the P–T model would be inappropriate even for a whale species of which the mature females each carry more than one live calf per year.

The May model would keep us just within "feasible" bounds since it would predict J% = 10 in these circumstances, but considering the optimism of the above assumptions, I do not think even its use can be recommended.

Possible courses of action in the second stage of screening would be (a) abandon the P–T model in favor of the Shepherd model, arbitrarily choosing appropriate parameter values; (b) abandon the testing of very low values of p_x; (c) bring the upper MSYR limit substantially lower than 4%. In this case, if use of the P–T model was continued, to examine low p_x, the high MSYR value could not reasonably be more than 1.8.

For second-stage screening involving variations or trends in carrying capacity, the stock-recruitment function used should, I think, be chosen to be well-behaved at high population levels, including $p > 1$. Neither the P–T nor the May functions are entirely suitable in this respect, especially with parameters that give high MSYL and/or MSYR, since they can lead to negative recruitment rates. In the P–T model, recruitment becomes negative when $p > [(1 + A) / A]^{1/z}$ (define threshold a p'). In the May model, the recruitment curve for $p_x > 0.5$ is inflected at $p = -1/a$ and $p' = (A + 1)/(A - a)$. The following table gives exemplary values.

		p'	
p_x	MSY%	P–T	May
0.5	4	1.63	1.63
	1	3.50	3.50
0.6	4	1.30	1.24
	1	1.88	1.50
0.8	4	1.07	1.03
	1	1.17	1.05

Chapter 26

Sensitivity of Management Procedures to Discrepencies between Catch Limits and Catches[1]

Testing of proposed algorithms for setting catch limits has, to date, involved the implicit assumption that catch limits and catches are identical. There are a number of ways in which that assumption may not be valid. Throughout the history of the management of commercial whaling, catches have more often than not been inconsistent with the prevailing scientific advice. In the great majority of these cases, the catches have been higher than those advised. The following situations can be identified.

A. Cases in which it is known that the catches differ from the advice:

1. The Commission sets a catch limit higher than consensus advice.
2. The Commission sets a limit which is between lower and higher numbers provided by the Scientific Committee.
   2.1. Committee consensus on a range
   2.2. Different views within the Committee
3. The Commission fails to decide on a catch limit, and catches subsequently taken differ from the advised catch limit.
4. The Commission decides on a catch limit, but excess catch is taken under objection to the decision.
5. Catches are taken under special permits for scientific purposes over and above the established catch limit.
6. Infractions are subsequently reported.
7. Catches are taken by non-Members or ex-Members of the Commission, and reported.
8. An agreed catch limit is not reached for operational reasons.

B. Cases in which it is not known that catches differ from the advice:

1. Excess catches are taken by Members of the Commission, but not discovered.
2. Catches are taken by non-Members of the Commission, but not reported.
3. Excess catches are taken, or thought to have been taken, by Members of the Commission, subsequently discovered, but not reported as infractions.

---

[1]Editors' note: written by Sidney Holt, Delegate for Seychelles, on May 21, 1990 as Document SC/J90/Mg4 for the 1990 meeting of the IWC Scientific Committee; it was never published in *Reports of the International Whaling Commission.*

I suggest that the Commission, in taking its eventual decisions regarding a Revised Management Procedure and its implementation, would benefit from an appraisal by the Scientific Committee of the sensitivity of the performance of any proposed procedure to such discrepancies in future. This could help the Commission decide, on a rational basis, on priorities for improvements in implementation.

When a new algorithm is incorporated in the Schedule, all other basic features of the 1946 Convention remain in place. It is, therefore, not reasonable to assume that catches will always conform with a single number derived from the perfect application of the algorithm. None of the above sources of discrepancy will, in principle, be eliminated by the acceptance by the Commission of revised procedures replacing Paragraph 10 (and 13) of the Schedule. Annual catch limit decisions will still be taken by consensus if possible, but if that cannot be reached, a three-quarters majority will be needed. The right of objection will remain, as well as the right to issue special permits. Whaling outside the Commission remains possible, and Members may still withdraw from membership. National inspection and even international observer schemes are likely to remain imperfect.

The possibility will remain that even if the Commission adopts a new algorithm by consensus, involving certain commitments to abide by it, this commitment will not subsequently be maintained; we have examples of that process both before and after the adoption of the New Management Procedure in 1974–1975. We must also accept that the possibility will remain of the Scientific Committee being unable to agree on an interpretation of new data, for example, from a sightings survey, no matter how much it strives for consensus.

My intention in raising this matter is not to be pessimistic about the likely effectiveness of any new algorithm, but to suggest that the Committee should recognize reality and try to give some guidance to the Commission as to the likely consequences of various scales, types and patterns of discrepancy. One feature of some of the processes identified above which lead to discrepancies is that the scale of discrepancy may be correlated with the pattern of change in catch limits. We have experienced one such case recently where a reduction in a catch limit was said to have led to a substantial increase in underreporting of catches.

I do not suggest that the Committee undertake a full set of simulations of various scales and patterns of discrepancy, but it seems that some testing will be necessary if we are to offer advice as to which processes leading to discrepancies may be more detrimental to the performance of a new procedure than others, and hence should receive greater attention from the Commission.

Chapter 27

Implementation of a Revised Management Procedure for Commercial Whaling[1]

Introduction

Since a global moratorium on commercial whaling was declared in 1982 by the International Whaling Commission (IWC), scientists have been attempting to devise a set of rules for determining annual catch limits for each species and population of whales, under which a cautious resumption of commercial whaling might be permitted. There appears to be a widespread perception that once an appropriate set of such rules has been recommended by the IWC Scientific Committee (SC), whaling could safely resume. However, it has been pointed out by governments and by other international organizations and non-governmental bodies that several other conditions would need to be met. One of these is scientific in nature: that the effects on marine ecosystems of removing large quantities of the top predators—the whales—is known, as set out in the World Conservation Strategy. The non-scientific conditions include: formal adoption of the set of rules by the Commission; improvements in enforcement of regulations; commitment to certain kinds and levels of research and monitoring; freer access by interested parties to data collected during whaling operations. Use of more humane hunting and killing methods has also been emphasized.[2]

These suggested conditions for any relaxation of the moratorium are based on long experience of inadequate monitoring and research, failures to enforce regulations, and hindrance of access by Commission Members to vital statistical information. Nevertheless, representatives of those few countries that are eager to resume commercial whaling have charged the majority with procrastination in the process of establishing revised management rules; two of them have threatened to withdraw from the IWC if they are not awarded "interim" catch limits either under the old rules or arbitrarily.

While all the above conditions are important, they are not sufficient to justify a hasty resumption of commercial whaling. One of the obvious features of the history of the IWC is the frequency with which the Commission has received the requested advice from its scientists and not acted in accordance with it. There are many ways in which this slippage at the

[1]Editors' note: written by Sidney J. Holt and Carole Carlson in 1991 for the International Fund for Animal Welfare and reproduced here with the permission of IFAW. A one-page list of acronyms and a four-page synopsis in the original version are omitted.

Authors' comment: we are grateful for ideas from several professional colleagues inside and outside IFAW, and also acknowledge with thanks editorial and production assistance from Mrs. Christine Watt, Information Services Executive, IFAW.

[2]See "Commercial Whaling Humane?" IFAW April 1991. Also "Why Whales?" WDSC 1991, and recent publications by the Royal Society for the Prevention of Cruelty to Animals (UK).

interface between science and decision-making can occur, and many reasons for it. Those reasons include absence of scientific consensus, ambiguous formulation of advice, intrinsic flaws in the set of rules which the scientists are supposed to follow, and the procedures by which the Commission comes to decisions by a sequence of votes.

There is, of course, also the right of Member States to object to decisions, provided they do so within 90 days. Ultimately, they may withdraw from the Commission if they do not like its decisions. More importantly, plausible *threats* to withdraw can be sufficient to weaken the will of other Members to resist efforts to depart from the scientific advice or from agreed procedural rules. This last has recently become important.

The IWC has always exercised its power to enact several kinds of conservation measures simultaneously, in addition to classifying species or stocks and setting catch limits for those left open to exploitation. There are, for example, long-standing rules against the killing of suckling calves and lactating females, against the killing of animals of all species, except minke whales *Balaenoptera acutorostrata* and *B. bonaerensis*, less than a certain minimum length, and against the operation of factory ships in certain regions, as well as rules for setting closed seasons and closed areas. In this paper, however, we concentrate entirely on the way in which catch limits have been determined, and the related matter of classifying species or stocks according to their assessed states.

The IWC has, three times, attempted to establish rigorous, scientifically based procedures for setting catch limits. The first was in 1960. Then, it undertook, by 1964, to bring Antarctic baleen catch limits into line with scientific evidence concerning sustainability, and it set in motion a process for obtaining the best available scientific advice. The failure of the IWC to do this, and its continuing failure throughout the 1960s, were instrumental in gaining support outside the Commission for an immediate ten-year moratorium on commercial whaling, which found expression in a resolution of the UN Conference on the Human Environment in 1972.

The second commitment to firm procedures was in 1974 when, in response to the 1972 resolution, the New Management Procedure (NMP) was launched, being formally adopted in 1975. The evident failure of that effort stimulated resumed pressure worldwide for a moratorium, which eventually found expression in the IWC's 1982 decision in favor of an indefinite moratorium with delayed implementation.

The third attempt to formulate workable rules is the current effort to devise a Revised Management Procedure (RMP) under which it might be safe to modify the 1982 decision.

It might seem that the events of the 1960s are old history of little relevance to current problems; it is, after all, well known that the IWC failed to manage Antarctic whaling in that period. However, those events illustrate well most of the problems which still exist with the IWC decision-making procedures. Because few participants in the IWC are familiar with them, we review them here in some detail. We follow that with a similar, though necessarily more compressed, analysis of the failed implementation of the NMP from 1977 to 1985. This last is illuminated by a review of the principal events in 1976 when the will of Member States to honor the compromise reached in 1974 was first tested.

The Commission passed the test then, but has failed it in most years since. The wording of Schedule paragraph 10, which has, since 1975, defined the rules for setting commercial catch limits, requires that the Commission *shall* take certain actions in certain circumstances. But, in practice, it has very frequently set *ad hoc* catch limits and agreed to ill-defined stock classifications with less and less regard to those rules. Such action has occasionally been facilitated when the SC has, in good faith, offered advice far beyond the scope of policy and procedural guidelines given to it by the Commission. In this connection, we should mention the disruptive effect on orderly assessment and management processes of the SC being called upon from time to time to answer specific "leading" questions—usually ambiguous ones—put to it by individual Commissioners from countries seeking to justify a particular *ad hoc* catch limit.

Any and all of these practices could be the rock on which a revised set of management rules will founder. We emphasize that we have not merely listed *theoretically possible* events, but rather cataloged only those types of situations for which we have found several examples. After reviewing relevant aspects of decision-making in the IWC, we offer suggestions as to how some improvement in that process could be effected by appropriate drafting of a replacement for paragraphs 9 and 10 of the Schedule, guided by a few basic principles.

The Transition from NMP to RMP

The approach now being taken by IWC scientists to the task of producing a set of rules for determining annual catch limits for each species and population of whales is roughly as follows. They first set up a simple idealized mathematical model of a whale population and then subject it to simulated exploitation for a century or more under a series of alternative management regimes. The performances of the regimes are evaluated against a set of policy objectives that has been determined by the IWC and which is derived from an interpretation of the purposes of the Commission, as laid out in the 1946 Convention for the Regulation of Whaling, the statute under which the IWC is established.

The specified objectives are the avoidance of possible extermination or excessive depletion of any population, the restoration of depleted populations, the attainment of the highest feasible sustainable yields in the long term, and the avoidance of sudden large changes in catch limits so that relative stability of the whaling industry can be assured. These were also the declared aims of the NMP.

The NMP was constructed by scientists who, at the time, were confident that knowledge about the dynamics of whale populations was adequate for its application. It soon became clear that they had been too optimistic. Furthermore, there turned out to be many circumstances to which the rules were not applicable. This procedure under which the SC was required to advise each year on catch limits for the following whaling season, for every whale "stock," had evidently been formulated with insufficient thought.

In 1985, Dr. William de la Mare proposed to the SC a new approach to the formulation and testing of decision rules for establishing catch limits.[3] He showed how the NMP was not efficient at meeting declared objectives even in circumstances to which it could, in principle, be applied, and he indicated how an efficient alternative might be developed. De la Mare's approach to this problem is now being followed by five groups of "procedures developers" within the SC. The testing arrangements have become very complex, the aim being to devise a set of rules which covers all conceivable circumstances and the efficiency of which does not depend critically on particular assumptions about dynamics of real populations as embodied in particular mathematical models.

[3]"Simulation studies on management procedures," Reports of the International Whaling Commission 36:429–450, 1986.

It is expected that the SC will, on the basis of these tests, be ready to propose to the Commission, by 1992, if not in 1991, an RMP. Meanwhile, the NMP remains in the Schedule, although it has effectively been put in abeyance by the moratorium decision. That decision was subject to review at the 1990 meeting of the IWC, which decided not to amend it at this time. However, three countries have expressed their wish to resume commercial whaling now, either under the flawed NMP rules or under "interim" catch limits set arbitrarily. The majority of Member States of the IWC have made clear their intention not to agree to any such catch limits until an RMP has been agreed and a formulation of it has been voted into the Schedule as a replacement for the NMP.

Reaching Decisions

We now describe the rules for setting annual catch limits, whether under the NMP or an RMP, in the context of the entire regulatory scheme available to the IWC under the 1946 Convention.

Catch limits are set by amending the Schedule in accordance with provisions in Articles III(2), V(1)(e), and V(2) of the 1946 Convention. Decisions involving amendment of the Schedule are binding on all Members unless they object (see below). Article V(1)(e) empowers the Commission to adopt regulations fixing the maximum catch of whales to be taken in any one season. Article III(2) says that any amendment requires a three-fourths majority of Members voting (abstentions not being counted). Article V(2) specifies conditions that must be met in principle by any amendment: (a) that it shall be in accordance with the objectives and purposes of the Convention; (b) that it shall be based on scientific findings; (c) that it shall not involve allocating specific quotas to any factory or ship or land station or any group of those: and (d) that it shall take into consideration the interests of the consumers of whale products and the whaling industry.

The Commission's practice with respect to Article V(2) does not normally involve specific consideration of these four conditions, unless a particular proposal for amendment is challenged by one or more Members. The 1982 moratorium decision was challenged primarily on grounds of not fulfilling conditions (b) and (d) and, in a loose sense, (a). However, the proposal was defended with to respect to (b) on two related grounds: some members of the SC had supported it, and it was argued that the absence of adequate and specified scientific knowledge was itself a relevant scientific finding. With respect to (d), it was pointed out that the Commission was supposed to serve the interests not only of present, but also of future consumers and industry, and that the Convention expressly recognized the need to safeguard whale stocks for future generations.

The letter of V(2)(c) has always been followed rigorously. But the way that the NMP was implemented has resulted in numerous *de facto* allocations to particular nations or groups of nations, so that it is now common for delegations from whaling countries to refer to potential catches accessible to them in the first person possessive: "our quota." This came about because of drastic reductions in factory ship operations and in the number of whaling nations, and by agreements being made effectively to divide the oceans, especially in the Northern Hemisphere, into "management areas" in such a way that generally there were active land stations of only one country within each such area. In few, if any, cases were the boundaries of these management areas based on solid biological information. One of the procedural difficulties in implementing the NMP has been seriously to consider revision of boundaries, or even to examine the implications of various options, in a situation in which they were originally adopted *provisionally* by consensus, but now need a three-fourths majority to amend. Replacement of the NMP by an RMP will involve a completely different approach to the stock boundary question.

Practically from the time of the second annual meeting of the IWC in 1950, two separate organs have considered the matter of catch limits: the SC or its precursors and the Technical Committee (TC). Although the SC formally reports to the Plenary of the Commission, its report is normally referred to the TC early in each annual or special meeting of the Commission.

The TC, now always presided over by the Vice-Chairman of the Commission, considers the advice offered by the SC and may do a number of things with it. The scientific advice may be accepted by consensus or acceptance may be determined by a vote. In this case, as in all proceedings of the TC, decision is by simple majority of Members present, excluding abstentions. It is not uncommon, however, for the scientific advice not to be adopted in a straightforward way. In earlier years, a scientific consensus often did not command a majority in the TC. Instead, a series of numbers were offered by delegations according to their particular economic interests, and from these would emerge either an agreement or a majority position. In recent years, the TC has more often than not accepted a scientific consensus if there was one, but there has frequently been no such consensus. The TC will usually then choose one of the alternative scientific views offered or decide on a compromise among these. Occasionally, the TC has been unable to come to any decision and so has passed the matter back to the Plenary.

In the Plenary, where a similar array of options may emerge, decision is usually more problematic, mainly because of the three-fourths majority requirement. The TC's majority recommendation, if there is one, may be accepted with or without a vote, but sometimes it will not attract sufficient votes. It may then be amended in such a way as to attract the necessary additional votes. Occasionally, no decision can be reached; we return later to the implications of this.

The IWC has its own written and unwritten rules about voting procedure. These have been fairly consistent over time, but they have varied in their application, sometimes at the whim of the current Chairman of the Commission. Proposals from the TC do not generally require secondment, but other proposals, including all proposals for amendments of proposals, must be seconded. In principle, debate follows the normal rules of formal international meetings. Amendments to proposals are voted upon in the reverse order to that in which they were made. The majorities required for adoption of an amendment, i.e., to incorporate it in the original proposal, have not been consistent; sometimes a simple majority has been treated as sufficient, but sometimes a three-fourths majority has been required. The amended proposal or the original, depending on the outcome of the vote, is then put forward for adoption by three-fourths majority.

Occasionally, it will be suggested that a proposal be voted upon paragraph by paragraph or section by section; if that suggestion is not generally accepted, it will itself be put to vote (simple majority). A proposal which has not been adopted may generally not be put forward a second time, except by a consensus agreement or by a Chairman's ruling which is not challenged and overturned by a simple majority vote.

The effects of there being no proposal adopted depend on the nature of the proposal and the way the relevant section of the Schedule to which amendment was being sought was written. For example, under the NMP, each stock of each species was supposed to be "classified" in one of three categories according to its assessed state. In fact, it soon became clear that there were, in fact, four categories—one category, in practice, being "unclassified." These classifications remained in place until changed by a proposal which obtained a three-fourths majority. But, catch limits were usually set for one year only, so that if one year no proposal obtained a three-fourths majority, no catch limit was set. This does not mean it became zero automatically, but rather that the whaling countries concerned were free to determine their own limits. An exception was that a "block quota," defining a total that may be taken over a certain period of years (usually also with a proviso that the catch in any one year may not exceed a certain number), would stand throughout that period unless altered by a three-fourths majority.

Before or after a vote, a delegate may express his country's "reservation" concerning a decision about to be taken or just taken. This expression of disapproval enters the record, but has no specific legal effect. However, amendments to the Schedule only become effective 90 days after they have been notified to governments, and before that time, any government may lodge a formal "objection," the effect of which is to exempt that government from the decision. If any such objection is lodged, a further 90 days will pass before the amendment comes into effect; this is to give the opportunity to more governments to object and thus maintain equity with the first objector. Countries that express reservations do not always object; sometimes countries that do not express reservations nevertheless subsequently object.

The Commission may, under Article VI of the Convention, make *non-binding* recommendations to any or all contracting governments on any matters which relate to whales or whaling and to the objectives and purposes of the Convention. Such recommendations require only a simple majority of Members present and voting. Their relevance here is that they can be precursors to binding decisions intended for consideration later, usually the following year.

The process of reaching decisions described here is set in train by the Report of the SC which, in future, is expected to contain advice according to an RMP. Where there is no consensus view—and, unfortunately, that has very commonly been the case—two or more scientific views, and alternative specific advice (e.g., on a catch limit or a stock classification), are expressed in the report. The SC does not vote on management advice. In principle, it could do so, on the basis of one vote per national delegation, but there is an understandably strong resistance to putting a scientific issue to a vote. The situation is complicated by the fact that some delegations to the SC consist of one person only, others may run to ten or more. Scientists present representing certain international organizations participate fully in the proceedings, but are expected to refrain from making specific management proposals. There are also always present a number of "invited participants" who are expected to contribute only to discussion of the subjects or items to which they were invited, though this understanding is interpreted rather liberally by the Committee.

Occasionally, a view is attributed by name[4] to one or more scientists, sometimes to one or more scientific delegations by nationality. Over the years, the SC has sought to express the relative strengths of views of members by using subjective phrases such as "most members believe," "some members believe ..., others believe ..., yet others ..." and so on. There have been efforts to attain more consistency in use of such expressions, but these have not been very successful. In recent years, it has, therefore, become more usual to express differing interpretations neutrally, and an opinion held by two or more members become identified with "some." If one or more members have strong reservations about what it has been agreed to say in the

[4]Normal scientific practice is, of course, always to identify the authors of hypotheses, analyses, and conclusions. Some members of the SC have, however, claimed that to follow that practice within the IWC would be equivalent to voting and is, therefore, unacceptable. Current practice is to identify names only if the individuals concerned wish that to be done; thus, the holders of some expressed opinions are concealed behind anonymity.

body of the report, they may append to it signed "minority statements." That term is, however, a hangover from past efforts to quantify various opinions; no vote has been taken, it simply means that those named persons hold a certain opinion and are exercising their right to bring it to the attention of the Commission in their own words. Minority statements are, however, scrutinized in draft by the SC as a whole to ensure they carry no improper implications for others.

The First Commitment to Management Rules: 1960–1964

At the time of its Twelfth meeting in 1960, the IWC was in crisis. Norway and The Netherlands had withdrawn from membership because of disagreements within the Commission on the total catch limit for baleen whales in the Antarctic and outside the Commission on the allocation of shares of that total among the five pelagic whaling countries.[5] In response to blandishments from the remaining Members to rejoin, Norway had set three conditions for remaining in the IWC, one being that The Netherlands would also rejoin. The Netherlands reentry was said to be conditional on "prior agreement being reached on the future level of the total catch limit (when) that was reimposed and on the establishment of national quotas for the countries engaged in pelagic whaling."

At the Twelfth meeting, the Commission suspended, for the 1960/61 and 1961/62 seasons, the existing continuing catch limit of 15,000 Blue Whale Units (BWU)[6] so that no limit would be in force. At that time, the Schedule was written in such a way that the limit adopted for the preceding season would remain in place unless changed. But Japan and USSR objected to the suspension decision so that, as far as they were concerned, the total was 15,000 BWU and it could be shared between them, while there would be no limits for The Netherlands, Norway, or UK.

The Twelfth meeting adopted a resolution asking the five countries to limit the sizes of their national catch quotas to levels not greater than each had imposed for the 1959/60 season. The responses to this resolution were: Norway and The Netherlands accepted it, in the sense that they self-imposed their previous allocations of 5,800 and 1,200 BWU, respectively; UK, having reduced the number of whaling factories from three to two, set itself a new limit of 1,800 instead of 2,500 BWU. Japan and USSR had, on the contrary, increased the number of their whaling expeditions[7] from six to seven and from two to three, respectively. Japan correspondingly increased its national limit from 5,100 to 5,980. The USSR announced that it would limit itself to 3,000 in the 1960/61 season.[8]

The Norwegian, Dutch, and British fleets had not reached their allocations during the 1959/60 season, but the total catch nevertheless exceeded the 15,000 ceiling by 512 BWU. The consequence of the 1960 decisions and events was that the total catch in the 1960/61 season was 16,433 BWU. Norway had rejoined before the season started, and the combined catch by the (now) four pelagic whaling Members exceeded the previous total; it was 15,422 BWU. The Dutch factory accounted for an additional 1,011 BWU.

In this disastrous international climate, the IWC adopted an important resolution. This defined tasks for its *Ad Hoc* SC and also established a Special Committee of Three Scientists

[5]Japan, The Netherlands, Norway, UK, and USSR.

[6]One blue whale *B. musculus* = 2 fin whales *B. physalus* = 2.5 humpback whales *Megaptera novaeangliae* = 6 sei whales *B. borealis* = 1 BWU.

[7]Each expedition consisting of a factory ship and a number of catchers.

[8]This last must be seen in the context of a failure by the five countries to agree to the proposition that 20% of the total BWU should be available to USSR and the remainder to the others to share among themselves.

"qualified in population dynamics or other appropriate science and drawn from countries not engaged in pelagic whaling in the Antarctic."[9] Their task was, working with the *Ad Hoc* SC, "to examine in the light of the advice and data available to the Commission the scientific evidence and to the extent justified by this evidence report ... to the Commission within one year of their appointment on the condition of the Antarctic [baleen] whale stocks, on the level of sustainable yield that can be supported by these stocks and on any conservation measures that would increase this sustainable yield."

The Commission, at the same time, formally declared its intention "... in setting up this special group of scientists that the Commission should, not later than 31st July, 1964, bring the Antarctic catch limit into line with the scientific findings, having regard to the provisions of paragraph 2 of Article V of the Convention."

These decisions contained three features new to the IWC. First, a specific recognition of a goal to set catch limits no higher than the *current* sustainable yield (CSY) of the stocks. Second, that by conservation measures—such as maintaining catch limits below the CSY—the stocks might be permitted to increase and thus, later, sustain a higher yield. Third, a commitment, in principle, to adopt catch limits in accordance with these concepts and scientific advice concerning them. This was also the first time that individual scientists had been brought into the advisory process to supplement the skills and knowledge of the members of the SC, who were then, as now, nominated by governments.

The Special Committee of Three presented its First Interim Report to the 1961 meeting of the Commission; it dealt only with procedural matters, particularly the assembly of data.[10] Its Second Interim Report was circulated several months before the 1963 meeting of the Commission; its final report was presented at the beginning of that meeting, and a supplement to that was circulated during the meeting.[11] The Commission also had numerous related reports from the SC. Meanwhile, the Commission, at its 1962 meeting, had taken a decision, in principle, "to extend the stock assessments to sperm whales [*Physeter macrocephalus*] in all regions and eventually to all other species of economic importance."

What, then, were the "scientific findings" with which the catch limits should be "brought into line" by July 1964 at the latest? In discussing them, it should be mentioned that the Committee of Three recommendations were all supported by the SC.

One was that the BWU should be abolished immediately because sustainable yields could only be maintained by quotas for each of the species. The Commission "noted this recommendation," but "felt for the present the BWU was the only practical method that could be administered."

Another was that the humpback whale *Megaptera novaeangliae* should be completely protected for a considerable number of years. This was accepted after the failure of an attempt by Australia to authorize continued catching from land stations north of 40°S.

A third was that the blue whale *B. musculus* should be protected "for a considerable number of years." On this, the TC made no recommendation. In the Plenary, Japan opposed the recommendation on the grounds that it wished to continue taking "pygmy" blue whales *B. m. brevicauda* in lower latitudes and that this subspecies, newly discovered by its scientists, was not endangered. The Commission accepted the Japanese view by deciding, by eight votes in

[9]D.G. Chapman (USA) and K. R. Allen (New Zealand), both of whom were later elected as Chairman of the Scientific Committee, and S. J. Holt, a British national but working then for FAO of the UN. They were appointed in March, 1961.
[10]Reports of the International Whaling Commission 13:36–38, 1962.
[11]Reports of the International Whaling Commission 14:35–38, 1964.

favor, none against, but five abstentions, to protect blue whales "in the waters south of 40°S latitude, except in the waters north of 55°S latitude, from 0° eastwards to 80°E longitude." The abstentions on this vote were significant; they indicate that countries which did not like the compromise were, nevertheless, not ready to oppose it for fear of losing the entire amendment.

On the primary question, the scientists had concluded that obtaining a sustained yield of fin whales *B. physalus* would involve reducing the catch each year to about 5,000 whales,[12] i.e., 2,500 BWU. They had been unable to make any assessment of the sei whales *B. borealis*, catching of which had only recently begun on a large scale.[13] They thought, however, that "... a very large increase in the present catch might be damaging ..." but that "... there is not enough information to suggest any but an arbitrary catch limit ..." They held out hope that a realistic assessment might be made in two or three years' time.

Acceptance of the 5,000 fin whale recommendation and maintenance of the sei whale catch at the 1962/63 level would, if the blue whale protection had been carried through, have implied a combined quota of about 3,400 BWU.

Having rejected the proposal to abolish the BWU, the TC discussed the BWU limit for the 1963/64 season. Proposals for 4,000, 10,000, and 12,000 BWU were all rejected, and discussion was returned to the Plenary. There, proposals for 4,000 and 10,000 were considered. The latter, by Japan and USSR, was adopted, seven in favor, one against, and five abstentions. Some of those who would have voted for 4,000 abstained on this vote because the other three pelagic whaling countries would find whaling uneconomic at the implied level and would, therefore, be expected to object within the statutory 90 days, so that the limit would revert automatically to 15,000 BWU.

The 1963/64 catch only reached 8,429 BWU because the UK dropped out of Antarctic pelagic whaling and because, despite still increasing catches of sei whales, other species were now so scarce that two of the four countries were unable to catch their allotted shares.

At the 1963 meeting, the Commission also considered the future of the special scientific investigations, especially with respect to obtaining advice about sei and sperm whales. According to the Chairman, "Some Members felt strongly that, as the findings and recommendations of the Committee of Three had not been met by appropriate action at this meeting, there was no reason for the continuance of work the results of which were not used." Eventually, however, a proposal by the USA to reaffirm the 1960 resolutions (by which appropriate action would be taken by 1964) was carried by a show of hands (ten in favor). A further motion was carried, six in favor, three against, with four abstentions: "requesting the Secretary to discuss with the Director-General of FAO the possibility of that organization undertaking the work thereafter if requested."

The Sixteenth meeting of the IWC, at which it had committed itself to bring regulations into line with scientific advice, was held in June 1964 in Sandefjord, Norway, the cradle of Antarctic pelagic whaling. It received a message from the Director-General of FAO who, while agreeing to the desired collaboration, in principle, said that such collaboration could not be expected "... if the Commission permitted the results of the scientific studies to be used merely for organization of the more efficient destruction of the resource for which it

---

[12]Fin whale catches had been as follows: 1959/60 – 26,415; 1960/61 – 27,374; 1961/62 – 26,438; 1962/63 – 18,668. Subsequently 1963/64 – 13,870.

[13]Sei whale catches had been: 1959/60 – 3,234; 1960/61 – 4,310; 1961/62 – 4,749; 1962/63 – 5,508. Subsequently 1963/64 – 8,286.

was responsible." The Commission had also received appeals from several non-governmental organizations, including the World Wildlife Fund, the International Union for the Conservation of Nature and Natural Resources, and the Fauna Preservation Society, urging action in accordance with the 1960 commitment.

The Committee of Three had been augmented, for its last year of work, by a fourth scientist.[14] This Committee of Four presented a report to the 1964 meeting,[15] though it was incomplete because the budget for the continued investigation had not come through until two months before the meeting. This report essentially reaffirmed the earlier recommendations, but contained three new features of interest. The first was a recommendation that special sightings surveys be conducted under the auspices of the Commission "so that some attempt could be made to measure the response of these stocks [primarily blue and humpback whales] to protection and to obtain more precise predictions when it will be feasible to resume catching them."

The second new feature was a recognition that there was no longer a complete consensus among scientists on the states of the stocks. Dissent came from two sides. T. Doi, S. Ohsumi, and T. Nemota had presented assessments for fin whales which differed substantially from those made by the Committee of Four and endorsed by most of the SC. The Japanese scientists agreed with estimates of a maximum possible sustained yield (MSY) of 20,000 fin whales from a recovered stock of 200,000, and that the population was currently much less than that giving the MSY. But, according to the Committees, the Japanese stock estimates were too high, their estimates of optimal stock too low, so they had overestimated the CSY and underestimated the time required to recover to optimal level.[16] Their estimates of CSY provided the basis for a Japanese/USSR proposal to the Commission.

This year, the Committee of Four had come forward with an estimate of the CSY of the sei whale. They gave the range 2,400 to 8,400 animals, but said they thought the true value was nearer the lower than the upper end of that range. On this basis, the <u>majority</u> of the SC concluded that "any catch of more than 5,000 (i.e., just below the central value of 5,400) is likely to be greater than the sustainable yield." The dissent from conclusion was by the senior USSR scientist, V. A. Tverianovitch, his view being that the assessments were insufficiently well founded.

A third feature of the 1964 SC meeting was that on the set of three formal recommendations to be made to the Commission, votes were taken (one vote per represented nation). These were: (i) to abolish the BWU [eight for, one abstention (USSR)]; (ii) to reduce Antarctic catches to below 4,000 fin and 5,000 sei whales [eight for, one against (USSR)] equivalent to 2,833 BWU; and (iii) to protect blue whales, including those in the "pygmy" area [seven for, two abstentions (USSR and Japan)].

At the 1964 Commission meeting, the complete protection of the blue whale in the Antarctic (i.e., south of 40°S) was at last achieved (UK/Norway proposal adopted twelve for, two against) following the defeat—by the same votes in reverse—of an effort by Japan and USSR to keep the pygmy blue whale area open. But that year, the Commission did not even consider the renewed request from scientists for the abolition of the BWU.

---

[14]J. A. Gulland of the UK.

[15]Reports of the International Whaling Commission 15:47–62, 1965.

[16]It was realized much later that, for several technical reasons, even the Committees' estimates of current and maximum sustainable yields were very much too high.

Next came consideration of the form of compliance with the 1960 commitment. The USA and Australia proposed a phasing down of the BWU limit: 4,000 in 1964/65, 3,000 in 1965/66, 2,000 in 1966/67. Japan and USSR proposed amending this to 8,500 in 1964/65. Norway and Iceland then proposed further amendment to 6,000 in 1964/65. The amendments were voted upon in succession. The Norway/Iceland and the Japan/USSR amendments were rejected; the original US proposal received ten votes in favor and four against and, therefore, was not adopted in Plenary. The four contrary votes were those of the four pelagic whaling countries.

Later in the meeting, a Netherlands/Norway proposal for the catch limit "not to exceed 8,000 BWU in any one season provided that no lower catch limit is established" failed: three for, ten against, one abstention.

The Chairman's report on this episode is revealing: "The resolution of the USA/Australia was compatible with the scientific advice which the Commission at their meeting in 1960 had undertaken to implement by July 1964." The case of those who voted against it, however, was based on Article V of the Convention, which states that regulations with respect to the conservation and utilization of whale resources "shall take into consideration the interests of the consumers of whale products and the whaling industry." Some of the opponents contended that "the condition of their whaling industries could not support the catch restrictions recommended."

After the meeting, the delegations of Japan, The Netherlands, Norway, and USSR conferred and agreed to recommend that their governments "declare jointly" an informal catch limit of 8,000 BWU for the 1964/65 season.[17]

Thus, in 1964, the Commission failed to honor the commitment made four years previously, but not for want of trying by a majority of its Members. Starting with scientific consensus, dissent began to appear just as the moment of truth arrived. Some scientific recommendations were deemed "impracticable." Others were contested as having implications contrary to selected passages of the 1946 Convention.

Starting the NMP: 1974–1976

The proceedings of the 1972 UN Conference on the Human Environment reveal that the governments of the countries engaged in commercial whaling were not each speaking with a single voice. None of them opposed the resolution calling for a moratorium. A few abstained on the vote, but most voted for it. One interpretation of this is that the ministers and departments representing governments at Stockholm were different from those represented in the IWC and that unified governmental policies had not been determined. Another interpretation is that those governments realized that so long as the UN recommendation was directed to an eventual action by the IWC, they would be able to continue to control events in that forum, in which they were a substantial rather than a tiny minority.

Proposals for implementation of the Stockholm resolution, which was subsequently endorsed by the UN General Assembly, were put to the IWC by various groups of Members at the meetings following the 1972 Conference. They attracted simple majorities, but not the necessary three-fourths for Schedule amendment. In 1974, a resolution was adopted which was described as a compromise; it expressed an intention to amend the Schedule the following year. This was done, and the NMP was thus launched.

[17]The Netherlands did not participate in the 1964/65 season, but instead sold its factory ship to Japan which thereby acquired the allocation that had been negotiated for The Netherlands. The actual catch (6,986 BWU) nevertheless fell far short of the informal catch limit.

After holding a special meeting, the SC gave advice in 1975 for the classification of all species/stocks and for catch limits for the 1975/76 Southern Hemisphere and 1976 Northern Hemisphere whaling seasons. This involved first identifying "management stocks" of species within oceans, and then defining their boundaries, and we need to say a little about these processes before passing to consideration of stock classifications and catch limits. Their catches revealed some differences in such characteristics as size, age, and sex compositions which may or may not have been indications of the existence of different populations. But the subdivisions were convenient for the presumed purposes of the NMP. Given these identities, it was thought desirable to establish geographical boundaries for them. Since stock assessments called for consideration of catch histories and current catches, the boundaries drawn were primarily related to existing statistical systems.

The only complication in this approach arose from the conduct of "small-type" pelagic whaling operations by Norwegian vessels throughout the North Atlantic. It has effectively been resolved by the withdrawal of these operations to Norwegian coastal waters and the Barents Sea. Until that happened, Norway was constrained to negotiate with Iceland and Denmark (Greenland) for an allocation of minke whales from the catch limits pertained to "their" regions.

The identity and boundary decisions were written into the Schedule in two stages, but, in all cases, by consensus, with the Commission following the SC's advice. At the time, it was clear that these decisions were to be regarded as "provisional." However, once in the Schedule, it then became necessary to obtain a new consensus or a three-fourths vote to change any of them.

The only times that changes have since been agreed—and there have been very few—they were made for the convenience of one or more whaling countries. In fact, even though there was little scientific basis originally for most of these boundaries, there has been strong resistance to subsequent proposals for change, or even for serious consideration of alternatives and their implications for assessments. The problem has been that suggested changes usually had little more certain scientific basis than the original decisions.

Resistance to change, or consideration of alternatives, was not of great importance in the early stock assessments under the NMP because of the particular assessment procedures used. They were based on catch series, estimates of numbers of whales from marking experiments, and series of catches per unit effort assumed to be indicative of changes in the abundance of whales. Thus, the assessments were really of *notional* stocks effectively available to particular fishing operations rather than of whales in certain bounded regions.

In recent years, the assessments have more often used the results of sightings surveys which, by their very nature, must be expressed in terms of geographical areas. But the other data remain what they have always been, i.e., defined in terms of the locations of whaling operations. The consequence of this change is that the results of assessments, especially as indicative of classifications, have become critically dependent on the precise stock boundaries adopted. Only during the process of developing an RMP has it become widely realized that a different approach must be taken to the stock identity and distribution question.

The NMP required that each stock be classified into one of three categories according to whether it was assessed as being significantly below its MSY level, close to MSY level, or substantially above it.[18] The first category was called *Protection Stock* (PS) and should automatically have a zero commercial catch limit. The second, *Sustained Management Stock*

(SMS), should have an annual catch limit at 90% of the estimated MSY, or 10% less than this for every 1% it was below the MSY level. The third category, *Initial Management Stock* (IMS), would have a catch limit either of 90% of estimated MSY or, "in the absence of positive evidence that a continuing higher percentage will not reduce the stock below MSY level," no more than 5% of the initial stock number, i.e., the number before exploitation. These alternative formulations arose from a dispute about the rate at which an IMS stock could be reduced to SMS category. Japan argued that this should be as fast as possible, otherwise potential productivity of the resource would be "wasted," and preferred an even faster rate than the first option which, in the normal applications, would be about 6% of the initial stock per year. Others emphasized the uncertain knowledge of the dynamics of—especially—lightly exploited stocks.

There were two supplementary rules: (i) we call SMS*bis*—if a "stock has remained at a stable level for a considerable period under a regime of approximately constant catches," it would be classified SMS in the absence of any positive evidence that it should be otherwise classified; and (ii) we call IMS*bis*—if a stock had not previously been exploited (and would, therefore, automatically be classified IMS), exploitation "should not commence until an estimate of stock size has been obtained which is satisfactory in the view of the Scientific Committee."

Early on in the application of these rules, the SC adopted some rules of thumb, necessitated by the fact that, in no case, could either the actual MSY level or the MSY be estimated. One such working rule was that for baleen whales, the MSY level would be assumed to be 60% of the initial size of the stock. Another, also for baleen whales, was that if there was no contrary evidence, the MSY would be taken as 4% of the size of the stock when it was at MSY level. For sperm whales, the interpretation was very much more complicated because of the special biological characteristics of that species, e.g., that males are larger than females and can, therefore, in many circumstances, be distinguished by gunners and killed selectively; and that the sexes are segregated for part of their lives; individual sexually dominant males are accompanied by and fertilize several females.

These rules offered even more scope for controversy among scientists than did the simple and less formal rules applied in the 1960s. All were in the form of imperatives in the Schedule, but the drafters had not envisaged that information would be insufficient for classification in other circumstances than those covered by SMS*bis*. Very quickly, a fourth category was recognized: *unclassified*. This term was used in SC advice and in tables of decisions in the Schedule, but never *defined* in the Schedule. The same term has always been applied to several quite different situations.

Classification, in many cases, required a precision in assessment that was unobtainable; the determination of whether a stock was in the range of 54–60% of its original number, and, if so, exactly where in that range, had an enormous effect on the catch limit decision.

In some cases, the SC could not identify whether a stock was in one class or an adjacent one, but *unclassified* would not be an appropriate designation, so it recommended that a stock be placed in one or other of the defined classes "provisionally" or "for one year only." But, stocks were, in any case, supposed, in principle, to be classified every year so these designations carried no concrete meaning, except when, as occasionally happened, the Commission

[18]A more complete description of the NMP is given in "Guide to review of the management of whaling," by S. J. Holt and N. M. Young, Center for Marine Conservation, Washington, DC, 1990. A revision of this guide will be published in April 1991.

itself used them and gave them some meaning. Such "provisional" classifications usually gave the benefit of doubt effectively to the whaling nation, so that SMS *provisional* was a more common designation than PS *provisional.*

Classifications were entered in the Schedule by the same principle as were the old BWU quotas. That is, they remained year after year unless changed by another three-fourths vote or by consensus. With the exception of stocks classed as SMS*bis*, catch limits were set annually. This meant that if there was no positive decision by the Commission in a particular year, there would be no limit for the following season. In the case of SMS*bis* stocks, however, it became usual to set so-called "block quotas," that is, an upper limit for the total catch over a specified number of years, usually with a provision that no more than a certain number could be caught in any one year.

The block quotas were set purely for the convenience of whalers who, if they had a bad season, had a chance to make up for it later. But block quotas had another, more important, consequence: even if was found in the middle of the defined period that earlier assessments had been in error, a three-fourths majority was required to make adjustments. A similar provision for the convenience of whalers was in what became known as "allowances." These were applied to pelagic minke whaling in the Southern Hemisphere; catch limits were set Area by Area, but any Area limit could be exceeded by a certain percentage provided that the grand total was not exceeded. Allowances were also made to permit overkill of one or the other sex of sperm whales.

In the first years of application of the NMP, the question arose as to what should be the catch limit for SMS*bis* stocks. Most commonly, it was decided to be the average of some past period of catches, the length of the period being variable. In addition, a "rolling quota" was suggested, this being the average adjusted each year to include the previous year's catch. Versions of this included cautious "rolling quotas" in which the catch limit for the next season was set below the recent catches. A similar, supposedly cautious, approach was sometimes taken to other advice which was not provided for under the NMP. An example would be the occasional application of a 10% "safety factor" (analogous to the similar safety factor in the NMP under which only 90% of an estimated MSY could be taken) when RY estimates (see below) were used as the basis for catch limits.

Another problem that had not been provided for explicitly in the original interpretation of the NMP by the SC arose from the fact that some baleen whaling operations catch more of one sex than the other (usually more females than males, the former being larger animals) and that, therefore, in an exploited stock, the reproductive potential will be affected by the exploitation-induced changes in the sex ratio. Eventually, the SC adopted the practice, in such cases, of classifying the stock according to the numbers of mature or recruited females relative to the initial numbers, on the assumption that the sex ratio in the unexploited stock had been 50:50.

In trying to apply the NMP to the Southern Hemisphere minke, a difficulty arose because some scientists thought the stocks had not been in equilibrium, but rather had been increasing, when intensive whaling began in 1971/72. This meant that they could not usually be classified because classification depended on an estimated relation between current stock size and the initial stock size assuming the latter had been constant for many years previously. It also meant that a stable MSY could not be calculated. In these, and some other cases where doubts prevailed (as happened when there was declining confidence among scientists that the "4% rule" for MSY was adequate), the scientists resorted to offering "provisional" advice about replacement yield (RY). This is an estimate of the catch which could be taken in any one year and leave the stock unchanged at the end of that year. It is not the same as CSY, and it depends critically on the recent history of exploitation. There is also no single agreed method of calculating RY. In some cases, a compromise was offered in which an average RY was calculated.

Evidently, the NMP suffered from being incomplete in the sense that there were many unforeseen circumstances in which it could not be applied. It was based on an assumed ability of scientists, which did not exist, to make estimations of certain variables. As a result, the SC had to resort to hastily conceived supplementary rules. And, it did not give guidance for action in the case of differences of scientific opinion. In retrospect, it is difficult to understand how such a scheme could have been agreed in 1974 or how the SC could have been confident that it could apply the rules, considering what was known about the scientific uncertainties of the 1960s. Its acceptance can perhaps only be explained by the anxiety of governments to escape from the deadlock within the Commission that the moratorium proposals had created.

Many of the SC's recommendations to the 1975 meeting were accepted with little debate. There were, however, some difficulties of principle. The application of the sliding scale between 54 and 60% of initial stock, to give reduced catch limits in the lower end of the SMS range, was not acceptable to Japan when applied to the Antarctic sei whales because it implied drastic reductions in catches. One suggestion by Japan was that the MSY level should be 50% rather than 60% of the initial stock; another was that the sliding scale should run from 48 to 60%. The latter was accepted as a compromise for all stocks except that in Area III, which was classed PS.

The IWC had hitherto treated Bryde's whales *B. edeni* and sei whales as the same species, and it was practically impossible to disentangle the historic catch data. The Commission agreed to treat them as separate species in the future.

The first attempts to classify and set catch limits for Bryde's whales ran into another difficulty. This species, in particular, had been exploited in all oceans by vessels flying flags of non-Member States; the catches were not publicly recorded and no other kinds of data were available. Then, it was realized that the same thing was happening with respect to some other species in various areas. The difficulties of setting catch limits in all such circumstances did not begin to be overcome until virtually all whaling countries had joined the IWC (by 1980) and companies using flags of convenience of non-Members were forced out of business by various measures.

Setting catch limits for the Southern Hemisphere minke proved to be difficult in several ways. There were few data since exploitation had only recently begun, though Area IV was already judged to be at SMS level. This classification, with all the other Areas being IMS, was accepted by the Commission. The SC applied the "5% of initial stock" rule to the five IMS Areas to establish catch limits. For Area IV, the SC proposed, on the basis of an obscure analogy with fin whales, a catch limit calculated as 7% of 60% of the estimated stock size just before exploitation began. This was accepted by the Commission, but there were strong differences of opinion both in the SC and in the Commission about whether the catches by Brazil in the northern part of Area II should be treated as coming from the same stock as the Antarctic pelagic catches in that area.[19] The SC's recommendation for a total of 6,710 minke whales for the entire Antarctic was adopted by the TC, but the differences of opinion were used to justify a number of proposed amendments in Plenary to the SC's majority recommendations, all aimed at increasing the catch limit for Area II. All were defeated, but the total was,

[19]This was subsequently resolved by the recapture off Brazil of marks placed in whales in the Antarctic.

nevertheless, raised in Plenary to 6,810 and in addition a 19% "allowance between Areas" was agreed.[20]

Treatment of Southern Hemisphere sperm whales was similar. The TC adopted the SC's recommendations, but a proposal was made in Plenary to increase the limits for males in Divisions 5 and 8.[21] This was not adopted, but a compromise—that the catch limit be divided between six Sectors instead of the nine Divisions—was agreed. Sector-to-Sector allowances were again made: 14% for males and 19% for females.

There were attempts in the Plenary to overturn the TC's adoption of the SC's advice on North Pacific sperm whales, but these were defeated. No assessments could be made of minke whales in the North Pacific; data for catches by Japan were not available, and the whales were known to be hunted also by non-Members. The Commission accepted the scientists' advice that the 1976 catch should not exceed the 1974 catch. Similar decisions were made for the North Pacific Bryde's whale and the North Atlantic minke whale. These decisions set a pattern for future treatment of numerous situations of complete uncertainty.

The 1975 meeting can be regarded as a "shakedown run" for the NMP. The first real test came in 1976.

Japan still did not like, in principle, the rule for setting catch limits for exploited IMS stocks, arguing that it allowed them only to be reduced to SMS level slowly and that this was unnecessarily wasteful of potential production. The question was referred back to the SC.

When the NMP was adopted, there had been no assessments of stocks in the North Atlantic, the catching of which had not previously been regulated. A special meeting was held for this purpose in April 1976 and its results were taken into account at the 1976 Annual Meeting. A proposal by Iceland that, in some cases, regulations could be based on restriction of whaling effort rather than by catch limits was rejected by the SC because of the impossibility of standardizing effort indices.

The first major difficulty arose because the SC did not agree whether the Area IV Southern Hemisphere minke stock was IMS or SMS and, therefore, offered two figures, 1,830 and 1,386. The TC chose the higher figure. In Plenary, Mexico proposed 1,386, but was defeated. However, the 1,830 alternative did not obtain three-fourths majority, either. The Commission approved a Norwegian compromise of 1600 (approximate average of 1,386 and 1,839). It also agreed to a Japanese proposal for a 10% Area-to-Area allowance; the question as to whether this was a matter on which the Scientific Committee should be expected to give advice was unresolved.

A second important topic of disagreement concerned the sei whales of the Southern Hemisphere, SMS in all Areas except III, which was PS. The SC had changed its assessment procedure from that used the previous year because "it no longer believed that it is appropriate to assume constant recruitment at a time when the parent stock has been substantially reduced from the original level." Using the new procedure, the SC advised a total catch limit of 1,863. Japanese and Soviet scientists could not accept the improved method and, using the method of the previous year, recommended 2,308. The scientific representatives of FAO [*Editor's note: Holt*] and UNEP disagreed with both, arguing that the procedure used by the Committee was still not conservative because it contained a hidden implication that recruitment had actually <u>increased</u> as the stock had declined. The SC acknowledged this and other problems with the

[20]The problem of negotiating national allocations of minke whales between USSR, Japan and Brazil impinged on the IWC total catch decisions until 1983.

[21] Southern Hemisphere sperm whales are divided into 9 management stocks, called Divisions and identified by Arabic numerals, while baleen stocks have all been assigned to 6 Areas,identified by upper case Roman numerals.

assessments, but could do nothing about them in the time available.

The argument about the Southern Hemisphere sei whales continued in the TC and was, therefore, not recorded in detail. The Commission simply adopted the majority scientific view, with only Japan opposing.

In 1976, the first block quota was set: 1,524 fin whales for Iceland over the period 1977–1982, with not more than 304 in any one year. These numbers came from an agreement by the scientists to propose, in the absence of any assessments, "a principle of not exceeding the average of the past five years' catches" together with the Commission's agreement with a 20% year-to-year allowance.

The impossibility of classifying some stocks, for lack of data, led the SC to consider two new categories, "vulnerable" and "indeterminate", and to suggest rules by which catch limits for these might be established.[22] The cases at issue that year were Spain–Portugal–British Isles fin whales (at that time exploited only by Spain, then not a Member of IWC), West Greenland fin whales, Iceland–Denmark Strait sei whales, and West Greenland minke whales (exploited by Denmark/Greenland and Norway). For the sei whales, the SC recommended that the annual catch be limited to the average catch over the past five years—132 animals. There were two recommendations of a similar kind for the West Greenland minke whales: one in the range 227–250, which represented the average Greenlandic landings in 1966–1975 calculated from incomplete and uncertain statistics; the other in the range 406–429, which was the average in the same period of the Norwegian and Greenlandic landings combined.

These proposals caused great difficulty in the TC and the Plenary. The Commission essentially ignored the suggestions for new categories and debated at length whether catch limits could be decided for unclassified stocks. One view was that a recommended limit should be agreed, but not inserted in the Schedule; another that such stocks should be treated as SMS (tentative) and the Scientific Committee's suggestions accepted; yet another view was that the Schedule should show a blank for "unclassified," but that agreed catch limits would be inserted. In two instances, the last was adopted by the Commission, but in later years, the ideas of "tentative" or "provisional" SMS classifications came up again without clear definition of what they meant or a decision on a rigorous and repeatable procedure for setting catch limits for them.

The third case, the West Greenland minke, proved more complicated. The TC opted for SMS tentative/provisional. As to the catch limit, the SC's divided opinion called for a decision one way or the other as a principle, but it was treated as if it were a range, and the Commission toyed with compromises. The TC proposed 400, but the Plenary adopted an amendment to 325.

The final test of the NMP approach came in the discussion of the catch limits for sperm whales in the Southern Hemisphere, especially for males in Divisions 2 and 8. The Chairman's report of this is succinct: the TC agreed with the SC's proposals (consensus except for dissent by Soviet scientists, but no specific proposals from them) for each Division and each sex, in some cases by consensus, in some by majority votes. Proposals by Japan and USSR in Plenary for higher limits for males in three Divisions were turned down, and the scientists' proposals were adopted, with a 10% allowance in each Division for each sex. Japan and USSR both reserved their positions on all these quotas, but neither subsequently lodged objections to them.[23]

---

[22]Reports of the International Whaling Commission 27:57, 1977.

[23]This was the first occasion, and the last, on which the Commission was presented with alternative catch limits calculated on the basis of seeking an eventual maximum sustainable catch by number of whales, or a maximum by weight (i.e., of products) as envisaged in the 1974 resolution. Movement towards the weight criterion would have involved larger initial reductions in catch limits and was, therefore, quite unacceptable in the intensely controversial atmosphere of the time.

This condensed account obscures a long and sometimes passionate, sometimes funny, debate in the Commission about principles behind the basic compromise in 1974 between countries favoring a moratorium and those opposed to it. The Commission came very close to arbitrary deals which would have undermined the NMP initiative at an early stage. It was saved from these mainly by the uncompromising stance taken by the Commissioner for Mexico and the tenacity of the US Commissioner. Much of the debate, which also revealed some of the weaknesses of the Commission's rules of debate, took place in Plenary sessions, of which verbatim records are kept. An extract of the parts most relevant to the subject of this article is given in Appendix 2; we have tried to retain some of the flavor of the debate as well as its substance.

The determining factor in the debate on the Divisions 2 and 8 male sperm whales was the insistence by Mexico and the USA that they would not go above the numbers proposed by most of the SC. A very different situation arose when the Northwest Pacific minke whale was discussed. The SC had recommended SMS classification and a catch limit of 541, which was passed on to the Plenary by the TC. There, Mexico/USA proposed amending this to 400. Japan pointedly asked if the Mexican Commissioner was now "refuting the recommendation of the Scientific Committee?" His response was that, in this case, "we do not consider the recommendation of the Scientific Committee as having been based on any scientific criteria whatsoever."

The Mexico/USA amendment was lost: five in favor, nine against, one abstention. The original proposal was then adopted: eleven in favor, one against, three abstentions. However, it is quite clear from the records that Mexico and USA, with their supporters France, New Zealand, Argentina were correct. The SC gave no arguments for its recommendation of SMS classification, which certainly did not fit either of the possibilities in the Schedule, and it departed from its own suggestions about how to deal with stocks it was unable to classify in accordance with the specified criteria. But, the catch limit recommendation had even less scientific basis; it was simply the highest catch ever previously taken by Japan.[24] Furthermore, the SC did not, in fact, recommend 541; it said the catch limit <u>should not exceed</u> 541. So, the Mexico/USA proposal could not reasonably be construed as "refuting" the SC's advice, regardless of whether that advice was "scientific" or based on the Schedule requirement.

This was the first of many occasions when the SC, usually when departing from the provisions of the NMP, suggested that catches should not exceed a certain number, but in which some delegations tried to insist (and they usually succeeded in this) that that number was a formal scientific recommendation and that any lower number would be contrary to the Committee's advice.

It is perhaps instructive to close this section by reverting to the proceedings of the Special Meeting of the SC in December 1974 at which the process of fleshing out the bones of the 1974 resolution was begun, in particular the drafting of criteria for defining stock categories. The SC agreed on a number of statements of which one was:

"A whale stock can only be brought to or maintained at a desired level by the application of a long-term continuing policy of management measures. It must be recognized that this long-term policy may be changed as a result of a change in objectives, of application by the SC of new data or concepts, or of adoption by the Commission of quotas or other

---

[24]A difficulty was that, although Japan had taken 541 in 1973, Republic of Korea, then not an IWC Member, had taken 882.

measures which differ from those of a proposed long-term policy. Except in these situations, however, **a long-term policy based on scientific evidence should not be changed once it has been adopted.**"[25] [*Editors' note: emphasis added by the authors*]

1977 to the Present

We have seen that flaws in the NMP appeared very quickly, as did some of procedural difficulties faced by the Commission. Our concern here is not with the NMP itself or whether the SC's advice was correct, truly scientific, or good, but rather with how the Commission dealt with that advice. So here, we do not continue with a detailed historical account, but rather identify the kinds of difficulties that emerged, and their frequency.

In 1976, the Commission had succeeded in setting the necessary catch limits, one way or another, and there were no subsequent objections to them. We have seen that problems arose even when scientific advice was unanimous or almost so, and even when it followed as strictly as possible the NMP principles. In the following years, however, the frequency with which there were differences of scientific opinion, and the frequency with which it was found that the NMP was not applicable, both increased. The character of the scientific differences also changed. Sometimes there was near consensus, but with one or two scientists or named national groups of scientists having a different opinion, but occasionally three or more different views emerged or differences were expressed as the opinions of "some" or "others" in the SC. Furthermore, as the SC increased in size, a declining portion of the participants took an active part in discussion or expressed their views about classifications or catch limits. Having no prior agreement on principle or written provisions in the NMP about how to deal with differences in conclusions reached on the basis of the same data, it is little wonder that the Commission found it more and more difficult to make unequivocal choices among the options offered.

Reports of the SC became more and more technical. They were usually written under great pressure and each year became less comprehensible to non-scientists or, indeed, even to scientists who had not participated in the SC meetings.

The SC attempted to offer useful *ad hoc* advice even when the requirements of the Schedule could not be met. This was usually where stocks could not be classified. The most common *ad hoc* advice involved the RY. At best, such numbers could only safely be applied for one year as an interim measure, but since the fundamental problems with the NMP and available data were not resolved, "one year only" turned into successions of years. There were also unresolved disputes within the SC about how the RY should be calculated, so that different numbers were sometimes offered even for that apparently straightforward quantity.

As assessment methods improved, it occasionally became possible to specify ranges within which a true number was thought to lie. Unfortunately, the SC has never standardized its specifications of these measures of uncertainty, so its presentations to the Commission are often incomprehensible to people not versed in statistical methodology. This has certainly not simplified the Commission's task of implementing the advice according to agreed rules as far as practicable.

However, when ranges were given for conclusions reached, the Commission rarely drew the fine, but important, distinctions between different kinds of alternatives. For example, a particular method of assessment might yield a "best" estimate of e.g., stock size near the center of a range expressed in terms of *confidence limits*, usually such that there was judged to be

---

[25]Reports of the International Whaling Commission 26:105–109, 1976.

a 19 in 20 probability that the true value lay within the range. In such a case, different delegations might consider it best to adopt catch limits, or even classifications, in accordance with the "best" (i.e., central) estimates, while others might wish to be more cautious and so would focus on the lower confidence limit. In some circumstances, some delegations preferred to look to the upper confidence limits especially if doing otherwise would lead to large reductions in quota from the previous year.

Even in the SC's way of expressing statistical ranges, there has been no consistency. Sometimes these have been given as confidence limits, sometimes as coefficients of variation, sometimes as standard errors. But, regardless of the manner in which they are expressed, they are quite different from higher and lower numbers derived from different interpretations of limited data by different groups of scientists, yet the Commission has commonly treated these two types of "ranges" identically.

The theory, usually, but not always, unspoken, was that if two scientific groups differed, then the correct answer must be somewhere in between. But, putting aside the matter of precision, in such cases either one or the other was correct, or both were wrong. But, if both were wrong, there is no reason, in principle, that the correct answer should lie *between* the two; it could be above the higher one or below the lower one.

Usually one of the scientific opinions was close to what whalers wanted for economic reasons. What was decided, in practice, depended simply on the voting strengths of the various groups of nations within the Commission, always with the possibility hanging over the negotiations and the votes that later there would be objections to uncomfortable decisions taken by majority votes. Thus, for a few years, the Commission sought consensus at almost any price. When a simple vote did not yield a decision, then attempts at consensus or to gain a requisite majority through compromise were inevitable.

Compromises did not always take the form of acceptance of intermediate numbers; they could become very much more complicated. Signifying a classification or a number as "provisional" and so on was a tactic of diminishing returns because these phrases quickly lost any meaning they might appear to have had in 1976. The Commission several times resorted to convening extraordinary sessions of the SC to try to clarify a controversy. Sometimes even deals were made (in closed informal meetings) which involved compromises on several stock and/or species quotas simultaneously, some of which could affect more than one country. There was a steep descent into the sort of political haggling that the NMP had been devised to eliminate, but, in one sense, worse than in the earlier years: worse because the various more-or-less scientifically based options became used as pawns in this bargaining process. Inevitably, this fed back, to some degree, to the SC itself.

One reason for differences of opinion among the scientists arose from genuine uncertainties and lack of relevant data, so that a wide range of interpretations could be legitimate. However, in certain cases, some data were in existence, but not available to the SC as such. Such restricted data, where they exist, are always in the hands of the whaling countries. The SC came to no agreement, in principle, on whether assessments made by scientists from the whaling countries which made use of such data could be acceptable for the purposes of implementing the NMP. In many cases, different views were differentiated by whether or not the various groups were relying for their opinion on the use of such restricted data. More recently, some of those data have been made available to the IWC for use by the SC during its meetings, but that only goes part way to resolving the problem because most serious scientific assessment work has to be done between sessions of the SC.

Freeing of previously restricted data does not mean that there will necessarily be fewer scientific opinions, at least not immediately. But, if existing data are all equally available, there is at least some chance of resolving genuine scientific differences. With respect to the issues addressed in this article, however, choice by the Commission among options presented by the SC has not been made easier by the fact that the exact reasons for the existence of several options have rarely been made clear to the Commission in straightforward language.

We have given this background to the SC's presentations of its views, or rather the views of groups of its members, so that one aspect of the discrepancies between scientific advice which is aligned as far as practicable with the NMP, and the subsequent actions by the Commission and by particular governments, can better be understood. We are now ready to try to categorize situations faced by the Commission since 1977.

Where the SC's advice about classifications and catch limits has been unequivocal, and has not implied any substantial changes in whaling operations, it has usually been adopted. That is hardly surprising. Although we had hoped to do so, we have found it impracticable to say how relatively frequent that has been. This is because there have been so many different types of advice, given in a great variety of circumstances: such as, for example, the continuation of an existing classification (absence of a specific recommendation implying continuation of the Schedule provision unchanged) that there was no time in a particular year to conduct or complete an assessment, and so on. We have not counted in our statistics the cases where, after the first decision for PS status, there has been no further advice offered or any discussion of the revision of that; these are principally the protections of fin and sei whales in the North Pacific and Southern Hemisphere and of blue whales everywhere. For the rest, our impression is that **the Commission's actions, or lack of actions, has departed from the scientific advice, or lack of advice, in about one-third of all cases**.

There is a very large number of theoretically possible sequences of events. Here, we only consider those which we have found actually to have occurred, at least once. The frequencies quoted are approximate because there are specific circumstances pertaining to many cases which make unambiguous categorization of them impossible. We have also not, for the purposes of the list, distinguished advice and decisions about stock classifications from those about catch limits. A problem in this respect is that advice from the SC about classification of stocks for which evidence is obviously quite insufficient does not, in retrospect, always stand up to logical evaluation.[26] The TC and the Plenary evidently cared little about classification alternatives except when the alternative was PS or either some form of SMS or unclassified; their interest was, not surprisingly, focused on the catch limit. Indeed, the Commission has occasionally changed a classification retrospectively in order to remove an apparent inconsistency with a compromise catch limit.

The types of cases were:

(1) SC made a recommendation without dissent, which was adopted by TC, but not by Plenary. [eight cases]

In four of the eight cases, the Plenary simply adopted a number substantially higher than that recommended by the TC in order to achieve a three-fourths majority; in one of those,

[26]In particular, the arguments used to justify advice that a classification, usually SMS, should be "provisional for one year only" (a phrase with no substantive meaning in most cases since assessments were supposedly reviewed annually) were not always consistent with those used to justify the proposed associated catch limit.

it also changed the classification. Two of the others involved agreement to permit an overlap from a block quota that the SC had recommended against; two involved compromises achieved by shifting stock boundaries in ways not reflected in the SC report.

(2) SC made a recommendation which was not accepted by the TC. [eight cases]

In two of the eight cases, the TC made a recommendation which was accepted in Plenary. In one of these, the catch limit was higher than recommended by the SC; in one case, it was lower.

In six cases, proposals by the TC were not adopted in Plenary. In two of these, a catch limit was set which was higher than proposed by the TC, but lower than that advised by the SC. In two cases, a proposal to modify or eliminate a block quota was rejected, so the block quota stayed in effect. In one case, no proposal was adopted, so no catch limit was set. In one case, a specific provision in the Schedule was deleted, having the effect of setting a catch limit higher than that recommended by the SC.

(3) SC was unable to make a numerical recommendation, but TC did make one. [eleven cases]

In four of the eleven cases, the TC recommendation was adopted by Plenary. In seven cases, the Plenary did not accept TC recommendation. In one of these, the Plenary adopted a more conservative classification than recommended by TC. In five of them, a catch limit was set higher than recommended by the TC, while in the other case, the Schedule was left unchanged because no action was agreed. One of the cases in this category ended in an objection.

(4) Neither SC nor TC were able to make a recommendation. [six cases]

In one of the six cases, the Plenary set a new catch limit. In two cases, both involving block quotas, the Schedule was left unchanged. In two cases, a footnote was added to the Schedule. In one case, no action was taken.

(5) SC offered two or more alternatives; TC adopted one of them. [twenty-seven cases]

In eleven of the twenty-seven cases, the Plenary adopted the TC choice; in one of these eleven, however, the decision was followed by an objection. In one case, the Plenary took no action. In two cases, the Plenary decided on a catch limit lower than proposed by the TC, but within the range offered by the SC. In the remaining thirteen cases, the Plenary adopted catch limits higher than proposed by the TC, but within the range offered by the SC.

Apart from the high number of cases in it, this category is important because of the high frequency with which a substantial number of delegations abstained in the decisive Plenary vote—in five cases, in one of which a majority of participating delegations abstained. Four cases in this category ended with objections.

(6) SC offered two or more alternatives; TC adopted none of these, but made its own recommendations. [four cases]

The Plenary did not adopt the TC recommendations in any of these four cases; all decisions reflected a desire to agree on some numbers rather than take the risk that any catch limits would be left open. In two of them, the Plenary rejected the TC proposals and simply adopted higher catch limits. In the other two cases, the TC adopted a "combination" proposal, i.e., it took various elements of the SC proposals. In one of these, there were alternative proposals from the SC for catch limits in each of two successive years; one was for SMS–220 in each year, the other for unclassified–143 in each year. The TC recommended unclassified–220 in the first year, 143 in the second, but the Plenary accepted only the recommendation for the first year. After failing to adopt any number for the second year—including rejecting a proposal for 240—a special closed meeting of Commissioners came up with a new proposal (440 for both years, but not more than 240 in either of them) which was adopted by the Plenary by consensus. The other case was similar, except that instead of juggling with two successive years, the Plenary juggled with the classifications and catch limits for the two sexes of sperm whales.

(7) SC offered advice which was accepted by TC and Plenary, but there was subsequently an objection to the decision. [two cases]

One by Spain in 1979: fin whales, SMS-143; the other by Chile in 1982: Bryde's whales, IMS-0. Other objections are included in categories (3) and (5).

We end this analysis by commenting on a precedent created by one of the objections referred to in category (5) above. This was the objection by Norway to the change of classification of Northeast Atlantic minke whales from *unclassified* to PS by decision of the 1985 Commission meeting: twenty-five in favor, one against, ten abstentions. Norway was dissatisfied both with the scientific basis for this decision and the procedure adopted in the Commission, during which it expressed a position of "not participating," i.e., present, but neither voting nor abstaining. (See Appendix 3 for an account of the precedent events.)

Norway subsequently objected to the PS classification; that objection still stands. But, it also created an *ad hoc* independent group of scientists, including some non-Norwegian nationals, none of whom had been involved before in whale stock assessments nor had attended IWC meetings. The group did not work with or consult the SC (as the Committees of Three and Four had been required to do), but it was given access to data that had been denied to the SC.[27] The report of this group to the Government of Norway was very critical of the competence and the procedures of the SC, as well as of the management objectives of the IWC. It did not support the SC's assessment and conclusions. Instead, it proposed an RY of about 200 as a basis for continued catching of minke whales, regardless of whether the stock was depleted to below optimal level.

---

[27]The SC obtained access to these data late in 1989 and used them at its 1990 meeting. Unfortunately, however, because of similar restrictions, the SC has not yet been able to use similar Icelandic data for North Atlantic fin whales in its attempt to assess them comprehensively.

Norway's dissatisfaction with the Commission's handling of the classification of the only stock in which it has a current interest was paralleled by its lack of enthusiasm for the development of a RMP[28] and a desire to continue to apply the NMP. This is consistent with Norway's standing objection to the 1982 moratorium decision because that decision effectively put the NMP into abeyance. The scientific controversy about this stock classification continues, both within and outside the IWC. It was not resolved at the 1990 SC meeting when, for the first time, all of the national catch and effort data were available, and has been complicated by revelations that some of those data had been falsified. The issue has now been put on the 1991 agendas of the Commission and the SC.

Meanwhile, the new practice by dissatisfied Member governments of taking controversial issues to other international bodies has been extended. Thus, the Governments of Norway and Iceland, in the context of their moves to create an alternative management organization than the IWC for North Atlantic whaling, have approached the International Council for the Exploration of the Sea (ICES) to determine whether that organization would be prepared to serve as provider of scientific advice to any such new organization.

These disputes, and consequent moves with respect to the future of the management of whaling in the North Atlantic, illustrate a difficulty that the IWC may now face in implementing any RMP. We have suggested that there might be ways of dealing with decision problems arising from differences of scientific opinion *within* the SC, in terms of the procedural rules, and so without resorting always to Plenary voting. But, it is difficult to see how the IWC can respond constructively and consistently to different scientific views expressed by its own advisory body and some external body, and especially if those groups are giving advice according to different policy objectives.

As to the resolution of scientific controversies, the priority should be to find ways of resolving those within the SC. But the SC has had limited success in doing this in the past decade; one reason is the SC is still mostly composed of scientists who are appointed by and are responsible to national governments with keen interest in the implications of its advice. In these circumstances, it should be acknowledged that efforts using other mechanisms, even outside the IWC, to contribute to resolution of such differences might sometimes be useful. But the creation of the Norwegian group did *not* resolve any scientific questions; rather, it exacerbated the scientific controversy as well as muddying the policy issue. If, nevertheless, some kind of "arbitration" or "second opinion" is eventually thought by the Commission to be needed to try to resolve particular controversies in the SC, then we suggest the Commission itself, or a responsible body such as an organ of the United Nations system, should make appropriate arrangements. However, any additional advice should remain strictly within the terms of the RMP, which is oriented to the Commission's statutory objectives.

In 1960, the IWC followed precisely this course by creating an "independent" Committee of Three and requiring it to work closely with the SC and in accordance with terms of reference which were strictly aligned to the implementation of an explicit policy decision. It would be unwise, however, to follow that path again blindly because the SC has since changed its character in two ways. Firstly, and mainly because of the events of the 1960s and early 1970s, the SC now includes advisers from other competent technical organizations: FAO, UNEP, and IUCN. These advisers have contributed substantially to the work of the Committee, although in recent years, their participation has been limited by reduced funds.

---

[28]For details, see "Guide to review of the management of whaling," Center for Marine Conservation 1990, pp. 14–17, covering the critical period 1982–1986.

Secondly, the SC also includes, on an *ad hoc* basis, a number of scientists invited by the Commission to participate, in their personal capacities, in defined areas of discussion. All the supernumerary members may contribute scientific studies, and all are treated as "accredited persons" with respect to access to data which are not public. The Commission also has a permanent secretariat in which scientific credentials are required.

Taken together, these changes since the 1960s mean that the SC structure should now, in principle, be adequate to provide for the flow of the best scientific advice for implementation of the IWC's management policies and the development of procedures to that end.

Conclusions and Suggestions

Throughout the periods we have analyzed, the ways that decisions have been reached were conditioned by a few critical factors. These were:

1) The nature of the reference in the Schedule to the current situation, i.e., if no decision were to be reached, would the previous catch limit or classification remain for another year or would there be no limit or classification?
2) The strength of the will to ensure that there would be some catch limit, and whether that was expressed by seeking limited compromises to secure sufficient votes for a three-fourths majority (by increasing the pro votes or the number of abstentions) or—in part of the period—by seeking consensus and thus avoiding voting at almost any cost.
3) The strength of the will to ensure the survival of the IWC and, hence, the nature of the compromises considered for that purpose.
4) The form of the scientific advice. In particular, whether or not the advice stemmed from the NMP, whether there was a scientific consensus or not, and, if not, how precisely the alternatives were spelled out and what weights were attached to them, and by which groups of scientists.
5) The voting relations between various "blocs." These appeared as follows: (a) some of the pelagic whaling countries versus the others (in the early years only); (b) pelagic whaling countries versus whaling countries only with land stations; (c) whaling countries versus non- and ex-whaling countries.[29] These relations changed several times as countries dropped out of pelagic whaling, countries dropped out of land station whaling, whaling countries that had not previously been Members joined the IWC, new non-whaling countries joined the IWC, newly ex-whaling countries rejoined the IWC. More recently, voting relations have changed again as countries have left the IWC for financial reasons or fallen into default and thus been denied their votes.

Controversy in the 1960s was caused essentially by some pelagic whaling countries seeking to disengage from the commitment made in 1960 to bring Antarctic baleen whale catches down to sustainable levels by a certain date. During the period following adoption of the NMP, the most difficult discussions arose from the insistence of some whaling countries that

[29]Other relations, of course, played their parts, especially e.g., members of EEC or Nordic or Latin American groups, trade and other economic links, involvement in aboriginal subsistence whaling.

it would be economically difficult to reduce catches rapidly in compliance with assessments made. In both periods, differences in scientific opinions were used as a tactical device, though much more frequently in the later years.

In 1960, a commitment was made to follow a certain future course of action. From 1964, the 1960s were characterized by efforts to disengage from that commitment. In 1974/1975, a new, broader, but more precisely defined, undertaking was given to follow certain principles in setting catch limits. This was a compromise involving a commitment to confine exploitation to stocks that were not depleted to below optimal levels and thus allow depleted stocks to recover to optimal levels as fast as nature would allow. We have seen that attempts to disengage from that commitment began already in 1976 and have continued since.[30] As one observer of discussion in recent IWC meetings about the minke whales in the Northeast Atlantic has commented: "One gains the impression that some countries are happy to champion a management procedure when it delivers the numbers they want, only to spurn it as soon as its dictates cease to chime with their wishes."

In adopting an RMP, a commitment to follow accepted rules needs to be renewed and reinforced. Before the IWC takes the bold and serious step of considering the reopening of commercial whaling under an RMP, it is necessary to be sure not only that the procedure itself is sound, but also that the countries interested in using it to go whaling are serious in their acceptance of it and prepared to abide by the revised procedure, not just when it suits them, but also when it doesn't.

Given the history of this matter, any such assurance will need to be convincing. Credibility might be achieved by embedding a formal commitment in the wording of the Schedule. Countries that do not like possible implications of the procedure would have a chance to object, but only one chance—that is, within 90 days of the adoption of the RMP. From then on, the catch limits should flow essentially automatically from the rules. The only future decisions that might require a vote, and provide an opportunity for further objection, would concern amendments to the RMP to take care of any flaws that are later found, or for choosing between defined options as specified below.

Testing by simulation of alternative RMPs has involved the assumption that catch limits defined by the procedure will, in fact, be adopted and those catches taken. The tested procedures are robust to small and temporary departures from those assumptions, but clearly, events of the kind associated with the implementation of the NMP, when even unanimously recommended catch limits were commonly not adopted, should not now be acceptable. The first requirement is obviously that the procedure itself must be as comprehensive as it can possibly be; each conceivable eventuality must be provided for in the Schedule or in an explanatory annex to it which carries the same force. Although the SC will not have to make stock assessments of the same kind as those demanded by the NMP, there will always remain the possibility of scientific disagreement, for example, over the interpretation of a survey result. It will be necessary for the RMP to contain the precise decision rules to be applied in such a case.

---

[30]Recently, the most common tactic for such disengagement has been to refer to an estimate of stock number and to argue that "no damage would be done" to that stock if a certain number of whales was killed the following year, regardless of its classification or uncertainties about its status. The most common way of implementing this tactic, within the Commission, has been to pose simplistic leading questions to the SC. Outside the Commission, especially in statements to the mass media, this has taken the form of affirming that the stock in question is not "endangered," thus confusing the public about the objectives of the IWC.

Some tentative suggestions for the guiding principles are as follows. They are offered in the expectation of an RMP embodying at least the elements contained in a seven-country paper submitted to the IWC 1989 meeting (Appendix 4):

(a) Management units for which any catch limits are to be set need to be defined in terms of populations of whales effectively accessible to particular commercial whaling operations or groups of operations, or by sufficiently small subareas.
(b) No catch may be taken from any stock for which there is no catch limit.
(c) The SC should be required to provide precisely formulated advice according to the rules, although if there is no consensus, this may take the form of several options so long as they are all expressed in terms of the rules.
(d) If the SC encounters a situation which is not provided for in the rules, it should bring this to the Commission's attention for appropriate action in amending the rules.
(e) All catch limits should be in strict accordance with the scientific advice based on the RMP. In cases where two or more scientific options are provided, the catch limit should be one or another of them. A choice between them would, if necessary, be made by vote. The right to object to the result of any such vote would remain, but then principle (b) above would apply.
(f) The number of future seasons for which a series of annual catch limits are set during a year in which the RMP is applied is determined in advance and in accordance with a requirement that a new evaluation of all relevant data be made not later than the end of that period.

Such principles could be expressed in appropriate binding wording in the Schedule without it being necessary to amend the Rule of Procedure of the Commission and without affecting the right of Member States to lodge objections to decisions.

The substantive wording would be a replacement for paragraph 10 of the current Schedule. However, it would also be necessary to replace paragraph 9 Classification of Areas and Divisions **by a short list of species/regions (e.g., North Atlantic fin whales; Southern Hemisphere minke whales) supplemented by a list of management units drawn up in accordance with the principle (a) above. The details of this would depend upon the precise formulation of the agreed RMP.**

Appendix 1

More Detailed History

Moment of Truth: 1964

Here, we recall some other features of the Commission's debates in 1964, other than or supplementary to, the specific consideration of the 1960 commitment. The Commission had been increasingly occupied with several issues other than the BWU in the Antarctic. We have noted that an important feature was the negotiation of national quotas for Antarctic pelagic whaling countries outside the Commission. Associated with this was the beginning of negotiation of an International Observer Scheme to supplement the national inspection arrangements specified in the Convention. This, too, was outside the Convention and was limited to the five countries that had been involved in Antarctic pelagic whaling.

The Agreement among the pelagic whaling countries, which carried implications for the Commission, was presented at the 1964 meeting, but it was not implemented until many years later, by which time the entire pattern of whaling operations had changed. International observers were needed so that the different whaling nations could keep their eyes on each other and ensure the national allocation agreements were honored. During the negotiations, some countries proposed that international observers should be of nationalities other than those engaged in the particular group of whaling operations and should be paid and appointed by IWC; this idea came to nothing ostensibly on ground of excessive cost to the Commission. But the scientists had also drawn attention to the fact that there was obvious and extensive cheating in the reporting by national inspectors of the lengths of whales,[31] and that this added to the difficulty of making assessments; it was supposed that an international observer scheme might cure that.

At this time, many countries were issuing large numbers of Special Scientific Permits, and the scientists began to press for an opportunity to review proposed permits before they were implemented.[32] Again, it would be many years before such prior review became a formal requirement. Some Members of the Commission were also increasingly concerned about whaling by non-Members, and thought was beginning to be given to the need to regulate catches from land stations; they had not previously been regulated.

While the arguments about Antarctic BWUs continued, the industry was moving on. Declines in blue, humpback, and fin whale catches in the Antarctic were accompanied by increases in sei and sperm whale catches there and a shift to the North Pacific. Such a sequence was to be replicated in the 1970s: while regulatory attention was focused on the protection of fin and sei whales in the North Pacific and Antarctic, the industry moved to Antarctic minke whaling, "pirate" whaling increased, and whaling from land stations in the North Atlantic and elsewhere also increased.

Some views expressed by the SC in 1964 concerning the situation then in the North Pacific are worth recalling; they were equally applicable to the Antarctic and other regions, and are still relevant, though the problem of dealing with uncertain scientific advice is a little more complicated than the Committee then thought:

> "Estimates of stock size and of sustainable yields are by their very nature statistical statements in which there is some margin of error; it is for this reason that each assessment is as far as possible presented in the form of possible and probable ranges, with a most likely value within the latter range.
>
> "It must be said that it is inherent in the nature of the scientific assessment of the condition of a whale stock that the most incontrovertible statements can be made when that stock has been seriously depleted and the history of this depletion has been well documented. It follows therefore that if conservation action by the Commission is delayed until 'better' assessments are available, such action will be too late—too late in the sense that the stock will have been depleted to a level such that it is ... [economically] very difficult to continue whaling and at the same time permit the stock to recover.

[31]See "Outlaw whalers 1990," Greenpeace International 1990 for more details of these matters.

[32]See "Scientific whalers? 1985," Greenpeace International 1985.

"Furthermore ... in considering action based on a value within the likely range of the sustainable yield it is in the long run better—better in the sense of getting higher cumulative yields, and with less total effort—to act on conservative rather than on too optimistic values."

Picking Up the Pieces: 1965–1968

A Special Meeting of IWC was convened in May 1965 at the request of several Members "in order to obtain some agreed total quota which would allow the restoration of the Antarctic whale resources." The scientific basis for discussion was provided by a reconstituted Committee of Four and analyses presented by FAO, but not involving the SC. Notwithstanding the strong position taken by FAO at the 1964 meeting, its D-G had agreed to continue cooperation with the IWC: "While FAO had expressed the wish not to be associated with a policy which could imply, ultimately, the effective destruction of the whale resources, ... there was nevertheless clearly an urgent need for an appraisal of the consequences of continued intensive pelagic whaling in the hope that at some future date it would prove possible for the interested countries to agree on measures of conservation and rational exploitation ..." This Special Meeting confirmed, by the presence of numerous observers, the increasing interest being taken in the activities of the IWC by those outside it. There was little change in the advice that emerged except that it was getting more complex and less generally comprehensible. This was largely due to the efforts made to evaluate the immediate and medium-term consequences of an increasing number of options.

The critical discussion took place in a Committee of Six countries set up by the Chairman: Norway, Japan, USSR, Canada, France, and USA. This discussion revolved simply around a series of more or less arbitrary numbers. The three remaining pelagic whaling countries made reduced demands as compared with their earlier ones, partly it may be supposed because the high catch limits were getting ever harder to reach.[33]

The USSR molded its proposal around its contention that evidence regarding the sei whale was "incomplete" and, therefore, should not be given great weight. The USSR also introduced a new concept: that regulation should include reference to whaling effort as well as catch limits. Its specific proposal was for a reduction in the number of expeditions by not less than 50%. This was too radical a move to attract other nations, and also the Committee of Four had some questions about its implementation, e.g., what if the numbers of catchers per expedition were increased?

Several countries suggested a phasing down of catches (in BWUs) over three seasons. None of these were found to be generally acceptable. The closest to agreement had been a second proposal from the USSR for a catch limit of 4,000 BWU in 1965/66. The Plenary adopted this wording for a resolution (not a Schedule amendment): four in favor, none against, nine abstentions. Evidently the considerable majority of nations did not like it; they would have preferred 3,000 phasing down to 2,000, but knew they did not have the necessary three-fourths majority to carry an amendment, but that even if they had, the whaling nations would almost certainly have objected to any such decision.

Eventually, there was unanimous acceptance in the Plenary of a resolution, the text of which is worth recalling:

---

[33]See Reports of the International Whaling Commission 16:25–49, 1966; pp. 33–34 for a useful review of policy options offered by the SC.

> "The Commission recognizes that this quota of 4,500 BWU will not in the 1965/66 season reduce the catch below the sustainable yield as determined on the basis of the best scientific evidence available, which is incomplete at least in regard to sei whales. This quota is agreed as a transitional limit to assist the pelagic whaling industries to adjust to the reductions required to begin rebuilding the whale herds. All Members agree that they will recommend to their governments that they support further reductions for the 1966/67 and 1967/68 seasons that will assure that the quota for the 1967/68 season will be less than the combined sustainable yields of the fin and sei whale stocks as determined on the basis of more precise scientific evidence."

The press statement from the Special Meeting ended: "There is for the first time a plan for the effective conservation of whales in the Antarctic."[34]

At the 1965 Annual Meeting, the wording of the resolution from the Special Meeting, including the promise for the future, was put into the Schedule, but as an imperative: 12–0 vote. This appeared to meet the concerns of the scientists who said they opposed any quota higher than 2,500 BWU unless there was a firm undertaking to reduce the quota to less than the combined sustainable yield by 1967/68. By going along with that, the Commission sidestepped another proposal from the scientists: that the Sanctuary in the Pacific Sector of the Antarctic, which had been opened to whaling in 1955, should again be closed.

Previously, the scientists had supported the whaling nations' pressure to open the Sanctuary and keep it open on the grounds that this would take the pressure off the declining stocks elsewhere. But, this was conditional on there being an effective and sufficiently restraining overall quota. The issue was not simply the numerical quota, but also the effectiveness of its implementation and that of other regulations. The scientists were worried about the fact that four nations—Japan, USSR, Norway, and UK—had objected to the blue whale protection decision, and that size limits were still not being enforced.

That year, the Committee also faced procedural difficulties. Its very distinguished Chairman, Prof. J. Ruud of Norway, had resigned in protest against the foot-dragging policies of the pelagic whaling countries, including especially—from his point of view—his own. The Committee members had been unable to elect a new Chairman. Instead, they appointed D. G. Chapman, Committee of Three/Four member, now on the US delegation, as Rapporteur. Other activities of the scientists included giving greater attention to sperm whales, pressing for arrangements to be made for assessment of all exploited species and stocks, and development of procedures for sightings surveys, to which the ICSU Special Committee on Antarctic Research (SCAR) was beginning to contribute.

In 1965, the SC also proposed to the Commission that it should make arrangements for the study of economic problems in parallel with the evolving arrangements for scientific assessments.[35] The reason for this proposal was that the scientists saw that undocumented economic arguments were being used to avoid or delay the implementation of their advice. All the Commissioners thought this interesting idea "should be borne in mind" for a year. Nothing came of that initiative. The Commission later passed the subject to FAO which, in turn, found

---

[34]Norsk Hvalfangst-Tidente 1965(5):113–114.

[35]See document "Economic assessment" by Committee of Four members Allen, Gulland, and Holt, Reports of the International Whaling Commission 16:62–63, 1966.

it could do nothing because economic information was not made available by the industry or by the governments possessing it.[36]

The report of the 1965 meeting of the SC is marked by increasing differences of opinion between scientists. For example, the FAO Stock Assessment group, which was continuing the work of the Committee of Four[37] concluded that the CSY of sei whales was about 4,500, while Japanese scientists estimated it as 7,400, even though the population estimates themselves were not incompatible. Other differences were emerging as well. Japanese scientists thought that their own estimates of the sei whales were too low because there were other areas "not adequately exploited." Conversely, a Norwegian scientist, Å. Jonsgard, thought that because sei whales were less mobile than the larger species, they were concentrated in a few areas. These areas were precisely the ones being now exploited and, consequently, the rather rapid decline in apparent abundance represented a real decline in the stocks as a whole. Japanese scientists also now contended that the estimates obtained from the study of Antarctic catches underestimated the SY because fin and sei whales also occurred elsewhere in the Southern Hemisphere.

The SC dealt with the difference of opinion on sei whale CSY by simply reporting that "the best estimate is 4,500–7,500." But, with respect to the fin whale, the Committee commended the suggestion by the FAO group that the fin whale should be protected. If BWU were to be continued, then the quota should be "sufficiently below the combined sustainable yield so that fin whale stocks can begin to rebuild. This is particularly important since the scientific evidence indicates that the fin whale stocks could be the most productive in terms of numbers and economic value."

The Commission, not unexpectedly, did not like the suggestion to protect the fin whale, nor did it move on the matter of abolishing the BWU. Instead, it calculated that, depending on whether the majority estimate or the Japanese estimate for sei whales was used, the BWU quota implied was either 3,000 or 3,500 BWU. It adopted 3,500 BWU for 1966/67 unanimously and reaffirmed "that the total catch for 1967/68 should be less than the combined sustainable yields of the fin and sei whale stocks as determined on the basis of more precise scientific evidence."

At the 1967 SC meeting, the controversy between the FAO group, the Japanese scientists, and Dr. Jonsgard about the sei whale continued. Once again, a wide range for CSY of this species was quoted (4,400–7,000, slightly lower than the previous year) and a consequent range 3,100–3,600 BWU. The SC again said that the BWU should be abolished, but passed on a new idea from the FAO group: if the Commission did not wish to protect the fin whale, it should consider giving it a lower value within the BWU formula. The SC also noted that the FAO and Japanese estimates for sei whales in Area II (Southwest Atlantic sector) agreed in that this species was below MSY level there and that consideration should be given to a complete closure of that Area. This was the first time the scientists made an explicit proposal to protect a species in a region because it was thought to be below optimal level; the suggestion was a precursor to the idea of Protection Stocks under the NMP.

Ignoring all the SC's advice except that pertaining to the BWU range, the Commission began the usual trading exercise. The TC, by a majority vote, recommended 3,100 BWU, and

---

[36]See "Economic studies of whaling regulations" in Reports of the International Whaling Commission 18:48–49, 1968.

[37]It now included L. Boerema, a Dutch scientist who had recently joined the FAO staff.

its Chairman, supported by Argentina and Mexico, proposed that to the Plenary. The UK and USA proposed amending the number to 3,200, seemingly because they had already come to the conclusion that the lower number, not being acceptable to the pelagic whaling countries, would not attract a three-fourths majority. Japan and USSR then proposed amending the number to 3300. Eventually, the UK/USA proposal was adopted: eleven in favor, two against, two abstentions.

When the SC met next in Tokyo in June 1968, it was confronted with the report of a meeting in Oslo, February–March, on age determination in whales. One of the findings was that the rings in ear plugs of fin whales should be interpreted as being laid down one, rather than two, each year. There had, for some years, been growing doubts about the interpretation on which all previous stock assessments had been based; now, it seemed that the whales in all samples were twice as old as had been thought. That meant that estimated mortality rates needed to be halved, but also that estimates of SY rates, as they were then being determined, also should be halved.

The SC and the FAO group agreed that in these circumstances, it would be necessary to go back to re-examine earlier conclusions. But, it was not possible to do that immediately, so for the time being, the figure of 5,000 BWU was put forward as an approximate estimate of CSY.

The Japanese scientists and the FAO group still did not agree about the condition of the sei whales.[38] But, now another element appeared: the SC carried out estimations for Area II during the meeting and came up with yet another number. The score for this Area was as follows: FAO—1,270; Japan—1,680; SC—1,900.

It was not possible to produce estimates for the other Areas during meeting. The Japanese and FAO estimates for Areas II–IV differed by about 30% on average; the Committee judged they "did not differ greatly." (If the Committee had had time to reassess these other Areas, they might well have produced very different numbers for any or all of them.) This year, however, instead of quoting the alternatives, the SC advised the Commission on the basis of the averages of the FAO and Japanese estimates, except for Area II where the Committee's own estimate was used. This led to a total estimate of 5,375 sei whales, rounded up to 5,400. This, together with an estimate of 5,000 fin whales, led to advice for a combined quota of 3,400 BWU or less.

In addition to this exercise, the SC tried to determine whether the stock of sei whales in each of the six Areas was above or below the level that would give MSY, again foreshadowing later efforts to classify stocks under the NMP by essentially the same criterion. Everyone agreed that the stock in Area I was still above MSY level. Japanese scientists maintained that the situation was the same in all other Areas. But, "the rest of the Committee" agreed among themselves that the stocks in Areas II, III, and IV were already at or below MSY level and that in Areas V and VI they might be at or just above MSY level.

The SC's new approach to the treatment of dissent and uncertainty—taking averages of different opinions—made the Commission's decision in 1968 easy. A quota of 3,200 BWU was set without argument for the 1968/69 season. The Commission had comfortably reduced the quota to below the estimated combined sustainable yields and apparently finally met its 1960 commitment. However, Norway dropped out from the competition, and Japan and USSR were only able to catch 2,470 BWU between them in the 1968/69 season.

---

[38]See Reports of the International Whaling Commission 19:29–38, 84–120, 131–137, 1969.

In retrospect, it must be admitted that the scientists were rather careless in their presentation of advice. In the years leading up to 1968, when dissent had appeared, the estimated CSYs of sei whales were presented as if they continuously spanned a range. But, in fact, only the two numbers at the end of this "range" had any meaning. They came from different analyses, one or both of which might be wrong. There was no reason at all to imply that greater weight could be given to any numbers between the extremes, but the Commission was induced by this presentation to compromise within a spurious range rather than to chose between the extremes on the basis of policy.

In 1968, the procedure of taking an average of opinions and presenting only that figure as advice to the Commission did not even offer the Commissioners any choice, and they were able to agree on the same quota as that in the preceding season. Yet, the SC had recognized that the sei whale stocks were probably in worse trouble than previously thought. The differences between the FAO and Japanese assessments were still substantial, and the SC had determined a preferred estimate for Area II which was quite different from both of those. There is no hint in the report of the SC how the new estimate for Area II was obtained or why it is quite different from the others. The way the average was obtained is clear in the SC's report, but it finds no reflection in the Chairman's report of the Commission meeting.

Trouble in 1969–1970

The Twentieth Report of the Commission gives a detailed account of the consequences of change in interpretation of rings in fin whale ear plugs.[39] It was clear that the Committee of Three had underestimated the total population size, but overestimated the net recruitment rate.[40] FAO continued what had by now become a routine updating of previous analyses, virtually an accounting exercise. But completely new analyses were presented by Japanese scientists T. Doi, S. Ohsumi, K. Nasu, and Y. Shimadzu, and by D. G. Chapman and by K. R. Allen.

The fin whale CSY estimates ranged from 1,000 to 5,600. "Most members" believed the best was "3,300 or less." The Japanese scientists concluded that the best estimate was in the range 4,900–5,600, "and that it is not possible to give a single estimate within this range." This time, the SC did not bother to average these incompatible figures, but simply handed them all to the Commission without comment.

Despite the discrepancies found the previous year, the SC was unable to do more than reiterate its conclusion that the sei whale CSY was about 5,000. No more was said about the matter of depletion in some Areas. The SC had practically given up its often reiterated request for abolition of the BWU.

L. Boerema, representing FAO, said now that the Antarctic quota had been reduced, and because national delegations had acquired sufficient expertise in population dynamics, his organization had decided to withdraw from its agreement to provide annual stock assessments.

The range of fin whale numbers given by the various assessments implied a quota of between 1,333 and 3,633 BWU. It was convenient for the Commission that the average of these was 2,483, and that the catch the previous season had only been 2,472. Adding a few more, a quota of 2,700 BWU for the 1969/70 season was unanimously agreed.

---

[39]Reports of the International Whaling Commission 20:33–96, 1970.

[40]The percentage number being applied to the population estimate to obtain the CSY.

A Special Meeting of the SC was held in Honolulu in March 1970 mainly to try to resolve the differences about fin whales. It failed in that respect.[41] At the Annual Meeting in June, "all members except Japan agreed that the recent level of fin whale catch (2,700 over the last five seasons) appears to be fairly close to the present sustainable yield." Japanese scientists, however, "believe the best estimate is 3,520 to 4,350." There were still no improved estimates for sei whales; by now, it had been almost forgotten that the 5,000 figure was a rough average between two discrepant assessments; recalculation had still not been made for Areas other than II.

As far as it concerned Antarctic pelagic baleen whaling, the Commission meeting was tranquil. The catch in 1969/70 had again been substantially below the quota: 2,470 BWU. The Commission retained the quota at 2,700 BWU for 1970/71, but said that this was done "on the understanding that catches would not be substantially increased above recent levels."

Soon after the 1970 IWC meeting, the US Secretary of the Interior placed eight species of whales, including the blue, fin, sei, sperm, and humpback, but not the minke, on the Endangered Species List under a new domestic law. This made it illegal to import parts or products of these species into the USA. At the same time, he announced that an international conference of leading whale experts would be convened, under the sponsorship of several Government agencies and non-governmental organizations "to consider the plight of the great whales and their biology as related to their conservation." The conference was held in June 1971.[42]

Meanwhile, the Japanese whaling industry, assisted by the Japan Fisheries Agency and its scientists, was quietly preparing its onslaught on the Antarctic minke whales, having made some surveys and trials. Arrangements were made to encourage the USSR to engage in similar operations by buying the minke whale meat produced from them.

Norway experimented in the 1969/70 Antarctic season with a small factory/catcher boat which killed four fin, twenty-two sei, and a number of minke whales, but was not an economic success. She became, however, the model for factory/catchers that soon began operating in the Atlantic, as "pirate" whalers, flying flags of convenience and killing fin, sei, and Bryde's whales.

Outside the Commission, moves were beginning to bring the whaling issue back to the United Nations. Fin and sei whales were protected in the Southern Hemisphere in the mid-1970s; since then, it has been found, from sightings surveys, that they and the humpback had been reduced to very much lower numbers than was thought in the 1960s or at the time of full protection.

Appendix 2

1976 Debate on Catch Limits for Southern Hemisphere Sperm Whales

[Debate starts on catch limit for males in Division 2, exploited by Japan and USSR. SC recommendation IMS, catch limit 764, adopted by TC by simple majority.]

PLENARY

Japan proposes amendment: 1,369. The basis for this is that by following the NMP, it would take a very long time for this stock to be reduced to SMS classification, and that more rapid re-

---

[41]Reports of the International Whaling Commission 21:34–39, 1971.

[42]"The whale problem: a status report", edited by W. E. Schevill, Harvard University Press, Cambridge, Mass, 1974.

duction is desirable to optimize use of the resource. Amendment lost four in favor, nine against, two abstentions. Vote on the original proposal: eleven in favor, four against, so not adopted.

Statement by Mexico, Dr. Rozental: The entire success of the New Management Procedure depends on its being applied scientifically and depends on its being applied according to certain parameters and criteria. ... The NMP does not admit, or should not admit, of the type of political bargaining which has characterized this Commission in the past. The fact that we have now taken two votes on two separate quotas, one on the quota recommended by the Scientific Committee and one on the quota the Japanese Commissioner desires, and that neither of these has received a three-quarters majority, as required by our rules of procedures, leaves us at the moment with no quota at all. It is the opinion of this delegation that we will not accept a single whale additional to the quota which has been suggested by the Scientific Committee and the Technical Committee. To do so would be to bury once and for all the NMP and ensure that what we have done so far goes down the drain. We would be prepared, and would favor and would be happier, to go home without any quota whatsoever than to start bargaining across the table, as has been the past custom of this Commission.

Chairman: My responsibility, and the responsibility of all Commissioners, is to leave this building with what is, in my opinion, a quota to ensure these stocks are not overfished.

Denmark, Mr. Lemche: [Disagrees with Mexico] In my opinion, this is not a question of taking some scientific advice based on pure science because the scientific advice was based on the wording in the Schedule, which is not done by science, but as a political thing [Agrees with Japanese views, and that the 1975 Schedule amendment defining the NMP obviously needs revising. Says only voted for the NMP in 1975 because the then Chairman said it would be reviewed in 1976. Denmark then proposes 1,050, which is the mean of the SC and the Japanese proposals (1,066) rounded down].

USSR, Dr. Nikonorov: [Supports Denmark and Japan] We can suppose that Japanese scientists are erroneous. Equally, other members of the Scientific Committee may be erroneous. The truth is somewhere in between.

[Put to the vote, the Japanese proposal is lost: seven in favor (all the whaling countries except Brazil, which did not catch this species), eight against.]

USA, Dr. White: [Explains vote against. Supports Mexican statement] I just want other delegations to know that the United States made its compromise with the whaling nations last year. We do not intend to compromise further than the compromise we made last year. We cannot keep manipulating these numbers to suit our needs each year. We have a procedure. We should follow it. What we are doing may be good politics, but it is bad science and it is bad conservation. The United States intends to stick with the recommendations of the Scientific Committee, even if it means leaving this meeting without a quota.

[UK, Iceland, Canada say they understand both points of view, but, for the time being, would support the SC's recommendation on the understanding that, in 1977, the whole procedure would be reviewed so as to allow a faster reduction of IMS stocks.]

South Africa, Mr. Kleu: As I said before, we fully subscribe to the scientific basis of management, but we think it should be applied in moderation and without going to extremes. [Objection to the procedure is that it allows too drastic changes in quotas with relatively small changes in model parameters.]

USSR: [Suggests he would exchange his position paper with that of Dr. White.] Perhaps the Commission could be divided into two parts, one of which could consist of whaling nations with an understanding of the difficulties (helped by another Scientific Committee of experts in economics), the other part could consist of the countries which sometimes do not realize what the matter is all about. [Proposed 1,040, but this is not seconded.]

USA: [Suggests move to another topic, "otherwise we will all be sitting here for the rest of the afternoon staring at each other."]

Chair: It does not worry me: they are all rather good-looking people, Dr. White, and I get a beautiful view from here!

[Debate passes to sperm whales in Division 8, with virtually identical procedure, but no speeches. TC proposes, by majority: 826. Japan proposes amendment: 3,188, which is lost, six in favor, nine against. Original proposal does not receive three-fourths majority; vote is ten in favor, four against, one abstention.]

[LATER, same Plenary]: Chair, as Australian Commissioner, proposes: 800 for Division 2, seconded by Iceland. Mexico announces it will not propose an amendment because the number it would propose had already been voted upon, and defeated. [Later in the session, there is a long and extremely confused procedural debate on whether or not a motion once lost can be proposed again. It is finally accepted that it can in the case that the vote did not lead to a decision; whether a revote can be taken in cases where the first vote did lead to a decision was left undecided, but not before the Mexican Commissioner had tried proposing a number one less than the SC recommendation and been chided by Japan for making a proposal not in accordance with the SC!]

[The Australia/Iceland proposal is put to vote and lost: six in favor, seven against, two abstentions.]

[It is then noted by the Chair and the USSR that although the SC did not actually recommend the number proposed by Japan, it had not repudiated it, either.]

[The Chair tries to close debate by moving to the next item, which would leave no quotas agreed for Divisions 2 and 8. He is overruled in a challenge: four in favor, eight against.]

ADJOURNMENT

[On resumption of the Plenary, South Africa proposes that quotas are set as in previous season: 990 in Division 2, 1,260 in Division 8; seconded by Norway. An additional remark is that, unless the SC comes up the following year with a different recommendation, its recommendation of this year will be adopted then. This is lost by a vote of two in favor, ten against, thre abstentions. The opposition was, in part, because such a compromise was unacceptable, in part, because it was unacceptable to some Members that the Commission would thus be committed that year to a certain Schedule change the following year.]

Mexico then makes its proposals, seconded by USA of 763 in Division 2, 825 in Division 8. This is not adopted: nine in favor, four against, two abstentions.

USA, seconded by Mexico, then again proposes: 764 in Division 2, 826 in Division 8. This is ruled out of order by the Chair. The Chair is challenged and by a simple majority vote, four to ten, is overruled. The USA/Mexico motion is then put to the vote and adopted, ten in favor, three against, three abstentions.

Appendix 3

Minke Whales in the Northeast Atlantic: Classification and Catch Limits

Under the NMP, the minke whales caught by Norwegian small-type commercial operations in the Barents Sea, along the coast of Norway, and in the North Sea were treated as belonging to the following management stocks:

1975 meeting, decisions for 1976 season: stock East of Cape Farewell, Greenland;
1976 meeting, decisions for 1977 season: East Atlantic stock;
1977 meeting, decisions for 1978 season: Svalbard–Norway–British Isles stock;
then unchanged until 1980 when renamed Northeastern stock;
unchanged thereafter.

In the following summary, we refer to the year in which each decision was made, to be applied to the following year.

The Northeastern stock was classified SMS in each season from 1977 to 1981, with a catch limit of 1,790 reaffirmed each year. This classification was based on the provision we have called in this article SMSbis, i.e a stock which "has remained at a stable level for a considerable period under a regime of approximately constant catches ..."

The classification was based on the Scientific Committee's inability to detect any trend since 1952 in crude catch per unit effort (CPUE). But each year, the SC expressed its limited confidence in the validity of the crude CPUE as a valid index of whale abundance, describing it as "problematical."

The catch limit was the mean annual catch 1967–1976. In 1975 and again in 1976, the Norwegian scientists expressed "their concern over the appropriateness of catch quota regulation for this fishery in view of the impossibility of introducing an effective control system ..."

At the 1982 meeting, the majority of the SC, while still describing the CPUE series as "problematical," recommended a classification as "provisional SMS," and a catch limit of 1,690, which was the mean catch in the ten years 1972–1981. A minority [Chapman, Holt, de la Mare], believing that there was "a real probability" that this stock had been declining during the period of regulation, proposed it should be *unclassified* with a catch limit "substantially lower than 1,690 and in any case not higher than 90% of that number (1,521)."

The TC defeated a proposal 1690 and adopted, by a majority, a proposal for 1,521. But the TC recommendation was not seconded in Plenary and was not, therefore, formally treated as a proposal [We have noted that the Commission's procedure with respect to secondment of a majority TC recommendation has not been consistent]. Then, Norway, seconded by Iceland and Denmark, proposed the majority recommendation of the SC: 1,690. An amendment to this was proposed by Seychelles, seconded by The Netherlands, Oman, Australia, and St Lucia; this was a compromise by which a 5%, rather than a 10%, safety factor would be applied, thus making the proposal 1,606. This amendment did not attract a three-fourths majority: fourteen in favor, five against, thirteen abstentions. The original Norwegian proposal was then voted upon and also failed: ten in favor, four against, eighteen abstentions.

Norway assured the Commission that, in the absence of an adopted catch limit, its catches in 1983 would not exceed 1,690.

Seychelles sought to reopen the discussion of a catch limit for this stock. This led to a lengthy procedural debate which the Chairman closed by ruling that "unless there is a unanimous decision to the contrary, when a subject is closed without any reservation to reopen it later, the matter cannot be reopened for discussion." There being no consensus, Seychelles withdrew its proposal. No catch limit was set despite the strong desire of almost all countries not to leave the meeting without a decision. The stock was entered in the Schedule as *unclassified*, although no specific decision to that effect was taken by the Plenary.

At the 1983 meeting, the SC made a "careful in-depth examination" of the CPUE data, concluded that "the reliability of the information on changes in stock abundance has been increased," and that "there is some evidence for a decreasing trend in the CPUE series, but no estimate of the present stock size relative to the initial size is available." The analysis on which these conclusions were based had been presented to the SC by Norwegian scientists who had been working with two British scientists [Beddington and Cooke]. The analysis had been made possible by the accumulation of data from an improved statistical system introduced by Norway in 1976. The SC, therefore, recommended that the stock be listed as *unclassified*, pending further data analyses by Norwegian scientists.

As there did exist estimates of current number, from marking experiments, and a best estimate of the rate of recent decline, from improved CPUE indices, the SC was able to calculate replacement yields. The values obtained from a variety of assumptions had a very wide overall range. A more likely range was considered, without dissent, to be 365 to 905, and the best estimate within that range, 635. This number was recommended as the *maximum* catch limit. The SC pointed out that this and the other RY estimates over the range were, in fact, average RYs during the period for which the data had been analyzed (1970–1982). The SC thought the current RY (i.e., pertaining to 1984) might be different from this average, but did not believe the difference to be large.

Australia, second by St Lucia, proposed the advised number—635, *unclassified*—to the TC, which adopted it. In the Plenary, Norway, seconded by Iceland and Japan, proposed an amendment to 885, i.e., 98% of the *upper* end of the most likely range. This amendment was described by its proponents as "an exceptional measure for a delay in accepting the SC's recommendation to ease the transition ... to the lower level proposed." The justification for this measure was the indication in the 1946 Convention of the purpose of "providing for an orderly development of the industry."

Several countries expressed support for the Norwegian position, but others said they could not go beyond the SC recommendations. The amendment was put to the vote and not adopted, it being ruled that a three-fourths majority was needed for adoption: fifteen in favor, seven against, thirteen abstentions. The original proposal was then adopted by consensus, but Norway, in accepting the consensus, recalled its right to lodge an objection. Norway did not subsequently object to this decision.

At the 1984 meeting, the SC made some corrections to the 1983 calculations and reassessed this stock. The corrections enhanced the downward trend in recent years and hence implied lower RYs than previously calculated. The SC could still not propose a classification, but now proposed an RY range of 300–747 (mid-range = 524). The majority statement by the SC was that "If the Commission chooses to set a catch limit near the upper end of this range, then a 10% reduction as a safety factor is appropriate," but two scientists [Gunnlaugsson and Ohsumi] dissented from the idea of applying a safety factor in this case. There was no other dissent in the SC as a whole, but in the Subcommittee which carried out the assessments, some members expressed their preference for one of the two CPUE series used; this gave an average RY of 708. Other members thought the evidence did not justify this choice as the sole basis for calculating RY.

In the TC, Norway, seconded by Japan, proposed applying a 10% safety factor to 708, giving 637. Australia, seconded by Sweden and New Zealand, supported by Sweden and Seychelles, proposed the lower end of the SC consensus proposal for the range: 300. The Australian proposal was adopted and transmitted to Plenary. There, F. R. Germany, seconded by USA, proposed an amendment to 525, the mid-range. On a vote, the amendment failed: eleven in favor, seven against, seventeen abstentions. The original motion for 300 then also failed: fifteen in favor, seven against, fourteen abstentions.

This lack of a decision was discussed at length. During the discussion, Norway proposed that the 1984 catch limit of 635 should be continued in 1985 and "remain unchanged as long as there is no proposal to the contrary ... because it is not in Commission's interest to have no quota set and to rely on national management of the stock." The Norwegian proposal was seconded by USSR and then adopted by consensus, but with a reservation by Sweden which continued to favor 300. Norwegian whalers killed 771 minke whales in their 1985 season.

At the 1985 meeting, the SC received from Holt a calculation of the initial number in this stock made by a new procedure which has since—in its computerized form—came to be known as the HITTER–FITTER procedure formulated by de la Mare. Holt's calculation used the 1984 results of the SC regarding current stock size and rate of recent decline, the population model which had been used by the SC on several occasions, together with the historic catch series, to calculate the ratio of current to initial stock. "Most members of the Committee, considering that these calculations were based directly on the analysis made in 1984 and agreed by the Committee, believed that the Commission should be advised that this stock should now be classified PS." However, those members also thought that, in consideration of a commitment by Norwegian scientists urgently to review and perhaps extend the CPUE series, "the Commission might consider postponing a decision on classification until 1985." In such circumstances, they said, they would strongly urge that the 1986 catch not exceed the lowest estimate of RY given by the new analysis, which was 306. Two scientists [Gunnlaugsson and Ikedal] dissented from this opinion on the grounds of questions about the new procedure. It is important to note that the majority of the SC did not *propose* postponement of a decision by the Commission, but rather an upper limit to the acceptable catch the following year, should the Commission take such a position.

The discussion of this assessment revealed that Norwegian scientists believed that there had been up to 30% underreporting of catches in 1984, when the catch limit had been sharply reduced and catch quotas had been introduced for individual ships under domestic law. They also "suspected that underreporting had occurred at an increasing rate since catch limits were introduced for North Atlantic minke whales in 1977." One scientist [Butterworth] pointed out that if underreporting had occurred at an increasing rate from 1977 to 1984, the rate of decline of the stock would be less than indicated by the official CPUE data. This effect would, however, be at least partially offset if estimates of the actual catches for those, rather than the reported catches, were used in the assessment.[43]

In the TC, Norway proposed postponement of consideration of classification. However, the new uncertainty about the validity of catch data, the fact that the assessments indicated that the stock was likely to be far below the MSY level, and the near consensus among the scientists, influenced many members of the TC. The majority agreed with a proposal by India that this stock be classified immediately as PS.

In Plenary, Finland, supported by India, Sweden, and Switzerland, seconded the TC proposal. Norway proposed, seconded by USA and supported by Japan, that consideration of this proposal be deferred until 1986. This led to an extended discussion about the relationship between the two proposals and which should be voted upon first. The Chairman ruled, and was not challenged, that there were two different proposals relating to the same question and that they should be voted on in the order in which they had been made. The proposal for PS classification was adopted: twenty-five in favor, one against, ten abstentions.

Norway registered its position as being present, but not participating, thus neither voting nor abstaining. Norway subsequently objected to this decision; its objection still stands. Norway announced it would set a *national* catch limit of 350 for 1986, i.e., a little less than the provisional figure that had been suggested by the SC. Norway was within its rights in doing this as it had also objected, in 1982, to the zero catch limits (commercial moratorium) which would come into effect as from the 1986 season.

The SC reassessed this stock at its meeting in 1986, the first year of the moratorium. This time it employed the HITTER–FITTER program developed by W. K. de la Mare. Two scientific views emerged: "Some members believed that the stock should remain a Protection Stock while others concluded that no classification could be recommended." In the Commission meeting, no formal proposals for change were made, so the stock remained classified PS. At the end of the general discussion, the UK asked Norway and other Members which had objected to the 1982 decision for a moratorium to withdraw those objections, and "urged Norway to consider its position very carefully in the light of the damage it may be doing to this particular stock and its national reputation." Norway responded "by recalling the deep divergence of scientific views over this stock" and said it "took its decisions with the greatest seriousness."

The reported Norwegian catch of minke whales in 1986 was 379. In July 1986, immediately after the IWC meeting, the Government of Norway decided to halt commercial whaling after the 1987 season. The catch reported for 1987 was 375.

[43]Dr. Lars Walløe and the small group of scientists appointed by the Norwegian Prime Minister in 1986 to "review the basis for Norway's harvesting of minke whales" reported "In recent years there have been rumours (sic) of insufficient reporting of catches, particularly for 1984–1986." Walløe organized a study of a sample of records of a few boats in each of the years 1980–1983, some of which had had an inspector on board and others not. He reported "there was no evidence of underreporting for these boats and years," but as the sample was small, "no firm conclusions can be made." Elsewhere in their report, the Walløe group said that apart from a published estimate by C. J. Rørvik of at least 6% underreporting each year in the period 1976–1983, it had "not found any evidence of substantial underreporting for that period." Since whalers had little if any motivation for substantial underreporting before boat quotas were introduced, the Walløe group concluded that underreporting "seems an unsatisfactory explanation for the decline in the Barents Sea CPUE index for 1976–1983." The Walløe group made no recommendations regarding the improvement of national inspection, nor have any proposed changes been announced with respect to any resumed commercial whaling.

The classification of the Northeast Atlantic minke stock was discussed at the 1990 meeting of the Commission in the context of the comprehensive assessment being conducted that year of the minkes in the North Atlantic as a whole. The scientific differences have not been resolved, nor did the Commission take any new decision. The question is on the agenda of the 1991 meetings.

The situation poses a dilemma for the Government of Norway. There is no legal reason why Norway should not resume commercial whaling at any time, as it has standing objections both to the classification of this particular stock and the moratorium. But, for various reasons, it is reluctant to do that. If it were to withdraw its objection to the latter decision, and if the moratorium were to be relaxed, then it would still find itself in the invidious position of resuming exploitation of what the IWC has determined to be a depleted stock. Thus, Norway now necessarily pursues a two-pronged fail-safe strategy: first to provide scientific evidence that the stock is not depleted to PS status in terms of the NMP, and thus obtain a reversal of the 1985 decision; and, second, to seek a catch limit under an RMP, but at the same time to work for the adoption of one of the RMP options which does not provide for an explicit stock protection level or, at least, one in which any protection level is much lower than the assumed 54% level which has been applied to the NMP.

Appendix 4

Preliminary Considerations on Management Procedures

[The following is an extract from Doc. IWC/41/42, 1989 with this title, submitted by the delegations of Australia, Federal Republic of Germany, New Zealand, Seychelles, Sweden, Switzerland, United Kingdom.]

"In the light of the review of the moratorium, it would be essential to develop, adopt and incorporate into the Schedule a revised management procedure which overcomes the deficiencies that have been revealed in the implementation of paragraph 10 (a), (b) and (c) of the Schedule, and which has been fully evaluated using the testing protocols that are being developed by the Scientific Committee.

"The primary objective of such a revised management procedure should be to ensure that currently abundant populations of whales are maintained at or above high target levels, and to ensure the recovery of depleted populations to these levels as soon as natural processes will permit.

"The revised procedure should ... in particular ... provide sufficient allowances for uncertainty, including uncertainty regarding stock boundaries, that lack of data does not prejudice achievement of the primary objective [Negligible risk of extinction].

"The revised procedure must retain the concept of a protection level, which shall be significantly high*."

[* This was subsequently clarified as meaning certainly no lower than the protection level adopted by the SC for implementing the NMP, with several countries having a strong preference for a level substantially higher than that.]

Postscript

Since this article was written, the Government of Iceland has proposed to the IWC that "some limited catch quotas be set for the 1991 season onwards, until the RMP (Revised Management Procedure) has been implemented."[44]

In considering this request, delegates to the 43rd IWC meeting might assume that, given the pace of development of an RMP by scientists, a *fully effective* RMP could be in existence by 1992. Such an assumption is, at present, unwarranted.

Analysis of past experience shows that the mere *existence* of an accepted set of rules does not guarantee that those rules will actually be applied by the Commission in setting catch limits or determining stock classifications. Past failures to apply rules rigorously have been numerous. The reasons for this include: finding that the rules do not cover certain important cases; the failure of the Commission to act on the advice given by its Scientific Committee in cases that the rules *did* cover; objections by whaling countries to IWC decisions.

The Commission has not yet given any attention to this matter in the context of an RMP, nor to the equally important matter of tightening up national and international inspection arrangements. Until it does so, the public cannot be confident that the IWC will soon be able to control any resumed whaling so as to ensure recovery of depleted whale stocks and avoid any further depletions.

When the IWC proclaimed an indefinite moratorium starting in 1985, it was expected to trigger the difficult management decisions needed to fulfill the Commission's mandate to look after the interests of future generations. Instead, governments have taken the easy way of tossing the ball to the scientists (as they have done twice before, in 1960 and 1974) and waiting for them to deliver a magic formula.

The justification offered by the Government of Iceland in asking for interim quotas is that they are needed "for social and cultural reasons to provide continuity" in the whaling industry. But, continuity of that industry was supposedly broken several years ago when Iceland decided not to object to the 1982 moratorium decision. If the industry is now permitted to resume under *ad hoc* quotas, it will be very difficult to ensure that they revert in future to such numbers as may be called for under an RMP, which may be smaller, or even zero. If a new "continuity" is thus re-established, renewed pressure for catch limits to confirm with vague "social and cultural factors," rather than with the requirements of resource conservation, is inevitable.

Requests for interim quotas effectively *prejudge* the revised rules, *steer* the revision process in a particular direction, and *subvert* the assembly of a management package in which scientifically tested rules, improved decision-making procedures within the Commission, and more credible enforcement are wrapped together.

There is no guarantee that a whaling country, once it has been allocated interim quotas, will later feel obliged to agree to a Revised Management Procedure, especially if it is one that could result in catch limits lower than the interim ones, or even the temporary protection of the stocks in which it has a commercial interest.

---

[44]Document IWC/43/2 of March 22, 1991, Annotated Provisional Agenda for the 43rd Annual Meeting, page 9, Agenda Item 10.3. This communication from Iceland raises another important question: the validity of that Government's claim that the 1982 decision establishing an indefinite moratorium on commercial whaling has automatically expired for any whale stock for which a "comprehensive" stock assessment had been made (regardless of whether such an assessment has been completed successfully and of its results). There is, of course, no legal basis for this claim.

Acquiescence by other nations in moves to placate one of intending whaling nations by allowing it an indeterminate period of *ad hoc* quotas would be to lose—probably irrevocably—hard-won gains made during 1980s towards reaffirming global responsibility for a great natural resource. When demands for interim quotas are accompanied by threats to walk out if they are not met, acquiescence will be seen as capitulation to short-term and parochial economic interests at the expense of the long-term interests of people of all nations.

Chapter 28

Fin Whales in the North Atlantic: A Comprehensive Assessment [1]

Introduction

The Scientific Committee (SC) of the International Whaling Commission (IWC) was told by the Commission to carry out *three* **comprehensive assessments** by 1990. The *first* was to determine the effects on whale stocks of the decision in 1982 for an indefinite pause in commercial whaling (called "the moratorium"). The *second* was to determine the effects on stocks of the adoption in 1983 of a new management scheme for aboriginal subsistence whaling.[2] The *third* comprehensive assessment, also agreed in 1983, was to be "an in-depth (and in-breadth) evaluation of the status of all whale stocks in the light of management objectives and procedures." This last was to include "the examination of current stock size, recent population trends, carrying capacity and productivity."

The first comprehensive assessment has so far been attempted only with respect to a few stocks of fin whales *Balaenoptera physalus* and minke whales in the North Atlantic (*B. acutorostrata*) and minke whales in the Southern Hemisphere (*B. bonaerensis*). It has not been successful for a simple reason: the moratorium has not been in force for long enough to detect, with the present rough methods of counting whales, any population changes, which are expected —from what little is known about the reproductive and natural death rates of whales— to be quite small. Furthermore, in no case have surveys yet been repeated after an interval, so no matter how fast whale populations recovered, or how accurate the surveys were, there are no figures that can be compared over time to measure any changes.

This article is my response to enquiries about one piece of the *third* comprehensive assessment. Apart from being far from "comprehensive," the report of the scientific meeting at which it was carried out[3] is inconclusive. It must also be virtually incomprehensible to anyone who did not participate in that meeting. It continues a "tradition" of opacity which began with reports of simi-

---

[1]Editors' note: written by Sidney Holt for the World Wide Fund for Nature, May 1991, and reproduced here (with minor edits) with their permission. Not included from the original is a page of acronyms and contractions as well as the author's summary.

[2]This has not yet been started. The aboriginal subsistence scheme was a simple modification of the scheme adopted for commercial whaling in 1975. Since the commercial scheme of 1975 was quickly found to have fatal flaws, it is not surprising that review of the 1983 aboriginal subsistence scheme should await replacement of the former, now underway.

[3]In Reykjavik, Iceland 25 February – 1 March 1991. IWC Doc. "Report of the Comprehensive Assessment Special Meeting on North Atlantic Fin Whales" published in *Forty-Second Report of the International Whaling Commission, 1992*. (Editors' note: Holt was a participant in this meeting as a representative of the Seychelles.) I refer to it as "SMR." First circulated by the IWC Secretariat as an unnumbered document. Some of the errors I have found may have since been corrected.

lar assessments of whale stocks that have been subject to commercial whaling which were carried out in 1990. And, perhaps more important, the analyses reported are less complete and, I think, presented more casually and carelessly, than any of the work done by the SC in recent years.

Such hastily written reports created confusion in the 1990 meeting of the IWC, and have been widely misunderstood by those other scientists, media people, politicians, and administrators who have been brave and diligent enough to read them. The conduct of comprehensive whale population assessments would, at best, be a scientifically difficult task. It has, however, been made more difficult than it need have been by the pressure from three countries for numbers to be generated which could provide a quasi-scientific basis for their resumption of commercial whaling before the Commission has agreed on a Revised Management Procedure (RMP) to replace the scheme adopted in 1975, and subsequently shown to be seriously flawed, called the New Management Procedure (NMP).[4] The NMP has been put in abeyance by the terms of the moratorium decision adopted by the IWC in 1982.

The partial assessments which have been attempted so far reveal, if the language in which they have been reported can be penetrated, that the areas of fundamental ignorance about these animals are vast. In no cases are the breeding areas known, nor are either the long-distance migrations nor the foraging movements understood. Knowledge about how many separate populations of each species live in each major ocean basin, how those populations are distributed, and how they intermingle either on feeding grounds, on breeding grounds, or both, is virtually non-existent. There are no estimates for any of them of a crucial number called the maximum sustainable yield rate [MSYR(%)].[5] That rate, if it could be measured, would, with knowledge of the current numbers and the historic catches, be the basis for setting annual catch limits under the NMP. The MSYR encapsulates in one number several of the basic features of the population dynamics of whales, particularly the expected *density-dependent responses* of rates of reproduction and natural death, to changes in the total numbers of animals of each sex in a population.

On the positive side, as a result of substantial research efforts in recent years, there now exist direct estimates, from systematic sightings surveys, of the numbers of each of the two species in the regions so far examined.

The idea of "comprehensive assessment" was introduced in 1982 as a dramatic gesture, a definition of at least one condition which, if fulfilled, would open the way to reconsideration of the present indefinite moratorium on commercial whaling. As use of the term expanded, it has become a cliché almost devoid of content.

What is an Assessment?

The **primary ingredients** of a whale stock assessment, as this is now understood by the IWC's Scientific Committee(SC), are:

1) As complete as possible **a series of annual catches**, with information about the locations of catch and the sexes of the animals caught. Some information about the ages and/or lengths of animals caught is also necessary.

---

[5] A non-technical account of the NMP and the revision process is given in "Guide to Review of the Management of Whaling," second edition, 1991, produced by the Center for Marine Conservation, Washington, DC for the 43rd Annual Meeting of the IWC. The first edition was distributed at the 42nd Annual Meeting.

[6] This is, roughly speaking, the rate at which a population would grow each year if it was not exploited and if it was at a certain optimal fraction (usually assumed to be 60%) of its pristine number.

2) One or more **estimates of the numbers of animals in a stock** in one or more recent years. These estimates may, in the case of the baleen whales, come from systematic scientific sightings surveys of large areas or from experiments in which animals have been marked with metal tags embedded in their skin, some of which have been recovered during whaling operations. Sometimes the results of both surveys and marking experiments are available.[6] Series of indices of whale abundance, as specified in 3) below, have also been used to estimate numbers indirectly.
3) A **series of annual "indices of relative abundance."** These are numbers that are thought to be proportional, or otherwise related in a definable way, to the numbers in the population. That is, a change in the index from one year to the next reflects a corresponding change in the numbers of whales. The index most commonly used is the "catch per unit of whaling effort (CPUE)", suitably adjusted in ways that are described below. Estimates obtained by scout boats attached to pelagic whaling expeditions have also occasionally been used. It is not necessary for the series of abundance indices to extend throughout the period of exploitation, and, in practice, it hardly ever does so. Indices for later years, when whaling operations were better documented, will usually be the most useful, but rough data for the early years can be important.
4) **A procedure for jointly interpreting the first three ingredients of data.** The procedure now used by the SC is called *HITTER/FITTER*.[7] When reliable sets of all three data ingredients exist, much can be deduced about the whale stock by using this procedure. If either of information types 2) or 3) are unavailable, reduced versions—*FITTER* and *HITTER*, respectively—can be used, but more, and essentially untestable, presumptions then have to be made or supplementary information provided.[8]

From this it should be clear that in trying to conduct a "comprehensive assessment," the SC does not set out to write down all that is known about certain populations of whales, but rather decides first on an analytical procedure for the assessment [ingredient 4)] and then ascertains the availability, and appraises the reliability, of the three data ingredients.

HITTER/FITTER

I presume that more readers of this report will be familiar with the operation of a bank savings account than with the methods of demography. So, I have given, in Appendix 1, an analogy of the H/F computer programs with the process of understanding a savings account, with limited knowledge about the initial deposit, the interest rates, and the bank charges.

[6]A new method of estimating numbers involves repeated recognition of individual whales from distinctive natural markings, but it has not been applied to the species of commercial interest which have been assessed so far.
[7]I used a crude form to assess minke whales in the North Atlantic in 1985 and 1986. As now used, it was developed by William de la Mare, who presented it to the SC in 1986. D. S. Butterworth has also written a computer program for this procedure: his and de la Mare's current improved version calculate confidence limits as well as "best" estimates for certain parameters, but they do this in different ways and with somewhat different results.
[8]The SC has used variable terminology concerning these procedures. Here, I adopt the following convention: HITTER/FITTER (H/F) is the program as a whole; HITTER-FITTER (H-F) is the mode in which that program is used with both one or more population targets and one or more CPUE series; HITTER (H) is its use only with a target; FITTER (F) is its use only with CPUE series, no target. What are described as "FITTER" runs in the SMR are what I have defined here as (H-F), but this is not obvious from the SMR itself—see later.

The output of H/F consists of an estimate of the numbers of animals of each sex in the stock when whaling began and an estimate of the current numbers. From these, a *depletion level* can be calculated. This is important for *classifying* the stock according to certain regulatory provisions; it is usually expressed as the ratio of the current number of mature females to the *initial* number, as a percentage. The programs also provide, in the form of a graph, a calculated population *trajectory* between the initial and current numbers.

H/F, using all three types of data, also gives, in principle, an estimate of the *maximum sustainable yield rate*, written as MSYR(%). This is, roughly, the annual rate at which a population would increase in number if it was not exploited and if it was at a certain *optimal* level of size. This optimum level is usually expressed as a fraction or a percentage of initial population size, written as MSYL or MSYL(%). It has not been possible to determine MSYL empirically for any whale population, but from theoretical considerations, the SC has usually assumed it to be 0.6 (60%). Some scientists, however, have said they think it may be higher, perhaps as high as 80%.

MSYR(%) cannot be estimated if there are no data of type 3). In such circumstances, it has become the practice to assume a range of values of this *parameter*. The range has been based on what is thought to be known about its value from studies of other whale species or the same species in other regions. Biological considerations such as the frequency of birth and the age at sexual maturity have been used to suggest upper limits, but they give values which are very much higher than what is thought to be a plausible range consistent with other information.

Strictly, MSYR(%) and MSYL must be defined not only in terms of numbers of whales, but also of the proportions of whales of various ages in the population. For a stock at MSYL to yield maximum sustainable catches, it must have a demographically stable age and sex composition. However, in the SC calculations, the "structure" of the stock considered—apart from the sex ratio—is limited to distinguishing between mature and immature and between recruited and non-recruited (prerecruit) animals.

The age at which recruitment occurs, i.e., the age at which animals begin to appear in catches, is determined by the interaction between the biological characteristics of the whales, principally their individual rates of growth and their migration pattern, and the seasonal timing and geographical distribution of the whaling operations. It may also be affected by any minimum legal size limits that may be in force and by the standing prohibition on killing lactating females, as judged by the presence near them of suckling calves.

An estimate of the age at recruitment is required by the H/F program, but it does not need to be precise nor be estimated every year. Assessment is made more complex if the age at recruitment changed during the period of exploitation (as it might have done for the fin whales near Iceland) or if—as in the case of the minke whales in the North Atlantic—it varies significantly from one location of whaling to another.

Measures or guesses of MSYR(%) have usually been expressed as percentages of the optimal number of mature females, but they can be expressed in other ways, e.g., as percentages of the recruited stock (including all mature animals and some immature ones), or even to the total stock including all weaned, but still immature, individuals.

MSYR(%), expressed in relation to recruited or mature stock, has been assumed to be in the range 1–4% for the purposes of testing revised management procedures. When this range was selected, it was thought to encompass a reasonable low and reasonable high rate, no more than that. The uncertainty about this parameter is, however, wider than this range suggests.

For purposes of fin whale assessments, the assumed range has been 0–10%. This was done because some scientists think the MSYR(%) *could* be less than 1%, while others think it *could* be substantially higher than 4%.

Application of H/F requires also an estimate of the annual natural mortality rate (expressed as M[9]) in the recruited population. There is evidence that adult natural mortality increases with age, as it does in all mammals for which there are data. There is also some evidence that it differs between the sexes, again as it does in other mammals, with the larger sex (females in the case of the baleen whales) having a higher death rate and, therefore, being somewhat shorter lived on average. However, computer simulations with H/F have shown that the results of assessments of the kind described here are not very sensitive to the exact value of M nor to its variations according to age or sex.

In fact, even with excellent data of all three types, the estimation of MSYR(%) cannot be precise. This, too, has been shown by computer simulations. The output which is least sensitive to the variability in data and to the specific assumptions made in using the program is the degree of depletion. Hence, the results of this assessment procedure may be useful in determining stock classification, but much less useful for estimating sustainable yields as a basis for setting any catch limits.[10]

In principle, a correct application of H/F gives estimates of current replacement yield (CRY). This is the catch that theoretically could be taken next year and exactly replaced by incoming recruitment at the beginning of the following year, adjusted by the natural mortality next year. CRY, or sometimes average recent RY (ARRY), has, in the past, been cited by the SC as a basis for interim catch limits when the existing management rules turn out not to be applicable.

The CRY of a stock that has been changing is not the same as current *sustainable* yield (CSY), precisely because it is not sustainable. Thus, if the *same* RY were to be taken as a catch the following (second) year, the stock at the beginning of the third year would not be the same as at the beginning of the second. This is because the age composition of the population will not be the same.

The CRY is very sensitive to the particular assumptions made and to data variability, as well as to the sizes of the most recent catches. If RY is to be used at all as a basis for calculating catch limits, then some calculated recent average is more appropriate than an estimate of CRY. Calculation of such averages is, however, rather arbitrary, since different answers can be obtained by taking different lengths of time for the average. Advice on possible interim catch limits could also, in theory, be given in terms of calculated average *future* RYs, but that calculation has the additional problem of depending upon the assumptions made as to what the future catches will actually be.

The particular form of population model used in the H/F makes a specific assumption about the way that the *net recruitment rate* (the difference between the rate of recruitment

[9]This is usually cited as an exponent, e.g., $M = 0.04$. When M is low, as it is for whales, this number can be envisaged as a percentage. In this example, about 4% die naturally per year.

[10]Just as one of the outputs from H-F is a value of MSYR(%), so an estimate of this quantity may be an input to H. In some of the reports of the SC, including that on the North Atlantic minke whales, estimated values of resilience A are given instead of MSYR(%). For the particular population model used by the SC, the resilience parameter can be converted to MSYR(%) by a simple formula if the value of M is specified. For example, with MSYL(%) = 60% and $M = 0.04$, the formula is: MSYR(%) = $A \times 2.8$; for $M = 0.1$, as it seems to be in the minke whale, then MSYR(%) = $A \times 7.0$.

and the rate of natural mortality), which is the same as the SY(%), changes with changes in the numbers of whales. It is assumed that in the unexploited stock, SY(%) is zero, and that as the stock declines, SY(%) continuously increases, reaching its highest value just as the stock approaches extinction. This change is said to reflect the *density-dependence* of these rates in the whale population. An equation called the *Pella–Tomlinson* function is usually used to describe this change algebraically.[11]

For the sake of orientation, suppose the MSYR(%) = 2% for MSYL(%) = 60%, then the Pella–Tomlinson function would tell us that the SY(%) in a very much reduced population would be about 2.8%. The sustainable yield itself is obtained by multiplying the SY(%) by the size of the population. The MSY is thus obtained by multiplying the MSY(%) by 60% of the original population number.

In evaluating the various runs presented in the SMR, it is worth recalling that any value of MSYR(%) higher than 1.2% implies an assumption (or a finding) that there is *super-compensation* of recruitment. This term describes a situation in which, as a stock is reduced below its initial numbers, not only does the net rate of recruitment increase, but also the absolute number of recruits.

Two examples illustrate this. If the MSYR(%) is 4%, then as the population is reduced to, say, about 80% of its initial number, the recruitment rate will have shown such strong density-dependence that the *absolute number* of recruits will be nearly 30% more than they were in the unexploited stock. Even if the stock were to be reduced to its MSYL, the number of recruits would still exceed the original number by about 20%. Not until the population was less than about 45% of its original number would the number of recruits become less than in the unexploited stock.

Let us look at the consequences of assuming a very high MSYR(%), such as the value 10% used in some of the SC calculations. Now, the number of recruits in a population reduced to its MSYL would be more than double the initial number, and still increasing as the stock is depleted further. In fact, the stock would have to be reduced to less than one-quarter of its original number before recruitment begins to be adversely affected. The consequences of assuming such an extreme density-dependence, resulting in *super*-super-compensation, is that the population becomes unstable. This is illustrated in some of the diagrams in the SMR which shows oscillating trajectories of increasing amplitude.

Some scientists think that, for reasons connected with the biology and evolution of whales, super-compensation is very unlikely to occur in the real world. Other scientists do not accept such theoretical reasoning, but nevertheless believe that in the absence of direct estimates of MSYR(%), it is risky to assume, for the purposes of assessments, that MSYR(%) could be so high as to result in super-compensation. Yet others think that some degree of super-compensation is possible; for support, they point to the rather high rates at which some species populations that were nearly exterminated have apparently been increasing under protection.

These rates of increase of small residual populations are greater than we should expect to find in a population that has recovered to its optimal number. Whether it is acceptable to estimate MSYR(%) roughly from the recovery rate of a stock which is still depleted is a matter of scientific controversy, especially if—as will usually be the case—the comparison is made between different whale species. But, *if* such comparison were acceptable, then the expected MSYR(%) should be no less than about one-half of the observed recovery rate.

---

[11]Several other functions could be used, but for the types of assessment normally undertaken by the SC, the particular function used does not matter very much.

So, much of the discussion of MSYR(%) is either about theory or about extrapolation from other species because it is now realized that methods of estimating it in the past were faulty. Those methods depended on examination of age compositions of samples of catches or on apparent density-dependent changes in the age at sexual maturity and/or the percentage of mature females in catches that were pregnant. But there are serious and unresolved practical difficulties in interpreting such data unambiguously. Furthermore, the recruitment rate depends not only on the rate of reproduction, but also on the juvenile death rate, which cannot be measured. Computer simulation, which is being used to test possible revised management procedures, has also revealed that many "traditional" analytical procedures are flawed.

I have discussed this aspect of whale stock assessments in some detail because they all depend critically on current ideas about the magnitude of MSYR(%). The two approaches—one looking at recovery rates of very depleted populations, the other at the output from H/F—both give very uncertain results. Difficulties with the former are outlined above. As for the latter, it is now clear that although, with good data, useful estimates of the status of a stock may be obtained, any estimates of the MSYR(%) are much less reliable. This means that there is little scientific basis for setting any catch limits according to the rules of the old New Management Procedure (NMP), as defined in paragraph 10 of the Schedule of the 1946 Convention.

With this general background, we are ready to look at the assessments of the North Atlantic fin whales that were attempted at the Special Meeting of the Scientific Committee. The first decisions that have to be made concern the identities of "stocks" of whales of a particular species.

Stock Identity

Scientists are virtually certain that populations of baleen whales (except possibly the subtropical Bryde's whale *B. edeni*) that feed and breed in the North Atlantic do not mix with populations of the same species in the southern hemisphere, and certainly not with those in the North Pacific. Very little is known, however, about the existence of distinct populations of a species within any major ocean region. For this reason, the SC decided, in 1989, to attempt comprehensive assessments of what it called groups of stocks of each species, but in doing so, it did not commit itself to any assumptions about the existence of only one or more than one population of any species in any region.

The migration routes and patterns of fin whales in the North Atlantic are largely unknown, as are the locations of breeding grounds, which are presumed to be in lower latitudes than the feeding areas, even in subtropical latitudes. Thus, it is not easy even to suggest plausible hypotheses about whether there is only one or there may be two or more distinct breeding populations, or how much whales from any subpopulations mix during their migrations or life cycles.

For the purpose of applying the NMP, the SC proposed, and the Commission accepted, in 1976, the existence of a number of "management stocks." It is important to recall the reason why the splitting up of entire ocean regions into smaller management units was considered by the SC to be so important. Until then, the Commission had been very reluctant to set catch limits by smaller units; this reluctance to split up the process of deciding on catch limits had extended for a decade to a refusal even to abandon the notorious blue whale unit (BWU). This reluctance had the same cause in both cases: inconvenience to whalers. Just as it was demonstrated that the BWU led to the successive depletion and near extinction of one species after another, so setting species catch limits for entire regions would lead to the successive depletion of one population after another of that species.

Thus, the introduction of the management subdivisions in the North Atlantic, and elsewhere, represented an advance in regulatory arrangements and was potentially conservative in its consequences. But, like all such decisions, whether it was really conservative in practice depended on exactly how it was implemented. The special arrangement for designating as a "Sustained Management Stock" any "stock" of which an assessment was impossible for lack of data, but which appeared not to have declined during an extended period of roughly constant catches, was certainly *not* conservative. Some other implementations of the NMP were, however, conservative in terms of the assessment procedures then used. However, as we shall see, regulatory decisions that are conservative if one assessment procedure is used to provide advice, may be the opposite if another assessment procedure is used.

A very simple, but real, example illustrates this. If, say, a catch limit is being determined by calculating a percentage of a population estimate, then a conservative catch limit would arise, other things being equal, from using a lower, rather than a higher, population estimate. However, if a recent average replacement yield is being calculated, for management advice, from a rate of decline of a stock and a level of catch that caused that decline, conservative advice will emerge from the adoption of relatively *higher* stock estimates. The latter case occurred when the SC assessed the Northeast Atlantic minke whale stock in 1983, leading to a large reduction in the catch limit for Norwegian whaling.[12]

Some of the decisions about stock identity and appropriate boundaries, made in the 1970s, are not at all conservative in their effects when the new assessment procedures are used. In particular, the extremely large management areas in mid-Atlantic, which have been adopted in the provision of advice on catch limits for Icelandic fin, sei *B. borealis*, and minke whaling, have dangerous consequences when H/F is used, and if the input includes sightings estimates of current numbers.

The management stocks in the North Atlantic were identified essentially in terms of distinct land-based whaling operations or groups of operations, some of which were thought to have been mainly responsible for depletions in limited areas during the early decades of "modern whaling" in the North Atlantic, and in a few cases more recently. Such operations included, for example, Norwegian land stations in north Norway and west Norway, stations in Faroes, Shetlands, and Orkneys, the Outer Hebrides, on the Canadian east coast (including Newfoundland), off the northwest coast of Spain, and around Iceland. It had been observed that the availability of whales had declined at different rates and in different periods in several of these locations, and it could be assumed, therefore, that even if there were not entirely separate populations near them, then at least the rate of mixing between the locations was rather slow.

Catchers operating from land stations had a limited operational range so their catches could be tabulated with little or no reference to predefined geographical areas. For assessment, it would then only be necessary to agree on how such sets of catch data should be

---

[12]Until then, Norwegian scientists had been producing higher estimates of stock numbers every year from recalculated results of marking experiments. In these experiments, as the years pass by with no more marks being recovered, the stock estimates keep rising! In 1983, with the acceptance, at last, that this stock had been declining for more than a decade so that RY needed to be calculated, lower stock estimates were in favor. The sudden cut in catch limit from 1,700 to 600 has been presented as the responsible acceptance by the government involved of a hard blow to its whaling industry. But, from another point of view, it was the retrieval of something from the wreckage of previous reckless regulations; if the slightly higher stock estimates had been used, the catch limit would have been cut, not to 600, but to 60.

grouped. But substantial catches of fin whales were also taken in the earlier years by pelagic operations. It seems reasonable to presume that these operations took whales which were also accessible to some of the land stations. In order to subdivide and allocate the pelagic catches to management "stocks," it was, therefore, necessary to draw boundaries far out to sea. The boundaries that were so drawn are shown in Figure 1.

These boundaries have no biological significance. Some of them embrace an extended area through which fin whales were presumed to migrate, judging mainly from the successive dates within each season at which the catches various operations peaked. The "British Isles, Spain, and Portugal" stock is an example of such a grouping, though there is no compelling reason for the Faroes, for example, to be excluded from it rather than included in the "West Norway" stock. Several of the boundaries were drawn to conform with existing international systems for compiling fisheries statistics or with the original Norwegian whaling statistical areas. The same process was followed in drawing "stock" boundaries for the minke whale and other species.

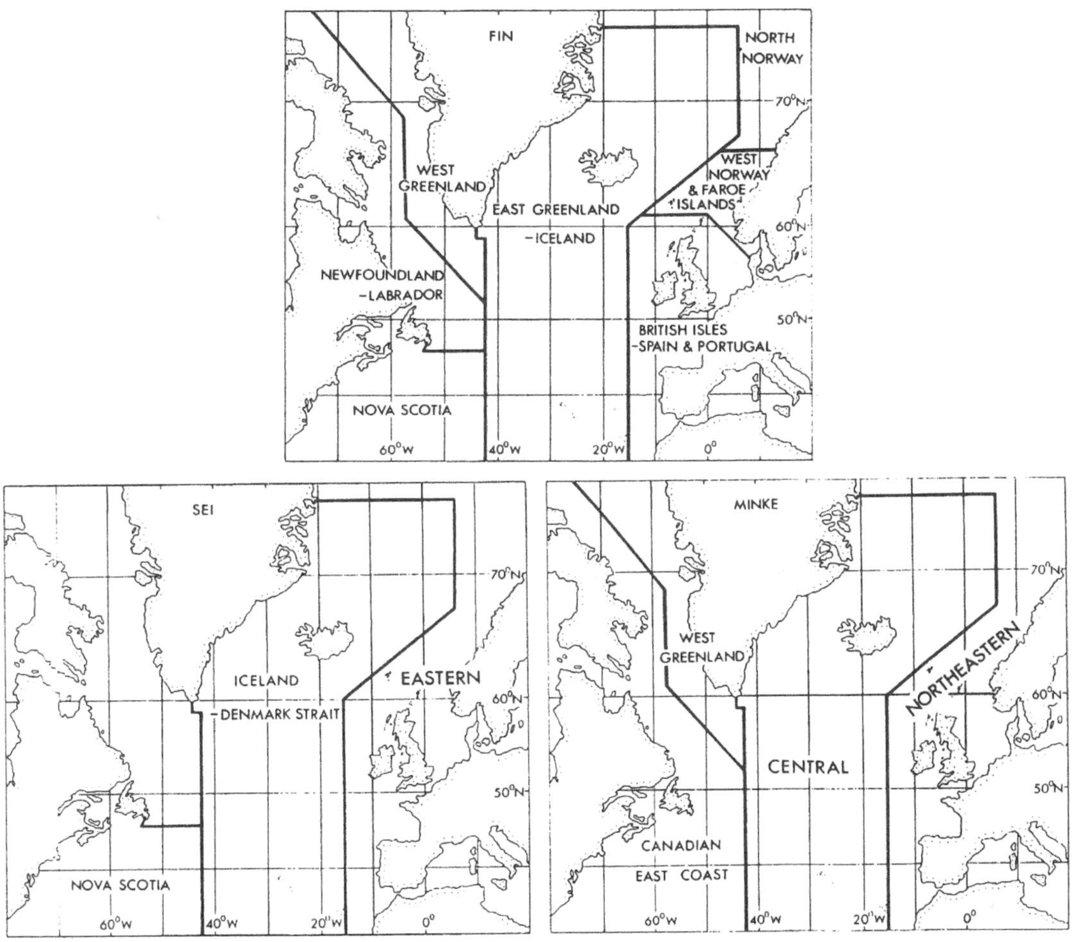

Figure 1. Schedule stock boundaries: usual Mercator projection. Shown here for three baleen species. Note that boundaries of the mid-ocean management areas are all identical and arbitrary; they have no substantive basis for any of the species.

It will be seen that the subdivisions of the North Atlantic basically run along the eastern and western rims of the ocean, leaving a large area in mid-ocean which was named "East Greenland—Iceland" in the case of fin whales and "Central" in the case of minke whales. Just how relatively large that central area is can better be seen on a map drawn on an equal area projection (Figure 2). It is so large because it is really a residual area, i.e., what is left after the rims areas have been split off.

Two kinds of data exist for examining mixing of whales between different locations, but the data for the North Atlantic fin whale are extremely sparse. One kind comes from marking experiments. Between 1950 and 1984, 685 marks were placed in fin whales and 97 of these were recovered. Whales were mostly marked in or near whaling areas and, of course, only recovered from places where whaling had continued, at least for a short while, after marking.

Of 304 marks placed in whales off the coasts of Canada, 38 were recovered, of which one was recovered off the southwest coast of Iceland. Of the 37 recovered by Canadian operations, 3 were found in management areas other than the one they were marked in. The Special Meeting concluded that this indicated sufficient mixing between the Newfoundland–Labrador management area and the Nova Scotia area to justify lumping these two areas together for any assessments.

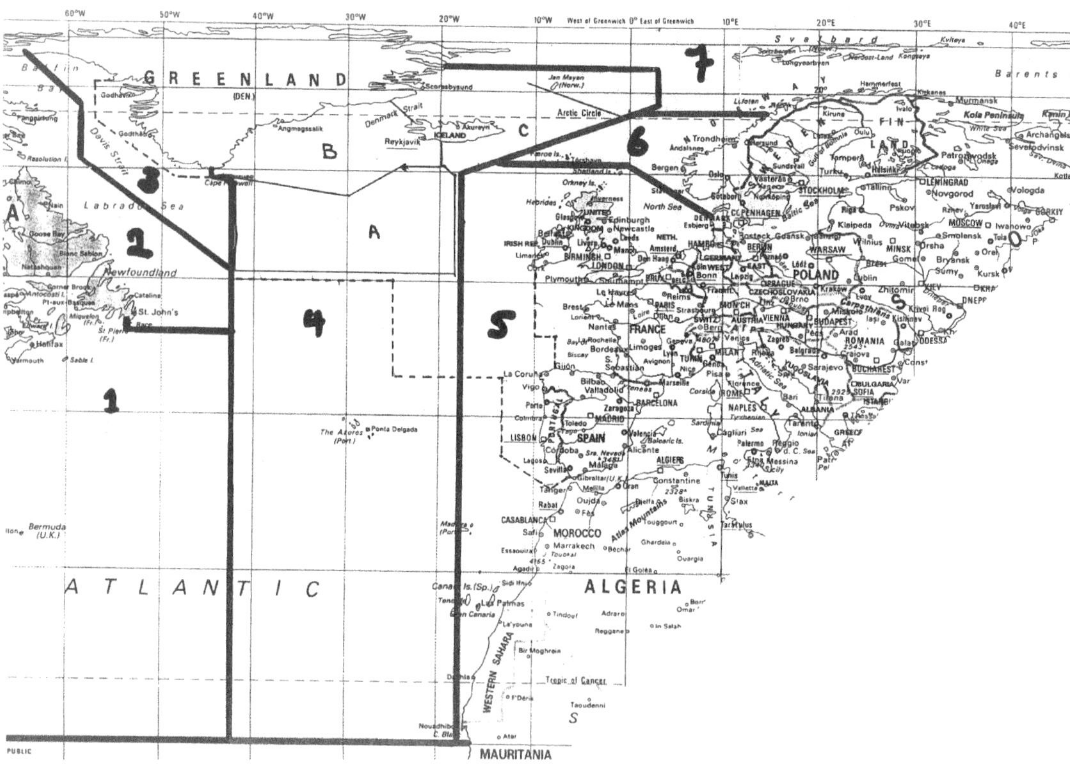

Figure 2. Schedule stock boundaries: equal area projection (Peter's). Dotted lines indicate southern and eastern NASS (North Atlantic Sightings Surveys) boundaries. Thin continuous lines indicate NASS blocks A, B, and C.

Some mixing was also observed between groups of whales off East Greenland and southwest Iceland, both being within the "East Greenland–Iceland" management area.

The one recovery of a "Canadian" mark off Iceland told something about the possible range of mixing rates; in fact, there being "only" a single recovery did not even mean that the mixing rate is particularly low. Further, when the number of marks is few, and especially if whaling is sporadic, the mere *absence* of recoveries of marks from one location of those placed in another does not indicate that mixing is low or non-existent. It is not procedurally correct, in such circumstances, if there are no mark recoveries indicating mixing, to act for management purposes as if there is no mixing. Yet, that has commonly been advocated by scientists who should know better. The correct procedure is to calculate the probabilities, using the *absence* of recoveries as relevant data, that mixing at a certain hypothetical rate has or has not occurred. The absence of "exchanges" of marks between, say, Iceland and the Northeast and East Atlantic is not inconsistent with there being considerable mixing between these parts of the entire region; but, it is equally not inconsistent with a hypothesis that there is very little mixing with fin whales that are seen near Iceland and some other adjacent areas to the north and/or the south.

One of the puzzling features about whales that is suggested by mark experiments is that they show, on the one hand, considerable ranges of movement on separated feeding grounds as well as during the feeding/breeding migrations, but on the other hand, marks are quite commonly recovered from *exactly* the same locations as they were fired several years previously.[13]

The other kind of relevant information about stock identity of North Atlantic fin whales came from a few recent genetic studies and from morphometrics (analyses of lengths of parts of the whale's body). These indicated that whales caught off the southwest coast of Iceland are different from those caught off northwest Spain.

On the basis of this extremely fragmentary information, some members of the SC concluded that "the available evidence was sufficient to suggest that the boundaries (in Figure 1) were suitable for the purposes of assessments..." except that Nova Scotia and Newfoundland–Labrador could conveniently be combined. Other members concluded that "...the limited information available meant that no definitive stock divisions or their boundaries could be proposed." Those members thought that the only reasonable approach would be to examine "a variety of divisions and boundaries, including small areas and broad regions."

The H/F runs that were eventually made looked at the North Atlantic as a whole, except for the southern part of the Spain–Portugal–British Isles management area and the Canadian–US coast area. Runs were also made for most of the seven "stocks" listed in the Schedule and defined in its paragraph 9(c), except that Nova Scotia and Newfoundland–Labrador were combined. However, no runs were made for the Spain–Portugal–British Isles management stock. Lastly, the entire NASS-surveyed area (see below) was cut up into the smallest portions that the plan of the NASS could accommodate. These portions were not, however, such as would have been selected if there had been a free choice; for example, one possible choice would have been the area encircling Iceland. This lack of freedom caused some problems when it came to deciding in what ways subsets of the smallest areas might be combined and

[13]This latter phenomenon has been noted with, for example, the marked minke whales recovered from northern Iceland; it is referred to in recent SC reports as "site specificity." Recoveries of marked fin whales from southwest Iceland have not yet been examined from this aspect. The term does not imply those whales are "resident" (although they might be, in some cases), but simply that they may return, in their migratory cycle, again and again to exactly the same spot.

in allocating some early catches to one small area or another. I shall describe later the way in which this problem was partially resolved.

In previous attempts to assess North Atlantic whale populations, the locations of the arbitrary boundaries between the designated management areas did not much matter. This is because most of the historic catches have been from land stations, and the outer limits of those operations were far from most of the boundaries; the problem of appropriate compilation of catch statistics was minimal. If CPUE data existed, they, like the catch data, were indicative of what was happening to a self-defined "stock," that is, a "population" of whales that was *effectively* exploited by a certain set of whaling operations, but not defined geographically except by reference to the locations of the land stations. Similarly, if there was a "population" estimate, it came from marking experiments, and the population thus estimated was, therefore, *also* defined in terms of the whaling operations, not of a geographical area. There was thus consistency among the datasets used.

In a few cases, a management area covered two or more operations physically separated by great distances; the example most relevant to the assessments made this year was the East Greenland–Iceland area, which included the recent land-based operations off southwest Iceland and early pelagic operations around Jan Mayen and elsewhere in this large area. Another example was the inclusion of pelagic operations near Gibraltar with the land-station operations in Shetland and the Hebrides. In other cases, operations not far apart happened to have been placed in different, though adjacent, management areas without there being any evidence that they were, in fact, exploiting different real populations. Examples are West Greenland coast/southwest Iceland, Faroes/Shetland and Hebrides, Faroes/southeast Iceland. Marking experiments have not helped to sort this matter out—far too few whales were marked in the eastern Atlantic, and none were recovered from there or from anywhere else.

Formally, the H/F program is only valid if it is applied to a single, biologically self-contained population. This is because it computes the incoming recruitment each year as a function of the numbers of mature female "parents" left in the population as potential mothers at a time a few years earlier. However, under certain conditions, for example, if immigration and emigration rates to the stock effectively exploited (strictly the *differences* between the immigration and emigration rates) are small, then the results of an H/F analysis are likely to be fairly indicative of the degree of depletion of the effective stock.

In recent assessments, the population estimates have more often come from sightings surveys. This is because there has been a general agreement that sightings estimates were "better" than marking estimates. In one sense, they are: the results do not depend on the vagaries of commercial recovery operations, the surveys can be properly planned, and special experiments can be carried out to examine uncertainties in the analytical procedures. But, surveys estimate *different entities* than do marking experiments; they are defined purely in geographical terms, not operational terms. For this reason, their results must be used with great care when applied in conjunction with operationally defined catch and CPUE data, as in computing with H/F programs.

If a great deal was known about whale movements within a large area, then the conditions for a proper application could perhaps be specified. But, without such information, the sightings estimates can be determined practically arbitrarily by adjusting boundaries, and this will commonly lead to inconclusive assessments. If the boundaries are maintained or adjusted with policy criteria in mind, the results of assessments may *appear* less inconclusive, but they will be biased towards fulfillment of policy.

Stock Estimates

Although the North Atlantic Sightings Surveys (NASS) in the summers of 1987, 1988, and 1989, primarily from ships, but some from aircraft, are collectively the most extensive surveys of whales ever carried out in the northern hemisphere, they were not complete. As can be seen from the rough sketches given in Annex D4 "Fin whale distribution from recent systematic surveys" in the SMR, they were confined, with the exception of the ice-free coasts of Greenland, to the region west of 42°W. Also, although the management stock areas for the North Atlantic are defined as extending southward to 20°N, the surveys, with the exception of that near the coast of Spain as far as the Strait of Gibraltar, did not extend further southward than 43°N. The North Sea was not surveyed; fin whales are now very rare there, as they are also in the Barents Sea, though they were once more common. In fact, one significant conclusion could be that the range of the fin whale in the North Atlantic has contracted as a result of earlier whaling on the eastern rim, and it has not yet expanded again since those whaling operations ceased.

The fact that surveys have not been carried out in the southern part of the management areas is not a practical disadvantage if it can be assumed *either* that the fin whales are there mainly during the winter and migrate in spring and autumn into and away from the main areas of feeding concentrations northward of about 40°N, *or* that the whales in the southern part are not migratory and so would be considered for assessment purposes as effectively in different stocks from those found more to the north. Nevertheless, if we are ever to know more about the basic distributions and migrations of fin whales, whether or not such knowledge would be used to manage future whaling, then it will be essential to conduct sightings surveys and make behavioral observations in the southerly parts of the range of the species.

The fact that the NASS surveys westward of 25°W extended southward only to 50°N, while those east of 25°W extended to about 43°N introduces an anomaly into the estimates, as can be seen by comparing the sketch map in Annex D4 of the SC report with Figure 3 in the body of that report and in this article. Fin whales were seen in the surveyed "rectangle" area bounded by 25°W50°N, 20°W50°N, 20°W43°N, and 25°W43°N. Some presumably may also be found in the unsurveyed area around and north of the Azores (at about 40°N30°W) and south of the subdivision "A" in Figure 3.

The overall picture is that there are thought to be very roughly 35,000 fin whales which feed in the North Atlantic northward of about 40°N. Of these, nearly one-half are to be found in the mid-Atlantic region called the East Greenland–Iceland management stock area, and nearly one-third on the western side of the North Atlantic; so, about one-sixth of the whales counted were in the surveyed part of the Atlantic eastward of the mid-Atlantic region.

The possible statistical errors are such that the total number can only be said to be probably in the approximate range 20,000–60,000.

Although no NASS surveys were conducted in the Newfoundland–Labrador–Nova Scotia combined area, sufficient marking data were available to give an approximate estimate of the stock that had effectively been exploited from the Canadian land stations. This was 10,800, with a 95% confidence range from 5,400 to 21,700. It was the only estimate that could be used to examine the state of this "stock," but it could not properly be combined with the sightings estimates in adjacent areas for methodological reasons; it could have included some whales that were actually to be found in the adjacent surveyed areas or, equally, might exclude some whales that would have been found in parts of the Canadian coast area if sightings surveys had been carried out in it.

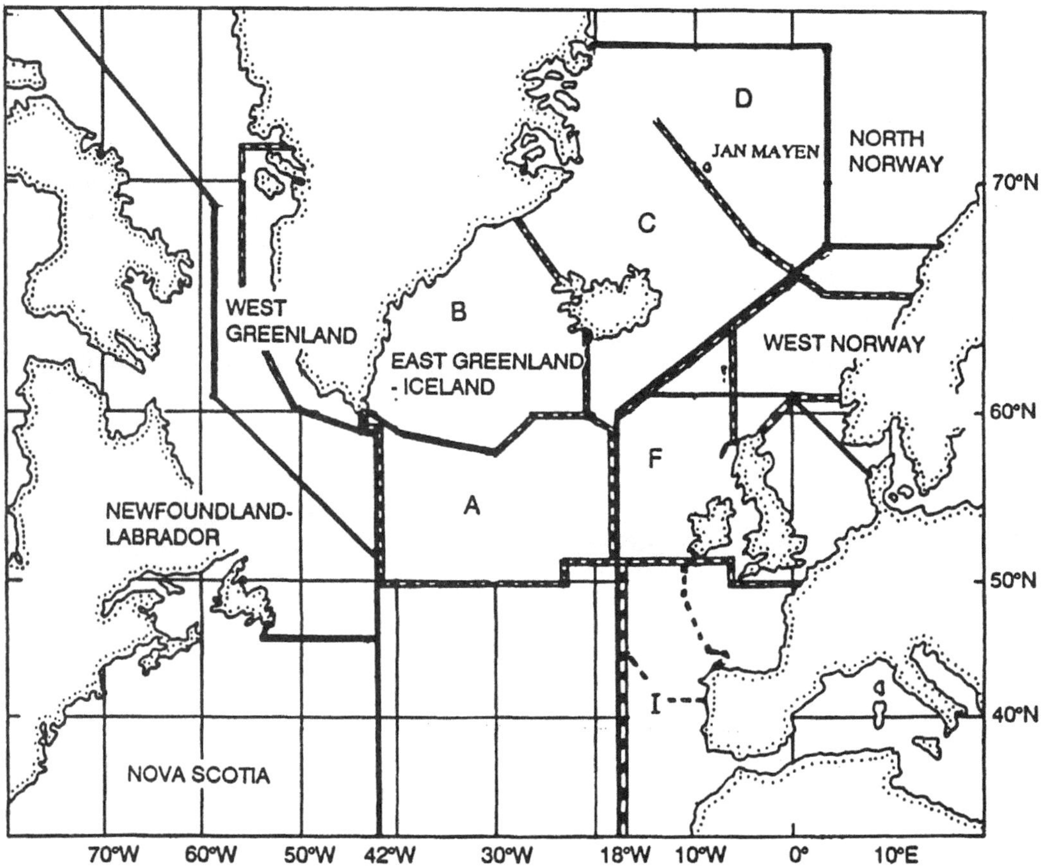

Figure 3. Areas used for H/F runs (as Figure 3 in SMR). Solid lines delineate Schedule management areas, as in Figure 1. Barred lines delineate blocks of the NASS surveys the estimates from which could be used in H/F runs, separately and combined in various ways.

The marking estimates for the Canadian operations are given in Table 1 of the SMR. But, similar estimates were also available for the Icelandic operations within the East Greenland–Iceland management area. These are given in Annex E3 of the SMR "Mark–Recapture Analysis" by J. Cooke. A number of values were obtained by slightly different procedures. Those used in H runs are given in the last table in Annex E3: one, 5,090, referred to the experimental period 1965–72; the other, 2,819, to the period 1979–89. The 95% confidence intervals for these (not given explicitly in the SMR) were, respectively, and approximately 1,400–18,000 and 1,400–6,000. Note that the confidence limits for the slightly longer, more recent period are substantially narrower than those for the earlier period.

All other stock estimates used came from sightings surveys carried out in 1987, 1988, or 1989 including, of course, the estimates for the whole and portions of the East Greenland–Iceland (EGI) management area. The various subdivisions adopted are shown in Figure 3 of the SMR, which is reproduced here also as Figure 3.

With respect to the EGI area, it may be seen that the sightings estimates for the subdivision in which recent operations have been conducted (labeled B in Figure 3) are quite close to, though somewhat higher than the range of the better and more recent marking estimates. One set of sightings estimates comes from the NASS-1989 survey, another from NASS-1987; in Table 1 of the SMR, they are labeled "East Greenland/Iceland B(i) and East Greenland/Iceland B(ii)" with, respectively, ranges of about 3,800–7,300 and 2,500–6,000. These are also given as rows 3 and 4 in Table A of Annex G "Results of Runs Using the HITTER/FITTER Programs."[14] It should be noted, however, that while the sightings estimates are taken to be of the total number of animals 1-year old or older, the marking estimates refer to the number of *recruited* animals.

Indices of Abundance

In past assessments of North Atlantic fin whales, crude early CPUE data were extensively used, as also were more recent data from Spanish operations. At the Special Meeting, only the recent data for the Icelandic operations were considered. But, the SC "... could not reach agreement on whether the Icelandic CPUE data could be used for assessment purposes (at the Special Meeting); some members believed this was appropriate, others did not." Here, I review how and why these differing views emerged, especially as Icelandic scientists had made considerable effort to extract and compile relevant data from ships' logs. It was also inevitable that special attention would be given by the SC to this matter because it was known that the Government of Iceland would be proposing to the Commission an immediate resumption of the catching of fin whales.

As part of its preparations for comprehensive assessments, the SC held, in 1987, a workshop "to review the use of CPUE and other relative abundance indices for identifying and estimating trends in population abundance and productivity." The workshop had concluded that if CPUE data were to be used, certain conditions should be met. These were (1) Member nations should provide detailed descriptions of the methods and strategy of the relevant whaling operations and other specified information particularly to reveal any changes in these with time, and (2) the SC should then develop models of the relationship between CPUE and abundance and carry out appropriate tests of the validity of the models through simulation and in other ways.

The necessary descriptions of the Icelandic operations were published in 1988. Condition (2), however, could not be met nor had the specified problems yet been adequately addressed by Icelandic scientists. According to the SMR, time constraints at the Special Meeting prevented either precise definition of the analyses that might have been made or agreement on what analyses should be conducted in the future (but see Appendix 2).

Icelandic scientists presented to the Special Meeting an update of their previous interpretations of CPUE data for the seasons 1962–1989. This series consisted of two parts: 1962 to 1978—effort measures had been extracted from the log books of the catchers operating in that period; 1979 to 1989—effort was from time–budget forms that had been introduced on 1 July 1979.[15] The basic unit of effort is now taken to be the time spent during a trip by a catcher on the whaling ground before it catches its first whale. A problem arises from the fact that in earlier years, this was a multispecies whaling operation, with sperm whales *Physeter macrocephalus* and sei whales also being caught, and some adjustments have to be made to try to account for biases that this introduces.

[14]This table contains errors in row 4; the column "Catch series used" is "EGI(B) —low". Also, the headings of the "95% conf. limits" columns should be interchanged: for "Upper" read "Lower," and vice versa.

[15]This late initiation of the collection of detailed effort data, five years after the adoption of the NMP, was one of the reasons why the SC has always had difficulty in assessing the state of the fin whale stock accessible to Iceland.

Five catcher boats have been operating for some parts or all of this period. They were of varying sizes and powers, so that either corrections have to be made for differences in their "whaling power" or a separate CPUE series has to be calculated for each of them. Two of the vessels, one of which began working in 1962 and the other in 1966, have continued to catch fin whales since 1986 under special scientific permits.[16] Although when Iceland's proposed research program was first evaluated by the SC, it was argued, in part justification of the program, that it was intended to continue operations in such a way as to continue the CPUE series, it is doubtful whether that purpose has been fulfilled. For example, the CPUE in 1989 was rather high, but Icelandic scientists explain this by observing that "a large proportion of the whales in this year was taken south of Reykjanes,[17] where very small catches were taken in earlier years. Only a limited number of whales can be found in this area, which results in more search hours needed per whale, but the shorter sailing time to and from the station more than compensates for this."

The matter of shifting operations from one whaling ground to another could be important. Two distinct grounds have been described (apart from the one visited in 1989). The *northern ground* is in the deep water between the continental shelves of western Iceland and south-eastern Greenland, a mixing area of the warm Irminger Current and the cold East Greenland Current. The *southwestern ground* is south of 65°N along the continental shelf as defined in this region by the 500-m isobath. Large differences have been reported in the average lengths of each sex of fin whale caught on the two grounds, and these vary also with the period of capture within the season. These observations indicate that the CPUE should be examined for the two grounds separately, but that has not yet been done.

The compiled whaling effort data from logbooks relate to the period 10 June to 20 July each year, and the CPUE series presented for the more recent years are also for that period. But changes in the average length of whales caught occur *within* that period. If, over the time-period of the entire CPUE series, there have been any shifts in the average date within the season that fin whales were caught (and this is quite possible, considering the change from a three-species to a one- or two-species operation), the CPUE will be distorted. Thus, an examination of the seasonal nature of the fin whaling operations, and adjustments if changes are detected, is also an essential prerequisite for the use of these data.

Although the effort data were not available, the length compositions of catches by year and sex could be examined. These showed there had been a declining trend (in median length) for both sexes over the period covered by CPUE data from commercial operations.[18] They also showed that both sexes of the fin whales taken under special permits from 1986 to 1989 were larger than had been taken since about 1970. Thus, this analysis cast doubt on the validity of including the years 1986–1989 in the CPUE series that might be used for assessment using H or F. But, it also confirmed that an analysis must be made of whether this change can be accounted for by shifts in the whaling operations, within-season changes in operational dates, changes in the population itself, perhaps as a consequence of exploitation, or any combination of these or other factors.

This study of sizes of whales in catches also casts doubt on the validity of using an assessment procedure in this case in which the age at recruitment is taken to be constant. It was

[16]Three catchers operated from 1983 to 1985, four from 1966 to 1984, and five from 1962 to 1965.

[17]The most southwesterly point of Iceland.

[18]The results of this analysis are given in Annex E1 of the SMR. The median length is the length at which half the whales in the catch are shorter and half are longer.

pointed out that "although the change in median length over time was small," it could be translated into about a 10% difference in the size of the recruited stock, but that this conclusion depended on the actual length compositions of the catches, which had not yet been fully analyzed.

These are among the problems that resulted in lack of agreement on whether the CPUE data should be used. Nevertheless, some members of the SC wished to try to use them, and their results are given in the SMR. I discuss these later.

Catch Histories

So-called "modern whaling," using an exploding harpoon fired by a cannon from a steam catchboat to kill fast-swimming rorquals, especially fin whales, began from land stations in northern Norway in the 1860s. From there, it spread to Iceland (1883), the Faroes, and Spitzbergen. By 1903, the technique reached the British Isles (Shetland, Outer Hebrides, and western Ireland—1903–1908). Modern whaling started on the west coast of Norway in 1918. Thus, in attempting a comprehensive assessment of North Atlantic fin whales, we are dealing with stocks that have been exploited for up to a century and intensively so in the early years in several areas.

The SMR is laconic on the matter of the catch data used at the Reykjavik meeting—two lines, citing three submitted documents, including a compilation by the Secretariat of what information had been computer coded. None of these documents cites any catch data earlier than 1920, except that the Secretariat paper mentions a catch of 13 whales on the north coast of Norway in 1903 and 178 whales caught in Shetland from 1906 to 1914. But more than 4,000 fin whales have been recorded as being caught from land stations on the British Isles alone between 1903 and 1914. Between 1883 and 1915, nearly 10,000 fin whales were caught off Iceland, as compared with 9,500 caught between 1936 (when whaling was resumed after a long ban) and 1989. These and other early catch data have been used in previous assessments; if they are not included in the catch series used for H/F analyses, then the depletions of stocks can be substantially underestimated.[19]

About 45,000 fin whales were killed in the North Atlantic between 1920 and 1989. Roughly 20,000 were killed before 1920. The total removals over a century have, therefore, been about double the best estimates of the whales now remaining. Since 1920, about one-quarter of the total killed have been taken in the geographically defined western Atlantic—10,500 on the Canadian coast and 700 off Greenland. However, if Icelandic catches are counted as "western," as the stock identity discussion indicated might be appropriate, then the western removals were not very much fewer than the eastern.

The sex ratio in the catches overall since 1920 is probably very close to 50:50; it cannot be stated precisely because in a few areas, notably the British Isles, only a small proportion of the total catch was identified by sex. There were clear and persistent differences in the sex ratios of catches from different locations. In particular, the catches landed in Iceland, Canada, and Greenland all show a preponderance of females, there being from 8 to 12% more females

[19] I suspect the statement in SMR about catch data is incorrect; certainly some of the trajectories shown for H runs start in the early years. It also seems possible that the set of catch data used for H-F runs for the East Greenland–Iceland management stock was not the same as that used for H runs. The former were carried out, not by the SC, but during the meeting by Dr. Butterworth's team in South Africa, who certainly had been using all the early data in their preparatory work.

caught than males. On the other hand, catches in western and northern Norway, Faroes, and off Spain, Portugal, Morocco, and Madeira all showed an excess catch of males over females by from 4 to 12%.

Whether these sex ratios indicate preponderances of males or females in populations, or selectivity of the whaling operations, or both, cannot be ascertained. But, imbalance of this magnitude, if it is due to selectivity, and if it persists through a long period—as it does in the cases cited—can affect assessments of the kinds now being carried out. It is, therefore, surprising that the SMR contains no discussion of this matter, and it does not, in fact, make clear how the H runs were organized in this respect, although it appears that the H-F runs for East Greenland–Iceland were carried out using the data subdivided by sex.

The Assessments

The use of H to produce hypothetical stock trajectories with an extreme range of assumed values of MSYR(%) has some interest as part of the research in this field in order to get a "feeling" for how the procedural model behaves. But its value in a presentation of results to a regulatory body such as the IWC is very limited and confusing when there is—as now—no consensus on even a reasonable hypothetical range of MSYR(%). Many members of the SC, including those most favorable to the H/F procedure as a tool for jointly evaluating datasets, are skeptical of the value of presentations such as those to be found in the SMR. No guidance is given to delegations, beyond the comments of groups of members of the SC (Section 9.3), as to how they might try to interpret the 22 pages of graphs and tables reproduced in Annex G. It is as if the Commission had asked the SC to bake a loaf of bread and had instead been handed several lumps of dough which had not yet been allowed to rise or even been thoroughly kneaded.

Nevertheless, those 22 pages are there, and an interpretation of the SMR would be incomplete without some guidance through them. I have tried to do that in Appendix 3. Here, I make some more general observations.

Notwithstanding the inconclusiveness of the SMR, an important breakthrough was achieved during the Reykjavik meeting. A breach was opened in the wall that has been constructed in the last few years to resist any assessments using geographically defined areas other than those determined in 1976, unless incontrovertible evidence can be produced by those who suggest change that other definitions are, in some mysterious way, "better." That, of course, cannot be done, and there has been a growing awareness that the correct approach, in the almost complete absence of factual knowledge, is to examine a wide range of possible alternatives. In this way, which is analogous to the "sensitivity analysis" used to decide what is important and what not in any mathematical calculation, it should be possible *either* to conclude that the answers do not depend much on the particular stock boundaries used, *or* to become aware that some answers do depend greatly on the stock boundaries chosen, and then to chose between them according to some explicit policy criteria.

One such criterion is the cautious one: to act on that which will be "safest" or best meet a set of declared objectives, as is being attempted by RMP developers. Another criterion involves choice of a stock area which will lead in the direction of a policy of reopening whaling soon, with substantial catch limits. That policy is helped also if CPUE series show little or no declines (or, under the NMP, if they are so variable that they show nothing), if historical catches are lower rather than higher, and if targets are higher rather than lower.

While targets are based on estimates from marking experiments, the stock boundary has only a secondary significance of defining the catch series to be used in a H/F analysis. But, when targets are based on sightings surveys, then the bigger the stock area chosen, the higher the target. However, extending the stock area beyond the range of one set of whaling operations can bring problems because it may, as it is further extended, begin to include other whaling operations, especially past ones. Then, the additional catches added to the catch series may more than counter-balance the consequent raising of the target, particularly if the whales were so reduced in the locality of the former whaling operations that the recent surveys find few whales.

This interplay of several factors can be illustrated with reference to the British Isles–Spain–Portugal stock of fin whales and to the East Greenland–Iceland stock. In the former case, there were recent operations towards the south, off northwest Spain, and there had been very intensive early operations in the north—Ireland, Hebrides, Shetland—which ceased long ago. High catches from these latter, and very low current numbers found in the NASS surveys, would combine to indicate that the entire stock is now greatly depleted. So, such a choice, whether it was right or wrong in terms of the biological reality, was not favorable to Spanish operators who might wish to avoid the consequences of the NMP, properly applied.

Now look at the opposite case concerning Iceland. Here, the bigger the area, the better, so long as it does not intrude to places where there has been intensive whaling (high past catches) and where there are now few whales. Thus, an extension, in this case, to include, say, Faroes or even the northern British Isles, would be disadvantageous. But, extension northward and southward to, for example, the blocks A and D, is highly advantageous. There has been no whaling in A, but fin whales are apparently there in substantial total numbers—it is a huge area (Figure 4). In D, which is quite small, whales are relatively dense near Jan Mayen (which is why Norwegian pelagic whalers went there), but the catches overall were not high. Division C is a little awkward because there were high catches in it in early years from both northwest and southeast Iceland, and some pelagic catches. But the NASS surveys showed there to be more whales there now than in any other block, so including C is acceptable. In any case, it would hardly look right to define a management area that was discontinuous.

Including West Greenland in the "mid-Atlantic" area would bring no advantage to anyone seeking to justify resumed fin whaling in Iceland; the area is small and the stock estimate is only a little less than in A, but there were also historical catches in it. Such inclusion would, in any case, bring a political disadvantage; it could open up the possible need in future to share a commercial catch limit with Greenlandic whalers, and the difficult negotiations that would involve.

Thus, the present boundary of the East Greenland–Iceland management area is the best possible one for Iceland's whaling ambitions (provided Greenlanders do not try to develop commercial whaling off their east coast), especially if either the NMP is followed or if *ad hoc* catch limits are set according to some arbitrary percentage of a stock estimate—even an apparently low percentage. It could hardly be better if it had been designed that way, and it is, therefore, not surprising that it is now vigorously defended.

I now return to the particular assessments carried out at the Reykjavik meeting. The assigned values of parameters used for H/F runs are listed in Table 2 of the SMR. The values of biologically determined parameters—natural mortality rate (4%) and age at first parturition (9.5 years median)—were simply those used in some previous assessments; there was no reanalysis.

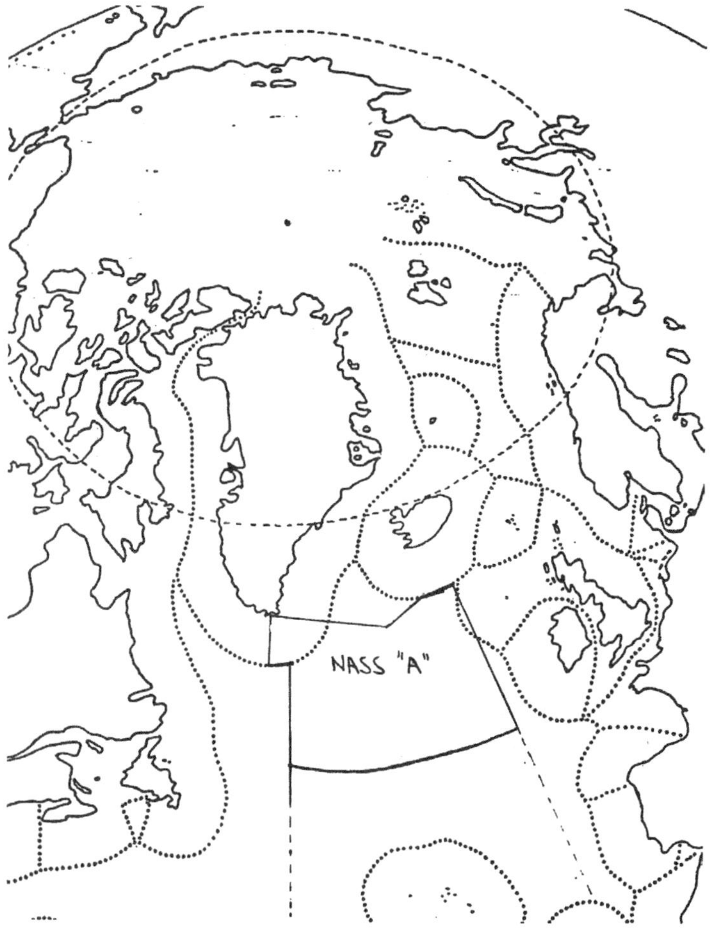

Figure 4. North Atlantic: polar projection. Dotted lines indicate approximate limits of 200-mile zones. Continuous lines enclose NASS block "A." This block, plus Jan Mayen (block "D"), account for nearly 40% of the fin whales estimated to be in the EGI Schedule area.

The parameter which is determined by a combination of biological and operational factors—age at recruitment in each sex—was given the same values everywhere. They were those used previously for assessments of fin whales in the East Greenland–Iceland management area: median 4 years for females and 5 years for males. This may be appropriate for the recent Icelandic operations in that area, but it may not be either for earlier operations there or, more particularly, for both land-based and pelagic operations elsewhere.

The age at recruitment for fin whales is affected by the minimum size limits in force; in theory, these differed for different operations, depending on whether or not land-station catches were for local consumption. But it also depends on the availability of whales of different sizes and ages at different locations. Some of these datasets are more reliable than others, and many may be biased by the frequent falsification of records of animals close to the minimum length, but they could be indicative of whether it really is appropriate to use a single pair of age-at-recruitment values in all assessments.

The results of applying the H procedure to a wide variety of subdivisions of the North Atlantic, using many different targets and assuming a wide range of values of MSYR(%), are given in Tables 1–16, and associated graphs, in Annex G of the SMR. The tables give results pertaining to numbers of *mature* animals in the "stocks", while the graphs plot the estimated *total* numbers of animals 1 year old or older. Care is, therefore, needed in comparing tables with figures and in attempting to evaluate the results of both.[20]

H runs were made for all "Schedule stocks" (i.e., management areas defined in the Schedule) except British Isles–Spain–Portugal. No explanation of this omission is given. In previous years, this management area has received more attention from the SC than any other except East Greenland–Iceland. Now, no doubt because the Government of Spain, showing no signs of wishing to resume commercial whaling in the near future, is not pressing for priority to be given to it, the area has been ignored in what is supposed to have been a comprehensive assessment of the North Atlantic as a whole. Nevertheless, it is worth giving it some attention now because it illustrates some important inconsistencies in the SMR, in particular the contrast between the tenacity with which some scientists insist that one of the existing management areas must remain unchanged, and the easy abandonment of another, which happens not to be of current commercial interest. In fact, because of the lack of pressure from commercial interests, the SC virtually ignored the entire eastern rim of the North Atlantic. This matter is examined in more detail in Appendix 3.

Assessments were not carried out for the Atlantic as a whole (as explained in Appendix 3), but H runs for a substantial part of the eastern half of the ocean were made—NNE in Table 11 of Annex G. These show that in order to avoid concluding that, if the NMP were applied, the entire region should be protected, it is necessary to make incautious assumptions: that the upper confidence limit should be taken as the best target and/or that MSYR(%) is higher than about 5%. If a cautious assumption about MSYR(%) is made, i.e., that assumed values implying very strong super-compensation of recruitment (2% or more) be excluded from consideration, then the fin whales in the region would be protected, even if the upper confidence limit for the stock target were adopted.

Now, in order to get some idea of the state of the whale population available to operations from southwest Iceland, we need to look first at a series of H runs, in the following order of the tables in Annex G; 12, 15 and 16, 3 and 4, 13 and 14. The necessity for pairs of tables, in all except the EGI Schedule stock case, is explained only in a footnote on page 13 of the SMR: that different assumptions had to be made to allocate to survey blocks early catches the positions of which were unknown.[21]

Taking as our criterion an RMP protection level the same as that adopted for the NMP, it is clear from Table 12 that if the EGI Schedule area is chosen as the assessment unit, the "stock" would be protected only if the best sightings estimate of the number in that area is used and the MSYR(%) is no higher than 1%; or if the lower confidence limit target is used, MSYR(%) must be no higher than about 2%.

---

[20]The graphs for the H-F runs also show the trajectories of total stocks, but so do the corresponding tables, although this is not so indicated. Tables 17–21 are, therefore, not comparable with Tables 3 and 4, 12, 15, and 16.

[21]I think this problem arose primarily because catches taken in early years off northwest Iceland, where there were a number of land stations close to the northern segment of the line separating blocks B and C in Figure 3, could not be located precisely. I also think "high" catches means "relatively high in block B," even when it is applied to assessments for blocks C + D.

The second most favorable configuration for the Icelandic whalers is EGI(A+B)+WG, given in Tables 15 and 16. Here, protection would be accorded if the best abundance estimate were taken and the MSYR(%) was about 2% or more.[22] However, if the lower confidence limit is taken as target, the "stock" would be protected so long as the MSYR(%) did not exceed 4%. Furthermore, if a cautious MSYR(%) value of 1.2% or less were adopted, the stock would be protected even for the upper confidence limit of number.

No runs were made for the combination EGI(A+B), but for the reasons explained, I do not think they would have differed much from the EGI(A+B)+WG runs.

I suspect these runs provide the rationalization of the proposal by the Commissioner for Iceland that interim catch limits be established at either 1% of the Schedule stock number or 2% of the "estimated part of the whale population accessible to the fishery." Both, by definition, imply the "stock" is close to MSYL, so that the appropriate percentage of that number would be the MSY. Table 12 shows that in the case, the "best" MSY value would be 100, going down to about 65 if the lower confidence limit of number is used. Tables 15 and 16 give slightly higher corresponding numbers: about 120 and 80. So, in offering to accept the lower of the two options he has specified, the Commissioner for Iceland will be implying that the existing management area (which also, of course, includes the large unexploited block A) gives the more conservative interim catch limits.

Furthermore, by arguing for *interim* catch limits, higher numbers can be conjured than if, for example, the NMP were being applied. This is because the NMP limits would progressively be reduced by 10% for every 1% that the stock was below 60% of the initial number, and an automatic safety limit cut of 10% would also be applied.

But, now look at Tables 3 and 4, which report the EGI(B) runs. Block B would apparently be a block which includes all the area accessible to catchers based in southwest Iceland. These runs show that the "stock" would be protected if the MSYR(%) was less than about 7%, using the best estimate of numbers. Furthermore, if the lower confidence limit were adopted, then the stock thus defined would be protected even for the highest hypothesized values of MSYR(%). Even if the upper confidence limit were used for the target number, the stock would be protected for all MSYR(%) values less than 4%. The "2% of accessible stock" proposed by the Government of Iceland as an interim catch limit would give limits of 56 or 41 whales. But the rationale for such numbers equally implies that such catches were being taken from a "stock" which is far below the NMP protection level, being between 20 and 45% of its initial number.

The marking estimate, for all its methodological weaknesses, does measure the number of whales effectively accessible to the Icelandic fishery based in southwest Iceland (Tables 13 and 14 of Annex G). But, again, the MSYR(%) would need to be higher than 6% to avoid protection status under the NMP rules or their equivalent under an RMP.

One way to escape the dilemma of there being no prior consensus on a value, or even on a reasonable range of values, of MSYR(%) is to attempt to estimate it by using CPUE in H-F runs. This is what D. S. Butterworth and team sought to do; their results are presented in Tables 17–21 of Annex G.

I discuss the results, and how they were obtained, in Appendix 4. A reasonable summary would be that they did not resolve any problems. Estimated MSYR(%) was high (between 3 and 10%) and unreliable. It was dependent entirely on the assumption that the recent CPUE

[22]The "high" and "low" catch runs do not differ very much, and I will ignore the differences.

series correctly represented an unchanging stock, whereas those series cannot, statistically speaking, distinguish between an increasing, a declining, or a stable stock and, furthermore, have yet to be analyzed in the appropriate ways that the SC has specified.

Some possible interpretations of both the H and H-F runs turn on the question of whether or not it is reasonable to include block A in any assessments pertaining to possible future land-station operations based in southwest Iceland. As Figure 2 makes clear, migrating fin whales passing through block A appear to be, *a priori*, as likely to travel to the south and east of Iceland as to the west through the Denmark Strait. If they did that, we would need to know how closely they approach the south and east coasts of Iceland to know whether such whales would be available to an Icelandic operation newly based in the southeast. This illustrates well how the virtually complete lack of basic biological knowledge about the North Atlantic fin whale, despite a century of exploitation, impedes serious assessment of this stock or stocks according to traditional norms.

Conclusions

I think the main conclusion I draw from participating in the Reykjavik meeting and subsequently reviewing the report of it is that the conduct of such meetings is not an appropriate way to carry out a serious "comprehensive assessment" of anything. The SC announced a few years ago that comprehensive assessments would have "greater depth and breadth" than previous efforts. The assessment of North Atlantic whales has, so far, been shallow and narrow.

A second conclusion is that just as a few years ago, the need for the SC to argue about scientific permit proposals impeded progress in preparing a comprehensive assessment, the pressure now to force either belated applications of the NMP or decisions on interim catch limits is seriously impeding its conduct. The Reykjavik meeting simply confirmed that the only feasible way to begin to resolve the problems facing the IWC and its SC is to focus all efforts on completing the development, testing, and application of a revised procedure. No more "management advice" should be asked for or offered until that is done. Meanwhile, the powerful analytical techniques that have been devised by a few members of the SC can be polished so that they will be ready for genuine assessments designed to "kick off" an RMP. At the same time, also, data such as CPUE series could either be analyzed in depth (as was originally decided when planning of the comprehensive assessments was begun) or abandoned as incurably flawed.

Some of the "management advice" offered from the Special Meeting by named individuals is astounding. If taken seriously, it would put the Commission back to its pre-1960s days. For example, Dr. Sigurjónsson, supported by Dr. Goto, brought the number 100 out of his hat as the sustainable yield of the EGI stock, with no rationale and certainly no scientific evidence.

Worse came from Dr. Schweder, whose argument was that since the Icelandic CPUE shows no obvious decline, the sustainable yield of that stock must be no less than the average catches taken from it in the past. Yet, that "logic" was shown to be quite false more than a decade ago, by Dr. D. G. Chapman and others, even if applied to CPUE series thought to be relatively reliable. It was because that sort of argument, as made in implementing the special SMS provision of the NMP, was fallible and incompatible with orderly and sustainable resource management, that the IWC decided to concentrate on a management procedure which, among other things, would exclude it.

A final observation that might be made is that although there were conversations about the matter in the Working Group that reported on Ecosystem and Habitat Considerations (SMR Annex H), the data from the research sampling of fin whales played no discernible part in the assessments. Nor did any ideas emerge from those conversations regarding concrete ways in which such data might be used in any future assessments.

Appendix 1

An Analogy

Here, I shall try to illustrate the basic principles of the calculations performed by the SC, on the assumption that more readers of this article will be familiar with the problems of trying to balance a bank account than with the methods of demography.

Imagine, then, that you have a savings account for which your bank manager can tell you the current balance (current stock size). You want to trace your dealings with the bank and have a record of all your withdrawals (catches) since you opened it. You would like to know what deposit you made to open the account (initial stock) so you can estimate your future regular income (SY) should you retire, but have lost that piece of information and so has the bank (it was accidentally wiped off the computer!). But, the bank can tell you the rate at which interest was paid each year (recruitment) and also the fixed scale of bank charges (natural mortality). You are a cautious gambler and think that historical knowledge will help you predict the future net yield from the account. Also, if the original deposit was an inheritance, it would be good to know how well or badly you have husbanded it.

Given that information, calculating the original deposit involves straightforward arithmetic. But, in a reasonably stable economy, under which you fortunately live, the interest rate does not change enormously from year to year, though it does depend on how much capital you have. Usually, the interest rate is higher the more capital one has, but your bank is perverse, or socially benign, depending on one's point of view: it pays interest at a higher rate on smaller savings, and none at all until you start withdrawing from the initial deposit, whatever that might have been (the recruitment rate increases as the population get smaller). If you know the average rate of interest or even the rate in a particular year—say, last year—and the way the bank determines its sliding interest scale, you can still, rather more laboriously, calculate what was the original deposit. If you had a copy of the HITTER program, you would be able to make calculation less laboriously.

Bank managers and accountants expect to be able to tell you to the nearest cent what money you now have. Here, our analogy begins to break down because we can only estimate stock numbers very roughly; most recent estimates are thought to be more or less correct only to within about 40 or 50% either way. By "more or less correct," we mean, usually, that we think there is a 19 to 1 chance that the true number is within a calculated range, which we call the *95% confidence limits*. The HITTER program makes all the requisite calculations so quickly that you could make a large number of trials to see how wrong your answers might be if you do not have exact figures (sensitivity analysis).

Your bank manager is one of those people who like number puzzles. To test your skill, he declines to tell you the bank charges and the interest rates directly, but he does let you know, for example, that your credit declined at about 2% a year from, say, 1984 to 1989. Armed with a copy of HITTER-FITTER, you would be able to solve his puzzle and calculate the original deposit and the interest rate or, to be more precise, the net difference between the interest rate and the bank charges.

The program would find a *trajectory* for your account which declined at about the same rate as had the account during those years and also hit the target: your current credit. It could, if you wished, project the trajectory forward a few years. Because of the way the interest rate declines as the credit increases, you will not be able to increase your income indefinitely simply by building up your capital. The H/F program could calculate for you the size of account you should maintain (MSYL) in order to maximize your future annual income (MSY).

Appendix 2

Availability of and Access to Data

Computer tapes containing Icelandic catch and corresponding effort data were sent to the IWC Secretariat in December 1990; the data were for use by accredited persons before, and in preparation for, the Special Meeting. Members of the SC were not informed that the data had been made available. When the possibility of carrying out analyses using these data was raised in the Special Meeting, it emerged that this database was not physically available to the meeting in a form that could be used.

The SMR is thus not entirely correct in asserting that lack of time was the reason for not conducting such analyses. In fact, on other occasions—for example, with respect to North Atlantic minke whales assessed at the 1990 Annual Meeting—it has been practicable very quickly to carry out analyses of CPUE data in accordance with the prescriptions of the CPUE workshop, provided data were available in the proper form. It is, however, indisputable that a working time of only 4 days, as in the case of the Special Meeting, is hardly sufficient to carry out serious analyses, and certainly not enough to justify calling any results obtained "comprehensive."

Attempts have been made by some SC members, since the Special Meeting, to rectify this failure to examine the Icelandic CPUE data by securing access to them in order to carry out analyses the results of which might be presented to this year's Annual Meeting. These attempts revealed two features of the data submission that had not previously been made public:

1) The Government of Iceland had placed an embargo on their use after the Special Meeting. When this embargo was eventually lifted, little time remained for analysis before the Annual Meeting. Furthermore, a condition was imposed that scientists using the data "clear" their results with Icelandic scientists before presenting them to the future meeting. Apart from reducing available time even further, a question evidently arises about the propriety of imposing such a condition on data analyses in the context of "comprehensive assessments" being carried out for the IWC in which members of the SC are expected to participate.

2) One part of a CPUE analysis is the examination of data for homogeneity, of which one measure is change in the size of the animals caught and another is change in the locations of whaling grounds. Studies during the Special Meeting revealed that there

> had been a prolonged downward trend in the sizes of whales in commercial catches. Studies had previously been published showing that the sizes of whales caught on the several whaling grounds accessible from the present land station varied according to the location within the accessible area. It is, therefore, essential to determine whether the overall decrease in size might have been caused by shifts in the whaling areas.

Such an analysis calls only for data which must be freely and promptly available in accordance with paragraph 27 of the Schedule, especially paragraph 27(c). This specifies that "particulars *shall* be given with respect to each whale treated in the ... land station ... as to the date and *approximate* latitude and longitude of taking, the species and sex of the whale, its length and, if it contains a fetus, the length and sex, if available, of the fetus."

However, in submitting this required "Schedule" information, the Government of Iceland apparently *omits* information, for each whale treated, about the latitude and longitude of taking. Knowing that fin whales taken in Iceland in recent years have all been treated at one land station, the Secretariat has adopted the practice of entering the exact latitude and longitude at which that station is located as the "approximate" location at which every whale was caught.[23]

The approximate locations of catches are, in fact, known; they are included in the "restricted" catch and effort dataset, wherein the position of each capture is indicated for every whale caught by small statistical "squares" with sides 1 degree of longitude by 30 minutes of latitude. Parts of that dataset have been made available for analysis by *selected* scientists from other countries for several years past.

It appears that examination of data which should be freely available under Schedule paragraph 27(c) is impeded by restrictions placed on a crucial element of those data.

Appendix 3

Interpretation of HITTER Runs, Except Those Directly Affecting Iceland

Among the "non-Schedule" H runs was one for a small area labeled "Iberia." This corresponded with one of the blocks of the NASS surveys. It is not clear what catch series was used for this. Figure 3 refers to the area "between enclosed dotted lines," which appear to embrace the operational area of recent whaling from northwestern Spain. But Table A in Annex G of the SMR says the catch series used was "Spain and Portugal." This reference to Portugal could only be to earlier catch data from the southern part of the area towards Gibraltar and presumably including the early pelagic whaling in this region. Judging from the trajectories shown in SMR Table 10 showing steep declines starting in about 1925, these early more southerly catches were probably included.

Table 10 and the three accompanying graphs are typical of many of the H runs. They show steep declines just after an early period of intense whaling, declines which were noticed at the time by the whalers who found it not worthwhile to continue. Then comes a long period of slow recovery when there was little or no whaling, followed by another decline when new operations started.

---

[23] I do not know if that practice was also followed by the BIWS when it was responsible to IWC for statistical compilations nor to what extent it may now be followed in the cases of land-station catches elsewhere.

In previous assessments of whales in this region, crude early CPUE series have been used, such as the catch per boat. It is not appropriate to use such series as the forcing data to produce trajectories using H-F, but they can be used to help appraise H runs. In fact, Icelandic scientists have used that approach in their studies of the operations near Iceland. In the Iberian case, the relatively low MSYR(%) assumptions (0–4%) imply that the "stock" would, in the early years, have declined catastrophically before the whaling operations ceased. The higher MSYR(%) values (6–10%) imply that the decline would have only been to between two-thirds and one-half of the initial number. Such a limited decline is simply at variance with the history of whaling in the area. For what they are worth, the Iberia H runs give support to the lower range of values of MSYR(%).

When the SC attempted, in 1976, to determine identities and working boundaries for the North Atlantic fin whale stocks, one of the few decisions based on factual evidence was that in which the whales off Spain were considered to be in the same population as those caught off Ireland, the Hebrides, and probably Shetland. The circumstantial evidence cited was mainly the sequential timing of observed migrations—and hence of whaling seasons—both northward and southward. Not conclusive evidence, to be sure, but indicative. Additional evidence was synchrony of the early depletions in different parts of the region, but this is not very convincing since all were subject to intensive operations.

One suggestive observation in this process of stock identification was anecdotal. Modern whaling started in Faroes a decade before it did in Scotland. When it did start in Scotland, whalers noticed that the fin whales there were not "shy" as they had become at Faroes, thus indicating little or slow exchanges between those two locations. Thus, in 1976, Faroes was lumped with West Norway, not with Scotland and Ireland.

Table 9 of the SMR gives the results of H runs labeled "Faroes"; this is really the division "F" of Figure 3, and the catch series used were for Faroes and Hebrides and Ireland, *but not Shetland*. This was done *solely* because the NASS block happened to have been designed that way. Shetland, however, was lumped with West Norway for the same reason, but no H runs were carried out for this Shetland–West Norway division.

The Faroes–Ireland–Hebrides H runs show much the same pattern as the Iberia runs in early years, but not, of course, the later decline following some recovery because there was no large-scale resumption of whaling in those areas as there had been in northern Spain. Runs for the Schedule management area West Norway (i.e., Faroes plus western Norway; Table 8 of the SMR) also show the same pattern, but with a continuing decline until the 1940s commensurate with continuing whaling in that area.

In all these cases, the general history of the operations is more closely represented by trajectories for the lower than those for the higher values of MSY(%).[24] Unfortunately, because no runs were made for the Schedule management stock British Isles–Spain–Portugal, we are not in a position to confirm this for that stock as a whole.

No runs were made for the North Atlantic as a whole, the nearest to that being NNE (Table 11 of Annex G). This excluded the Canadian coast area because it was not considered desirable to add a stock estimate from marking to those from NASS. It also excluded "Iberia." The reason for this exclusion is not explained in the SMR, but the argument ran more or less as follows: "It has been agreed that samples from southwest Iceland and from northwest

[24]The wavy lines in several figures (7 and 10, for example), representing the MSYR(%) = 10% trajectory, and which do not pass near the target, can be ignored. Strong oscillations can arise from use of an unreasonably high MSYR(%), but in these cases, the computer program is possibly not handling these high rates correctly.

Spain show some genetic differences. We want to include the huge East Greenland–Iceland area, from one corner of which the sample came, in any 'greater North Atlantic' assessment, but if we leave Iberia in also, we shall be treating as one 'stock' what we know to contain at least two genetically separated groups. Therefore, we will exclude the smallest definable area including northwest Spain which, in terms of sightings surveys, is the Iberia division."

This is half of an argument around, in the context of the entire North Atlantic, one fragmentary piece of evidence. It would have been equally valid, for example, to have defined an NNE as the entire ocean *except* the southern Denmark Strait (division B in Figure 3), West Greenland, and the Canadian coast. Choice of both combinations would at least have the advantage of symmetry. The unfortunate fact is that combinations cannot be rationally selected without much more geographically extensive knowledge of the genetic differences and similarities in and between many localities within the region as a whole. Until such time as that exists, we shall have to make do with a number of possible choices all equally compatible with limited data. One of these would need to be the North Atlantic as a whole, to examine the hypothesis that there is only one biologically distinct population there, but with slow mixing throughout.

The NNE runs show the expected ambiguity.[25] If MSYR(%) is no more than about 5% and the "best" stock estimates are used, then the entire Northeast Atlantic would be protected if the protection level of the NMP were to be applied. If caution rules (using lower confidence limits for the stock target), then the entire area would be protected even if the MSYR(%) was as high as about 11%. Then, if caution still rules, but is expressed by holding MSY(%) to an upper value that excludes super-compensation (1.2%), or that confines super-compensation to the first stages of stock reduction (about 2%), then even if the upper confidence limit for stock target is used, the entire Northeast Atlantic would be protected.

Now, let us look at the Canadian coast stock, where a marking estimate was used as the target. Here, the target is an estimate of the number of whales that were available to the Canadian operations, regardless of the boundaries of a geographical area they were supposed to be inhabiting. The graphs in Table 1 of Annex G show that for lower MSYR(%), the stock around the time of the marking estimate (where all the lines cross, 1969) would have begun to decline steeply. For higher MSYR(%), it hardly declined at all, or continued oscillation, with increasing amplitude. Unfortunately, the CPUE records from the Canadian operations at that time are very poor and do not directly help us discriminate between high and low MSYR(%). But, it has been said that the Canadian east coast operations were shut down in the 1970s because of the declining availability of whales.

Appendix 4

EGI H-F Runs

SMR Annex G: "B. FITTER runs," says that the method used was practically the same as that used in Document SC/F91/F7. This was a paper provided to the Special Meeting by Butterworth and his colleague A. E. Punt. In that document, only the Schedule-defined EGI management area was examined, whereas during the Special Meeting, EGI(B) and

[25]In the case shown in Table 11 of Annex G, the program is fitting the 10% MSYR(%) option correctly to pass through the target. For the "best" and for the higher confidence level of target, this high MSYR(%) implies population oscillations. There is no oscillatory pattern if the lower target is selected because that implies that the population was more depleted in the early years.

EGI(A+B)+WG were also explored. There are as yet unresolved problems with the way that the Butterworth–Punt program calculates confidence limits of estimates, and it is advisable for the time being to ignore the standard errors given in the tables in italics.

Some explanation is needed concerning the CPUE series used; it is given very cryptically in the SMR and could not be understood at all without reference to the Butterworth–Punt document.

In Annex G, five CPUE series have been used. Four are recent ones: they were not made available to the Reykjavik meeting, but are contained in the Butterworth–Punt document. They use as effort the time on the whaling ground until the first whale is caught, and one series is calculated for each of the four catcher boats operating after 1962. Calculating these separately avoids the need to calibrate the effort in terms of relative efficiencies of the four boats. The series were those used by the SC at its 1988 meeting, and so ran from 1962 (one of them, in fact, starts only in 1966) only to 1987 (not to 1989 as did the single series presented to the Special Meeting by Icelandic scientists). These same truncated series were used by Butterworth and Punt during the Special Meeting. The Icelandic scientists had warned that they had doubts about the 1989 value. But, if these series are to be used at all, they would better be terminated in 1985, since the analyses of the length compositions of catches indicated that the values obtained from research whaling are probably anomalous.

The fifth series referred to was also used at the 1988 meeting of the SC. It is catch per boat-day (in multispecies operations) for the period 1901–1915, and is thus a very crude indicator of stock changes. This series suggests a fall in stock during 12 years to about one-quarter of the original number.

In one set of runs, all five series were used, and in another set, only the recent four were used. A range of assumptions was also made about the form of the relationship between the CPUE index and the abundance of whales (the so-called "non-linearity coefficient").

In all attempts to use these series together, including those made this year, a difficulty has arisen because the trajectory determined by the recent series is not compatible with that determined by the earlier series. This is not surprising, considering the nature of all the series; the Icelandic scientists discussed this problem rather thoroughly in their paper to the 1988 meeting. When all series are used together, the computer program gives them all equal "weight," which is also not very reasonable.

In their paper, Butterworth and Punt tried to resolve the conflict by assuming that the natural carrying capacity of the mid-North Atlantic for fin whales increased in 1910 by 50% with respect to 1890. That means that if whaling had stopped then, it would be expected that the population would eventually increase to 50% more than its size before whaling began. Of course, there is no evidence that this happened, but equally, there is no evidence that it did *not* happen. To examine such assumptions in an attempt to reconcile datasets is not an unreasonable procedure, but it might be questioned whether it is really worth doing when the datasets are of such poor quality.

In any case, the reconciliation attempt failed. Butterworth and Punt concluded that "...it is impossible to find a trajectory which shows all the qualitative features of a marked decline from 1901 to 1915, little trend between 1962 and 1987, and a current total abundance greater than some 10,000 whales."

What is surprising is that they neither looked at the possibility that the EGI stock area definition was inappropriate and that the current abundance in an appropriate area might be much less than 10,000 ([EGI(B) is half of that), nor questioned whether the "features" they were trying to reproduce were real. There was also a fourth feature which they did not mention in their summary conclusion here quoted, but which they accepted and which has recently been repeated in several non-scientific publications by the Icelandic Government: "that after the Althing banned land-station whaling in 1915, the stock 'showed a reasonable degree of recovery' by 1929 when whaling was resumed."

Let us look at two of these features, starting with the fourth. This supposition is based entirely on the observation by Icelandic scientists that after 1929 "annual catches approaching 220 fin whales were frequently achieved." These catches were taken both by renewed land stations and by pelagic operations, the efficiency and economics of both of which were quite different from those pertaining to the pre-1915 land stations. The fact that with a new type of operation (pelagic and/or better boats) it is possible to catch whales offshore some years after it had become uneconomic to catch them from shore is not, in itself, *evidence* of any recovery, even if it suggests a promising line of investigation. Yet, this myth, that it has actually been demonstrated, even by circumstantial evidence, that the stock had recovered by 1930, has become embodied in several documents distributed by Icelandic authorities and in several brochures and popular articles authored by government scientists. Butterworth and Punt have probably been trying to fit a calculated trajectory to a mirage!

The other dubious "feature" that they and Icelandic scientists have been trying to reproduce is "little trend between 1962 and 1987." But, the recent CPUE series have such high year-to-year variability that, statistically speaking, they are not incompatible either with an increasing trend or with a declining one over the period. Since these CPUE series have not yet been analyzed in the ways that the SC, for good reasons, has specified, in this case, too, an attempt is probably being made to account for a feature (no trend) which has no substance.

In the H-F runs conducted during the Special Meeting, a number of stock area definitions, and hence targets, were evaluated. These runs gave very variable results. No attempt is made in the SMR to evaluate these, and I will not attempt that here. The exploration of the issues was far from complete (for example, the upper and lower confidence limits of targets were not examined) and at least some members of the SC doubt whether fragmentary assessments done in this way are appropriate undertakings for four-day international meetings.

One feature common to all runs is a high MSYR(%) estimate ranging from 3 to 10% and mostly around 7 or 8%. The upper limit is an artifact of the computer program, which does not permit itself to predict a higher rate; if it could have done it, it would have done it. But these estimates arise entirely because the trajectory is essentially being constrained to reproduce the "no recent trend" scenario. This can be modeled only if the stock is calculated to be very slightly depleted, if at all, and for that reason providing a sustainable yield less than the theoretical maximum. But this requires an MSYR(%) so high that for higher stock targets, violent oscillations are produced, as shown in Figures 17 and 18 of Annex G.

Chapter 29

Minority Statement Advice on Selection of a Revised Management Procedure for Commercial Baleen Whaling[1]

Introduction

The Commission is expecting to receive this year from the Scientific Committee *either* a single complete proposal for a Revised Management Procedure (RMP), *or* more than one to choose from, given they have clear descriptions of the significant differences between them.

Overall, the five proposed RMPs[2] now perform rather similarly on single-stock trials. In reality, all cases in which it is envisaged that an RMP would be applied in the near future are in areas in which it is not clear how many biological populations of a species exist and, if there are indications for more than one population, it remains unknown where to draw natural boundaries between them. Unfortunately, not all the trials for multiple-stock situations have been completed, in particular those for coastal and mixed coastal–pelagic operations are still to be done.

It is understandable that scientists hesitate to present proposals for an RMP to the Commission for adoption when they have not been absolutely fully tested. But, while the process of testing is very nearly done, the Committee has been presented with and requested to advise on new, so-called "interim" or "safe" procedures, which have not and cannot be tested in much less time than was required to test the five draft RMPs.

In our judgment, there is now sufficient understanding about the behavior of the five draft RMPs to justify presenting them to the Commission for adoption of one procedure at its 43rd Annual Meeting. After adoption, the Commission could immediately pursue its task of preparing an appropriate amendment to the Schedule.

Thus, we believe that it is essential that the Commission this year be enabled to select a complete procedure. We also believe that the risks in doing so for whale populations will be minimal, provided that the procedure contains sufficient safeguards and that the testing process in the Scientific Committee will continue in parallel with the drafting process in the Commission.

---

[1]Editors' note: written by J. G. Cooke, W. K. de la Mare, S. J. Holt, K. Lankester, and T. Lyrholm as Annex Q Minority statement Q4 in the 1991 Report of the Scientific Committee and published in Reports of the International Whaling Commission 42:267–269, 1992.

[2]Editors' note: the five proposed RMPs are: Cooke procedure (C), de la Mare procedure (dlM), Magnússon–Stefánsson procedure (M–S), Punt–Butterworth procedure (P–B), and Sakuramoto–Tanaka procedure (S–T).

The Objectives of an RMP

The three management objectives which the Commission accepted in 1987 (Reports of the International Whaling Commission, 38:36) may be summarized as (i) relatively small year-to-year changes in catch limits, (ii) acceptable risk that a stock not be depleted below some chosen fraction of its carrying capacity, and (iii) making possible the highest continuing yield.

At the 1989 Annual Meeting, all delegations that spoke considered that objective (ii) should have the highest priority. In addition, various delegations in the Commission expressed as their primary objective (a) negligible risk of extinction, (b) assured recovery of depleted stocks and avoidance of further depletion of those already depleted, and (c) stability of catches as most important. No delegation gave highest continuing yield as their first objective.

Seven countries presented their choice of objective (ii) and relevant explicit features to be incorporated into a future RMP (IWC/41/42):

1) The primary objective of an RMP ... is to ensure that currently abundant populations are maintained at or above high target levels, and that recovery is ensured of depleted stocks to these [high target] levels as soon as natural processes permit;
2) The RMP, insofar as it is intended to address other objectives, should not jeopardize achievement of the primary objective;
3) The RMP should be so designed that it guarantees the achievement of the primary objective with respect to each population of whales ... and provide sufficient allowances for uncertainty ... regarding stock boundaries that lack of data does not prejudice achievement of the primary objective;
4) The RMP must retain the concept of a protection level which is significantly high.

This last phrase was clarified in the presentation as intending "certainly not lower than that used in implementing the NMP" and "actually achieved, not merely aimed at."

Selection of a Proposed RMP

The five procedures which have been developed all have attempted to accommodate the set of defined objectives. Since these cannot be optimally achieved simultaneously, there must be a certain degree of trade-off between them. [To give a simplified example, a procedure which allows for more catches will generally lead to a higher risk for the population.] Selection of an RMP is not solely a scientific process; it is also a policy matter for the Commission because it involves taking due account of the preferences for primary objectives which have been expressed by Commissioners in response to the Scientific Committee's questions. Recognizing that the Commission may select a proposal for adoption this year, we consider it appropriate to present to the Commission our preference for one of the procedures, while not restricting the Commission's options.

Arguments for the Target Adjusting Procedure (D-Procedure)

We believe that the Target Adjusting Procedure (the TAP or D-Procedure)[3] is the only currently viable candidate for selection of an RMP. A complete and detailed description

[3]Editors' note: the Target Adjusting Procedure (the TAP or D-Procedure) is part of the dlM.

of the TAP is given in Document SC/43/O 18-Annex, presented to this meeting. While all procedures appear to meet objective (ii), they do so with different margins of safety. Although the other procedures have also been tested for robustness in all single-stock trials, the TAP is the only one complete in all major respects. It is now fully specified and ready for application. In addition, it has a number of advantages beyond the aspects reflected in the performance indices in the screening tests. It is specific where essential and flexible where possible. It distinguishes between the responsibilities of the Scientific Committee where scientific judgment can be routinely employed versus those where high standards of evidence are required. Furthermore, its catch limit rules have already been designed for coastal and mixed coastal/pelagic operations, and these will not need to be altered for the coastal trials.

Our preference must be clarified on two items:

1) It has been suggested by other scientists that, in view of the different testing stages for multiple stocks "pelagic" versus "coastal," a procedure can be proposed for adoption now for pelagic operations, leaving coastal operations until later. We consider this to be impractical because (a) the Commission cannot be expected to adopt two separate procedures at different times or to extensively amend a procedure adopted for pelagic whaling in order for it to be accommodated for coastal operations as well, (b) an important case is a mixed coastal–pelagic operation, and (c) whaling operations could change in relatively few years from coastal to pelagic or vice versa.
2) It is essential that an RMP contains a rule with three geographical scales, to be adopted universally, and not only a two-scale scheme as preferred by some other members in the Committee. In the three-scale scheme, the top level is unambiguously defined as an ocean basin: the North Atlantic, the North Pacific, or the Southern Hemisphere. This might only require amendment if commercial whaling of some Bryde's whales *Balaenoptera edeni* stocks is reconsidered. The intermediate level divides the ocean basin in survey areas, which each would normally be surveyed in any one year over a six-year period. The third level divides each area in relatively small subareas or sectors, necessary for the designation of catch limits in areas where natural stock areas are unknown. If the first level (ocean basin) would be abolished, the definition of areas could be endlessly controversial (like the "stock areas" as presently in the Schedule), thus impeding the implementation of the RMP. In a three-level scheme, this controversy could be bypassed when there is no agreement where to draw the area boundaries.

To support our preference, we put forward the following characteristics of the TAP, which are not shared by any of the other four proposals:

a) The primary objectives are the restoration and conservation of whale stocks; subject to this, it aims at obtaining the highest possible continuing catch and at limiting yearly fluctuations in catch limits necessary for the orderly development of the whaling industry;
b) It specifies a Protection Level at 55% of the population size before exploitation occurred, which it realizes with a 95% probability that populations not yet depleted remain above that level;

c) It ensures restoration of already depleted stocks. In all robustness trials, there was not a single realization in which such further depletion occurs. We consider this important since real life may be represented by *any single* simulation, and not necessarily by the mean;

d) It sufficiently allows for uncertainty by ensuring that stocks estimated to be above the protection level may only be exploited if there is also a 95% probability that the stock is above 40%;

e) In conclusion, considering that, in the TAP, the catch-limits decision rules seem so intrinsically cautious (conservative), it is notable that it has proven possible to devise a scheme with satisfactory, if not spectacular, performance in providing high sustained catches in the future, and minimal disruption to an industry once it has been permitted to resume. A part of the price paid for this performance is that a depleted stock may be protected longer than would be the case with the other procedures before such permission would be given.

Addendum Produced in Response to Annex J

The term "realized protection level" has been used in two different ways in this annex and Annex J. The purpose of this addendum is to avoid the possibility of confusion by clarifying the difference in terminology.

The "indicator of realized protection" (IRP) is defined in Annex J as:

i) in development cases (populations with little previous exploitation), the IRP is the lower 5th percentile of the distribution of lowest population size over the 100-year period;

ii) in rehabilitation cases (populations depleted prior to the management period), the IRP is the lower 5th percentile of the distribution of final population size at the end of the 100-year period.

The concept of "realized protection level" (RPL) employed in this Annex is the level below which the stock is protected with 95% certainty; that is, the level such that there is less than a 5% risk that the population will be inadvertently exploited below this level during the management period.

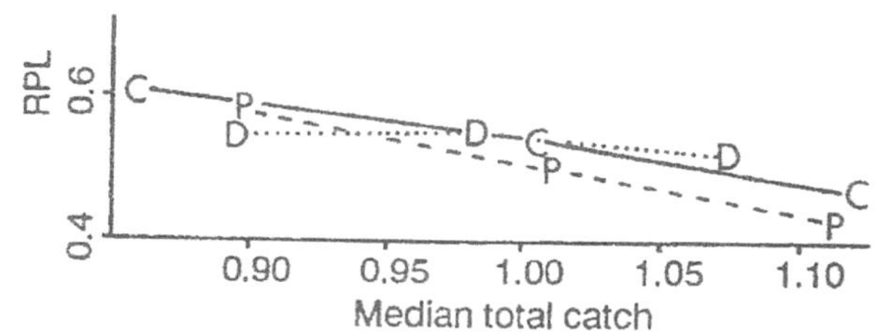

Figure 1. Realized Protection Level as a function of median total catch in the D1 scenario for the P–B, C, and dlM procedures. For each procedure, the three points plotted refer to the "low," "middle," and "high" tunings, respectively.

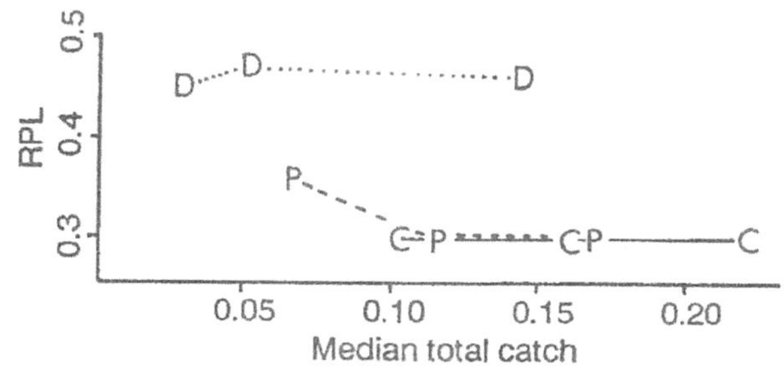

Figure 2. Realized Protection Level as a function of median total catch in the R1 scenario for P–B, C, and dIM procedures. For each procedure, the three points plotted refer to the "low," "middle," and "high" tunings, respectively.

Table 1. Comparison of IRP and RPL for the P–B, C, and dIM procedures at three tunings.

	Procedure	Tuning 0.60	0.66	0.72
Development case				
IRP and RLP	P–B	0.43	0.50	0.58
	C	0.47	0.54	0.61
	dIM	0.52	0.55	0.55
Rehabilitation case				
IRP	P–B	0.39	0.44	0.50
	C	0.44	0.49	0.55
	dIM	0.53	0.55	0.52
RPL	P–B	0.30	0.30	0.36
	C	0.30	0.30	0.30
	dIM	0.46	0.47	0.45

The two definitions are equivalent in the development case, but have different consequences in the rehabilitation case. The figures for the realized protection level using the two definitions are given in the table below for the three procedures compared in Annex J (i.e., C, P–B, and dIM) for development and rehabilitation scenarios for MSYR = 1%.

The relationship between realized protection level (RPL as used in this Annex) and median total catch across tunings are shown in Figures 1 and 2 for the development and rehabilitation scenarios, respectively, for the three procedures.

[*Editors' note: In 1992, the Scientific Committee pointed out in its progress report to the Commission* (Report of the Scientific Committee, Reports of the International Whaling Commission 42:51–82, 1992) *"that no procedure performed uniformly best over all trials and all tunings." It agreed that, of the five proposals considered, the primary candidates for recommendation to the Commission were the P–B and C procedures. Expecting "that the Commission would wish to place highest emphasis on risk minimization," the Committee preferred the C procedure "since it achieves better performance in indicator of realized protection level and in catch than the P–B procedure at all tunings for MSYR = 1% (the least productive case)." The Committee, therefore, recommended the C procedure to the Commission as the "best" single stock procedure. At its 1992 meeting, the Commission accepted the Committee's recommendation.*]

Chapter 30

Selection of an RMP[1]

At the 1989 meeting of the Commission, delegations expressed their views about the desirable features of an RMP. These views pertained to what might be called "design criteria" and, within those, to *priorities* among objectives; only at a second level of evaluation was reference made to "balance" among them. These views are expressed in (a) document IWC/41/42 presented by seven delegations; (b) the Chairman's report of the meeting; and (c) the verbatim report of the Plenary discussion.

The Scientific Committee's judgment of five suggested RMPs should take account of whether suggested design criteria have been met and whether, within the criteria, the primary objectives are attained as well as can be expected.

As to *primary* objectives espoused by various delegations in the debate, these were expressed as (a) negligible risk of extinctions; (b) assured recovery of depleted stocks and avoidance of depletion of those not already depleted; and (c) stability of catches. No delegation gave possible attainment of high cumulative catches as their primary objective.

All procedures appear to meet (a) sufficiently well, which does not, therefore, provide a basis of choice.

One approach to choice is to see which RMP gives a notional balance to (b) and (c) simultaneously, while being adequate with respect to catches. Another approach is to see which meets a given choice (b or c) and is adequate both with respect to the other choice (c or b) and to catches.

I prefer the second approach, but realize that the Commission may also wish to have an opportunity to follow the first.

The explicit features specified by seven countries were, in their words:

"The *primary* objective of an RMP ... is to ensure that currently abundant populations are maintained at or above high target levels, and that recovery is ensured of depleted populations to these [high target] levels as soon as natural process will permit." (*Holt's emphasis*)

"The RMP, insofar as it is intended to address other objectives, should not jeopardize achievement of the primary objective."

"The RMP should be so designed that it guarantees the achievement of the primary objective with respect to each population of whales ... and provide sufficient allowances for uncertainty ... regarding stock boundaries that lack of data does not prejudice achievement of the primary objective."

---

[1]Editors' note: written by S. J. Holt in 1991 as Appendix 12 for the Report of the Sub-Committee on Management Procedures and published in Reports of the International Whaling Commission 42:134–135, 1992.

"The RMP must retain the concept of a protection level which is sufficiently high."

This last phrase was clarified, in the presentation, as intending "certainly not lower than that used in implementing the NMP" and "actually achieved, not merely aimed at."

The dlM procedure conforms explicitly with these last cited features; its relative inflexibility is a consequence of that conformity and is not, therefore, a "fault" in it. The only other procedure with explicit features is S–T, which conforms with the primary objective expressed by Japan, which was stability of a resumed industry. Thus, both the S–T and dlM procedures should be forwarded to the Commission. However, while dlM has performance which is adequate, if not optimal, with respect to the criterion of low interannual catch variability (especially if the deficiency of AAV as an index of that is remembered), the S–T does not and, as far as I can see, cannot fulfill the criteria which dlM is designed to fulfill. Therefore, my choice of one procedure must be for that of dlM.

It has been the choice, it seems, of other developers than dlM, to ignore or undervalue the criteria specified both by Japan and by the seven proposers of document IWC/41/42. They achieve, in various ways, certain mixes of subjectively weighted objectives, but they do that, at least in part, precisely *because* they do not conform with the specifications, especially for a high and realized protection level. One or other of them could be considered as a fallback in case the Commission will not adopt the dlM procedure, either because the latter pays too much attention to the specifications in document IWC/41/42 or because some delegations may have changed their minds since 1989. If, for one of those reasons, P–B, M–S, or C is to be considered, the choice would, I suggest, go to that one, the performance of which is closest to the primary objective defined in IWC/41/42. That seems to me to be the C procedure.

In summary, the dlM procedure adheres to design criteria and a primary objective specified by several Members of the Commission, while not denying the requirements of other Members. P–B, M–S, and C achieve only slightly higher performance levels on one or more objectives, and they do so by ignoring those criteria. If the high-catch objective, which was not given as a first objective by any country, is given relatively low weight, then the C procedure comes out marginally as the best of those three.

Chapter 31

Scientific Advice on the Management of Commercial Whaling[1]

Just as in 1991, the Scientific Committee was faced with the problem of choosing between five possible core RMPs, some of which, at least, had rather similar performances, so this year, it is likely to be considering at least two types of multistock rules: "capping" and "cascading." It might be useful to discuss, *a priori*, exactly what judgment criteria are to be applied. In offering the following suggestions, it should be understood that I do not intend to revisit the process followed last year (though I do remain convinced that the majority advice, as presented to the Commission, was misguided and the cause of considerable difficulties to the Commission); the C procedure has been adopted by the Commission, but with specified provisos. But, I do now suggest that, on the basis of the 1991–1992 experience, the Committee should try to be more consistent in its practices.

In 1991, a minority of Committee members (Australia, New Zealand, Netherlands, Switzerland, Seychelles, IUCN) considered that, while the performances of two or three of the five candidate core procedures were rather similar with respect to the three quantified objectives of the Commission, as specified to the Committee, they differed among themselves sufficiently with respect to certain "features" of their implementation to warrant two or more of them being forwarded to the Commission with the Committee's comments on these and perhaps other features.

The features specifically mentioned were (a) the "protection levels" actually realized in various circumstances and (b) the general pattern of annual catch limits set over an extended time; in some cases, these generally tended to increase, in others to decrease even where the overall performance statistics were virtually identical.

The majority of the Committee held that the Committee had agreed to provide advice on the basis only of comparisons among the three sets of summary statistics and that consideration of such additional statistics should not be involved in the formulation of advice. However, in doing so, the Committee provided the Commission with only three "tuning" options, so that the Commission was placed in a situation of either accepting one of the Committee's arbitrary suggestions or asking the Committee to conduct further trials and thereby delay the final approval of a core procedure. Furthermore, the Committee did not provide the Commission with sufficient information for it to make its own choices.

However, at the Special Meeting in Copenhagen, several members of the Committee who, last year, opposed consideration of additional statistics that I and others thought to be relevant, in effect (it seemed to me) began to give considerable weight to another "feature" that is <u>not</u> based on any general instruction from the Commission and which has very considerable policy

[1]Editors' note: written by Sidney Holt in 1992 and presented to the IWC Scientific Committee as Document SC/44/O 6, but was not published in *Reports of the International Whaling Commission*.

implications. This is the level of catches when whaling is initially reopened on any stock. That is naturally of political interest to those countries which are anxious to resume whaling immediately with economically viable catch limits, just as the question of realized protection levels was—and surely remains—of interest to those countries who are not willing to see what they regard as advances in conservation and management principles under the NMP to be lost or compromised under an RMP.

The reason I gave at the 1991 meeting for wishing to advise the Commission about realized protection levels (and also about time-trends in catch limits) was that a number of countries were on record as having expressed concern about this and, despite the fact that the Commission as a whole had not at that time made any recommendations on the matter, we could not properly ignore it. However, I can find nowhere in the records of the Commission any indication, even from a few countries, that the levels of initial catches should be among the criteria used for choosing what scientific advice should be offered and, specifically, for selecting certain rules or preferring one set of rules over another.

It will perhaps turn out that trials of the various "capping" and "cascading" rules will indicate clear advantages of one over the other(s) with respect to performance as judged strictly by the agreed three sets of statistics which provide measures of long-term cumulative catch, variability of catch limits from year to year, and probabilities of depletion/extinction. If, however, no such clear discrimination is possible from such considerations, then I suggest it will be necessary to present alternative proposals to the Commission and describe clearly the various performance characteristics of these, in addition to the three sets of agreed statistics, on the basis of which the Commission itself may make policy decisions.

I am also concerned that some Committee members are anxious to change the parameter values and assumptions in the adopted core procedure, particularly in such a way that it will be less "conservative." If that is taken seriously, then I think it calls for strong positive evidence that changes are needed, and a review of the simulation trials that led to the selection by a majority of the Committee of one out of five proposed core procedures. The substantive aspects of suggested modifications to the core procedure are discussed in IWC/SC/44/O 3.

Another concern I have about the development process is taken up in another paper I have provided to this meeting. It is that the robustness trials of core procedures regarding possible changes in carrying capacity seem to me not to have been modeled correctly, and that the error is in a direction such that the procedure appears to be more robust to this possibility than it really is. Another concern is that the criteria applied to the determination of the subdivisions for North Atlantic multistock trials are unclear and, furthermore, the selections made do not seem to test very stringently the robustness of any of the suggested multistock rules to specification errors of stock divisions; perhaps this concern will be allayed by the report that Greg Donovan was asked to provide on consultations among members on this matter.

Lastly, I regret that I was prevented by an accident from attending the last two days of the Copenhagen meeting. If I had been able to do so, I would certainly have withheld my agreement to the appended draft of the full RMP. This is because I believe it implies far greater discretion to the Scientific Committee to decide exactly what multistock rules will be used to calculate catch limits in any particular case than is warranted, and with no agreed criteria—or even suggestions as to which criteria—to be applied in making such choices. Such a provision would virtually guarantee continuing difficulties in the Committee with regard to providing agreed future advice on catch limits: every single instance would give opportunities for political wrangling under the guise of scientific debate, something the Committee has been trying to move away from for the past four years.

Chapter 32

Thoughts about MSY Rates and Levels[1]

An MSY rate cannot reasonably be regarded as an intrinsic "property" of a whale population; it is the manifestation of an interaction between a population and its environment. Thus, it is not a constant in a fixed function relating recruitment to population size or density. This is evident from consideration of whether it is reasonable to assume either that the same population can be modeled with a recruitment curve unchanging from one period to another, even if, say, food supplies change drastically, or that different populations of a species occupying different areas will have the same population growth patterns. Any assumption that there are universal family-wide constant parameters of a recruitment function is even less plausible. Prediction of a value of MSYR from a population growth rate estimate for that population at a level other than MSYL requires knowledge of, or specific assumptions about, the shape of the recruitment curve under constant environmental conditions. There is no reason to assume that this must be an uninflected curve, the values of which decline monotonously with increasing population. Furthermore, even if such a form was appropriate in relation to particular elements of the recruitment function (such as the density-dependence of the rate of live births), it does not follow that the recruitment function itself will have such a form. And, even if the general form and scale of the recruitment curve could be reasonably postulated, any interpolated value for MSYR depends critically on an assumption about MSYL. These assumptions cannot be reliably tested against any existing empirical data. For example, if a recruitment curve appeared to be linear (corresponding with a simple logistic expression for population growth), it could not be distinguished from a slightly non-linear function (whether inflected or not inflected) having an MSYL far from 0.5K. A short discussion and a detailed appendix show that attempts by Fowler[2] to find empirical confirmation of his theory that large mammal populations must have relatively high MSYLs are unconvincing, even if the theory itself is plausible. Fowler's more recent attempt to argue not only that recruitment curves of such species must be strongly convex (upward), but that they cannot have inflections, is supported neither by data nor theory. In fact, in presenting his case in "Reviews" to the Scientific Committee in the past two years, Fowler has unfortunately chosen to ignore valid criticism of his previous analyses, to repeat serious methodological errors, and to continue to present misleading graphics. It has been shown that non-monotonic and inflected recruitment curves can theoretically arise both from consideration of the various components of recruitment showing density-dependence

---

[1]Editors' note: written by Sidney J. Holt, UNEP, as Document SC/44/O 3 for the 1992 meeting of the IWC Scientific Committee and published (abstract only) in Reports of the International Whaling Commission 43:534, 1993. It was decided to include only the abstract in light of the large (43 pp) size of the original document.

[2]Editors' note: Fowler and Baker (1991) is a good source of relevant references on the issues raised here.

and from consideration of spatial heterogeneities. In the virtually complete absence of empirical date for any whale population, "best" values of MSYR or of any universal lower bound to that quantity cannot validly be inferred from either intra- or interspecific analogies or extrapolations.

References

Fowler, C. W., and J. D. Baker. 1991. A review of animal population dynamics at extremely reduced population levels. Reports of the International Whaling Commission 41:545–554.

Chapter 33

Protocol for the Scientific Evaluation of Proposals to Cull Marine Mammals[1]

Background

The Planning and Coordinating Committee (PCC) of the UNEP Marine Mammal Action Plan (MMAP) has, as its members, a number of intergovernmental and non-governmental organizations which share an interest in marine mammals. Among its activities is the establishment of an *ad hoc* Scientific Advisory Committee (SAC) to advise on current issues as the need arises. In 1992, in response to an intensifying debate on the interactions between fisheries and marine mammals, a SAC was convened to consider the question of culling marine mammals for the purpose of protecting fisheries resources.

The task given to the SAC was twofold:

- compile a list of historical and current culling programs and situations in which proposals have been made for culls for the intended benefit of fisheries resources, and
- draw up guidelines for the scientific evaluation of culling proposals, including a specification of the information which is needed to evaluate a cull proposal.

The SAC held two meetings to fulfill this mandate, in November 1992 (Liège, Belgium) and August 1994 (Crowborough, United Kingdom). The reports of the meetings are available from the MMAP Secretariat at UNEP. The present protocol was drafted by a subcommittee of

[1]Editors' note: this is a Report of the Scientific Advisory Committee of the Marine Mammals Action Plan (MMAP), October 1999, financially supported by Greenpeace, World Wide Fund for Nature (WWF), International Fund for Animal Welfare (IFAW), and the United Nations Environment Programme (UNEP). The MMAP was a unique organization in that its members consisted of both intergovernmental organizations (UNEP, IWC, IUCN, IATTC, FAO) and NGOs (Greenpeace, WWF, IFAW), all on equal footing, able to discuss, vote, etc. Sidney Holt represented IFAW at the MMAP and, in 1992, persuaded the members of the MMAP of the need to develop a protocol for the scientific evaluation of proposals for the culling of marine mammals for their alleged benefits for fisheries. This was a topic of considerable controversy at the time, with various hunts of marine mammals being justified by claims that fisheries yields would improve as a direct consequence. Such claims were, and still are, largely based on hand-waving arguments, with the notable exception of a proposed cull of Cape fur seals (*Arctocephalus pusillus*) in South Africa (which was not conducted due to a scientific evaluation of the probable consequences). The protocol was the fruit of two workshops (1992, 1994) attended by scientists from 11 countries. In addition to conceiving the idea for the protocol, Sidney was an active participant in both workshops and member of the drafting group which produced the Report. The protocol was subsequently used in a study conducted for the European Parliament on the impact of culling grey seals in the North Sea (Harwood and Walton 2002). The protocol remains pertinent as culling proposals continue to appear that are justified by sophistry rather than science. On occasion, such proposals still get implemented, though, as far as the editors can determine, no cull has ever been carried out having been justified by a scientific assessment of its probable impact. The Report is reproduced here with the permission of UNEP.

the SAC and then sent to all members of the Committee for review. Annex 1 lists the members of the Committee who attended the two meetings and commented on the draft protocol. The final version, incorporating comments received, was then submitted to UNEP and the other member organizations of the PCC.

Introduction

Concerns are often expressed over the quantities of fish consumed by seals, dolphins, and other marine mammals and the possibility that they are affecting the size and availability of the fish stocks and thus the viability of fishing industries. Consequently, demands are made for culls to reduce the number of marine mammals in the belief that fewer marine mammals will consume less fish, which will leave more fish available for fisheries to catch, with resulting benefits for the fishing industry and the livelihood of fishers. During this century, many control programs aimed at reducing or limiting the numbers of seals or cetaceans have been implemented around the world. These activities have taken various forms, from bounties and culls lasting for several decades [e.g., a bounty on harbor seals (*Phoca vitulina*) in eastern and western Canada] to one-off hunts sponsored by governments [an open season on New Zealand fur seals (*Arctocephalus forsteri*) in 1946]. Culls can also be implemented through existing commercial hunts, for example, where the authorities subsidize a hunt to encourage it to take more animals than it otherwise would or where a marine mammal harvest is managed with the aim of reducing, rather than merely maintaining, the marine mammal population. Table 1 lists the instances known to the SAC where culls or control programs for marine mammals have been carried out for the apparent purpose of benefiting fisheries.

Given the depleted status of many fish stocks and fisheries, it is not expected that demands for culling programs will be reduced in the near future. According to a review by the Food and Agriculture Organization (FAO 1993), one in four of the world's fish stocks for which an assessment is available is classified as overexploited, depleted, or slowly recovering from depletion. A further 44% are fully or heavily exploited, leaving fewer than one stock in three which is lightly or moderately exploited and which allows some scope for expansion. As the fishing industry attempts to maintain or increase catches from generally dwindling stocks, marine mammals are liable to be perceived increasingly as competitors to fisheries.

The limited number of cases that have been examined in depth to date reveal that interactions between marine mammals and fisheries are not always as simple as they might at first sight appear. Simplistic assumptions, such as that reductions in the abundance of a seal population will necessarily be reflected by a concomitant increase in fishery yields, can yield erroneous conclusions about the likely effects of a cull (see below on South Africa). Furthermore, experience shows that the issue of culling marine mammals can lead to emotional and polarized debates which may interfere with a rational evaluation of the potential merits and detriments of a specific proposal. For these reasons, an objective and scientific approach to the evaluation of culling proposals is important. Equally, all interested parties should have easy access to all the relevant information regarding any proposal to cull marine mammals. The SAC considered that a standard protocol for evaluation of culling proposals would assist this process by helping to ensure that all major relevant factors are taken into account and by providing a suitable conceptual framework for making use of the available information. Use of a common protocol can also facilitate the comparison of cull proposals which, in turn, can facilitate the application of experience gained in one situation to another. In addition to the

Table 1. Instances of current or historical culls or bounty programs for marine mammals due to perceived ecological interactions with fisheries.

Country	Marine mammal	Cull or bounty program
United Kingdom	Grey seals *Halichoerus grypus*	Various control programs from 1934–1982
Ireland	Grey seals	Bounty until 1977
Ireland	Harbor seals	Bounty until 1977
Norway	Grey seals	Cull from 1980–1989
Norway	Harbor seals	Cull from 1980–1989
Norway	Killer whale *Orcinus orca*	Expanded commercial hunt from 1978–1982
Baltic Sea states	Seals	Bounties for various periods from 1889–1976
Sweden	Grey seals	Experimental cull 1997
Finland	Grey seals	Experimental cull 1997
Iceland	Grey seals	Bounty from 1982–??
Iceland	Harbor seals	Bounty from 1982–1990
Iceland	Killer whales	US Naval operation in 1976
Greenland	Killer whales	Bounty from 1960–1975
Eastern Canada	Harbor seals	Bounty from 1927–1976
Eastern Canada	Grey seals	Cull from 1967–1984, bounty from 1976–1990
Eastern Canada	Harp seals *Pagophilus groenlandicus*	Seal hunt variously portrayed as "sustainable harvest" or "population control mechanism"
Eastern Canada	Beluga whales *Delphinapterus leucas*	Various control programs during 1920s and 1930s
British Columbia	Steller sea lions *Eumetopias jubatus*	Various control programs from 1912–1968
British Columbia	Harbor seals	Bounty from 1914–1964, recent localized culls
Alaska	Harbor seals	Bounty from 1920–1967
Western United States	Harbor seals	Various bounties from 1920s–1972
New Zealand	New Zealand fur seals	Open season in 1946
Australia	Australian sea lions *Neophoca cinerea*	Open season in 1920
Australia	New Zealand fur seals	Open season in 1920
Australia	Australian fur seals *Arctocephalus pusillus doriferus*	Open season in 1948/49
Japan	Small cetaceans	Various control operations until 1991
Japan	Steller sea lions	Bounty
South Africa	Bottlenose dolphins *Tursiops truncatus*	Official killing in early 1900s
Namibia	Cape fur seals	Fur seal hunt portrayed as "sustainable harvest" or "population control mechanism to protect fisheries"

essential scientific evaluation of culling proposals, there are economic and social implications of culls which should also be evaluated.

Since the last major reviews of the problems of interactions between marine mammals and fisheries (Northridge 1984, 1991; Beddington et al. 1985), there have been useful scientific developments relevant to the analysis of culling questions. Understanding of the properties of foodwebs has advanced, including methods of analyzing foodwebs involving many species, but with only limited data on each one (Yodzis 1989, 1994, 1998, 2000). Other important developments include approaches to the management of marine mammal and fish stocks which involve simulation of their management under a range of scenarios (Cooke 1995; Punt and Butterworth 1995); this method provides a flexible means of making use of the available data and provides a consistent vertical integration of the various levels of the process from the biological through to the management levels, and is gradually becoming more widely applied.

The SAC reviewed recent scientific developments and experience relevant to the analysis of culling questions in particular and to the management of multispecies fishery situations in general and used these to develop an evaluation protocol which aims to combine practicality in terms of the requirements for data and analysis, with a satisfactory degree of ecological and operational realism.

Overview of the Protocol

The scientific evaluation of a proposal to cull marine mammals for the purpose of benefiting one or more commercial fisheries is not a trivial exercise as it must consider the complexity of ecological interactions among the marine mammal population(s), the relevant fish stocks, and the fishery/fisheries which catch them. The protocol developed by the SAC consists of two parts. The first (Section 2) defines in detail the information which must be provided with the culling proposal which will allow a scientific evaluation of the biological and ecological aspects, including:

i. a brief description of the components of the interaction (the marine mammal population involved, the fish species and the fishery or fisheries which take the fish and are thought to be in competition with the marine mammals)—Section 2.1;
ii. the objectives of the proposed cull and the expected benefits and risks—Section 2.2;
iii. ecological information on the marine mammals (distribution, population size, per capita food consumption and diet, the total food consumption of the population, and their population dynamics)—Section 2.3.1;
iv. ecological information on each of the target fish species (stock assessments, the consumption of these fish by their significant predators, and the principal prey species of the target fish stocks)—Section 2.3.2;
v. ecological information on other major components of the ecosystem—Section 2.3.3;
vi. information on the fisheries (total catches including, where significant, discards, and the method by which the fishery is managed)—Sections 2.4.1 and 2.4.2;
vii. basic economic information such as relative prices for the commercially important species and subsidies—Section 2.4.3;
viii. details of the culling program (number of animals to be killed, their sex and age, the target population size for the marine mammal, and the methods to be used to monitor the population size) —Section 2.5;
ix. provisions for monitoring the effects of the cull in order to ascertain whether it achieves its objectives and the procedure for managing cull—Section 2.6.

Section 3 provides a description of the procedure for evaluating the likely effects of the proposed cull. The evaluation is performed using a particular type of ecological modeling known as scenario modeling. This technique has been shown to be a useful and versatile way to address fishery management problems which is able to incorporate a large amount of information while taking into account the uncertainties in our understanding of ecological relationships.

The steps involved in the evaluation exercise are:

i. Summarize the available information and verify that requirements of Section 2 are fulfilled.

ii. Identify those species (fish, marine mammal), fisheries, and other components which must be included to create a "minimal realistic" model of the ecosystem. Increasing the number of species makes the model more realistic, but also more complex and difficult to interpret. The SAC considered that a reasonable approach was to include enough species to account for most (i.e., at least 80%) of the natural predation for the commercial species of concern.
iii. Construct the simulation model which includes all of the relevant components and the interactions among them.
iv. Select realistic values and make realistic assumptions for the model. For instance, it is important that the model be constructed in such a way that the marine mammal population will not increase indefinitely if they are not culled. The uncertainties in either the parameter values or the model structure must be incorporated by, for example, using a range of values or alternative plausible model structures.
v. Choose the appropriate statistics which will be generated from the model to be used in assessing whether or not the cull will meet its objectives (performance measures). For instance, if increases in the catches of certain fish stocks are the objective, then the predicted catches following the cull would be a relevant performance measure. The appropriate comparison would be between catches following the cull and catches in the absence of a cull.
vi. Run the model as constructed under the various scenarios.
vii. Interpret the results: if the results tend to show either clear benefits or clear detriments from the proposed cull, then this would be evidence either for or against the cull, respectively. If the results are less clear cut, with some results showing gains and others suggesting disadvantages, further research may reduce the ambiguity.

An approach similar to that outlined here was used in South Africa to study the effects of a proposed cull of Cape fur seals (*A. pusillus pusillus*) on the hake (*Merluccius capensis*) fishery. A "minimal realistic" model was constructed that included the hake fishery, the fur seal population, the cull, the hake stock (in at least two age/size classes and including cannibalism), and a lumped component representing "other predatory fish" (Anonymous 1991). The basic model (Punt 1994) predicted that a seal cull would benefit the fishery, although the size of the predicted effect was fairly small. A more realistic version, with two species of hake, one of which ate the other, predicted that a seal cull would have negligible benefits or a possible detrimental effect on the fishery. Subsequently, a decision was made not to proceed with the cull.

While the SAC developed this protocol using a particular modeling technique—scenario modeling—other approaches may also be valid. They would need to include the same interactions among the components (marine mammal, culling program, fish stocks, fishery, management regime) as well as the uncertainties relating to both parameter values (fish stock abundance, catches by the fishery) and the structure of the model (ecological relationships).

References

Anonymous. 1991. Report on the Benguela Ecology Programme workshop on seal-fishery biological interactions. Reports of the Benguela Ecology Programme of South Africa 22.

Beddington, J. R., R. J. H. Beverton, and D. M. Lavigne. Editors. 1985. Marine mammals and fisheries. George Allen and Unwin, London.

Cooke, J. G. 1995. The International Whaling Commission's Revised Management Procedure as an example of a new approach to fishery management. Pages 647–657 *in* A. S. Blix, L. Walløe, and Ø. Ulltang, editors. Whales, seals, fish and man. Elsevier, Amsterdam.

Food and Agriculture Organization. 1993. Marine fisheries and the law of the sea: a decade of change. FAO Fisheries Circular 853.

Harwood, J., and M. Walton. 2002. Interactions between seals and commercial fisheries in the North-East Atlantic. Final Report to the Committee on Fisheries of the European Parliament, January 2002. European Parliament Directorate-General for Research, Fisheries Series, FISH 110 EN.

Northridge, S. P. 1984. World review of interactions between marine mammals and fisheries. FAO Fisheries Technical Paper 251.

Northridge, S. P. 1991. An updated world review of interactions between marine mammals and fisheries. FAO Fisheries Technical Paper 251, Supplement 1.

Punt, A. E., and D. S. Butterworth. 1995. The effects of future consumption by the Cape fur seal on catches and catch rates of the Cape hakes. 4. Modelling the biological interaction between Cape fur seals *Arctocephalus pusillus* and the Cape hake *Merluccius capensis* and *Merluccius paradoxus*. South African Journal of Marine Science 16:255–285.

Yodzis, P. 1989. Introduction to theoretical ecology. Harper and Row, New York.

Yodzis, P. 1994. Predator-prey theory and management of multispecies fisheries. Ecological Applications 4:51–58.

Yodzis, P. 1998. Local trophodynamics and the interaction of marine mammals and fisheries in the Benguela ecosystem. Journal of Animal Ecology 67:635–658.

Yodzis, P. 2000. Diffuse effects in food webs. Ecology 81(1):261–266.

Protocol for the Scientific Evaluation of Proposals to Cull Marine Mammals

1. Scope

The protocol covers the case where marine mammals are perceived as ecological competitors with a fishery, and a cull of the marine mammals is proposed for the purpose of benefiting the fishery. It does not address other situations in which a marine mammal cull may be proposed, such as: (i) operational interactions (e.g., damage to fishing gear or raiding of nets by mammals); (ii) transmission of parasites from mammals to fish; (iii) actual or perceived damage to habitats by marine mammals.

The protocol focuses on the biological and ecological aspects of the evaluation of cull proposals. For this purpose, the economic objectives of the proposed cull should be noted, so that the evaluation can focus on the most relevant aspects of the expected effects of a cull, but a full evaluation of the economic consequences of a proposed cull is beyond the scope of the current protocol.

There are two main parts to the protocol:

- specification of the information required for the scientific evaluation of culling proposal; and
- outline of the evaluation process for cull proposals.

2. Information required for the scientific evaluation of culling proposals

In order to evaluate scientifically a proposal to cull marine mammals to benefit fisheries, a certain minimum amount of information is needed. To facilitate the evaluation, it is preferable that cull proposals be made in written form and include the information specified below, with references to the original sources of the information.

The potential amount and scope of ecological data that is relevant to the evaluation of a cull proposal is virtually unlimited. It is recognized that ecological data can be costly to collect and that long time-series may be required for them to be useable. This protocol focuses on specifying the minimum set of information that is required for a scientifically meaningful evaluation of a cull proposal. It does not attempt to provide an exhaustive list of additional data that, if available, could be used to improve the evaluation of the cull proposal.

In situations in which some of the data listed are not available, attempts should be made to obtain them. If too much information is missing, an evaluation of the proposal may not be feasible. In such cases, the cull could not be justified on scientific grounds because it would not be possible to judge whether it would be likely to achieve its stated objectives or, alternatively, have detrimental and/or unexpected impacts on the target fish stocks or other species in the system. A summary list of the data required is given in Table 2.

2.1. Basic information on the marine mammal/fishery interaction

The following should be specified:

- the marine mammal species and population(s) involved, their geographical extent, and seasonal movements;
- the target fish species which are thought to be the subject of competition between marine mammals and fisheries;
- the fishery or fisheries which take the target species, their area and season of operation, gear used, and details of other species taken in significant quantities;
- the geographical area where the competition is believed to occur.

2.2. Objectives of the proposed cull

The objectives of the cull, and the expected benefits, e.g., biological or economic, need to be specified clearly so that an evaluation can be made of the likelihood of achieving them. Any potential environmental costs and risks should also be noted.

The objectives and expected benefits of a cull could be, for example, biological (e.g., increases in the abundance of certain fish stocks), economic (e.g., increase in the profitability of a fishery) or catch-related (e.g., increases in realized or permitted catches, or realized catch rates of certain fish species by certain fisheries).

If scientific analyses have been conducted to estimate the expected risks or benefits of the proposed cull, these should be fully documented, including a specification of the data used, the assumptions made, and the details of any models used.

2.3. Ecological Information

Certain ecological data are required for each of the marine mammal population(s), the fishery or fisheries, and the target fish stocks which are believed to be involved in the interaction. The information listed below is not an exhaustive list of all the relevant information, but represents the minimum required for a meaningful evaluation of the cull proposal.

Table 2. Data required for the scientific evaluation of cull proposals.

1. Basic information: - the marine mammal population, range, and migration - the target fish species - the fisheries involved and the species they take - the geographical area of concern
2. Cull objectives: - biological - operational - economic
3. Ecological information:
(i) Marine mammal: - distribution, population size, and population structure - *per capita* food/energy consumption - diet composition, including methods of sampling and estimation - total food consumption - demographic parameters
(ii) Target fish species: - distribution and migration - demographic parameters (weight at age, age at spawning, etc) - details of current stock assessment (if any), including: ○ data used (e.g., catch at age, research survey abundance indices, commercial catch per unit effort) ○ details of assessment model and results
(iii) Other predators and prey of the target fish species: - abundance, amounts consumed, details of stock assessments if any
(iv) Other components of the ecosystem: - 2-way matrix of 'who eats whom' with estimated or guessed annual consumptions - estimated abundance by species
4. Fisheries: - catches, including bycatches, of the target and other species - management system in place, if any, including procedure, if any, for determining TAC - basic economic data, where relevant to objectives
5. Culling program: - envisaged duration - annual numbers to be killed or details of other measures, e.g., bounties - measures to estimate the numbers killed - target population size or other reduction/control targets, if any - provisions for monitoring marine mammal population - provisions, if any, for monitoring the effects of the cull.

2.3.1. The marine mammals

Estimates are needed of the total consumption by the marine mammals both of the target fish species and of all other species that form a significant component of their diet, since these other species may be involved in direct or indirect interactions with the target species. Obtaining these estimates will require the following information:

2.3.1.1. Distribution

In order to identify which marine mammals are actually involved in the interaction, information is necessary on the distribution of the population, including any seasonal migrations, so that the fraction of the population in the area of interest at each time of year can be estimated. In many species, males and females or adults and juveniles are found in different areas.

2.3.1.2. Population size

An estimate is needed at least of the current size of the marine mammal population, and preferably also estimates of the population size in previous years and the current trend in abundance. All estimates should specify the component of the population to which they refer and have associated confidence limits (e.g., 525,000 ± 48,500 animals aged one year and older). The method of estimation should be specified and documented. Where possible, information on the age and sex composition of the population should be provided. The latter is especially important in species where males and females differ greatly in size.

2.3.1.3. Per capita food consumption and diet

The *per capita* food consumption of the marine mammal population, partitioned by prey type, can be estimated either by (i) direct measurements of the absolute amounts consumed, e.g., from stomach content or fecal analysis; or (ii) by converting an estimate of relative diet composition to an estimate of total consumption using estimates of the energy requirements of the mammal and the energy content of the food species. Method (i) is only feasible in some cases, and even when it can be used, method (ii) should also be applied as a cross check. Application of method (ii) requires:

- estimation of the annual energy requirements of the marine mammal;
- estimation of the composition of the marine mammal's diet;
- estimates of the energy content of each prey species.

The *per capita* energy requirement estimates should be for each age class and sex. Where specific data are not available, generic formulae for mammalian food consumption by body size and life stage may be used. They should be for the period of the year in which the marine mammal is in the area of interest.

Estimation of the composition of the diet, by food species, requires care. Diet composition can vary widely: by time of day, between seasons, from region to region, from year to year and, often, between males and females and among animals of different ages and sizes. It is necessary to ensure that dietary information is representative. Stomach contents or observed feeding behavior of animals taken in or near fishing nets are unlikely to be representative. Likewise, the diets of animals at times and places where they can readily be seen during breeding or hauling out may not be representative of their diet through the year and their range. Since all of the techniques currently available to estimate diet composition are biased in one way or another, the methodology used should be noted and the results interpreted cautiously.

The energy content of prey items can also vary seasonally in some species, as fish are often richer in their energy content just prior to spawning. Data on these variations, which can have a large impact on estimates of total consumption, are needed for the most important prey species. Information on the size of the fish which are consumed, especially of the target species, may also be necessary to evaluate the cull in some cases.

2.3.1.4. Total food consumption by the marine mammal population

The above information on diet, distribution, and abundance can then be combined to give estimates of how much fish and other prey of each species is consumed by the marine mammal population in the area of interest.

2.3.1.5. Population dynamics of the marine mammal

An essential part of the evaluation of the ecological effects of the proposed cull is an estimate of the impact of the cull on the abundance of the marine mammal. For this purpose, estimates of basic population parameters for the marine mammal are required: fecundity, age at maturity, and natural mortality by age. Where available, data on changes in any of these parameters over time should be given. Natural mortality rates can be difficult to estimate: if assumed values are used, an explanation and brief justification should be given. Using this information, predictions should be made of the impact of the proposed cull on the marine mammal population. The assumptions and model used for this purpose should be detailed.

There will usually be considerable uncertainty about some key population processes, such as the presence and nature of density-dependent effects in the marine mammal population dynamics. In this case, a range of alternative assumptions should be made to span a plausible reasonable range for the unknown effects. Models which imply that the marine mammal population would, in the absence of a cull, increase without limit, are not realistic.

2.3.2. The target fish species

2.3.2.1. Stock assessments

Full details of the current stock assessment of each of the target fish species should be provided. These include:

- assumed and measured values of biological parameters, including weight at age, age at first spawning, natural mortality rates;
- data used for the assessment: e.g., time-series of annual catches by age or size, time-series of abundance data from research surveys, if available, time-series of catches per unit effort by fishery;
- details of the assessment model used and the results.

If no stock assessment has been performed, the above data should be provided to enable this to be done.

2.3.2.2. Other predators and prey of the target fish species

Estimates should be provided of the annual consumption of the target fish species by each significant predator species in the area. Although it is difficult to give firm guidelines as to what constitutes a significant predator, the following guideline has been found to provide a reasonable balance between inclusiveness and practicality: the major predators of the target species should be listed in decreasing order of estimated consumption and enough predators included to account for most (e.g., 80% or more) of the total estimated predation mortality on each age class of the target fish species in the area. As much of the life history of the target fish species should be included as possible, but it may be necessary to ignore predation on the earliest life stages due to lack of reliable data. Estimates of predation mortality should include cannibalism (predation of larger fish on smaller fish of the same species) where this occurs.

The main prey species of the target fish species should be identified, and estimates of consumption of each prey species made. This is especially important when the target species prey upon their own species or each other, or on species which are also preyed upon by the marine mammals. In some cases, it may be necessary to include one or more additional prey species which are of particular interest to managers, even though they would not normally be included on the basis of their consumption by the marine mammal in question.

2.3.3. Other components of the ecosystem

Conceivably, other species in the system, besides those already discussed in Sections 2.3.1 and 2.3.2, could play a significant role in the response to a cull. For instance, if other predators of the target species are significant, then changes in the abundance of those other predators might also be significant, and the model would need to predict such changes. This might require detailed information about the predators and the prey of those other predators of the target species. By the same reasoning, one might need detailed information about the predators and prey of the predators and prey of those other predators of the target species, and so on; indeed, one might need to know all the interactions in the whole system. However, full information of this kind (say, accounting for at least 80% of consumption of all species in the system) will rarely, if ever, be available. Therefore, the SAC agreed that such multispecies issues could be addressed in the following way.

All the major taxonomic groups in the ecosystem (i.e., species feeding, or being preyed upon, in the area of interest for part or all of the year) should be identified, as well as the trophic interactions between them (who eats whom). Initially, this information should be assembled in qualitative terms: i.e., include all significant components and trophic interactions without quantifying them. The information is best expressed as a two-way table, with rows for prey and columns for predators. When this has been done, available data should be searched for a range of estimates of abundance, in biomass, of as many components as possible. Estimates of the total food consumption and the consumption of individual prey species should be entered where available. Gaps in the table should then be filled with reasonable guesses, in such a way that the "books balance." The distinction between estimated and guessed values should be kept clear. For estimated values, the source of data should be listed, while for guessed values, a brief explanation of the rationale should be provided.

Estimation of food requirements based on general allometric relationships relating energy requirements to body size for different groups of organisms should be used to derive a plausible range of estimates where specific data are lacking.

This table will be used for deciding what species should provisionally be taken into account in an evaluation of the likely effects of the cull. Taxonomically similar species should not be lumped together if there are substantial differences in their distribution and diet, such as between species occurring inshore and offshore.

2.4. Information on the fisheries

2.4.1. Catches

For each fishery which exploits the target fish species, reliable data should be provided on total quantities of fish caught (including those which are discarded), their size distribution, and the location and season of the catches. Some fisheries may need to be included on the basis of their bycatches alone if these are significant. Data are also required on the catches of other species, including both commercially valuable species and bycatches, by each of these fisheries.

2.4.2. Management

In most instances, a cull of marine mammals would be intended to increase the abundance of fish stocks on the assumption that this would lead to greater catches and profits for the fishing industry. However, the extent to which potential increases in a fish stock are reflected in increased catches can depend on the way the fishery is managed.

In order to evaluate the possible effects of a cull on a fishery, it is necessary to specify the way in which the fishery is to be managed during and following the cull. Typically this is accomplished by setting management measures, such as effort controls or a total allowable catch (TAC), on an annual basis, using an assessment of the state of the stocks at that time, which is, in turn, based on a series of catch and other data collected up to that point. The method by which the annual TAC is determined or calculated should be specified, so that this can be included in simulations of the effect of a cull under various scenarios.

If there is no clearly defined procedure for the determination of management measures (e.g., TACs), the main objectives of the fishery management regime should be specified.

2.4.3. Economics

A full evaluation of the economic effects of a cull is beyond the scope of this protocol. Nevertheless, the economic objectives of the cull, if any, should be stated, since these help to guide the choice of biological and technical indices (such as the predicted level of catch for a given period into the future or the level of fishing effort required to achieve a given catch) used to summarize the predicted effects of a cull, and to ensure that they are relevant.

A full economic analysis of the effects of a cull on a fishery would require information on operating costs, capitalization, market conditions, subsidies, taxation, etc. However, even in the absence of such an analysis, certain basic economic information should be provided where appropriate. For example, if the main fishery of interest catches several species of commercial value, data on the relative prices of each should be given, so that predictions of the level of catch by species can be converted to a total catch value. This may be particularly relevant in cases where the ecological modeling analyses suggest that the marine mammal cull may increase the abundance of some commercial fish species, but reduce the abundance of others.

It should be emphasized that increases in the potential yield of a fish stock do not necessarily translate into an increase in the net economic yield of the fishery. In the case of fisheries which are inadequately managed and dependent on high levels of subsidies, enhancing the potential yield of the fish stock, through a marine mammal cull or other means, can even exacerbate the net economic losses made by the fishery. However, these issues are beyond the scope of this protocol.

2.5. Details of the proposed cull or culling program

There are several types of control programs which can be used to reduce marine mammal populations, such as culls and bounties maintained for several years or events limited to one or two years. The effects of the control program will depend upon its extent and the methods chosen.

The number of marine mammals to be killed each year should be specified, along with details of how this number has been determined, the predicted age and sex of the animals, where and when they would be taken, and how long the culling program is expected to continue. The proposal should also note the precise timing in relation to the annual cycle of the marine mammal population, such as a cull during or immediately after the reproductive period or an opportunistic hunt throughout the year. If the number to be killed is not fixed in advance, but is dependent on data to be collected, such as a fixed proportion of the estimated pup production in each year, then the formula for determining the number and class of animals to be culled should be specified, along with details of the data which will be used to implement the formula.

If the number to be culled is not fixed or based on a fixed formula, for example, if it is based on a bounty program to provide an incentive to people to kill the marine mammals opportunistically, then the nature of the incentives should be specified, along with a rough assessment of its likely effect in terms of expected numbers killed, plus any relevant data from past experience.

The intended effects of the cull on the marine mammal population, such as to reduce it to a specific target level, should be specified.

Details should also be given of how the cull is to be implemented and what measures are proposed to verify or estimate the numbers of animals which are actually killed, as well as the scientific information to be obtained from the animals killed and its intended use.

The intended procedures for monitoring the marine mammal population (such as annual surveys of abundance) during and after the cull period should be specified.

2.6. Provisions for monitoring the effects of the cull

The provisions, if any, for monitoring the effects of the cull, with a view to ascertaining whether it had the intended effect, should be specified. Specific monitoring provisions should be undertaken with respect to each of the major elements in the objectives of the cull and the expected effects of the cull on the fishery and the marine mammal population. Management decisions to continue, stop, or modify a cull may be linked to the results of the monitoring program.

3. Evaluation of cull proposals

3.1. Introduction

This section describes the procedure for evaluating, in advance, the likely effects of a proposed cull, especially in terms of the likelihood of achieving its objectives. The specified evaluation should be completed before a cull is undertaken.

Although the focus of this protocol is on the prior evaluation of a proposed cull, the approach can, in principle, also be used to evaluate the effects of a cull that has already taken place. However, delaying the evaluation until after the cull has taken place is not recommended for two reasons: (i) the prior analysis might indicate that the cull has a high probability of being detrimental to the stated objectives, and hence it would disadvantageous for the cull to take place; (ii) the prior analysis might reveal that in order to enhance the reliability of the assessment of the likely effects of the cull, certain data need to be collected from the pre-cull system. Without these data, it may be considerably harder to determine whether the cull had the intended effect.

The prior evaluation should also include an evaluation of the chances that the planned means of monitoring the effects of the cull will yield reliable results.

The recommended method of evaluation involves the ecological modeling exercise detailed in the following sections. The modeling approach used—scenario modeling—has been shown to be a useful and versatile tool in addressing fishery management questions. The exercise will not necessarily yield a definite verdict as to whether the proposed cull will be beneficial or detrimental, but it may yield an indication of the relative likelihood or plausibility of the cull having positive or negative effects with respect to its objectives. The evaluation exercise may also pinpoint those key aspects of the system on which more information would enable the effect of the cull to be predicted and monitored with more confidence.

It is recognized that, in some cases, the available modeling expertise may not be available, or there may be features of the situation that make the modeling exercise described below unduly difficult, in which case the evaluation may be performed in a less formal way. In such cases, it is, nevertheless, recommended that the following sections be studied and that: (i) the data that would be required to implement a model of the type described are assembled; (ii) all the major features which are recommended for inclusion in the modeling exercise should be assessed for their relevance and taken into account in the evaluation to the extent possible. Table 3 lists the main steps involved in the evaluation process.

3.2. Structure of the evaluation

The evaluation exercise involves the following steps:

(1) Summarizing the available information, including a determination of whether the information specified above is complete and of sufficient quality to conduct an evaluation of the possible effects of the cull. In particular, the information needs to be sufficient to construct a reasonable simulation model of the system, as described below.

Table 3. Procedure for the evaluation of the cull proposal.

Verify that data listed in Table 2 are complete and adequate
Identify species, fisheries and other components to be included in a minimal realistic model of the system
Specify quantitative performance measures relevant to cull objectives
Construct simulation model of the system
Specify a range of alternative scenarios, covering the range of plausible parameter values and model structures
Run multiple replicate simulations of each scenario and compute performance measures and other relevant statistics
Interpret and synthesize results
Evaluate power of indices proposed to monitor effects of cull

(2) Identification of those components of the system which need to be taken into account for at least a minimal realistic evaluation of the likely effects of the proposed cull: these will include the marine mammal population, the cull itself, the target fish stock or stocks, other major predators of the target fish stocks, and, where appropriate the fishery, the management of the fishery, and other relevant components of the ecosystem.
(3) Construction of a simulation model of the system which includes as submodels each of the components identified under (2) and all significant interactions between these components.
(4) Specification of quantitative performance measures and other model outputs which are used to summarize the predicted performance of the cull relative to its objectives, and to throw light on other relevant aspects of the predicted effects of the cull. For example, if one of the objectives is to increase catches from a given fishery, then the predicted average annual catch by this fishery over the next 10 or 20 years would be a relevant performance measure. Other relevant output statistics include the extent to which the marine mammal population is predicted to be reduced by the cull.
(5) In general the available information will be insufficient to specify uniquely the most appropriate choice of models and parameter values. Hence, a variety of different scenarios should be constructed to span the range of plausible alternatives with regard to model assumptions and parameter values.
(6) In most cases, some important components of the system, such as annual recruitment to the fish stock, will be subject to unpredictable random variability. The model should explicitly include this random variability, and multiple replicate simulations (typically 100 or more) of each scenario should be run so as to indicate the range of probable outcomes.
(7) Each replicate of each scenario should be run with and without the proposed cull so that the differential effect of the cull on the predicted outcome can be analyzed. The performance measures and other relevant output statistics should be summarized in a form to assist with understanding of the results and, if possible, to reach an overall verdict on the relative merits or demerits of the proposed cull or, failing that, to pinpoint those aspects of the system where more data are needed to reach a conclusion.

Some of these steps are elaborated in more detail in the following sections.

3.3. Construction of the simulation model

A simulation model should be constructed which contains the following submodels to simulate the components of the system:

3.3.1. Population dynamics of the marine mammal

This submodel should normally simulate the marine mammal population by age and sex over time and be constructed so as to make use of available information on population parameters of the marine mammal (reproduction and mortality, age at maturation, weight at age by sex, etc). The submodel should contain some density-dependence to avoid the unrealistic prediction that the population will expand indefinitely in the absence of a cull. Since density-dependence is hard to measure in populations directly, it may be necessary to make a range of assumptions about density-dependent processes, such as a range of values for the carrying capacity of the environment. For mammals dependent on a highly variable environment, whose populations are liable to fluctuate even in the absence of culling, at least some of the simulated scenarios should reflect this variability; an example would be the effects of El Niño on certain pinnipeds.

If the diet of the marine mammal is sufficiently varied (within or between years) or if the fish stocks of interest constitute only a minor portion of the marine mammal's annual diet, it is probably not necessary to model explicitly the dependence of the marine mammal on its food resources. Otherwise, a range of assumptions may have to be made relating the population carrying capacity to the available food.

3.3.2. The proposed cull

The submodel should allow the cull to be simulated by removing given numbers of marine mammals by age and sex from the simulated population.

The proposed cull may not always specify a fixed number of animals to be taken, but may involve a variable number to be taken over time depending on estimates of the remaining numbers of animals. In such cases, the formula needs to be specified for determining the number of animals to be culled, the type of data on which this calculation will be based, and the frequency and precision with which it will be collected (for example, annual pup counts with a given coefficient of variation; see also Section 3.3.7—Data collection submodels).

3.3.3. Food consumption of the marine mammal

This submodel should allow data on the diet composition of the marine mammal, preferably broken down by population component, such as mature and immature males and females, to be used to predict the total consumption of the species of interest by the simulated mammal population as a function of the simulated numbers of mammals by age and sex at any time in the future.

For this purpose, it will be necessary to make assumptions about how diet composition might vary as the availability of the various food items changes. It will not usually be realistic to assume that the diet composition is independent of the relative abundance of food items, although this assumption might be used to construct one extreme scenario to bracket one end of the range of possibilities. A suitable intermediate assumption is that the relative proportions of the different food items vary in proportion to the ambient abundance of the respective items, i.e., relative food preferences do not vary. If the food species under explicit consideration constitute only a minor part of the total diet of the mammal, this assumption can be approximated by the assumption that the mammal exerts a fixed mortality rate on the species of interest.

An alternative scenario to include as a sensitivity test could be that the mammals focus on the more abundant food items, provided that this is consistent with the available data.

3.3.4. Dynamics of the target fish stock or stocks

In the case of fish stocks subject to an existing program of assessment and management, it is preferable that the submodel for the dynamics of the target fish stock or stocks is equivalent to that used for the routine assessment and management of the fishery. However, if the model used for the routine assessment of the fishery is purely a stock reconstruction model, without predictive power, it will be necessary to extend it accordingly.

Given the variety of fishery assessment models in current use and the need to match the choice of model to the specific characteristics of the fish stocks or fisheries in question, it is not appropriate to prescribe the details of the model here. However, the model does need to be of a form to enable predation on the stock by the marine mammals and other predators to be modeled explicitly. Since different predators, including fishers, often focus on different sizes of fish, it will normally be necessary to use a size- or age-structured model. Prediction of the future trends in abundance of the fish stock requires assumptions about recruitment to the stock in the form of a stock–recruitment relationship or the assumption of stock-independent recruitment levels. Since recruitment to most fish stocks is subject to considerable unpredictable variability, the assumed recruitment function should include an appropriate level of random variation. This will necessitate simulating multiple replicates of each scenario, whose outcomes will differ according to the different random patterns of annual recruitment that occur in each.

3.3.5. Other species

If an attempt to predict the outcome of a cull were to consider only the commercial fish species and the marine mammal thought to be one of its important predators and ignored all other species in the ecosystem, then it is almost inevitable that the analysis would conclude that a cull would lead to increases in the size of the fish stock. However, the limited experience to date in examining interactions between marine mammals and commercial fisheries suggests that restricting the scope of the analysis to only these components of the system is liable to yield misleading predictions of the effects of a cull.

It is desirable to include as many participants in the interaction as possible in models which are used to investigate the consequences of culls. However, it is recognized that increasing the number of species in the simulation model to make it more realistic also rapidly increases its complexity and decreases the ease and confidence with which the results can be interpreted. There is a trade-off to be made between these two tendencies, which makes it difficult to give a firm guideline for the minimum adequate number of components to be included in the system.

The question of whether reliable predictions can be obtained by singling out a few key species and ignoring the others is still the subject of ongoing research from which general conclusions have not yet emerged. Nevertheless, a model which contains so few components that only one kind of outcome is possible regardless of the input data, is clearly insufficient.

At a minimum, it is recommended that the models include enough components so that: (i) more than one type of behavior is possible, depending on the input data; and (ii) the species included account for most (e.g., typically 80%+) of the predation mortality (including cannibalism, where this occurs) experienced by each age class of the commercial fish species of concern (data for the earliest life stages may not be available).

Judgment also needs to be exercised as to which interactions between the species in the model need to be included. To keep the model simple, it can be assumed that if the species in the model make up only a small part of the total diet of another species in the model, then the dynamics of the latter will be independent of the species modeled. For example, if the fish species included in the model make up only a small part of the diet of the marine mammal, then the predation impact of the marine mammal on the fish needs to be included; however, the impact of prey abundance on the population dynamics of the marine mammal need not be considered explicitly.

In the early stages of model formulation, information on other species (Section 2.3.3—Other components of the ecosystem) can be incorporated selectively in order to achieve a balance between workability and completeness of the model. Models with differing levels of inclusiveness of other species in the system may be a part of the variety of different modeling scenarios considered (Section 3.2—Structure of the evaluation). However, care should be taken not to permit too great a difference between models in the quality of data utilized. In many cases, this requirement will limit the extent to which information about other species can be included.

3.3.6. The fishery and its management

The extent to which it is appropriate or feasible to model the fishery and its management will depend on the objectives of the cull, the nature of the fishery, and the arrangements for its management, if any. Three main types of situation can occur:

(1) In the case of fisheries not subject to quantitative management restrictions, the aim of the proposed cull may be to enhance the abundance of fish and thereby provide increased catches to the fishery at the existing level of fishing effort. In this case, it may be sufficient to model the relationship between the abundance of the fish stock and the level of catch for one or more given levels of effort on the assumption that the cull will have no effect on the level of fishing effort.

(2) In the case of fisheries subject to quantitative management restrictions, such as TACs, these will usually be based on some form of assessment of the fish stock. A cull of marine mammals may affect the abundance of fish, and hence the data on which the assessment is based, and thereby affect future TACs, which will, in turn, affect the future abundance of fish, and so on. Modeling the effect of the cull on the fish stock, therefore, requires modeling the process of assessment of the fish stock and the calculation of TACs. This will be possible provided that the procedure for setting the TAC is sufficiently well defined. It will also involve modeling the data on which the fish stock assessments are based (Section 3.3.7—Data collection submodels).

(3) In the case of fish stocks subject to TACs or similar management restrictions, but where the procedures for setting TACs or other management measures are not sufficiently well defined for them to be meaningfully included in a model, it will not be possible to model the direct impacts of a marine mammal cull on fish catches. In this case, all that can be done is to model the impact of a cull under a range of reasonable assumptions about catch levels over a limited period to ascertain the effect of the cull on the final level of the fish stock. The assumption is that the higher the level of the fish stock at the end of the period, the greater the potential future catch levels, even though the relationship between the two cannot be quantified. An assumption of this approach is that levels of TAC over this period are independent of whether the cull takes place. This assumption will tend only to be realistic for relatively short periods, say 5–10 years.

3.3.7. Data collection submodels

Some of the cases mentioned above involve simulations in which management actions are regularly taken on the basis of the data collected to date, such as the number of mammals to be culled or the TAC to be set for the fishery. In such cases, the type of data to be collected needs to be specified, along with their precision and the method by which they will be used to determine the management action. All such data are subject to a certain degree of random error. For each type of data, a submodel is required to specify the nature of the random error (e.g., lognormal distribution) and its extent (e.g., coefficient of variation). It is well established that the predicted performance of a fishery management strategy based on data subjected to realistic levels of error can be very different from the performance that would be predicted on the basis of perfect data. This aspect of the modeling exercise can be one of the most important.

3.4. Choice of parameter values

The parameter values for each of the submodels should be chosen to be consistent with what is known about the various components. There are two main types of uncertainty in parameter values or model structure:

i. *Uncertainty that is not readily quantifiable*—for example, a choice between alternative model structures, or the setting of a parameter value for which little or no quantitative data are available. A common example of this kind of uncertainty is the relationship between predator diet and prey availability (functional response).
ii. *Quantified uncertainty*—for example, where an estimate, with an associated variance, is available for a parameter value. Examples of this kind of uncertainty include the current abundance of the marine mammal and the fish stocks.

Uncertainty which is not readily quantifiable can be handled by choosing a reasonable intermediate model or parameter value for the base case scenario, together with one value or model on either side of this, chosen so as to span the possible or plausible range for the parameter value or model. Uncertainty which is quantifiable can be handled in a similar way by taking, say, the 95% upper and lower confidence limits of the parameter value as the end of the range of likely values.

The advantage of this approach is its transparency and conceptual simplicity; a disadvantage is that it requires constructing two additional scenarios for each uncertain parameter value, thus increasing the overall number of scenarios which will later have to be taken into account in reaching an overall verdict.

Uncertainty which is quantifiable can alternatively be handled by defining a probability distribution for the value of the uncertain parameter (for example, log-normally distributed about the point estimate of the parameter). For each random replicate of each scenario, a value for this parameter will be chosen randomly from the defined probability distribution.

Since it will normally be necessary to run multiple replicates of each scenario anyway (because of the presence of inherently random or unpredictable factors, such as fish stock recruitment), the second method of handling parameter uncertainty requires no additional scenarios nor even additional replicates of each scenario. A disadvantage is that it will not be readily transparent which source of uncertainty is contributing substantially to the overall uncertainty in the final result. Hence it will normally be necessary to employ a combination of both methods.

More sophisticated approaches for handling parameter uncertainty are available, including Bayesian conditioning, which can be used provided that the required expertise is readily available and provided that this does not result in the analyses becoming bogged down in the complexities of these methods.

Plausible ranges for some unknown parameters, such as the potential growth rates of different species can be estimated from general allometric relationships where specific data are lacking.

3.5. Performance measures and other output statistics

The simulation model will generate a large number of results for each replicate of each scenario. In order to interpret and use the results, it is necessary to focus on a few well-defined output statistics.

The predicted range of possible outcomes can often be quite broad, reflecting the level of uncertainty in the model. Hence, it is useful to focus on the differences between the results of the cull and no-cull variants of each scenario. Even if the overall range of outcomes is very broad, useful conclusions can nevertheless be drawn when, for example, the effects of a cull are all in a consistent direction. Three main types of output statistics need to be identified and defined:

i. For each objective of the cull, specific, numerical performance measures need to be defined for use in quantifying the merits or demerits of the proposed cull relative to those objectives. For example, if an objective of the cull is to increase the level of catch from a given fishery, a suitable performance statistic would be the difference (between the cull and no-cull scenarios) in the mean annual level of catch from that fishery over a period of 10 or 20 years following the initiation of the proposed cull or, in the case of a single event, the year of proposed cull. The main statistic for each scenario could be the median value of this difference over the replicates, but, in addition, the upper and lower 5th percentile of the difference should be given, to reflect the level of uncertainty.
ii. Performance measures should also be specified to reflect subsidiary or implied objectives of the cull or qualifications of the main objectives. For example, if the cull is intended to reduce, but not endanger, the marine mammal population, a relevant performance statistic to reflect the latter consideration would be the lower 5th percentile of the minimum size of the marine mammal population over the cull period.
iii. Additional output statistics are required to enhance the transparency of the results, even when they are not directly related to the objectives of the cull. For example, if an objective of the cull is to increase catches from the fishery, then, in addition to the mean level of catch from the fishery, the mean abundance of fish would be a useful supplementary performance measure. If the model predictions indicate, for example, that the cull would not appreciably improve fish catches, then it is important to see whether this result is because the model predicts that there will be little effect of the cull on the abundance of the fish stock or whether it predicts that the fishery fails to take advantage of an enhancement of the fish stock.

3.6. Running the model

The simulation analysis will consist of a range of scenarios: multiple replicates (typically 100) of each scenario are simulated to take account of the random variation inherent in many aspects of each scenario. Each replicate of each scenario should be run with and without the proposed cull, so that the differential benefits or losses of the cull can be discerned.

In an initial analysis, it is most convenient to define a base case scenario which contains what is considered to be the set of most likely or intermediate choices of parameter values and models. Each additional scenario differs from the base case in only one aspect: one parameter takes a value near an extreme of the plausible range.

After initial analyses have identified those parameters which appear to have substantial effects, scenarios involving combinations of alternative values of the parameters can be constructed to explore interactions. In many cases, it may emerge that the results are sensitive to only a few parameters and even fewer interactions. Where this is not the case, a statistically more systematic approach to the design of the simulation "experiment" may be called for.

In programming the model, the same sets of random numbers should be used for each scenario with and without the cull, so that the differences between scenarios and between the cases of a cull and no cull are not affected by the random differences between replicates.

3.7. Interpreting the results

It is important that models be structured, parameterized, and conditioned in such a way as to enhance the transparency of the results. Where possible, it is important to understand why the model produces the results it does and which factors, if any, are dominant in determining the results. If no single interaction or small set of interactions is dominating the results, then it is important to recognize this. This will help to identify priorities for further data collection.

The effect of a marine mammal cull on fishery yields is the outcome of three main factors:

i. the effect of the cull on the marine mammal population itself
ii. the effect of fishing on the fish population
iii. the "substitutability" of fish removed by marine mammals and fish removed by the fishery.

"Substitutability" here denotes the extent to which a given reduction in consumption of fish (whether of a given species or in total) by the marine mammals achieved by a cull can be reflected in increases in catches by a fishery without depleting the fish stock further or increasing the risk of stock collapse. The substitutability can also refer to the converse situation, in which increases in consumption by growing populations of marine mammals result in decreases in fishery catches. It can be useful to distinguish further between the pure ecological substitutability that could be achieved, given perfect knowledge of the fish stocks, and the degree of substitution that is achievable in practice, given the imperfections of fisheries assessment and management.

Model outputs should be presented in a way which enables these three factors to be separately discernible.

In cases where some of the parameters of the model are not input, but are functions of other parameters, the values that the model gives to these parameters should be produced as output so that their biological feasibility can be assessed (in some cases, there may exist more direct estimates or observations of the parameters against which the values can be compared).

Outputs should also be presented in ways which are meaningful to fishery scientists in terms of the ways in which they normally assess fish stocks. For example, the single-species dynamics for the fish stock of interest implied by a multispecies model (in essence the projection of that model onto one dimension) should be presented so that fisheries scientists can relate the behavior of the model to that of the models which they normally use for stock assessment.

3.8. Conclusions from the modeling exercise

The conclusions to be drawn from the modeling exercise depend on the pattern of results. If the results tend to go one way, this would constitute evidence in favor or against the cull, depending on the direction of the results.

For example, if the results indicate a substantial benefit from culling, in terms of the main performance statistics, in some scenarios, while in other scenarios, they indicate small benefits or detriments for culling, but indicate no substantial detriments from culling in any scenario, then this would constitute evidence in favor of the cull.

If the results are less clear cut, for example, if they indicate substantial benefits from culling in some scenarios and substantial detriments in others (and possibly no substantial effects in further scenarios), then two courses of action offer themselves:

(i) Closer analysis of the available data may enable the range of feasible scenarios to be narrowed in some respects.
(ii) Through inspection of the results and possibly additional sensitivity studies, it may be possible to identify one or more key parameters whose values determine whether the cull is predicted to be beneficial or detrimental. A program to collect further data for estimation of this or these parameters could then be designed.

A further possibility is that even within scenarios, the range of results from individual replicate runs includes both outcomes indicating substantial benefits of culling and outcomes indicating substantial detriments from culling. This situation is more likely to occur if the second method of handling parameter uncertainty described in Section 3.4 is used. In such cases, the conclusions to be drawn will depend on an assessment of the relative risks of culling and not culling.

3.9. Monitoring the effects of a cull

The proposed means of monitoring the effects of a cull should be evaluated along with the cull proposal. For each scenario and replicate, the model should produce as output the simulated values of the index which has been proposed to monitor the success of the cull. The simulated values should take account of the type of data on which the index is to be based, the statistical properties of these data, and the way the index is to be calculated from them.

The results of the simulation model can then be used to estimate the probability that the proposed indices will show significant effects of the cull and, if they do, the probability that the indicated effects will be in the same direction (i.e., beneficial or detrimental) as the true effects.

A finding that the planned means of monitoring the effects of the cull are unlikely to yield reliable results does not necessarily imply that the cull should not go ahead. There may be cases where, while it is possible to predict with reasonable confidence that a cull is likely to be beneficial, there are nevertheless inherent reasons why it will not be feasible to verify subsequently that the predicted benefits have indeed accrued. This is particularly liable to be the case in ecosystems subject to high levels of unexplained natural variability.

In some circumstances, it may be desirable to develop adaptive management plans, such that subsequent management actions (e.g., to continue, stop or modify the cull) are dependent on predetermined events or effects being detected by the monitoring program.

Annex I: Members of the Scientific Advisory Committee

Name	Country	1992	1994	Commented on protocol
Arnoldus Schytte Blix	Norway	x		
Monica Borobia	UNEP	x	x	
Justin Cooke	Germany	x	x	x
Enrique Alberto Crespo	Argentina	x	x	x
Michael Earle	Belgium	x	x	x
John Harwood	United Kingdom	x	x	x
Sidney Holt	Italy	x	x	x
Toshio Kasuya	Japan	x		
David Lavigne	Canada	x	x	x
Andrew Read	United States	x		
Jean-Paul Roux	Namibia		x	
Keith Sainsbury	Australia	x		x
Mark Simmonds	United Kingdom	x		x
Tony Sinclair	Canada	x		x
Peter Yodzis	Canada	x	x	x

Annex II: Glossary

Allometric relationship: mathematical relationship describing the relative increase in a part of an organism or a measure of its physiology or behavior in relation to some other measure, usually its overall size.

Density-dependent: those factors which influence a varying proportion of organisms in a population, depending on population density.

Ecological competition: use or defense of a resource by one individual that reduces the availability of that resource for other individuals, whether of the same species (intraspecific competition) or of another species (interspecific competition).

Functional response: change in the rate of exploitation of prey by an individual predator as a result of a change in prey density.

Stock–recruitment relationship: relationship between the spawning stock biomass and the rate of recruitment of their progeny to the population.

Chapter 34

ICES Involvement in Whaling and Whale Conservation, and Implications of IWC Actions[1]

The Early Years: a Potted History of "Modern" Whaling

"So remorseless a havoc..." was Herman Melville's famous description of the worldwide activities of Yankee hunters of sperm whales *Physeter macrocephalus* during the 19th century. But this was as nothing to the havoc wrought mainly by "modern" British and Norwegian whalers in the Antarctic during the 20th century. However, that living marine resources were finite, that they could possibly be exterminated in pursuit of profit, but might also be used and conserved, occurred first to those whalers—or rather to scientists from those same countries accompanying and studying them. The theory, if not the practice, of sustainable use was first applied to a few very large species of whalebone (baleen) whales—the rorquals—which feed in Antarctic waters and breed in the warmer waters of the Southern Hemisphere.

What was called "modern whaling" by Norwegian historians began with the invention, and deployment in 1864 from steam vessels, of an invention by Svend Foyn: a harpoon fired from a cannon and tipped with an explosive grenade (Tønnessen and Johnsen 1982). This combination provided, for the first time, a means of hunting the fast-swimming rorquals: the four species of baleen whales—humpback *Megaptera novaengliae*, blue *Balaenoptera musculus*, fin *B. physalus*, and sei *B. borealis*—the catches of which were later lumped as Blue Whale Units (BWU), and also the minke whale *B. acutorostrata* and *B. bonaerensis* and, later, the sei-like Bryde's whale *B. edeni*. It also greatly facilitated hunting for sperm whales which, with the right whales *Eubalaena* spp, had been the prime target of "Yankee-style" pelagic whaling during the 19th century.

Modern whaling began off the northeast coast of Norway from shore stations. As the stocks of blue and humpback whales there were depleted, whalers moved further afield to Bear Island. Shore operations also moved westward to coastal areas important for cod (*Gadus morhua*) fishing. This aroused the opposition of fishermen and led, by 1903, to the prohibition by the Norwegian government of whaling in such areas. But before that, these whaling operations had extended to Iceland (1893), Korea (1889), Sakhalin (1897), Faroes (1894), Newfoundland (1898), Shetland (1903), Svalbard (also 1903), then the Hebrides and Ireland [see Tønnessen and Johnsen (1982), the classic study which is the main source of the historical information in this section, but see also Chatterton (1931) and Ellis (1992)]. The old whaling industry of Japan switched to the Norwegian method at the turn of the 20th century.

---

[1]Editors' note: written by Sidney J. Holt, presented at the ICES Symposium "100 Years of Science under ICES" held in Helsinki 1–4 August 2000, and published in the Symposium proceedings (Anderson 2002) as Holt (2002). Permission to reproduce this article, with minor editing, has been granted by ICES.

ICES was established in 1902, following the recommendation of the 1895 International Geological Congress for "a scheme of international cooperation in marine scientific research in which scientific investigation would be accompanied by a practical exposé of the steps to be taken in order to bring exploitation of the sea fishery more in accord with the natural conditions regulating the growth and increase of fish" (Hjort et al. 1931). The following year, the expedition vessel "Antarctica," in the charge of a Norwegian whaler and sealer, C. A. Larsen, was crushed in the Antarctic ice. He and his crew were rescued by an Argentinean warship and taken to Buenos Aires. By December 1904, Larsen was in South Georgia, with a crew of Norwegians, heading an Argentine company, Cia Argentina de Pesca. Antarctic whaling had begun. By 1912, ICES was discussing the collection of statistics about catches of seals and whales in that region.

Shore-based whaling operations in the North Atlantic had virtually ceased by 1909, seven years after the formation of ICES, and so received little attention by the Council. But before the end of the 19th century, many expeditions had visited the western Antarctic looking for seals and right whales. Few right whales were found, and the seals near the sub-Antarctic islands were soon depleted (Ellis 1992). However, during these expeditions, large numbers of fin whales were noticed. By 1905/1906, after Larsen had begun at South Georgia, similar activities were in full swing at South Shetland by Chilean and Newfoundland companies. In the following years, such whaling began elsewhere in the Southern Hemisphere at Durban (1908) and (in 1909) in Angola, Kerguelen, and on the coast of Chile. These stations required, however, spacious sites with assured access (i.e., not blocked by ice) and ample and continuous supplies of freshwater, both of which were rare in Antarctica.

The next move was, therefore, to the use of floating factories which could be moved from place to place according to the availability of whales and which could carry—and later produce—freshwater. At first, these were moored in sheltered harbors, the whales were tied alongside for flensing, and the flesh and blubber hauled aboard for processing. Then, in the early 1920s, factory ships with stern ramps began to be used; the whales could be hauled aboard the ship for processing, and so the factories could operate far from shore.

Throughout this development, although other countries sought involvement, the industry was firmly in the hands of Norway and the United Kingdom (UK), and the former, in particular, long had a virtual monopoly on skills (especially the all-important gunners). This monopoly was reinforced by its national laws. Experience in the North Atlantic and elsewhere had shown that whale stocks were not inexhaustible, and by the end of the 1920s, local depletions were evident in the Antarctic region (Holt 2000a). Given the nationalities involved, and the Atlantic experience, as well as the fact that the pelagic expeditions were mainly based in Europe and were bringing products to European markets, it is understandable that ICES would be moved to take an initiative, despite the remoteness of the whaling operations.

Few people now closely associate ICES with research on whales and the regulation of whaling. Yet, the present arrangements for these activities, through the International Whaling Commission (IWC), were set in train by ICES in 1926, at its meeting that year, by the creation of a Comité International pour la Protection de la Baleine (CIPB), subsequently referred to in the *Rapports et Procès-Verbaux des Réunions* as the "Whaling Committee," with Johan Hjort (Norway) as Chair (ICES 1927a).

Statistics

The ICES Whaling Committee met in Paris in April 1927 (ICES 1927b). The lead was taken by the representatives there of the French Interminsterial Committee for the Protection of Whales who argued that "whaling as carried out was proving disastrous to the stock" and called for international investigations and cooperation. At this meeting, the Committee laid out a comprehensive plan for research and collection of statistical materials from whaling stations. One of its main recommendations was that "the Council request the Norwegian Government to organise a central institution to collect statistics from the whaling industry throughout the world, on the understanding that a report be laid before the Committee yearly." These statistics should "be collected from every country, also from such as are not members of ICES." The Committee said that efforts should be made to compile statistics for previous years. A further specification was for the annual compilation of a list of all operating whaling companies and the number of catcher boats employed.

The Norwegian government acted promptly. It established a Committee for Whaling Statistics (CWS) comprising the Director of the Norwegian Statistical Central Bureau (Gunnar Jahn), the Secretary of the Norwegian Whalers' Association (NWA) (Sigurd Risting), and Professor Hjort (Risting 1928; Hjort et al. 1930). This Committee's original idea was to try to build an international statistical system on official returns from countries, but soon agreed it would be better to develop, worldwide, the long-standing NWA system of soliciting data from the whaling industry. This led to the successful long period of annual publications by the CWS's instrument, the Bureau of International Whaling Statistics (BIWS), which continued until the IWC took over the task in the 1970s.

Data about whaling companies and ships were partially published in *Norsk Hvalfangertidende* (*Norwegian Whaling Gazette*) and partly in the BIWS reports *International Whaling Statistics*, which also contained some information about the production of baleen oil ("whale oil") and oil from sperm whales, as well as meal, blubber, and other "by-products." A severe limitation of these published data, however, was that information on operations and yields by individual expeditions and catchers was withheld from scientists and regulators as a commercial secret, even though it was available within BIWS. This was not corrected until the 1960s.

An unusual feature of the arrangement with BIWS was that, in later years, it was made responsible for administering some regulatory decisions. Thus, from 1949 through to the mid-1970s, pelagic expeditions telegraphed their catch results to the BIWS headquarters at Sandefjord where they were compiled in what we would now call "real time." From these reports, BIWS predicted when the agreed total allowable catch (TAC) would be reached and then sent back messages of closure of the Antarctic whaling season.

From the first years of Antarctic whaling, Norwegian domestic law provided for the presence of national inspectors on each factory ship, and the other big whaling country at the time—UK—did likewise, voluntarily. With the formation of the IWC in 1946, this became mandatory for all IWC Member States. In theory, this ensured that statistics were provided honestly and that regulations (such as prohibition of the catching of calves, nursing mothers, and certain protected whale species, and those on species size limits) were honored. In practice, although this was better than nothing, scientists concluded from circumstantial evidence (such as discontinuous length distributions) that many "infractions" were not reported.

Another weakness in the BIWS operations was a failure to move with the times. Norwegian-style industrial whaling was almost entirely concentrated on oil production. Japanese whaling was always primarily for production of meat for human consumption, but these statistics were entirely lacking from BIWS publications.

The ICES Whaling Committee

Throughout the pre-World War II years, the ICES Whaling Committee (WC) was very active in coordinating and evaluating research, monitoring the industry, and intervening in diplomatic and political issues. In 1928, it advised on the draft whaling bill proposed by the Norwegian government (ICES 1928). The same year, it assisted ICES in establishing relations with the League of Nations and pointed out the important differences between the whaling issues—which were global—and other fisheries issues which could and should be dealt with regionally. But ICES did add that "it was not at present disposed to advance any concrete proposals for the regulation of whaling for embodiment in an international convention of general application."

In 1929, the WC issued a very significant statement (ICES 1929):

> "While fully admitting that it is not likely that any definitive results can be derived from the scientific researches now in progress until a considerable time has elapsed, the Committee feels strongly that the enormous expansion of the whaling industry in recent years constitutes a real menace to the maintenance of the stocks of whales, and that if the expansion continues at the present rate there is a real risk of those stocks being so reduced as to cause serious detriment to the industry.
>
> While admitting that until the scientific researches have reached a definite conclusion it will be impossible to devise any measures of protection of a permanent nature, the Committee is of opinion that the Governments of the countries interested in whaling should, as a matter of urgency, give serious consideration to the question of taking immediately temporary measures for dealing with the situation."

The statement went on to detail such possible measures. There could hardly have been a clearer enunciation of what we would now call "the precautionary principle."

The 1930 and 1931 WC meetings found that the Norwegian and French laws were in harmony, but did nothing more because of the crisis in the industry, connected with the world economic situation, involving huge unsold stockpiles of oil. In 1930, they were waiting for the results of the recently held conference in Berlin under the Economic Committee of the League (ICES 1930), and in 1931 gave their endorsement to the draft Geneva Convention for the Protection of Whales and the Regulation of Whaling (ICES 1931), which followed closely the ICES proposals of 1927. By the 1933 meeting (ICES 1933), the WC members were becoming pessimistic "since the industry was concerned more with finding markets than with the whale stock." Then, and in the succeeding years, the WC followed the ratification process of the Convention and noted the failure of the hoped-for collaboration between Norway and the UK. Warnings were given each year, both to industry and to governments, that trouble lay ahead, but to no avail. By 1938, the British scientists were suggesting, quaintly, "a year's holiday" for the humpback whale. The problem that most preoccupied the WC by 1937 was the difficulty of involving Germany and Japan in regulatory and conservation measures. Pessimism ruled through to 1939, there having been little progress in intergovernmental negotiations, but scientific results at least were beginning to be significant, particularly from marking experiments. It was now apparent that the fin whale, as well as the blue whale, was in decline (ICES 1939).

The WC met next in October 1945 (ICES 1946), but did little business, in expectation of a conference to be convened by the United States government the following year. At the August 1946 ICES meeting (ICES 1947), the eminent Norwegian scientist, J. T. Ruud, announced success in determining the ages of blue and fin whales from ridges on their baleen plates. By 1946, the International Convention for the Regulation of Whaling (which created the IWC) had been signed. WC members were concerned that the current working limit of 16,000 BWU for pelagic catches of baleen whales in the Antarctic was too high, and they agreed to pursue scientific work that would be pertinent to the new IWC when it came into being. This continued until the WC's 1949 meeting (ICES 1950) when—the IWC having met earlier that year—the question arose as to whether the WC should continue to exist. Meanwhile, it was agreed to send scientific results such as Ruud's age distributions to the IWC Secretary. In 1950 (ICES 1951), there was a majority against abolition of the WC, but it was agreed to suspend the hitherto regular meetings, the then Chair, N. A. Mackintosh, being asked to convene future meetings as and when he thought necessary. An interesting point at this meeting was the recognition that the small minke whale had now acquired economic importance.

Eventually, a meeting of what was then called the ICES Whaling Sub-Committee was held in 1953 (ICES 1954). Apparently this remained inactive until 1963 when its status was again reviewed by the Council. No changes were made at that time. The WC did meet again—now with J. T. Ruud as the acting Chair—in October 1966 to consider the transition agreed by an extraordinary meeting of ICES Delegates in May of that year to transform it into a Marine Mammals Committee (MMC). From then on, the MMC met annually through 1997, after which the Committee became part of a new Living Resources Committee. An examination of the MMC's proceedings through to 1978 indicates that, although there was some reference in these meetings to research on the large whales, and especially to marking with Discovery tags, there is, remarkably, no trace of the fact that the IWC and the whaling industry had been passing through one crisis after another, that all the earlier concerns about the future of whales and whaling had been more than justified, and that there had been a renaissance of science within the IWC beginning in the 1960s. Even in October 1972, there is no reference to the fact that the United Nations (UN) Conference on the Human Environment, held in Stockholm in June of that year, had discussed the whaling issue as one of its most important topics, just as had the League of Nations 40 years previously.

Similarly, in 1974, 1975, and 1976, there is no reference to the fact that, in 1974, the IWC had embarked on a new, science-based approach to regulating whaling, seeking a balance between sustainable use and protection. While IWC and ICES were exchanging observers (usually not scientists) at their annual meetings, there was no apparent substantive flow of scientific information between them. ICES responded negatively to a request by the Food and Agriculture Organization (FAO) of the UN that it collaborate in an FAO/UNEP (UN Environment Programme) project to review the status of all marine mammals. (This project led to a landmark international conference in Bergen, Norway, in 1976 attended by 400 scientists and whose proceedings were published by FAO in four volumes entitled *Mammals in the Seas* in the following years.)

Law

Apart from arranging for statistical collection, compilation, and publication, the most important action by the ICES Whaling Committee was to bring the need for regulation of Antarctic whaling to the attention of the League of Nations. With the support of ICES, the League convened a meeting of experts in April 1930 attended by representatives from France, Germany, Japan, Norway, Portugal, the UK, and the USA. They proposed an international regulatory agreement following the principles of the Norwegian law of 1929. The draft proposed by ICES was re-examined by another committee—the Committee of Experts for the Progressive Codification of International Law (CEPCIL)—in Geneva in September 1931, the report of which was adopted the same month by delegates from 26 governments (Birnie 1985).

This "Geneva Convention" was the basis for all future international agreements, both among governments and among whaling companies. The Rapporteur of CEPCIL was an Argentine lawyer, José Leon Suarez, a man far ahead of his time insofar as the legal principles of managing industries based on marine mammals were concerned. In particular, Suarez proposed the designation of a substantial part of the Southern Ocean as a "sanctuary" for whales (Holt 2000a), an idea that was partially implemented by the IWC from 1946 to 1955, and then fully from 1994.

The Geneva Convention included a ban on the catching of right whales, calves, lactating females, and sexually immature animals. It required full utilization of carcasses, that whaling could only be undertaken on the basis of governmental licenses, that statistics be submitted to BIWS, and that accurate logbooks be maintained. Most importantly for the future of all whaling management, the Geneva Convention provided that bonuses to gunners would be based on size, species, and oil yield of the whale, and not merely on the numbers of whales. The latter, favored by the British companies, was in practice, of course, very wasteful of whales. Because of this disagreement, the UK did not ratify the Convention until the end of 1934. This was perhaps the first of many occasions when such prolonged delays impeded actions promoting the conservation of whales and the sustainability of the industry. In the longer run, the complex BWU formula, based roughly on the relative oil yields from blue, fin, humpback, and sei whales, came to dominate regulatory considerations for 40 years.

During the 1930s, ICES involvement was essentially a "watching brief." Scientific advances were being made, but scientific input to the diplomatic negotiations was negligible. These negotiations, and related activities, are described in detail by Holt (2001).

Sustainability and Catch Limits

The American demographer R. Pearl hypothesized that population growth rates would slow as a linear function of population size or, rather, of population density (numbers per unit area) (Pearl and Reed 1923; Lotka 1925; Lewontin 1969), which leads to the well known symmetrical *sigmoid* growth curve, or *logistic*. The party of *continuity*, insofar as it was interested in wild animals, was perhaps best exemplified in the works of the Italian scientist, Y. Volterra, published in the 1920s by ICES (Volterra 1928; see also Volterra 1938; Holt 2000b). The first application of Volterra's ideas in the field of fisheries was to whales, by Hjort, Jahn, and Ottestad, in 1933. These authors at first considered that the catch per catcher boat would give an index of the abundance of whales. However, because the length of the whaling season was regulated by governments and/or the industry, they later switched to the "catch per catcher's day's work" as the fundamental index. Noticing that this catch per unit effort (CPUE) for blue whales had declined rather quickly in the Antarctic, they fitted the logistic curve in the opposite direction to the usual one of population growth, and pointed out that if a whale population could be held near the inflexion of that sigmoid, or allowed to increase to it, a *maximum (optimal)* catch could be obtained in perpetuity.

Hjort and his collaborators had additional information about whales which was pertinent to their conclusions (Hjort et al. 1931). Studies of dead animals on factory ships and land stations had revealed the relationship between size and sexual maturity. They had noted declines in the average sizes of caught whales and had the results of tagging experiments that had been launched (unsuccessfully) by Hjort and (successfully) by the British Discovery Committee in 1932.

A major obstacle was lack of an ability to determine the ages of whales. In 1929, Mackintosh and Wheeler concluded that the number of years elapsed after sexual maturity of females could be estimated by counting the corpora lutea in their ovaries. To those numbers had to be added, of course, the average age at maturity. A rough idea of this had come from studies of growth over short periods of time, including from tagging experiments. Then in 1940, as already mentioned, Johan Ruud published a method based on counting baleen ridges (Ruud 1942). Subsequently, a more reliable method of age determination, from laminations in the earplugs of fin and other baleen whales, was developed by P. E. Purves and M. O. Mountford and published in 1959. These techniques came too late, however, to be of much practical use before the Antarctic whaling industry effectively collapsed as a result of overfishing.

The logistic curve approach to management was applied by M. Graham (1939) (see also Russell 1931, and Ricker 1946) to cod and other groundfishes and received much attention in ICES circles. In the post-World War II years, it was replaced in most European fisheries applications, and hence in ICES publications, by age-structured models which had been pioneered by the Russian mathematician T. I. Baranov as early as 1918, but largely forgotten, and developed some decades later, especially by R. J. H. Beverton and the present author (Beverton and Holt 1957). In the United States, however, the logistic curve methodology was applied especially to the Pacific tuna fisheries (there being no means then of determining the ages of tunas), and its development is associated especially with M. B. Schaefer (1954).

For essentially the same reason, the same approach, in only slightly modified form, persisted until the late 1970s in the IWC's efforts to regulate whaling on a scientific basis. The modification was little more than the use of a non-linear (power or polynomial) function, as proposed by J. J. Pella and P. K. Tomlinson (1969), who also studied tuna. This produced an asymmetrical sigmoid population growth curve which was applied by the IWC's Scientific Committee with improved methods of using historical data to estimate some of the parameters of the model (de la Mare 1987). Thus, the approach pioneered by ICES in the 1930s persisted for nearly half a century. The application of age-structured models to whales was also impeded by the interest of the IWC only in "optimizing" the numbers of whales caught rather than the total weight of the catches (and hence their value). This was, in some sense, a degrading of the approach during the 1930s when the management focus was primarily on the quantity of the main product.

Main Developments in the IWC

Details of the following account may be found in the annual Reports of the IWC (London and Cambridge, UK) and summarized and analyzed from different perspectives in Birnie (1985) and Holt (2000a).

When it began work in 1949, the IWC took over the overall catch limit of 16 000 BWU for Antarctic pelagic whaling south of 40°S. This limit—about half of the high catch of 1937/1938—had no scientific foundation, but was agreed on the basis of three considerations:

1) the assumption/hope that it would be sustainable;
2) a presumption that the whales had increased substantially in number as a result of the reduction in whaling during the war years;

3) the expectation that exploitation pressure would thereby be removed from the blue whale.

The decision was wrong on all counts.

As catch rates declined through the 1950s, some countries (particularly the UK and Norway, and also the USA) made strenuous efforts to obtain agreement on minor reductions in this catch limit and to give more protection to the most depleted species (humpback and blue) by adjusting seasonal opening and closing dates by species and, eventually, by designating as "protected" those species the very existence of which was perceived to be endangered. However, by the 1957/1958 season, the only agreement was to bring the limit down to 14,500 BWU, and it was put up again to 15,000 for 1958/1959. For the next three seasons, no limit was adopted by the IWC, and the five pelagic whaling countries—the UK, Norway, The Netherlands, Japan, and the USSR—agreed among themselves to keep catches no higher than 17,500 BWU.

One problem was that, in their efforts to maintain or increase their shares of an overall total, countries steadily increased the sizes of factory ships, the tonnages and horsepower of catcher boats, and the numbers of catchers attached to each expedition. The consequence was called "the whaling Olympics." There were other problems connected with the efforts by other countries—especially non-Members of IWC—to get into this profitable business. In 1958, the UK government suggested that the five countries should get together to negotiate equitable shares of the overall catch limit. Such negotiations continued right through the 1960s and were inextricably linked with negotiations to establish an international observer scheme so that the five could keep their eyes on each other (Holt 2001).

By 1960, the situation was so bad that the IWC decided to invite three "independent" fisheries scientists not specialized in whale assessment and from countries not engaged in Antarctic whaling to provide advice on the current sustainable yields from the stocks being exploited in the Antarctic. They were D. G. Chapman (USA, spokesman for the so-called "three wise men"), K. R. Allen (Canada/New Zealand), and the present author (British, but working for FAO and, therefore, acceptably neutral). There was a vague commitment by countries eventually to act on the recommendations of this Committee of Three, and unprecedented arrangements were made for their access to data. Their analysis provided the first instance of electronic computers being used in the sphere of whaling management.

By that time, the Antarctic catches consisted primarily of fin whales. During the Committee of Three's work, the catching of the smaller sei whale was greatly increased, but no data were available for making meaningful assessments. The blue whale was not yet protected, but its catches were small and rapidly declining. Whaling for the smallest rorqual, the minke, had not yet begun; since 1972/1973, it has dominated the residual Antarctic whaling.

The appointment of the Committee of Three effectively brought the IWC back into the arena of scientific fisheries management from which numerically competent scientists had been effectively disengaged since ICES stepped back. The Committee was required to report back by June 1964, but was so worried by its interim findings that it reported early in an effort to get the global catch limit drastically reduced for the 1963/1964 whaling season. It recommended that catching be immediately reduced from 15,000 to about 4,500 BWU and that limits be set by species, applied also to Antarctic land station catches, and eventually to the entire Southern Hemisphere. Further, it was considered that eventually for catches to be sustainable and for stocks to be allowed to recover towards optimal levels, catches could be no more that 4,000 fin whales (2,000 BWU) and 5,000 sei (<1,000 BWU), with the blue whale being completely protected.

Through the next 5–6 years, the IWC haggled over these numbers, but the limit was brought down to 2,300 by 1971/1972. The work of the Committee of Three was not based strongly on any particular population model, but rather an empirical approach was taken, resting heavily on CPUE data, but using as far as practicable the gathering results of tagging programs and age determinations. For the first time, corrections were made to catch-per-catcher-day data (following methods used by ICES in its studies of bottom trawling) in an attempt to calibrate for increases in efficiency with sizes and powers of catchers, to look separately at the performances of the various national fleets, and to allow for disparities in the balance of factory handling capacity and numbers of catchers per factory.

When the Committee of Three had finished its work, by 1965, continuation of increasingly routine assessments was made the responsibility, for a few years, of a small group of scientists within FAO, who happened to have been closely associated with ICES: Luit Boerema and John Gulland.

Although progress was being made in the 1960s in terms of reducing the pressure of whaling, the reluctance of the IWC to act responsibly in accordance with its mandate and to effect adequate controls, while some species, especially the blue whale, seemed to be driven towards extinction, led to a demand at the 1972 UN Conference on the Human Environment held in Stockholm that the IWC enact a 10-year moratorium on all commercial whaling. The USSR was not represented at Stockholm, but the other four Antarctic whaling countries did not oppose the idea. This was evidently a tactical decision because when the debate moved to the IWC in that and following years, Norway, Japan, and the USSR strongly opposed the proposal. Since a binding decision by IWC requires a three-fourths majority vote (not counting abstentions) and any such decision could, in any case, be the subject of an "objection" by any country, there was clearly no prospect of enacting a moratorium. Following the Stockholm meeting, the IWC did, however, agree to set catch limits by species for each and every whale "stock." It also appointed a full-time Secretariat for the first time (taking it a step towards the demonstrably successful ICES structure) and launched a greatly expanded program of research.

In these circumstances, Australia—then still a shore-based whaling country—proposed in 1974 what it saw as a compromise between the proponents of a moratorium (notably the USA, the UK, and France) and the "business-as-before" countries. In this proposal, which was adopted in 1975, implemented for the 1975/1976 and 1976 seasons, and called the New Management Procedure (NMP), stocks deemed by the Scientific Committee to have been reduced to less than 54% of their "virgin" number were to be protected, while the rest would be subject to catch limits at 90% of their estimated maximum sustainable yields (MSY). This led quickly to the closure of whaling in the Southern Hemisphere for the four "BWU" species, but it was not useful—because of lack of data—in regulating the new industry for minke whales or the continuing hunt for sperm whales. Furthermore, operations recently set up in the Northern Hemisphere in Canada, Spain, Iceland, and Korea to provide meat to the Japanese market were difficult to assess. Naturally, the burden of proof of overexploitation remained with those seeking more conservative regulation. The small "safety factors," such as the provision for taking "only" 90% of the estimated MSY, were wholly inadequate.

The assessments made for applying the NMP were based on a modified logistic, employing the Pella–Tomlinson model, with parameters adjusted so that the MSY would be available when the stock was at 60% of its carrying capacity. There was a limited age structure in that the model was applied to the "recruited" (essentially the mature) stock and all density dependence (of natural mortality and reproduction combined) was assigned to the pre-recruit phase. The application relied heavily on very questionable estimates of natural mortality rates, on long historical series of catch data, and on one or more estimates of the absolute number of whales (from, for example, tagging experiments and. later, direct visual surveys). What constituted a "stock" was arbitrary—as it still largely is—since very little was known about the identity and degree of mixing/separation of biologically distinct populations. A computer simulation submitted to the IWC in 1985 (de la Mare 1986) finally brought an end to attempts to implement the NMP. De la Mare demonstrated that the NMP must lead to the continued depletion of stocks even if parameters of the model used were well estimated and if the model accurately reflected the real dynamics of a whale population. But in 1982, the Commission had at last adopted a proposal for an indefinite moratorium on all commercial whaling, which would come into effect in 1986.

This decision was essentially for a sufficiently prolonged pause in commercial whaling. However, some of those governments supporting the idea, and many nongovernmental organizations, came to regard the moratorium as a step towards a permanent closure of commercial operations (there were already moratoria in place on all catching of sperm whales and on pelagic whaling except for minke whales), some on ethical grounds and others in consideration of pessimism—in the harsh light of history—that high seas whaling, especially pelagic whaling, could ever in practice be rigorously regulated.

De la Mare's simulations and subsequent studies by J. G. Cooke (Cooke 1995) pointed to a radically different way of providing scientific advice for management. The same approach that had demonstrated the inevitable inadequacy of the NMP could be used to test the consequences of applying any other algorithms for the calculation of catch limits. In 1992, the IWC adopted the algorithm devised by Cooke under the name of Revised Management Procedure (RMP). This has, however, not yet been implemented, pending completion of negotiations for effective international inspection and other elements of a proposed Revised Management Scheme (RMS).

Here I briefly review the essential features of the RMP, in case this approach might be useful in situations of direct concern to ICES:

1) The IWC Scientific Committee presided over a competition among five groups of scientists to create the most effective algorithm, having established precise performance criteria.
2) The Committee induced the Commission itself to clarify the specific objectives of regulation and to establish its priorities among these. The top priority objective was to avoid the unintentional depletion of a stock and to provide for its potential recovery to high, but not predetermined, levels. It would thus embody the "precautionary principle" in a formally quantified manner. The second priority was to assure high cumulative catches in the long term (simulations were undertaken to the then maximum feasibly computable period of 100 years). The third objective was to ensure that there would be few unnecessarily wild changes in annual catch limits which could destabilize the industry.

3) The catch-limit algorithm had to be robust to the consequences of poor estimation of variables and to failure of any population model to reflect the real-world dynamics in order to allow adequately for known and unexpected uncertainties in these, and to abrupt or any long-term environmental (carrying capacity) changes which could include the consequences of interspecific interactions.
4) The procedure would not depend critically on knowledge of population trajectories, such as from abundance indices like CPUE. In fact, the adopted procedure depends only on reasonably adequate knowledge of catch histories and on single estimates of whale abundance from periodic, scientifically planned, conducted, and analyzed surveys together with robust estimates of sampling errors in those surveys.
5) The procedure must be tested to have been robust to systematic errors in reporting of catch statistics.
6) Most importantly, the procedure must be applicable to groups of notionally separate but insufficiently specified and identified "stocks." This is potentially the greatest weakness in the revised system.

The Scientific Committee of the IWC is fundamentally composed of scientists nominated by Member Governments, although representatives of FAO, IUCN (International Union for the Conservation of Nature), and, more recently, UNEP, have full but non-voting status in it. The IWC has been plagued throughout its existence with suspicion that members of the Committee are frequently less than objective in the provision of scientific advice on management questions. There is much circumstantial and some direct evidence for this in terms of the known support by some individuals for policies of their governments. The IWC has, however, for many years, made provision for the appointment, on a year-to-year basis, of "Invited Participants" taking part in their individual capacities, though barred from comment on the formulation of specific management advice. This arrangement has, I think, ensured a higher quality of scientific scrutiny than would otherwise have pertained. Despite this, it has often proven impossible for the IWC's scientific advisers to reach consensus on advice which has, from time to time, led to external proposals that reference should be made to supposedly "independent" advisers. This did work once, in 1960–1964; the task given was a strictly defined and limited one, and the membership of the IWC's own Committee at the time was manifestly not technically competent to conduct the kinds of analyses that the Committee of Three undertook. Subsequent informal efforts to make alternative analyses other than those conducted by the Scientific Committee have rarely been fruitful; the reality is that the scientific methodology has become so technical and specialized that it is increasingly difficult for "outsiders" to contribute convincingly.

Since the IWC completed this part of developing a Revised Management System, the approach devised by its Scientific Committee has begun to be looked at by groups concerned with the management of fisheries for tunas, cod, and some other species. Those who are accustomed to the procedures followed in fisheries research and management for the past 50 or so years find some difficulty in understanding a "theory of fishing" which does not rely on the calculation of parameter values in a population model used for assessment. I believe that a new paradigm has been created by the IWC in the special circumstances provided by a moratorium, and the lessons from that exercise may influence the future work of ICES and other fisheries scientific and management organizations.

Summary and Conclusions

Beginning in 1927, ICES engaged actively and successfully in matters of whaling regulation—especially in the new Antarctic industry—and of research on whales. It initiated the first international treaty, through the League of Nations, and followed closely, through the 1930s and in the immediate post-war years, the succession of follow-up agreements and the development of the relevant science.

As early as 1929, the Council recognized that current science was inadequate for providing reliable advice about managing whaling, and that it would be many years before it might be in a much better position. At that time, the ICES Whaling Committee enunciated what we would now call a "precautionary principle."

During the 1930s, ICES led the way in providing a theoretical framework for considering sustainability. The original studies by Volterra were applied by Hjort and his collaborators to the blue whale. In those years, and in the post-war period, ICES was advanced in developing and trying to apply a "theory of fishing." Scientists—particularly the Norwegians and the British—working through ICES made important original scientific contributions. The Whaling Committee closely followed the whaling story through to the first meeting of the IWC in 1949.

Thereafter, however, the contacts between the two intergovernmental organizations became largely symbolic. ICES observers followed the work of the IWC from its second meeting in 1950 until 1996 (except 1984, 1989, and 1994), while the IWC frequently nominated observers to ICES meetings. ICES observers at IWC meetings have nearly always been from Nordic countries, particularly Norway, but are not invited to attend IWC Scientific Committee meetings, unlike representatives from FAO, UNEP, and IUCN.

The lack of substantive contact between ICES and the IWC was unfortunate because the IWC, which had been operating in its first decade virtually without relevant scientific advice, entered a crisis in 1960, to the resolution of which ICES scientific input might have contributed. The new surge of advice had instead to come from fisheries scientists from the USA, Canada, and FAO. Thereafter, ICES remained disengaged from the renewed attempts at scientific management begun in 1974 after the UN had pressed the IWC to enact a 10-year moratorium on all commercial whaling. When a moratorium was, in fact, declared in 1982, a new wave of scientific studies related to management, within the IWC's Scientific Committee, led to the construction of a fundamentally new advisory paradigm (de la Mare 1986; Cooke 1995). Three decades ago, an opportunity was lost for the ICES scientific expertise to continue to assist in the conservation of whales and the regulation of whaling. Now, ICES might, in its fisheries role, itself benefit from consideration of the new approaches pioneered by the IWC. The IWC probably suffered scientifically over the years from being rather isolated from other major international fisheries bodies, especially ICES. Now, following many years of intensive original studies by the IWC Scientific Committee, the reverse might be true.

Acknowledgements

My thanks are owed to Olga Nève for her help in assembling reference material, to numerous participants in the Helsinki Symposium, and to Emory Anderson and the referees who offered comments on the first and second drafts of this paper.

References

Anderson, E. D. Editor. 2002. 100 years of science under ICES. A symposium held in Helsinki, 1–4 August 2000. ICES Marine Science Symposia 215.

Baranov, E. L. 1918. K voprosy o biologicheskikh osnovaniyahk rybnogo khozyaistva (On the question of the biological basis of fisheries). Nauchnyi Issledovatelskii Ikhtiologicheskii Institut Izvestiya (Report of the Division of Fish Management and Scientific Study of the Fishing Industry) 1:81–128. (In Russian).

Beverton, R. J. H., and S. J. Holt. 1957. On the dynamics of exploited fish populations. Fishery Investigations Series II, 19. HMSO, London.

Birnie, R W. 1985. International regulation of whaling: from conservation of whaling to conservation of whales and regulation of whale watching. Two volumes. Oceana Publications, London.

Chatterton, E. K. 1931. Whalers and whaling. W. F. Payson, New York.

Cooke, J. G. 1995. The International Whaling Commission's Revised Management Procedure as an example of a new approach to fishery management. Pages 647–657 *in* A. S. Blix, L. Walløe, and Ø. Ulltang, editors. Developments in marine biology 4. Whales, seals, fish and man. Proceedings of the International Symposium on the Biology of Marine Mammals in the North East Atlantic, Tromsø, Norway, 29 November–1 December, 1994. Elsevier Science, Amsterdam.

de la Mare, W. K. 1986. Simulation studies on management procedures. Reports of the International Whaling Commission 36:429–450.

de la Mare, W. K. 1987. Fitting population models to combinations of CPUE series and absolute abundance estimates. Reports of the International Whaling Commission 37:379–381.

Ellis. R. 1992. Men and whales. Robert Hale, London.

Graham, M. 1939. The sigmoid curve and the overfishing problem. Rapports et Procès-Verbaux des Réunions du Conseil Permanent International pour l'Exploration de la Mer 110:17–20.

Hjort, J., G. Jahn, and P. Ottestad. 1933. The optimum catch. Hvalrådets Skrifter 7:92–127.

Hjort, J., G. Jahn, and S. Risting. 1930. International Whaling Statistics 1:1–4. Committee for Whaling Statistics, Bureau of International Whaling Statistics, Sandefjord, Norway.

Hjort, J., G. Jahn, and S. Risting. 1931. The development of modern whaling. International Whaling Statistics 2:4–17. Committee for Whaling Statistics, Bureau of International Whaling Statistics, Sandefjord, Norway.

Holt, S. J. 2000a. Whales and whaling. Pages 73–88 *in* C. R. C. Sheppard, editor. Seas at the millennium: an environmental evaluation. Volume III. Global Issues and Processes. Elsevier Science, Amsterdam.

Holt, S. J. 2000b. Ethical aspects of the sustainable use of marine living resources, especially of whales. Contribution to the Fifth World Congress on Bioethics.

Holt, S. J. 2001. Sharing the catches of whales in the southern hemisphere. Pages 322–373 *in* R. Shotton, editor. Case studies on the allocation of transferable quota rights in fisheries. FAO Fisheries Technical Paper 411.

Holt, S. J. 2002. ICES involvement in whaling and whale conservation, and implications of IWC actions. ICES Marine Science Symposia 215:464–473.

ICES. 1927a. Report of the 19th meeting of the Council. Rapports et Procés-Verbaux des Réunions du Conseil Permanent International pour l 'Exploration de la Mer 41:9–209.

ICES. 1927b. Whaling Committee. Meeting in Paris, 7th–9th April 1927. Rapports et Procès-Verbaux des Réunions du Conseil Permanent International pour l 'Exploration de la Mer 45:84–98.

ICES. 1928. Whaling Committee. Rapports et Procès-Verbaux des Réunions du Conseil Permanent International pour l 'Exploration de la Mer 49:84–101.

ICES. 1929. Whaling Committee. Rapports et Procès-Verbaux des Réunions du Conseil Permanent International pour l'Exploration de la Mer 60:62–64.

ICES. 1930. Whaling Committee. Rapports et Procès-Verbaux des Réunions du Conseil Permanent International pour l'Exploration de la Mer 66:78–79.
ICES. 1931. Whaling Committee. Rapports et Procès-Verbaux des Réunions du Conseil Permanent International pour l'Exploration de la Mer 74:78–93.
ICES. 1933. Whaling Committee. Rapports et Procès-Verbaux des Réunions du Conseil Permanent International pour l'Exploration de la Mer 85(1):41–42.
ICES. 1939. Whaling Committee. Rapports et Procès-Verbaux des Réunions du Conseil Permanent International pour l'Exploration de la Mer 109(1):46–47.
ICES. 1946. Whaling Committee. Rapports et Procès-Verbaux des Réunions du Conseil Permanent International pour l'Exploration de la Mer 116(1):38.
ICES. 1947. Whaling Committee. Rapports et Procès-Verbaux des Réunions du Conseil Permanent International pour l'Exploration de la Mer 117(1):43.
ICES. 1950. Whaling Committee. Rapports et Procès-Verbaux des Réunions du Conseil Permanent International pour l'Exploration de la Mer 127(1):48–49.
ICES. 1951. Whaling Committee. Rapports et Procès-Verbaux des Réunions du Conseil Permanent International pour l'Exploration de la Mer 129(1):44–45.
ICES. 1954. Whaling Sub-Committee. Rapports et Procès-Verbaux des Réunions du Conseil Permanent International pour l'Exploration de la Mer 135(1):33.
Lewontin, R. C. 1969. On population growth in a randomly varying environment. Proceedings of the National Academy of Sciences, USA 62:1056–1060.
Lotka, A. J. 1925. Elements of physical biology. Williams and Wilkins, Baltimore, Maryland.
Mackintosh, N. A., and J. F. G. Wheeler. 1929. Southern blue and fin whales. Discovery Report 1:257–540.
Pearl, R., and L. J. Reed. 1923. On the mathematical theory of population growth. Metron 3(1):6–19.
Pella, J. J., and P. K. Tomlinson. 1969. A generalized stock production model. Bulletin of the Inter-American Tropical Tuna Commission 13:419–496.
Purves, P. E,, and M. O. Mountford. 1959. Ear plug laminations in relation to the age composition of a population of fin whales. Bulletin of the Natural History Museum. Zoology 5:125–154.
Ricker, W. E. 1946. Production and utilization of fish populations. Ecological Monographs 16:373–391.
Risting, S. 1928. Whales and whale fetuses. Statistics of catch and measurements collected from the Norwegian Whalers' Association 1922–25. Rapports et Procès-Verbaux des Réunions du Conseil Permanent International pour l'Exploration de la Mer 50.
Russell, E. S. 1931. Some theoretical considerations on the "overfishing" problem. Journal du Conseil Permanent International pour l'Exploration de la Mer 6:3–20.
Ruud, J. T. 1942. A review of the investigations on whales and whaling in recent years. International Whaling Statistics 16:72–83. Committee for Whaling Statistics, Bureau of International Whaling Statistics, Sandefjord, Norway.
Schaefer, M. B. 1954. Some aspects of the dynamics of populations important to the management of the commercial marine fisheries. Bulletin of the Inter-American Tropical Tuna Commission 1(2):26–56.
Tønnessen, J. N., and A. O. Johnsen. 1982. The history of modem whaling. C. Hurst and Co., London.
Volterra, V. 1928. Variations and fluctuations of the numbers of individuals in animal species living together. Journal du Conseil Permanent International pour l'Exploration de la Mer 3(1):3–51.
Volterra, V 1938. Population growth, equilibria and extinction under specified breeding conditions. A development and extension of the logistic curve. Human Biology 10(1):1–11.

Chapter 35

Opening Statement by Dr. Sidney Holt on Behalf of NGOs Attending the Intersessional Meeting of the IWC in Rome in March 2009 and as the Observer for the Antarctic and Southern Ocean Coalition (ASOC)[1]

My name is Sidney Holt. I have been asked by a number of NGOs to speak about conservation.

When I first addressed this Commission, in 1961, I was one of three scientists called in to help resolve the IWC's first crisis. But accomplishing our given task of bringing Antarctic baleen whale catches down to sustainable levels, which was supposed to have been done within three years, eventually took sixteen. That delay led to the near extermination of blue *Balaenoptera musculus*, fin *B. physalus*, and humpback *Megaptera novaengliae* whales.

The second crisis erupted in 1973. Conflict had been triggered by the call for a ten-year pause in commercial whaling—a moratorium. All whaling countries except Japan had supported that in the UN General Assembly the previous year. From it emerged a stunted "chimera" called the Amended Moratorium or New Management Procedure (Paragraph 10a, b, and c in the Schedule).

The Commission survived both crises, but it was sick. The whaling industry suffered a double blow, first the Antarctic, then globally.

Surgery was prescribed for the IWC. In the following years, I participated in three attempts to negotiate improvements in the ICRW (International Convention for the Regulation of Whaling) or to replace it through conferences in Copenhagen, Reykjavik, and finally Lisbon. All failed dismally. Problems of legal competence were not resolved. Not one of the critical issues, such as the absence of instruments to ensure compliance with binding decisions, was resolved.

We did learn, then, that short-term "provisional solutions" could lead to nasty long-term consequences.

What crisis management really needed was for governments to have the will to change and to act in good faith. But promises to act definitively within a certain, specified time-frame were repeatedly broken.

---

[1]Editors' note: written by Sidney Holt and presented on March 10, 2009 on behalf of, and as an observer to the meeting in question, for the Antarctic and Southern Ocean Coalition (ASOC) and reproduced here, with minor edits, with the permission of the ASOC.

In those days, Commissioners for whaling nations sometimes said: "We've just invested in new ships and equipment, and we're even spending lots of money on research. Conserving renewable resources, and sustainable use of them, may be nice ideas, but we have bank loans to pay off and men to keep employed." It was so often forgotten that whaling is primarily a business, and follows business rules.

The IWC is now in its third crisis. If it is dysfunctional, as some say—though I don't agree—that will not be resolved by disgruntled governments walking away from it. And the last thing needed is a so-called "safety net" with an uncomfortably large mesh size.

The real need is to get the conservation of whales and the management of any and all whaling back under IWC control, in accord with the enhanced authority it has gained from the Convention on the Law of the Sea and the related United Nations document, *Agenda 21*, to which practically all nations have adhered.

Those hard and soft laws apply, like the ICRW, to the ocean as a whole. They point to the interests of future human generations. They define whales as highly migratory species to which special rules are applied, throughout their ranges, by appropriate intergovernmental organizations (Article 64). They apply other special rules uniquely to marine mammals, including confirming the rights of competent intergovernmental organizations and states to prohibit their exploitation, where that's desired for conservation or any other good reasons (Articles 65 and 120). And, by the way, UNCLOS says that fisheries must be managed in such a way as to leave enough food for dependent species (such as cetaceans)—not the other way around [Articles 61.4 and 119.1(b)].

Some common terms are not defined in the ICRW. One—"coastal whaling"—is dangerously ambiguous. Colloquially, it means "near the shore," but some governments seem to think it could mean "within 200-mile wide Exclusive Economic Zones, or even further." That's practically what Aristotle called a *reductio ad absurdum*, making whale habitat from the Barents Sea to the coast of Labrador a zone of "coastal whaling." Further confusion comes from something called "Small-Type Coastal Whaling" (S-TCW), which is just an English translation of a Japanese administrative category by which catchers of less than 48 tons displacement are allowed to hunt small whales on one-day trips: that's about 50 miles from base.

The restoration of functionality requires the withdrawal of all "objections," with a commitment by states to a binding timetable for that. International trade in commodities from whales is another—and uncontrollable—threat to the conservation of whales and the management of any whaling, so the same applies to "reservations" to CITES Appendix I listings. And, of course, the cessation of using Article VIII of the ICRW as a substitute for commercial whaling under Article V.

As for science, we welcome the initiative of the Government of Australia in launching the Southern Ocean Research Partnership (SORP), which will hold its first workshop this month.

I see that as a late start in producing a coherent management plan for the Southern Ocean Whale Sanctuary.

Supporting Organizations

Antarctic and Southern Ocean Coalition
Animal Welfare Institute
Centro de Conservacion Cetacea
Cetacean Society International

The Cousteau Society
Dolphin and Whale Action Network
Eastern Caribbean Coalition for Environmental Awareness
Environmental Investigation Agency
Humane Society International
Institute de Conservacion de Ballenas—Argentina
International Fund for Animal Welfare
LegaSeas International
Norwegian Society for the Protection of Animals
Pew Environment Group
Society for the Conservation of Marine Mammals, Denmark
Whale and Dolphin Conservation Society
World Society for the Protection of Animals
WWF

Chapter 36

Science is Rooted in Conversations[1]

First, let me say what a pleasure it is to be meeting you all, and in the place where I spent so much time as a youngish fisheries scientist. I'm looking forward to the rest of this meeting, and thank Jean Collins and Joan Parker for inviting me.

My revised title,[2] and my theme, come from the memoirs of Werner Heisenberg, quoted by Neil Belton (2005) in his biography of Schrodinger, *A Game with Sharpened Knives*. I shall talk about my needs for information in various stages of my scientific life in the hope that it might be of some use to you. Of course, it represents a sample of just one, and as a sort of amateur statistician, I would be wary of drawing any conclusions from that. You have been warned!

When I arrived on the scene of my first job, in 1947, at the Fisheries Laboratory in Lowestoft, Suffolk, England, I was greeted by the then Director of the lab, Mr. Michael Graham, with an oral outline of my main task and a scruffy blue envelope. He explained the nature of the problem: to come up with a theory of fishing that would be applicable at least to North Sea trawling, particularly for plaice *Pleuronectes platessa* and haddock *Melanogrammus aeglefinus*. The writing on the envelope, he said, came from an RAF colleague, a mathematician/physicist, Mr. H. R. Hulme, who had written it while sitting with him on a tank turret in Normandy, plotting the statistical distributions of shells and bombs. It bore a short differential equation and a suggested integration of that. I said "But I know hardly any maths!" "Then go learn some," was the answer. So off I went to find "teach yourself" books and, later, to a part-time course in statistics at University College, London, being offered to civil servants at His Majesty's expense. At the same time, I was analyzing, with a colleague, Dick Margetts, the historical data that would demonstrate the expected recovery of the plaice stocks as a result of the "rest" given them by the restriction of North Sea fishing during the Second World War (Margetts and Holt 1947).

Note the two "Mr."s plus my own making three. At that time, few of us thought PhDs were worth having. Most were anxious to begin useful work. All of us then bore the title "Naturalist." And most of us were influenced by the recently ended war; scientists should be useful, and many of us admired the contributions to the outcome of the conflict made by those scientists—mostly physicists, biologists, and mathematicians—who had been engaged in "operational research" (Operations research, to the Americans).

---

[1]Editors' note: presented by Sidney Holt on October 8, 2005 at the 31st IAMSLIC (International Association of Aquatic and Marine Science Libraries and Information Centers) Conference: Information for Responsible Fisheries: Libraries as Mediators, held in Rome, Italy October 10–14, 2005, and published in the conference proceedings (Anderson and Thiery 2006) as Holt (2006a). Permission to use this article was given by Kristen L. Anderson, Director, Health Sciences Library, University of Hawaii, Honolulu, HI, who was co-editor of the conference proceedings.

[2]My original, rather obscure, title, "Scaling Fisheries," came from a book about the history of fisheries science by Tim D. Smith.

What emerged from that second meeting, three years later (the first had been a pre-appointment interview, that had also involved the previous Director, E. S. Russell,[3] in a tiny, dark and musty cubby-hole in the Ministry of Agriculture, Fisheries and Food, in Horseferry Road, Westminster), was the text of a book (Beverton and Holt 1957), written with my late colleague, Ray Beverton. It took His/Her Majesty's Stationery Office seven more years to publish it, but that's another story. But the relevance of that bit of history to this conference is that on my second or third day at Lowestoft, I told my boss I would be spending some time in the library reading the literature on this subject, to which he replied: "Do the work first, then read 'the literature' to see whether anyone else got there first. Reading will only confuse and distract you. By the way, here are a couple of papers by some Americans who seem to have tried hard but I think got it a bit wrong. And here's a book I wrote during the war."—it was *The Fish Gate* (Graham 1943).

The advice not to read the literature didn't worry me at the time. My education at Reading—a "Red Brick" university, as we used to call such newcomers to academia—was weighted more towards practice than literature reading, and my personal interest was really in solving puzzles without bothering whether someone else had solved them before me. Lowestoft had—and I think still has—a good general ecology and fisheries library, and I certainly read voraciously when I was there. But the papers and books that came to my attention and helped me in my task came mainly from colleagues, through contacts abroad via Graham, and by reading general periodicals such as *Science* and *Nature*.

Things changed when I came to FAO, here in Rome, in December 1953. My taskmaster here was then Dr. Geoffrey Kesteven, an Australian, Chief of the Biology Branch of the Fisheries Division. His Branch then comprised four people, including him and myself. The idea that we were doing "biology" was rather strange to me. In Lowestoft, we were Naturalists; we all—including the Director—went to sea regularly and frequently on both research ships and commercial vessels, thought of ourselves as partway to being fishermen (women were not allowed on the ships), and encouraged to think of ourselves as doing operations research, in the post-war fashion.

Anyway, my first task in FAO was to analyze a report by a Chinese consultant, Shao Wen Ling, a fine painter of prawns, about the sweet-water fisheries of the Mekong River system and the extraordinarily productive variable lake, Tonle Sap, and what French engineers under UN auspices were planning to do to them. So, back to the drawing board. My thesis at Reading had been on the anatomy of a European cyprinid, the tench (*Tinca tinca*), but that didn't seem to be very relevant to the multispecies fresh- and brackish-water fisheries of Southeast Asia. So, now I really did have to seek out such literature as was available in the embryonic FAO library, and rather more in the Vatican and in the private library of an Italian (Roman) retired oceanographer and well-known gourmet, Professor Giorgio Bini, whom Kesteven had befriended. Among other things, Bini was the author of a six-volume treatise on the fishes of the Mediterranean. With my newfound expertise, I concluded that the great plans for hydroelectric dams along the Mekong were not a good thing, at least not for the hundreds of thousands of families living from the biological productivity of this great river system.

Bini had spent his latest years—apart from collecting ancient armor, and cooking and eating—in compiling an annual *Bibliografica Oceanografica*. Kesteven's idea was that we would help him internationalize it, broaden its scope, and publish it in English as well as in

[3]Russell was the author of an influential book, *The Overfishing Problem*, published during the war, in 1942. He had published his original contribution 11 years previously, in the context of ICES discussion of this problem.

Italian. Bini was also much in arrears, and we thought the value of the *B.O.* would be enormously enhanced if we could get it up to date and then keep it current. One week, I think in 1954 or 1955, I went by train to Paris with Kesteven, Bini, and two of his Italian colleagues. Our idea was to visit UNESCO and also talk to some French marine scientists. What I mainly remember of the trip was learning, on the train, how good was Valpolicella—a red wine of the Lake Garda region—a subject about which I was then entirely ignorant; and, from the Italians, how impossible it was to find anything decent to eat in Paris!

The environment in FAO at that time was conducive for a junior staff member to continue to exercise his professional skills, and I did a great deal of "reading the literature" because a few of us were having to deal with a vast array of issues. We began our excursion into the realm of information rather than the solving of scientific and technical problems by becoming "high tech." We typed bibliographic references on cards that I designed, with punched holes and/or slots around their edges, sorted with knitting needles, we made multiple carbon copies of 3 x 5 note cards, and we invested in Flexiwriters, like the ticker-tape machines one sees in old American movies, and so noisy the operator had to work enclosed in a glass cabin. Titles were insufficiently informative, published abstracts too long to copy and also often not very informative, so we decided on annotations, continuing the system begun by Giorgio Bini. Kesteven thought that all the staff of the Branch should regularly contribute annotations, and to that end, incoming publications were circulated throughout the Branch. This was onerous, but did ensure that we all kept in touch with our disciplines and did not become complete bureaucrats or even aid-giving technocrats. This approach was familiar to me: like Graham's insistence that everyone spend time at sea.

Another technique introduced at the time involved punching small holes all over special cards, then holding batches of them up to the light to detect coincidences. This was, of course, a "low tech" version of the famous Hollerith/IBM punched cards that could be sorted electrically. The story of my later involvement with the latter remains vivid. A Committee of Three Scientists established in 1960 by the International Whaling Commission (IWC), chaired by Professor Douglas Chapman of the University of Washington, Seattle, had arranged for all existing biological data collected by scientists aboard whaling vessels in the Antarctic to be compiled on punched cards. Nearly two decades later, when the Scientific Committee of the IWC was engaged on what was rather misleadingly called a Comprehensive Assessment of all whale stocks, Chapman was asked to retrieve the old cards. A year later, he was asked to report back. He said that the cards had unfortunately been ruined by damp and nibbled by mice, but all was OK, he had found the holes in heaps on the floor!

But I digress. The development of a current periodic bibliography, and then the Current Contents in Aquatic Sciences and Fisheries,[4] led us to create related tools, such as a list of periodicals and their short names, publishers' addresses, and so on, directories of research institutions, and an associated list of marine scientists. Kesteven also had the idea of creating

[4]The decision we made that the most appropriate mini-universe to map would be the envelope "aquatic sciences and fisheries" was not an easy one, and several alternative definitions of the scope of our embryonic information services were considered. The final choice reflected our belief at the time that fisheries science and oceanography (or "oceanology," and the Russians correctly insisted) were particularly closely related, as were marine science and limnology. It also reflected the fact that the Division, and hence such powers of decision as we had, were organized that way. And, finally, that we had developed fruitful working relations—in the days before the UNESCO Marine Sciences Division and the Intergovernmental Oceanographic Commission—with the very active Scientific Committee on Oceanic Research (SCOR) of the International Council of Scientific Unions (ICSU).

what we called Thesauruses: one for compilations of data on Species and Stocks of fishes, shellfishes, etc. (SAST), the other for information on Methods and Subjects (MAST). These evolved, respectively, into the long series of FAO Fisheries Technical Papers on species and species groups, and the Manuals on methods of stock assessment and the like, issued sequentially in the same series, as well as technical material in FAO Fisheries Circulars. I was interested to see that these series have now been supplemented by the attractive, smaller-format FAO Technical Guidelines for Responsible Fisheries.

Our encyclopedic approach also spawned a series of comprehensive reviews of specific knowledge areas of fisheries interest. These were a series of books on such matters as fishing vessel design; fishing gear design, construction, and operation; aquaculture: shrimp and prawn fisheries and cultivation; marine pollution; and marine mammals—this last with UNEP. In each case, the books were the outcome of world conferences on the subject, with hundreds of participants. But these were each prepared during one, two, or even three years beforehand through consultations with as many specialists as we could persuade to cooperate, and conduct of a series of special "workshops." In all, thousands of specialists contributed their knowledge and time.[5]

I should say that although our idea was to provide services to others, we were driven by our own needs to plan, supervise, and evaluate fisheries assistance projects, to conduct training courses and centers, and to participate constructively in the work of international fisheries organizations, including the older bodies such as ICES, CIESM, ICNAF (now NAFO), and NEAFC, as well as the newer regional commissions and councils being established under the FAO umbrella, such as the GFCM and the IPFC.[6]

From about 1960 to 1972, I was engaged essentially in administrative, managerial, and political stuff in FAO, UNESCO, and the UN, but then returned to be engaged in the population dynamics of marine mammals, backed by the FAO and UNEP infrastructures and the human resources offered by membership of the Scientific Committee of the IWC. After my early retirement from the UN system in 1980, I continued to do scientific work, but with little supporting infrastructure, although I was associated with, and partially financially supported by, a number of international nongovernmental organizations (INGOs) campaigning on environmental and animal welfare issues,[7] as well as with the environmental branches of some governments, especially of the Republic of Seychelles. In the first decade of the Third Millennium, I have gradually moved back to a focus on broader issues of fisheries science (though still including marine mammals) and have had to work to bring myself up to date with what has happened in that field during the past two decades.[8] In doing that, I should say that the FAO Fisheries Library and Jean Collins have been most helpful.

---

[5]Not all of these were primarily concerned with science, of course; the first two mentioned were mainly about naval, mechanical, and systems engineering. In recognition of the relations of these professional disciplines with scientific research, the Division of the Department of Fisheries, of which I was for a time Director, subsumed both the original Biology Branch and the Technology Branch except for the Section concerned with fish processing.

[6]International Council for the Exploration of the Sea, International Commission for the Scientific Exploration of the Mediterranean Sea, International Commission for the Northwest Atlantic Fisheries, North Atlantic Fisheries Organization, North-East Atlantic Fisheries Commission, General Fisheries Council for the Mediterranean, IndoPacific Fisheries Council.

[7]Especially the Threshold Foundation, World Wide Fund for Nature (WWF), Greenpeace International, International Fund for Animal Welfare (IFAW), Friends of the Earth (FoE), Humane Society International (HSI), and International Union for Conservation of Nature (IUCN).

[8]One result of that exercise is Holt (2004), my Foreword to the 2004 printing of Beverton and Holt (1957).

I have also discovered the joys and some of the sorrows of the Internet and the World Wide Web. Practicing science outside any institution, even working in Britain or Italy, and living on a pension of diminishing value, has some features in common with practicing in a so-called developing country. I am on e-mail, have a working computer, and think twice about buying a newer one. I use the telephone sparingly and as yet do not have access to broadband. Printer cartridges and paper are costly, as is postage, and increasingly so relative to income. Travel to relevant meetings is also increasingly expensive and, with advancing age, physically more difficult. Subscriptions to scientific periodicals are now mostly beyond my means.

So, where are the solutions to the communication and access problems posed by these realities? First, I should say that friendly or at least cordial personal contacts with other scientists are crucial. When I come across a reference to a paper that looks relevant and interesting, my first thought is: who among my professional contacts is likely to have a copy of it and be able to find it and willing to go to the trouble and expense of sending it to me, as hard-copy or digitally? Clearly, to have a wide range of such friends, particularly well-organized ones, is an enormous advantage. Obviously, next best is friends who themselves have ready access to good local libraries.

Second, while ability to comfortably read the texts of papers is evidently important, the lists of references appended to them or in foot- or endnotes are often more important. Again and again, I find that the paper I thought would be interesting is not, but its reference list leads me to something that is. In this connection, I might say that the practice of some scientific periodicals, such as *Science*, in truncating the references—usually by omitting titles, and sometimes by failing to register both the first and the last page of articles—is, to say the least, unhelpful. It makes it much more difficult to judge whether a reference might be useful, and to ask someone for it or to order it.

Third, creative science calls for what the Maltese philosopher, Edward de Bono, labeled "lateral thinking," and on a broad scale. I mentioned that in the 1940s and 1950s, some of us thought of ourselves as being engaged in operations research in fisheries, rather than in the specific disciplines in which we were trained (mine were zoology with also botany and chemistry). In the first couple of years engaged on my first research project, I "discovered" the theory of searching for submarines in the Atlantic (classified stuff then); the theory of games and economic behavior (von Neumann and Morgenstern); cybernetics and servomechanisms (Norbert Weiner, A. Porter); mathematical applications to biological problems as advanced by Verhulst, Volterra, Lotka, Rashevsky, Kostitzin; systems theory (especially Ludwig von Bertalanffy); animal physiology, especially metabolic processes (e.g., S. Brodie et al.); the nature of the biosphere (Vladimir Vernadsky, Charles Elton); animal behavior (Lorenz, Tinbergen et al.); human behavior (umpteen authors). In case you wonder, this last theme relates to Holt's law that for every regulation imposed by authority, fishermen find a behavioral response tending to nullify its intended effect—much like Newton's law that to every action there is an equal and opposite reaction. Most, if not all, of these discoveries came from browsing, not in libraries, but in bookshops in London and Cambridge and in other people's houses.

And, of course, I have made other "discoveries" in recent years, such as the relevance of population genetics to problems of sustainable utilization and conservation of natural resources (Lars Witting), the beauty and utility of fractals, vibrations, and chaos (Mandelbrot, The Beach Boys, Robert May, et al.), the joy of computer simulation (my colleagues

William de la Mare and Justin Cooke), and the awe of history (my friend Mary Carmel Finley and others).[9]

So, from that limited experiential trajectory, what suggestions can I offer to an assembly of librarians and information providers associated with major institutions, such as you?

Perhaps first: people are overwhelmingly important. Such institutions should position themselves to be able to facilitate personal contacts, as well as provide "the literature," maintain open-access, continuously updated address lists. Help disadvantaged researchers be in contact with each other.

Second: bibliographic references are, of course, important. Current Contents (CCASF) and compilations of abstracts such as ASFA are useful, but I have found that the most useful tool for my way of working is a citation index. I think such indexes are still only available at a price—by subscription. I wish they were "subsidized" and free. To help lateral thinking, particularly by people with limited access to bookshops and big general libraries, such indexes have to be very broad, not limited to references from a relatively small selection of periodicals. They should also include books, critical reviews of books (which often contain substantive material, including corrections of the author's mistakes), and the so-called gray literature such as submissions to conferences, workshops, and symposia. However, beware the warped sense of humor of some writers. It's common knowledge that a few years ago, a couple of scientists submitted a completely fake article to a primary journal that got past peer reviewers and was published. Ray Beverton and I put a fake reference in the bibliography of our book, for fun, and as far as I know, no one has detected it in nearly fifty years!

Third—and related to the first and second—encourage publishers and editors of journals always to provide information about the coordinates of the authors they publish, at least at the time of going to press. And, if annotations to references are not to be provided, at least encourage authors and their editors to provide meaningful, comprehensible abstracts that will help others to decide whether they really should go to the trouble of getting hold of the paper itself. Such abstracts should be written not in the jargon of the subject, but with an eye to attracting the attention of people working outside that subject. That is, the abstract should be a vehicle for facilitating interaction across fields and disciplines.

Fourth, remember the importance of state and local archives. We are working in an area often of political and economic sensitivity in which disinformation is rife. In recent years, I have been fascinated by revelations, from the declassification of previously secret papers, concerning the international negotiations of the mid-1950s on the law of the sea and, especially, fisheries development and management, and in the 1960s concerning the future of Antarctic whaling, in both of which I was a rather innocent participant.

I suppose these suggestions might look trivial to you. I could step outside my own experience and offer more general suggestions about access to information, but shall refrain now from doing so. I would end, however, by reference to a subject of great interest to me: it is the communication of science to wider publics. A few years ago, I undertook to review that problem, in the field of oceanology and marine affairs, as a member of the Independent World Commission on the Ocean (IWCO), set up and chaired by Dr. Mario Soares, ex-President of Portugal, but actually the brainchild of Mario Ruivo, my successor as Director of the FAO

[9]The history bit has found expression in my chapter "On the notion of sustainability" (Holt 2006b) in a forthcoming book edited by David Lavigne, published by University of Limerick Press and the International Fund for Animal Welfare, on the sustainable use of wild living resources (Lavigne 2006).

Fisheries Resources, Operations and Environment Division. We came up with the idea of an Ocean Observatory, using that term not to refer to a building housing a telescope, but rather a system of comprehensively organized information—in our case, about the sea, in all its aspects. As with the earlier notions of MAST and SAST, the imagination far outstripped the means, but at least now the fantastic advances in digital technology meant that the horizon could be more clearly seen. In this connection, I was excited by reading just this week about the Internet Archive and the Wayback Machine developed by Brewster Kahle in "an attempt to achieve what the ancient Greeks and Egyptians tried at the library of Alexandria: to make a permanent record of all human knowledge."[10] (Naturally, I have some doubts about the permanency bit, considering the piles of punch card holes and the cases of erasures and deterioration of magnetic media.)

Such efforts raise important questions of, for instance, language translation, but my concern here is more the channels and ways of communication of ideas and concepts to people in other disciplines—such as lawyers, economists, engineers, and artists of various species—and to administrators, politicians, media people, and the general public too. This has always interested me, I suppose because the results of fisheries research are of little lasting interest unless they are communicated to, understood by, and enthusiastically acted upon by those who may be influential in this respect. I long ago came to the conclusion that the research scientists themselves should take some responsibility in trying to do that, though they often need help both in writing and speaking each their own language.

But constructive feedback from the audience/readership is also of great importance. In the 1990s, I had an interesting experience in that regard. A small so-called Development Group of scientists (mostly mathematical biologists with a sprinkling of engineers) were charged by the IWC to come up with a new and better way of setting management rules for any future commercial whaling. They worked at this for many years, taking advantage of the prolonged pause in such whaling beginning in 1986. They triggered, I think, a minor paradigmatic revolution à la Thomas Kuhn in this field (Kuhn 1970) that is now spreading broadly into fisheries generally (see Cooke 1999).

This group had to explain the new things they were doing—essentially computer simulation of the entire management process—both to other scientists in the IWC's Scientific Committee (which turned out to be more difficult than expected) and to the IWC Commissioners and members of their national delegations. This latter was not, I think, quite so difficult, except with respect to the few such Commissioners who had themselves been exposed to training in science. However, it turned out that the Development Group could not advance far without more precise, quantitative specification of management goals and time-frames by the Commissioners representing their governments, and, through those, the various interest groups—"stakeholders"—in their countries.

That interactive process was difficult, but eventually fruitful. The specialized scientists had to modify their language; those empowered with management had to struggle to define, quantitatively, their political and operational objectives and—this was new—a time-frame for fulfilling them. All had to read materials they had previously ignored. Their archivists and documentation specialists and systems helped. And although the IWC worked formally only in English, translations were essential to true understanding. I think all of us who were engaged one way or another in this exercise learned from it.

[10] "Brewster Kahle" by Betty Hogge in the *New Statesman*, October 17, 2005, p. 26.

References

Anderson, K. L., and C. Thiery, editors. 2006. Information for responsible fisheries: libraries as mediators. Proceedings of the 31st Annual Conference of the International Association of Aquatic and Marine Science Libraries and Information Centers, Rome, Italy, 10–14 October 2005. IAMSLIC, Fort Pierce, Florida.

Belton, N. 2005. A game with sharpened knives. Weidenfeld & Nicolson, London.

Beverton, R. J. H., and S. J. Holt. 1957. On the dynamics of exploited fish populations. Her Majesty's Stationery Office, UK.

Cooke, J. G. 1999. Improvement of fishery-management advice through simulation testing of harvest algorithms. ICES Journal of Marine Science 56:797–810.

Graham, M. 1943. The fish gate. Faber and Faber, London.

Holt, S. J. 2004. Foreword to the 2004 printing. Pages i–xxiii *in* R. J. H. Beverton, and S. J. Holt. 1957. On the dynamics of exploited fish populations. Blackburn Press, New Jersey.

Holt, S. J. 2006a. Science is rooted in conversations. Pages 137–145 *in* K. L. Anderson, and C. Thiery. Information for responsible fisheries: libraries as mediators. Proceedings of the 31st Annual Conference of the International Association of Aquatic and Marine Science Libraries and Information Centers, 10–14 October, 2005, Rome, Italy. IAMSLIC, Fort Pierce, Florida. https://hdl.handle.net/1912/1480

Holt, S. J. 2006b. The notion of sustainability. Pages 43–82 *in* D. M. Lavigne, editor. Gaining ground: in pursuit of ecological sustainability. International Fund for Animal Welfare, Guelph, Canada and the University of Limerick, Limerick, Ireland.

Kuhn, T. S. 1970. The structure of scientific revolutions, Second edition. University of Chicago Press, Chicago.

Lavigne, D. M. 2006. Gaining ground: in pursuit of ecological sustainability. International Fund for Animal Welfare, Guelph, Canada and the University of Limerick, Limerick, Ireland.

Margetts, A. R., and S. J. Holt. 1947. The effect of the 1939–1945 war on the English North Sea Trawl Fisheries. Rapports et Procès-Verbaux des Réunions du Conseil Permanent International pour l'Exploration de la Mer 122:26–46.

Russell, E. S. 1942. The overfishing problem. Cambridge University Press, Cambridge, UK.

Smith, T. D. 1994. Scaling fisheries: the science of measuring the effects of fishing, 1855–1955. Cambridge University Press, Cambridge.

Part III

Getting Back to First Principles

Foreword to Part III: Getting Back to First Principles

By the early 2000s, fisheries in the European Union (EU) were in a parlous state. Many stocks were overfished, though the EU had no formal definition of what that meant, and the economics of the fishing industry were not promising. Historically, annual total allowable catches (TACs) had been set in marathon meetings of EU Council of Fisheries Ministers following proposals for individual stocks from the European Commission, which, in turn, were based on advice from the International Council for the Exploration of the Sea (ICES)—though the Commission did not always relay ICES advice. Such a process favored short-term political compromises on TACs, with many being set at higher levels than either the Commission's proposals or ICES advice. For example, in 2004, the Council adopted a total of 158 TACs, of which 31% (49 stocks) were set greater than the proposal of the Commission, including 19 TACs that were 40% or more greater.

Following the reform of the Common Fisheries Policy (CFP) of 2002, the EU had begun to adopt recovery plans, which included simple rules to set TACs based on the status of the stocks—this was designed to reduce the politicization of setting TACs.

By 2006, such plans had been adopted for a handful of stocks, most notably for Cod *Gadus morhua* and Hake *Merluccius merluccius*. The objective of these plans was for stocks to increase to, and subsequently be maintained within, "safe biological limits" which were, in turn, defined as a "low risk of transgressing certain limit reference points."

The Commission was well aware that this lack of clarity in objectives was unhelpful and, in 2006, published a pair of discussion papers (Commission 2006a, 2006b) proposing that fisheries management in the EU should seek to move towards pursuit of maximum sustainable yield (MSY). In its formal response a year later, the Council agreed to move towards an MSY approach, accompanied by an environmental, social, and economic impact assessment as well as support measures for the transition period (Council 2007).

At the time, one of us (Earle) was adviser on fisheries for the Greens/European Free Alliance (Greens/EFA), a political group in the European Parliament. He had known Sidney for years and contacted him for advice; this led to the latter's involvement for a decade in revising certain components of the CFP. Over that period, Sidney produced a large number of documents, most of them directly targeting the debates raging around the adoption, and implementation, of MSY as a management objective for the CFP. These ranged from lengthy and detailed policy papers for the

Michael Earle and Sidney in conversation. Photo by Tim Holt.

Greens to submissions for public consultations convened by the Commission; there were many short notes aiming to explain a particular matter being discussed in the ongoing negotiations for a given regulation or to answer a question put to him by Earle. Along the way, Sidney made his views on MSY more widely known, outside the inwardly-focused EU institutions caught up in the day-to-day machinations of the CFP, via a number of presentations to his peers, both fisheries scientists and fisheries historians, some of which were subsequently published. Over half of these chapters were done for the Greens/EFA Group in the European Parliament, either under formal contract, as short notes, including ones to Earle, or as public presentations such as to the Fisheries Committee of the European Parliament. The Greens/EFA Group has given permission for their use. The origins of other chapters are provided as necessary.

The chapters in this section comprise many of these works, beginning with his comments on the 2006 Commission documents that began the entire process, through the revision of the CFP "basic regulation" (EU 2013), and up to the first detailed, quantitative incorporation of the concept of MSY into a fisheries management plan for the Baltic Sea in 2016 (EU 2016).

A minimal understanding of the political and legal context of the CFP and the discussions and negotiations occurring at the time is instructive to better understand the motivation and argumentation of these papers by Sidney. Further detail can be found in Earle (2021).

As noted above, decisions on TACs were, and still are, made solely by the Council, which, prior to the management plans, had a virtually free hand to follow the convenience of political pressures for higher quotas. Since the objectives of the CFP were numerous, inconsistent, and often contradictory, it was easy to find a justification for a particular quota, usually on social or economic grounds. The opening move by the Commission to rein in the Council was the aforementioned papers from 2006. Shortly thereafter, in 2009, came the Green Paper (Commission 2009), a discussion document to begin debate over the reform of the CFP, scheduled for 2012,[1] in which the Commission argued in favor of pursuing a policy of managing fisheries "within MSY."

Sidney relaxing with his olive trees and vineyard in Paciano, Italy in 2010. Photo by Tim Holt.

The legislative proposal that emerged (Commission 2011) formally proposed "*that exploitation of living marine biological resources restores and maintains populations of harvested species above levels which can produce the maximum sustainable yield.*" (Article 2). Since 2009, all regulations of the CFP must be adopted by both the Council and the European Parliament (with the notable exception of TACs, which remain the responsibility of the Council alone). A battle royale thus ensued in which the European Parliament supported the idea of setting fishing mortality rates to allow stocks to recover to levels above those capable of producing MSY, whereas the Council advocated fishing mortality levels that produce MSY.

What emerged (EU 2013; Article 2.2) was a political compromise which satisfied nobody because the primary question remained open—could fishing mortality be equivalent to F_{MSY},

[1]Contrary to popular belief, a decadal reform of the CFP is not legally required. A provision in the basic regulation to allow Member States to govern access to the 12-mile limit is limited to ten years, at which point it must be renewed or amended. The Commission uses that requirement to propose other amendments to the CFP that it judges worthwhile.

with stocks fluctuating around MSY levels, or was it required to be less than F_{MSY} so as to allow stocks to be above MSY levels? Since this was a framework regulation, ambiguity was accepted.

Sidney relaxing at home in Paciano in 2010. Photo by David Lavigne.

It was only in 2016, with the adoption of the specific management plan for the Baltic Sea (EU 2016), that the question was decided. That regulation, as adopted, allows fishing mortality rates to be set well above the point estimate of F_{MSY} provided by ICES. The upper bound is the F-value on the right-hand slope of the yield curve that would produce 95% of MSY. Such F-values are greater than F_{MSY}, sometimes substantially so. Subsequent plans have followed a similar approach (Earle, 2021 describes this decade of negotiation).

Throughout these developments, the thrust of Sidney's writings was to advocate for fishing at levels significantly less than F_{MSY} for economic and environmental reasons. He pointed out that substantial savings in terms of fishing effort could be realized by accepting marginally lower catches than MSY. By basing his work on basic life history characteristics (the ratio of natural mortality to growth rate—M/K), Sidney was able to arrive at conclusions that are applicable across a very wide range of teleost taxa. The essence of these conclusions is contained in Chapter 46, his 2012 policy brief for the Greens. While much of his work was directed at the EU's CFP, his conclusions and recommendations are relevant elsewhere. Indeed, he made a number of presentations to various ICES theme sessions/symposia between 2012 and 2015 (Chapters 49, 50, 53, and 57).

Sidney taking time for a morning cappuccino in 2015. Photo by Tim Holt.

The evidence he marshalled to support his arguments demonstrates clearly his wide-ranging and long-standing interest and personal involvement in fisheries science and management, and marine conservation generally, since the late 1940s. He discusses the evolution of surplus production models and their weaknesses, explains the processes that led the International Whaling Commission (IWC) to abandon such models for a revolutionary new approach that eschews reference points entirely and led to modern concepts of management strategy evaluation, and attempts to answer the question of how far below F_{MSY} would be a reasonable target for fisheries management.

Spread throughout his work are fascinating anecdotes about the negotiations among the great powers after World War II that led to the 1958 Convention on the High Seas, his role in the development of the IWC's Revised Management procedure (RMP), and our changing views on what is meant by "sustainability." Even a cursory reading of some of his historical musings makes clear the debt he felt to his mentor, Michael Graham. Reading Graham's own work, in turn (e.g., Graham 1939), demonstrates the intellectual link between the two men.

The papers included here vary widely in length and scope. Some were written and polished over a matter of months; others were dashed off very quickly and never revisited. They thus vary in the depth of their analysis and the clarity of their presentation. The negotiations on the CFP were dynamic and complex and did not often leave time for detailed revision. As they were produced for different audiences at different times, there is a certain degree of repetition of some of the basic information. Additionally, a few ambiguous statements, odd turns of phrase, or questionable numbers in tables appear occasionally that we were unable to resolve and which were left unchanged, as we did not want to second-guess what Sidney intended. Though some of the details may be uncertain—Sidney was not infallible—the essence of his writings here is coherent and powerful. These papers offer a unique insight into the thinking of a great scientist after a career spanning seven decades, as he reviews his own original work and its place in the history of fisheries science and ruminates on how it has been used and, at times, misunderstood or worse. Comparison of the ideas presented in this part of the book with his earlier work shows that while his thinking deepened and evolved, he remained unwaveringly committed to rational fisheries management that leads to sustainable and profitable fishing—exactly the same arguments he made in the 1950s.

Emory D. Anderson and Michael Earle
Editors

References

Commission. 2006a. Communication from the Commission to the Council and the European Parliament. Implementing sustainability in EU fisheries through maximum sustainable yield. COM(2006) 360. Brussels.

Commission. 2006b. Commission Staff Working Document Accompanying the Communication from the Commission to the Council and the European Parliament. Implementing sustainability in EU fisheries through maximum sustainable yield. SEC(2006) 868. Brussels.

Commission. 2009. Green Paper. Reform of the Common Fisheries Policy. COM(2009)163. Brussels.

Commission. 2011. Proposal for a Regulation of the European Parliament and of the Council on the Common Fisheries Policy. COM(2011) 425. Brussels.

Council. 2007. Council Conclusions—Communication from the Commission [COM(2006)360 final]. Implementing sustainability in EU fisheries through maximum sustainable yield. 2841st Agriculture and Fisheries Council meeting. Brussels, 17 December 2007.

Earle, M. 2021. Maximum sustainable yield in the EU's Common Fisheries Policy—a political history. ICES Journal of Marine Science, doi:10.1093/icesjms/fsab037.

EU. 2013. Regulation (EU) 1380/2013 of the European Parliament and of the Council of 11 December 2013 on the Common Fisheries Policy.

EU. 2016. Regulation (EU) 2016/1139 of the European Parliament and of the Council of 6 July 2016 establishing a multiannual plan for the stocks of cod, herring and sprat in the Baltic Sea and the fisheries exploiting those stocks.

Graham, M. 1939. The sigmoid curve and the overfishing problem. Rapports et Procès-Verbaux des Réunions du Conseil Permanent International pour l'Exploration de la Mer 119:15–20.

Chapter 37

Commentary on "Implementing Sustainability in EU Fisheries through Maximum Sustainable Yield"[1]

The following is an interim comment, prepared for an Experts meeting of the Commission's Advisory Committee on Fisheries and Aquaculture (ACFA) in Brussels, September 15, 2006. It should be regarded as a template for a more comprehensive ***Commentary****, in preparation, in which the arguments are elaborated and sources fully referenced. They refer only to the scientific and procedural aspects of the* ***Communication.*** *(Comment by Sidney J. Holt)*

The *Communication* and the accompanying *Commission Staff Working Document*—"Technical Background"—provide, in the Communication's Sections 1, 2, and 3, admirable accounts of the straightforward scientific approach to management for sustainability that has evolved in the scientific community over the past 50 years. It is, quite understandably, now anchored to the Implementation Plan adopted at the World Summit on Sustainable Development, Johannesburg, 2002, with particular reference to its Section 31. This called, *inter alia*, for actions to "*maintain or restore stocks to levels that can produce the maximum sustainable* ***yield*** *with the aim of achieving these goals for depleted stocks on an urgent basis and where possible not later than 2015.*" (*Holt's emphasis*)

That statement is derived from the principle enshrined in Chapter 17 of Agenda 21 from the UN Conference on Environment and Development, Rio de Janeiro, 1982. States therein committed themselves "*to the conservation and sustainable use of marine living resources*" and specifically to "*maintain or restore populations of marine species at* ***levels*** *that can produce the maximum sustainable yield ...*" (*Holt's emphasis*) That, in turn, was derived from provisions in the UN Convention on the Law of the Sea, 1982 (UNCLOS), but there the central objective—"*to maintain or restore populations of harvested species at levels which can produce MSY*"—was "*qualified by relevant environmental and economic factors*" and by "*taking into account fishing patterns, the interdependence of stocks and any generally recommended international minimum standards*." The authoritative UNCLOS formulation gives us a little more leeway in interpretation of the MSY imperative than do subsequent contractions of the rule.

Here, I think we should concern ourselves with a qualification necessitated by scientific and practical considerations. The notion of the existence of an MSY is not an empirical one,

[1]Editors' note: this commentary by Sidney Holt, dated September 10, 2006, was on a Communication from the Commission to the Council and the European Parliament Document COM(2006) 360 final, Brussels, July 4, 2006 and was done for The Greens/European Free Alliance in the European Parliament, who have given permission to use this document.

but rather a theoretical one derived from ideas about the population dynamics of wild animals dating from the 1930s and even before. The approach is quite simple, even, I would say, simplistic. A catch cannot be taken repeatedly from a fish population without changing it, in two ways—reducing somewhat its numbers and changing its age composition. So, the sustainable yield from a "pristine" population must be zero. Likewise, the yield from a population that has been reduced to tiny proportions, possibly close to extinction, must also be zero. So, somewhere in between, there should be a maximum, a peak in a curve of sustainable yield (SY) against population size, and, by extension, in the curve of SY against the intensity of fishing.

The MSY notion—and also the concept of biological sustainability and hence SY—have been strongly criticized by numerous scientists and others from several points of view, some questioning their existence or reality, some the practical possibility of estimating them, others their suitability as a management objectives and targets. Here, I focus on the matter of *practicability*, and the question of whether there is an approach to management that will get us into the situation in which we want eventually to be—basically, depleted stocks recovered to a better state, in which they can provide higher, steady, and continuing yields in the long-term, without undue risk of renewed depletion. This is the spirit of UNCLOS Articles 61.2, 61.3, and 119.1 which require States and responsible intergovernmental organizations to "*ensure ... that the maintenance of the living resources ... is not endangered by overexploitation*."

First, it must be understood that although theoretical models and discussions about sustainability usually treat it as "forever," in practice, **a limited time-frame** has to be defined; in the modern approach, that is determined by at least four factors: (a) the normal lifespans of the animals concerned; (b) the life/work spans of the humans involved; (c) the expected (minimum) duration of any specific management instrument and system, including arrangements for continuing monitoring; and (d) computational limitations.

Second, management objectives and rules have to recognize and take account of the fact that any qualified sustainable yield is dependent not only on the **level** of the exploited population, but also on, among other things, **its composition by ages and hence body size of the fish, and by sex ratios and possibly other biologically defined features**.

Third, any management procedure that relies on fitting mathematical models to population data in order to calculate TACs and the like, without fully recognizing the implications of great uncertainty as to how realistic are the structures and inevitable assumptions of those models, as well as the uncertainty of estimates for values of their parameters, is doomed to catastrophic failure. Independent general studies, using exhaustive computer simulations, have shown that failure—especially as a result of attempts to find and attain MSY—is very likely, even if those assumptions and estimates were virtually perfect.

This third consideration is crucial. Happily, a way through the maze is now available to formulate a robust and precautionary management regime, and it comes mainly from intensive work over many years by five competing teams of scientists associated with the International Whaling Commission (IWC), all of which adopted the same general approach (formulated by the IWC's Scientific Committee), one of which—the so-called C-Procedure—gained most points for **robustness** and **efficiency** in meeting the IWC's explicit management objectives.

This development followed more than a decade of failed attempts to implement responsibly an MSY-focused approach to management adopted by the IWC in 1975. Notwithstanding the labors of many competent scientists, neither the supposed MSY level nor the MSY rate could be determined empirically, so it was "guestimated" on the basis of hypothetical population models without sound scientific bases. This is also the situation with regard to most depleted fish stocks. By the early 1980s, the IWC's catch limits were mostly being set quite arbitrarily, the MSY approach having been found simply not to be feasible in practice in most situations. As it happened, the declaration by the IWC in 1982 of an indefinite moratorium on all commercial whaling released the scientific community from the onerous responsibility of advising every year on catch regulations for every species and stock, and so provided a unique opportunity to attend to finding fundamentally new ways of regulating any future whaling.

The IWC's new approach had two features. One was to get agreement from those responsible for management policy on a small number of specific, precisely defined objectives, all of which were to some degree mutually incompatible. The other was to test, by agreed methods, the ***performance*** of alternative calculation procedures (called ***catch limit algorithms—CLAs***) applied to artificial "data" generated from plausible population models, in attaining the array of objectives, to which relative priorities also had to be assigned by the managerial authority—the IWC itself.

The three objectives identified by the Commission, from quantifiable suggestions made by the scientists collectively, were, in order of priority, (1) provide **the highest cumulative catch** over the selected management period (in this case, it was 100 years), that is compatible with (2) a specified **low risk that a stock would be depleted unintentionally** during that period (the probabilities were specified and also the acceptable frequency in which some depletion might accidentally occur), and (3) to **keep year-to-year changes in catch limits to the minimum** necessary, in order to help stabilize the industry.

The population models used to generate catch "data" all featured an MSY and MSY stock level. However, the testing was carried out with a model in which the parameter values were not well known except for their possible ranges. Even more important, "data" were generated from several structurally different models. An acceptable CLA had to be **demonstrably robust** to all those uncertainties, and also to possibly large, sudden changes or trends in the biological productivity of the environment (including the abundance of the whales' foods) and to possible interactions with competing species. These possibilities were all tested to the breaking point. Application of the CLA within a Revised Management Procedure (RMP) called for little more than reasonably reliable series of past catch histories—especially of the recent past—and estimates of the current abundance of the whales, with some information about the sex and age structure of the population. The RMP incorporates, in addition to the CLA, a procedure for applying the CLA efficiently and safely in situations where there may be more than one population of the species in adjacent and/or overlapping areas, but where the stock boundaries and mixing rates are not well-known.

The RMP developed by the IWC scientists was formally adopted by the Commission, but has not yet been implemented, and perhaps never will be. This has nothing to do with the validity of the suggested Procedure, but everything to do with the difficulty the IWC has in reaching agreement on measures for monitoring, control, surveillance, and other elements of **compliance** with a Revised Management System (RMS) incorporating the approved RMP/CLA. Additionally, in this special case, many, perhaps most, States evidently do not now wish commercial whaling to be resumed, for reasons other than the problem of managing it.

Nevertheless, the approach described above is open for consideration as a means to improve fisheries management generally, and some efforts to apply it to certain finfish fisheries are, I believe, now in hand. It does not seek MSY explicitly, in recognition that the conditions for attaining that are unknown and possibly unknowable. But, it does seek sustainability, within a defined time-frame, and, at the same time, due precaution and respect to the particular needs of industry. It offers the enormous advantage of avoiding endless largely pseudo-scientific debate about where exactly does the MSY level lie, what is the maximum sustainable rate of yield, whether these are changing, and to what extent are they naturally variable and possibly unpredictable.

An additional word of caution is relevant here. Fisheries scientists and managers are well aware that in many fisheries, especially but not only for small pelagic species, management for sustainability has somehow to cope with large and unpredictable *fluctuations*, especially of recruitment. Also, the possible existence of *oscillations* in stocks, arising from environmental oscillations, and/or from predator–prey dynamics, is now widely recognized. Research in recent years has, however, indicated that wild animal populations might have *intrinsic* oscillatory patterns of long wavelength, and we now need to study how best to deal with these in management. The abstract notion of sustainability, as well as theoretically covering infinite time, is closely connected with the idea of the existence of natural, and in general naturally stable, equilibria (and related notions such as "steady state" and "carrying capacity"), and models that describe transitions from one equilibrium to another. It is, I think, time to grasp the nettle of change in the fundamental ideas of population dynamics.

The new approach, exemplified by the IWC's RMP, is, I suggest, a modern version of the "philosophy" of fisheries management expounded by Michael Graham (Director of the English fisheries laboratory in Lowestoft, Suffolk in the 1940–1950s), and by some other European colleagues of that time, that I would paraphrase as "Don't look for universal 'maxima,' 'optima,' or other grand illusions of paradise, but identify a direction in which to move, to improve matters, and move that way and monitor carefully and continually what happens." Graham insisted, I think rightly, that controlling fishing effort was the better way to go than setting physical catch limits, and in his last writing, published posthumously, I think my old colleague Ray Beverton clinched that argument. Stocks that have recovered to higher levels, whether by reducing fishing effort directly, or indirectly by adherence to appropriate catch limits, provide both bigger sustainable catches, with bigger average sizes of fish, and higher rates of catch per effort. What more can we ask?

Chapter 38

Commentary on Commission Staff Working Document[1]

The following is an interim comment, prepared for an Experts meeting of the Commission's Advisory Committee on Fisheries and Aquaculture (ACFA) in Brussels, September 15, 2006. It should be regarded as an adjunct to my Commentary dated September 10, 2006 [Chapter 37 in this book] *on the original* ***Communication*** *from the Commission. It refers only to the scientific and procedural aspects of the* ***Working Document****. (Comment by Sidney J. Holt)*

I consider the Staff Working Document to be a basically correct exposition of the simple theory of fishing in language that should be easily understood by non-specialists. Its author has evidently worked hard to achieve such clarity. However, I differ from him in a few important respects, and that is the subject of this commentary.

In his Paragraph 2.2, the author writes: "... the stock size that can produce the maximum sustainable yield ... is often difficult to measure because of the variability of the marine environment," and this appears to be the only difficulty he recognizes. While his statement is correct, it is far from complete. There are major and fundamental difficulties with such measurement, and even more with estimation of the level (MSYL) of stock from which MSY might be obtained. As the author points out for some types of stocks and modes of exploitation, the shape of the expected curve of sustainable yield against stock size is a ***plateau-topped*** dome. This means that the MSYL is, for practical purposes, almost indeterminate, although the rough estimation of the MSY itself would appear to be immune to that phenomenon.

The simple theory goes that when a stock is reduced by fishing or by other processes (natural or related to human activities), the rates of natural mortality and/or of effective reproduction ("recruitment") change. The process is called **density-dependence**, the existence of which is a practically universal assumption in ecological science, although it has often proven very difficult to pin it down in the real world. It is generally assumed that with declining density or population number, the natural mortality, if it changes, will decrease, and the reproductive rate, if that changes, will increase. It is such changes that, in certain circumstances, generate a sustainable yield. But, the precise circumstances are critical, and it is the nature of that criticality that makes the estimation of both MSY and MSYL so difficult in practice.

---

[1]Editors' note: this commentary by Sidney Holt, dated September 11, 2006, was on a Commission Staff Working Document SEC(2006) 868 that provided technical background to the Commission's Communication "Implementing sus-tainability in EU fisheries through maximum sustainable yield: a strategy for growth and employment" and was done for The Greens/European Free Alliance in the European Parliament, who have given permission to use this document.

In the idealized pristine stock, at equilibrium, the natural mortality rate and the rate of reproduction balance each other: the stock neither increases nor declines, although as we know, there may be very strong fluctuations in short-lived species, due mainly to the extremely variable mortality of the eggs, larvae, and prerecruit juveniles. For the moment, let us simplify the discussion and think only about numbers of animals, not body growth and biomass—they add complications, but do not affect the basic argument. If either or both the natural mortality and recruitment change in such manner that there arises **a positive numerical difference between** them, that **difference** is what generates a sustainable yield, which will be available if the **fishing effort** that generated the **fishing mortality rate** that caused the original population to decline is stabilized, and for long enough for the population to move from its original equilibrium to a new one.

The **sustainable yield** (*SY*) is calculated by multiplying this difference, which is commonly called the **net rate of growth** or **of increase** in the population number, by the size of the population (it is called that because it is the rate at which the population would increase if the fishing pressure were to be suddenly removed). Normally, this rate cannot be measured directly because a growing population is also changing its age/size and possibly sex composition. Also, the estimation of the natural mortality rate and reproductive rate separately presents problems. It is very difficult to estimate the mortality rate itself, but estimating small changes in it as the population changes size is practically impossible. Likewise, although the effective reproductive rate is usually easier to estimate than the mortality rate, there is the problem that it is, in most fishes, very variable from one year to the next, so detection of changes in it requires a long series of observations. But, when one has to deal with very long time-series, the possibility arises that other factors have changed during the period of observation.

According to the simple theory of fishing, **this difference** between the natural mortality rate and the effective reproductive rate—the net increase rate—**will itself increase as the population declines**; this process is called **compensation** in the scientific literature. So, the changes in the population and its net increase rate are in opposite directions; it is that circumstance that produces a domed-shaped curve with a maximum (which may be described as a "peak," but is often more like a plateau)—the **MSY**. However, exactly **where that maximum lies within the population range** from "pristine" to "virtually extinct" is determined by **the *rate* at which the difference changes**, that is—in the language of the mathematical calculus—by the second derivative. Identifying the location of the maximum from data is thus intrinsically doubly difficult. Identification of all reference points related to the maximum, such as $F_{0.1}$, is equally difficult.

It has been demonstrated that, in fisheries, the intensities of which are regulated, determining both the MSYs and the locations of MSYL (there is no reason to believe that this location would be the same for all species and stocks) is less difficult when regulation is by **limitation of fishing effort** than it is when control is by **limitation of catches** (such as by TACs). But, the great technical difficulty is still there.

An additional point should be made about the process of compensation, which is not dealt with in the Staff Working Document. A basic assumption in virtually all mathematical models used to make fishery assessments is that the net rate of population increase increases continuously as a population declines, becoming at its highest when the population is tiny; its value then is called ***the initial net rate of population increase*** and this is a parameter that repeatedly appears in the theory of population dynamics, symbolized as ***i***. (It is called "initial" because it is the rate at which an initial very small population would begin to grow; it is not related to the effects of initiating fishing on the pristine stock.) However, there is evidence that, in some species or in some circumstances, **the net rate of increase begins to *decrease* at "small" population levels**. This process is usually called **depensation**, but alternatively may be referred to as **negative compensation** or **the Allee effect**, named after the American ecologist (Allee 1931) who first hypothesized and described the phenomenon.

Depensation, when it exists (and it is not easy either to detect or to reject it as a hypothesis) has very important consequences for depleted species. First, they reach a point at which they continue to decline even if fishing pressure is relaxed; and they may even continue to decline if fishing ceases. This latter is called **critical depensation**; suddenly, the natural processes are seen to be **irreversible**, contrary to the usual assumption. Second, if there is depensation, but it is not critical, then the stock will begin to increase (to "recover") when its exploitation ceases, **but it will recover at a slower rate than would be expected from standard theory**. Thus, we see a phenomenon sometimes called "**take-off**" in which a small population begins to increase rapidly after a prolonged period of relative inertia. The biological features that can produce depensation are varied and include, for example, situations in which adults prey on the young, as does the cod.

The problem for fisheries management is this: beyond the detection of depensation, we really need to know at what point on the sliding scale of population size it manifests itself, so that management measures can be taken to reduce the probability that a population will ever be unintentionally reduced to that level and below. In the IWC scientists' approach described in my Commentary on the Commission's Communication (*Editors' note: Chapter 37*), it was assumed, by mutual agreement, but without strong evidence, that the MSYL was, for baleen whales, near to 60%. It was also assumed that if there was depensation, it would manifest itself at stock levels below about 30%, at the most. So, a Procedure that ensured stocks would never, or hardly ever, be unintentionally reduced below that seemed "safe." But, in most models used in finfish fisheries assessments, the MSYL is much lower, commonly around 30%, and never higher than 50%. Thus, a management procedure that seeks MSY by maintaining stocks at MSYL is far more likely to bring them, unintentionally, into a depensation zone. The problem is compounded by the fact that in the limited scientific literature analyzing this question, statistical tests have been applied in which the *null hypothesis* is usually taken to be "no depensation" and the test is to see whether that assumption is rejected by the data. But, there is no reason to prefer that null hypothesis over its opposite, and testing the other way produces a higher likelihood that the data *do* reveal depensation.

My last point concerns the notion of **density-dependence**. **Density** is the number (or biomass, or other such measure) per unit area or volume in a specified part of the geographic range of the population. The **average density** is the number of animals divided by the area of its **range**. Density-dependence is thought to arise from competition for food or other resources needed for life within the population. If each individual animal is considered in isolation, then its dependence will be related to the available resources *per capita* in its local space. If the individuals are shoaling or in other ways relating to each other *en masse*, the situation is different and the likelihood of natural death or reproductive success is likely to be related more closely to the average density than to the local density. In trying to find the form of the relationship between the net rate of increase and the size of the population, the range of which does not change, it does not matter whether one specifies population size or the average density. But, if the range changes as the population size changes, then the situation is different. Range does generally change in some way as a population either grows or declines.

If the population is close to its pristine level, changes in its range will be small and usually negligible. But, the range of a depleted population can change much more with changes in the degree of depletion, so special care is needed in making assessments. This is one of the phenomena that have plagued fisheries scientists for decades in attempting to use *catch per unit effort* as a valid index of population size.

The Commission's **Communication** and the **Staff Working Paper** are focused especially on overfished, i.e., depleted, stocks. So, the matter of contraction or expansion of the range can be important. To illustrate this, consider an example in which the stock, although depleted, is continuing to decline, but there is no indication of depensation. We want to know what net rate of increase to expect in the smaller stock. If we have been looking at data relating the net rate of increase to the population size, then we would expect to see it increasing at a faster rate, and consequently would feel "safe." But, if the range has also been contracting, the density in the area still inhabited by the species will not decline as fast as the population as a whole, or even at all. So, the actual change in the net growth rate in the now-inhabited area will be less "beneficial" than thought if the range contraction were ignored, and there would be a greater probability of continued overfishing. In any case, the fish are not uniformly distributed in the first place; they will tend to be less dense in less biologically productive or otherwise poor areas of their distribution, and it is those areas that will be "evacuated" as the population declines. The theory of all this is fairly well worked out, but seldom applied in practice.

I have not commented specifically on the sections of either the Communication or the Staff Working Paper that refer to the process of moving from an overfished state to a better one, and its social and economic consequences, but at least one thing needs to be said from the scientific point of view since it is in that transition that some loss of catch has to be endured in the short-term in order to get higher catches later without again depleting the stock. In that period of transition, we may be aiming at a new and better equilibrium state as far as the stock and the continuing catches are concerned, but we are not normally starting the process from an equilibrium. Hence, knowledge of the current age composition of the stock and calculation of the **transients**—that is, the details of the recovery process as a stock that has an unbalanced or non-equilibrium composition increases—is essential if an economically and socially acceptable transition to a better long-term state is to be achieved.

References

Allee, W. C. 1931. Animal aggregations: a study in general sociology. University of Chicago Press, Chicago.

Chapter 39

New Policy Objectives and Management Procedures for EU Fisheries: a Commentary and Suggestions to the Greens/European Free Alliance in the European Parliament[1]

Introduction

The European Commission has proposed that sustainability in EU fisheries be implemented through a policy of seeking maximum sustainable yield (MSY)[2] and has provided accounts of the approach to management for sustainability that has evolved in the scientific community through the application of mathematical population models. The Commission anchored its proposal to the Implementation Plan adopted at the World Summit on Sustainable Development (Johannesburg, 2002) This called, *inter alia*, for actions to "*maintain or restore stocks to levels that can produce the* ***maximum sustainable yield*** *with the aim of achieving these goals for depleted stocks on an urgent basis and where possible not later than 2015.*" (*Holt's emphasis*) That statement is derived from the principle enshrined in Agenda 21 from the UN Conference on Environment and Development (Rio de Janeiro, 1992). States therein committed themselves "*to the conservation and sustainable use of marine living resources*" and specifically to "*maintain or restore populations of marine species at* ***levels*** *that can produce the maximum sustainable yield ...*" (*Holt's emphasis*) And **that** was derived from the UN Convention on the Law of the Sea, 1982 (UNCLOS), but there the objective—"*to maintain or restore populations of harvested species at levels which can produce MSY*"—was "*qualified by relevant environmental and economic factors*" and by "*taking into account fishing patterns, the interdependence of stocks and any generally recommended international minimum standards.*" UNCLOS provides the more authoritative "hard" law and gives more leeway in operational interpretation of the MSY imperative.[3]

---

[1]Editors' note: with minor edits, this is a briefing document prepared Sidney Holt for the Greens/European Free Alliance in the European Parliament, February 8, 2007, and reproduced here with their permission. It was sponsored by Ms. Marie-Hélène Aubert, MEP.

[2]Proposed in two documents circulated by the European Commission to the Council and the European Parliament, and presented to an experts meeting of the Commission's Advisory Committee on Fisheries and Aquaculture (ACFA) on September 15, 2006— [COM (2006) 360 final, Brussels 4.7.2006, "Implementing sustainability in EU fisheries through maximum sustainable yield" and an accompanying Commission Staff Working Document SEC (2006) 868 "Technical Background"]. The author offered two "Commentaries" on these (September 10, 2006 on the "Communication" and September 11, 2006 on the "Technical background") that were noted by ACFA. Editors' note: these are included as Chapters 37 and 38.

[3]Unfortunately, the "legal" documents refer only to levels of stocks, whereas the compositions (by age, body size, sex, and other features) of a biological population are as important as their relative sizes in consideration of sustainability. For the present purpose, we may mentally always write "and composition" after "level."

History

European fisheries management practices since the end of World War II resulted in depletion of most fisheries resources in what is now the EU Exclusive Economic Zone (EEZ), and the endangerment of some of the most valuable of them.[4] Some of those that had been highly productive for centuries have been so reduced that it has been necessary to impose moratoria on their exploitation. The primary cause among several other secondary causes is, without any doubt, excessive and inappropriate deployment of fishing power and effort. So the new proposal by the Commission that the Common Fisheries Policy be amended to provide explicitly for the core objective of management to be identified as the restoration of fish stocks to levels and states in which they are capable of providing maximum sustainable yields, and maintenance of those stocks at or above those levels, is welcome.

However, attainment of MSY as a valid management objective, and the question of whether it is a real biological feature of exploitable wild populations and not merely a construct from a mathematical model, have long been contested by scientists and economists.[5] Some efforts to implement it have failed, partly because fisheries authorities have not acted on the relevant scientific advice and partly because it has not always been very good advice. Such implementations have commonly involved the construction of plausible mathematical models to mimic the dynamics of exploited fish stocks, research to estimate the values of their parameters, and calculation from them of total allowable catches (TACs) that should be sustainable, and/or of the levels and patterns of fishing effort that would be necessary to prevent further depletions of stocks and allow their eventual recovery to higher and more productive levels. In recent years, there have been efforts to include a precautionary element in such calculations, but the measures offered have been essentially arbitrary. The lack of rigor in formulating scientific advice, due principally to continuing uncertainties about the dynamics of stocks, has diminished the force of recommendations to cut back *present* fishing effort in order to gain *future* benefits from increased catches. Additional benefits would include more valuable catches at less cost, a more stable industry, and, hopefully, more good protein food at lower prices to consumers.[6]

The "classical" population models provide curves of sustainable yield against equilibrium population size and against the rate of mortality due to fishing (see Figure 1). Such curves, of course, pass very close to zero at very small populations and to zero at high population sizes (so-called "carrying capacity") when there is no fishing. The curves in which fishing mortality is plotted along the horizontal axis show extinction of the stock at some finite level of fishing intensity if such intensity is sustained. The curves are all peaked or domed, and the highest point on them shows the greatest yield that can be obtained sustainably, referred to as maximum sustainable yield or MSY, at a population level MSYL and generated by a fishing mortality rate.

---

[4]It is perhaps worthy of record here that many resources that had been depleted before the war recovered during it when fishing practically ceased. Records showed increases in stock biomass up to fourfold after five or six years "rest" (see ICES 1948).

[5]The best known of such criticisms is that by Larkin (1977). Larkin made, however, a partial retraction of his "epitaph" a few years later (Larkin 1993). See also Kesteven (1997) for a thoughtful discussion.

[6]An interesting discussion of the concept of "sustainability" as rhetoric rather than as science, in the context of contrasts between "sustainable utilization" and "sustainable ecosystems," as seen by social scientists, is given by Massey and Rees (2004).

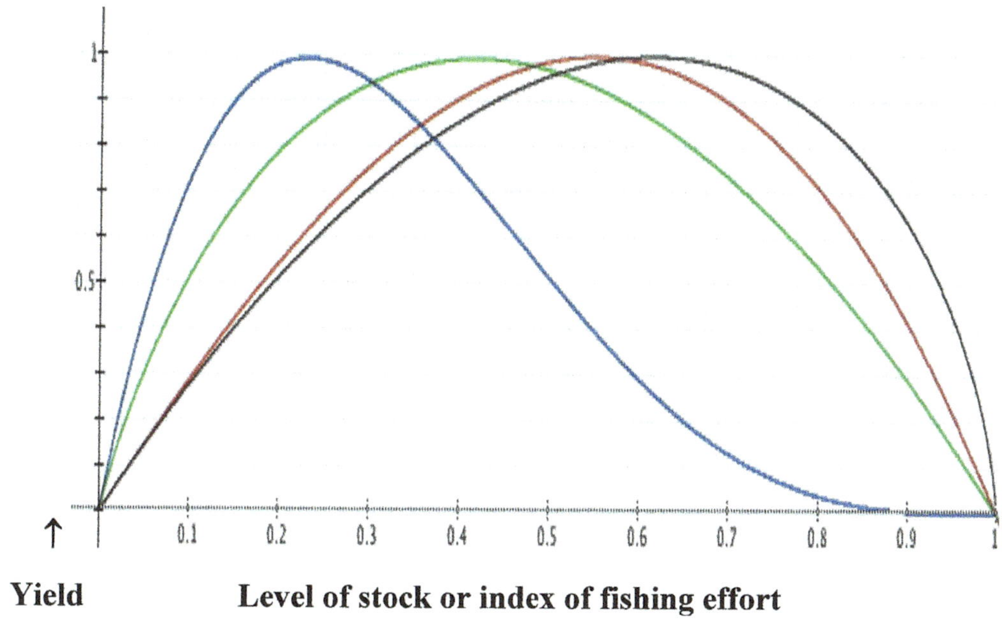

Figure 1. Curves of sustainable yield against stock level and fishing effort.

Explanation: Two pairs of curves are shown for sustainable yield plotted against stock level and level of fishing effort. All are scaled on the vertical y-axis to MSY =1. One pair—black and red lines—comes from the modified logistic model with parameter values such that MSYLevel is at 0.55 (55%) of the carrying capacity; the other—blue and green lines—comes from the same model, but with MSYL at 0.40 (40%) of carrying capacity.

The red and green curves are of yield plotted against stock level relative to carrying capacity; the black and blue curves show yield plotted against an index of fishing effort relative to the level of effort that, if maintained, would result in stock extinction (this effort is assumed to be standardized to be proportional to the fishing mortality rate it causes).

Notice that while the two curves of yield against stock level are similar in shape, those of yield against fishing effort are distinctly different from each other. The latter curve (black), in the case of MSYL = 55%, has a peak at 63% of the extinction effort. It is also relatively flatter-topped than the corresponding curve (blue) for MSYL = 40%. This last has, however, a peak of sustainable yield against fishing effort much lower—at 23% of extinction effort—than that of yield against stock level (green). This curve is also less "domed," more "peaky." These differences mean that if the MSYLevel is low (below 50% of carrying capacity), the "optimal" fishing effort is far below the "danger" level of extinction, but also that a relatively small error in estimating its location leads quickly to catches substantially less than the maximum.

Note that these are features of the model, not necessarily of the real world.

A first difficult, and commonly insoluble, problem is to know what is the stock size (MSYL) that would produce MSY in relation, on the one hand, to carrying capacity and, on the other, to *present* population size. A related difficulty is knowing what the current level of fishing mortality is compared to the level that would produce MSY—above it or below it—and to the level that would drive the stock to extinction. A second major difficulty is to determine whether the sustainable yield curves are sharply peaked or flattish-topped and also asymmetrical. In such case, it is particularly irrational and inefficient to attempt to manage a fishery by aiming to bring the stock to MSYL, and catches to MSY, because catches close to MSY could be obtained with very much less effort, hence less cost, and always with less fear of undesirable consequences. In practice, it is also necessary to pursue such modeling to deal with the considerable annual fluctuations we observe in many fisheries and to calculate the trajectories of stock and catch as changes are generated by new management objectives and procedures.[7]

A New Approach

Bringing European fishing activities back to sustainability and eventually higher productivity will inevitably be economically painful to producers, but not as painful as the disruption caused by taking very strong measures—such as fishery closures—haphazardly and too late. It is, therefore, necessary to move, and move fast, beyond the promulgation of ideological concepts—some would say obscure, even mythological concepts such as MSY targets, as they have, until now, been defined—toward operational procedures that, at the same time, allow recoveries of depleted stocks, control overly rapid expansions of new fisheries, and do not disrupt the industry even when they impose short-term strains on it.

The inadequacy of MSY as a rational and feasible management objective has been the subject of many scientific studies, resulting in suggestions for choice of targets—commonly referred to as *reference points* other than MSY and/or MSYL (e.g., Caddy and Mahon 1995). Those are, however, mostly the coordinates of other points on curves of sustainable yields against stock abundance or fishing effort, curves that lie barely visible in a fog of uncertainty. Adopting them will not solve our problem.[8] Even more to the point is the demonstration by Australian scientist William de la Mare twenty years ago, by simulation, that a simplistic approach to achieving sustainability and higher productivity or stock restoration will fail—in the sense that further depletion is likely to occur—even if the adopted model properly mimics the real population dynamics and the values of its parameters are known exactly (de la Mare 1986).

Among the most dramatic failures of an effort to implement an MSY approach to management was that undertaken by the International Whaling Commission (IWC) from the mid-1970s to the mid-1980s. That failure gave impetus to the decision to declare a global moratorium on all commercial whaling in 1982–1986. It also freed some of the most brilliant scientists associated with the IWC to develop a completely new approach to providing advice on TACs and related regulatory measures—it took an eight-person Development Group of the IWC's Scientific Committee about four years to do that.[9] Their approach has since been embraced by other scientific groups elsewhere, for a variety of fisheries, although the IWC's

---

[7]The models used to construct the graphs in Figure 1 are examples of a class of what are generally called surplus production models in which age and size structure of the population is ignored. The most accessible account of these and other models commonly applied in fishery assessments, requiring only school-level mathematics, is Cadima (2003). Cadima does not, however, discuss depensation or self-regenerating models (see later).

[8]A more sophisticated view of the value of "reference points" in the context of ecosystem-based management is expressed by Greenstreet and Rogers (2006).

[9]This group was led by G. P. Kirkwood (an Australian now at Imperial College, London) who also had the difficult task of explaining to Commissioners what the group had done. Members of the group were scientists from Australia, Iceland, Japan, South Africa, and UK/Germany. The work was organized as a competition, by

Revised Management Procedure (RMP) is undoubtedly the most fully developed and rigorously tested so far. It is time and opportune to apply the same approach in a Revised Common Fisheries Policy for the European Union. But, to do so calls for a radical rethink of its objectives, in an operational sense, and of the arrangements for its execution.

The first necessary reorientation is in the definition of MSY itself, not as a theoretical point on a curve, but as a real quantity. That quantity could conveniently be the biggest possible ***cumulative*** catch obtainable from each exploited stock, over a predefined period, a period that is not too long, not too short—50 years, perhaps. The second necessary reorientation is in the idea of the time-frame of ***sustainability***. Few official or popular documents mention this; theory and philosophy imply it is infinity.[10] But, we all know that nothing lasts forever, and we know that the natural world continues to change in unexpected and unpredictable ways. So, we might settle on the same duration as the redefined "MSY"—50 years? We should be satisfied if we can expect to have stable and productive fisheries at the end of such time, and notable improvements during it, once we are through the painful beginning. Putting these two definitions together, we should contemplate a ***management period*** of the same duration; that is to say, a management regime that we intend to keep in place for long enough to achieve the desired results. Naturally, amendments might be needed from time to time to take into account new events, findings, and happenings, but if we do engage in such a scheme, we need to make a commitment to seeing it through, applying new rules for long enough to make a real difference.

So, some details of how this might work. First, conditions—secondary management objectives—have to be attached to the new aim of securing the biggest cumulative catches. The *first* and most obvious of these is that the management Procedure[11] must be such that there is only a very low probability that the stock will be depleted unintentionally at any time during the management period to below some predefined level or, more precisely, driven into a zone of low productivity and possible irreversible decline.

The *second* is that the stock must, insofar as its natural rate of growth permits, be abundant (also to be predefined quantitatively) at the end of the period—and not, of course, to be wiped out at that time in order to enhance the cumulative catch!

These are the conditions as they relate to the catches. Other conditions relate to the industry. The management Procedure should be such that allowable catches, or the permissible fishing effort, are not unduly changed from year to year, i.e., more than is necessary for the other conditions to be met, except to respond to entirely unexpected events or discoveries. And, as far as possible, catches should not be brought down to levels that threaten the closure of the industry for economic or social reasons.[12]

---

mutually agreed rules, between several subgroups—five initially, whittled down to three—devising possible efficient management procedures separately. As far as I know, this was a unique scientific development process. The winner was Dr. Justin Cooke—by acclaim of the entire Scientific Committee (although, remarkably, when it came to a vote, he expressed preference for de la Mare's auxiliary algorithm to deal with multiple substocks). In the IWC publications, his solution is referred to as *The C-Procedure*. Dr. de la Mare was the runner up, with Drs. Butterworth and Punt a close third. Interestingly, all three were funded from non-governmental sources, not by their governments. This author's rôle was to think of plausible scenarios that might cause the CLAs to fail to perform as desired. I was singularly unsuccessful.

[10]For discussion, see Holt (2006).

[11]This term is used, and capitalized, here to mean the application of a tested algorithm to define regulatory measures that are not merely advised, but are acted upon without amendment for political or other arbitrary reasons.

[12]The above set of management objectives does not conflict with, but, rather, makes operational the more general objectives listed in the 1995 FAO Code of Conduct for Responsible Fisheries.

What is new in the approach pioneered by the IWC is that TACs are set by applying a ***catch limit algorithm*** (CLA)[13] that has been tested for its performance in meeting defined management targets, for its efficiency, and for its robustness, by computer simulation, using "data" generated by a mathematical model that is thought reasonably to represent the dynamics of the real exploited fish population. Random variables have to be inserted in the model to represent the statistical variability of the central values of each parameter in it which describe the patterns of reproduction, growth, mortality, and other relevant features of the dynamics of the population so that very many simulations are calculated for each situation to provide probability distributions. Each simulation begins with an estimated population (size and structure), and a time-series of previous catches, calculates a TAC and applies it to the population, then calculates the population the following year, calculates a new catch limit, and applies it, repeatedly, until the end of the management period.

The next stage is to repeat this process with the same model, but with different central parameter values. Typically, these would represent different assumptions about what precisely are the shape and scales of the curves of sustainable yield created by the model. If necessary, the starting algorithm is then modified to ensure it is robust for all the chosen parameter sets for this model. A further step is to repeat the entire process with other different and plausible population models (i.e., models that are not incompatible with whatever relevant data are available or with what is known generally about the dynamics of animal populations) and again make any necessary modifications to the CLA., with the intention that the algorithm eventually adopted will not be dependent on the particular population model favored. This is a crucial difference from the "traditional" approach to providing scientific advice on TACs.

Further tests, following the same procedure, are made to determine whether and in what way the consequences of applying the algorithm might change as a result of external environmental changes, such as by pollution, habitat loss, ocean climate change, and changes in food supply and abundance of predators. In doing so, changes of any degree, kind, and direction can, in principle, be modeled. Adjustments to the algorithm may again be needed to ensure it is still robust and, if applied, would both avoid stock collapse and ensure high long-term cumulative catches. "The environment," in this sense, includes not simply the physical, chemical, and structural properties of the habitat of the stock, but also possible dynamic interactions between the stock and its biological competitors, predators, and prey.[14] Thus, without moving into questionable complex models of such interactions and even of ecosystem dynamics, such a Procedure takes us an important step towards the ideal of ecosystem-based management.

The real data needed for the application of such a Procedure are a series of historical catch statistics, which are as complete as possible, and, later, the catches under the new management

[13]"A finite ***Procedure***, written in a fixed symbolic vocabulary, governed by precise instructions, moving in discrete steps 1, 2, 3…, whose execution requires no insight, cleverness, intuition, intelligence or perspicuity, and that sooner or later comes to an end" (definition by Berlinski 2000).

[14]There have been numerous attempts to deal with this matter by the construction of multispecies population models. However, as Caddy and Mahon (1995) remark, such models are extremely data-intensive and are still beyond the practical reach even of assessments in well-studied regions such as the Northwest Atlantic. However, it must be remembered that according to the UN Fish Stocks Agreement of 1995 (which, while specifically directed to the management of fishing on straddling stocks and highly migratory species, is of general application), "the absence of adequate scientific information shall not be used as a reason for postponing or failing to take conservation and management measures."

regime, together with periodic estimation of the size and composition of the stock—starting with its size and composition at the beginning of the management period—and the results of monitoring what are judged to be the interacting species. In examining the robustness of the CLA, prior tests will have been done to ensure that it will not be completely invalidated either by historic data being falsified or otherwise unreliable or by errors in the data obtained under management—such as poor catch statistics and the inevitable statistical errors of estimating population abundance. **A precautionary approach is not, however, explicit in the CLA algorithm, but rather is defined in the choice of management objectives the algorithm is designed to achieve** (e.g., Cooke 1989; Sainsbury et al. 2000).

The conditions that must be defined to make operational a precautionary approach would need to be complemented by conditions to ensure, as far as is practicable, given the conservation status of each stock, that continuity of the industry is feasible within the constraints of resource conservation and restoration. Such conditions could also be incorporated in the creative process of constructing and testing the CLA. And, finally, the CLA has to be tested in the context of uncertainty about whether or not there are substocks of the basic species stock and, if so, what are their geographic boundaries and what mixing occurs between them. This requires the invention of a second algorithm that determines precisely how the CLA will be applied to a multiple-stock situation. That second algorithm and the CLA together are what the IWC scientists termed a Revised Management Procedure (RMP),[15] and is the basis for a form of *adaptive management.*

A few studies have recently been made, or at least begun, involving the simulation of management Procedures in fisheries generally; the idea has begun to be broadly accepted. A case in point is that by L. Kell and his colleagues concerning the Plaice *Pleuronectes platessa* stock of the North Sea (Kell et al. 1999) and an ongoing international project led by Kell and funded by the EU Research Directorate and others.[16] None of these has, however, yet progressed—as far as I know—to the point reached by the IWC scientists some years ago.

More Work Needed

There are some other practical and theoretical problems to be confronted in developing, by simulation, an adaptive management Procedure for particular fisheries and circumstances, but it is unnecessary to elaborate all of them here. Three should, however, be mentioned. The first is that the simple population models used for the IWC simulations dealt with *numbers* of animals in populations and catches (the IWC still being committed to regulating whaling by setting numerical limits), whereas, of course, in finfish and shellfish fisheries, we necessarily deal normally with *weights* of catches and *biomasses* of populations, with their *compositions* by age, size, and sometimes sex, and with highly variable recruitment that is rarely closely correlated with the size of the sexually mature part of the stock. Population models suitable for this task are, however, available, one being called the *self-regenerating model* published by the late R. J. H. Beverton and the author in the 1950s and further explored by Beverton (1994).[17] An example of the result of applying such a model is given in Figure 2.

[15]The RMP is also accompanied by precise specifications for how abundance surveys are to be conducted.

[16]This project is called Framework for Management Strategies (FEMS), one element of which is creation of an open-source software framework (code name FLR). Kell et al. (2007) describes this process and gives references to applications of this approach elsewhere.

[17]Holt (2004) comments further on the self-regenerating model.

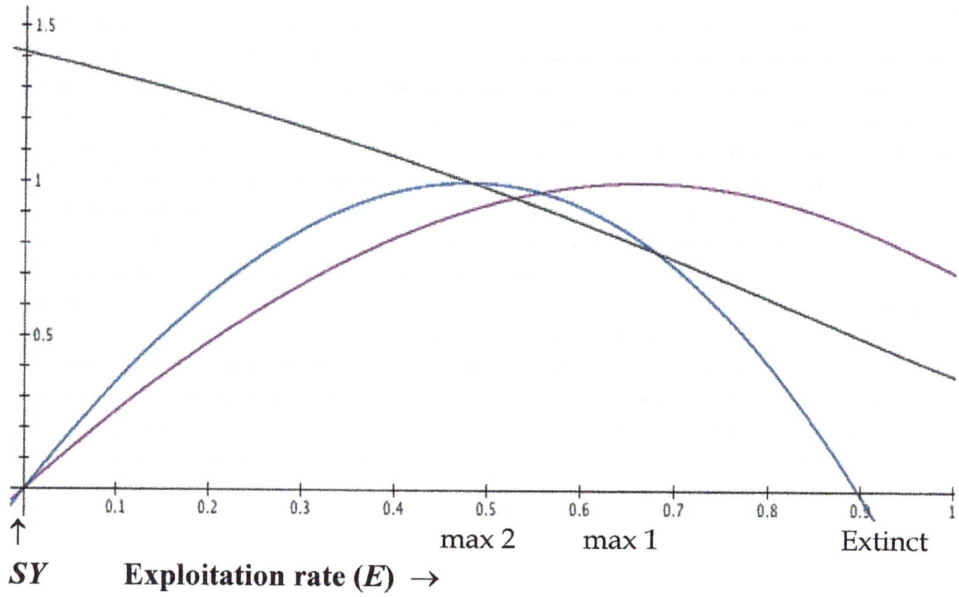

Figure 2. Example of self-regenerating model for sustainable yield.

Explanation: The purple line is an example of a Beverton–Holt yield-per-recruit (Y/R) model—i.e., sustainable yield with constant recruitment—with parameters chosen arbitrarily to provide an MSY exploitation rate of 0.67 (max 1). The blue line is the same model, but with recruitment dependent on spawning-stock biomass according to the B–H stock–recruitment relationship. The vertical axis is scaled for both so that MSY = 1. The dependence of recruitment on spawning stock causes the MSY exploitation rate to be pushed back to about 0.5 (max 2). The black line shows the mean weight of fish in the catch; it is scaled, simply for convenience, to have an indicative weight of 1 at max 2. If larger fish are more valuable per kg than smaller ones, it is advantageous to reduce the exploitation rate even possibly to be a little lower than that producing MSY. The magnitude of any benefit depends, of course, on the model parameter values; the choice made of parameter values for this example is such to make substantial changes in the average size of caught fish. Interpretation of such averages must be made with care; with high E, most of the catch will consist of fish little bigger than the size at recruitment, but with lower E, there will be a wider range of sizes in the catch from the small post-recruits up to significant numbers of much larger fishes. The parameters for this illustration were set such that the maximum weight to which the fish grows is about eightfold the size at recruitment. The average weight of fish in the catch at very low exploitation rate is up to a factor of three and one-half more than the weight of recruits, or rather less than one-third of the theoretical maximum weight. At the MSYL of the self-regenerating curve the average weight is about a factor of two and one-half more than the weight at recruitment.

If a fishery was assessed to be more-or-less stabilized at $E = 0.55$, it might be concluded, from a Y/R assessment, that increased effort would bring increased total yield, though the average size of fish caught would be smaller. But, assessment using a self-regenerating model would indicate that the same increase in sustainable yield would come from a roughly equivalent reduction in fishing effort, and the average size of fish caught would be bigger.

The second problem is that none of the models used so far in simulation studies have been designed to exhibit *depensation* or the "Allee effect" (named after the American ecologist who first studied it; Allee 1931). This phenomenon can arise at low population levels where, as the stock is further depleted—e.g., by continued intense fishing—it becomes unstable and can be driven to extinction or, if fishing pressure is relaxed, it may begin to recover, but much more slowly than expected from the "classical" models (Figures 3 and 4).[18] A slow "take-off" is becoming a familiar observation in the discipline of "restoration biology." Again, models do exist embodying depensation, but they just have not much been applied.[19] This is because the effect is not easy to detect from fisheries data, and it has been too easy simply to assume it is not there.[20] But, it becomes critically important when we are attempting to allow depleted resources to recover and, at the same time, continuing to exploit them, even—we may think—cautiously and lightly.[21] Lastly, we are used to thinking, in the classical theory, that wild biological populations tend to move to and between intrinsically stable steady states. We regard shifts from such states as "perturbations" and try to calculate the transient situation—the trajectory—from one quasi-equilibrium to another. Deviations from this simple situation are fairly well understood—they mostly involve interactions of the population with predators or prey, or delays in the life cycle in attainment of sexual maturity, both of which can lead to cycles ("oscillations") or even chaotic—hence utterly unpredictable—population behavior. Recent theoretical research in population genetics has, however, led some scientists to conclude that natural populations may have an inherent tendency to long-term oscillations, with wavelength of several generations.[22]

This phenomenon has immense implications in considering sustainability and the setting of sustainable TACs, as has been demonstrated by some data analyses of marine and other populations. No simulations of possible management Procedures have yet been carried out in which such behavior is manifest; it is desirable that it be done in order to judge whether it has to be taken into account specifically in devising a robust and efficient CLA.[23]

---

[18]Hutchings (2000) has reviewed the fisheries literature and demonstrated that slower recovery of depleted stocks than expected is common.

[19]See Kapitza (1992) who described human population growth throughout history. Kapitza's model possesses a property appreciated by mathematically inclined scientists of being "self similar," meaning its characteristics do not alter with changes of scale.

[20]However, Liermann and Hilborn (1997) demonstrated that data for many species, from numerous taxonomic groups in a huge database "were consistent with moderate levels of depensation."

[21]It is possible this phenomenon was manifest in the Northern Cod *Gadus morhua* situation off Newfoundland and eastern Canada. When this stock had been greatly depleted, the Canadian authorities closed down the commercial fishery, but allowed a continued subsistence take. Recovery was not as expected, and the latter had also to be closed. Possibly the most spectacular failures of premature reopening of collapsed fisheries, and recovery rates being slower than expected from "classical" population models, are provided by the history of the Peruvian Anchoveta *Engraulis ringens* fishery.

[22]This theory has been elaborated by Witting (1997), which also contains a lucid exposition of the "classical" theory of population dynamics as a preliminary to Witting's critique and fundamental reworking of that. See also Witting (1999).

[23]Witting's most convincing application is to the gray whale *Eschrichtius robustus* of the North Pacific, now recovered from near extinction to numbers very much higher than before commercial whaling began in the 19th century, despite loss of much of its breeding habitat. Conventional assessments lead to the conclusion that the sustainable yield of that population is now several hundred animals per year. But, it is now near the peak of a Witting cycle; it will soon begin to decline naturally, and so the sustainable yield is zero. Fromentin and Kell (2007) have carried out some management simulations concerning Atlantic Bluefin Tuna *Thunnus thynnus* in which long-term oscillations occur, although they did not follow Witting's lead regarding the underlying theory.

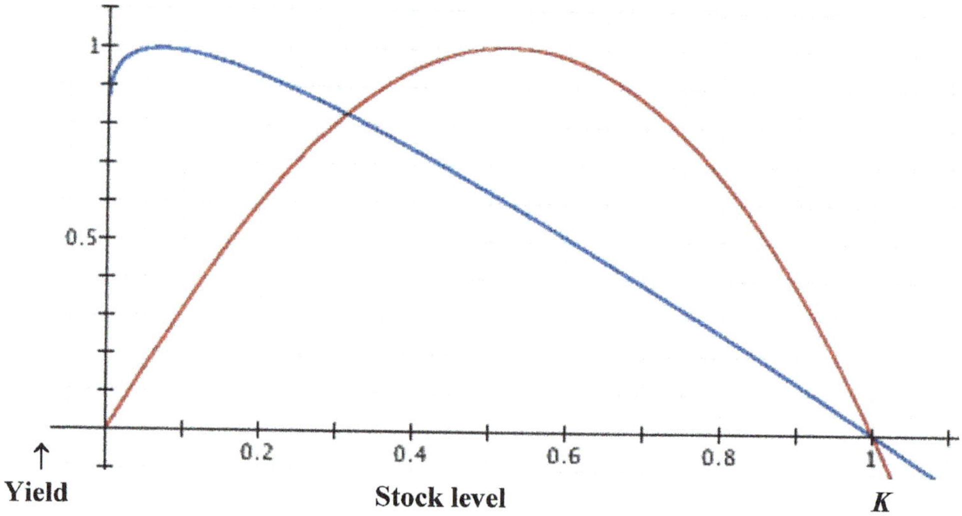

Figure 3. Illustrating the "Allee effect"—depensation.

Explanation: The simple logistic model, with MSYLevel = 0.5 (50%), gives identical curves of sustainable yield against stock level and fishing effort that are symmetrical domes (actually, parabolas), mirror images of each other. This is not illustrated here, but rather a closely similar model, but with a depensation element added. The depensatory process exists throughout the stock and effort ranges, but becomes dominant and obvious only at low stock levels and high effort levels. In this example (only the curve of sustainable yield against stock level is shown here—red line), the generally weak depensatory effect has shifted the MSYL from 50% to 55%. But, on the left-hand side, this red curve has a very slight, barely visible *inflection* (S-shape) at about SYLevel = 0.1. In this case, the stock becomes unstable (not viable) if depleted to below 10% of its original (pre-exploitation) abundance.

The blue line maps the ***slope*** of the red yield against stock size curve. It is, thus, the curve of relative rate of increase, i.e., the difference between the rate of reproduction and the rate of (total) mortality. Both curves are scaled so that their peak values are 1. The peak of the blue line marks the stock level below which depensatory processes dominate. If the depensatory processes were to be removed, the red line would revert to a symmetrical parabola and the blue line would become straight, passing from a maximum relative rate of increase (= 1) at zero stock size to zero relative rate of increase at carrying capacity (stock = 1).

Beyond TACs

The period of devising and testing a precautionary and adaptive management system of the kind suggested here offers an opportunity for reconsideration of alternatives to setting TACs or at least supplementing them by regulations aimed at limiting and appropriately adjusting the intensity of fishing directly, as measured by the fishing effort exerted and/or the value of the fishing mortality that effort generates. There are considerable advantages, particularly in terms of stability of the fishery in the medium- and long-term, of moving away from a purely TAC-regulated system. These become obvious from analyses of self-regenerating population models in which the usual uncertainty about the numerical relationship of annual recruits

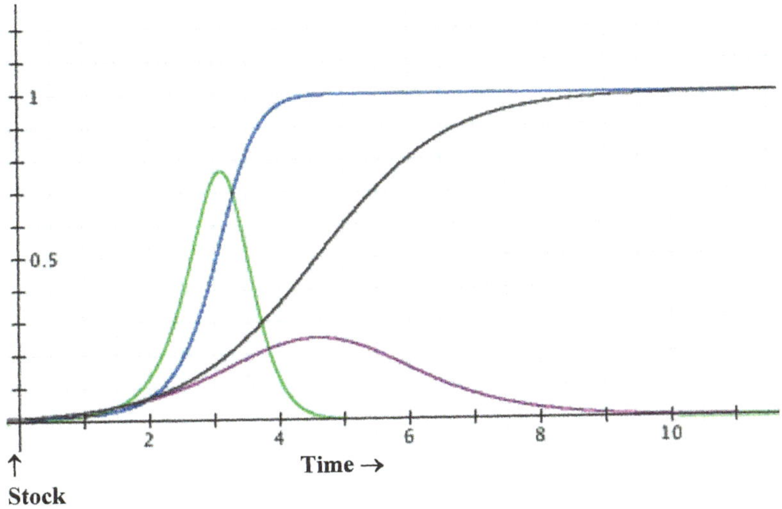

Figure 4a. Depensation and stock recovery: growth and yield.

Explanation: The black curve illustrates simple logistic population growth. The blue curve shows the corresponding growth curve of the model with depensation illustrated in blue in Figure 3. Parameter values are set, for illustrative purposes, such that MSYL = 0.5 (50%) of carrying capacity for both models, but the MSY in the blue model is about threefold the MSY in the black model. This can be seen from the green and purple curves. These graph the slopes of the growth curves, that is, the sustainable yield as it changes over time because the population is growing. Notice that, for the logistic (black and purple), the curve of SY against time is bell-shaped, in contrast with the parabolic shape of the SY against stock size. The non-logistic model with depensation (blue, green) is similarly bell-shaped, though more peaked, less spread. So, although the two growth curves seem to be of similar shapes, their patterns are distinctly different, over and above the difference in the scales of MSY.

to the size of the mature "parent" stock that spawned them is taken into account (Beverton 1994). Implementation of the ideas sketched here would call for three changes in present practice by the Commission and its scientific advisers. First, the scientific effort to create, test, and evaluate the necessary algorithms would be substantial, and new arrangements would be needed to ensure this. Second, this approach requires that the managers, industry operators, and other stakeholders agree to specify quite clearly the set of secondary management objectives to be sought (presuming all agree, *a priori*, on the primary objective of securing a cumulative catch over the management period, which is as big as possible within the constraints of a very low probability of inadvertent stock depletion). The scientific advisors should be able to assist in this process by suggesting what might be feasible options and reasonable probabilities to specify as well as outlining the scientific aspects of the choice of the management period, which coincides with the temporal limits of the definition of sustainability. Thirdly, during the implementation of the Procedure, the management authority and the providers of scientific advice would have to work much more closely together than seems to have been the case hitherto, and similarly both of those entities with the operators of the fishery.

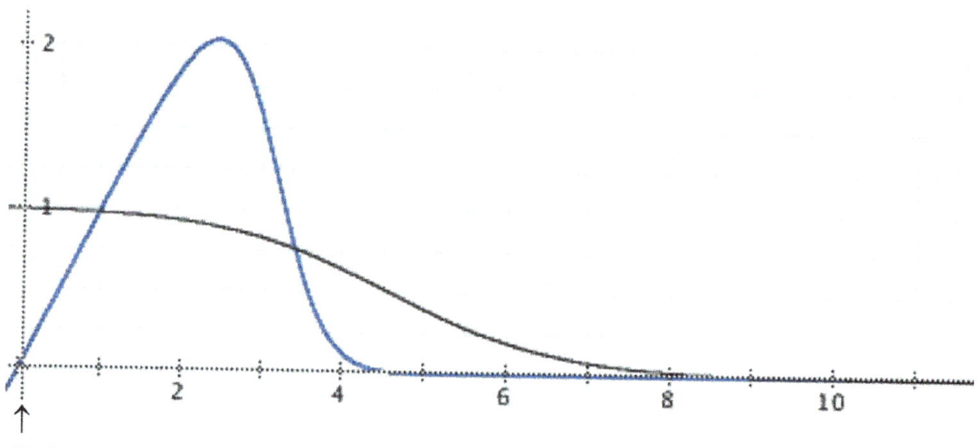

Figure 4b. Depensation and stock recovery: relative growth rates.

Explanation: The two graphs are from the same models and same parameter values as Figure 4a, but show the ***relative*** rate of growth, i.e., the slope (sustainable yield) divided by the population size as a function of elapsed time. If plotted against population size rather than time, the simple logistic (black) would be a straight line from unity near zero population to zero near carrying capacity. The superficially very similar non-logistic model (blue) has a completely different pattern of relative growth rate against time. The rate of increase is close to zero at very small population sizes. As the population slowly increases, the rate of increase accelerates, reaching a peak after an elapse of about 2.5 arbitrary time-units, but declines rapidly thereafter, approaching zero, of course, as the population approaches carrying capacity.

Conclusion

Progressive reduction in fishing effort, resulting from the institution of a "computer-aided" management system as described here, consistently implemented throughout a predetermined management period, would bring with it advantages beyond simply bigger, sustainable catches at much lower cost. In many cases, the age and size compositions of the catches would change in such a way as to give greater stability of catch from year to year (as good and bad recruitments have less effect on the variability of catches when fish from several cohorts and so of a range of ages and sizes are well represented in the population) and an increase in unit value that may result from increases in the average size of fish caught. Additional benefits would be substantial reductions in the incidental kill of other species, such as birds and small cetaceans, and in the volumes of discards of both unmarketable fish species and undersized individuals of marketable species. Furthermore, the UNCLOS imperative that fisheries should be regulated so as to leave adequate sustenance for dependent species would be met. These consequences of reduced effort would mark an important step towards the generally desired, but still elusive, management and the preservation of biological diversity.

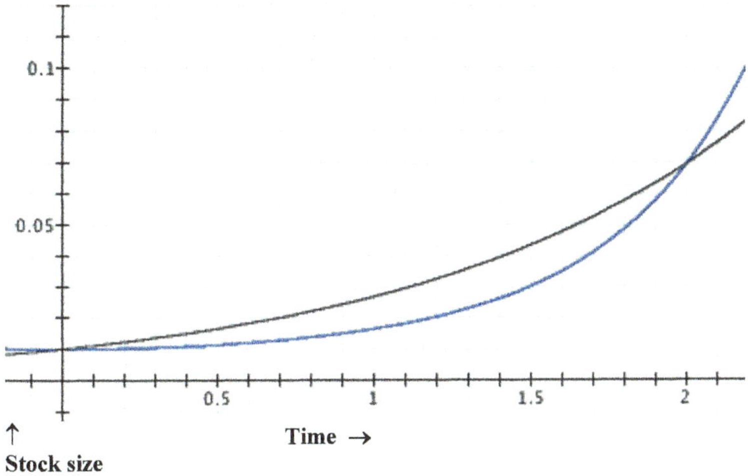

Figure 4c. Pattern of stock recovery at low population levels.

Explanation: Here, we look closely at the lower ends of the two population growth curves in Figure 4a, but with the stock-size scale adjusted so that the figure represents the situation in which the stock barely survives, having been reduced by fishing to only 1% of its original size (carrying capacity). The logistic (black) shows the usual "classical" feature of expected recovery, after the cessation of fishing, with relative growth rate highest when the population is extremely small. By contrast, the non-logistic with depensation (blue) shows a very slow beginning of recovery, but accelerating, though not reaching the same level as the logistic until two arbitrary time-units have passed and the stock has recovered to about 7% of carrying capacity; this might be regarded as the limit of the preponderance of depensatory processes. Thereafter, it continues steeply upward until it reaches the much higher MSY slope and then approaches carrying capacity more rapidly than does the logistic model.

This particular graph shows the general features of these alternative models, but naturally both the logistic and non-logistic parameters can be adjusted so that, for example, non-logistic MSY is no higher than that of the logistic model. Both can be generalized to have MSYL higher or lower than 0.5 (50% of carrying capacity). In recent stock assessments, the logistic model and its generalized family have rarely been used in this simplistic way; normally now, there would at least be an effort to incorporate age and size distributions, time-lags to sexual maturity, and specific functions relating stock to recruitment. But, the consequences of ignoring possible depensation when predicting recovery trajectories of greatly depleted stocks remain essentially the same.

References

Allee, W. C. 1931. Animal aggregations: a study in general sociology. University of Chicago Press, Chicago.

Berlinski, D. 2000. The advent of the algorithm: the idea that rules the world. Harcourt, San Diego.

Beverton, R. J. H. 1994. The state of fisheries science. Pages 25–54 *in* C. W. Voitlander, editor. The state of the world's fisheries resources. International Science Publisher, Lebanon, New Hampshire.

Caddy, J. F., and R. Mahon. 1995. Reference points for fisheries management. FAO Fisheries Technical Paper 347.

Cadima, E. L. 2003. Fish stock assessment manual. FAO Fisheries Technical Paper 393.

Commission of the European Communities (COM). 2006. Communication from the Commission to the Council and the European Parliament: implementing sustainability in EU fisheries through maximum sustainable yield. COM (2006) 360 final. 4.7.2006.

Cooke, J. G. 1989. Improvement of fishery management advice through simulation testing of harvest algorithms. ICES Journal of Marine Science 56:797–810.

de la Mare, W. K. 1986. Simulation studies on management procedures. Reports of the International Whaling Commission 36:429–450.

Fromentin, J. M., and L. T. Kell. T. 2007. Consequences of variations in carrying capacity or migration for the perception of Atlantic Bluefin Tuna (*Thunnus thynnus*) population dynamics. Canadian Journal of Fisheries and Aquatic Sciences 64:827–836.

Greenstreet, S. P. R., and S. L. Rogers. 2006. Indicators of the health of the North Sea fish community: identifying reference levels for an ecosystem approach to management. ICES Journal of Marine Science 63:573–593.

Holt, S. J. 2004. Foreword to the 2004 printing. Pages i–xxiii *in* R. J. H. Beverton and S. J. Holt. 1957. On the Dynamics of Exploited Fish Populations. Blackburn Press, Caldwell, New Jersey.

Holt, S. J. 2006. The Notion of Sustainability. Pages 43–82 *in* D. M. Lavigne, editor. Gaining ground: in pursuit of ecological sustainability. International Fund for Animal Welfare, Guelph, Canada and the University of Limerick, Limerick, Ireland.

Hutchings, J. A. 2000. Collapse and recovery of marine fishes. Nature 406:882–885.

ICES. 1948. Report of the Scientific Meeting on the Effect of the War on the Stocks of Commercial Food Fishes, 24 October 1947. Contributions to Special Scientific Meetings 1947. Rapports et Procès-Verbaux des Réunions du Conseil Permanent International pour l'Exploration de la Mer 122.

Kapitza, S. P. 1992. A mathematical model for global population growth. Mathematical Modelling 4(6):65–79.

Kell, L. T., I. Mosqueira, P. Grosjean, J-M. Fromentin, D. Garcia, R. Hillary, E. Jardim, S. Mardle, M. A. Pastoors, J. J. Poos, F. Scott, and R. D. Scott. 2007. FLR: an open-source framework for the evaluation and development of management strategies. ICES Journal of Marine Science 64:640–646.

Kell, L. T., C. M. O'Brien, M. T. Smith, T. K. Stokes, and B. D. Rackham. 1999. An evaluation of management procedures for implementing a precautionary approach in the ICES context for North Sea plaice (*Pleuronectes platessa* L.). ICES Journal of Marine Science 56:834–845.

Kesteven, G. L. 1997. MSY revisited. Marine Policy 21(1):73–82.

Larkin, P. A. 1977. An epitaph for the concept of maximum sustainable yield. Transactions of the American Fisheries Society 106:1–11.

Larkin, P. A. 1993. The year 2000: will we be ready technically? We're ready already. Fisheries 18(3):6–11.

Liermann, M., and R. Hilborn. 1997. Depensation in fish stocks: a hierarchic Bayesian meta-analysis. Canadian Journal of Fisheries and Aquatic Sciences 54(9):1976–1984.

Massey, E., and E. Rees. 2004. Sustainable utilisation of fisheries as governmentality: constraining the potential for ecosystem-based management? New Zealand Geographer 60(1):25–35.

Sainsbury, K. J., A. E. Punt, and A. D. M. Smith. 2000. Design of operational management strategies for achieving fishery ecosystem objectives. ICES Journal of Marine Science 57:731–741.

Witting, L. 1997. A general theory of evolution: by means of selection by density dependent competitive interactions. Peregrine Publisher, Århus, Denmark.

Witting, L. 1999. Optimization of management procedures with control on uncertainty risk. ICES Journal of Marine Science 56:876–883.

Chapter 40

Statement to Fisheries Committee of the European Parliament[1]

It is an honor to be invited to address this Committee, and I shall try to be helpful.

I prepared two papers (*Editors' note: Chapters 37 and 38*) concerning the proposal by the EU Commission that a revised fisheries policy for the waters under EU jurisdiction should be focused on the target of securing maximum sustained yield (MSY). These were considered by the Commission's Committee on Fisheries and Aquaculture. Subsequently, I prepared more extensive technical briefs on the subject for the Green Alliance in Parliament (*Editors' note: Chapter 39*) and the European Policy Office of the World Wide Fund for Nature (WWF) (Holt 2007). All four documents have already been made available to the rapporteur, chairman, and secretariat of this Committee, as well as to some others; copies are, of course, available for other Committee Members on request from the Secretariat. Copies of a book entitled *Gaining Ground*, containing a chapter by me on "The notion of sustainability" have also been provided to some participants, and further copies can be obtained, gratis, from Michael Earle, Advisor to the Green Alliance, for anyone interested in following the relevant history of this topic. I shall assume, however, that few of you will have yet had time to study all or even any of these materials.

First, let me say that as a scientist and as someone engaged for most of his working life on questions concerning international fisheries management and the conservation of marine living resources, I very much welcome the Commission's proposals and hope they will attract your support and will be implemented. In the present state of many of the fisheries resources in this region and elsewhere, attainment of their **sustainable use** is not a sufficient objective of policy; **the aim should be to ensure recovery of depleted stocks to high productive levels and at the same time ensure that stocks and species not yet depleted do not become so**. In support of its proposals, the Commission has cited the UN-sponsored conferences in Rio de Janeiro and Johannesburg, but I suggest that a stronger—and at the same time more flexible—basis for support is in the 1982 UN Convention on the Law of the Sea, and I would refer you to that text. Not only does it begin to define the MSY approach to management in a realistic way, but requires implementation of some aspects of what is now called ecosystem-based management, for example, the requirement that sufficient quantities of smaller prey species be left for the adequate nutrition and reproduction of larger, dependent, and often more valuable predatory species.

MSY targets have, over the years, been designated for management of many fisheries. The practical results have frequently been such as to discredit the whole idea. At the same time, the concept has received the critical attention of many competent scientists, with more negative than positive views being expressed by them. There are good technical reasons for this, and I shall

---

[1]Editors' note: delivered by Sidney Holt on April 11, 2007.

not try to explain them here. But, the basic problem is that the concept has rarely been expressed in operational form, and in its "traditional" form makes assumptions about the properties of animal populations, especially fish populations, that are simplistic, that cannot be verified with hard data, and that depend too critically on a precision of scientific knowledge that does not exist and is probably not attainable. In fact, most of the non-technical discussion of the matter has been based on outdated concepts of the dynamics of animal populations.

There have been attempts to overcome or circumvent some of the deficiencies of MSY management targets following the fashionable idea of alternative **reference points**, but personally, I do not consider these to be very helpful; indeed, they are distracting and can be misleading, all of them being linked—like the usual view of MSY—to a particular mathematical population model.

The precautionary approach to management and conservation of resources tells us to avoid doing nothing until more and more science gives us incontrovertible answers. Instead, I advocate the development of management procedures that will, with high probability, provide the desired results even when biological and ecological knowledge is very uncertain.

Scientists associated with the International Whaling Commission (IWC) were faced sharply with this situation when, in the early 1980s, they found themselves freed from the task of advising on total allowable catches ("catch limits" in IWC-speak) every year for every species and stock of whales. This ended seven or eight years of ultimately failed attempts to apply a "traditional" MSY target procedure of regulation to the residual stocks of baleen and sperm whales (*Physeter macrocephalus*). They consequently gave their full attention, for several years, to devising a management procedure that, unlike the previously existing procedures, would actually achieve its purpose. Their consensus proposals were adopted by the IWC, but have not been implemented, partly because they should be complemented by compliance arrangements that have not yet been agreed. Nevertheless, the approach developed by the IWC Scientific Committee has set a totally new pattern for fisheries management that is already being studied in several places, including—under research contracts—within the EU. I will try to explain to you very briefly the essence of this new approach.

First, it is based on computer simulations of fisheries situations, using only data that can certainly and necessarily be available, such as series of historical annual catches and at least one good recent estimate of the size and composition of the stock. The purpose of the simulations is to find an **algorithm** (that is, a set of calculation rules) that will work with maximum independence from assumptions about the dynamics of the fish populations in general and the managed one in particular. That is to say, it should be independent of particular models of population dynamics and of specific suites of values of the parameters in those models.

(When I began my career in fisheries research and management in the 1940s, I heard from some quarters that mathematics had no place in fisheries management. Now, we wonder how we ever managed to do anything without mathematics. But, we do now face a different, but still difficult, conceptual problem; the approach I want to introduce to you here also presents an intellectual challenge of considerable difficulty, and not only to lay persons. We are *not* talking merely about simply devising better mathematical models and analytical procedures to replace the ones that have failed us—that is often a first presumption, and I'm afraid an erroneous one. One of my closest colleagues, someone who has been instrumental in developing the new methods, has said we are not talking about science as such; we are talking about the more-or-less scientific basis of engineering.)

Second, the notion of MSY has to be **reinterpreted** in a new and **operationally meaningful** way. Sustainability itself has to be defined in a specified, real time-frame, not as an abstract concept to be applied to infinite time, forever and ever. So, MSY is to be regarded as the objective of obtaining the **largest possible cumulative catch** over a predefined and limited time (for the whale scientists, that was 100 years, but for practical fisheries practice, it would be much shorter than that, perhaps a decade for shorter-lived species, a couple of decades for longer-lived ones). Further, it is important that the regulatory mechanism has the same time-frame as the time-limited "sustainability" and revised "MSY-like" definitions. That is to say, a clear, workable procedure has to be agreed at the outset for setting TACs and/or other measures such as fishing effort/power/capacity limitations, and such an agreement must be honored throughout the period, not subject to endless annual renegotiation and compromise, unless circumstances arise that make adjustments absolutely necessary.

Third, obtaining the highest cumulative catch has, of course, to be conditional on other conditions being met. One obvious condition is that it is not permissible to "top-up" at the end of the regulatory period in order to boost the catch; the stock must be left in a good productive state at the end. Another condition is that there must be a low, predefined probability that the stock will be depleted to or below some predefined level, accidentally, at any time during the management period. Another condition is that the procedure must be robust to possible unpredicted and unpredictable external changes, such as in the productivity of the environment or noticeable biological interactions among the species whose exploitation is regulated. Furthermore, the minimal acceptable stock level is determined, of course, to remove any likelihood of extinctions and unwise reductions in spawning stock biomass, but more importantly to the need to retain high biological productivity—which can mean reducing catch rates long before there is any threat to the stock's continued existence. At this point, I should perhaps emphasize that recent scientific research in population dynamics has shown that we should not expect stocks necessarily to "bounce" back from depletion when they are relieved of fishing pressure. Slow "take-off" may be expected, and even failure to recover at all. We have some practical examples of such behavior, such as the Cod *Gadus morhua* in the Northwest Atlantic. Such advances have to be taken into account in the development of workable safe procedures for management.

Lastly, but certainly not least, the procedure should be such, if possible, as to avoid the "extinction" of any fishery unless, of course, such is desired for economic and/or political reasons and is acceptable to stakeholders.

Unfortunately, the various statements about MSY in international political documents all concentrate on the so-called **stock level**, which is commonly interpreted as meaning some number or biomass, possibly related to the original size of the stock before intense fishing began ("carrying capacity"). But, the productivity of a fish population depends as much on its composition by ages and body sizes of the fish, and the sex ratio, etc., as on crude measures of stock size or level. A new procedure must take that properly into account. One of the implications of this is that catches—especially of the relatively longer-lived species—will be more stable from year to year as they are taken from stocks with better balanced compositions comprising more age groups.

I finish this presentation with three points:

1) The steady reductions in fishing over-capacity that a highest feasible cumulative catch policy implies will result in direct reductions in incidental deaths of other species such as seabirds, turtles, and non-target fishes, including small species and undersized individuals that must be discarded under the current regime.

2) In the medium term, fish should be more plentiful in the market and less costly because caught with less deployment of fishing capacity and effort.
3) Fish has long been, and should in future become again, a prime source of high quality animal protein for the populations of Europe. At present, about one-half of the fish consumed here is imported. Europeans cannot safely assume that others, outside the Union, will always take good care of *their* fisheries resources. **Restoration of depleted stocks in the European EEZ and ensuring they all stay highly productive is, therefore, in the longer term, a matter of European food security**.

Thank you for listening to what I had to say.

References

Holt, S. J. 2007. New policy objectives and management procedures for EU fisheries. A commentary and suggestions. A briefing paper prepared for the European Policy Office of the WWF. Unpublished. 67 pp.

Chapter 41

Does History Serve Conservation?[1]

It is sometimes said that we should never believe a scientific theory until it is verified by experience. But … also … we should never believe an observation until it is confirmed by a theory.
Sir Arthur Eddington, British cosmologist

I have come here from the 11th conference of the North Atlantic Fisheries History Association (NAFHA) in Bergen, Norway, the first I have attended. I was pleased to learn that the answer to my title question was "yes." But, the historians there were looking at sequences of events, and it seemed that the main lesson was that there had been many failures from which we might learn what not to do in the future. But, in any case, my theme today is not so much *events* as *ideas*.

In his majestic science fiction saga, Last and First Men (1930), Olaf Stapledon, English novelist and philosopher, wrote: "*Theories, theories, myriads upon myriads of them, streamed over me like windborne leaves, like the contents of some titanic paper-factory flung aloft by the storm, like dust-clouds in the hurricane advance of the mind. Gasping in this vast whirling aridity, I almost forgot that in every mote of it lay some few spores of the organic truth, most often parched and dead, but sometimes living, pregnant, significant.*" Stapledon hypothesized the trajectory of human evolution through two billion years to the ultimate burnout of the solar system, though he did not look backwards to before 1930. Later writers, of course, offered us escape routes to planets in other star systems, lengthening our lifespan, and even to other universes, *à la* Philip Pullman's trilogy *His Dark Materials*.

I think I was 16 years old when I first read this. I had just decided to become a biologist, though I had little interest in natural history. More to the point, the branch of the British Museum in the part of northwest London where I grew up, which housed the Museum's newspaper collection, had just been bombed. The Luftwaffe was trying to disable the old aerodrome nearby where new RAF fighter pilots were being trained, a field too small for modern aircraft and where, later, Metropolitan Police drivers, in what were called the Flying Squads, being polished at Hendon Police College, learned how to survive high speed skids.

We in Colindale awoke after the air raid to find our streets and gardens littered with tattered papers. That was my first introduction to the gray literature, although we did not know it by that name. For some years, my home museum contained a few torn scraps of old newspa-

[1]Editors' note: presented by Sidney Holt on October 8, 2007 at the 33rd IAMSLIC (International Association of Aquatic and Marine Science Libraries and Information Centers) Conference: Changes on the Horizon held in Sarasota, Florida October 7–11, 2007 and published in the conference proceedings (Anderson 2008) as Holt (2008). Permission to use this article was given by Kristen L. Anderson, Director, Health Sciences Library, University of Hawaii, Honolulu, HI, who was editor of the conference proceedings.

pers along with a piece of true British anti-aircraft shrapnel from our loft, along with a piece of German incendiary bomb—that failed to explode.

Some of these gray particles, Stapledon's dust clouds, could coalesce, the condensate being called "the published literature," to be reviewed, revised, retrieved, indexed, and filed, and in the traditional scientific manner, metaphorically torn apart, demolished, and proven wrong. Nothing—not even theories and ideas—is permanent, yet we strive for continuity within change. I suppose it is in the turbulent teens that worries about survival and mortality, evolution and stability are seeded, some eventually blossoming into notions such as conservation and sustainability. I believe the history of such notions is in the gray literature; the published forms tend to obscure it.

An Iranian-American, Amil Imani, recently wrote, in the online magazine *Global Politician*: "All things on earth are subject to a limited life span, be they bacteria, trees, mountains, humans or ideas—including religions." That last is perhaps a counterpoint to Stapledon's later book, *Star Maker* (1937), which introduced ideas remarkably like the modern *Intelligent Design*, and I suppose it is no coincidence that I found the Imani quotation in the Newsletter of the (British) National Secular Society—more gray literature! In *Star Maker*, Stapledon undertakes—according to Wikipedia—"the immense task of describing the entire history of life in the universe. He tackles philosophical themes such as the essence of life, of birth, decay, and death, and the relationship between creation and creator." I intend today to be considerably less ambitious.

Anyway, we see in Imani the idea of *lifespan* as something looking forward, towards an end, and that is a common interpretation. But, a "span" has a beginning as well as an end; it is finite—think bridges. (When I Googled the word, I was led, unexpectedly and with delight, to an organization with that name, listed as a Self Publishing and Independent Publishing Trade Association. Most appropriate!) I liked Amil Imani's attribution of mortality to ideas as well as things since I have, since I spoke at the IAMSLIC conference in Rome two years ago, immersed myself in the history of some of the **ideas** associated with conservation of natural resources, especially marine living resources, as distinct from the sequences of **events** in the real world such as those explored by Jared Diamond in his *Collapse* and by Brian Fagan in his *Floods, Famines and Emperors: El Niño and the Fate of Civilizations*. I supposed it would be worth asking if such exercises are useful as well as interesting to an eclectic few. And, since that immersion turned out to be largely in the gray literature, it seemed appropriate to talk about it here, where tomorrow's theme is ***Capturing the Shadows***. But, I'll bear in mind the overall theme of this conference—***Changes on the Horizon***—which conjures simultaneously the interlocked concepts both of space and time. As with lifespan, we usually, most of us, tend to sit watching the setting sun, not the dawn. But here, we have to deal with ***cycles***—I prefer the term ***good vibrations***. I'll have a few words to say about that later.

The excursion on which I spent the longest time tracing myths was to the origins of the modern ideas of sustainability, sustainable use of resources, and sustainable development, this last having been described fairly convincingly as a monstrous oxymoron. It has been reviewed critically elsewhere by others and I'll just refer this audience to the contributions by William E. Rees and Sharon Beder to *Gaining Ground* (see below), respectively: "Why conventional economic logic won't protect biodiversity" (Rees 2006) and "The changing face of conservation: commodification, privatization and the free market" (Beder 2006). The ideas I want to talk about concern the first two mentioned, and within those, the notions of ***density dependence, depensation, maximum and optimum sustainable yields (MSY, OSY)***, as well as the gray (including secret, confidential, and private materials) and published sources concerning them. The relatively new idea of the now much-touted—and, by some, feared—***precautionary principle*** is already worthy of historical analysis, but it is probably too soon to consider whether such a study would serve conservation. A brave beginning was made in 2001 by a team assembled by the European Environment Agency ("Late lessons from early warnings: the precautionary principle 1896–2000") (Harremoës et al. 2001), and the summary of it—"Twelve late lessons" —by the Editorial Team is a thoughtful progress report. It has a chapter on "Fisheries: taking stock," by Malcolm MacGarvin, but the case studies therein are geographically limited and, in my opinion, do not include some of the most revealing situations and are especially thin on the European experience.

The idea that ***biodiversity***, at genetic, taxonomic, and ecosystem levels, is a good thing is also relatively new and fashionable, but perhaps not yet quite ripe for critical historical analysis. It may or may not be closely related to the even more fashionable idea of taking an ***ecosystem approach to management*** of the use of ocean space, natural resources, or whatever. The "may or may not" phrase is there because this much-talked-about approach, notwithstanding an enormous gray literature referring to it, has about as many quasi-definitions as there are buns in a baker's dozen. This has been well brought out in a recent article by Joji Morishita (Japan's Alternate Commissioner to the International Whaling Commission—IWC) "What is the ecosystem approach for fisheries management?" (Morishita 2008). What is clear is that, apart from the basic problem of definition, there is remarkably little depth to the science about it. I propose to leave it alone until the fog clears.

So, let me begin with "The Notion of Sustainability," the title of my chapter in a book published in 2006 (Lavigne 2006). I can recommend reading several chapters—other than my own, of course—especially the first and the last by Lavigne and some of his collaborators and the aforementioned ones by Rees and Beder.

To understand the theory of sustainable use of renewable living resources (the theory is a bit different for non-living resources), we go back to the start of the 19th century, when the Rev. Thomas Malthus (1798) postulated that human and other animal populations tended to increase *geometrically*—2, 4, 8, 16, 32, etc.—or, as mathematicians expressed it, *exponentially*. Something had to give to prevent the planet from filling up, and Malthus noted that occasional catastrophes knocked off the "surplus" from time to time, but, more generally, that the contrast between a tendency for exponential growth and the presumed reality of only linear increase in access to resources (that is proportional to population number) led to periodic collapses and new beginnings, and hence to cycles. These were associated with the class structure of society—many poor, few rich—so the Reverend was more or less in line with Karl Marx and Frederick Engels, both of whom also had scientific/mathematical credentials. However, the mid-19th century, what with the turbulence and revolutions of 1848 and thereabouts, was a time when the European establishments wanted—and needed—stability, so a series of publications by a Belgian mathematician, P-F. Verhulst (1838, 1845, 1847) was welcome. Verhulst postulated that the geometric rate of population increase would itself be reduced as a population got larger. To make the algebra simple, he further postulated that the relative rate of increase (that is, the absolute rate divided by the size of the population at that time) would be a linear function of population size. This leads to an S-shaped curve of population size plotted against time, reaching an equilibrium final size eventually, and with the fastest increase rate occurring when the population has reached about 50% of its final size. The phenomenon of declining rate of increase as population size increases has come to be called ***compensation*** or (negative) ***density-dependence***.

Successive generations of scientists have sought, with limited success, to identify the assumed phenomenon as the result of a recognizable biological *process* or complex of processes, and some have advised the use of this ***logistic*** S-curve for the management of major sea fisheries, most notably the tuna fisheries of the tropical eastern Pacific. Even economists have clutched it like a straw for insertion in econometric models, thinking—mistakenly—that it is a scientifically established theory in the biological sciences. The S-curve or ogive, especially a symmetrical one, has had a mystical appeal to some: Michael Graham—one of the "greats" of fisheries science—was particularly attracted to it and elaborated that in his classic book *The Fish Gate* (Graham 1943). But, direct evidence of density-dependence is weak, and evidence that it is linear with population size is non-existent (despite the misleading entry in Wikipedia on the subject). Verhulst's simplistic postulate has been conceptually useful, but can be disastrous in unthinking application. A later variation of the idea has been that the dependence is not linear, but more like a parabola or even a polynomial function of a higher order. Such modifications, nevertheless, retain two important features of the simple logistic. One is that the supposed equilibrium upper limit to population size (sometimes called "carrying capacity" or, in fishing and whaling circles, "virgin stock") is only reached after infinite time, and that applies to any restoration process after a resource has been exploited, but subsequently protected, as well as to the attainment of sustained exploitation at any chosen level of population and yield. As playwright Tom Stoppard once wrote: "Eternity's a terrible thought. I mean, where's it all going to end?" For more about eternity, I recommend John D. Barrow's *The Infinite Book* (2005), and for a view from the other end of the scale *The Book of Nothing* by the same author (Barrow 2001). And while we are on the subject of good books, I only recently discovered the writings of Nassim Nicholas Taleb, especially *Fooled by Randomness: The Hidden Role of Chance in Life and in the Markets* (2004) and *The Black Swan: The Impact of the Highly Improbable* (2007). Taleb, who is a market trader, does not like the normal/Gaussian form of a bell-shaped curve, considering it to have little empirical justification except in trivial pursuits like measuring the heights of people and calculating means and the spread of differences. I'm inclined to agree with him and was rather pleased to find myself not entirely alone on a limb of heresy or, worse, of apostasy. Taleb has two gripes. One, that the normal curve is symmetrical, while real life is asymmetrical. The other, that the tails of the normal curve seriously underestimate the likelihood of extreme events. Fair enough. And the paucity of empirical substantiation of a neat theory is exactly what worries me about the famous logistic curve of population growth, at the heart of the shaky discipline of population dynamics, *my* discipline.

An important recent shift in ideas about managing the use of renewable resources is the realization that *eternity* has to be abolished, and replaced by the notion of sustainability as referring to a finite, and not necessarily very long, period of time. In practical terms, this was first implemented by scientists working in the IWC who devised the Commission's Revised Management Procedure (RMP) in the 1990s, but its application is spreading to fisheries management advice more generally.

The other feature of the logistic and its relatives is that the highest rate of population increase occurs when the population is tiny. The opposite phenomenon, in which the rate of increase of small populations *increases* as population *increases* is called ***depensation***. Its existence results in, among other things, the possibility that a population which has been overexploited continues to move towards extinction even if exploitation ceases. If that extreme and irreversible event does not occur, because the strength of depensation is less than "critical," it will result in a slower recovery than expected, speeding up only when the population has grown considerably bigger. This is characteristic of the dynamics of many "initial" animal populations: a slow start, then take-off. So, in that little detail of the bend in a curve lies the devil of nasty behavior. Fisheries managers had better watch their backs.

Fisheries research and management have a history of it being assumed that there is no depensation unless data demonstrate that it exists, and classical statistical analysis has difficulty in producing such evidence. That approach is evidently a misuse of Occam's razor. William of Occam was a 14th century Franciscan friar who enunciated "the principle that the explanation of any phenomenon should involve as few assumptions as possible, eliminating those that make no difference in the observable predictions of the explanatory hypothesis or theory" (thanks to Wikipedia, which tells us that this is a heuristic maxim that advises economy, parsimony, or simplicity in scientific theories). The precautionary principle, on the other hand, would advise us that the safer option is to assume that depensation **does** exist and to formulate our statistical tests accordingly, taking care correctly to balance as well as seek to minimize both type I errors (false positives) and type II errors (false negatives). In any case, the idea that non-existence of depensation is the simpler assumption is entirely illusory; no more parameters are needed in a population model to include than needed to exclude it.

Sorry, yet another diversion. While making sure that I got Bill Occam's date and pronouncement right, I found, to my delight, a website dedicated to Occam's Toothbrush. What's that? "Making easy things hopelessly complex." With the byline: "If you're going to multiply propositions needlessly, you're going to need good teeth." Hurrah for the World Wide Web!

I think this notion of depensation is a clear case of where the study of the history of an idea and of its consequences serves conservation. So now, let's move to my second example: the ideas of maximum and optimum sustainable yields as objectives in managing the desired sustainable use of fisheries and wildlife resources. The "family" of logistic population models, and others, predict that the highest continuing (i.e., "sustainable") yield can be obtained by holding the population at some intermediate level between the "carrying capacity" and the low level at which depensation might come into play. There are many economic and other practical reasons why targeting on the maximum might not be a good idea, including computer simulations showing that management procedures intended to strike that target do not work well. The idea is biologically dangerous, too, because it puts all the emphasis on the size of the population and ignores its structure—by age, sex, and so on. Naïve efforts have been made in recent years to resolve the practical problems by defining other "reference points" than the maximum on a curve of sustainable yield against population size, but all of these depend—as does MSY itself—on knowing the shape and scale of the curve and where the population now is on it, and also having the ability to aim at a chosen reference point efficiently.

Despite all these serious problems, the MSY target is still present in many fisheries management plans, is at the heart of the UN Convention on the Law of the Sea, and was endorsed by the UN environment conferences in Rio de Janeiro and Johannesburg. And, the European Union is now considering replacing its so far disastrous Common Fisheries Policy (CFP) by one incorporating the MSY target. So, how has this situation come about, that a management objective which scientists know to be an unreachable mirage (and which would be undesirable even if it was real) has become virtually cast in stone in international treaties? The answer is in politics—surprised?

The period 1955–1958 was a very active one with respect to development of the Law of the Sea, partly as a consequence of the Truman administration claiming, a decade before, administrative and jurisdiction rights over the resources of the seas adjacent to the US coast, up to 200 nautical miles, an act which was swiftly followed by the more general territorial 200-mile claims of a number of Latin American states. The US delegations came to the various UN meetings determined that MSY should be the general objective for international fisheries management. Most of the Europeans—then the other major powers in fishing, apart from Japan and the USSR—were more inclined to looser, negotiable objectives and were also more ready than the USA to accept restrictions—in the cause of conservation—on fishing effort rather than merely on catches. The USA argued that a universal, numerically determined, physical catch limit objective was essential and that no international agreement was feasible on economically or socially defined reference points. It was probably correct in this, but it was not the real reason for the US insistence on the MSY target. To understand why, we have to look back a decade to the post-war negotiation of a peace treaty between Japan and the USA.

I was a staff member of the Food and Agriculture Organization of the UN (FAO) from 1953, but had also been involved—from the side of the UK government—in fisheries treaty negotiations with the USA, Canada, and other European states in 1949. So, I had some idea of what might be going on. But, I have also been fortunate in the last few years in meeting Dr. Mary Carmel Finley who has just completed her doctoral thesis on the history of Japan-US fishery relations, of which she has kindly given me a copy. It is entitled "The Tragedy of Enclosure: Fish, Fisheries Science, and U.S. Foreign Policy, 1920–1960" (Finley 2007). Dr. Finley reached me by coming across my name in US State Department files recently declassified under the 50-year rule, and getting in touch first with FAO and then with the IWC. So, what with as-yet unpublished theses and no-longer-secret cables from Rome to Washington DC, we are looking at a gray literature *par excellence*.

The American occupying power was sympathetic to the Japanese wish to re-engage in global high seas fishing and to that end, among other things, encouraged the resumption of Antarctic whaling. But at that time, by far the most valuable fisheries on the North American seaboard were for Pacific salmon *Oncorhynchus* spp., traditionally taken by US and Canadian fishermen in rivers and coastal waters. Japan wished to catch those salmon on the high seas; North Americans contested this—they said they had originally "developed" the fishery and were spending money and foregoing other benefits to keep the rivers good for spawning salmon. So, their problem was to halt the Japanese. To that end, the very influential fishery advisers in the US State Department—all of them biologists, incidentally—invented the "principle" of ***abstention***. The idea was that if a coastal state was fully utilizing high seas resources adjacent to its shores and managing that utilization on a sound scientific basis, then other states should refrain from entering the fishery. Pursuit of that principle involved defining "fully utilizing, etc.," and—Hey, presto! —the MSY target-notion provided the definition. Japan's acceptance of the abstention principle in the International North Pacific Fisheries Convention (INPFC) was one requirement for completion of the peace treaty; another was that Japan—which had always refused to participate in various conservation management measures adopted by the dominant Europeans—would join the IWC and abide by its rules.

One consequence of the adoption of the abstention principle was a gross distortion of scientific research. Under the INPFC, a line, roughly north–south, had to be drawn through the North Pacific Ocean to mark the separation of the salmon stocks that spawned in Asia from those spawning in North America and, therefore, subject to Japanese abstention. For years, great efforts were made to define the line, though it became pretty clear quite early on that there was no such fixed line. I have since then come across legal and administrative rules that have similarly distorted scientific efforts; the most recent, that I know best, is the corruption of research on whale stocks by Japanese efforts to justify whaling on commercial scale and with commercial intent by the issue of special permits for scientific purposes, and especially now to provide "evidence" that the fact that some whales eat some fish of commercial interest is a threat to fisheries and to the food security of humans.

When I was involved in the negotiations in Washington, DC for the creation of an International Convention for the Northwest Atlantic Fisheries (ICNAF, now transmuted to NAFO), being faced with pressure for insertion of the MSY target in the proposed new treaty, we in the UK delegation asked—among ourselves—why? One suspicion was that the USA, and possibly also Canada, wanted, one day, to be able to exclude the Scots and other Europeans from the then-great fisheries off their Atlantic coasts. To counter this, we began to talk about "historic rights," something the Japanese could not plausibly claim in the Northeast Pacific. So, the European Northwest Atlantic fisheries were "saved"—until they were practically destroyed by overfishing (despite ICNAF) and eventually by the declaration of 200-mile exclusive economic zones (EEZs). But, "historic rights" subsequently found their way into the negotiations that resulted in the UNCLOS.

Soon after the INPFC "saved" the salmon—and also the valuable Halibut *Hippoglossus stenolepis*—fisheries for the North Americans, the predominant US economic interest switched to Pacific tunas—which were also, of course, of great interest to the Japanese. Now US political priorities were reversed—US operators wanted to fish freely off the coasts of Latin American and other states, not only for the tunas themselves, but also for small bait fishes, which mostly lived close to shore. This development naturally divided, in terms of policy, the US west coast fishing industry, but the tuna interests prevailed, and the US government spent many years opposing—ultimately unsuccessfully—the "enclosures" being increasingly claimed by Latin Americans and others. Further big changes came when the tuna fishermen switched from baited lines to purse seines and no longer needed to venture close inshore off Central America and northwestern South America.

Returning to the matter of the revision of EU fishery policy, when I was invited to suggest specific changes, I faced a dilemma. Clearly, the European Commission wanted to ensure depleted stocks were permitted to increase to more biologically productive levels, as a major step towards renewal of fisheries in the European Union's EEZ, and I believed good management for long-term sustainability—and, more urgently, providing for recovery of severely depleted fish stocks—would call for abandoning the unreachable and inappropriate MSY target. However, a confrontation over that, in the face of the UNCLOS and the Rio and Johannesburg words, would, at this stage, probably not be successful, and I advised instead an objective based on the work of the IWC's Scientific Committee. This could, I thought, be regarded essentially as involving ***redefining*** MSY, not as the highest possible sustainable annual catch, but as a maximum **cumulative catch** (i.e., the sum of all catches) over a defined management period, consonant with meeting other objectives, principally a minimal probability that a stock will be unintentionally reduced below a defined threshold in any year during the period. A management algorithm to achieve this can be found by computer simulation and, if necessary, in such a way that it is effective and robust in being—for practical purposes—independent of any particular population model. This proposal and various responses to it are all in the grey literature of the European Commission, The Greens/EFA Group in the European Parliament, the European Policy Office of WWF, and in Parliamentary documents. Whether this little egg will hatch—and, if so, produce a monster—depends on the complex relations between the Commission (generally favoring the long-term view, and hence conservation); the Council of Ministers (which tends to favor short-term commercial views that please most of the present generation of fishermen); Parliament (rather unpredictable) and the scientific establishment (which tends to be schizophrenic, with some not wanting to admit the weakness of past methods and others hoping to build their careers on new approaches).

Choosing the duration of the defined management period is not a trivial matter. The IWC scientists chose 100 years. This was, at the time, the longest time for which available computers could practically carry out trial simulations. It happens to be also of the same order of magnitude as the lifespans of whales and humans, but is surely longer than a management instrument and system are likely to endure. For most fisheries, a period of 20 to, say, 50 years might be more appropriate. But, Norwegian government scientists associated with the whaling industry have recently suggested a longer period—two, or even three, centuries. Present computer power can cope with simulating that, but it is, of course, quite unrealistic to expect management instruments/arrangements/procedures to persist for so long; we are not living in Byzantium and running the Eastern Roman Empire for a millennium! But, there is method in Norwegian apparent madness. With a longer period, current and near-future whale catches could be much higher—which is naturally what the whalers want—but after a couple of centuries the then-depleted stocks could be subject to a much lower exploitation rate, and so be expected to recover to high levels by the end of the third century—and who is there to care now what happens to whales then?

Let me conclude by asserting that understanding of the birth and evolution of key ideas, notions, concepts, unproven theories, untested hypotheses, and glorious principles can help us invent better ones and thus serve conservation if that is our purpose. But, possibly more important: that history can help us distinguish between, and remember, what is a hypothesis, what is an observed set of facts, what is truly a principle, what is a mere assumption, and so on. In particular, to be aware of the process by which what is convenient assumption (like easing the application of the calculus, in Verhulst's case) becomes believed by nearly everyone, after a generation or so, to be reliable and incontrovertible fact on which can be based the management of marine ecosystems. I have learned that one should not underestimate the passage of time since an idea is first promulgated and when it may be lost, then reinvented. In fisheries research, a classic example of this was the rediscovery, after World War II, of a mathematical model by a Russian scientist F. I. Baranov (1918) which described a relationship between sustainable yield and fishing effort. I have only very recently come across an even older reference by a Professor G. Chrystal who, on 30 November 1901 wrote, in a letter to Mr. Noel Paton, an equation for the exponential decline of a fish stock, without recruitment, under conditions of steadily increasing efficiency of fishing operations (remarkably, if efficiency increases linearly with time, the trajectory of the stock is the right-hand side of a normal curve). Chrystal's letter was published in 1902 as an Appendix III to a report of the British House of Commons on fisheries (Chrystal 1902). I liked Prof. Chrystal's closing paragraph: "There are many questions to be solved, but they do not seem to be beyond the reach of statistics combined with patient and intelligent observation controlled by well-directed special experiments. Hasty conclusions and premature application of formulae are to be earnestly deprecated. All that mathematics can do at this stage is merely to assist the imagination." I don't know whether such reports count as gray literature, but they are pretty obscure as to their location; this one was found for me in the excellently catalogued library of the CEFAS Fisheries Laboratory in Lowestoft, Suffolk, England, by Andrew Payne, Editor-in-Chief of the *ICES Journal of Marine Science*.

So, as to the theme of this conference as a whole, my experience has been that such historical studies can only be pursued through access to the gray literature of all kinds. I have been making such studies on the whaling issue—particularly concerning Japan's policy—which would have been practically impossible if the IWC had not, until recently, kept verbatim records of its plenary sessions and—after some early resistance—been forced to put them in the public domain. Much of the crucial scientific work on which conservation actions are based remains in the gray documents of committees, not all being published eventually, and some never. These are immensely important because they usually contain detail that gets omitted from later publications. In looking at whaling history, the logbooks of the whalers are among the best sources of hard information; they are steadily being scrutinized, but finding funds for such laborious pursuits is difficult. And state archives are, as Dr. Finley found, treasure houses of revelations. But, with those, one must take care and be more than usually skeptical. She came across cables from Rome to Washington, during the Law of the Sea meetings in the 1950s, saying that I had recruited a Dutch biologist—a good friend and colleague, now deceased—to go on a prolonged mission to Panama, ostensibly to help the local shrimping industry, but actually to spy upon, and possibly interfere with, the burgeoning US tuna-fishing operations in the region. Bizarrely untrue! I suspect this happened because I had been spotted, during the UN technical meeting in Rome, being unusually friendly with the Soviet scientists who were experiencing their first excursion outside the USSR; I was helping edit their documents, written in not-very-good English!

Now, as promised, and finally, I return to the Rev. Thomas Malthus and exponential growth. His argument began, as did countless subsequent arguments, with an assumption that population growth is essentially geometric, and his successors made further assumptions about what factors would modify the rate of growth, and how. However, a Danish mathematical geneticist living in Greenland, Dr. Lars Witting, has demonstrated that the simple exponential assumption is incompatible with evolutionary processes as now understood. His idea is conceptually simple, but mathematically difficult; I won't bother you with the latter (Witting 1997). But as to concept, suppose each individual in an initial population has a tendency to reproduce periodically or even only once, and also a probability of dying one way or another, that the tendencies vary from one individual to another, and that the individual's tendencies have some genetic element, that is, they are at least partially genetically determined. Witting has shown that the population must evolve in such a manner that the population increase rate will not be exponential in form, but what he calls *hyper-exponential*. This has immense implications for population dynamics. In particular, it introduces long-term cycles in population size with different properties from those with which we have been familiar, arising from interactions between populations, such as predators and prey. One application of Witting's theory explains why, for instance, eastern Pacific gray whales *Eschrichtius robustus*, which were nearly exterminated by American commercial whaling in the 19th century, have not only recovered under decades of protection, but are now far more abundant than they ever were, despite the elimination of a large part of their breeding habitat along the coasts of California and Baja California, and possibly also by some destruction of their food supply by bottom trawling. This application also tells us that these whales are reaching the peak of a cycle, are about to decline naturally whether or not whaling is resumed, and hence—notwithstanding their abundance—now have zero or negative sustainable yield. This is in stark contrast with assessments made with the orthodox population dynamic models which say that catches of several hundred animals annually would be sustainable, i.e., would not cause the population to decline.

However, another paper has very recently been published by three other geneticists on the history of the gray whale; they have really put a cat among the pigeons. They claim that gray whales were enormously more abundant before 19th century commercial whaling than they are now. Whom to believe? It's not gray literature; it's in the august *Proceedings of the US National Academy of Science* (Alter 2007). How is such a discrepancy to be resolved? Not, I think, in the Scientific Committee of the IWC.

To summarize, I have given a few examples where critical examination of the history of certain ideas—exponential growth, density-dependent limitation of growth, sustainability, and so on—reveals that what begins as a hypothesis or an assumption can turn, over time, in peoples' minds, into a "theory" or an empirically validated fact, or even—almost—into an ideology. My thesis is that these are not isolated cases and that such study *can* serve conservation.

I close by acknowledging the value of conversations over the years with many friends and colleagues, some alive and some deceased, among them Sir Peter Scott, Dr. David Lavigne, Prof. Peter Jewell, Dr. Philip Thomas, Ms. Melanie Salmon, Dr. Mario Ruivo, Ms. Leslie Busby, Dr. Mary Carmel Finley, Dr. Geoffrey Kesteven, Dr. Vassili Papastavrou, Mr. David McTaggart, Dr. Justin Cooke, Dr. William de la Mare, Dr. Lyall Watson, Lord May of Oxford, Dr. Philip Clapham, Mr. Maxwell Bruce, Dr. Kees Lankester, Mme Elisabeth Mann Borgese, and Dr. Salvino Busuttil. None of them are, in the least, to blame for my errors and misunderstandings.

References

Alter, S. E., E. Rynes, and S. R. Palumbi. 2007. DNA evidence for historic population size and past ecosystem impacts of gray whales. Proceedings of the National Academy of Sciences of the United States of America 104(38):15162–15167.

Anderson, K. L. Editor. 2008. Changes on the Horizon. Proceedings of the 33rd Annual Conference of the International Association of Aquatic and Marine Science Libraries and Information Centers, 7–11 October, 2007, Sarasota, Florida. IAMSLIC, Fort Pierce, FL. https://hdl.handle.net/1912/2946.

Baranov, T. I. 1918. On the question of the biological basis of fisheries. Nauchnyi issledovatelskii ikhtiologischeskii Institut Isvestia 1:81–128.

Barrow, J. D. 2005. The infinite book: a short guide to the boundless, timeless and endless. Pantheon, New York.

Barrow, J. D. 2001. The book of nothing: vacuums, voids, and the latest ideas about the origins of the universe. Pantheon, New York.

Beder, S. 2006. The changing face of conservation: commodification, privatisation and the free market. Pages 83–97 *in* D. M. Lavigne, editor. Gaining ground: in pursuit of ecological sustainability. International Fund for Animal Welfare, Guelph, Canada and the University of Limerick, Limerick, Ireland.

Chrystal, G. 1902. Appendix III to the Report of the Committee (of the British House of Commons) Appointed to inquire into and Report as to the Best Means by which the State or Local Authorities can Assist Scientific Research as applied to problems affecting the Fisheries of Great Britain and Ireland, with Minutes of Evidence, Appendix and Index, pp. 56–57; 136, HMSO, London.

Finley, M. C. 2007. The tragedy of enclosure: fish, fisheries science, and U.S. foreign policy, 1920–1960. PhD Thesis in History (Science Studies), University of California, San Diego.

Graham, M. 1943. The fish gate. Faber and Faber, London.

Harremoës, P., D. Gee, M. MacGarvin, A. Stirling, J. Keys, B. Wynne, and S. Guedes Vaz. Editors. 2001. Late lessons from early warnings: the precautionary principle 1896–2000. Environmental Issue Report 22. European Environment Agency.

Holt, S. J. 2006. The notion of sustainability. Pages 43–82 *in* D. M. Lavigne, editor. Gaining ground: in pursuit of ecological sustainability. International Fund for Animal Welfare, Guelph, Canada and the University of Limerick, Limerick, Ireland.

Holt, S. J. 2008. Does history serve conservation? Pages 3–14 *in* K. L. Anderson, editor. Changes on the horizon. Proceedings of the 33rd Annual Conference of the International Association of Aquatic and Marine Science Libraries and Information Centers, 7–11 October, 2007, Sarasota, Florida. IAMSLIC, Fort Pierce, FL. https://hdl.handle.net/1912/2946.

Lavigne, D. M. 2006. Gaining ground: in pursuit of ecological sustainability. International Fund for Animal Welfare, Guelph, Canada and the University of Limerick, Limerick, Ireland.

Malthus, T. R. 1798. An essay on the principle of population. J. Johnson, London.

Morishita, J. 2008. What is the ecosystem approach for fisheries management? Marine Policy 32(1):19–26.

Rees, W. E. 2006. Why conventional economic logic won't protect biodiversity. Pages 207–226 *in* D. M. Lavigne, editor. Gaining ground: in pursuit of ecological sustainability. International Fund for Animal Welfare, Guelph, Canada and the University of Limerick, Limerick, Ireland.

Taleb, N. N. 2004. Fooled by randomness: the hidden role of chance in life and in the markets. Cengage South-Western, Boston.

Taleb, N. N. 2007. The black swan: the impact of the highly improbable. Random House, New York.

Verhulst, P-F. 1838. Notice sur la loi que la population suit dans son accroissement. Correspondance mathématique et physique 10:113–121.

Verhulst, P-F. 1845. Recherches mathématiques sur la loi d'accroissement de la population. Nouveaux Mémoires de l'Académie Royale des Sciences et Belles-Lettres de Bruxelles 18:1–42.

Verhulst, P-F. 1847. Deuxième mémoire sur la loi d'accroissement de la population. Mémoires de l'Académie Royale des Sciences, des Lettres et des Beaux-Arts de Belgique 20:1–32.

Witting, L. 1997. A general theory of evolution: by means of selection by density dependent competitive interactions. Peregrine Publisher, Århus, Denmark.

Chapter 42

Greed Enthroned: Shall Future Generations Eat Fish or Whales?[1]

I confess that when I was invited to give this lecture, I knew little about Gresham College except that it had once been the home of the Royal Society. Accordingly, I plunged into the magical digital world to learn more. I soon found myself immersed in esoteric accounts of Freemasonry, alchemy, Rosicrucianism, British Israelism, the Knights Templar, the ideas of reconstructing London after the Great Fire as the New Jerusalem, and much, much more.[2] Down-to-earth websites told me the college was founded more than four centuries ago under the bequest of Sir Thomas Gresham and was closely connected with the Mercers' livery company (Numero uno in the hierarchy of the livery companies) as well as the Corporation of London. Another Worshipful City company, Haberdashers', was an offshoot of Mercers' (It's now number eight in the hierarchy), and I was educated at Haberdashers' Aske's Hampstead School. We usually beat Mercers' School at rugby, and I was sorry to read that it closed in 1959. Then, in the published history of the college, I read that Sir Thomas's uncle, Sir John Gresham, had founded Gresham School in Holt, Norfolk, which still functions. There might be my family's origins, which we think was in East Anglia, "holt" being the Norse word for a small woodland, or copse. Then, I read that the Worshipful Company of Fishmongers (fourth in the hierarchy) has been the Trustee of that school since 1555. My first job (1947) was at the Government's Fisheries Laboratory, in Lowestoft, and some of us used to visit Billingsgate from time to time and pay homage to Fish Hall at London Bridge. Which brings me to my subject.

You will see that I have modified the title of my talk. That is because, although I shall focus on fishes and whales, I think my subject could be of wider interest. The day after I decided on the change, the *naufrage* of the American Insurance Group (AIG) erupted in the media and much talk began about the greed of bankers, insurers, CEOs bearing golden parachutes, shareholders, and investors. I know very little about such matters, but it seemed to come down to the question of an imperative—said to be embedded in the current version of globalized capitalism—to maximize profits by any legal means and to ensure, when necessary, changes in law to facilitate that. And that is conceptually close to what the management of fishing and whaling, and the conservation of fishes and whales, are about. A few days later—September 28, 2008—the US House of Representatives was engaged in revising the Administration's proposal for "bailing out" the financial system, and then voting to reject the revised version.

---

[1]Editors' note: a lecture by Sidney Holt at Gresham College, London, November 19, 2008. Permission to reproduce, with minimal editing, this lecture was given by Professor Michael Mainelli and by the International Fund for Animal Welfare who encouraged and facilitated his participation in this lecture. Audio and video versions of the presentation are available at https://www.gresham.ac.uk/lectures-and-events/shall-future-generations-eat-fish-or-whales.
[2]Thanks largely to the WWW and Gilbert (2003).

Greed was rapidly becoming the flavor of the year. The theories of the Chicago Gang of Free Marketeers, led by Professor Milton Friedman, so warmly embraced by the chemist Margaret Thatcher and popular actor Ronald Reagan, seemed to be in trouble. Exactly four decades earlier, Professor Garrett Hardin had published his controversial *The Tragedy of the Commons* (Hardin 1968) that put a powerful phrase into our language, describing the fate of unregulated use of limited resources. It now seemed that the scarcely regulated financial system of 21st century capitalism constituted another, larger, Commons.

In pursuit of my revised title and thesis, I propose to spout three parables and fill the interstices with examples and attempted explanations. The first is a parable of connectedness. The second is a story of illusion; the third is one of the relationship between greed and hegemony and empire.

On 28 September [2008], the *New York Times* published an editorial by Eduardo Porter that I cannot resist quoting. Porter wrote: "Of course, it's all Gordon Gekko's fault![3] 'Greed and irresponsibility,' blasted Barack Obama. 'Greed and excess and corruption,' charged John McCain. President Luiz Inácio Lula da Silva could tell from as far away as Brazil that the 'boundless greed of a few' blew up the American financial system. With the Eureka moment behind us, I would suggest that this insight offers a way to try to restore the abused financial markets to health. If greed is to blame, the question is whether we can line up a reasonable array of alternative incentives—and disincentives—to do away with greed for good."[4]

In an earlier period of financial chaos—the Great Depression of the 1930s—President Franklin D. Roosevelt said, in his second inauguration address: "We have always known that heedless self-interest was bad morals; we know now that it is bad economics."

Early Christians designated greed as the second of Pope Gregory the Great's and Dante's seven cardinal sins. *Avaritia male est.* Looking further back, a 21st century blogger reminded we Web-surfers that Zeus had decided to sink the idyllic island of Atlantis because its people had become too greedy. I'm addicted to the excellent website *Wordsmith* that I peruse every morning over coffee. A few weeks ago, it introduced me to **pleonexia**, meaning "excessive or insatiable covetousness," or, as the Christian activist Don McClanen put it: "an insatiable need for what I already have."[5]

Eduardo Porter went on to write "Capitalism, in all its cleverness, decided that what you can't beat, you should use. It worked to harness greed. To be able to discuss it in polite company, economists renamed it 'maximization of utility,' and built a theory of the world that everyone benefits when we seek to maximize our own individual welfare." He still had hopes, he wrote, "Because there is a crucial brake that has been missing from the edifice of high-tech financial capitalism—a counterbalance at the other end of the scale from where utility gets maximized. That brake is *fear*. My suggestion, then, is to put fear back in the picture."

---

[3]Gekko is a fictional character from the 1987 film *Wall Street*, portrayed by actor-producer Michael Douglas.

[4]Porter continued: "This will be no easy task. From populist opprobrium to elitist disdain—standard social behavioral devices have proved unable to dent humanity's greedy nature. ... Even religion's not insubstantial powers of persuasion (think Hell) and coercion (think Inquisition) have proved insufficient to blot out this insidious sin. The free market, it should be obvious by now, hasn't been up to the task either. Greed reached its zenith in the 1980s which brought supply-side economics and its bedrock belief that the path to prosperity for all required removing every obstacle to utility maximization, including most regulations and taxes. Then financial markets crashed. The Rev. Jesse Jackson lambasted America's greedy corporations. And one survey found that 83% of Americans blamed 'unmitigated greed' for the financial crisis. A few years later the markets were again soaring; greed was back in style."

[5]See Jaye Scholl's article in Barrons (New York), September 18, 2000: "Don McClaren offers the wealthy a different kind of freedom."

But fear of what, exactly? Fear of the nasty consequences of greed? Or fear of the consequences of not exercising one's greed, such as the fear of a fisher who, refraining from taking a fish now rather than later (when it has grown bigger or reproduced), worries that someone else will come and take it, or that it will die one way or another—perhaps by being eaten by a whale—before he gets back to it.

It used to be said that in not caring for the conservation of finite resources we are borrowing from our children with little intention of arranging for them to be paid back. By the standard of maximizing present utility, it may be that the debts thereby incurred will never be repaid. Will Nature repossess the planet? Is that the inevitable consequence of the rhetorical question "What have future generations ever done for me?"

Another *New York Times* editorialist, Barbara Ehrenreich, wrote on September 23 [2008] "Greed—and its crafty sibling, speculation—are the designated culprits for the financial crisis. But another, much admired, habit of mind should get its share of the blame: the delusional optimism of mainstream, all-American, *positive thinking*." She continued, brilliantly, to advocate *The Power of Negative Thinking*. Vice President Dan Quayle once defined positive thinking well: "The future will be better tomorrow." I would assert that in many situations "positive thinking" sends us into a state of denial of facts and real processes and, in the case of the depletion of living resources, impedes the implementation of the precautionary approach or principle adopted by UN conferences on the environment and accepted, in principle, by most governments.

The corresponding virtue is—at least for Roman Catholics—charity. For the agnostics among us, frugality or perhaps parsimony might be better words.[6] For those of you who enjoy Latin tags, here's another one relevant to this talk, at least insofar as it concerns whales and whaling: *Crudelitatis mater est avaritia*, Greed is the mother of cruelty.

Awareness that the needs of others are denied is essential to the concept of greed. The demon who tempts to greed is Mammon, and although Mammon is for us usually associated with money, he is interested also in other sorts of wealth. And, significantly for us, Mammon was also personified in the Middle Ages as the Demon of Injustice, and his ideas brought into the 16th and 17th centuries by Edmund Spencer and John Milton.[7] Later occultist writings such as Collin De Plancy's *Dictionnaire Infernal* (Paris, 1818, three years after the battle of Waterloo) describe Mammon as Hell's ambassador to England, so that should put this audience on guard.[8]

Now, why am I going on so about greed? Primarily because I shall be talking mainly about injustice to future generations of humans, but the revolutionary notion of injustice to other species should also be brought into the picture. The work of preparing for this talk was eased by the publication in October of a study by the Agriculture and Development Department of the World Bank, assisted by staff of the Fisheries and Aquaculture Department of the Food and Agriculture Organization of the United Nations (FAO): *The Sunken Billions: The Economic Justification for Fisheries Reform* (World Bank and FAO 2009). Here are what I think are its most important findings:

> Long before the fuel price increases of 2008, the economic health of the world's marine capture fisheries has been in decline. The build-up of redundant fishing fleet

[6] The best non-technical account of this idea of which I know is Muddling towards Frugality: A Blueprint for Survival in the 1980s (Johnson (1979).

[7] Educated, respectively, I'm glad to say, in the City of London at Merchant Taylors' School and St Paul's School, which was founded by Sir Henry Colet, Lord Mayor and also a distinguished member of the Mercers' Company.

[8] De Plancy was a free-thinker, influenced by Voltaire.

> capacity, deployment of increasingly powerful fishing technologies and increasing pollution and habitat loss have depleted fish stocks worldwide. Despite the increased fishing effort, the global marine catch has been stagnant for over a decade, at about 80 to 85 million tonnes, while the natural fish capital has declined. At the same time the margin has narrowed between the global costs of catching and the value of the catch. In many cases the catching operations are buoyed by subsidies, so that the global fishery economy to the point of landing is in deficit. The cumulative economic loss to the global economy 1974 to 2007 has been about two trillion US dollars. The difference between the potential and actual net economic benefits from marine fisheries is conservatively estimated as about $50 billion per year.[9] Improved governance of marine fisheries could capture a substantial part of this annual economic loss. Reform of the fisheries sector could generate additional economic growth and alternative livelihoods, both in the marine economy and in other sectors.

Half of the marine capture fish stocks monitored by the FAO are said to be *fully exploited*, producing at or close to their maximum sustainable yield (MSY). Another 25% of the stocks are assessed as "overexploited," "depleted," or "recovering from depletion" and are rendering less than their MSY. The remaining 25% of the stocks are "under-" or "moderately exploited" and, while this implies that more could be produced from them, many are of low-value species or species for which harvesting may be uneconomical. Globally, the proportion of fully exploited and overexploited, depleted, or recovering stocks has increased from just over 50% of all those assessed in the mid-1970s to about 75% in 2005. So, in economic terms, more than 75% of the world's marine capture fisheries are underperforming or subject to economic overfishing. In 1974, about 40% of the assessed stocks were rated as underexploited or moderately exploited; this had fallen to 25% by 2005.

According to the World Bank, "...global production of seafood from wild stocks is at or close to its long run biological maximum." I do not believe this, and I am publishing elsewhere the technical reasons for my opinion (Holt 2009). While the Bank's conclusions are, I think, qualitatively correct, the numbers and the methodology of calculating them, are questionable. The scale of losses is uncertain, but probably wildly underestimated by the Bank. Possibly the most important conclusion [*Holt's paraphrasing*] from the World Bank study is this:

> If the overall MSY is about 95 million tonnes—as is assumed in the study—the current (2004) recorded catch being 86 million tonnes, worth about $80 billion, then the "optimum economic yield" would be just over 80 million tonnes, i.e., 6% less than the current catch by weight and 15% less than the Bank's assumed MSY. The difference in total market value of the catches would be less than that because the optimal catch would contain more of the larger, more valuable species than the current catch, and also the average sizes of individuals of the valuable species would be higher and generally more valuable per kilo. The optimal catch would, however, be taken with a fishing "effort" (size of fleet, time operating) roughly half as much as the current effort. The optimal catch would be taken from an overall biomass of fish in the ocean

[9]The authors' 95% confidence interval for the lost economic benefits was between $26 and $72 billion for the base year, 2004. The cumulative period begins in 1974 because that was the reference year of FAO's first global marine fisheries resource assessment.

two- to threefold the current biomass (the difference depends on which of two theoretical fish population models you believe), attained by allowing overfished stocks to increase naturally.

This general conclusion is not really new. In the 1940s, the English naturalist and later Director of Fisheries Research, Michael Graham, showed that by reducing trawling effort on North Sea bottomfish stocks, the same catch could be obtained with about half the effort. My then colleague Ray Beverton and I demonstrated the same effect for Plaice *Pleuronectes platessa* and Haddock *Melanogrammus aeglefinus* stocks in the North Sea, Graham's assumptions having been confirmed by observation of how much the stocks had increased during World War II because of the sharp reduction of fishing activities. What *is* new is the fact that this phenomenon is now global and applies to the total of marine fisheries.[10]

However, the World Bank report does not do justice to the fact that reaching any optimal state from where we are now will be painful. It would require acceptance of a possibly prolonged reduction of catches while depleted stocks recover. Some fisheries would, as a result, become unprofitable during the recovery period (if they were not so already), and the only sensible thing might be to subject them to a moratorium, as has already happened with a few fish stocks and, as we shall see, with whaling. This is how our ability to control greed will be tested.

Now, although I am not qualified to question the economic assessments in *Sunken Billions*, I do question some of the biological science quoted. Too many numbers referred to as "estimates" are, in fact, assumptions or guesses. The crucial assumption made that the combined MSY of all fished stocks is a little more than the current reported annual catch is particularly questionable, yet that number is the main determinant of the study's specific conclusions. It is taken to be 95 million tonnes, but available data could easily allow it to be at least double that. The report's discussion of whether the MSY might or might not be as high as 101 million tonnes is bizarre scholasticism. So, while we can accept that the economic losses from lack of, or bad, management have been, and continue to be, enormous, we do not need to fix our sights on particular figures. Furthermore, the methodology adopted—based on treating all species in the global commercial catches as one huge collective lump of "biomass" and fitting to it an assumed curve of the type used in the 19th century to describe the expansion and ultimate limitation of human populations—was not even minimally examined as to whether it is mathematically or biologically defensible.

That being said, I propose to outline a few of the reasons why we have arrived at the deplorable situation described in the World Bank report and in terms not commonly applied to this issue.

Ecologists and ecologically informed campaigners, and some political scientists, are fond of saying "Everything is connected." Here is an example of that truism. During the summer of this year, there were many news accounts of piracy in the Indian Ocean, particularly off the coast of Somalia. Certain TV reporters managed to talk to the pirates who had taken a Ukrainian ship full of heavy armaments said to be on their way to the Sudan via Kenya. Many of the pirates were Somali fishermen. They had taken up their new profession

---

[10]A useful adjunct to the Bank report, in plain language, would be Professor Elmer A. Keen's little book *Ownership and Productivity of Marine Fishery Resources: An Essay on the Resolution of Conflict in the Use of the Ocean Pastures* (Keen 1988). Of course, the global picture was not then so dire as now.

because they could no longer make a living from fishing. They said this was because large fishing vessels were operating in those waters, coming from Europe and East Asian countries, and scouring the sea of fish. Those vessels operated without hindrance within 200 miles of the Somali coast because Somalia has no government and, so, there has, in recent years, been no control of any kind over its 200-mile-wide Exclusive Economic Zone (EEZ). So why is there no government? The reasons are multiple and complex, but one is that the land was caught up in the cold war. When the British colony comprised of the Seychelles, Mauritius, and the Chagos Archipelago became independent, it was broken three ways into two new independent states and the British Indian Ocean Territory, the largest island of which is Diego Garcia. Between 1967 and 1971, the long-term Chagos residents—whose very existence had, at first, been denied by the British Government—were expelled so that the largest island, Diego Garcia, could be handed over to make a naval and air base from which the United States could, if circumstances required, bombard the Middle East and Southern Asia, and, as it later turned out, use as a destination for rendition flights. The USSR wanted a corresponding military base in the Indian Ocean and sought an arrangement with the newly independent Republic of Seychelles. But, Seychelles, guided by President Albert René, opted for non-alignment and neutrality.[11] The next option was Somalia, the government of which, under the presidency of Mohamed Siad Barre, was Soviet-friendly at the time. I was working in FAO then and hired a well-known fisheries scientist, Dr. Mikhail Edelman, a Soviet citizen, to go and work there, setting up a fisheries administration and trying to introduce marine research. It was going well, and I was sending Mikhail, who was passionately interested in classical music, packages of recordings unobtainable in Somalia. The Barre regime went pear-shaped in 1991, with, it is widely believed, a little help from the USA via Ethiopia; since then, Chaos has reigned.

So, one fisheries problem is the propensity to go elsewhere, to fish in someone else's waters—maybe with a license from the coastal state, maybe not—when one has depleted the resources in one's own waters. Or, go to the high seas where control is nominal. Consider this: more than half of the fish now being consumed in the European Union are caught outside the waters governed by the EU's Common Fisheries Policy, caught by vessels flying European flags or others. This situation is not unique to Europe: over 80% of the seafood consumed in the USA is now imported, causing a trade deficit in this sector of $9 billion annually.[12] A large part of the global fish catch—variously estimated as between one-third and two-thirds of the total—is taken in what is now designated as IUU fishing—illegal, unregulated, and unreported—much of it under flags of convenience (FoC), and the practice of "flag-hopping," to circumvent regional conservation and management measures, is rife.[13]

---

[11]Some of us on the Seychelles delegation to the International Whaling Commission (IWC) in the early 1980s—particularly Dr. Lyall Watson—had a special interest in the Horn of Africa—specifically the low-lying Afar Triangle region that abuts Djibouti (ex-French Somaliland), as the favored place for exploring human origins, especially our possible aquatic antecedents. I'm rather glad that funding for our little expedition didn't come through!

[12]See Rubino (2008). Marine aquaculture enthusiasts assert that this industry "has the potential, if it is properly managed and profitable (aye, there's the rub *sjh*) to supplement the currently diminishing wild fisheries and to provide more of our seafood." Also see Halvorson and Dulf (2008), which I have no time to discuss, but one major problem is the supply of feed for caged fish, which generally must include small wild-caught fishes or products from them.

[13]The international trade in IUU-caught fish was estimated to have been worth between 3 and 10 billion Euros in 2004. The legal landings by the EU fleet were worth just under 7 billion Euros.

Now, for another excursion into connectedness. The World Bank report bases calculations on maximum sustainable yield. This notion is embedded in the UN Convention on the Law of the Sea (UNCLOS); attaining it is deemed to be the universal aim of fisheries management, though hedged by various weasel phrases. Scientists have, for 50 years, been very critical of this idea for a variety of reasons that I do not have time to explain here. The World Bank at least recognized one big fault of the notion, which I was writing about in the late 1940s with Ray Beverton, that mathematical economists such as Prof. Colin W. Clark at UBC, Vancouver have since analyzed in depth (Clark 2006). What matters is not the absolute size of the catch, but the difference between its value and what it cost to take it—the net economic yield (NEY). However, when it comes to international regulation of fishing, a maximum NEY is not easy to agree as a target, so practically all international management is based on regulated catch sizes, commonly expressed as TACs—total allowable catches—and the requirement that these be sustainable.

Now, I beg your indulgence for what American educators might call Population Dynamics 101. The notion of sustainability—that universal buzzword—in managing the exploitation of wild animals such as fishes and whales goes back to a Belgian mathematician, Pierre-François Verhulst, who published, in 1838, an equation that sought to explain why the Rev. Thomas Malthus was wrong in thinking that human and animal populations would grow indefinitely, that is geometrically (exponentially) until they collapsed from lack of basic resources (Malthus 1798). The essence of the equation was what ecologists call density-dependence or compensation, meaning that the relative rate of population increase (that is, the absolute rate—e.g., numbers per year divided by the size of the population at any given time) decreases as the population gets bigger until eventually the population ceases to grow. This process would be recognized now as one of negative feedback, a process that in nature tends to stop runaway change and promote peace (equilibrium).

Verhulst's equation leads to a curve of population growth through time, which is S-shaped, called in the scientific literature a sigmoid or ogive. As a population grows, the *absolute* rate of increase at first increases, then decreases. Where it bends, at some intermediate size—what we call the *inflexion*—is the maximum rate of growth. If we can, by fishing, adjust our effort (number of ships and their size, amount of time spent fishing, and so on) so that the population size is held at that inflexion, then we can forever continue to take the maximum sustainable yield, year after year. If a population is reduced, by too much fishing effort, to below the level of the inflexion, it is said to be overfished. That is the essence of the simple theory of fishing and much of the practice of what Michael Graham called, in 1943, *rational fishing*.[14] I'll not worry you here with the details—such as that there will be variations from year to year and perhaps cyclic changes and other forms of disequilibrium.

Verhulst had no computer. He did what paper-and-pencil mathematicians are inclined to do: he supposed that density-dependence was linear; that is, for every increment of population number the relative rate of increase would diminish by the same amount. This gives a *symmetrical* sigmoid curve of population against time, in which the inflexion occurs when the population has attained one-half of its final number called, in ecology, the carrying capacity. Such a curve is called a logistic. As often happens with such things, Verhulst's simple as-

[14]The classic reference is his *The Fish Gate* (Graham 1943). Graham—and Beverton and I—also looked at another type of overfishing called "growth overfishing," in which too many fish are caught before they have each sufficiently grown in size. Most overfishing is actually a combination of both kinds.

sumption of linearity, made for algebraic simplicity, came to be thought of, as time passed, as a scientific finding, an observed fact. There was no empirical evidence for it then, and there is not now.[15] Yet, the World Bank study used it, as many others have done before them, as the given, biologically endorsed wisdom.[16]

MSY as a universal objective of management dates to the immediate post-World War II period when the United States was negotiating a peace treaty with defeated Japan. Japanese people were starving, in need of fats especially (as was Europe), and were traditionally seafood eaters. General MacArthur arranged for them to return to factory-ship whaling in the Antarctic, on which they had embarked in the mid-1930s.[17] The Japanese fishing industry wanted to operate again practically world-wide, and that was OK with the American occupiers so long as Japan kept out of the Northeast Pacific where the United States' most important fishing industry was located—the Pacific salmon *Oncorhynchus* spp., shared, of course, with Canada. The US negotiators invented a new "principle," the principle of abstention. This was that if a coastal state was *fully utilizing* a marine fishery resource in waters adjacent to its shores, other states should refrain from exploiting that resource. So, what did "fully utilizing" mean? US scientific advisers came up with the idea that it meant the resource was being managed for MSY. From that point, this objective was pursued relentlessly by the US fisheries diplomats everywhere, being introduced to the global UN fisheries negotiations of 1955 and 1958.[18] It later proved somewhat embarrassing; the abstention principle was dropped, although MSY unfortunately survived. This was because the salmon industry declined and the tuna industry grew to dominate the US West coast and Washington's ongoing fisheries negotiations. The tuna industry in the early years needed bait fishes, and these were best obtained from the coastal waters of Pacific Latin American countries; roles were thus reversed. In the UNCLOS, the old abstention principle has effectively mutated; it now says that if a coastal state is *not* fully utilizing the fish resources in its waters, it should license others to take the "surplus."

In 1986, the Australian engineer William de la Mare performed a computer simulation of a procedure intended to bring an exploited stock of whales to MSY level by setting TACs, which the whaling fraternity calls catch limits. He showed that the best procedures would fail and lead to overexploitation even when there were excellent data and the biological parameters perfectly known (de la Mare 1986). We might say that enthroned, "managed" greed, institutionalized in MSY management theory, breeds destruction. De la Mare was working in the context of the effort by scientists associated with the IWC to develop a safe management system for any renewal of commercial whaling. That could bring me to my second topic—*The Whaling Problem*. But first, let me offer you a brief look at how official bodies have, in prac-

[15]For those who would like to read more about all this, I recommend my own chapter, "The Notion of Sustainability" (Holt 2006), in a book entitled *Gaining Ground: in Pursuit of Ecological Sustainability* (Lavigne 2006). Several other chapters of that book would also amply repay your attention. And, again blowing my own trumpet, my chapter "Sustainable Use of Wild Living Marine Resources: Notion or Myth?" (Holt 2008) in *Foundations of Environmental Sustainability* (Rockwood et al. 2008).

[16]The authors did also use another equation, closely related to the logistic, in which the maximum sustainable yield comes when the exploited population is at 36%, rather than 50%, of the carrying capacity. There is precious little empirical evidence for that either.

[17]The Japanese excursions in the 1930s were not, as current propaganda would have it, to obtain food for people, but indirectly to acquire fuel oil for the military machine. Thereby hangs another tale.

[18]This political process has been investigated thoroughly by a historian, Mary Carmel Finley. She is writing it up as a book, but meanwhile, her study is available at the University of California San Diego as a 2007 doctoral thesis entitled "The Tragedy of Enclosure: Fish, Fisheries Science, and Foreign Policy, 1920–1960."

tice, been dealing with the matter of stabilizing fisheries, preventing or curing overfishing, and bringing them to a state of sustainability, high productivity and—maybe—profitability.

Where better to do that than by reference to the EU's Common Fisheries Policy, which is supposedly dedicated to sustainability and also to the recovery of depleted resources. Advice on TACs is provided to the EU by the International Council for the Exploration of the Sea (ICES), an intergovernmental body dedicated to science, established early in the 20th century. ICES advice is transmitted to the European Commission, which advises the decision-making body of the EU—the Council of Ministers. The most common scenario is that the best estimate of a sustainable TAC offered by ICES is topped up by the Commission and topped up further by the Council of Ministers to satisfy the short-term interests of the fishing industries. To make matters worse, the actual kill is often much bigger than the TAC for two reasons—huge quantities of undersized or unsalable fish are discarded at sea, and the continuing blight of IUU fishing. It is almost as if it is thought possible to negotiate with Nature, who actually decides how many fish will be born and how many will die naturally, and hence what might be any "surplus" available to humans. ICES and other scientists can make mistakes—sometimes big ones—and sometimes they are overoptimistic and sometimes too pessimistic, but they are nevertheless better at sounding out Nature than the others, even than the fishermen themselves.[19]

I'll give you just one example of this process: the Anchovy *Engraulis encrasicholus* fishery in the Northeast Atlantic. ICES advice for a sustainable catch in 2003 was 12,500 tonnes. The Commission advice to the Ministers was 19,800 tonnes. The Ministers settled on 33,000 tonnes. The industry could not find all those anchovies; the reported catch was 10,600 tonnes. The following year, the ICES advice was 11,000 tonnes, and this was accepted by the Commission. The Ministers, however, stayed with 33,000 tonnes. The reported catch was 16,400 tonnes. This upward swing was not surprising as anchovies are short-lived fish with highly variable reproduction. Nevertheless, ICES found it could advise only 5,000 tonnes for 2005. The Commission accepted this, and the Ministers generously reduced the TAC from 33,000 to 30,000! The actual catch that year was 1,100 tonnes, this from a fishery that had yielded an average of 30,000–40,000 tonnes per year from 1990 to 2001.

I could also quote you examples for Cod *Gadus morhua*, Herring *Clupea harengus*, and other staples among the historic fisheries of the Northeast Atlantic, for tunas in several regions, and for the relatively new fisheries of the Antarctic zone of the great Southern Ocean.

So, at last, to the whales. First, I note that the International Convention for the Regulation of Whaling 1946 (ICRW), which provides the umbrella for the IWC, was the first international treaty to refer explicitly to the interests of future human generations. The first paragraph of its Preamble reads: "Recognizing the interest of the nations of the world in safeguarding for future generations the great natural resources represented by the whale stocks." In view of what actually happened in the following 40 years, as these "great resources" were practically exterminated by vast fleets from Europe, Japan, and the USSR, we might exclaim—as Vice-President-Elect Joe Biden did when asked by an interviewer whether "Spread the Wealth" Obama was a Marxist who believed in "from each according to his ability; to each according to his need"—"Are you joking?!"

---

[19]There was a time, in the IWC, when the scientists used to top up their own advice in the belief that it was unwise to suggest scientifically determined sustainable catch limits so low that decision-makers would reject them out of hand.

Now, a look back in history. "Modern whaling,"[20] by the Norwegian method—that is, using powered catcher boats, cannons firing harpoons tipped with explosive grenades, and with air pumps to keep dead whales afloat—was, from the beginning, closely connected with militarism and imperialism, especially their manifestations in the Antarctic. Whaling was a reason for British tenacity in holding the Falklands Islands and the Dependency—the Islands of South Georgia, the South Shetlands, the South Orkneys, and the South Sandwich Islands—all of them, at one time, whaling bases. The Norwegian whalers paid license fees to operate land stations on South Georgia before the First World War, although they eventually escaped by inventing the whaling factory ship with a stern ramp up which dead whales could be hauled. This new technology also opened the entire Southern Ocean to modern whaling. Before and during that war, the British threatened to cut coal supplies to Norway, on which Norwegian industry depended for all its energy, unless its trade in whale oil with Germany ceased; this was because the oil from baleen ("whalebone") whales—called "whale oil" in the trade and statistics—was the raw material from which glycerin was made, in turn the raw material for explosives.

After the war, the relationship between Britain and Norway was two-sided. On the one hand, the companies in the two nations were fiercely competitive, but on the other, they collaborated in efforts to sustain an Anglo-Norse monopoly of Antarctic whaling. The special relationship is illustrated by the fact that the biggest whaling company anywhere, ever, was Salvesen of Leith, Scotland, of Norwegian descent. The company's history illustrates how commercial whaling can be biologically unsustainable, yet economically "sustainable." Salvesen, which began in 1851 as ship brokers and timber merchants, went into whaling in the Arctic at the beginning of the 20th century, shifted later to the far more profitable Antarctic, made so much money that when they pulled out of whaling in 1963, they transformed into one of Europe's biggest transport and logistics firms after experimenting with, and profiting from, fishing and cold stores.[21]

During the depression years of the 1930s, Germany and Japan both decided to break the British–Norwegian monopoly and sent factory ships to the Antarctic. Although the two totalitarian powers formed an Axis (also with Italy), they had somewhat different, though connected, interests. "Guns not butter" and "First guns, then butter" were Nazi slogans—variously attributed to Hermann Goering and Joseph Goebbels—and although they had a broad metaphorical meaning related to the drive for national self-sufficiency, they also had a specific one—whale oil was used to make a substitute for butter—margarine—by a chemical process, *hydrogenation*, developed in Germany. Japan, on the other hand, desperately needed fuel oil (it had plenty of coal after occupying Manchuria), especially for military use, which, at the time, could best be purchased from the USA. Japan's industrialists devised a brilliant way of securing this. They built whaling factories, the oil produced by which was transported, in the factories, to Rotterdam for off-loading to the German market, at a time when the League

---

[20]This is the term given to the method by the Norwegian historians J. N. Tønnessen and A. O. Johnsen in their classic *The History of Modern Whaling* (Tønnessen and Johnsen 1982). Where catcher boats operate from and tow their catches to shore for processing there, it is called land-station whaling; if they bring their catches to a mother ship with a stern ramp, or if the catchers are large enough and fitted to haul whales onto their decks and process them onboard, it is called pelagic whaling.

[21]The Norwegian pelagic whaling companies, based at Sandefjord in southern Norway, had similar histories and, in fact, that thriving city is virtually built on blue whales *Balaenoptera musculus*. One of Salvesen's catcher boats is now in Sandefjord's whaling museum.

of Nations, led by Britain and France, was applying sanctions to Germany. The factory ships then crossed the Atlantic—cleaning their tanks on the way—passed into the Pacific via the Panama Canal, loaded mineral oil at west coast US ports, and carried it across the Pacific to Japan. In those years, most of the meat was thrown back into the sea or converted to meal and used as fertilizer. Although Japan's authorities repeatedly claim that Japanese whalers have always produced food for human consumption, while Europeans did not (as if margarine was not edible!), the fact is that the market for whale meat depended on the post-war development and application of the new technology of freezing at sea.[22]

Just before the Second World War broke out, Norwegian entrepreneurs and diplomats noted that Japan and Germany presented a threat to their hegemony. By the end of the war, all the Japanese factory ships had been destroyed. Nevertheless, General MacArthur, commanding the US occupation of Japan, encouraged the resumption of Antarctic whaling on humanitarian grounds; the Japanese diet was short of both fats and proteins. A few fibs were told by the US Administration, such as that the deal was only for one year; in fact, it was open-ended, and no limits were imposed on the Japanese whaling fleet, except, initially, with respect to where it could operate within the Antarctic.

Immediately after the war, several countries—especially Argentina and Italy—wanted to get into the Antarctic whaling business, which was set to become very profitable again. One way and another, the British and Norwegians put a stop to that. Occupied Germany also wanted to revive its industry and had as good a reason to do so as Japan, considering the shortage of fats in the diet of a near-starving population. But Germany was unlucky; the occupied region of the Baltic ports, Hamburg and Bremerhaven, that had served the German whalers previously, was administered by the British, who were not prepared to allow the resurgence of whaling competitors. However, German interests and expertise had a back door—they were important in the creation of a whaling fleet, the *Olympic Challenger*, under the Panamanian flag, owned by Aristotle Onassis and managed by a company board that was based in the USA and almost wholly composed of American nationals. This fleet caused untold trouble to the IWC, was arrested by the Peruvian authorities and released with a fine, and was finally purchased by a Japanese company. Another fleet that caused great post-war grief to the other whalers flew the Dutch flag—the *Willem Barendsz I*, later replaced by *Willem Barendsz II*. But, through the 1950s and 1960s, Japan steadily increased its fleet, in the last stages by buying the fleets of the UK, Netherlands, and Norway as they went out of business due to declining catches. The vendors got reasonable prices mainly because these factories carried with them negotiated percentages of the overall IWC catch limits. In the 1960s, the one-factory-ship Soviet fleet—the *Slava*, a vessel acquired as war reparations—was augmented by several very large new factories.[23] These were eventually producing frozen whale meat only for the

[22]Other myths are still regularly propagated by the Japanese whaling lobby. One is that whaling in the Antarctic, entirely from factory ships—except for the brief post-World War II renting of a station on South Georgia from the British—is an ancient national tradition that must not be lost. In fact, Japan was one of the last nations to engage in that industry.

[23]See how the whaling ship trade went: the 13,000-ton *Slava* began in 1929 as the British-built and flagged *Vikingen*. From 1939 to 1945, it was renamed Wikinger and flew the German flag. In 1945, it became British again, renamed *Empire Venture*, and was then handed to the USSR in 1946 and worked in the Antarctic until 1965/1966. The new Soviet factory ships, coming in the 1960s, were nearly threefold the size of the *Slava*. The *Slava* was the main source of large-scale falsifications of data about species, locations, and quantities of Antarctic whale catches that were revealed by Russian scientists and operatives—ex-national inspectors—after the collapse of the USSR.

Japanese market, but had as a secondary purpose the large-scale catching of sperm whales *Physeter macrocephalus*, which are practically inedible, but yield a special industrial oil as a strategic material.[24] The Soviet sperm oil did not enter the global market, and the equally important strategic stockpiles desired by the USA were derived from numerous sperm whaling operations by a group of client states.

Most whale populations were depleted by "modern whaling," earlier than most of the fishes. That was not because they were less abundant (in terms of biomass), but because they were even more valuable. It is not widely known how big were the catches of the large whales in the 1930s and again in 1940s to the mid-1960s. This is partly because catches were always quoted as numbers, with fish being quoted in tonnes, and production was recorded in barrels of oil, not weight. The tonnage of whale catches from the Antarctic—from which most of them came—was between 15 and 20% of the total world fish catch, by weight, and considerably more in market value. But, you may ask, why are you fussing about that now, when commercial whaling has been suspended for a couple of decades? Certainly, commercial whaling continues in the guise of "scientific research," but its scale is relatively small, although increasing and diversifying.

Of course, there had been no hint of sustainable use of the resources in these operations of the 1930s and 1950–1960s; they were mining the resources. There were, from the early 1930s, regulations limiting the catches from the Antarctic by the infamous and disastrous *blue whale unit* (BWU), but these were initially imposed in an effort to stabilize the price of whale oil, especially by preventing overproduction—similarly to the actions of OPEC regarding mineral oil in the late 20th century. After World War II, there were efforts to use the BWU limitation as a conservation measure, but with little success. Scientists called for its abolition almost every year in the first decades of the existence of the IWC, but this did not happen until 1973, in the wake of a UN Resolution of 1972 calling for a ten-year moratorium on all commercial whaling.

At least some of the great whales are now doing what they were supposed to do after they were protected from whaling—increasing in number. We have good data for only some of them—particularly the spectacularly watchable singing humpbacks *Megaptera novaeangliae*. We don't have good reason for thinking the others are *not* recovering. If, as I suppose, most of them are, there are now a lot more fin whales *B. physalus*—by far the most valuable species—than there were when they were protected in the Antarctic in the 1970s and everywhere else in 1986 when the so-called commercial moratorium (voted in 1982) came into force. I believe the Japanese whaling and whale-meat lobbies think the same way and, having gained a complete monopoly of factory ship whaling when the USSR dropped out of the business in 1987, are now executing a long-term plan which—if successful—will one day bring them great profit.

Japan lodged an "objection" to the 1982 moratorium, as the IWC's rules allow, so was not bound by the decision to put all legal catch limits to zero, indefinitely—to allow stock recovery and give time for development of new safe procedures for limiting catches to sustainable levels and for allowing depleted stocks to recover to high, much more productive levels. Japan, however, withdrew its objection on the promise that its fishing vessels would

---

[24]Their tertiary purpose, along with the huge Soviet fleet of very large fishing vessels—stern factory trawlers, mainly, derived from the Salvesen design of whaling factories—was as a globally extensive source of military and economic intelligence.

be licensed to operate in the new 200-mile wide fishing limits of the USA in the Pacific. A change in US policy soon thereafter cancelled that permission, and Japan decided to exploit a loophole in the IWC's rules, which allows any country to award its whalers, unilaterally, any number of whales, of any species, to be killed for "scientific purposes," with a requirement, moreover, that the carcasses be "processed." The beauty of this for a whaling operation is that the "scientific whales" can be killed anywhere (including in sanctuaries and other protected areas), at any convenient time, and without regard for long-standing rules that protect calves and nursing mothers, protected species, and undersized individuals; and avoiding the minimal rules that aim to reduce the inhumanity of hunting and killing methods. The requirement that the dead whales be processed, of course, practically ensures that the meat and other commodities from them will be marketed, although some of it may be given away, or sold cheaply, to schools, prisons, hospitals, and the like, and to help promote consumption generally. The original idea of the processing requirement was to avoid waste, and there were other such well-meaning regulations, such as setting a limit to the number of dead whales that could be used as fenders on the side of the catchers and as buoys to which other dead whales, awaiting processing, could be moored.

At first, the "scientific whaling" program involved the killing only of minke whales *B. acutorostrata* and *B. bonaerensis*, a "small" species (adult individuals each weigh "only" about 6–8 tonnes) that had been killed by the Soviet and Japanese whalers in tens of thousands in the last days of legal commercial whaling: from 1970 until the moratorium. It is not profitable to catch and market a relatively small number—a few hundred rather than thousands—only of this species, and the research expeditions, therefore, have to be highly subsidized. The "research program" has been moved a bit towards profitability by steadily increasing the numbers of minke whales caught, operating also in the North Pacific during the southern winter and, recently, adding other much larger species to the list of "scientific whales." The excuse for this expansion is that it is necessary to study how the different species of whales might be competing with each other for food. The excuse for increasing the numbers of whales killed is that they might be competing with humans for commercially valuable fish and this, too, needed study that can, it is claimed, only been done by looking in their stomachs. In fact, the "problems" that are supposed to have been solved by scientific whaling have changed throughout the 20 years, and it is of interest that the IWC's scientists reported last year that not one of the specified research problems has, in fact, been resolved.

The expansion of the "research" operations has put some strain on the relatively small (6,000 tons displacement, 2,000-tonne hold capacity) elderly factory ship—*Nisshin Maru*—and its associated catcher boats that conduct them. The increase in Antarctic catches has led to an operational need to use a transport vessel to bring frozen meat from the Antarctic to Japan in mid-season, there being insufficient hold and freezer space in the ship's factory. The transport also serves to refuel the whaling fleet, an operation that violates IMO rules regarding pollution dangers in the vulnerable protected seas surrounding the continent. This Japanese-owned auxiliary—*Oriental Bluebird*—flies or rather flew the Panamanian flag of convenience. As I write this (October 29, 2008), news has come through that the owners have deflagged this vessel after the Panamanian maritime authorities had fined them a ridiculously small sum when it was found to be "in violation of domestic and international regulations relating to its permissible use, the safety of human life and the preservation of the marine environment." It is now (November 15) unclear whether the auxiliary will accompany the Japanese fleet this coming season and, if so, what flag it will fly and what will be the nationalities of its captain and crew.

In the most recent season, the research subsidy covered about one-quarter of the expenses of the supposedly not-for-profit operation. Another quarter was covered by an interest-free loan from the government, and it has been reported in the Japanese press that the Institute for Cetacean Research (ICR)—which is responsible for the scientific whaling operations—has defaulted on that loan. Meat sales covered the remaining 50% of costs. Evidently, there are financial problems that probably cannot be resolved by trying to catch and process more whales using the old factory ship, and it could be that the Japanese authorities will not be content to provide subsidies and erase debts on an increasing scale, indefinitely. So, there is now talk of commissioning a new, larger and technically more efficient factory ship with better on-board processing facilities. If that happens, it will, I think, be virtually impossible to rein in and in any way internationally regulate the "scientific" whaling and prevent an expansion to more of the larger whales, as they—presumably—recover.[25] If that happens, the answer to the second part of the question I posed in my original title will likely be "Yes, but mostly only by Japanese consumers, although a few Norwegians might continue to eat Northeast Atlantic minke whale *B. acutorostrata* meat, and some South Koreans might continue to consume, as well as to export to Japan, the numerous minke whales they claim to catch 'accidentally' in their fishing nets."[26]

There is another scenario. It is made plausible by the fact that the whale-meat dealers are finding it difficult to sell their produce to a younger generation of consumers, who have grown up in a period when whale meat was available in small quantities and was very expensive. This scenario is that Japanese whaling interests may be considering—or might even have already decided on—an exit strategy, notwithstanding the big investment they have made for two decades in a fake research program. They have to contend, too, with escalating fuel costs for very long voyages. If that is so, they will certainly conceal their reasons for such a decision, pretend they might get out of Antarctic whaling as a goodwill gesture to the rest of the world, and seek a suitable *quid pro quo*. Think about it.

During the scientific whaling period, the IWC's scientists have devised a completely new kind of procedure for regulating any future legitimized commercial whaling. It is exceedingly conservative—they think—and precautionary in principle. The MSY objective has been abandoned. The Revised Management Procedure (RMP) aims at a fairly large, but not maximum, *cumulative* catch over a predefined period (100 years) during which the annual catch limits set by the procedure would be adhered to strictly—that is, no haggling. But, the *prime* objective in the RMP is that there should be an extremely low probability that whaling would *accidentally* lead in any year to depletion of the exploited stock to below about half the original number of whales in each population. The IWC decided that, for any such procedure to be implemented, there must be a water-tight scheme to ensure compliance by States and their whalers with the rules. It should come as no surprise that the Japanese—and also the Norwegian—authorities have consistently made difficulties in reaching an agreement, which would necessarily include a severe restriction on scientific whaling. Meanwhile, more and more of the countries that previously were engaged in commercial whaling—for instance UK, Aus-

[25]In recent years, the US Administration has been moving away from its previous leading position inside and outside the IWC in support of conservation of whales, as it has become less committed to multilateral actions, in favor of bilateral politics. We await signals that the new Administration will correct course.

[26]Published research by Prof. Scott Baker and his colleagues has shown, by DNA evidence, that these numbers are very much higher now than those reported in Korea's official statistics submitted to the IWC.

tralia, New Zealand, The Netherlands, South Africa, Chile, Spain, Portugal, Brazil—as well as many that were not, at least not recently—Argentina, Mexico, Ireland, India, Germany, for example—have come to the conclusion that "the moratorium" should stay in place for a long time to come, and perhaps be permanent.[27] One idea that is being discussed is to declare the entire Southern Hemisphere as a whaling-free region; that seems to be what the southern coastal states want. Another is that whaling should not again be permitted on the high seas—which includes the ocean surrounding the Antarctic continent. The idea is growing that future human generations should be left to decide for themselves whether they want or need to eat whales, and that the overwhelming priority now is to give the chance for the Southern Ocean and other marine ecosystems, in which whales were once very abundant, to recover, and that will take many more decades of protection.

In recent years, the Government of Japan has repeatedly threatened to leave the IWC if it does not get its way, which is mainly the lifting of the moratorium and the abolition of whale sanctuaries. I doubt they would do that. Why should they? If they withdrew, they would attract far more international diplomatic, legal, and popular trouble for themselves and would forego the practical advantages of continuing scientific whaling without any international regulation. Furthermore, the RMP is so cautious that many of the whale stocks they would eventually like to exploit would receive zero catch limits for many years to come. No, those threats are bluffs.

In recent years, the Government of Japan has been making the utterly bizarre claim—in such places as FAO's Committee on Fisheries and its General Conference and in the UN General Assembly—that an immediate renewal of commercial whaling is a necessary part of ***world food security***. A better case might be made for it having a role in the fairly distant future, when stocks, especially in the Antarctic, have—hopefully—recovered under prolonged protection. But even then, a ***sustainable*** catch of large whales would contribute rather little to the total sustainable production of wild sea fish and shellfish, and real security comes from continuity, not one-off living resource "mining."[28] It will obviously be up to future human generations, not us, to decide what to do about the whales, whether to renew exploitation of them, and whether to eat them.

Lurking behind all this debate is the fact that no way has been found to make whale killing acceptably humane and, frankly, the remaining whalers are not doing anything much to make it so—and, naturally, are doing everything possible to prevent anyone else seeing, and recording, what they are up to.

Let me close by returning to the matter of ***greed***. In developing the RMP for baleen whales, the IWC scientists abandoned both the MSY target and the strict definition of sustainability as unattainable. Instead, they invented, and tested by computer simulation, a target which is the highest feasible *cumulative* catch over an extended period (they chose 100 years, for reasons connected with the lifespans of whales, the maximum likely life

---

[27]The EU is currently reviewing its policy on commercial whaling, and the strong indications are that it will, as a whole—with a possible opt-out by Denmark—unanimously support the prolonged continuation of the moratorium.

[28]Another equally bizarre aspect of this issue is that Japan is routinely supported in the international fora mentioned, and in others, by a claque of representatives of mostly small, poor, developing countries to most of which Japan has provided special aid packages, and which have in recent years joined the IWC and regularly vote with Japan. See Japan's "Vote Consolidation" Operation at the International Whaling Commission. Third Millennium Foundation, Paciano, Italy, May 2007. 96 pp.
(http://opinion-former-resources.politics.co.uk/microsites2/364355/graphics/iwc.pdf).

of any management institution and—most definitively—the computing power available to them at the time—the early 1990s—to carry out the necessary simulations). This maximum cumulative catch is, however, only the second priority of management; the first is to have negligible risk that the stock will be *accidentally* reduced by whaling in any one of the 100 years to less than some low, but presumed safe, minimum threshold.[29] Furthermore, once a procedure had been agreed, in detail, allowable catches would be determined each year automatically, avoiding the usual annual negotiations, of which I have given fisheries examples.[30]

What is not clear is whether pelagic whaling in the Antarctic, under RMP regulations, would ever be *economically* sustainable. The history of it does not help us, because those operations have always been mining, not for sustainability. The limited-scale minke whaling conducted by Japanese interests for 20 years is not economically sustainable; it has always been heavily subsidized. An increase in catches, which has been happening slowly, but steadily, tends to reduce the price received, even though the Japanese market is "socialized," i.e., the wholesale price is determined by the producer, the ICR. As we have seen, a further increase would entail investment in a new, bigger factory ship, which would be costly. It is of interest that the new plans for Special Permit "scientific whaling" includes substantial numbers of the (presumably) recovering fin whales. Although the excuse for this is the need to look at their stomachs to determine whether they are competing with other whale species and even with fisheries (although the fin whale diet is almost completely confined to plankton), I think that Japan's planners are fully aware that if there is an economic future for Antarctic whaling, then it rests mainly with the fin whale stocks, which have historically yielded far more than any other species, including the blue whale, which is twice as large, but was less numerous.

So, whether biologically sustainable whaling, other than by very limited land-based operations in a few places, can ever be profitable—especially in the Antarctic, where most of the whales originally fed and will, we hope, do so again one day—is an open question. I think the answer is that it would not be profitable. It is made even less likely by two new factors: one is the fact that warming in the Antarctic zone is reducing the length of the ice edge, which provides the special habitat needed for krill *Euphausia* spp.—the planktonic food of the baleen whales and other Southern Ocean resources. The other is more human greed: the current drive to exploit the Antarctic krill *E. superba* to provide feedstock, mainly for feeding cultured freshwater and marine fishes, as well as some terrestrial livestock such as poultry.

Attentive listeners/readers will notice that I have completely failed to identify how greed can be cancelled by fear, as advocated by Eduardo Porter. The appropriate epiphany is elusive.

---

[29]A third criterion, that need not concern us here, was that there should not be such large year-to-year changes in allowable catches as to make the logistics of whaling impossible, with a loophole there for emergency action.

[30]The European Commission is currently reviewing its fisheries management policy and procedures. I have suggested that the IWC approach points the way forward, within the limitations now set by the UN Convention on the Law of the Sea and other related legal instruments, by redefining MSY not as a point on a graph, but as the highest cumulative catch, over a defined time-period, that is compatible with conservation considerations such as taking care not to reduce stocks to below thresholds that risk reproductive failure and possible extinctions (*New Policy Objectives and Management Procedures for EU Fisheries: A Commentary and Suggestions*. A briefing paper prepared for the European Policy Office of the Worldwide Fund for Nature, Brussels, January 27, 2007; *Editors' note: this is a longer version of Chapter 39*).

Thanks

I thank Professor Michael Mainelli and the Gresham College Academic Board for inviting me to give this lecture, and the International Fund for Animal Welfare, especially Fred O'Regan, its CEO, and Patrick Ramage for encouraging and facilitating my participation. I thank Mary Carmel Finley and Michael Earle especially for information they generously provided, and Tim Holt for his frank comments on an early draft.

References

Clark, C. W. 2006. The worldwide crisis in fisheries: economic models and human behavior. Cambridge University Press, Cambridge.

de la Mare, W. K. 1986. Simulation studies on management procedures. Reports of the International Whaling Commission 36:429–450.

Gilbert, A. 2003. The new Jerusalem: the extraordinary true story of how a secret society rebuilt London. Corgi Books, London.

Graham, M. 1943. The fish gate. Faber and Faber, London.

Halvorson, H. O., and J. Dulf. 2008. Offshore aquaculture legislation designed to balance production and protection. Marine Pollution Bulletin 56:1673–1675.

Hardin, G. 1968. The tragedy of the commons. Science, New Series 162(3859):1243–1248.

Holt, S. J. 2006. The notion of sustainability. Pages 43–82 in D. M. Lavigne, editor. Gaining ground: in pursuit of ecological sustainability. International Fund for Animal Welfare, Guelph, Canada and the University of Limerick, Limerick, Ireland.

Holt, S. J. 2008. Sustainable use of wild marine living resources: notion or Myth? Pages 185–201 *in* L. L. Rockwood, R. E. Stewart, and T. Dietz, editors. Foundations of environmental sustainability: the coevolution of science and policy. Oxford University Press, New York.

Holt, S. J. 2009. Sunken billions—but how many? Fisheries Research 97:2–10.

Johnson, W. 1979. Muddling towards frugality: a blueprint for survival in the 1980s. Shambhala Publications, Boulder, Colorado.

Keen, E. A. 1988. Ownership and productivity of marine fishery resources: an essay on the resolution of conflict in the use of the ocean pastures. McDonald and Woodward, Blacksburg, Virginia.

Lavigne, D. M. Editor. 2006. Gaining ground: in pursuit of ecological sustainability. International Fund for Animal Welfare, Guelph, Canada and the University of Limerick, Limerick, Ireland.

Malthus, T. R. 1798. An essay on the principle of population. J. Johnson, London.

Rockwood, L. L., R. E. Stewart, and T. Dietz. Editors. 2008. Foundations of environmental sustainability: the coevolution of science and policy. Oxford University Press, New York.

Rubino, M. Editor. 2008. Offshore aquaculture in the United States: economic considerations, implications & opportunities. NOAA Technical Memorandum NMFS F/SPO-103.

Tønnessen. J. N., and A. O. Johnsen. 1982. The history of modern whaling. C. Hurst & Company, London.

Verhulst, P. F. 1838. Notice sur la loi que la population suit dans son accroissement. Correspondance Mathématique et Physique 10:113–121.

World Bank and FAO (The World Bank and Food and Agriculture Organization). 2009. The Sunken billions—the economic justification for fisheries reform. The World Bank, Washington, DC.

Chapter 43

Comments by Sidney J. Holt to the Commission of the European Communities on The Green Paper: Reform of the Common Fisheries Policy[1]

"Since resources do not serendipitously start out at MSY conditions and, further, stochastic events continually displace resources from equilibrium conditions, MSY solutions are not in themselves the most appropriate policies for managing [the use of] resources. Rather, they provide a reference point for the design of dynamic harvesting policies, preferably feedback or adaptive policies, that can respond to unforeseen changes in the state of the resource." (Getz and Haight 1989)

Prologue

I begin with the proposition that the primary objectives of managing a fishing industry are to: (i) ensure continuing supplies of high-value fish, (ii) ensure profitability, and (iii) ensure employment in fishing and associated industries and to sustain fishing communities.

In parallel with this, we can define primary objectives of managing the exploitation of fish stocks as to: (i) seek a high average catch over a predetermined time-period, (ii) ensure that the risk is very small of accidentally depleting a fish stock to a level (and to a stock composition) substantially less than a designated optimum, and a practically zero risk of any accidental depletion to a level where the continued existence of the stock is threatened, and (iii) maintain stability from annual season to season insofar as that is possible in the light of natural variations in the fish stocks.

There are common problems of uncertainty about the dynamics of the systems being managed—some intrinsic, some arising simply from inadequacies of data—and others arising from external sources of unpredictable variability. For these reasons, the two triads inevitably contain partial incompatibilities among different primary objectives. For example, in the latter triad—which is the subject of this comment—it is impossible simultaneously to maximize the short-, medium-, or long-term catches and to minimize the likelihood of accidental depletion

[1]Editors' note: this paper was submitted by Sidney Holt on November 20, 2009 in the form of comments to the Commission of the EU on The Green Paper: Reform of the Common Fisheries Policy (Document COM(2009)163 final, 22.4.2009). See https://ec.europa.eu/fisheries/sites/fisheries/files/docs/body/sidney_holt_en.pdf. Parts of the original paper have been omitted for the sake of brevity; these include two annexes, one a short biography of Holt and the other a report by Ms. Hélène Bours containing a review of the history of the use of ideas about fishing capacity and effort by the European Union. In the original, the Summary and Conclusions were at the beginning of the paper; here, they have been moved to the end. The preparation of this paper by Holt was funded by the Pew Environmental Group, which has given permission to use the paper in this book. Copyright © 2017 The Pew Charitable Trusts. All Rights Reserved. Reproduced with permission. Any use without the express written consent of The Pew Charitable Trusts is prohibited.

of the managed fish stock. Thus, a management procedure seeking to meet multiple objectives must be such as to provide a working compromise, defined in terms of a suite of acceptable and pre-defined probabilities and priorities. To these must be added an overall decision about the desired level of precaution to be built into the procedure.

In 2007, on the occasion of the Commission's decision to try to achieve compatibility of a revised Common Fisheries Policy (CFP) with the decisions of two major UN conferences concerning maximum sustainable yield as a primary management objective, I provided two documents for consideration by the European Union—a long, technical paper for the Commission's Scientific, Technical, and Economic Committee for Fisheries (STECF) (Holt 2007a) and the other, a shorter and less technical one, for The Greens/European Free Alliance in the European Parliament (Holt 2007b). These, and the text of my presentation of the second one to the Fisheries Committee of Parliament, are referenced below.[2]

These papers drew largely on the experience of the International Whaling Commission (IWC), which was the first international organization—and virtually the only one at that time—to have developed and exhaustively tested, by computer simulation, a new approach to stock management.[3] Now, let me try to explain briefly and non-technically the essence of the new approach; it will be familiar to modern engineers. First, to be clear, the IWC was prepared only to continue regulating whaling by setting annual catch limits, by numbers of whales, for each species and putative biological stock, rather than the more common practice in fisheries of referring to weight of catch, as a proxy for commodity production and value. This meant that any population model used need give little, if any, attention to the age structure of the exploited population; the only structural features in the models used were the separation of the sexes and the distinction between mature and immature (hence older and younger) females.

Since 1975, when the IWC formally adopted the prime objectives of seeking MSY and allowing depleted populations to recover to abundant levels, the Commission's Scientific Committee had been using what is usually called a ***surplus production*** model—or, rather, a modified one called BALEEN, to take the structural elements into account—to calculate supposedly "safe" catch limits. The properties of the model had largely to be assumed. The level of population giving MSY, for example, relative to the so-called "virgin" (unexploited) stock was assumed to be 60% although there was little empirical evidence to justify that, just as there is virtually no such evidence to justify various percentages that have been used—and are still being used—in fisheries management of finfish and shellfish fisheries. Similarly, although there was some evidence, for some species, of natural mortality and reproductive rates, there was practically none about the density-dependence of those rates, which is what must be known to estimate sustainable yield rates. The result was that although the New Management Procedure (NMP), as the 1975 decision was called, led to the closure of commercial whaling on greatly depleted stocks, it became evident, in just a few years, that it was almost useless for the management of whaling on stocks that were not yet depleted.

---

[2] "New Policy Objectives and Management Procedures for EU Fisheries: A Commentary and Suggestions," a briefing paper for the European Policy Office of the WWF, February 13, 2007, 60 pp. "New Policy Objectives and Management Procedures for EU Fisheries: A Commentary and Suggestions to The Greens/European Free Alliance in the Parliament," February 8, 2007, 21 pp. (*Chapter 39*). Statement by Sidney Holt to Fisheries Committee of the European Parliament, April 11, 2007, 4 pp. (*Chapter 40*).

[3] Since then, the scientific development of this approach to management has been pursued, and examples are given, but the stage has not yet been reached elsewhere, as in the IWC, of having been formally approved (if not implemented) by management authorities.

The NMP was rooted in the idea that each stock could be "classified" according to its assessed conservation status—Protection Stocks would be a few percentage points below the assumed MSY level; Sustained Management Stocks would be fairly close to the MSY level, and Initial Management Stocks would be somewhat above the MSY level. It turned out that this sort of classification called for knowledge that far exceeded what was actually available, so there was endless argument at both scientific and administrative levels about the catch limits for stocks that were close to, either above or below, a classification boundary, with a corresponding temptation to distort science for essentially political reasons. It is important to ensure that the EU, in revising the CFP, does not fall into that trap.

Other unforeseen difficulties arose, but these illustrate the main ones. In applying the NMP, the Scientific Committee was obliged to spend almost all its time and energy in trying to calculate catch limits for every whale stock every year. This, of course, meant the assessments were not as good as they could have been, even with limited data. ICES, in providing advice to the EU Commission on annual TACs, has encountered the same problem, on an even larger scale. As a side effect of the declaration in 1982 of an indefinite moratorium on all commercial whaling, the IWC's Scientific Committee was at last released to put its skills to devising a "Revised" Management Procedure (RMP). This was indeed one of the conditions that, when attained, would open the door to possibly modifying the moratorium; the other condition was that depleted stocks would have substantially recovered. The approach taken by the IWC scientists was to use population models (with BALEEN as the default) to generate artificial "data" that would be used in simulations of candidate procedures for calculating catch limits, to test them for efficiency and robustness, and to match data requirements with practical reality. A prime need was for a procedure that was not dependent on either the parameter values of a particular model or on which model was used of all conceivable models. The philosophy was the engineers' "testing to destruction." It soon became obvious that procedures that called, for example, for classification of stocks as "overexploited," "underexploited," or "optimally exploited" were unsuitable, at least in those simplistic versions.

A second realization was that sustainable yields could only be discussed usefully in a finite, pre-established time-frame. A stock does not, at a particular point in time (during a year?), have a sustainable yield; that had been made clear in many scientific papers. A sustainable yield can only be thought of as an average (or cumulative) yield over a certain time—and not an infinite time. Hence, the suggestions for a revised procedure were intended to maximize the cumulative yield—in the whaling case, this was in a period of 100 years; I will discuss later the criteria for deciding the length of the period—the management horizon. In the IWC case, this was determined by a combination of computing power and the lifespans of whales and humans, having in mind also the likely lifespan of any social instrument charged with implementing the procedure. This last is important because the use of regulatory methods, such as that devised by the IWC scientists, depends on an initial agreement by managers, after which catch limits are set automatically, without annual negotiation. Of course, there has to be some kind of fail-safe to deal with very exceptional situations, but that too should be specified in advance. These details are explained further in the account of the RMP by Dr. Justin Cooke, who was the "winner" of a prolonged and strongly refereed competition among different development groups in the IWC encompassing very severe tests of efficiency and robustness to unknowns and unforeseeable events, including strong environmental changes (Cooke 1999).

The problem of avoiding accidental depletion during the management period came down to specifying a lower bound and the acceptable probability that the stock would be accidentally pushed below that bound at least once in the period. The IWC scientists chose—with the agreement of the Commission—not to define a lower bound related to any supposed dangerously low stock where extinction might be threatened, but a much higher level, at 54% of the estimated virgin stock, that is, at what had been the boundary between Protection Stocks and Sustained Management Stocks in the old system. The assigned probability for testing by simulation was 99% in 100 years, that is, no more than one time in the management period. The problem of compromise between conflicting management objectives is easily seen here; if either or both the thresholds was set lower, or the acceptable probability of depletion higher, bigger short-term catches would be obtained at the expense of a higher risk of depletion. (It was later realized that tests must be made also of alternative distance of the management horizon; if the period were to be made longer, then bigger catches could be made in the short-term, but recovery of depleted stocks would only be achieved after a long delay). I should say, too, that the end-game has to be specified; it is not allowed, as the horizon is approached, to catch most of the recovered stock in order to boost the long-term average! Naturally, the more depleted is the target stock, or all the stocks in a management area, the longer will the recovery period necessarily be. It would seem to be commonsensical that the time-horizon should extend to, and probably beyond, the time when recovery will have been more or less completed, which depends on both the degree of depletion and the lifespan and general dynamics of the target species.

In my two 2007 advisory documents, I sought to explain how the IWC approach might be applied to fishery situations in which the aim is to maximize the average catch during a specified period (I wrote usually about cumulative catch, but one can, of course, equally consider mean catch). The main change is to use multicohort models, with an age/size structure to provide test "data" for simulation runs. This is obviously necessary when one is trying to maximize average catch in weight, where fish growth has to be taken fully into account. But, there is an additional essential reason for this approach. It is that even if the simplistic idea that there exists, for a stock, a peaked curve of sustainable yield against fishing mortality (fishing effort) or population size (biomass), the location and height of that peak depend crucially on the ages to which the fishing mortality is applied—which vary with, for example, mesh size in trawls, hook size in lines, location and season of fishing for practically all methods. In fact, if there is an MSY, it would be obtained—approximately—by catching all the surviving fish in every cohort (year class) in a very short time when they reach an optimum age. This would, however, normally require a practically infinite fishing effort, extremely narrowly focused, and is entirely impracticable except perhaps in a few special cases (some anadromous species, perhaps).

I suggested, for a start, the model I proposed with my colleague Raymond Beverton in our 1957 book (Beverton and Holt 1957) that we called a self-regenerating model (S-RM).[4] This is a combination of a model that expresses the sustainable yield-per-recruit as a function of fishing mortality rate (fishing effort) with specified selectivity or, equally, as a function of stock biomass (the curves sometimes being peaked, but not necessarily so, depending on the values of the parameters specifying individual growth and natural mortality rates), with a model specifying the expected numbers of recruits arising from a parent stock of specified

[4]The fourth printing of this, dated 2004 and published by Blackburn Press of New Jersey, contains a historical résumé by me (Holt 2004).

size. The yield curves of such self-regenerating models are always peaked, but the location of the peak varies greatly with the population parameter values, and the curves can be "kinked," a feature the importance of which we shall see later. The stock–recruitment relationship used was the familiar Beverton and Holt asymptotic model, and I illustrated this for the two papers to Parliament and the STECF in a simplified form in which the age of first liability to capture by the trawl-type gear in use is the same as the median age at sexual maturity, with males and females having the same parameter values.

In future studies, a more flexible stock–recruitment relationship should be used. There are several published candidates for this, and the well-known model by John Shepherd (1982) is perhaps the most suitable. Another that has been used in some applications is that by William Ricker (1954); this is particularly applicable if it is believed that at very high population densities recruitment begins to decline with increasing abundance of spawners/larvae/eggs. Shepherd also found an easier way to compute the self-regenerating models than that originally used by Beverton and me. More recently, further advances have been made in modeling by Kell et al. (1999a, 1999b) and also by A. J. Constable and W. K. de la Mare (1996), the latter authors having published a General Yield Model (GYM) of apparently universal application, although it was developed in the context of managing fishing in the Southern Ocean.

Almost all population models that have been used in fish stock assessments as a basis for managing fishing contain a usually unstated assumption that the fastest rate of population growth occurs when the stock has been reduced close to extinction. It has also commonly been assumed that the peak sustainable yield can be taken when the stock is well below one-half its pre-exploitation size, usually at about 30–35% of it. It is, therefore, not surprising that, given the huge uncertainties in assessment methods, and the assumption about high increase rates at low stock levels possibly not being correct, many fish stocks have been rapidly reduced to levels at which their survival might be in doubt. (Of course, such problems have been exacerbated by the unfortunately common practice of administrations to set TACs higher than the best available scientific advice would support—what I call trying to negotiate with Nature.) Attempts to avoid such situations have been made, involving setting a reference point at a level below which the spawning stock is considered not to be viable. But, such reference points, for minimal viable spawning stocks, have no empirical basis and, like the MSY reference point—whatever it might be—are entirely arbitrary, though sometimes presented as "best scientific opinions."

The phenomenon of rates of increase beginning to decline as the stock is reduced is called depensation. Early studies indicated that, where the few available data had any meaning, there was no evidence of the existence of depensation. However, that conclusion of "no significance" is an artifact of the statistical methods used. In more recent studies, it has become clear that depensation is not at all uncommon; in nearly one-half the cases where useful data exist, it has been concluded that the presence of depensation cannot be ruled out (e.g., Hutchings 2000). This suggests that, in simulation models, the possibility of depensation must be acknowledged, and that if a reference point for minimum spawning stock size is to be defined (with regulatory implications such as shutting down the fishery), it must be set much higher than has been customary.

Another hidden assumption in most models is that the number of recruits depends on some measure of the size of the spawning stock from which those recruits were derived each year. This is, superficially, reasonable, but at least two developments in population dynamics suggest that it might not be. One has been described by Ginzburg and Colyvan (2004) and called the ***Maternal Effect Model***. In this model, the survival of young (in fisheries the precursor of the recruitment process) is dependent not only on the numbers and "quality" of parents, but, in part, on the quality of the previous generation, especially through the maternal line which transmits the genes to the egg that contribute to good or poor survival probabilities. The maternal effect induces *inertial population growth*, which leads to long-duration population cycles. The likely existence of such cycles has, of course, profound implications for the notion of sustainable exploitation.

Another type of inertial growth leading to long-term cycles has been presented by Lars Witting (1997). It rests on the idea, related to that of Ginzburg and Colyvan, that there are genetic differences in the survival probabilities of individual eggs and larvae and of young animals which impel evolution to an intrinsic hyper-exponential growth rather than the intrinsic exponential (geometric) growth that has dominated population dynamics theory since proposed by Malthus (1798) and Verhulst (1845). Witting's models also predict cyclic population change, especially in long-lived species, and his application to, for example, the gray whale *Eschrichtius robustus* has made sense of otherwise inexplicable population recovery data.[5]

Other well known features of interactions between different species, especially predator–prey interactions, are responsible for population cycles, and these ultimately must be taken into account in any multispecies management scheme or "ecosystem approach to management." A special predator–prey interaction is common in the form of cannibalism in the bony fishes—teleosts—between different stages of the same species (e.g., adult Cod *Gadus morhua* are major predators on their own young) (Longhurst 2006). Longhurst is on record as claiming "There is plenty of support for Walter Nellen's suggestion that a transfer of energy occurs along a food chain from small to large individuals of many teleost species. For these, we might conclude that the function of large incoming year classes is not to enable rapid population growth, but rather to contribute to the nourishment of the entire adult population by transforming food particles that are too small to be useful to adults—even smaller fish, euphausiids, hyperiids, etc.—into items of food large enough to be consumed profitably. Perhaps the main function of young gadoids in high latitudes is to nourish their elders and their betters!" Longhurst's reference here to "their betters" is to the finding that, in many species, the contribution to the next generation by older and bigger spawners is high, beyond being merely proportional to their weight. This adds strength to my belief that assessments and regulatory procedures must take special care to maintain the proportion of older fish in the stock, beyond the fact that they are usually more valuable in the market than smaller ones.

In the development of the IWC's RMP, the candidate catch limit algorithms (CLAs) that would be used to calculate annual catch limits were tested against "data" generated by a variety of models and suites of parameter values. But, in fact, the choice of alternative models (made in the effort to ensure that the performance and robustness of a selected CLA would be as free as possible from dependence on the data-generation model) was somewhat limited, but because the accepted management criteria (MSY level; acceptably low probability of accidental depletion; lower cut-off level) were extremely precautionary, such limitation has probably not undermined the usefulness of the adopted Procedure. But, in applying this process to finfish fisheries management, a wider range of different population models would have to be tested.

---

[5]In a October 29, 2009 contribution to an open-ended *Nature Network Forum* on Theoretical Population Dynamics, Dr. Martin Chester contributed a theorem demonstrating that cycles should be expected in population size without necessarily the involvement of interspecific interactions, time-lags, or particular genetic considerations.

In seeking a way of managing the exploitation of individual stocks and species, the IWC scientists had to confront the problem of interactions between the target and other whale species, and other—competitive—predators on the same food resource as well as unexpected environmental changes. The approach to the latter problem was quite straightforward—simply conduct simulations in which hypothetical gross changes in the environment, short- and long-term, were assumed. The problem with species interactions is more complicated. There is, I think, a near consensus among scientists engaged in conservation and management research that we are still very far from being able to model such multispecies interactions or having adequate empirical data for safe management purposes, and even further from practical means for "ecosystem management," no matter how desirable those may be in principle. In these circumstances, those scientific groups agreed that a single-stock approach—what has been called, pejoratively, "extreme reductionism"—was the only practicable one at this time. Hence, there is a need for continuous monitoring of the other elements of the ecosystem, especially those known to have a direct biological connection with the target species, so that adjustments can be made from time to time to the details of the management decisions so as to deal with any detected changes.

Lastly, the CLA, which was devised for managing the exploitation of a single population of a species, had to be integrated into an overall procedure by which possible—but not well defined or understood—groups of populations of the species occupied an extended geographical region. This danger—of applying catch limits mistakenly to a large area in which one population was presumed to range when, in fact, there might be several populations, and applying a single catch limit would possibly lead to the overfishing or even the extermination of one or more of them—has for years plagued fishery managers as well as managers of whaling. The IWC scientists tackled this by dividing large management regions into numerous "small areas"—arbitrarily—and then testing, again by simulations, the consequences of various ways of applying the CLAs to all those areas. A similar approach could—and I think should—be taken to management of fisheries generally. But, a warning about this problem of managing the exploitation of separate, but possibly geographically overlapping stocks is timely. Reiss et al. (2009) have recently demonstrated that, in fishes such as Herring *Clupea harengus*, Cod, Hake *Merluccius merluccius*, and Haddock *Melanogrammus aeglefinus*, there is a mismatch between biological and fisheries management units. They emphasize the need for a revision of these units and question the appropriateness of current management measures. They conclude that "the implementation of complex and dynamic population structures into novel and less static management procedures should be a primary task for future fisheries management approaches."

I end this Prologue by a brief consideration of the means of ensuring the third objective of the fishing management system, namely maintenance of stable catches from year to year insofar as the natural variability of the stock permits. This was not achieved in the IWC case, although the problem was recognized. Since then, other assessment and regulatory agencies have approached it by making another decision rule that catch shall not be allowed to change—except in specified unusual circumstances—by more than a certain amount or percentage from one season to the next (or at the times of regular surveys and reassessments—five-year intervals in the IWC case) by providing that catch limits shall not change by more than a certain amount from one year to the next. ICES has followed a similar course in some of its advice to the EU Commission. There is, however, a danger in such a simple solution; it could be that the unexpected change in the stock, or to the assessment of its state, is so large that a simple proportional limit to the permitted change in a catch limit will maintain catches bigger than they should be to avoid continued reduction in the stock. This appears to have happened in past actions under the CFP. So, a supplementary rule is needed to prevent the stability rule from leading to catch limits higher than the calculated sustainable limits. But, further, if the management objective is to obtain the biggest practicable average catch over an extended period, as advocated here, then the supplementary rule must take that into account and cannot properly take the form of a simple adjustment of the current catch limit.

General Advice

The Commission should follow the path pioneered by the IWC Scientific Committee by defining MSY as the maximum feasible *average physical catch over a predetermined period*—the *management horizon*—that is compatible with a defined probability of any accidental depletion to a predetermined lower bound of the biological stock. The length of that period should not—as it was a decade ago in the IWC case—be affected by computing limitations; the relevant factors are the lifespans of the target fishes, reasonable appraisal of the economically and politically determined duration of the management institution, and assessment of the likely time required for substantial recovery of depleted stocks.[6] Alternative values for all of these decisions—probabilities of depletion, threshold reference points, duration of the horizon—should all be tested by simulations. Criteria for evaluating the results of those simulations must be agreed before they are undertaken. Simulations are needed also to decide whether those decisions should be made for each target species or a universal set of rules devised that would be applicable, but probably not optimal, to several or all species. Similar consideration must be given to alternative properties of the data-generation models, such as whether they allow the possibilities of depensation and cycles, and what is the range of putative MSY levels to be examined. But, the important expansion of the simulations like those undertaken by the IWC scientists is that a range of selectivity of fishing gears and operations has to be chosen.

In the Prologue, I mentioned several types of "data"-generation models: Beverton and Holt with alternative stock–recruitment functions, Shepherd, Constable and de la Mare, and there are perhaps others. It probably does not much matter which is used—convenience and flexibility are the key features. However, the output will necessarily include not only the size of the average catch by weight, but also some measure of the size of fish in it, such as the mean length or weight and also an appropriate measure of the statistical dispersion of the size distribution. This should make possible the calculation of an index of the relative value of the average catch as well as its size.

[6]William de la Mare has offered the opinion (personal communication, November 17, 2009) that, for short-lived species, the horizon needs to be at least 30 years and, for long-lived species, considerably longer. He has suggested a starting rule might be 10 divided by the natural mortality rate M, which would imply about 50 years for many commercial fishes, and up to two centuries for whales. Additional rules would have to ensure that no tricky operating decisions were made that would allow unreasonably large catches in the early years, with the thought that that bad behavior could be made up for in the very distant future. It is, nevertheless, clear that the horizon has to be far enough to accommodate more than one population cycle if the data and models have such a property.

The Stock Management Algorithm

Generation of this algorithm is, by far, the scientifically most creative part of the process of developing an operationally satisfactory stock management procedure, and advice on how to go about that is best given by one or more of those experts who have themselves done something similar. The output from the algorithm is naturally the defined primary aim of management, which I propose to be either an average catch over a predefined period or such an average weighted by the market or other relative values of fish of different sizes or other qualities. However, the regulatory data for applying the algorithm (and the input for the simulation tests) are not necessarily of the same kind. In particular, we should not necessarily be inputting putative annual catch limits (TACs) in order to predict average catches; in fact, I shall suggest that is possibly the worst way of limiting fishing in order to prevent overfishing or allow recoveries of depleted stocks, even though it is, by far, the most common approach to regulation today.

The main, and far better, alternative is the limitation of ***fishing effort***, and the arguments favoring that have been made in many scientific studies. It is quite possible to create an economically very inefficient industry by limiting fishing effort, but not limiting ***fishing power***, or ***fishing capacity***, as it has been called, though defined rather loosely. Documents of the European Commission and others frequently refer to this notion of capacity. Unfortunately, it has been defined by FAO as "*The amount of fish (or fishing effort) that can be produced within a period of time (e.g., a year or a fishing season) by a vessel or a fleet if fully utilised and for a given resource condition*." To use such an important term alternatively as a quantity of fish (output) or an amount of fishing effort (input) introduces counter-productive ambiguity into discussions of management. I avoid it, preferring to use well-defined terms in the scientific literature of fisheries management. But, if "capacity" is to be used at all, it should probably be as a quasi-synonym for "power." The fishing power of *a fishing unit* is the fishing mortality rate it would cause in the target population(s); it is not to be confused with the engine power of fishing vessels: hp or kW. Thus, the fishing power of the fleet should, by management, be adjusted so as to cause, when normally deployed, the desired fishing mortality rate that will ensure sustainability according to the management algorithm. In this case, the mortality rate must be defined as an exponential (not a percentage), so that the rates exerted by the various units are additive with regard to the effective power of the fleet. Fishing effort and fishing power must be properly matched, otherwise vessels are kept idle as and when it becomes necessary to reduce effort in order to reduce the fishing mortality rate. The aim of regulation is to generate an appropriate level and pattern of fishing mortality rate in the stock. TACs seem to provide an indirect way of doing that; fishing power control is a direct way, and for that reason alone is preferable, in principle.

Let me further clarify definitions. Fishing power requires <u>calibration</u> to deal with differences among vessels in the fleet and with changes (usually improvements in efficiency) over time. The methods for doing this when only one type of unit is being deployed (e.g., beam trawl or longline) are fairly well established; two methods can be used—comparative fishing experiments and analysis of catch and effort statistics—and both are generally desirable. Determining relative fishing powers of different types of fishing unit employed in the same fishery is more complicated, but possible. The main difficulties arise from the different selectivities (e.g., by fish size) of the different types of unit—remembering that the unit is defined with respect to the vessel, the gear, <u>and the mode of operation, including the location and less tangible skill factors, especially coming from skippers' experience</u>.

Article 3(n) of Council Regulation (EC) No 2371/2002 on the conservation and sustainable exploitation of fisheries resources under the Common Fisheries Policy[7] defines fishing capacity as "a vessel's tonnage in GT and its power in kW, as defined in Council Regulation (EC) No 2930/86. For certain types of fishing activity, capacity may be defined by the Council using, for example, the amount and/or size of a vessel's fishing gear." Elsewhere in that Regulation, e.g., Article 11, the term is used almost always to refer to the capacity of a fleet, and presumably that is the arithmetic sum of the capacities of all licensed fishing units.[8]

In 1983, the European Commission launched a series of Multiannual Guidance Programmes (MAGPs) concerned with fleet structure and especially the matter of reducing excess capacity and hence excessive potential fishing effort. A report by the European Court of Auditors[9] summarized the effect of the series of MAGPs: "*... the combined effects of over-modest objectives, management problems, and the use of structural funds for shipbuilding and modernisation resulted in the failure of these programmes. This led the Commission to abandon this approach in 2003.*"

Ms. Bours (*see Editors' note in footnote 8*) concluded her review thus: "*As mentioned in this report, the EU has abandoned programmes to decrease capacity. Excess fishing capacity has been identified as one of the major drivers of over-fishing, legal or not. The FAO has adopted an International Plan of Action on the Management of Capacity which, even though not very forceful, is not implemented. Even worse, there is strong resistance from Member States to have such programmes imposed again. Various management tools such as TACs and quotas, technical conservation measures, etc are bound to fail or be undermined as long as the amount of fishing capacity is far superior to what is needed to exploit fish stocks sustainably.*"

These past actions regarding limitation of fleet capacity began with moves to limit the operations of new entrants to the EU and then started to address the problem of comprehensive overcapacity. Here, I am proposing a far more sophisticated use of the notion of capacity, translated into effort, as the prime means of regulating fishing to attain high sustainable yields, reduce the probability of accidental depletion of the stocks, and facilitate the controlled recovery of depleted stocks. But, before going further into that matter, we should look at some other definitions in the Council Regulation of 2002 cited above.

Paragraph 3(h) defined *fishing effort* as "the product of the capacity and the **activity** of a fishing vessel; for a group of vessels, it is the sum of the fishing effort of all vessels in the group" [*Holt's emphasis*]. That is essentially the same as my usage if the capacity of a vessel is taken to mean the calibrated fishing power of a fishing unit.

Paragraph 3(f) defined *fishing mortality* rate as "the catches of a stock over a given period as a proportion of the average stock available to the fishery in that period." This is a quantity more commonly referred to in fisheries technical literature as the *harvest rate*. For regulatory purposes, the fishing mortality is better defined as an *exponential* rate F, not as a proportion,

---

[7]Editors' note: Holt wrote this report prior to the reform of the CFP in 2012, so these comments refer to the legislation in effect then. The current regime is established by Regulation (EU) No 1380/2013 of the European Parliament and the Council and is different in significant aspects.

[8]Editors' note: the original version of this report by Holt included an annex containing a review by Ms. Hélène Bours for the PEW Environment Group of the history of the use of ideas about fishing capacity and effort by the European Commission. See the Acknowledgement at the end of this chapter for Holt's mention of this report. In the interest of conserving space in this book, that annex is not included.

[9]Special Report No. 7/2007 on the control, inspection, and sanction systems relating to the rules on conservation of Community fisheries resources. It can be found at http://eca.europa.eu/portal/pls/portal/docs/1/673627.PDF.

so that the sum of the rates induced by each fishing unit is equal to the overall rate induced by all vessels on the stock (A common source of confusion in this matter is lack of clarity about whether one is talking about fishing mortality applied to the entire stock or to that part of it that is "available" to the fishery).

Lastly, Paragraph 3(g) defined a *stock* as "*a living aquatic resource that occurs in a given management area.*" This purely geographical concept is only partially satisfactory for management purposes since a management algorithm is basically applicable to a single biological population, and there might be several of the same species in any large management area. The IWC scientists have shown us how that problem may be approached by defining many small areas within the large one, more or less arbitrarily if there is no empirical information available about boundaries and mixing of populations and applying the algorithm to them individually. This could be more difficult if the algorithm concerns effort limits rather than catch limits.

Evidently, the efficient way to use direct control of the desired fishing mortality rate by limitation of the fleet fishing effort is through the limitation of capacity supplemented by such time-to-time limitation of its deployment as found to be necessary. The capacity of the fleet is not something that can be changed from year to year except as part of a process of transitional and continuing renewal, but the degree to which it is deployed, particularly in terms of time at sea or on fishing grounds or operating, can, in principle, be adjusted. The common mistake in some fisheries in other regions has been to adjust time by designating fishing seasons without limiting capacity. This not only forces inefficiency, but also provides opportunities for evading regulations and practically encourages such behavior.

It is reasonable to ask how efficient, in terms of catches, would be a management system aimed at high average catches with very low probability of accidental depletion during the defined horizon. There is no universal answer to that, naturally, but Dr. Justin Cooke, the inventor of the catch limit algorithm with the best overall performance devised by the IWC, has written that, in extensive simulations with parameter sets appropriate for Cod, Herring, Orange Roughy *Hoplostethus atlanticus*, Bluefin Tuna *Thunnus thynnus*, and Capelin *Mallotus villosus*: "*The average catches achieved range from 27–81% of the theoretical MSY. On the face of it, this does not appear to correspond to full utilization of the resources. However, this is achieved with a management procedure that ensures negligible risk of stock collapse, keeps stock biomass close to natural levels, and provides catches that are relatively stable within the constraints imposed by the level of available data and the natural variability of the stock. Additionally, fishing costs are kept low by maintaining the stock at levels which provide a high catch per unit effort. Currently managed fisheries typically do not realize a greater proportion of the theoretical MSY, because most of the stocks are overfished.*"

We shall look a little later at the other advantages of such an outcome. Cooke's suggestion, which I think is reasonable, is that the aim should be to try to keep fish stocks in the region of 75–80% of their unexploited level, and that this would be achieved by keeping the fishing mortality rate at roughly one-half of that which would theoretically provide MSY.[10] It is interesting, and more than a coincidence, that the authors of a booklet published by the World Bank and FAO (2009) concluded that a bit less than the relatively stationary global

[10]J. G. Cooke, "A precautionary approach to fishing," March 1995, unpublished manuscript, 25 pp. This document was written for Greenpeace International for the negotiations of the United Nations Fish Stocks Agreement of 1995. It provides, as an annex, the results of simulations of a catch limit algorithm similar to that used in the IWC's RMP fitting the suggested criteria.

catches could be obtained with about one-half the present fishing effort and would yield substantial profit rather than overall economic loss, countered only by subsidies.[11]

I should perhaps explain here that, in my opinion, management to seek high average net economic yield rather than high average physical yield is entirely in conformity with the requirement of the UN Convention on the Law of the Sea (UNCLOS) in which Articles 62.2 and 61.3 plus 119.1(a) seek, through an MSY objective, to: "... *ensure through proper conservation and management measures that the maintenance of the living resources in the exclusive economic zone/on the high seas is not endangered by over-exploitation,* [and] *designed to maintain or restore populations of harvested species at levels which can produce the maximum sustainable yield, as qualified by relevant environmental and economic factors* ..." (*Holt's emphasis*). Furthermore, sustainable exploitation at the lower fishing mortality/effort level implied by submaximization of the difference between the value of the catch and the cost of taking it resulting in higher catches-per-unit fishing effort, must be expected to be more economically beneficial to the individual fishing unit.

Why Regulate Input Rather than Output?

The idea that catch limitation is a simpler and more reliable instrument of governance than limitation of calibrated fishing power (capacity) and effort is an illusion. It stems primarily from the apparent simplicity and universality of collecting catch statistics. In fact, even the veracity of normal official records is illusory. How many users realize that, for example, several of the statistical series archived by FAO on the basis of submissions from governments have been constructed by applying conversion factors (multipliers, actually) to records of a derived commodity such as fishmeal or fillets? More important, however, is the prevalence and scale of illegal, unregulated, and unreported fishing (IUU). But, there are more fundamental reasons for preferring power/effort regulation, and they have to do mainly with the scale and types of uncertainties, both in acquired knowledge about fishes from relevant scientific research and from a vast assortment of sources of natural variation, as well as the relation between these and human activities.

Getz and Haight (1989), in their discussion of how the stochastic reality of fish populations (that is, the fact that growth, mortality, and reproduction parameters are subject to random variation as well as possible trends over time and uncertainties in estimating them) may detract from the efficiency of regulation based on single-species deterministic models,[12] characterize three general approaches as follows: "1. A *fixed-escapement* policy [that is, one designed to permit the catching of all the stock except a residual needed to ensure future recruitment *SJH note*] is designed to stabilize the stock but only at the expense of the catch, possibly even closing down the fishery in years when the stock is particularly weak. 2. A *constant-catch* policy requires that we allow the stock to fluctuate and, in the long run, and if a constant-catch policy does not respond to severe weaknesses in the stock, it can destroy the fishery altogether. 3. A *constant-effort* policy has the advantage of stabilizing labor and

[11]In my published critique of that document (Holt 2009), I have concluded that their data show that the economic benefits of reducing fishing capacity would be substantially greater than they claim and that the sustainable catches from recovered stocks would be higher than the present global catch.

[12]Several other titles in the literature of the science of fisheries management refer to the advantages of limiting fishing power or effort. Beverton (1998) did so in a posthumously published lecture. Getz and Haight (1989) cite the study by Pikitch (1987) of a US West coast demersal fishery as a successful example.

capital invested in the fisheries operation, but at the expense of transferring variability from the stock to the catch. Results from various case studies … indicate that constant-effort policies exhibit less yield variability than fixed escapement policies, and greater stock stability than constant-catch policies."

Essentially the same resultant characteristics arise from corresponding policies that permit regulatory actions to change from one year or period to the next. The present CFP may be regarded as a blend of Getz and Haight's policies 1 and 2 in variable form. Thus, the setting of annual TACs is a form of policy 2, and the change of decision (which can be as drastic as temporary closure of a fishery) when a stock is below a threshold of minimum spawning-stock biomass is a form of policy 1. We need to look more closely at this latter aspect of current EU governance.

The threshold level of the spawning biomass (its composition is hardly ever mentioned beyond the obvious fact that it consists of sexually mature or about-to-become-mature fish, possibly adjusted to the sex ratio) is a pseudo-scientifically determined quantity. That is to say, it is one proposed by scientists, but having practically no basis in empirical studies, by which I mean "information gained by means of observation, experience, or experiment." Two questions may be asked about it—does a threshold exist? And, if so, where is it? The present CFP[13] presumes its existence and sets special rules for management decisions when a stock is found to be—or more usually suspected to be—below the threshold. But, it goes further. In Paragraph 3(l), the threshold is defined in terms of ***safe biological limits***, meaning "*indicators of the state of a stock or of its exploitation inside which there is a low risk of transgressing certain limit reference points.*" Article 5 deals with Recovery Plans, and in its Paragraph 2, we see that they "*shall be to ensure the recovery of stocks to within safe biological limits*"; in the original Commission proposal for this regulation, there was a substantively different formulation.[14] Paragraph 3 requires that Management Plans shall be multiannual, "*be drawn up on the basis of the precautionary approach to fisheries management,*" "*take account of limit reference points recommended by relevant scientific bodies,*" and "*indicate the expected time frame for reaching the targets established.*" Paragraph 2 provides that "*Management plans shall include conservation reference points such as targets against which the maintenance of stocks within such limits shall be assessed.*"

ICES, as the scientific advisory body to the Commission, has given much attention to reference points, and specifically to that defining the *safe biological limit* threshold.

Although the correct approach is to look at the properties of appropriate self-regulatory models and their behavior at low population abundance, it is common to consider simply the relationship between recruitment and spawning biomass. Whichever of these is the focus of attention, one natural lower threshold of stability and continuity of the population is provided by a point of inflection in the relevant curve of sustainable yield against stock size, indicating the existence of depensation (positive density-dependent feedback). This reference point, if it can be found, identifies a threshold below which the population will proceed to extinction even if fishing pressure is reduced or ceases. But, in practice, it is usual to try to fit data for

---

[13]Editors' note: as noted above, Holt's text refers to the previous version of the CFP—Council Regulation 2371/2002. The current legislation (Regulation 1380/2013) defines, and uses, safe biological limits and management plans in a different manner—see, in that Regulation, Art. 4.1.18 for safe biological limits and Art. 9 and 10 on management plans.

[14]As follows: "… *for stocks outside safe biological limits, ensure their* ***rapid*** *return within those limits, and for stocks at or within safe biological limits, maintain them within those limits.*" (Holt's emphasis)

recruitment and stock size to a simple continuous function, of which the most common are those by Ricker (1954), Beverton and Holt (1957), or Shepherd (1982). Except for curves with certain parameter values in Shepherd's model signaling depensation, these are practically indistinguishable for small population sizes. They do not, in themselves, give any clear indication of where a practical threshold might lie.

C. M. O'Brien and D. L. Maxwell have proposed an ingenious way, called segmented regression, of fitting such data to a discontinuous graph, in which for lower stock abundance the recruitment is directly proportional to the stock size and for higher abundance the recruitment does not change with stock size (ICES 2002). The breakpoint between the two regressions suggested to those authors a possible threshold for a safe biological limit to be adopted for management purposes. In 2002, they applied their method to several species-stocks for which there were fairly good data—Anchovy *Engraulis encrasicolus* in the Bay of Biscay; Mackerel *Scomber scombrus* in the Northeast Atlantic; Herring in the Thames Estuary and in ICES Subarea IV and Divisions IIIa and VIId; Plaice *Pleuronectes platessa* in Division IIIa; Cod in Divisions VIIc–k, VIIa, VIa, Subarea IV, Divisions IIIa and VIId, and Subareas I and II; Northern Hake; and Saithe *Pollachius virens* in Subareas I and II. What was evident was that where a break could be identified statistically, its location on the stock-size axis varied enormously as between species and stocks. In some cases, it was at around 20–25% of the largest stock size historically encountered; in others, well over one-half of that stock size (O'Brien and Maxwell 2002).

One problem with the segmented regression approach is that if the break is taken to be the threshold where the management rule changes, then one is already too near to a collapse point, since movement below it will set in train—in theory—a steady decline of the stock, because the recruitment will decline proportionately, which will only happen with a continuous function when the stock has been reduced to extremely low levels. Such considerations have led scientists to try to define "precautionary reference points" which are safer, more conservative, perhaps by a certain fractional value than the basic ones.[15] In fact, the IWC scientists tried to do just that when the IWC adopted, in 1975, its New Management Procedure—so-called safety factors were that the permitted catch would never be set at more than 90% of the estimated MSY, although at the same time, the risky procedure was used to allow catch limits to be non-zero even when the stock was judged to be up to 10% less than the MSY level. These simplistic subsidiary rules were disastrous, not least because the scientists' ability to estimate stock size and biological productivity was nowhere near to what would be required properly to implement such "safety" and "risk" factors.

In some management practices, it has been assumed that the cut-off point might be about 20% of the unexploited size. Such evidence as there is suggests it should be higher than that (the IWC scientists assumed a threshold between regulatory stock classifications at 54%, that is, 10% below the assumed MSY level of 60%). From where came the "traditional" 20% is obscure, but apparently from another assumption not supported by empirical evidence, that the MSY level for fish populations—or at least teleost (bony) fish populations—is at about 40% of the unexploited stock size and that "safe" fishing operations would keep the stock at rather more than one-half that. This, however, is virtually "scientific hand-waving" and

[15]William de la Mare has reminded me that, in this connection, he devised a decision rule which used probability criteria for setting regulations based on the probabilities of being above or below "reference points." This is described in Logan et al. (2005).

is reduced to absurdity if statements are then built on it about the expected probabilities of excessive accidental depletion.

The approach by the IWC scientists is by far preferable: to set a threshold at which catch limits become zero (if TACs are the means of management) or where a moratorium is called for under a fishing effort limitation regime, which is not so much a dangerous minimum, but a desirable high level not far, if at all, below the presumed MSY. But, preferable to such provisions for "classification" of stock "conservation status" is to adopt algorithms that provide continuous fall in risk if the stock declines (or falls into a less productive state by change in composition).

Maintenance of stocks at high levels, and with compositions not too far from their unexploited compositions, simply by limiting fishing mortality rates by fishing capacity limitation and effort control, brings more benefits (catch rates, profits, and substantial long-term catches) than do the lower catches obtained by taking a larger percentage of a smaller biomass every year. One of those benefits that has long been recognized is the ***buffering effect*** of exploiting a stock with more age classes surviving for longer than would be found in intensely fished stocks. And, in addition, such management would automatically greatly reduce the problem of discards of target fish species. If TACs are not used to regulate catches, then there is little point in forcing the discarding of truly accidental "excess" catches. Control of the undesirable catching of juvenile fish is better done by regulating the gear type and the locations and times of fishing than by discarding. The incidental kills of non-target animals, such as marine mammals, turtles, and sea-birds, as well as of practically and unintentionally endangered species, such as skates, are likely to decline, at least in proportion to the reduction of fishing effort to meet revised objectives—a major saving. Similarly, the inevitable habitat destruction caused by heavy gears like bottom trawls should decrease in proportion to the reduction of effort.

At this point, it is appropriate to quote *in extenso* Justin Cooke's advisory paper of 1995 cited in footnote 8: "*The mainstay of precautionary fishery management should be to keep fishing effort under control, so as to limit the impact of fishing on the target species and the ecosystem. Limitation of fishing effort brings several advantages in the longer term. As fishing intensity increases, the target species will make up a smaller proportion of the total catch. By-catches of other species of fish and wildlife will increase, in both absolute and relative terms. According to conventional ecological models, the biodiversity of the marine ecosystem can be expected to decrease with increasing fishing intensity. In the shorter term, the loss in biodiversity is mainly quantitative, as reflected in standard indices of species diversity (an increasing proportion of the standing biomass consists of a decreasing number of species), without there necessarily being any actual loss of species. In the longer term, prolonged heavy fishing would eventually be expected to lead to a reduced diversity of species.*

In the case of bottom trawling, higher fishing effort means more frequent disturbance of the benthos, which changes the structure of benthic communities, reduces benthic biodiversity, and can make conditions less favorable for the reproduction of commercially important fish species, including those which are not themselves fished demersally. A further advantage of keeping fishing effort low and maintaining stocks at high levels is that the annual catch can easily be obtained by fishing in only a limited part of the area inhabited by the fish stocks. Thus, extensive closed areas can be established covering 50% or more of the stock area with little or no loss to the fishery. This is in contrast to the current situation in some overfished grounds where virtually the entire grounds have to be swept, sometimes several times in a season, in order to attain a TAC. The designation of protected areas (no fishing zones) will protect natural benthic communities or, in the case of previously disturbed grounds, enable them to reestablish themselves. The limited evidence available suggests that benthic communities can recover quite rapidly provided they receive protection in the near future while the constituent species are still present in the environment."

Transition

Finding ways of moving from an overfished situation to a good sustainable high-productivity one relatively painlessly, in economic and social terms, is as important as deciding on better management objectives and procedures. And, in some respects, it can be more difficult. Declaration of a species or species group or area moratorium can make things easier, provided there are alternative uses for the temporarily unemployed fishing units. This is a problem that should be tackled by conducting simulations of alternative rates of change (mostly of reduction, considering the number of already overfished stocks), either of TACs or, as recommended here, of fishing capacity and effort. As far as I know, no such simulations have yet been made for any fishery, but perhaps some are, somewhere, being undertaken. The reduction of capacity and effort is probably easier than that of TACs, though, in both cases, catches will inevitably fall until stocks have begun to recover. But, in the case of catch reduction, which involves dealing reasonably with annual variations in recruitment and stock availability, the economic disruption is likely to be greater.

The Commission's present recovery plans are specifically intended only to move depleted stocks out of the danger zone bounded by the zone of *safe biological limits*. This will take a shorter time than needed for real recovery to the high level and new more stable composition of stocks demanded by recovery towards highest feasible average sustainable catch; recovery to economically optimal state will take rather longer. Thus, any temptation to reopen or increase a fishery prematurely by upgrading the fleet capacity would have to be resisted. Just as attainment of the desired modified MSY target depends on getting agreement on a tested procedure and then refraining from negotiation either of annual TACs or changes in fleet capacity, there would have to be advance agreement on any ways in which, and reasons for, any adjustment of regulations and fleet limits and deployment would be negotiated—in exceptional circumstances—during the recovery period of the management horizon.

If the suggestions in this paper were to be accepted, then there would be two, presumably simultaneous, transitions: one from management merely for sustainability to management for high average physical yields, and the other from TAC-based regulation to fishing effort and capacity regulation as the primary management process. This would, of course, make simulation of various options and scenarios more complicated, but still feasible.

Summary and Conclusions

More than one-half the consumption of marine fish products in the European Union now comes from outside waters under direct EU jurisdiction, either from EU-flagged fishing vessels operating outside the EU EEZ or as imports from non-EU producers. This situation makes European consumers vulnerable in the long-term, and perhaps in the medium-term, to shortages arising from limitations exercised by others as a result of overfishing outside the EU EEZ or for other reasons related to local sustainable development and the global market for fish products. That fact alone would make the Commission's proposal that the Common Fisheries Policy should shift from mere sustainability and recovery of seriously depleted stocks from that dangerous status towards seeking maximum sustainable yields, admirable and timely. In addition, the proposed shift would eventually and inevitably lead to higher catch rates (catches per effort) and hence greater potential profitability. This is notwithstanding the fact that modern scientific and technical opinion is that the concept of MSY is illusory, and even if it were not, is unreachable, and certainly not economically optimal. It is, however, a useful concept that can be rendered operational by redefining it, and also the fundamental idea of sustainability, and seeking scenarios to approach it through exhaustive simulations of possible management procedures.

The necessary redefinition is **a high optimal average catch over an appropriately designated time horizon**; sustainability is not like diamonds—forever. Interestingly, this approach was pioneered by the IWC when it embarked, in the 1980s, on designing a RMP for regulating the catching of baleen whales. The core of the RMP was an algorithm that had been shown, by simulation, to result in a high annual average catch while avoiding, with high probability, any accidental depletion of stocks to below a predefined minimum threshold level, and which—as a third objective—ensured that, as far as reasonably possible, there would not be very large changes in catches from one year to the next as a result of the procedure itself.

I advocate the replacement of the present EU management system, involving the application of annual TACs for every stock, by a system based on limitation of fishing effort, mediated by a limitation of fleet capacities to ensure that regulation to reduce effort does not generally lead to inefficient operations. That is, regulating by input rather than output—to maintain fishing mortality rates at, or below, levels that are expected to provide highest feasible average physical yields over an appropriate predefined period. This would more effectively stabilize the European fisheries, but would also carry other advantages, in particular, reduction of discards, especially of non-target species, and release of scientific effort for more constructive activity than annual calculation of hundreds of TACs.

These proposals would call for an acceptance that, once Member States had agreed to a procedure and committed themselves to applying it, there would be no further political negotiation of details of annual or short-period regulations; that is, the current faux-negotiations with Nature (involving second-guessing of the best available scientific advice) would cease.

Any serious move towards some form of maximum yield policy will inevitably have costs in the short-term while depleted stocks are allowed to recover to more productive levels and compositions. These are likely, however, to be offset by savings from reductions in fleet capacities and in deployed fishing effort. The possible scenarios for less "painful" transitions from overfishing to optimal fishing, and from TAC management to effort and capacity management, should be examined by computer simulations of alternatives, such as of faster or slower transitions.

Finally, attention should be paid to the recent suggestions from the World Bank and FAO (2009) that, both globally and regionally, there are strong economic advantages to maintaining fishing mortality rates, hence fishing effort, significantly below the levels giving optimal sustainable physical yield.

The simple fact that if fish are allowed to grow up, get fat, and produce lots of young, instead of being killed as juveniles, the consequences of our taking some of them will be better and more profitable has been known for seventy years or more. A defining date was perhaps 1943, when the remarkable little book entitled *The Fish Gate*, written by Michael Graham

(England's first post-war Director of Fisheries Research) was published in London (Graham 1943). A few years later, when the war had ended and fishing had resumed in the North Sea, his and others' predictions of recoveries were vindicated: plaice and sole had increased more than threefold, haddock and cod more, during the respite that Graham called the ***Second Great Fishing Experiment*** (the First had been the Great War, during which similar stock increases occurred, but not so fully documented).

The European Union now has the chance to conduct the Third Great Fishing Experiment by deliberate, systematic phased reduction of fishing capacity and effort to gain more productive yield levels. That opportunity must be grasped.

Acknowledgments

Although the views expressed here are my own, I must acknowledge the suggestions and encouragement I have had from several colleagues in recent years, especially from Michael Earle, Justin Cooke, Hélène Bours, and William de la Mare. In particular, Ms. Bours' unpublished document entitled "Evaluating options for strengthening the European Union (EU) Common Fisheries Policy (CFP)," dated March 29, 2008, has been a most useful reference. I am further indebted to the European Marine Program of the PEW Environment Group for financial support and also encouragement.

References

Beverton, R. J. H. 1998. Fish, fact and fantasy: a long view. Reviews in Fish Biology and Fisheries 8:229–249.

Beverton, R. J. H., and S. J. Holt. 1957. On the dynamics of exploited fish populations. Her Majesty's Stationery Office, UK.

Constable, A. J., and W. K. de la Mare. 1996. A generalised model for evaluating yield and the long-term status of fish stocks under conditions of uncertainty. CCAMLR Science 3:31–54.

Cooke, J. G. 1999. Improvement of fishery-management advice through simulation testing of harvest algorithms. Pages 797–810 in A. I. L. Payne, editor. Confronting uncertainty in the evaluation and implementation of fisheries-management systems. ICES Journal of Marine Science 56(6).

Getz, W. M., and R. G. Haight. 1989. Population harvesting: demographic models of fish, forest and animal resources. Princeton University Press, New Jersey.

Ginzburg, L., and M. Colyvan. 2004. Ecological orbits: how planets move and populations grow. Oxford University Press, Oxford.

Graham, M. 1943. The fish gate. Faber and Faber, London.

Holt, S. J. 2004. Foreword to the 2004 printing. Pages i–xxiii *in* R. J. H. Beverton and S. J. Holt. 1957. On the dynamics of exploited fish populations. Blackburn Press, Caldwell, New Jersey.

Holt, S. J. 2007a. New policy objectives and management procedures for EU fisheries. A commentary and suggestions. A briefing paper prepared for the European Policy Office of the WWF. 67 pp. Unpublished.

Holt, S. J. 2007b. New policy objectives and management procedures for EU Fisheries. A commentary and suggestions to The Greens/European Free Alliance in the European Parliament. A study sponsored by Ms. Marie-Hélène Aubert, MEP. 23 pp. Unpublished.

Holt, S. J. 2009. Sunken billions—but how many? Fisheries Research 97:2–10.

Hutchings, J. A. 2000. Collapse and recovery of marine fishes. Nature 406:882–885.

ICES. 2002. Report of the Study Group on the Further Development of the Precautionary Approach to Fishery Management. Lisbon, Portugal, 4–8 March 2002. ICES Document CM 2002/ACFM:10.

Kell, L. T., J-M. Fromentin, F. Gauthiez, and V. R. Restrepo. 1999b. A simulation framework to evaluate management strategies for Atlantic tunas: a preliminary example based on East Atlantic Bluefin Tuna. ICCAT Working Document SCRS/99/11 2095-116.

Kell, L. T., C. M. O'Brien, M. T. Smith, T. K. Stokes, and B. D. Rackham. 1999a. An evaluation of management procedures for implementing a precautionary approach in the ICES context for North Sea Plaice (*Pleuronectes platessa* L.). ICES Journal of Marine Science 56:834–845.

Logan, G., W. de la Mare, J. King, and D. Haggarty. 2005. Management Framework for Strait of Georgia Lingcod. Canadian Science Advisory Secretariat Research Document. 2005/048.

Longhurst, A. 2006. The sustainability myth. Fisheries Research 81:107–112.

Malthus, T. R. 1798. An essay on the principle of population. J. Johnson, London.

O'Brien, C. M, and D. L. Maxwell. 2002. Towards an operational implementation of the Precautionary Approach within ICES—biomass reference points. Annex 2, Working Document 8. Pages 87–92 *in* Report of the Study Group on the Further Development of the Precautionary Approach to Fishery Management. Lisbon, Portugal, 4–8 March 2002. ICES Document CM 2002/ACFM:10.

Pikitch, E. K. 1987. Use of a mixed species yield-per-recruit model to explore the consequences of various management policies for the Oregon flatfish fishery. Canadian Journal of Fisheries and Aquatic Sciences 44(Suppl. 2):349–359.

Reiss, H., G. Hoarau, M. Dickey-Collas, and W. J. Wolff. 2009. Genetic population structure of marine fish: mismatch between biological and fisheries management units. Fish and Fisheries 10:361–395.

Ricker, W. E. 1954. Stock and recruitment. Journal of the Fisheries Research Board of Canada 11(5):559–623.

Shepherd, J. G. 1982. A versatile new stock-recruitment relationship for fisheries and the construction of sustainable yield curves. Journal du Conseil International pour l'Exploration de la Mer 40:67–75.

Verhulst, P-F. 1845. Recherches mathématiques sur la loi d'accroissement de la population. Nouveaux Mémoires de l'Académie Royale des Sciences et Belles-Lettres de Bruxelles 18:1–42.

Witting, L. 1997. A general theory of evolution: by means of selection by density dependent competitive interactions. Peregrine Publisher, Århus, Denmark.

(The) World Bank and Food and Agriculture Organization (World Bank and FAO). 2009. The sunken billions. The economic justification for fisheries reform. The World Bank, Washington, DC.

Chapter 44

Wherever is MSY?[1]

Whosoever seeks the truth will not proceed by studying the writings of his predecessors and by simply accepting his own good opinion of them.
Whosoever studies works of science must, if he wants to find the truth, transform himself into a critic of everything he reads.
He must examine tests and explanations with the greatest precision and question them from all angles and aspects.

Hasan Ibn al-Haytham
(10th century, scientific adviser to the ruler in Cairo of the Fatimid Ismaili caliphate, al-Hakim)

Introduction

I cannot aspire to emulate Hasan, but this paper is indeed about questioning, especially customary practices and given truths.

The UN Law of the Sea Convention 1982 (UNCLOS) came into force in November 1994 and, as of March 2010, had been ratified by 160 nations, but not yet by the USA. Its Article 61 defines the management objective for fisheries in Exclusive Economic Zones (EEZs) in the following terms (*slightly edited and contracted by Holt*):

> "Coastal State(s), taking into account the best scientific evidence available to it/them, shall ensure through proper conservation and management measures that the maintenance of the living resources in the exclusive economic zone is not endangered by overexploitation ... **Such measures shall also be designed to maintain or restore populations of harvested species at levels which can produce the maximum sustainable yield [MSY], as qualified by relevant environmental and economic factors** ... taking into account fishing patterns, the interdependence of stocks and any generally recommended international minimum standards... In taking such measures, the coastal State(s) shall take into consideration the effects on species associated with or dependent upon harvested species with a view to maintaining or restoring populations of such associated or dependent species above levels at which their reproduction may become seriously threatened." (*Holt's emphasis*)

[1]Editors' note: written by Sidney Holt in 2010; an earlier version was submitted to Fisheries Research for publication, but was declined pending revision and resubmission, which, unfortunately, Holt never did. The editors, nonetheless, consider it to contain valuable insights by Holt and so include it in this anthology. In the original, an Abstract was included at the beginning of the paper; here, it has been removed.

Article 119 of UNCLOS defines the management objective for fisheries on the high seas with respect to determination of allowable catches and establishing other conservation measures in identical language. The same provisions apply to fisheries for catadromous fishes (those that breed in the sea and feed in fresh- and brackish waters), but the catching of which on the high seas is prohibited under Article 67. A somewhat different regime is specified for anadromous fishes (that breed in inland waters and feed at sea).

Although the USA has not ratified UNCLOS, it was the main driver during the long negotiations that resulted in the UNCLOS specifications for conservation of marine living resources and especially in pressing for inclusion of the MSY target. We may, therefore, presume that the above words express a universally accepted feature of the Law of the Sea. Elsewhere, I have detailed their international legal origins in the insistence of the USA, during the UN Technical Conference held in Rome in 1955 and at the First UN Conference on the Law of the Sea held in Geneva in 1958, which adopted a Convention on Fishing and Conservation of the Living Resources of the High Seas, that the main objective of "conservation" is "the aggregate of the measures rendering possible the ***optimum sustainable yield*** from those resources so as ***to secure*** *a maximum supply of food and other marine products*" (*Holt's emphasis*). The historian M. C. Finley (2007) revealed that this "optimum/maximum" idea was embedded in the International North Pacific Fisheries Convention (INPFC) of 1952 as a device for excluding Japanese fishing vessels from taking salmon on the high seas in the Northeast Pacific. Finley described her study as "a political history of maximum sustainable yield." This so-called ***abstention principle*** did not get into the 1958 and 1982 Conventions, but reared its head again in the 1995 UN Straddling Fish Stocks Agreement, which came into effect in December 2001. Yasuko Tsuru (2004) has written that "States parties to the Agreement enjoy a right to access to high seas fishery resources, but on the condition that they respect existing regional conservation and management measures. This Agreement also introduces the concept of a precautionary approach. The end result is that the 1995 Agreement fulfilled the intent of the original 1952 principle of abstention—that is, it places some kinds of restrictions on high seas fishing. The degree of restrictions depends on the activities and policies of new regional fishery organizations like WCPFC [Western and Central Pacific Fisheries Commission] and so on. Even if the new version of the principle of abstention succeeds in excluding distant fishing states from high seas, it will do little to solve the larger problems facing today's ocean fishery resources."

So, it seems we are stuck with this idea both for optimizing fishing activities with a view to the future and as a means, in certain circumstances, of deciding which people and how they may fish in what is left of the high seas.

The general MSY concept—but not yet using that phrase—was first addressed by the Norwegian biologists Hjort, Jahn, and Ottestad (1933) in relation to the dynamics of blue whales *Balaenoptera musculus* in the Southern Hemisphere, and formalized notably by the American scientist M. B. Schaefer (1953, 1954), especially with respect to Yellowfin Tuna *Thunnus albacares* fisheries in the Eastern Central Pacific. These applications to organisms that, at the time, could not be aged led later to the idea of "recruitment overfishing" in which a stock has been reduced to a <u>level</u> less than that at which MSY can be obtained. It is important to note that the UNCLOS text always refers to "levels" of stock, with no consideration of their structure, such as their size, age, and sex composition.

Another "tradition" is associated especially with the British naturalist Michael Graham and colleagues connected with ICES (see especially Graham 1939, 1943). They were dealing mainly with fishes for which there already existed age-determination methods (scale and otolith rings, especially, and the result of tagging experiments) and concluded that it was necessary to limit fishing effort so that more fish had a chance to survive to grow bigger. This approach led to the idea of "growth overfishing," which was the focus of my studies with R. J. H. Beverton in the 1940s–1950s (Beverton and Holt 1957; Holt 2004). I have discussed in detail the two approaches to conservation in Holt (2006). It has been asserted (e.g., by Quinn and Deriso 1999) that Graham's name should be linked with that of Schaefer because he referred to "sigmoid curves" (which he considered to be practically "magic") and occasionally to the "logistic," the particular function adopted by Schaefer, but in fact, Graham's logic had little connection with Schaefer's method, and in an exchange of letters with Schaefer, Graham was very skeptical of the validity of Schaefer's approach.[2]

Does Location Matter?

Because of the current near universality of fisheries management by setting periodic catch limits—***total allowable catches*** (TACs) —most attention has been given to the scale of MSY, with very little to the stock conditions from which it may be obtained. Where attention has been given, it has focused almost always on the size—***level***—of stock biomass, especially as the UNCLOS and other legal criteria virtually all refer to "level" and not to its composition by age, sex, or body size nor to the fishing effort that will generate the fishing mortality rate required to produce the MSY. When practical matters are discussed, it is also nearly always presumed that scientists are either able to estimate the relevant level or number or make a reasonable educated guess. In fisheries management discourse, "guesstimates" about MSY location are the rule.

The simple logistic model predicts—or rather assumes—that MSY will be obtained when stock biomass is one-half the size of the unexploited stock, and this assumption has been made in many published fish stock assessments, although the exploited "biomass" in question is variously defined—either the total of the population, of the adults or spawners, or of the "recruited" or "exploited" stock. It is a very limiting assumption made originally solely for mathematical simplicity—the density-dependence of the local rate of population increase being linearly related to population size or density, as assumed by P. F. Verhulst (1838) in his original formulation. This simple formulation has been used, with no empirical justification, in countless published fish stock assessments, most recently by the authors of an important World Bank—FAO publication reviewing the state of global fisheries (Holt 2009).

Schaefer's fitting of his Yellowfin Tuna data to a parabola (Verhulst's original equation), implying that MSY was to be found at a stock level one-half of the level before exploitation began (MSYL= 0.5, or 50%), could have given an equally good statistical fit, or an even better one, by using a function implying an MSYL other than—indeed far from—50%. Several such functions have been put forward in the literature of fisheries science and of general ecology and population dynamics. In 1969, J. J. Pella and P. K. Tomlinson—who worked with Schaefer at the Inter-American Tropical Tuna Commission (IATTC)—published a modification of the logistic—in fact, a simple power function with one extra parameter that allowed a wide range of MSY levels. This was, in fact, used by the Scientific Committee of the International

---

[2] I am indebted to Dr. Finley for providing me with copies of that correspondence which she had located in various now-public archives.

Whaling Commission (IWC) throughout the 1970s to make assessments of baleen whale populations for which it was assumed that the MSY level was at 60% of the unexploited—"virgin"—stock size,[3] although, in that case, for policy reasons, it referred to numbers of animals rather than to biomass. The Pella–Tomlinson (1969) model has, however, one formal technical disadvantage, which is that it predicts that if MSYL is set to less than 37% ($1/e$ = 0.37), it would be impossible to exterminate the stock even by applying an infinite fishing mortality; for this reason, it is unrealistic in modeling low MSY levels (Quinn and Deriso 1999).

Schaefer and Beverton (1963) examined the question of where the MSYL for fishes would most likely lie and concluded it was probably lower than 50% of unexploited stock. The modified logistic of Pella and Tomlinson has a worrying mathematical characteristic which has rarely been noticed by those that have applied it—least of all by scientists, including myself, who conducted assessments of whale populations for the IWC. It is that if a parameter set is used that sets the MSYL high (as a fraction of unexploited biomass)—implying that the optimum value of fishing mortality F is low, then only a small increase in F above the MSY value will, if sustained, drive the stock to extinction—a counter-intuitive result well illustrated by Figure 2.7 in Quinn and Deriso (1999). De la Mare (1986) also showed by a simulation of a management procedure developed in the whaling context that even if the model was correctly specified and the values of its parameters precisely known, it would be impossible to home-in on the MSY by setting total allowable catches (TACs) if there was any stochastic variability in the system. His study opened the route for the IWC scientists to explore a completely different approach to the setting of catch limits—the Catch Limit Algorithm (CLA) that forms the core of the Commission's Revised Management Procedure (RMP) (Cooke 1995).

Worm et al. (2009) state that in their opinion "fishing for MSY results in a stock biomass that is substantially lower than the unfished biomass" and is typically 25–50% of that. In the same paper, they describe this as "the traditional" objective of management, but they unfortunately provide no references from which this "tradition" comes. Certainly, it comes more from theory than from empirical practice, and very loose theory at that. The upper figure is entirely circumstantial from the simple logistic function. The study presented here suggests that the theory of population dynamics guides us towards the lower figure.

Before I move to a deeper analysis of the MSY location question, I should illustrate why and how much it matters. For that, I chose two hypothetical situations, using as my guide the Pella–Tomlinson (P–T) population model. Stripped of scaling parameters, this is very simple:

$$g_p = 1 - p^n$$

where p is population size (usually biomass) expressed as a fraction of the size, in steady-state conditions, of the population before exploitation began (sometimes referred to as "carrying capacity"); g_p is the population increase (growth) rate, expressed as a fraction of the maxi-

[3]That was a political choice (the scientists were under pressure to be "conservative," and there had to be a consensus because without it, the New Management Procedure they had devised, and the Commission had adopted in 1975, could not be implemented. In the debates of the time, it was argued that because whales were thought to have low natural mortality rates and certainly had low rates of reproduction, this justified the assumption of a high MSY level. However, that level does not depend on the absolute values of the two contrasting rates, but rather on the density-dependence of the difference between them, and specifically on the nature of the non-linearity of that dependence.

mum rate g_i attained when the population is very small (sometimes referred to as the intrinsic rate of natural increase i); g_p is sometimes called the sustainable yield rate (SYR) from a population of size p (the product of SYR and p is the *sustainable yield* SY, and the value of g in a population at the level of MSYL, in steady state, is the MSYR); and n is a coefficient that determines the skewness of a curve of sustainable yield against population size or density (when $n = 1$, the P–T equation reduces to the simple symmetrical logistic with linear density-dependence, when n is greater than unity, the curve of SY against population size is skewed to the right, with MSYL being higher than 50%, and vice versa if n is less than one).[4]

My first example is the case where a population has been reduced to a very small size—say 10% of carrying capacity—by overfishing, then protected, and after a time, surveys show it is increasing. Suppose it has doubled—to 20%—and there is pressure to reopen the fishery, setting a TAC somewhat lower than an estimate of the current SY, so that the population will be expected to continue to increase, though at a slower rate than would be expected if the fishery were to remain closed. So, what is the SY? We need to look at the SYR at the 20% population level and think of it as a fraction of the maximum SYR, i, in a very small population. If we assume the SY–P curve is a simple logistic, then SYR/i is 0.8. But, if MSYL is not 50%, but is 40% (n = about 0.3), then SYR/i is 0.4, a twofold difference. If a TAC were to be set based on the assumption that MSYL is 50% when it is actually 40%, then it will be twice as high and lead to renewed stock depletion. It seems that, in such circumstances, it is more precautionary to assume a low MSYL. On the other hand, however, the lower MSYL really is closer to the current level, and so it is more likely that any overestimate of the current SY will result in further depletion, which, if continued, might be irreversible.

A current specific example of this problem is to be found in the IWC Scientific Committee's reconsideration of the RMP that it devised more than a decade ago, which the Commission adopted, but has not yet implemented. This requires, as input, a presumed range of values of the MSY rate, that is, the number which, multiplied by the estimated stock size at MSY level, gives the MSY. One way of getting an idea of this is to look at the rate of increase—obtained from repeated surveys—of a greatly reduced population under long-term protection. This would not be in a steady state as far as its age composition is concerned, but it is, nevertheless, commonly assumed that the increase rate is an approximation to i. If, as the IWC has assumed for decades, the MSYL for whale populations is at 60% of the carrying capacity, the Pella–Tomlinson formula would give an MSYR of $0.52i$. But, if the real MSYL is 40%, then the MSYR estimate would be $0.29i$. The erroneous MSYL assumption would give an 80% overestimate of MSYR and, if it were to be used to provide a figure of MSY, as under the IWC's earlier New Management Procedure (NMP)—still on its books—the consequence of the MSYL error would be a nearly threefold overestimate of MSY itself.

Now, let us look at another example. Suppose we have a previously unexploited or only lightly exploited stock and wish to allow exploitation, but with a precautionary TAC. A real example is in the IWC where the current management procedure sets zero TACs until such time as at least a scientifically acceptable estimate of the population size has been obtained. When that happens, the IWC rules say the permitted catch will not be more than 90% of the

---

[4]The P–T model reduces to the inflexible Gompertz-Fox model (Fox 1970) when n is zero, MSY level = 36.8% level. Fletcher (1982) achieved flexibility similar to that of P–T by using a polynomial (specifically, a quadratic) instead of the P–T power function.

presumed MSY. There was, before the general moratorium on commercial whaling was enacted, great pressure from whaling countries to follow this course, with the argument that not to do so would be a scandalous "waste" of natural resources. Let us suppose a situation in which it is agreed that this will be allowed and, IWC's assumption being that MSYL is 60% for all large whale species, it is decided, as an interim caution, to set a limit such that the stock will not be brought down below 80% of carrying capacity. What is the catch limit to be to hold the stock at 80%? We know, further, that the MSYL = 60% value (P–T $n = 1.39$) has little, if any, scientific support; the value was chosen by the IWC Scientific Committee in 1975 because it had been instructed by the Commission to be cautious—hence higher than 50%—and, furthermore, the new Procedure was called a Moratorium Compromise because when stocks were 10% or more below the presumed MSYL, catch limits would be zero.

After 1975, the IWC Scientific Committee used P–T-type models to provide catch limit advice. Suppose the real MSYL for whales is less than 60%; at the time, some scientists argued that it was nearer 50% (this was a political position taken by those from whaling countries—it meant it would be less likely that, when the stocks in which their country had an interest were assessed, the catch limits of interest would be zero). Others thought whales are not really different from fish, so MSYL could perhaps be as low as 40% of carrying capacity. In this situation, calculating catch limits assuming MSYL is 50% would lead to advice on a sustainable catch limit to be much higher than 90% of the real MSY if MSYL was truly 50 or 40% (by twofold or sixfold, respectively) and, of course, lead to rapid unintended depletion of a previously "healthy" population.

Evidently, if catch limits are to be calculated from a flexible population model, such as Pella–Tomlinson, Fletcher quadratic, or from another, inflexible one, such as Schaefer (logistic) or Gompertz-Fox (P–T with MSYL at 37%), then the chosen MSYL location matters very much indeed. Furthermore, this warning does not depend at all on the validity of the estimations or guesses as to the values of the other parameters in the model used.

A lesson from this exercise is that ***what is precautionary and what is not depends on the circumstances***.

I conclude this section by noting that values of MSY itself, as would be calculated from these models, are themselves wide ranging. For example, for the P–T model, the expected MSY values, scaled to 1.0 for the symmetrical logistic, would be 2.4 for MSYL = 0.7, 1.7 for MSYL = 0.6, and 0.4 for MSYL = 0.4.

Properties of Age-Structured Population Models

Those scientists who work mostly with logistic-type ("surplus production") models usually think of MSY in terms of equilibrium biomasses expressed as percentages of unexploited/virgin/carrying capacity biomasses. Those using age-structured models tend to think more about the level of fishing mortality that will stabilize a stock around its MSY level. It may, however, be useful to look at both quantities together, especially in consideration of alternative approaches to fishery regulation than by the currently dominant method of setting TACs, such as by the limitation of fishing powers, fishing efforts, and fishing intensities of fleets.

I use those terms here in the following senses: the fishing power of a fishing vessel and its gear (or other fishing unit) is the fishing mortality that would be generated by the full-time deployment of that unit for a season to a specified stock or stocks. The fishing power of a fleet is the sum of the powers of its components. Fishing effort is the product of the fishing power of a unit and the fraction of available time in which it operates; the fishing effort of a fleet is the sum of all those efforts. I use the term fishing intensity to mean fishing effort per unit area of fishing grounds. It is necessary to refer to intensity if the fleet distribution changes, the distribution of the fish changes, or, indeed, if there is any change within the whole fished area in the distribution of the fleet relative to the distribution of the fish. Although these are general questions, my focus is on the recent decision by the European Union to shift from a policy of precautionary sustainability to one of "achieving MSY for EU fish stocks by 2015, in line with the EU commitment made in Johannesburg in 2002" (see The Fisheries Secretariat, a non-profit organization based in Stockholm—27 September 2010).

In 1958, (see Beverton and Holt 1959), I published a transformed version of the simple steady-state yield equation that Beverton and I had developed a few years before, having realized that its properties depended not so much on the values of the natural mortality rate M and the parameters of a von Bertalanffy growth curve, K and W_∞, as on the ratio M/K, with the asymptotic body weight (a proxy of the general size of the species in question) playing no part in the shape of the yield curve, only the scale of its y-axis. There is some evidence that M and K are correlated, so their ratio varies among species less than either of its components. For example, Beverton and I found it to be close to unity for both Plaice *Pleuronectes platessa* and Haddock *Melanogrammus aeglefinus* in the North Sea, although both M and K for Haddock were double the values found for Plaice.

Similarly, another critical ratio was $c = l_R/l_\infty$, the ratio of the length at recruitment to final asymptotic length of the fish. Beverton and I repeated my earlier analysis, carrying it a bit further, and tabulated the steady-state yield-per-recruit for wide ranges of the two ratios, expressing the results in terms of the exploitation rate, defined as the variable:

$$E = F/(F + M) = F/Z \quad \text{(Beverton and Holt 1966).}$$

If one prefers to think in terms of the fishing mortality rate, it is sufficient to remember that:

$$F = M \times E/(1 - E)$$

Note that Worm et al. (2009) and some others have used the term "exploitation rate" for, roughly, the ratio of the fishing mortality rate to the biomass of the recruited population or possibly of the total population.

These tables allow us now to examine, without further computation, the properties of the model relating to MSY, both in terms of the fishing mortality rate and the stock biomass. For our present purposes, it is convenient to deal with $w = c^3$, that is, the ratio of the weight of the fish at recruitment to the asymptotic size. It is worth noting that the weight of a fish at the inflexion of the growth curve is always $W_I = 0.3 \times W_\infty$.

The simple yield-per-recruit equation is about a group of distinct cohorts of fish, each comprised of fish all born at the same time. It is well known that, in theory, the maximum catch could be taken by exerting an infinite fishing effort at the age at which the cohort attains its maximum weight through the interaction of the natural mortality and the intrinsic growth rate K. I called that the ***potential yield*** (Holt 1958; Chapter 4) and gave the expression for the ***potential yield-per-recruit*** Y^* as a fraction of the asymptotic body weight in the growth function. The body weight of fish in the potential yield (expressed as a percentage of the asymptotic weight) is a function only of M/K. Table 1 below gives some examples.

Table 1. Body weight of fish in the potential yield-per-recruit.

M/K	Body weight/W_∞ (%)	Y^*/W_∞ (%)
0.25	80.4	41.4
0.50	63.6	23.8
1.00	43.0	10.5
1.25	37.3	7.6
1.50	30.1	5.7
2.00	22.1	3.5
3.00	12.9	1.6
4.00	8.2	0.8

The maximum steady-state catch that will be obtained by otherwise unselective fishing from any given age or size at recruitment will always be less than that theoretical maximum—the potential yield y^*—often much less. Take as an example the case where M/K = 1.0. Each cohort attains its maximum weight at an age when the individual body weights are 43% of the asymptotic weight. The yield-per-recruit, expressed as a percentage of the asymptotic weight W_∞, is 10.5%. Now, suppose it is decided to exploit this stock using gears, such as trawls with an appropriate codend mesh size, that will retain mainly mature fish and that maturity is attained on average at 30% of the asymptotic size (c = 0.67). The tables show that the maximum steady-state yield-per-recruit will be obtained with F/M = 6.7, and its value as a percentage of the asymptotic body weight would be 10.2%, which is 97% of the potential.

Now, suppose a smaller mesh were to be used, so that fish began to be subject to capture at only 20% of their asymptotic size. In this case, obtaining the maximum yield-per-recruit would require F/M to be only 3.0 (calling for less than one-half the fishing effort), yet the catch would still be as much as 91% of the potential. If we now drop the maximum criterion and reduce fishing effort so that F/M = 2, then we would still obtain as much as 85% of the potential yield and 90% of the maximum obtainable with that mesh size.

Two conclusions should be obvious from this and similar examples. First, that it does not make sense to aim at the maximum if a substantial reduction in the fishing effort will reduce the sustainable catch only a little, while greatly reducing the cost of taking it. Second, the lower fishing effort, and consequent higher biomass of each cohort, mean it will be less likely that the stock, and hence the biomass of spawners, will be reduced by fishing to a degree that might cause recruitment to be reduced.

Potential yield is not itself a useful metric for practical management, but it can serve as a scale against which the overall efficiency of more realistic management objectives and procedures could be judged. It is, in this connection, worth noting that the age at which animals might be expected to become mature, commonly close to the inflection weight in the von Bertalanffy growth curve, depends only on the value of K, whereas the weight of a fish in the potential yield depends on both M and K, but is not fixed by the ratio M/K. Table 2 below illustrates this. Notice that for M/K =1.5, the age and hence size at sexual maturity and in the potential yield are the same. This means, as a rough rule, that the potential yield for $M/K \geq 1.5$ is not a sustainable quantity because all fish caught would be immature. This is a warning that high fishing intensities imposed on populations with relatively high M/K ratios might lead to depletion and even extermination if continued.

Table 2. Values of $t(0.3)$ and $t(p)$ for different levels of M and K, where $t(0.3)$ is the age, in years, at which the fraction 0.3 of the asymptotic body weight is attained, the inflexion of the growth curve and putative age at maturity, and $t(p)$ is the age at which the cohort attains its greatest biomass, i.e., the putative potential yield.

M	K	M/K	$t(0.3)$	$t(p)$
0.05	0.1	0.5	11.0	19.5
0.2	0.4	0.5	2.7	4.9
0.05	0.05	1	22.0	27.7
0.1	0.1	1	11.0	13.9
0.15	0.15	1	7.3	9.2
0.2	0.2	1	5.5	6.9
0.15	0.1	1.5	11.0	11.0
0.3	0.2	1.5	5.5	5.5
0.6	0.4	1.5	2.7	2.7
0.1	0.05	2	22.0	18.3
0.2	0.1	2	11.0	9.2
0.4	0.2	2	5.5	4.5

As is well-known, what I have said about cohorts here is applicable to the population as a whole, annual catches from the population being comparable with life-time catches from a cohort, provided we are concentrating on steady-state conditions and there is no systemic change in recruitment.

The body weights at which the potential yield would be taken are directly comparable with the average weights of fishes in the catch when the fishing intensity is finite, and it is of interest to compare them with the average weights when the fishing intensity is such as to hold the yield-per-recruit curves at their peaks. It is also of interest to compare those peak yields with the values of potential yield. With M/K values substantially higher than 1.5, the catches would be all of mature fishes and whether they would be sustainable, depends, of course, on how much higher the critical body weight is than the size at sexual maturity.

The yield-per-recruit curves usually displayed are domed.[5] The peak of the dome has sometimes been labeled as MSY and its location on the biomass or fishing mortality axis as MSYL or F_{MSY}—but not by Beverton or me. Such a label would only be justified when the average number of recruits is not sensibly affected by reduction of the numbers of fish, by fishing, close to the region of the peak. If, however, the recruitment is sensibly reduced as a stock is reduced by fishing to a level close to the peak of the dome, then the actual MSY, as it would be given by a model in which the recruitment changes with stock numbers or biomass, would be at a lower F value and higher stock biomass than that given by the constant recruitment model. Thus, the constant recruitment model can, in a way, serve as setting a lower bound to the location of MSY in terms of stock biomass and an upper bound to the location of MSY in terms of the fishing mortality rate.

[5]Quinn and Deriso (1999) write that the cohort biomass curve against time "is almost always domed," with its peak at what they call the critical age t^*. This is an inevitable consequence of, and they give the equations for, this and the corresponding body weight.

Table 3 below summarizes my extractions from the yield tables. The range of the ratio *M/K* was wide—0.25 to 4.0; of *E* from 0.05 to unity, and similarly *w* from zero to 0.94 (*c* = 0.98, with intervals of *c* at 0.02). Scale changes made to provide results mainly to two significant figures as a whole number percentage are indicated in the legend to the table. In Table I of Beverton and Holt (1966), the values of *Y/R* are given to six significant figures,

Table 3. Values of E_{MSY}*, F_{MSY}*/*M*, MSY*L, and MSY* for *M/K* ranging from 0.25 to 4 and *w* ranging from 40 to 1. E_{MSY}* is the values of the exploitation rate *E* giving the peak steady-state yield-per-recruit. These are, of course, decimal fractions less than unity, but the decimal points are omitted to reduce clutter. F_{MSY}*/*M* is the ratio E_{MSY}*/(1 – E_{MSY}*). MSY*L is the biomass of the recruited stock at the MSY*L expressed as a fraction of the biomass of the unexploited recruited stock. MSY* is the height of the peak of the yield-per-recruit curve expressed in units of 1000 x W_∞ . This arbitrary unit allows easy comparison of expected steady-state yields across the ranges of parameter values given in the table.

	M/K						
w	**0.25**	**0.5**	**1**	**1.5**	**2**	**3**	**4**
			E*MSY				
40	80	85	100				
30	75	75	85	100	100		
20	70	70	75	85	95	100	
10	65	65	65	65	75	90	100
4	65	60	55	55	60	70	80
2	60	55	50	50	50	55	60
1	60	55	50	45	45	50	50
			F*MSY*/*M				
40	4.0	5.7					
30	3.0	3.0	5.7				
20	2.3	2.3	3.0	5.7	19.0		
10	1.9	1.9	1.9	1.9	3.0	9.0	
4	1.9	1.5	1.2	1.2	1.5	2.3	4.0
2	1.5	1.2	1.0	1.0	1.0	1.2	1.5
1	1.5	1.2	1.0	0.8	0.8	1.0	1.0
			MSY*L				
40	15	10					
30	17	17	9				
20	21	20	14	8	3		
10	24	21	19	19	12	5	
4	23	23	25	24	20	14	9
2	26	26	27	25	24	21	18
1	26	25	25	29	28	23	30
			MSY*				
40	365	224					
30	341	209	102				
20	319	194	95	55	34		
10	292	172	83	49	32	15	
4	275	158	74	43	28	14	8
2	260	146	66	38	24	12	7
1	255	142	62	36	23	11	6

with indications of the locations of maxima, while Table IV gives corresponding values for the biomass of the recruited stock, appropriately scaled. This latter presents a problem. If w is much increased, e.g., in trawling with a much wider codend mesh size, or a shift of operations to locations where bigger, older fish are more likely to be found, then it would perhaps be more appropriate to identify the MSY points on an axis of total biomass rather than only of the biomass of the recruited phase. That can be done with rather more calculation. But, it is not very likely or common for unselective trawl fisheries to be conducted in such a way as to subject to capture only the big fish, say those already past the growth curve inflexion. I have, therefore, for purposes of illustration, confined my table to values of $w \leq 40\%$. I have omitted the line for $w = 0$ because it is not realistic to consider the von Bertalanffy growth curve as applying to values of w less than 1 or 2% of W_∞.

Interpretation

Perhaps the first thing to note is that steady-state yield-per-recruit curves against fishing mortality of stock biomass are not always peaked. Above a certain value of the size of first liability to capture relative to asymptotic size, they are asymptotic, with the critical value of w depending on the value of M/K, for $M/K \geq 0.5$.

The second point is that for a wide range of values of M/K for unselective fisheries (w = 1 or 2%), the MSY* levels are remarkably similar, of the order of 20–30% of the unexploited recruited stock, but MSY*L diminishes rapidly as fishing becomes more selective (i.e., as w is increased). Third, for the entire range of M/K, the ratio of fishing mortality rate to natural mortality is remarkably constant for relatively unselective fisheries. Fourth, over the wide selectivity range from w = 1% to 20 or 30%, the MSY is relatively little changed, but it diminishes drastically for higher M/K.

Comparison of Table 3 with Table 4 indicates how far short of the theoretical maximum steady-state yield-per-recruit the peak catches may fall when the size at first liability to capture is different from (usually less than) the body size at which each cohort attains its maximum total weight.

Another metric of interest in interpreting the yield-per-recruit curves is the average body size of fish in the steady-state catch. It is well known that the first indication that fishing is significantly affecting the stock size is often a reduction in the average size of fish caught. This is commonly observed as an average length, but that can be misleading because the length decline is roughly the cube root of the weight decline, which is what really matters. This failure to realize the difference between changes in length and changes in weight is classically seen in the history of the management of whaling. Whales are, of course, measured, never weighed.

Table 4. Values of w_{MSY}* and MSY*max for M/K ranging from 0.25 to 4. w_{MSY}* is the body size, relative to asymptotic size, at which the weight of the catch is maximized. MSY*max is the weight of the cohort, per recruit, in units of 1,000 x W_∞ when the individuals in it attain the relative body weight w_{MSY}. Of course, obtaining that calls for exertion of infinite fishing effort to generate infinite fishing mortality.

M/K	0.25	0.5	1	1.5	2	3	4
w_{MSY}*	78	64	40	28	22	12	7
MSY*max	414	238	105	57	36	16	8

In the early years of the Antarctic whaling industry, the decline in size was noticed and fully documented, and it signaled considerable reductions in potential oil production per whale (but obscured for years by continuing changes in the efficiency of oil extraction processes[6]), and in 1974, the IWC agreed that future regulation should eventually be such as to optimize total weight yield instead of the traditional numbers. But, that wise decision was set aside and never implemented, the reason being that it would have called for a longer period of protection of depleted stocks to provide for optimization. If it had been implemented, it would have led to lower catch limits in order to generate a lower fishing mortality rate, because in surplus yield models, attainment of MSY by weight calls for a lower fishing effort than does MSY by number.

The change in average weight of fish in steady-state catches with change in fishing mortality is very considerable, For example, Beverton and Holt (1957) calculated a yield-per-recruit curve for Plaice in the North Sea for a 70-mm codend mesh size, corresponding with $w_R = 0.043$ and an estimate of $M/K = 1.05$. This stock was severely overfished in the 1930s, with $F/M = 7.3$. The peak yield-per-recruit was estimated to be at $F/M = 2.2$. Reduction of the fishing effort to that "optimal" level would have led to an almost doubling of the average weight of fish in the catch. This has two specific implications. One is that, in some markets, the value per unit weight of a fish increases with its size, at least over a certain range. The second is a biological one that comes from the findings analyzed by Alan Longhurst (2010) that, although egg production is roughly proportional to body size (and hence cohort egg production to the total weight of its sexually mature members), the eggs and larvae of larger/older females tend to be more viable than those from smaller/younger ones and hence contribute relatively more to recruitment to the next generation and eventually to the stock as a whole.

The variations in the average weight of fish in the catches by fishing intensities yielding peak catches, the age at first liability to capture, with various values of the ratio M/K, are of some interest and are summarized in Table 5 below.

Table 5. Average weight of fish in the catches as a fraction of the asymptotic weight, as a function of the weight at first liability to capture (w) as a fraction of the asymptotic weight, and selected values of M/K.

	M/K			
w	**0.25**	**0.5**	**1**	**2**
0.40	0.64	0.52		
0.30	0.60	0.48	0.35	
0.20	0.58	0.43	0.30	0.20
0.10	0.52	0.36	0.24	0.15
0.05	0.47	0.33	0.21	0.11
0.02	0.47	0.31	0.18	0.09
0.01	0.45	0.29	0.16	0.08
0.00	0.42	0.25	0.13	0.05

[6]Not all of the oil in a baleen whale is in its blubber; nearly one-half may be in the bones, organs, and muscles, though the distribution changes dramatically during the several months of intensive feeding while fat is being stored. These were sometimes processed to extract the oil, sometimes not, and especially not when frozen meat became the most valuable product.

As expected, the average size increases with the age—and hence size—at first liability to capture, but not over such a wide range as the latter. More striking is the degree to which a relatively high value of M/K, which demands a relatively high fishing intensity for peak steady-state catches, leads to a relatively low average size when first liability to capture occurs at a young age/small size.

Another metric of interest in connection with such assessments is the average body size at sexual maturity. One form of this, called by Jensen (1996) one of the "Beverton and Holt invariants" (see also Munro 2011 and Mangel 2005 for more on invariants and dimensionless ratios) is the ratio of the body weight at sexual maturity to the asymptotic weight: $w_m \approx 0.3$, a value remarkably close to the inflexion of the von Bertalanffy growth curve. This ratio is not truly invariant, but it does perhaps offer a guide to judging whether yield-per-recruit calculations are being examined in terms of the mature stock $w \geq w_m$ or the whole, including juveniles.

The Next Step

Since our aim is to see whether population dynamics theory can help us set bounds to the possible location of MSY with respect to fishing mortality and/or stock biomass, it is useful to look at a ***self-regenerating model*** in which a yield-per-recruit model is combined with a function relating reproduction as spawning biomass or egg production to recruitment, such as that derived by Beverton and Holt (1957) and occasionally used since then. I have shown how solutions to this can be found easily when the size at recruitment coincides with the size at maturity, that is, when the recruited stock is the same as the mature stock (Holt 2007, Chapter 39), and Shepherd (1982) showed how the tedious iterative solutions used by Beverton and Holt could be avoided. Other simple models have been proposed; most are asymptotic, like those of Beverton and Holt, whereas two others by Ricker (1954) and Shepherd, can predict declines in recruitment with increasing spawner abundance, but that feature is of little interest in the present context. More importantly, Shepherd's model, with certain parameter values, exhibits depensation at low spawner abundance (and some others, all reviewed by Quinn and Deriso 1999, also have that property) and that might have more relevance to our quest for MSYL. But, for my present purpose, the Beverton and Holt model suffices, and I shall return later to the depensation issue.

The Beverton and Holt stock–recruitment equation, which is usually portrayed as having two constants, can be reduced to just one by expressing recruitment R at any spawner abundance S as a fraction or percentage of the asymptotic recruitment R_∞ arising theoretically from an infinite spawner abundance, thus:

$$r = R/R_\infty = S/(b + S)$$

where b merely determines the scale of the S axis, and its value does not change the shape of the S–R curve. This reduced formula has a convenient property that $r = 50\%$ when $S = b$.

Since $r^* = dr/dS = b/(b + S)^2$, it is evident that the slope of the R–S curve tends to unity at small S and, of course, approaches zero as S increases; the slope approximates to $1/S^2$ at high S.

When this expression is married with that for yield-per-mature-recruit, the unexploited population settles to a steady state with $r < 1$, which I shall identify as r_v. Several workers have given special significance to the average slope of the S–R curve between S_V and $0.2S_V$, labeling the ratio of r at the smaller spawning stock size to r_V as ***steepness*** *h*. (Mace and Doonan 1988). But Haddon (2001) attributes this innovation to Francis (1992). For further clarification, see Mangel et al. (2010).[7] The choice of 20% of the recruitment to the virgin stock is arbitrary, but appears to have been motivated by an assumption that this represents a "safe" lower limit—in management terms—to the size of the spawning stock, perhaps a limit above which it is reasonable to assume there are no depensatory processes operating that could threaten to exterminate the population. Some writers also appear to believe that h is, in some way, a property of the R–S function determined by values of the constants a and b (or their ratio) in the usual formulation, but it really is not, since the location of r_V is fully determined by the properties of the yield-per-recruit function, which arise from the value of the ratio M/K and the age/size at maturity, together with the overall natural mortality of eggs, larvae, and juveniles, pre-recruits generally. Then, *steepness* is readily shown to be:

$$h = 0.2/(1 - 0.8\, r_v)$$

where r is the ratio of recruitment at a given value of S to the asymptotic recruitment, and r_v is the value of that ratio in the unexploited population. An alternative, and one I would prefer, is to focus on the slope of the S–R curve at r_v, which is, with S_v being set, for convenience, at unity, simply:

$$dr/dS = r_v\,(1 - r_v)$$

The next step is to identify, from the yield-per-recruit function, the value of F/M that takes the steady-state stock to the level MSY*L, from which can be calculated a value for S at that stock level, and thus the negative slope of the R–S curve at that stock size. Clearly, then, if the stock were to be permitted to increase a little by slightly relaxing the fishing mortality, the steady-state catch would increase slightly rather than decrease, because the recruitment would have increased slightly. Thus, we can see that the MSY occurs at a lower fishing mortality and a correspondingly bigger stock size than MSY*. The question is "By how much should we expect?"

At a few points in the published literature, some calculations of the self-regenerating yield equation have been reported. Here, I shall simply refer as an example to Figure 18.6 in Beverton and Holt (1957) which illustrates the steady-state yields of North Sea Haddock, as given by various values of the scaling parameter of the Beverton and Holt equation for recruitment as a function of egg production or spawning biomass. When, instead of being assumed to be constant on the average, the recruitment varies with fishing mortality or stock size near to the peaks of the yield-per-constant-recruit curve, the resulting maxima occur, as they must, at lower F values/higher stock size. But, what was (to me) unexpected was how small that shift is. Over a tenfold range of b, the MSYF barely changes from $F/M \approx 1.5$ ($M/K \approx 1.0$), although the peaks, the MSY values, are considerably higher than the peak of the yield-per-constant-

[7]There is some confusion in the literature about this definition, some authors apparently thinking that the recruitment to the virgin stock is the same as the asymptotic value of the B–H or any similar R–S function. I also find this terminology confusing because h is not as one might expect the slope of the R–S curve, but an inverse of it. If we, instead, define "steepness *bis*" as the average slope between S_V and $0.2S_V$, symbolized by h', then $h' = 1/(1 + 1.2S_V + 0.2S_V^2/b)$, which tends to zero for large S_V and to unity for small S_V.

recruit. The stocks are, in some cases, exterminated at F values not much higher than the prewar fishing mortality ($F/M \approx 5.0$). The ratio of F/M for extermination to that for MSY ranges, for the examples given (for various b values), from about 3 to 8.

A rather different picture emerges when calculations are made with both recruitment and body growth (W_∞) being density-dependent, that is, with growth rate increasing as abundance falls.[8] Now, the values of F/M for MSY are substantially higher (hence the stock abundance lower relative to the abundance of the unexploited stock) than for the maximum for constant recruitment; in the examples given, more than 50% higher and with substantially higher MSYs.[9]

I think these figures warn of the dangers of aiming at MSY by setting catch limits, rather than by regulating fishing effort, when recruitment is variable and especially if body growth and/or adult natural mortality are possibly density-dependent. This because uncertainty in presuming a low MSYL plus related errors in the MSY rate can so easily lead to serious overestimation of MSY and thus further unintended depletion of the fish stock and even its endangerment.

Worm et al. (2009) suggested that "traditional" approaches to better management would include, among other things, use of more selective gear. I have not discussed here the corresponding behaviors of sustained yield models used to assess the consequences of changes in the ages and sizes of liability to capture, as with mesh or hook size changes. But, one general feature is clear: although yield-per-recruit calculations will often indicate that improvement is to be expected from, for example, increases in codend mesh size, attempts to optimize this will call for increases in fishing effort directed to the older animals and thus may result in a disproportionate reduction in the abundance of the biologically most productive individuals.[10]

In trying to define the location of MSY for management purposes, we are not entirely constrained by consideration of the peaks of curves of steady-state yields against fishing mortality or stock biomass. The Scientific Committee of the IWC, in devising its proposed RMP, chose to find an algorithm that would lead to the greatest possible average catch over a considerable, but specified, time-period that would be consistent with a very low specified probability of accidental depletion of the stock to a specified dangerously low level at any time during that period, at which its capacity to rebound when relieved of fishing pressure, or even to survive, may be sharply reduced. It might be reasonable to describe that highest such average catch as "MSY," though it is not exactly what the drafters of UNCLOS were thinking about. I should mention that the IWC scientists' algorithm was tested so as to be as far as possible independent of the specific population model used to generate artificial "catches" to use in simulations, and particularly to be robust to unsure assumptions about the location of MSY and the MSY rate.

---

[8]In Figure 18.17 in Beverton and Holt (1957), density-dependence is assumed to affect only the asymptotic weight, but the results are essentially the same if it assumed that K changes, or both.

[9]The shapes of curves of sustainable yield against fishing mortality and abundance change from the "standard" skew parabolas when growth and/or natural mortality in the recruited phase are density-dependent and recruitment is strongly dependent on spawning biomass (low h, high h'), including indications of depensation at low total or spawning biomass, even though this feature is not explicitly provided for in the R–S relationship.

[10]Several recent authors have suggested that the increasingly observed phenomenon of relative absence of large fish in catches is a consequence of selective fishing for the bigger individuals. This might be so in certain situations, but it is far more often simply the result of relatively unselective, but intense, fishing reducing the abundance of smaller individuals that would, if they had been allowed to live, have grown into bigger ones. This illustrates well the important difference between static and dynamic thinking.

In the UNCLOS, the following "weasel words" regarding the basic management and conservation objective offer a way out of strict adherence to the physical MSY reference point: "…**as qualified by relevant environmental and economic factors**" (*Holt's emphasis*).

It has long been understood that a target of slightly less than physical MSY, from a lower fishing effort, is economically advantageous by increasing profitability (in terms of the difference between market value of the catch and the cost of taking it), and that an economic optimum (MEY) might thus be defined, although its location relative to MSYL depends on several factors, including the discount rate (e.g., Clark 1976). A striking example of this has been provided in a recent publication by the World Bank and FAO (2009) (see Holt 2009), which argues that a roughly 50% reduction in global fleet fishing power would change the economic situation from loss-making and dependence on subsidy to one of high net profit. The Bank's calculations are based on two alternative assumptions about the location of MSY. One of these is that of the simple logistic, giving an MSY biomass stock level of 50%; the other uses the Pella–Tomlinson model, with the skewness parameter set to put MSY at 37% of the unexploited stock biomass. The conclusions are quantitatively different, although qualitatively similar, but they do depend on a basic—and I think dubious—assumption that, overall, the global fisheries are currently producing catches in the order of the total MSY of all currently exploited stocks. This would mean that F_{MEY} is about 50% of F_{MSY}.

The UNCLOS requirement that fisheries be managed sustainably, but in such a manner as to ensure that the productivity of "dependent species" is not hampered, does not, in itself, provide quantitative guidance as to how that can be done, but it seems likely that it would imply that stocks of exploited prey species, at least, should be maintained at levels higher than those providing physical MSY, hence subject to lower fishing mortality. The more general problem of "taking into account the interdependence of stocks" introduces more complex questions, to which there is probably no equivocal answer, and I shall not pursue that here.

Conclusions

If MSY is real, it is certain that it is not a unique metric for any fish population, with the possible exception of *potential yield*, as the weight of a cohort of fish at the age that it attains its greatest total weight. But, for fisheries in the real world that are necessarily more or less selective and for which the fishing intensity is never infinite, the notion of MSY depends critically on the selectivity of fishing as well as on the basic population parameters. Provided that selectivity is taken into account, the approach pioneered by the Scientific Committee of the IWC, that is, seeking the highest cumulative catch over a finite predetermined management time-period as is compatible with other objectives, such as avoiding unintended depletion and other adverse consequences, offers a way forward.

That being said, the properties of the generalized Beverton and Holt steady-state yield-per-recruit model suggest that maximum sustainable yields from single stocks may be obtained from stocks at equilibrium approximately of the order of 20–30% of their unexploited biomass, and probably much smaller than that if exploitation is confined to the large/older individuals in the populations. Over quite a wide range of parameter values, this corresponds with a fishing mortality rate around double the natural mortality rate. The physical MSY biomass level, thus, might be relatively lower than has sometimes been assumed, although it is likely to be above levels at which depensation might occur, with danger of inadvertent depletion and even extinction. Examination of a self-regulating model, in which recruitment changes with the biomass of the adult or recruited phase of the population, indicates that, while this can change drastically one's predictions of MSY, it may have little effect on the location of MSY with respect to fishing intensity and stock biomass. It is also clear that it will generally be beneficial to hold the stock size at a higher level than the physical optimum in order to ensure economic benefit without much loss of catch. The scale of that benefit depends critically, however, on how flat-topped are the curves of sustainable yield against stock biomass or fishing mortality.

There are several ways in which management by regulation of fishing power and effort has advantages over regulation by TACs, but what consideration of the properties of the self-regenerating, age-structured model shows is that even without the usual annual variations in recruitment so commonly encountered, the catches from stocks at an imprecisely estimated or assumed MSY level are far less predictable than the level of fishing effort that will reliably take them. A further consideration is that management by TACs depends on catch data in two distinct ways: their use in making stock assessments, and the setting and enforcement of the TACs themselves. Many authors have pointed out that official catch statistics are commonly quite unreliable for both purposes; recent revelations of the total inadequacy of Bluefin Tuna *Thunnus thynnus* catch statistics in the Mediterranean are a striking example. To put the matter crudely, it is much easier to falsify catch statistics than to hide fishing vessels.

Many criticisms of MSY as a management procedure have been published by fisheries scientists, the best known being that by Peter Larkin (1977). But, those criticisms cover a wide range of issues; an obvious one is that if attained, it would still be far from economically or socially optimal. But, I remember William Herrington, the US representative, saying to me in Rome in 1955 that—for political reasons—there was no hope of our being able to designate maximum economic yield as an international target; he was probably right at the time. But, other criticisms, relevant to the discussion in this paper, have been focused on where MSY is located, whether it can be estimated empirically, and whether the common methods of setting TACs can track it efficiently. Then there are difficult questions about its application to multispecies fisheries and even to ecosystems that I have not dared to touch here. Furthermore, as de la Mare so clearly showed, realistic management has to take stochasticity fully into account. I have, however, here sought to show that a useful step could be to look at the general properties of more or less realistic single-stock, deterministic, age-structured models of population dynamics. That might help us on the road to a useful redefinition of the MSY criterion in UNCLOS and now formally adopted by the EU, as the IWC has done in its RMP, by aiming at maximizing the average annual catch over a specified management horizon in such a way as to avoid undesirable unexpected consequences, including economic losses. Given the uncertainties in locating MSY—which could apparently involve reducing stocks to very low levels relative to their unexploited abundance—as well as in predicting future recruitment, only a precautionary management procedure that aims well above those theoretical levels and is implemented by limiting fishing power and effort rather than by setting TACs, can ensure the future productivity and profitability of our sea fisheries. The high uncertainties in locating MSY and the heavy price of error in both estimating and tracking it by management, means that to ensure long-term sustainability without nasty surprises, the target must be set much lower than the theoretical maximum. I hope this examination of the properties of general population models will begin to indicate just how much lower that might have to be.

References

Beverton, R. J. H., and S. J. Holt. 1957. On the dynamics of exploited fish populations. Her Majesty's Stationery Office, UK.

Beverton, R. J. H., and S. J. Holt. 1959. A review of the lifespans and mortality rates of fish in nature and their relation to growth and other physiological characteristics. Pages 142–180 *in* G. E. W. Wolstenholme and M. O'Connor, editors. Ciba Foundation Colloquium on Aging. Volume 5: The lifespan of animals. J. & A. Churchill, London.

Beverton, R. J. H., and S. J. Holt. 1966. Manual of methods for fish stock assessment. Part II. Tables of yield functions. FAO Fisheries Technical Paper 38 (Revision 1).

Clark, C. W. 1976. Mathematical bioeconomics: the optimal management of renewable resources. John Wiley, New York.

Cooke, J. G. 1995. The International Whaling Commission's revised management procedure as an example of a new approach to fisheries management. Pages 647–657 *in* A. S. Blix, L. Walløe, and Ø. Ulltang, editors. Whales, seals, fish and man. Elsevier Science, Amsterdam.

de la Mare, W. K. 1986. Simulation studies on management procedures. Reports of the International Whaling Commission 36:429–450.

Finley, M. C. 2007. The tragedy of enclosure: fish, fisheries science, and U.S. foreign policy, 1920–1960. PhD Thesis in History (Science Studies), University of California, San Diego.

(The) Fisheries Secretariat. 2010. Editorial by the Director, Niki Sporrong, Latest News, 27 September 2010.

Fletcher, R. I. 1982. A class of non-linear productivity equations from fishery science and a new formulation. Mathematical Biosciences 61:279–293.

Fox, W. W. 1970. An exponential yield model for optimizing exploited fish populations. Transactions of the American Fisheries Society 99:80–88.

Francis, R. I. C. C. 1992. Use of risk analysis to assess fisheries management strategies: a case study using Orange Roughy (*Hoplostemthus atlanticus*) on the Chatham Rise, New Zealand. Canadian Journal of Fisheries and Aquatic Sciences 49:922–930.

Graham, M. 1939. The sigmoid curve and the overfishing problem. Rapports et Procès-Verbaux des Réunions du Conseil Permanent International pour l'Exploration de la Mer 110:17–20.

Graham, M. 1943. The fish gate. Faber and Faber, London.

Haddon, M. 2001. Modelling and quantitative methods in fisheries. Chapman and Hall/CRC, Boca Raton, Florida.

Hjort, J., G. Jahn, and P. Ottestad. 1933. The optimum catch. Hvalrådets Skrifter 7:92–127.

Holt, S. J. 1958. The evaluation of fisheries resources by the dynamic analysis of stocks, and notes on the time factors involved. Pages 77–95 *in* Some problems for biological fisheries survey and techniques for their solution. ICNAF Special Publication 1.

Holt, S. J. 2004. Foreword to the 2004 printing. Pages i–xxiii *in* R. J. H. Beverton and S. J. Holt. 1957. On the dynamics of exploited fish populations. Blackburn Press, Caldwell, New Jersey.

Holt, S. J. 2006. The notion of sustainability. Pages 43–82 *in* D. M. Lavigne, editor. Gaining ground: in pursuit of ecological sustainability. International Fund for Animal Welfare, Guelph, Canada and the University of Limerick, Limerick, Ireland.

Holt, S. J. 2007. New policy objectives and management procedures for EU fisheries: a commentary and suggestions to The Greens/European Free Alliance in the European Parliament. The Greens/European Free Alliance, 8 February 2007.

Holt, S. J. 2009. Sunken billions—but how many? Fisheries Research 97:2–10.

Jensen, A. L. 1996. Beverton and Holt life history invariants result from optimal trade-off of reproduction and survival. Canadian Journal of Fisheries and Aquatic Sciences 53:820–822.

Larkin, P. A. 1977. An epitaph for the concept of MSY. Transactions of the American Fisheries Society 106:1–11.

Longhurst, A. 2010. Mismanagement of marine fisheries. Cambridge University Press, Cambridge.

Mace, P. M., and I. J. Doonan. 1988. A generalised bioeconomic simulation model for fish population dynamics. New Zealand Fisheries Assessment Research Document 88/4. Fisheries Research Centre, MAFFish, New Zealand Ministry of Agriculture and Fisheries.

Mangel, M. 2005. Invariant ratios vs dimensionless ratios. Science 310:1426–1427.

Mangel, M., J. Brodziak, and G. DiNardo. 2010. Reproductive ecology and scientific inference of steepness: a fundamental metric of population dynamics and strategic fisheries management. Fish and Fisheries 11:89–104.

Munro, J. L. 2011. Assessment of exploited stocks of tropical fishes. An overview. Pages 145–170 *in* V. Christensen, and J. MacLean, editors. Ecosystem approaches to fisheries: a global perspective. Cambridge University Press, Cambridge.

Pella, J. J., and P. K. Tomlinson. 1969. A generalized stock production model. Bulletin of the Inter-American Tropical Tuna Commission 14:421–496.

Quinn, T. J., and R. B. Deriso. 1999. Quantitative fish dynamics. Oxford University Press, Cambridge.

Ricker, W. E. 1954. Stock and recruitment. Journal of the Fisheries Research Board of Canada 11(5):559–623.

Schaefer, M. B. 1953. Fisheries dynamics and the concept of maximum equilibrium yield. Pages 53–64 *in* C. P. Idyll, editor. Proceedings of the Gulf and Caribbean Fisheries Institute, Sixth Annual Session, Miami, Florida, November 1953. University of Miami, Coral Gables, Florida.

Schaefer, M. B. 1954. Some aspects of the dynamics of populations important to the management of commercial marine fisheries. Bulletin of the Inter-American Tropical Tuna Commission 1:25–56.

Schaefer, M. B., and R. J. H. Beverton. 1963. Fishery dynamics: their analysis and interpretation. Pages 464–483 *in* M. N. Hill, editor. The sea, Volume 2: Composition of seawater. Comparative and descriptive oceanography. John Wiley, New York.

Shepherd, J. G. 1982. A versatile new stock-recruitment relationship for fisheries and the construction of sustainable yield curves. Journal du Conseil International pour l'Exploration de la Mer 40:67–75.

Tsuro, Y. 2004. Rethinking the principle of abstention: the North Pacific and beyond. Marine Policy 28(8):541–552.

Verhulst, P. F. 1838. Notice sur la loi que le population suit dans son accroissement. Correspondance Mathématique et Physique 10:113–121.

(The) World Bank and Food and Agriculture Organization (World Bank and FAO). 2009. The sunken billions. The economic justification for fisheries reform. The World Bank, Washington, DC.

Worm, B., R. Hilborn, J. K. Baum, T. A. Branch, J. S. Collie, C. Costello, M. J. Fogarty, E. A. Fulton, J. A. Hutchings, S. Jennings, O. P. Jensen, H. K. Lotze, P. M. Mace, T. R. McClanahan, C. Minto, S. R. Palumbi, A. M. Parma, D. Ricard, A. A. Rosenberg, R. Watson, and D. Zeller. 2009. Rebuilding global fisheries. Science 325:578–585.

Chapter 45

Some Simple Thoughts about the Science of Fisheries Management[1]

In general, control of complex systems is better effected by regulation of input than of output. Hence, these notes are directed more to the primary regulation of fisheries by controls on inputs (fishing power, operational rules) than by output (TACs, desired catches).

The dynamics of an exploited fish population are determined principally by the relationships among three intrinsic variables and one extrinsic one. The three are measures of the rate of growth of an individual in size, which we'll call K; the natural mortality rate M, and the reproductive rate R. The fourth is the fishing mortality rate F. The important relationships are:

- the ratio of fishing to natural mortality: F/M,
- the ratio of growth rate to natural mortality: K/M, and
- the dependence on population abundance/density on the difference $(R - M)$.

It is important to remember that abundance and density changes may differ. Thus, commonly, the effective geographical range of the population contracts as its abundance is reduced by fishing. This is because density (abundance per square kilometer) varies from place to place, and fishers will always prefer to operate, if possible, in places where natural or residual densities are highest. But, fish will be sooner eliminated from localities where natural densities are lower—i.e., where the habitat is less favorable. These factors lead to what has been called "the basin model" of assessment, and they explain that methods of estimation of stock trends from changes over time in "catch per unit effort" (CPUE) almost always lead to underestimation of the rate at which a stock is declining under the pressure of fishing. Another common cause of underestimation of stock decline is unnoticed technical or operational improvements in fishing methods.

The relationship between the two types of mortality is sometimes expressed as:

$$\text{Exploitation rate} = E = F/(F + M)$$

which ranges from 0 to 1.

The biological optimum is commonly in the vicinity of $E = 0.7$ or less, i.e., fishing mortality is not more than twice the natural mortality rate. This is only a very rough guide.

Evidently, for fishes with high natural mortality rates, fishing has to be relatively intense in order to optimize sustainable catches—fish have to be caught before most of them

[1]Editors' note: a short note written by Sidney Holt to Michael Earle dated September 16, 2010 for discussion at a "retreat" by The Green Members of the Fisheries Committee. The Greens/EFA Group has given permission for its reproduction here.

die naturally—and vice versa for naturally long-lived species; for those species, it is best to have low exploitation rates which allow the relatively long-lived fish to grow.

The details depend on the selectivity of the fishery, that is, whether fish of all sizes or a wide range of sizes are subject to fishing mortality, or only a certain limited range of sizes/ages. In general, the more selective the fishery, for whatever reason—specification of gear, locality, season of permitted deployment, or other—the higher will be the fishing mortality rate required to optimize sustainable yields.

The value of F (hence E) is determined primarily by the amount of fishing effort appropriately calibrated. Fishing effort is the product of the total fishing power of the entire fleet and the average deployment of that power i.e., days at sea, hours spent searching and fishing, etc, again appropriately calibrated, expressed as a proportion of available time. The fishing power of the fleet is simply the sum of all the powers of individual fleet units (vessels), also appropriately calibrated.

Immediately after the Second World War, the dominant idea in Europe was to limit the growth of fleets to something below their sizes before the war. For various reasons, this idea became replaced by the idea of catch limits, partly because, in the international sphere, the USA, very active in developing international regulation, saw fleet limitation as an undesirable limitation of the universal right to go fishing. But, effort limitation also had a bad reputation, especially from experience in regulating the Pacific Halibut *Hippoglossus stenolepis* fishery, because although deployed effort was limited, the fleet power was not; this led inevitably to an extremely inefficient fishery because ships were eventually only allowed to operate a few days per year.

The simplicity and directness of limiting fisheries by setting TACs are illusory. There are so many ways and such strong temptations to conceal the fact that illegal catches have been taken, unmarketable catch being discarded, and so on. In addition, the "correct" setting of a TAC depends on reliable prediction of future stock abundance, particularly of next year's recruitment. This is one of the most difficult things to do—and often expensive—in fisheries research and assessment practice. Fishing power and deployment are, on the other hand, relatively easy to keep track of, both in terms of presence in ports and, nowadays, with available tracking technology.

Computer simulations of managed fisheries have also shown that they are far more stable in the medium and long-terms when the primary control is limitation of the deployment of available fishing power rather than the limitation of catches. Economic stability is also best ensured by the proper balance between available power and percentage deployment of that power over time. Regulation by power and effort control do, however, call for much more attention to be given to the various necessary calibrations, but the time-frame for these is much longer than that needed to determine annual TACs; the efficiency of a certain trawler, for example, does not change wildly from one year to the next, nor do new methods with enhanced efficiency appear every year. A proper power regulatory system does, however, require that technological enhancements to fishing methods and other operational changes be properly accounted for in a timely manner.

Lastly, I should mention that more than 20 years ago, the Australian scientist/engineer William de la Mare showed by computer simulations that efforts to hold a stock at MSY level by setting catch limits must always fail, even when the population model is perfectly specified and all its parameters precisely estimated. De la Mare's study was made in the context of the management of whaling, but his results and conclusions are generally applicable.

Chapter 46

Brief for the Greens/EFA Group in the European Parliament on Reform of the EU Common Fisheries Policy (CFP): How to Achieve Sustainable and Profitable Fishing[1]

Contents

[1]Editors' note: as part of its work on the reform of the Common Fisheries Policy, The Greens/EFA Group asked Sidney Holt for advice on the wisdom and benefits of maintaining fish stocks at levels above those capable of producing maximum sustainable yield. This document is a revision of the version made available to The Greens/EFA Group in February 2012 and is reproduced here with their permission. It was little changed except to incorporate new material subsequently prepared by Holt as substantive addenda. He noted that the result was not entirely satisfactory and required further editing, but that was never done.

Summary

The European Commission's proposal that "maximum sustainable yield" (MSY) be at the core of fisheries management policy has rightly been welcomed as a signal of intention to allow the recovery of depleted fish stocks by imposing restraints on fishing. This brief extends the advice I provided earlier to The Greens/EFA Group in the European Parliament (Holt 2007; Chapter 39) supporting the proposal by the Commission to replace the old aim of the CFP, which was simply attainment of sustained yields, by a policy of seeking *MSY*. This brief is, however, neither a revision nor a supplement of that. It is a stand-alone document.

But, a strict MSY policy also formalizes and legitimizes greed and gives excessive status to an otherwise undesirable objective that may or may not be a reality and will impoverish the search party. An MSY policy is an economically very inefficient manner of managing renewable resources such as fish stocks for long-term profitability and social value. It is, therefore, very encouraging that the European Commission has declared that "Multiannual plans providing for conservation measures to **maintain or restore fish stocks above levels capable of producing maximum sustainable yield** shall be established as a priority" (Commission 2011). (*Holt's emphasis*)

Here, I give many reasons, with supporting calculations and arguments, for acceptance of the proposal that the aim should be to maintain fish stocks bigger than would be needed to provide maximal sustainable catches by reducing the proportion of the population that is annually removed from the sea by fishing, a process we shall call the application of a restrained fishing mortality rate. The big questions are why that is not only desirable, but is necessary, and by how much should management aim away from the notion of MSY? I give reasons why fishing rates must be much lower than is required to provide maximum sustainable physical yield in order to ensure the future economic viability of fishing, the stability of the industry, the well-being of coastal communities dependent on fishing, and continued supply of food and other commodities from marine living resources.

Answers to these questions come from the ideas and findings of the branch of science called population dynamics and from consideration of other desirable aims of management, particularly the sustained profitability of fishing as an industry, the integrity and well-being of fishing communities, and, equally, the well-being of the natural marine environment. What comes from consideration of general population dynamics—the main thread in this brief—is that all fish species' populations have essentially the same characteristics (so, generalizations are possible), and the desirable difference between what some still see as an ideal—MSY—and a very much better state is far greater than might be have been expected, would not demand much "sacrifice" in terms of physical catches, and would provide catches of higher quality and stability (so, more valuable) for much less effort and, hence, be both more profitable and otherwise socially beneficial.

The specific aims depend very much on the selectivity of fishing operations as well as on their intensity. By this I mean the sizes—hence the ages—at which fish become liable to capture, resulting from the type and details of the gears used and the locations and seasons of fishing. This selectivity can be such as to range from targeting juveniles, to mainly catching sexually mature fish, and allowing individuals to grow bigger and catching them just before they die naturally. Such varied fishing strategies have very different costs, hazards, and consequences which have somehow to be balanced.

Since the 1950s, there has been a cultural—but fundamentally irrational—concentration on exercising management through control of the output of each fishery by setting catch limits (TACs) supplemented by market controls such as the prohibition of landing fish determined for some reason to be "undersized" or even biologically "endangered." This has had the unintended result of causing huge waste by "discards," doing little to "save endangered species," or to reduce the mostly negative impacts of intense fishing on the natural environment. A more rational approach begins with reversion to the traditional principles of engineering—optimal control of an "open" system principally by managing the energy and other inputs to it, only secondarily—for "fine tuning"—engaging in limiting the system's output.

This brief deals only with the basic management objective related to maximum sustainable yield and particularly to the proposal that the objective should be to maintain the rate of exploitation of all fish stocks such that the stocks are held at, or permitted to return to, larger and more productive levels than are needed to yield the biggest possible sustained catches.

I have used, in several places, the ill-defined terms "optimal" and "maximal," as well as "overfishing," and I hope my meanings will be clear in context. Although "optimal" literally means "best," I have here used it as it is still commonly used in discourse about fisheries management, referring to the value of the fishing mortality rate, or the fishing effort generating it, that is expected to provide maximum sustainable catches in given circumstances.[2]

There is now, I think, a consensus that the aims of fisheries management are, primarily, to ensure that total catches are sustainable, preferably high—maybe maximal—and also profitable. Unfortunately, these aims are not necessarily compatible with each other, nor does attaining one of them ensure the other. Large catches can be profitable, but not sustainable. Catches can be sustainable, but not profitable. Maximal sustainable yields (MSY) would inevitably be unprofitable and, in fact, are unattainable for both technical and economic reasons. Another management aim is, generally, that the managed fishery be as stable as an inherently variable and capricious Nature will allow.

Note by Sidney. This document would have been clearer if it could have been illustrated with about 15 custom-made graphs, but time was not available to prepare those. Instead, a few existing published diagrams have been provided as annexes, each with an explanation of its meaning in the present context.

The Basics

A ***cohort*** of fish in an unexploited population (usually a "year" class) decreases in number, by natural causes, from the time it is spawned, but the survivors each increase in individual body weight. In fishes, *growth is indeterminate*—that is, it continues throughout life, as in most if not all marine vertebrates—but the rate of growth slows over time. Because the curve of numerical survival against time is exponential, roughly the shape of a reversed J (see Annex Figure 1, p. 547) and the curve of increase in body weight is S-shaped (*sigmoid*), approaching a maximum (asymptotic) weight, the total weight of each cohort reaches a maximum at some time. That time—sometimes referred to as the ***critical age***—and the size of the fish in the cohort when its total maximum weight is reached—are determined not by the rates

[2]These various usages were discussed in historical detail in my 1957 book written with R. J. H. Beverton (pp. 238+ in all four printings and editions). Little has changed in such usage since then except for matters to have become even less clear.

of mortality and growth in themselves, but by the relation between them. That relationship can be expressed as the ratio of the natural mortality rate—*M*—in the fish after they have been "recruited" into the "fishable population" (that is, the part of the population from which fish may be liable to capture) and a coefficient of growth rate—*K*. Because mortality and growth tend to be positively correlated as between species, the ratio *M/K* has a rather narrow range of real values. That simplifies our problem of making valid generalizations about all species from the management point of view. Here, I give results mostly for the *M/K* range 0.25 to 2.0.

We have a limited practical interest in the cohort from the moments of its being spawned, hatched, and undergoing larval development, except for one special reason that we shall explore later. But, there comes a time in the cohort's history at which the fishes in it are of potential economic interest. Those survivors that reach that age we call ***recruits***, constituting a ***stock***. But, we do not necessarily begin to catch fish as soon as they are recruited; we might decide to leave them to grow rather bigger in the hope of eventually getting bigger total catches, which may be possible so long as the rate at which the fish increase in size exceeds the rate at which their numbers diminish by natural death.

So, we can talk about the ***recruited stock*** (sometimes referred to as the ***exploitable*** stock) and the ***exploited stock***, which is that part of the recruited stock that is liable to capture by the gear in use, given the other specifics of the operation (location, depth, season, etc.). For some purposes—when we want to ensure sustainability so that future generations of humans can also catch fish in abundance—our focus can be in the mature or spawning stock.

That's enough definitions for the time being. The biggest catch from a cohort of recruits would be obtained by waiting until the cohort has reached its maximum total weight and then catching all the survivors immediately (see Annex Figure 1, p. 547). If we could do that to all the many cohorts in a population as they each attained their maximum weight, we would obtain maximum sustainable yield (MSY). We cannot do that with wild fish for two reasons. First, no fishing method is or can be so completely selective, by size or age. Second, it would require a virtually infinite fishing effort, and even trying to approach it by catching most of the fish near that "optimum" age would be impossibly costly. Our solution is to begin catching some fish before that time, and possibly to continue doing so after it. Our problem then becomes to determine at what size or age to begin catching them, at what rate (i.e., what rate of fishing mortality *F* to impose), and how much that might cost in relation to the value of the catch obtained. Our measurement scale could be what that catch would be as a percentage of the theoretical, but impossible, MSY.

A word about the ***fishing mortality rate***. It is easy to think of that, and also the ***natural mortality***, as an annual percentage of the stock. However, we can better express these as ***coefficients***, exponential coefficients with a property common to logarithms: they can be combined by adding them together, which cannot be done with percentages. (If the natural mortality rate alone is 60%—exponential rate 0.9—and the fishing mortality rate is also 60%, what is the combined rate? It cannot be more than 100%!) Another reason for measuring the effect of fishing as an exponential coefficient is that *F* is to a first approximation directly proportional to the amount of fishing effort (double the effort to double *F*, for example) expressed by figures such as hours trawling, days at sea, number of hook-hours, number of seine sets per year, and so on.

Further, just as we found that the ratio *M/K* tells us more about the dynamics of fish populations than do *M* and *K* separately, we find it is convenient to use a relative index of the fishing rate, *F/M*, calling that *f*. Here, we shall generally use that index because it is roughly proportional to the fishing effort exerted—assuming M is a constant. There are other indices, the most usual of which is exploitation rate (with possible values between 0 and 1) viz. $E \equiv F/(F + M)$. (For readers not familiar with mathematical symbols, ≡ signifies "identical to.") E is the proportion of a cohort that is caught throughout its life, so $1 - E$ is the proportion that die from natural or other causes, including any that may be induced by human activities other than fishing. The relation between the two indices is: $F/M = E/(1 - E)$.

Fishing affects the size of the exploited stock, and that size determines the ***catch-per-unit-effort*** (CPUE) which, together with the market price, is the main determinant of profitability. Fishing also affects the ***size of the spawning stock*** and hence future sustainable catches. It also affects the ***size of the recruited*** or ***exploitable stock*** and hence the calculations of the wisdom or otherwise of altering the selectivity of the fishery, such as by changing the meshes in the codends of trawls or the gauges of hooks. "Size" is, however, an ambiguous term. What really matters is the ***biological productivity*** of the recruited, mature, or exploited stock, and that depends not only on its weight (***biomass***), but also on its composition—by size, age, and sex of the fish in it. There is much confusion on this matter in the published "literature," and not only in "secondary" and "popular" writings and statements.

A common error—and it appears in fish stock assessments applying so-called ***surplus production*** methods—is a presumption that productivity, and hence potential sustainable yield, depends only on the biomass of the whole stock or some part of it (such as spawners or females). In fact, two stocks of the same species, with the same biomasses, can have very different productivities, depending on their size, age, and sex compositions.[3] For that and other reasons, we shall here ignore such methods (except to comment on them where they give particularly poor or misleading results) and focus on assessment methods that rely on age-structured models of fish populations. In particular, our comments and examples will be based on one set of models put forward by R. J. H. Beverton and myself in the 1940s–1950s (with a later modification) and that have been very widely used by fisheries scientists since then (see *Technical Notes* on page 536).

The growth of a fish can be described by an algebraic expression, due to Ludwig von Bertalanffy (LVB) (1938), having two basic parameters, one of which is K, which determines the *curvatures* of the growth curve. The other is W, the weight that the fish will theoretically attain if it lives forever, that we call the ***asymptotic*** or ***final weight***. (W is a theoretical average; it is not the largest size that a particular individual fish of the species can attain.) So long as W does not change, it does not affect any of the conclusions about management that we shall consider here, though later we shall take a brief look at the consequences that follow if it *does* change—most likely because of a changed food supply. The exponent K will also normally be constant; it would not be expected to change with food supply, but it might with water temperature which directly affects metabolic rates, especially of cold-blooded species.

The important properties for us of the LVB growth expression concern the effects of the value of K, even though they are mostly expressed through the M/K ratio, which, from here,

---

[3]These methods, usually associated with the names of researchers Schaefer, Pella and Tomlinson, and Fox, among others, are mentioned later in this brief, but there is no space here to go into detail about them. They have frequently been cited in advice about fisheries management, but are entirely unsuitable for the kinds of calculations and assessment made here. The degree of curvature of "sustainable-catch" curves based on those methods is unrealistic (and cannot be determined empirically), both near where MSY is supposed to be found and at the extremes of unfished biomass and near extermination.

we refer to as a single parameter m.[4] With the same convention, we designate the ratio F/M as f. The ***age*** at which a growing fish will reach a certain size depends on K, but not at all on W. It is convenient for us to speak about the size of a fish as a fraction or percentage of W, which we shall refer to as w. The steepest part of the curve of growth in weight against time or age is at the point of ***inflexion***, where the curve changes from being concave to being convex (as seen from above). In the LVB curves, this is always when $w = 0.3$, regardless of the value of K. This point of maximum absolute growth rate is also the size around which fish become sexually mature; where necessary, we'll call this size, relative to W, w_s. (Their age at maturity t_s depends—of course inversely—on the value of K. It is $t_s = 0.36/K$.) In a similar way, we can label the size at which fish are recruited as w_r and the size at which they first become liable to capture by the fishing operation practiced and the gear used—w_c [the corresponding ages are: $t_c = -\ln(1 - w_c)/K$, where ln is the natural logarithm (log base e)]; w_c can be smaller or bigger than w_s. We shall also label as w_{MSY} the size at which MSY is the (maximum) weight of the unexploited cohort: it depends only on m (see Appendix 3, p. 545); w_c could, in principle, be bigger than w_{MSY}, but it does not make much sense to arrange for it to be so because by the time a fish reaches w_{MSY}, its growth rate has considerably slowed; the cohort is declining steadily in number because of natural mortality, but the size of the surviving fish is increasing more slowly, even at a snail's pace. It is easy to think about this if we think not about ***absolute*** growth rates, but, rather, ***specific growth rate***, that is, the increment in weight in a certain time (perhaps a year) divided by the average of the weight of the fish at the beginning of that time and that at the end of it (sometimes called ***relative growth rate***, but that can be an ambiguous term, referring perhaps to comparisons between stocks or species). The maximum of the curve of cohort weight against age occurs when the specific growth rate becomes equal (and, naturally, opposite to) the corresponding specific mortality rate.

Sustainable yields are usually referred to in the scientific literature as Y, but we shall label them as SY. SY is generally directly proportional both to W and to R, the ***annual number of recruits***. If we can presume these both stay constant—or at least at a constant average—it is convenient to look at $sy = \mathrm{SY}/(W \times R)$, so sy will always be a decimal number less than 1. Our consideration of this will be by using what is usually called a "yield-per-recruit" model, in contrast with *self-regenerating* models, which incorporate an expression relating the number of recruits to the numerical size and other characteristics of a previous generation that spawned them (see Annex Figures 2 and 3, p. 547 and 548, respectively). That expression can never be a proportional one, but, rather, describes the ***density-dependence*** of the natural mortality of the "prerecruit" fish, in such a way that the mortality rate increases as the number of prerecruits increases. It is also possible that with an extremely low prerecruit abundance (caused most likely by extreme overfishing of the parent generation), there can be a reversed density-dependence causing what is called the ***Allee effect*** or ***depensation*** (Allee 1931). This is a dangerous condition that can lead to irreversible changes in the population—such as continuing declines or even extermination, even if fishing pressure is relaxed or ceases. However, although that possibility must be borne in mind in considering management of fisheries in general, here—focused as we are on management to maintain relatively large and productive stocks—the Allee effect need only worry us in connection with management to avoid the possibility of ***unintended*** stock depletions.

[4]This ratio is referred to by T. J. Quinn II and R. B. Deriso in their textbook *Quantitative Fish Dynamics* (Quinn and Deriso 1999) as "a fundamental life-history parameter," and referred to by others (e.g., M. Mangel) as one of the Beverton–Holt invariants, although it is not strictly so.

Possible changes in R resulting from changes in stock size and composition can be very important, and we shall have to look at those carefully later. However, such changes will generally occur when a stock has been much reduced by intense, prolonged fishing (or other catastrophes). Our interest here is mainly in what happens when we manage fishing in such a way that a stock is maintained at a relatively high level—or allowed to recover to such a level—so as to get as close as practicable to MSY. In that region of "parameter space," there may or may not be any or much change in average recruitment, even though the natural variability of R may be—and commonly is—very great. That is one reason, as we shall see, why trying to approach MSY by setting catch limits (such as TACs) is—at best—inefficient and frequently fails, sometimes catastrophically; this is that stocks brought to, or near to, MSY—and *msy*—levels (see later) will necessarily be much smaller than they were before they were exploited.

Applicable Theory

We have said the maximum sustainable yield (MSY) in weight from an exploited stock of fish is obtainable only by catching all the surviving individuals of each cohort (year class) when the cohort attains its maximum total weight, and that that is impracticable for wild populations since it would require an infinite fishing intensity; even getting close to it would always be impracticable. In the real world, fishing must be less selective, though the selectivity actually imposed makes an immense difference to rational management decisions. Here, selectivity is defined as the size of fish w_c in a population when the fish first becomes liable to capture, expressed as a fraction or percentage of the theoretical asymptotic size. This percentage, or fraction, we call w_c. (Low w_c means relatively unselective, bigger w_c means more selective. Examples of some of these numerical options are given in the appendices.)

Given a stock having a certain value of m, we can calculate values of sy for each possible w_c and for a wide range of values for $f = F/M$. From our point of view, given the fairly narrow observed range of m, a fin whale *Balaenoptera physalus* would be just a gigantic Haddock *Melanogrammus aeglefinus* or Anchovy *Engraulis encrasicolus* were it not for the scale and density-dependence of recruitment! In general, a graph of sy against f (sometimes called a sustainable catch curve) has a maximum—a peak value, designated as *msy*—when w_c is relatively small, but when w_c is relatively large, the curve has no peak—it approaches an upper asymptote. The boundary between "relatively smaller" and what is "relatively larger" varies with m, being at bigger w_c when m is higher (see Table 1 below).

Note: the approach to fish stock assessment and fisheries management taken in this brief is that the central questions concern **the amount of fishing** (in terms of fishing effort) and **the mortality rate of fish induced by that, and the selectivity of that effort** with respect to sizes and ages of fish caught. I am, however, aware that many choose to focus attention on the sizes of the stocks—their ***biomasses***—and the effects of fishing on the stocks.

Biomass: a Matter of Size and Shape

(Shape because I am thinking of the varied reversed J shapes of the age and size distributions of catches and the populations from which they have been taken.)

Fishing changes a fish population in many ways, but primarily by changing its size in numbers of individuals and in their total weight (biomass), and also by changing its composition by age and size of fish. And sometimes by sex ratio.

Fishing itself has many dimensions, but primarily two:

- its intensity; and
- its selectivity (here I refer exclusively to intraspecies population, not interspecific selection).

Here, we focus mainly on the consequences of both dimensions of fishing to features of the biomass. These features relate primarily to three definitions of biomass with different interests to different stakeholders in fishing: the operators, the fisheries scientists, and ecologists and managers of the uses of "ocean space."

The interests of the first group are essentially in the biomass of the ***exploited phase*** of the population, which determines their *catch rate*, as expressed in indices of *catch–per-unit-effort* (***CPUE***) and hence expected profitability. The second group of stakeholders has an interest in all three definitions, but mostly centered on the sector comprised of sexually mature fish, commonly called the **spawning** or **mature stock**. Changes in this may determine the recruitment into the next generation of the population, the pending cohorts. With the focus on reproduction, this definition is of particular interest—but not only—to those scientists who practice a method of fish stock assessment based on **surplus production theory**. The third biomass definition is of the **recruited stock** consisting of all the juvenile fish and adults that might be of interest to humans as a potentially exploitable resource. For our purposes, it would usually be comprised of all the fish above an age at which it can reasonably be assumed the parameters of growth and natural mortality do not change much, at least to a first approximation.

The difference between the recruited and exploited stock arises from the selectivity of fishing.

A fourth definition of biomass is of interest to some ecologists: that of the total population, including eggs, larvae, and fry as well as juveniles and adults. Since pre- and post-recruits are usually occupying different **ecological niches** (defined by such features as different available foods, shelter from predators and other threats), those interested in the roles of the fish species in the marine ecosystem in which it is an element or in the multiple consequences of fishing on that ecosystem (to form a basis for an **ecosystem approach to management**) are usually focused on this fourth definition.

In this section, I focus on the first three: "exploited," "mature," and "recruited." For this purpose, I have to choose a specific definition of the size (hence age) at recruitment, and I have chosen the size $w_r = 0.05$, at which the fish are 5% of the estimated asymptotic weight, a size that they would theoretically attain if they lived long enough, i.e., forever. I assume that fish become mature when they have grown to $w_s = 0.3$.

The indices of biomass are given by the Beverton–Holt yield-per-recruit expression with the fishing rate index, $f = F/M$, set to zero.

Given the fixed values for w_r and w_s, the indices of biomass (biomass-per-recruit-per-unit-asymptotic-weight) depend only on the parameter $m = M/K$. The results for most of the m-values used are given in Table 1.

The unexploited biomasses B_r and B_s are relatively small for higher m, reflecting the consequences of a relatively high natural mortality rate, so that few survive to grow very much, and particularly not to the age and size of maturity. This differential also naturally accounts for big differences in the percentages of the unexploited recruited population that are sexually mature.

Table 1. Indices of recruited (B_r) and mature (B_s) biomass. B_r is the biomass of the unexploited recruited stock, excluding eggs, larvae, fry, and juveniles living elsewhere.

m	B_r	B_s	$B_s/B_r(\%)$	w_{MSY}	MSY
0.25	0.655	0.633	97	0.79	0.414
0.5	0.455	0.421	92	0.63	0.238
1.0	0.346	0.200	81	0.42	0.105
1.5	0.147	0.101	69	***0.30***	***0.057***
2.0	0.094	0.053	57	0.22	0.035
3.0	0.043	0.016	33	*0.13*	0.016

Table 2. Percentage depletion of three components of the stock at varying levels of fishing intensity and selectivity.

				% depletion by weight		
m	w_c	*msy/RW*	*fmsy*	**Recruited**	**Exploited**	**Mature**
0.25	0.4	0.365	4.3	18.2	13.7	20.8
	0.3	**0.344**	**3.3**	**18.9**	**16.5**	**16.5**
	0.2	0.331	2.8	19.4	18.3	18.3
	0.1	0.294	2.1	21.8	21.5	21.5
0.5	0.4	0.224	5.2	21.8	11.3	25.8
	0.3	**0.211**	**3.4**	**20.8**	**14.8**	**14.8**
	0.2	0.184	2.4	20.7	17.5	17.5
	0.1	0.173	1.7	23.2	22.6	22.6
1.0	0.4	0.105	Inf.			
	0.3	**0.103**	**6.3**	**24.7**	**8.2**	**8.2**
	0.2	0.096	3.0	22.1	14.5	14.5
	0.1	0.084	1.7	22.4	20.7	20.7
1.5	**0.3**	**0.056**	**Inf.**			
	0.2	0.055	5.3	24.0	8.6	8.6
	0.1	0.049	2.1	20.5	16.8	16.8
2.0	**0.3**	**0.033**	**Inf.**			
	0.2	0.034	19	27.9	2.5	2.5
	0.1	0.032	2.8	19.8	13.4	13.4

We are now in a position to examine the effects of fishing. Note that the mature sector of the population can be partly within the pre-exploited stock or entirely within the exploited stock; it is excluded that any mature fish are in the prerecruit part of the population. To illustrate the consequences, I have chosen w_c values 0.1, 0.2, 0.4, and the "boundary" 0.3. I have calculated the depletion of the unexploited recruited stock and its parts—the spawning stock and the exploited stock—when the fishing intensities are those producing *msy* for each of the chosen selectivities (Table 2). "% depletion" is the ratio of the biomass at *msy*-level to that in the corresponding sector of the unexploited stock. The table is truncated at higher values of m and w_c—in the space where *msy* is obtainable only with infinite fishing effort.

Evidently the exploited stock is always depleted more deeply at *msy*-level than the entire recruited stock, and especially when m is higher than 1.0, and for the higher values of w_c. Over a wide range, the depletion is to about 20%. The depletion of the biomass of mature—and presumably sexually active—fish contributing to future recruitment is the same as that of the exploited stock for w_c up to 0.3 because all mature fish are within the fishery. But when exploitation does not begin until well after fish have matured ($w_c = 0.4$ or bigger, so some of the smaller mature fish remain unexploited for a while), then, of course, the degree of depletion of the mature is somewhat less. It is, nevertheless, always substantial, which raises the possibility that when the exploited stock is at its *msy* level and composition, the recruitment will have been reduced by the effects of fishing, and the catches and catch-per-unit-effort would have been correspondingly less than expected from consideration solely of a yield-per-recruit assessment model.

About MSY

For any given recruited stock of fish, the *maximum sustainable yield* (MSY) is theoretically obtained by catching each year all the survivors of a cohort (year class) when they attain their ***critical size*** at which the cohort reaches its maximum total weight.

A convenient expression of that critical size is its ratio (w_{MSY}) to the theoretical maximum weight (W) which a fish of that species, growing in that environment, is expected to attain if it lives forever. That critical weight is determined only by the parameter m, the ratio of the natural mortality rate (M) in the recruited population to an exponential coefficient of body growth (K). It is given by the expression:

$$w_{MSY} = [3/(3 + m)]^3$$

and

$$\text{MSY} = w_{MSY} \times [m/(3 + m)]^m$$

The range of observed values of m is narrower than the ranges in nature of M and K.

w_{MSY} is always bigger than the average weight at which fish become sexually mature (w_s) and begin to contribute to the reproduction of the population. The value of w_s is generally close to 0.3—at least in the bony fishes—and the age at maturity is close to the reciprocal of K, about $1.1/K$.

It is, of course, impracticable to obtain MSY; that would require the exertion of an infinite fishing effort. Thus, in practice, a lower policy target must be identified. This can be investigated from two points of view: what is a practicable selectivity (w_c) less than w_{MSY} and what are a desirable fishing mortality rate and a consequent sustainable catch and catch-per-unit-effort (CPUE). For the purpose of this brief, it is convenient to define a parameter f as the ratio of the fishing mortality rate (F) to the natural mortality rate (M). f will be proportional to the fishing effort if the latter is properly calibrated.

There are several ways of expressing such choices. If the feasible and chosen w_c is only a little less than w_{MSY}, a sustainable catch close to MSY could theoretically be taken, but only with enormous fishing effort and, therefore, at exorbitant cost. For any chosen w_c, there is a value of f (which, in some cases, will be finite, but in others, infinite like that required to yield MSY itself) that would give a "***local***" maximum sustainable yield (*msy*) and hence a CPUE index calculated as *msy/f*. It is also useful to express that index as a fraction (percentage) of the CPUE that would be obtained from exploiting a near-virgin stock with that selectivity. What is "near virgin" may be determined empirically or otherwise chosen arbitrarily; for this brief, I chose $f = 0.05$.

Another approach to policy choice is to seek maximum sustainable economic yield (*msey*) defined as the yield (and corresponding f and *sy/f* values) for which the market value minus the cost of taking it is maximized. That, as before, would be calculated for each selectivity (w_c). This will always be obtained with a fishing effort very much less than that required to take any chosen local *msy* and at a correspondingly much higher catch rate (*sy/f*).

These findings are based on a general population model in which it is assumed that the annual recruitment is not reduced as a consequence of the reduction of the overall size of the population by fishing. However, this is reasonable for the situations and policy options considered here. The options aim to avoid the possibility of any life phases of the stock—especially the mature fish—becoming much depleted, since reductions in recruitment—at least in bony fishes—which can lead to fishery collapses have been observed only in severely and persistently overfished, hence grossly depleted, situations.

Other features of sustainable catches and of population sizes (biomasses) and compositions can be examined in the same way. The feature most directly related to economic and social questions is the average sizes of fish in the catches, since fish of a certain species, but of different sizes, sometimes fetch a range of prices in the market. It is straightforward to calculate scenarios incorporating that feature: the *sy* in weight is simply divided by the *sy* in number given by the same model. The same method serves for calculation of the average size of fish in the recruited population which may have, for example, implication for consideration of the effect of fishing on the diet of the fish population (and hence its growth and biological productivity) in both its exploited and unexploited phases. Whereas the biomass of the exploited phase of the population is of particular interest to the industry—as that determines the catch rate and hence profitability—the biomasses of the mature sector of the population, and of the entire recruited population, are important in consideration of ecosystem approaches to managing human use of the ocean.

An important practical issue that is not within the scope of this brief is how a transition from an existing situation in a fishery to a better sustainable state is to be managed with respect to the time-scale, the short-term and medium-term gains and sacrifices, and the long-term cumulative benefits. This requires rather more complicated calculations (but not differing in principle from those applied here to examine steady states and sustainable scenarios), but does call for examination in each particular case.

More about Catch Rates

So, for every stock, there is an infinity of *msy*s obtainable with finite fishing effort, and depending on the selectivity. Each is to be obtained by exerting a certain intensity of fishing effort, and the *msy* varies for each selectivity. The intensity of fishing that generates an *msy* increases with increasing selectivity, but the *msy* itself also generally increases as selectivity is increased (and so becomes closer to MSY) except—as in the case for $m = 2.0$ or higher and $w_c = 0.3$—where it is somewhat higher in those cases than w_{MSY} (see Table 3 below).

For each value of *m*, the more selective the fishing operation, the bigger is the *msy*, but obtained only with a substantially higher fishing mortality and with a lower catch rate. In Table 3, the catch rates (last column) are given as percentages of the initial catch rate in the virtually unexploited stock *sy/finit%*, taken to be at $f = 0.01$. In the range of *m* from 0.25 to 1.5, the catch rates at *msy* with unselective fishing ($w_c = 0.1$) are about one-fifth of the rate when the stock is virtually unexploited (right-hand column). If the fish become first liable to capture as they become mature ($w_c = 0.3$), we find that for the lower range of *m* values (0.25–1.0), the catch rate (at *msy*) ranges from 8 to 17% of that obtained from the unexploited stock.

Table 3. *msy*, f_{msy}, CPUE.

$m = M/K$	w_c	*msy/RW*	*msy*/MSY%	f_{msy}	CPUE(*msy/f*)/(*sy/finit*)%
0.25	0.5	0.383	92	5.8	11
	0.4	0.365	88	4.3	14
	0.3	**0.344**	**83**	**3.3**	**17**
	0.2	0.331	80	2.8	18
	0.1	0.294	71	2.1	22
	0.05				
0.5	0.5	0.233	98	9.5	7
	0.4	0.224	94	5.2	11
	0.3	**0.211**	**89**	**3.4**	**15**
	0.2	0.184	82	2.4	19
	0.1	0.173	73	1.7	23
	0.05				
1.0	0.4	0.105	100	**infinite**	
	0.3	**0.103**	**97**	**6.3**	**8**
	0.2	0.096	91	3.0	15
	0.1	0.084	80	1.7	21
	0.05				
1.5	**0.3**	**0.056**	**99**	**Infinite**	
	0.2	0.055	97	5.3	15
	0.1	0.049	87	2.1	18
	0.05				
2.0	**0.3**	**0.033**	**95**	**infinite**	
	0.2	0.034	100	19	5
	0.1	0.032	93	2.8	14
	0.05				

Now, what else do these tables of the properties of the age-structured yield-per-recruit model tell us?

First, that in the range of realistic values for *m*, the fishing effort to obtain any *msy* is, with lower selectivity, such that the fishing mortality rate is around double the natural mortality rate (see fifth column f_{msy}), but that the sustainable catch varies six- to eightfold (see third column *msy/RW*), depending on the exact value of *m*. That seems one good reason for management aiming at a desired fishing mortality rate rather than a biomass and an associated catch, which suggests that aiming at fixing a desirable fishing effort will be more reliable than aiming at a certain catch limit.

Second, with low selectivity, the *msy* will be at least three-quarters of the theoretical (see fourth column *msy*/MSY%), but unattainable, MSY and more as selectivity increases, but at a rapidly increasing cost.

In interpreting the figures for the catch rate—relative to that obtained from the very lightly exploited stock—we must bear in mind that with $w_c = 0.1$, we are looking at practically the entire recruited stock, but with higher w_c (above 0.3), the effect of fishing on the entire recruited and the spawning stock is considerably less than is indicated by CPUE figures, which, of course, relate to only the actually exploited part of the stock.

As fishing intensity increases, the size (biomass and number) of fish in the exploited stock, the mature stock, and the recruited stock are reduced continuously to some quantifiable percentage or fraction of its original pre-exploitation size. Consequently, the rate of catch by individual fishing vessels or enterprises declines correspondingly, though the true rate of stock decline might not always be evident because of simultaneous increases in technical and operational efficiency and/or the reduced geographic range of a reduced stock, becoming concentrated in the part of its original range in which density was highest. As fishing intensity increases, so also does the average age, and hence size, of fish in the catch decrease, with likely significant economic consequences.

In unrestrained fisheries, the fishing intensity tends to increase until the fishery becomes unprofitable, overall, commonly past the level of intensity that would provide *msy*. In such cases, we experience ***overfishing*** (more specifically referred to sometimes as ***growth overfishing***, although that term is rather misleading.)

Even in a fishery regulated to hold the fishing effort at the intensity required for taking *msy*, the fishery is not necessarily profitable, with total value of catch exceeding costs. Profitability may be restored or achieved by reducing fishing intensity and hence increasing the catch rate as the stock recovers towards its original size, the so-called *carrying capacity*—another unfortunately mislabeled and misleading term! There is a level of intensity, for a given selectivity, in which the gross profit would be maximized. We call the sustainable yield *sy* at that point the ***maximum sustainable economic yield*** (*msey*). The fishing intensity for *msey* (f_{msey}) is always less than required to obtain *msy* (f_{msy})—and can be substantially less—with correspondingly higher catch rates.[5]

Because the curves of *sy* against fishing intensity tend to be rather flat-topped, it is almost always possible to obtain sustainable catches only <u>marginally</u> less than *msy* with fishing intensities <u>substantially</u> less than those needed to take *msy*.[6] This means that *msy* is also not necessarily an optimal choice as a management target. But, the asymmetry of the *sy* fishing intensity curves means that it can be dangerous to allow populations to be driven to levels of biomass and to age/size compositions, down past the *msy* points; that can induce ***recruitment overfishing*** (in which the annual recruitment declines significantly as a result of lack of spawners) such that population collapse and possibly irreversible change can occur.

A particularly dangerous situation arises in cases where a selective fishery (just for mature fish, say), perhaps linked to fishing for food for human consumption, is succeeded, for market and technical reasons, by a less selective one, involving fishing to provide industrial products. The North Sea Herring *Clupea harengus* fishery is an example, as explained later.

It is now natural to ask what would happen to the stock and the sustainable catch if we were to depart from an *msy* target, one way or the other? The fixed shape of a curve of sustainable yield against stock biomass or fishing mortality rate given by the commonly used "surplus production" assessment models due to Fox (1970), to Schaefer (1954), and to Pella and Tomlinson (1969) gives an inflexible answer to that question, whatever MSY catches they may predict. For example, the Schaefer version, with MSY available when the entire stock or the spawning stock is reduced to one-half its unexploited level, predicts that if the fishing effort were to be reduced or increased by, say, 10% from that yielding MSY, the sustainable catch would be reduced by 1%, but, as we shall see, the consequences of such change as

---

[5]It is important to remember that this refers to the sustainability of the stock exploitation, not economic sustainability; *msey* may or may not be itself sustainable, although it probably usually would be.

[6]A measure of the *flat-toppedness* or *peakedness* of a curve or distribution is called its *kurtosis*. It is not a feature of a sustainable yield–fishing effort curve that is much referred to in discussing fish population dynamics and management, but in the context of this brief, it is crucial. See *Technical Notes*, p. 536.

predicted by the age-structured yield-per-recruit model used to generate the figures in Table 3 are very different. That is essentially because the latter are much more flat-topped than the former[7] and also because they are not symmetrical—the left-hand limb (decreasing fishing mortality rate) descends more steeply than the right-hand limb (increasing fishing mortality). One consequence of this is that the catches continue to hold up well after a state of growth overfishing has been reached, though they come at an increasingly higher cost, consist of more and more smaller fish, and risk sudden collapse as recruitment overfishing sets in. (In fact, the curves of sustainable catch against fishing intensity are more nearly symmetrical over a wide range when values of *sy/RW* are plotted against the logarithm of *f*.)

We are more interested here in what would be expected to happen if the fishing mortality were to be moved towards the left of the curve, reduced mortality rate yielding less than *msy*, but providing a higher catch rate, hence usually being more profitable. Here, the flat-toppedness of the general age-structured model, as advocated here—especially to the left of the yield-per-recruit curve—is crucial. The catch curves are more like the profile of the Grand Massif of France, or the Cotswolds of England, with gently sloping plateaus than like the Alps or the Dolomites.

To examine that, we can look at a few selected scenarios. I have given the result for three such situations in Table 4: the consequences of reducing the fishing mortality (hence *f*) 1%, 5%, or 10% below that required to take *msy*, which I refer to as f_{msy}. There are two ways of looking at this. One is to choose a possible reduction in fishing effort and see how much of the sustainable catch is "sacrificed"; the other is to choose a level of "sacrifice" and see what reduction in fishing effort will be needed to bring about that situation. I have chosen to illustrate the latter in Table 4.

In this table, f' is the value that will yield 99, 95, or 90% of the relevant *msy*, which, in each case, is given as a percentage of f_{msy}. It is striking that for all the given values of m and w_c, and even though f_{msy} varies over a wide range, 99% of any *msy* can be secured by exerting a fishing rate between about 60 to 80% of that needed to take *msy*. If we were prepared to be less greedy, and "sacrifice" as much as 10% of any *msy*, we could take the other 90% with only about 20–40% of the fishing effort, with corresponding increases in stock biomass, spawning biomass, catch rates, and—presumably—profitability. In addition, the average sizes of the caught fish will be substantially bigger than they would be if *msy* was the target.

In the light of the findings summarized in Table 4, I looked at the consequences of bigger reductions of the fishing effort below the *msy* levels. The results were rather surprising and of interest, I thought, with respect to the current Green Group proposal. I summarize them here; they refer to the full ranges of m from 0.25 to 2.0 and w_c from 0.25 to 0.5.

As f is further reduced below each f_{msy} value, the sustainable catch, of course, is reduced, at first slowly, then more rapidly, and the relative CPUE increases correspondingly. If the f value is reduced to as little as one-half of f_{msy}, the sustainable catch, for all parameter sets, is never reduced by more than 10% of *msy*, with corresponding increases in the catch rate. The size of the exploited stock is at least 40% of the unexploited size, as compared with the general *msy* level around 20%. This last difference also makes it much less likely that the spawning stock will be so reduced by fishing to substantially affect the recruitment.

[7]This is because the curves produced by the age-structured model derived their shape from the mortality and, especially, body growth functions, whereas the shapes of the published surplus production models are simple arbitrary algebraic functions with no biological basis.

Table 4. Fishing mortality less than f_{msy}.

$m = M/K$	w_c	f_{msy}	$f'/f_{msy}\%$		
			99% *msy*	**95%**	**90%**
0.25	**0.5**	**5.81**	68.5	45.8	34.1
	0.4	**4.25**	73.2	51.3	39.3
	0.3	**3.29**	76.3	55.0	42.9
	0.2	**2.61**	78.5	58.2	46.4
	0.1	**2.06**	80.6	61.7	51.9
0.5	**0.5**	**9.49**	55.5	32.5	23.0
	0.4	**5.15**	67.2	43.9	32.8
	0.3	**3.39**	73.2	51.3	39.8
	0.2	**2.39**	77.4	56.9	45.2
	0.1	**1.72**	80.2	61.0	49.4
1.0	**0.4**	**20+**			
	0.3	**6.26**	59.1	35.9	25.9
	0.2	**3.03**	71.6	52.8	38.0
	0.1	**1.72**	78.5	58.1	46.5
1.5	**0.2**	**5.34**	59.0	26.3	26.4
	0.1	**2.06**	75.2	53.9	42.2
2.0	**0.2**	**20+**			
	0.1	**2.76**	**60.9**	**47.5**	**36.3**

I think it is worth noting here that a greatly reduced fishing effort leads to a bigger average size of fish in the catches. It leads also to less variability in annual catches because they are coming from stocks in which more cohorts contribute significantly to the population and catches. When stocks are very depleted, the catches depend more and more on the number of recruits and, further, there is also less "buffering" of the natural variability of the recruitment. A reduced fishing effort does not necessarily lead to a corresponding reduction in the sizes and numbers of coastal fishing communities—as seems sometimes to be feared—because appropriate social arrangements can ensure that each fisher or fishing unit can work less for much greater economic benefit.

The Sizes of Fish in Catches and Stocks

Fishing alters the size and age distributions of fish in the recruited and the exploited population. These changes can be illustrated most simply by calculating the average weight of fish in steady-state catches as a function of m, w_c, and F, or $f = F/M$. The steady-state catch in number of fish is given by the expression: $syn = f \times (1 - w^{1/3})^m / (1 + f)$, so the average weight of fish in the catch is: $w_{av} = syn/sy$.

With respect to w_c, it is convenient to express each average weight as a fraction of the average weight in the virtually unexploited stock. Table 5 below gives a selection of the results of such calculations, taking the "virtually unexploited **recruited**" stock as one subject only to a fishing intensity of $f = 0.05$, with $w_c = w_r = 0.05$ and when the value of f is that yielding the local *msy*. The table also gives the corresponding ratios to the virtually unexploited **exploited** stock. With respect to the latter series, the irregularities in the numbers reflect the fact that the average size is affected in one direction by the fishing and in the other by the selectivity.

Table 5. Average weights of fish in stock and catches.

m	Av Wt f & $w_r = 0.05$	w_c	$\%w_r$ A	$\%w_c$ B
0.25	0.73	0.4	87	74
		0.3	83	72
		0.2	77	71
		0.1	73	70
0.5	0.56	0.4	94	69
		0.3	87	67
		0.2	79	67
		0.1	72	67
1.0	0.38	0.3	99	64
		0,2	89	73
		0.1	78	68
1.5	0.29	0.2	102	66
		0.1	87	77
2.0	0.23	0.1	101	79

Explanation of column headings: Av Wt—the average weight of fish in recruited stock, $w_r = 0.05$, as a fraction of the asymptotic weight of fish; A—average weight of fish in the *msy* catch as % of the average weight in the unexploited recruited stock; B—average weight of fish in the *msy* catch as % of the average weight of fish in the very lightly ($f = 0.05$) exploited stock.

Perhaps the most evident feature of this table is that the average weight of fish in the catch (exploited phase of the population, column B) at *msy* "level" is always in the narrow range from about 65 to 80% of that in the very lightly exploited population at $f = 0.05$. It is notable that the reduction of average size as fishing intensifies means that to approach *msy*-level catches, the numbers caught must be increased substantially, hence the high values of f for *msy*.

Self-regeneration

When the size of a stock is reduced by fishing, it is reasonable to expect that, at some point, the average annual number of recruits to it will be significantly diminished. Our experience is, however, that this only becomes evident when the stock is very substantially reduced, at least in the bony fishes with very high egg and larval production rates. Any such density-dependence of recruitment is very important when considering depletion and possible stock collapses, even exterminations, but here we are discussing stocks that are to be maintained at biomass and numerical levels exceeding those that would provide maximum yields, and especially actions intended to allow stocks to increase above such levels. In this case, any density-dependence that leads to larger numbers of recruits coming from bigger numbers and biomasses of parents is obviously beneficial. The question is how much would this affect our conclusions, if at all?

The most common trial calculations for these circumstances have applied an expression derived by Beverton and Holt in the 1940s–1950s. This fixes a curve, convex as viewed from above, that reveals an extremely weak relationship when the spawning stock is large, and a strong, almost proportional, relationship when the spawning stock is very small. (Several other proposed expressions, such as that proposed by W. Ricker (1954), have basically the same properties, especially for middle-sized and much reduced populations, but differ slightly in quantitative detail.) One question that has to be addressed in fish population analysis—though not usually in these terms—is: where on this curve does an unexploited population "settle" in a steady-state, sometimes called—but I think misleadingly in this context—"*carrying capacity*?" (Misleadingly, I think, because that term focuses on a supposed property of the environment in which the population lives, whereas the attainment of such a state is a function of the dynamics of the population, within its habitat.)

The scale of the consequences of reducing fishing intensity to less than that yielding *msy* depends on the original location of the initial steady state on the stock–recruitment curve, the steepness of the curve at that location, and, even more, on its steepness in the region of the *msy*s on the yield-per-recruit curves, which might be anywhere between one-half and, say, one-tenth of the size of the unexploited population.[8] The effects of the stock reduction are, however, qualitatively sure: the estimate of *msy* stock level will be *increased*, and the fishing mortality needed to secure *msy decreased*, although only slightly. But, the expected *msy* catches will be bigger. How much bigger depends critically on a precise knowledge of the initial steady-state point. This has a major practical implication: we can determine the desirable fishing mortality rate within a limited range, but be able to predict the consequent average catches with much less precision. The implications are that while the desired fishing rate could be regulated by limiting the fishing effort, the desired state cannot be ensured by regulating catches on the basis of predictions of sustainable yield. Appendix 2 (p. 544) gives, as an example, the North Sea Haddock situation as it was in pre-World War II years.

Economic and Social Considerations

In his classic 1943 book *The Fish Gate* (Graham, 1943) (and, subsequently, in 1955, to a Conference of the United Nations), Michael Graham expressed his Great Law of Fishing: "**Fisheries that are unlimited become inefficient and unprofitable**."

And, while we're about it, some other basic principles can be mentioned:

- On ***Conservation***: HM Inspector of Fisheries, Mr. Ernest Holt, reporting to Parliament in 1895 "**...it is desirable that fish should have a chance of reproducing their species at least once before they are destroyed**."
- On ***Restraint***: Sidney Holt's Second Law of Fishing (1958): "**Fishers and fishing enterprises will do everything possible—preferably legally—to respond to regulative restraints in such a way as to tend to nullify their intended benefits, particularly their immediate effects**." (That is not just about "sheer cussedness," but rather an expression of the realities of continuing making a living, daily, while restraints take time—commonly several years—to provide material benefits, during which time they will usually—inevitably—cause temporary economic and social stress, which could be ameliorated only by correct social policies. Indeed, the fishers' responses can be, unintentionally, actually *to enhance the deficiencies* that the restraints were intended to correct, for example, by increasing their efficiency.)

[8]Note that this value of the recruitment to the unexploited stock is not the asymptote of the curve, which would only be attained—theoretically—from an infinitely big mature stock.

- On ***Innovation*** and ***originality***: Craig Tanimoto and Apple Computer (1987) "**Think different**." See also Cherry Thomas and Edward de Bono (1970), "**Lateral thinking**." Here, the main difficulty may be to free ourselves from ways of thinking about management methods, as well as about aims, that have become entrenched during recent decades.

Graham's *Great Law* generally comes into operation when fishing effort has gone above—sometimes far above—the level required for taking *msy*. This is usually because the expansion is practically always a sort of mining rather than sustainable resource use. The level of non-profitability could generally only be reached before "growth overfishing" begins if demand for the products of that fishery is, or becomes, quite limited. With modern forms of fish processing, transport, and global trade, that is now rare except, occasionally, when a product from fish—other than food for direct human consumption—is substituted in the market by a similar, but cheaper, or otherwise more desirable product from another source. Here, we can only deal with the case of a stable market for an unchanging product.

We can examine the possible patterns of maximum sustainable economic yield (*msey*) using the yield equations more or less as we have done in examining physical yields, assuming that parameters, such as the mortality–growth coefficient ratio (m), stay constant as a marker. Now, suppose we write the market value of the total catch (for any given value of f) as $sy \times p$ (p is price) and the cost of taking it as $f \times q$ (q is cost of exerting unit fishing). Then, the gross profit is $(p \times sy) - (q \times f)$, and we want to find where is the maximum of that difference, and at what value of f (call that f_{msey}), particularly in relation to f_{msy} and to the value of f when gross profit is zero (call that f_0). The simple geometry of the curves of sy against f tells us that f_{msey} lies at the point where the tangent to the curve (in mathematical terms, its first derivative) has the same slope as the value of sy/f at f_0. In Table 6, we look, for orientation, at some examples, applying the calculations from which Tables 1, 3, and 4 were derived. In our examples, we first assume that the gross profit is zero at *msy*, f_{msy}, and, alternatively, at values of f of 66.7% and 150% of f_{msy}, i.e., with zero profit being reached before growth overfishing occurs, and after. (We have chosen these two ratios, bearing in mind the general asymmetry of the sy-f curves, to make some symmetry of "above" and "below": 150/100 and 100/150.)

In Table 6, the values of *msy* and f_{msy} are the same as in Table 3. The right-hand column, ***rate of profit***, is the gross profit (value of the catch minus the cost of taking it) divided by that cost, expressed as a percentage.

(The last column of Table 6 has deliberately been left incomplete because the selected figures shown are indicative only of the scale of the possible profit rate from drastic reductions in fishing effort, and to avoid the temptation to make unwarranted comparisons between the results for different m and w_c values.)

There are, I think, three remarkable features in Table 6. The first is that for all values of m and w_c, except the highest of both, the ratios of *msey* to *msy*, and of f_{msey} to f_{msy} are similar. (This apparent anomaly is related to the fact that the calculations for $m = 2.0$ and $w_c = 0.2$ bring us close to the parameter space where the maxima in the *msy* curve are close to the asymptote.)

The second is that the effort needed to acquire the maximum economic yield is much less than (always well under half) that needed for *msy*.

And, third, that the economically maximum sustainable catch is not much less than the *msy*.

Table 6. Sustainable economic yields ($f_0 = f_{msy}$).

$m = M/K$	w_c	f_{msey}	f_{msey}/f_{msy} (%)	*msey*	*msey/msy* (%)	**Profit rate (%)**
0.25	0.5	1.48	25.5	0.318	83.2	225
	0.4	1.12	26.4	0.290	79.7	171
	0.3	1.00	30.4	0.277	80.4	165
	0.2	0.85	32.6	0.256	79.7	143
	0.1	0.71	34.5	0.231	78.7	128
	0.05	0.63	35.0	0.215	77.8	**c.100**
0.5	0.5	1.90	20.0	0.206	88.6	
	0.4	1.30	25.2	0.188	84.0	
	0.3	0.98	28.9	0.173	81.8	
	0.2	0.76	31.8	0.155	80.0	
	0.1	0.59	34.3	0.136	78.5	
	0.05	0.51	35.9	0.124	78.0	
1	0.3	1.36	21.7	0.086	86.4	
	0.2	0.86	28.4	0.079	82.6	
	0.1	0.57	33.1	0.067	79.7	
	0.05	0.45	35.2	0.058	78.2	
1.5	0.2	1.30	24.3	0.032	86.4	
	0.1	0.63	30.6	0.040	81.0	
	0.05	0.46	34.1	0.035	79.1	
2	0.2	2.30	12.1	0.032	92.7	
	0.1	0.76	27.5	0.027	83.2	
	0.05	0.50	32.9	0.023	80.3	

(It is worth recalling something that Beverton and I "discovered," in the 1950s, that an *msey* can always be found even when there is no *msy* except with infinite fishing intensity. I have not calculated such situations for these illustrative examples.)

The quantitative conclusions in Table 6 do depend on the elasticity of the price of fish, the presumption being that, over the supply range from *msy* to *msey*, the price will not change much —if at all— (elasticity zero or low) as a consequence of the change in supply. If elasticity is *not* low or zero, but price increases with reduced production—as might be expected in some circumstances—then f_{msey} will be <u>less</u> than the table indicates. These calculations can easily be modified if the price elasticity is known empirically or can safely be assumed on the basis of experience and the actual situation of a particular fishery.

An example of such calculations, for the bottom-trawl fishery for Plaice *Pleuronectes platessa* in the southern North Sea is given in Appendix 1 (p. 543).

Our next task is to see how much these results depend critically on the assumption of zero profit at f_{msy}. To explore that, we can look at some arbitrarily-chosen case where zero profit may be found when f is above or below the value giving *msy*. It is not necessary here to explore the entire range examined for the yield-per-recruit analysis. I have chosen to look mainly at the mid-range case of $m = 1.0$ for various selectivities and assume that f is one-third less or 50% more than f_{msy} (Table 7).

Again, we see that the value of f_{msey} is always—perhaps unexpectedly—low (and similar for all m and w_c values) and, for the $m = 1.0$ example, the *msey* catches are all in the range of about 70% to 90% of *msy*. There are similarly narrow ranges for other values of m.

Table 7. Effects of different levels of unprofitability.

$m = M/K$	w_c	$f_0 = 1.5f_{msy}$			$f_0 = 0.67f_{msy}$		
		f_{msey}	f_{msey}/f_{msy} (%)	*msey/msy* (%)	f_{msey}	f_{msey}/f_{msy} (%)	*msey/msy* (%)
0.25	0.5	1.87	32.2	88.7	0.68	11.7	59.7
	0.4	1.53	36.0	88.0	0.91	21.4	73.2
	0.3	1.29	39.2	87.7	0.47	14.3	55.0
	0.2	1.10	42.2	87.5	0.63	24.1	69.4
	0.1	0.92	44.7	87.1	0.52	25.2	67.5
	0.05	0.85	47.22	87.7	0.45	25.0	65.3
0.5	0.4	1.30	25.24	84.0	1.01	19.6	77.1
	0.3	0.98	28.91	81.8	0.75	22.1	73.5
	0.2	0.98	41.00	87.6	0.57	23.0	70.2
	0.1	0.78	45.35	87.6	0.44	25.6	67.8
	0.05	0.67	47.18	87.2	0.37	26.1	65.8
1.0	0.3	1.68	26.84	90.6	1.10	17.6	81.2
	0.3	1.08	35.64	88.5	0.66	21.8	74.5
	0.1	0.66	38.37	84.4	0.43	25.0	69.7
	0.05	0.59	46.09	86.8	0.33	25.8	66.6
2.0	0.1	0.95	34.42	88.9	0.59	21.4	76.6
	0.05	0.64	42.11	87.8	0.37	24.3	69.7

It is not always sufficient to make individual operations restrained to sustainability, attractive and profitable. Thus, there can be an argument for restraining fishing intensity to a level different from that required for *msey*. How different (and in which direction) obviously depends on the prevailing economic conditions (presuming that the expected rate of profit will not be so different from prevailing conditions in other industries as either to make fishing unattractive for investment, or so attractive that there will be irresistible pressure for investments to increase beyond what may be desirable for conservation and social reasons). Similarly, social policy decisions may dictate how big an industry, and product supply from it, is desired. For each fisher, vessel, or enterprise, the greatest benefit might come from having as few competitors for the fish as is possible, although account has to be taken of certain advantages from cooperation between several fishing units.

It should not be assumed, however, that a low fishing intensity must always imply a small fishing fleet. Our discussion of the desirable scale of *f* relates to **deployed** effort, not simply to available fishing power. An early example of successful management by fishing effort regulation illustrated this—that of the Halibut *Hippoglossus stenolepis* fishery off the Pacific coast of North America. Regulation by setting specific catch limits or limiting the number of participating vessels was not available to the managers of the US-Canada International Pacific Halibut Commission, and recovery of a fishery in crisis was essentially achieved by strictly limiting the number of days in each season that vessels could operate. The resulting recovery of the stock, yielding much higher catch rates, was sufficiently profitable for fishers to convert to multispecies operations for the rest of the year as well as to make necessary improvements in vessels and gear.

A policy decision to seek to minimize the impact of the fishery on the marine ecosystem of which it is part (including reducing incidental catches of non-target species), as well as to reduce the consumption of fossil fuels and other input resources, would push the regulated fishing intensity down to even less than is required for *msey*. On the other hand, there can be social reasons for maintaining fishing communities at certain viable levels, ensuring stable infrastructures, even military reasons for maintaining a potential reserve of experienced seamen, convertible vessels, and shipbuilding and processing facilities and so favoring fishing intensities somewhat higher than f_{msey}.

Many considerations point to the desirability of regulating fishing primarily by control of the input to it (fishing intensity and selectivity) rather than of the output (catch limits, minimum size limits, and so on). The most important of these are that:

1. the likely catch in the immediate future is one of the most difficult numbers to predict because of the wide annual variability of recruitment in most fishes as well as shifts in availability of the already recruited animals;
2. in the already recruited population, the biological productivity of the stock, which determines the sustainable yield, is far more stable than the stock's immediate size; and
3. errors in the estimates of model parameters (by which I mean differences between the estimates and their real values in Nature) can lead to very large errors in predicted yield, so poorly calculated catch limits can lead to serious overexploitation, requiring rapid and vigorous "corrections" in each succeeding year. This is particularly so when the intensity of fishing is such that it is likely that recruitment will decline as a result of diminishing female spawner numbers and biomass.

It is, finally, worth recalling that staff and consultants of FAO and the World Bank recently concluded that the global fish catch is now in the region of "MSY" and the fisheries necessarily heavily subsidized, and that large reductions in global fishing effort would restore overall profitability, with some sacrifice of total catch (World Bank and FAO 2009). In my opinion, their conclusion is qualitatively correct, but I have doubts concerning its quantitative aspects, in particular, my belief that the method they used considerably ***overestimated*** the sacrifice of total catch that would follow substantial reduction in global fishing power and deployed effort, and so underestimated its likely economic and social advantages (Holt 2009). This was because they used the—I think—very unreliable surplus production models (see below).

There follow some final thoughts about the theory and practice of fisheries management.

Policy Objectives

The policy objectives discussed below may all be desirable, but are not entirely compatible with each other. Compromises will usually have to be sought, on the basis of priorities assigned *a priori* among them.

By **sustainable catches**, we mean average annual catches which, if taken, will leave the *size and the composition* of the exploited population unchanged the following year, apart from the inevitable natural fluctuations in recruitment. Sustainability is often thought of as implying "forever"—like diamonds!—but in practice, the concept can only be useful when a limited, if long, time-frame is pre-specified, so that—among other things—computer simulations can be undertaken to test the likely consequences of proposed actions to promote sustainability and optimization. Sustainability is now accepted widely as a desirable state for wise use of a renewable resource, especially if its future ability to provide continuing yield is highly valued. It is a moral, economic, and social notion, not a scientific one, though science can serve to suggest what will ensure it and what endanger it. The fact that catching operations that are merely technically sustainable will not necessarily be profitable, and hence cannot become economically sustainable, is frequently overlooked in the enactment of fisheries policy, as it is in other industries.

In some fisheries, the sizes of the fishes caught—the proportions in the catch of smaller (possibly immature), medium, and larger—matter in that they attract different market prices, may even enter different markets, and are suitable for different kinds of processing. The average size depends on two factors operating in opposite directions. One is obviously the selectivity—highly selective fishing should normally yield a higher proportion of older, larger fish and hence a larger average size. On the other hand, seeking to optimize selective catching always demands more effort and hence a higher fishing mortality rate on the age classes being exploited. This leads to a change in age and size composition of the populations such that the average size of fish caught will tend to be smaller, other things being equal. So overall, optimization of the distribution of size in the catches calls for calculations for each particular case. Such calculations would look not only at the average sizes of the fish in catches, but at the distribution of size or at least at the categories defined by the market, for example, small, medium, large.

One important policy issue that should be mentioned here is the need for industry to have as much continuity from year to year in what regulations permit it to do. This has been a continuing problem with the annual setting of TACs—sometimes hundreds of them—and a problem for the scientific community too, who must regularly perform huge numbers of routine assessments, thus limiting their ability to make the deeper studies that will refine and otherwise improve such methods. The decision by the European Commission to engage in multiannual plans is a huge step forward, but might not in itself sufficiently stabilize regulatory measures and release sufficient research time. The approach to this matter taken some years ago by the International Whaling Commission by including a regulation-stability requirement in its Revised Management Procedure (RMP) as a specific management objective, and setting predefined limits to the amendment, timing, data requirements, and circumstances that could warrant changes, was helpful. The most difficult kinds of regulations to be treated in this way are catch limits such as TACs. Regulation of inputs such as fleet power and deployed effort are easier to stabilize in the medium term.

Now, let us take a quick look at the Second Law. This is not the place to discuss in detail the innumerable ways of nullifying the intended consequences of regulation of catches, fishing effort, and procedures, but one aspect of normal economic behavior demands our attention. This is the basic and continuing effort by fishers both to increase their efficiency—ideally at as low a cost as possible, of course—and to respond profitably to changes in fish location and abundance. This can have unintended consequences for management, especially for certain kinds of regulation. An example of this was the case of the Georges Bank trawl fishery for Haddock in the Northwest Atlantic where well-intended, but inappropriate, limited measures to increase sustainable yield by increasing codend mesh size eventually contributed to the collapse of the fishery. Here, I give an illustrative theoretical example.

We make our "default" assumption (based on a real case) that we are dealing with a fish stock with $m = 1.0$, with low selectivity trawling at, say, $w_c = 0.064$ (6.4% of W). This has an *msy* of $0.77 \times RW$ obtained with $f = 1.86$. This situation is relatively slight growth overfishing. A decrease in fishing effort to the *msy* value of $f = 1.5$ would give us 1.5% more sustainable yield for only 80% of the cost. But possibly, we have no authority to constrain the fishing effort even though the fishery was, prior to regulation, making zero profit. We decide that a higher *msy* can be obtained by increasing mesh size such that $w_c = 0.074$. Calculations tell us that this would provide a sustainable catch of 0.790, an increase of just 3% in both total catch and catch rate. A bigger increase of selectivity, say to $w_c = 0.26$, would reward us with a much bigger increase in sustainable catch—by 24%. Still higher selectivity would, however, lead to decreased catch, and at 50% selectivity, back to where we started. Now, we decide to be cautious, sticking to the original small increase in mesh. The increase in fish abundance and hence in catch rate, though small, will attract more effort to this fishery. We can expect such an effort increase to continue until the catch rate is the same as before regulation. If the effort were to increase by 25%, we should get the same catch as we started with, *but now at a much lower catch rate*. But, zero profit will have been reached again long before that. We shall end up with about the same catch as before, a slightly lower catch rate, and done nothing to "cure" either overfishing or its lack of profitability. In the real world, it would be rather more complicated; for example, larger-meshed codends are sometimes more efficient than smaller-meshed ones because of increased water flow and reduced need for towing power. Against that, we shall certainly face higher costs in monitoring and enforcing the new net regulations, not to mention the costs of research to ensure the new mesh size is doing to the population what it is supposed to do.

The point of this discourse is to illustrate that in considering such matters as seeking *msy* or some other related target or "reference point," it is essential to take into account the inevitable reactions of the industry and the fishers to the regulations. This example would naturally have been made much more complicated if we had examined the transition period between one steady state and the other, at the outset of which the catches would have been smaller than the original ones because of the escape of smaller fish which would have been retained by the original smaller mesh. That would have illustrated the truism that there can be no improvement without some pain.

Now for some other considerations. One of them concerns the R_∞ parameter of the Beverton and Holt or other stock–recruitment function (see Appendix 4, p. 545). In general, an empirical or assumption error in this can lead to a very wide range of possible *msy* values, or of corresponding *sy* values for other reference points connected with *msy*. However, the f values for *msy* or the other reference points are much more stable, shifting very little for substantial errors in r (see *Technical Notes*, p. 536). The direction of shift is towards *reducing* the "optimal" f value a little further as the expected steady-state catch increases. This again points to the regulation of fishing effort rather than limitation of catch as a better primary regulatory measure.

In addition to this source of systematic error, the annual catch is always affected by the essentially unpredictable natural variation of recruitment, in some species very wide, such as in the small, relatively short-lived clupeids. This means that seeking to limit f appropriately by setting catch limits is hazardous and can lead even to oscillatory changes. As we have seen from yield-per-recruit models, **the value of f required to meet policy goals does not depend on the number of recruits**. And, if fishing effort is restrained—as it should be—there is little point in making great efforts to predict the eminently unpredictable number of recruits, unless perhaps to prepare onshore processing facilities to handle highly variable landings and such like operational arrangements.

The asymmetry of curves of *sy* against f is such that *msy*s can be dangerously close to fishing intensities that will lead to irreversible unintended stock declines, especially as the density-dependence of recruitment begins to bite. This is especially true in situations where the function defining the relation between spawning stock and recruitment has been misjudged in such a way as to predict too-high *msy*s. Furthermore, stochastic simulations have shown that efforts to target *msy*, whether from an overfished or an underfished situation, by setting sequential catch limits, are not efficient, commonly overshooting the target and causing unintended further depletion of the stock.

The example given above of an attempt to improve a fishery by modifying a single input parameter—in that case, selectivity—shows that too can be counter-productive. The same can be said about regulation through control of fishing effort while leaving selectivity to change freely, but generally the consequences would not be so negative. On the other hand, attempts to reach a selected target only by controlling one parameter of the output—the catch—will almost certainly have unintended and deleterious consequence of a variety of kinds. That remains true even if an additional output parameter—e.g., the size of fish caught and landed—is regulated by setting minimum legal lengths. In this case, since fishers are technically—and operationally—unable to ensure that they do not unintentionally catch undersized animals, an unnecessary waste of discarded fish is inevitable. Similar wastage occurs when, for whatever reasons, catch limits are exceeded, if only temporarily.

Most fisheries have unintended catches of non-target species, including vulnerable or even threatened ones. A policy that aims at keeping fishing intensity well below that which is expected to provide the highest sustainable yields can be expected, in itself, to reduce the unintended catches proportionately and so contribute to the conservation of non-target species.

In recent decades, changes in markets and technology have sometimes led to dramatic—and, occasionally, catastrophic—changes in the selectivity of a species-fishery. The most well-known is that for Northeast Atlantic Herring which were exploited for high yields, almost certainly sustainably, for over a century by the capture of migrating spawners in driftnets for direct human consumption, then in later years—after World War II—by seines for fishmeal and oil, with no lower limit to size. A change in selectivity, from about 30% ($w_c = 0.3$) to 10% ($w_c = 0.1$) or less, quickly led to a shift in the fishing mortality from <u>less</u> than the old *msy* level to well <u>above</u> the *msy* level pertinent to the new selectivity. As we have noted, the f value for *msy* diminishes quite rapidly as selectivity is decreased, whatever the value of m. Thus, the likelihood of overfishing and collapse resulting from a shift from fishing for human consumption to fishing for industrial products is caused as much by a change in the selectivity of an unregulated fishery as by an increasing fishing effort inspired by the new and practically unlimited market.

The current arguments about the huge Jack Mackerel *Trachurus murphyi* fishery in the Southeast Pacific illustrate an early stage of a similar developing situation.

It is of some interest that the definition of management objectives in the 1958 UN Convention on Fishing and Conservation of the Living Resources of the High Seas emphasized the sustainable production of food for human consumption: "*measures rendering possible the optimum sustainable yield from those resources so as to secure a maximum supply of food and other marine products. Conservation programmes should be formulated with a view to securing in the first place a supply of food for human consumption.*" The idea of priority for food for human consumption was eliminated from the UNCLOS of 1982, possibly because, in the interval, more countries in need of better food supplies had taken up industrial fishing in the expectation of profits from the export of the products being used to pay for other imported foods. Efforts by FAO to develop interest in the production of food from fishmeal for direct human consumption were not successful.

The Global Picture

The World Bank and FAO (2009) recently published "The Sunken Billions: The Economic Justification for Fisheries Reform" a study by staff of, and consultants to, the two international organizations which concluded that the global catch is now in the region of "MSY," that costs of catching far exceed the market value of the catch—thus requiring massive subsidies—and that a large reduction in fishing power and deployed effort would lead to profitability and limited sacrifice of total catch. The economic analysis is, I think, pretty sound, but the biological treatment is poor because the consultants applied "traditional" surplus yield models crudely—and, I think, improperly—with alternative "MSY levels" at 50% and 37% of the unexploited fish biomass—whatever that was! I published a critique of this (Holt 2009), but consider the unofficial Bank–FAO conclusion to be qualitatively, if not quantitatively, correct. In my opinion, the study's assumption that current catches represent a global MSY is erroneous, and also these surplus yield models, giving rather peaked curves of *sy* against stock size (and also taking no account of selectivity), greatly <u>overestimated</u> the likely sacrifice of yield with reduced effort. Nevertheless, the World Bank–FAO report was a brave and useful attempt to step back from the world-wide fisheries crisis and provide a global, economic, and social perspective with great long-term consequences.

Conclusions

There are evidently numerous substantial economic, social, and ecological advantages to be gained by managing fishing to keep fish stocks substantially above the population sizes—"levels" (biomasses and numbers)—that are thought to be necessary to generate maximum sustainable yields. That requires an intensity of fishing substantially less than for the "maximum" requirement. This is always true regardless of the fact that the *msy* "levels" (and the fishing intensities required to sustain them) vary greatly. That variation depends principally on the selectivity of the fishery, by size, and also—but to a lesser degree—on the biological characteristics of the exploited species. Those are signified mainly by the ratio of natural mortality to the body growth exponent, and by the effective fecundity of mature females and the density-dependent natural mortality rate of eggs, larvae, and juvenile "prerecruits" that together determine the level of the annual recruitment.

Such management is required to:

- ensure sustainability and profitability;
- enhance overall profitability in terms of the difference between the market value of the total catch and the cost of taking it;
- keep the catch rates of fishing units high enough for them to be profitable as well as again playing a substantial role in feeding human populations;
- minimize the risk of unintended depletion of the resource by excessive fishing intensity and uncertain stock assessments and lack of knowledge about the dynamics of the stocks and the ecosystem of which they are a biologically important component;
- minimize the impact of fishing on the integrity and productivity of the marine ecosystems as a whole in which the managed species live;

- provide that the fish size and sex compositions of the catches are economically advantageous; and
- minimize non-target catches that generate enormous discards of unwanted fish of both target- and non-target species.

All of these benefits can be secured by appropriate regulation of the inputs to (fishing effort, gear selectivity) and outputs by (catches) the industry. Of these, regulation primarily of the **inputs** is more effective and efficient, for numerous reasons. These arise from uncertainties about near-future catches, especially due to highly variable recruitment and unpredictable changes in the availability of fish, and also from uncertainties in understanding the relationship between recruitment and parent stock in particular species and situations. All theory shows that it is preferable to set targets for fishing mortality rate than for catch, partly because of the general shapes (flat-topped, asymmetrical) of curves of sustainable yield against fishing effort and selectivity. Thus, there can be big changes in profitability arising from major changes in fishing effort, but with little consequent change in average catch.

It should be said here that even if—as strongly recommended in this brief—limitation of inputs were to be the primary regulatory action, it would, in some circumstances, be beneficial to designate secondary rules, such as specific size limits and catch limits.

Technical Notes

There is nothing so practical as a good theory
Kurt Lewin (psychologist) and Richard Feynman (physicist)

The purpose of models is not to fit data, but to sharpen thinking
Sam Karlin (mathematician)

The assertions about fish populations and management made in this brief are mostly derived from a combination of a vast body of the literature of fisheries science and the properties of a generalized model of population dynamics. This latter is a form of the model for steady-state yield-per-recruit as a function of fishing effort and selectivity of fishing developed in the 1940s–1950s mainly by R. J. H. Beverton and myself and used extensively by very many others since then. It is combined with our simple parallel model relating annual recruitment into a fish stock to the fecundity of the mature part of the stock and the prerecruit density-dependent mortality rate of youngsters ("prerecruits"). Together, these constitute what we called a *self-regenerating model*.

The model contains only four parameters which determine its properties, *viz*.

1. M/K (referred to here as m), the ratio of the rate of natural mortality M of recruited animals (that is, those that could be liable to capture and of possible interest to fishers) to a growth parameter K that defines the rate at which the increase in weight of a growing fish slows down as it approaches a final maximum size in very old age, if it survives. This growth parameter—an exponent—is considered to be specific to each kind of fish and has a narrow range, mainly between 0.5 and 2.0, notwithstanding a 100,000-fold plus range of size of animals to which it can apply, from anchovies and the Common Sole *Solea solea* to the gigantic blue *B. musculus* and fin whales of the Southern Hemisphere. It might change, however, with the ambient water temperature;

2. F/M (here referred to as f), the ratio of the fishing mortality rate F to M, which can be proportional to the intensity of fishing—expressed as ***deployed fishing effort per year***, appropriately calibrated;
3. A selectivity coefficient w_c, which is the weight at which fish first become liable to capture by the operations and gears in use as determined by technical features such as mesh or hook size used, limitations such as minimum legal size, the preferences of the market, as well as the times, seasons, depth, and locations of operation; w_c is expressed as a fraction or percentage of the weight W that the fish will finally approach if it lives long enough;
4. A number r, between 1 and 0, which is the fraction that the average number of recruits to the unexploited ("virgin") population is of the maximum number that there could be, theoretically, if the number or biomass of spawners was infinite. r is closely related to another occasionally used parameter *steepness* (h) that expresses the fraction by which the recruitment would decline if the spawning biomass were to be reduced (essentially by fishing) to one-fifth (20%) of its initial size: $h = 0.2/(1 - 0.8r)$. The choice of 20% is arbitrary, but appears to have been motivated by an assumption that this represents a "safe" lower limit —in management terms—to the size of the spawning stock, perhaps a limit above which it is reasonable to assume there are no *depensatory* processes (also referred to as "Allee effects"; Allee 1931) operating that could threaten to exterminate the population.

Two other quantities—R, the average (and often extremely variable from year-to-year and essentially unpredictable) annual number of recruits to the fishable population, and W—do not, in themselves, affect the interpretation of the dynamics of the exploitable/exploited population. Sustainable yield sy is represented here as an index, a decimal number less than 1, which provides the actual yield when multiplied by the product $R \times W$, in whatever units W is expressed.

The original sustainable yield-per-recruit, and self-regenerating models, published by Beverton and Holt (1956a, 1957), dealt with the natural mortality in the recruited phase of the population and the growth coefficient K as separate parameters. However, Holt (1958; Chapter 4) showed that the dynamics of the population depended only on the ratio of those two numbers. Furthermore, subsequent research by us and other scientists revealed that the two components are generally correlated as among species so that the ratio $m = M/K$ has a narrower range of interspecific variability than either M or K.

Beverton and Holt (1966) subsequently compiled tables of the modified yield-per-recruit model for wide ranges of values of w_c and m, and a series of f values from 0.05 to 20.0. Many of the conclusions presented in this brief can be drawn from a close examination of those tables, but I have also recalculated them with finer parameter intervals and an extended range.

The values of W and K together determine the absolute rate of growth of a fish at any time in its life. That rate is greatest when the weight of the fish is about 30% of W ($w_c = 0.3$), and that approximates to the average size at sexual maturity, although, in reality, maturity is usually spread over a range of age and size. Given that constant, the value of K determines the age at which the fish is most likely to become mature. In the growth model by L. von Bertalanffy (1938) used here, the size of maturity is reached in $1.1/K$ years, i.e., it is practically equal to the reciprocal of K. Because at the inflexion of the growth curve at $w_c = 0.3$, the rate of growth changes smoothly, calculations of sustainable yield for selectivities less than or greater than $w_c = 0.3$ exhibit a gradually changing pattern, with yield peaks being replaced by asymptotes.

Another frequently used growth function is that due to B. Gompertz (1825) which provides an asymptotic sigmoid curve of growth in body weight with an inflexion at $w = 1/e = 0.37$. This inflexion weight is considerably larger than that generally found empirically to be the size at maturity.

The Gompertz growth model, applied to populations rather than to individual growth, and a specific form of it due to W. W. Fox (1970), has been seen as an alternative to the simple logistic used by Schaefer (1954) and others. It places "*msy*" at 37% of the unexploited stock size, in contrast to the 50% of the logistic. A modification of the logistic due to J. J. Pella and P. K. Tomlinson (1969) provides, by inserting an additional parameter, a theoretical *msy* target range from 37% to near 100% of the unexploited biomass. An undesirable feature of these "surplus production" models is that when the MSY or *msy* "level" is assumed to be about 40% or less, even an infinite fishing intensity cannot exterminate the stock. They are thus in that sense—and in other ways, such as involving arbitrary assumptions about the shapes of the curves of *sy* against *F*—entirely unrealistic.

For the analyses summarized in this brief, it has been assumed that *M*, *K* (or, strictly, the ratio $m = M/K$), and *W* are constants. In reality, any or all of them may be variables. In particular, *M* might vary with the age of the fish and also be dependent on the size of the population, and *W* may vary with the amount of food consumed by the population as a whole and by the individuals composing it, and so also be density-dependent. *K* might conceivably also be a density-dependent variable, but there is little or no evidence for that. Extensive trial calculations have shown generally that although such variability and dependencies will have quantitative consequences, many of those will be small, interactive among themselves and with negligible qualitative consequences for the population properties considered here.

By far the most difficult parameter to estimate is *r* in the stock–recruitment function and hence for the self-regenerating model. In many fisheries, recruitment is not seen to decline much, if at all, as fishing intensifies and the abundance of spawners is sharply reduced, until the original "virgin" stock is extremely depleted. This is evidence that the value of *r* is not much less than 1, but the eventual decline of recruitment will depend as much on the selectivity of fishing as on its intensity. In the large whales, *r* is surely small, and it is possibly in mid-range among the sharks and rays which are much less fecund than most of the bony fishes (*teleosts*) and thus more vulnerable to unrestrained fishing.

The shape of the Beverton and Holt stock–recruitment curve is fixed and derives from the equation $R = S/(S + b)$ (*S* is the size of the spawning stock, and *b* a scaling factor). J. G. Shepherd (1982) provided a more flexible model by adding one parameter, an exponent, so that $R = S^n/(S^n + b)$. This becomes the Beverton and Holt model when $n = 1$. With $n > 1$, this can display depensation—the *Allee effect*. Use of the Shepherd model instead of Beverton and Holt would make little, if any, difference to the general conclusions drawn in this brief.

The self-regenerative models all predict population extermination by a finite fishing intensity if that intensity persists despite evident depletion. While stability and regenerative capacity of a fish population are due mainly to the density-dependence of the prerecruit mortality rate that provides a negative feedback controlling population growth, there is increasing evidence that, at very low population size or density, the direction of density-dependence changes and creates a negative, destabilizing feedback that can lead to extermination even if fishing intensity is relaxed or ceases. The self-regenerating model can, with certain parameter values, exhibit depensation as an intrinsic property, but it seems likely that it must be provided for more generally and explicitly in future models. However, keeping fishing intensity well below that which will yield *msy* will normally keep the population far from the level at which depensation might "kick-in," but care would be needed for highly selective fisheries in which very high fishing intensities are required to obtain high or maximum sustainable yields.

The main variable, the consequences of which are computed, is the fishing mortality rate F. This is usually expressed as an exponential coefficient, but its value is, like that of M, related to the mortality rate expressed as percentages. Because exponential coefficients are *additive*, the *total* mortality rate in an exploited population is $F + M = Z$, but such simple addition is not valid for percentage rates of death (simple percentage rates or their sums obviously cannot add up to more than 100%, but either or both F and M can take values greater than 1). F is proportional to the deployed fishing power (effort) that generates it, properly calibrated. A unit of effort is that which generates $F = 1$ or some other arbitrarily chosen scale; thus the calibration is particular to the vulnerability of the individuals or shoals of a target species as well as to a fishing method.

It is convenient, for reaching generalized conclusions, as in this brief, to express the fishing mortality as a ratio of F to the natural mortality rate, hence F/M. ***Fishing power*** expresses the ability of a vessel or other fishing unit to generate a contribution to the total fishing mortality rate, so it also is specific. The fishing powers of the several vessels in a fleet or fleets are additive if they are calibrated. Rough calibrations can sometimes be made from consideration of the horsepowers of ships' engines, the sizes of the vessels (lengths, tonnages), and so on, but proper calibration requires a combination of experiments and examination of detailed statistics of catches.

Fishing effort is the deployment of fishing power. The degree of deployment varies from 0—the vessel never leaves port!—to full use which would usually mean operation for as many days of the year and hours of the day as are technically feasible, having regard to vessel maintenance schedules, occasional repairs, crew changes and rest periods, general weather conditions, and the like. Properly calibrated and summed, the deployed effort would be proportional to the fishing mortality generated. However, where the effort is deployed over a wide area in which the densities (local abundance) of the target species vary from place to place (or from time to time, or both) a *weighted* sum, called the ***effective effort*** must be calculated by methods that have been determined.

Fishing intensity has a technical sense of the concentration of fishing effort in space or volume, analogous to "density", but here I have used it more loosely as a generic term meaning "amount of fishing"—with the corollary that overfishing is sometimes "defined" as "too much fishing." However, it can have a special importance in situations where a high fishing level may have led to contraction of the geographical range of the population (see McCall 1990).

Some confusion has arisen because T. D. Smith, in his popular 1994 book, labeled the yield-per-recruit model "the Hulme model". He apparently did this because a mathematical colleague, H. R. Hulme, was the lead author, with co-authors Beverton and Holt (in collaboration with M. Graham), of a 1947 letter to *Nature* announcing the preliminary results of our joint work (Hulme et al. 1947). Hulme was, of course, important in the development of the general theory, but the simple model in the *Nature* letter had very different properties from the eventual one used here because it incorporated an unrealistic function for growth of individual fishes in weight, as did earlier models by T. I. Baranov (1918), W. E. Ricker (1946), and others. The version with the sigmoid body growth function was published first in 1949 by the Challenger Society as short contributions by Beverton (1949) and Holt (1949), and elaborated in our chapter entitled "The Theory of Fishing" (Beverton and Holt 1956b) in a book edited by Graham in 1956. It is in that book that the idea of "***rational fishing***," attributed to Graham and his mentor, E. S. Russell, as well as to Johan Hjort, is further explored.

Kurtosis—a measure of the flat-toppedness or peakiness of a distribution that has a maximum *y* at an intermediate value of *x*. A common measure compares this with the shape of the normal (Gaussian) bell curve, which takes the kurtosis value 0. Positive kurtosis applies to a curve that is peakier than the normal; a negative value describes a curve that is more flat-topped. It is not, in general, particularly relevant to discussion of fisheries management such as in this brief, but here it is important in evaluating the benefits of lower rather than higher rates of fishing. But, we are not so much interested in the overall kurtosis (which is a function—KURT—in the lexicon of, for example, Microsoft EXCEL), but rather in the shape of the curve in the upper region of yield close to *msy* and, especially somewhat below it, that is, values of *f* substantially lower than that required to take *msy*, thus to the left-hand side of the curve of *sy* against *f* as it is usually plotted. For that purpose, I have used a simpler index: it is the percentage reduction in *sy* that would occur with unit percentage reduction in *f*. (When we come to looking at the economics of sustainable fishing, a feature of equal relevance is the steepness—slopes—of the *sy-f* curve, especially in the mid-range of the left-hand limb. In that range will usually lie the point at which a further reduction in fishing effort reduces the overall cost less than the reduction in the value of the total catch, so reducing the net profit.)

A sister property of kurtosis is ***skewness*** or ***skew***, usually designated by the third lower-case Greek "gamma" γ and calculated as the third moment about the mean. A negatively skewed curve "leans" to the right, a positive one to the left, having sometimes a longish right-hand "tail." (*sy-f* curves rarely have left-hand "tails", but the calculations of the index of skewness nevertheless apply.) The skew of a curve of sustainable yield against fishing rate can be skewed positively or negatively and can be affected by the degree of density-dependence of certain parameters, especially of recruitment, on the size of the parent population.

There are very important differences between the curves commonly applied in fisheries "surplus yield" assessments and those—as used in this brief—based on age- and size-structured population models. Those differences lead to substantially different conclusions about management objectives. The former are essentially inflexible, arbitrary functions. One that is still frequently used—due to M. B. Schaefer (1954)—is symmetrical, and distinctly "peaked" in comparison with the latter types, which are based on actual observations of the birth, growth and death of fishes in exploited populations. Pella and Tomlinson (1969) introduced some flexibility by adding another parameter; their curves are asymmetric—can be positively or negatively skewed, although the symmetrical Schaefer function is a special case—but are still relatively peaked—even more so than the symmetric Schaefer ("logistic") version. **This means that use of surplus production models leads to quite serious underestimation of the benefits of reducing fishing rates below those needed to obtain *msy*.** This is one reason why the global benefits of reducing fishing pressure estimated by the World Bank and FAO (2009) publication "Sunken Billions"—which used the Schaefer model and a negatively skewed form of Pella–Tomlinson, due to W. W. Fox—are unreliable. (A negatively skewed curve of yield against fishing rate becomes a positively skewed curve when *sy* is plotted against population abundance, as is commonly done in presentations of surplus-yield assessments.)

The generally positive skew of the age-structured model provides a partial explanation of how it is possible for overfishing to continue and, in certain circumstances, worsen because catches do not begin to decline sharply until fishing effort has increased to substantially above the f_{msy} level.

Finally, given below is the full expression used here for calculating lifetime catch from a cohort. The expression is:

$$\frac{C}{RW} = \sum_{n=0}^{3} U_n (1 - w_c^{1/3})^{n+m} / (1 + 1/f + n/mf)$$

where $U_0 = +1$, $U_1 = -3$, $U_2 = +3$, and $U_3 = -1$. (Note that $m = M/K$, $f = F/M$, so $mf = F/K$. The ratios f and F/K can alternatively be used as proxies for F or fishing effort or intensity, the choice being made according to assumptions as to the constancy of M or K.)

Other References, Sources, and Further Reading

Holt (2004) is the Foreword to the 4th printing of Beverton and Holt (1957). The Foreword contains the citations and explanations of most of the references made in this brief to relevant publications by the authors and other scientists since the original version went to press in 1954. A few others are listed below.

Quinn and Deriso (1999) provides accounts of the derivation and properties of most of the population models referred to here.

Myers et al. (1999) compiled evidence of depensation in fish populations.

Pamela Mace has published numerous papers relating to the "steepness" h of the stock–recruitment function, most of which are cited in Mace (2001). In some of these, a related parameter was used to define the shape of the curve in a chosen space: tau = $(1 - h)/4h$. W. G. Clark (1991) popularized yet another parameterization of "steepness": $A = 1 - h$.

Holt (2002, 2006) provide summary accounts of the development of population dynamic theory associated with Verhulst, Malthus, Volterra, Pearl, Hjort, and Ottestad and others leading to the "surplus yield" approach to fisheries management associated especially with Schaefer.

The World Bank and FAO staff leaders of the project that culminated with the publication of (World Bank and FAO 2009) were, respectively, Kieran Kelleher and Rolf Willmann. The lead consultant economist was Prof. Ragnar Arnason of the University of Iceland, who "developed the theory and modeling underpinning the study and undertook the economic rent-loss calculations." Nicole Franz, of FAO, assisted with the statistical analyses, and numerous other listed persons and organizations were involved. My critique of this is published as (Holt 2009).

Readers interested in a lighthearted account of the conduct of fisheries research in post-World War II England might enjoy reading "Three Lumps of Coal" (Holt 2020), which is an address I gave to a gathering of scientists in Lowestoft, 2008, on the occasion of publication of *Advances in Fisheries Science: 50 Years on From Beverton and Holt* (Payne et al. 2008) celebrating (a trifle late) a half century since publication of Beverton and Holt (1957). Some readers might also be interested in reading the Foreword I wrote for this book (Holt 2008).

My Lowestoft talk ended with a celebrated statement—my real reason for mentioning this article—by Michael Graham in his seminal 1943 book *The Fish Gate* quoted at the UN/FAO Technical Conference in Rome in 1955 preparatory to the Second UN Conference on the Law of the Sea, Geneva 1958: "The trail of fishery science is strewn with opinions of those who, while partly right, were wholly wrong." Graham's *bons mots* are as valid today as they were then.

Acknowledgments

I particularly wish to thank Michael Earle and Tim Holt for patiently reading, commenting on, and helping edit this brief. Residual errors are still mine.

References

Allee, W. C. 1931. Animal aggregations: a study in general sociology. University of Chicago Press, Chicago.

Baranov, T. I. 1918. On the question of the biological basis of fisheries. Nauchnyi issledovatelskii ikhtiologischeskii Institut Isvestia 1:81–128.

Beverton, R. J. H. 1949. Assessments of yield in the plaice population of the southern North Sea. Challenger Society 3:10–11.

Beverton, R. J. H., and S. J. Holt. 1956a. A review of methods for estimating mortality rates in exploited fish populations, with special reference to sources of bias in catch sampling. Rapports et Procès-Verbaux des Réunions du Conseil Permanent International pour l'Exploration de la Mer 140:67–83.

Beverton, R. J. H., and S. J. Holt. 1956b. The theory of fishing. *In* M. Graham, editor. Sea fisheries: their investigation in the United Kingdom. Edward Arnold, London.

Beverton, R. J. H., and S. J. Holt. 1957. On the dynamics of exploited fish populations. Her Majesty's Stationery Office, UK.

Beverton, R. J. H., and S. J. Holt. 1966. Manual of methods for fish stock assessment. Part II. Tables of yield functions. FAO Fisheries Technical Paper 38 (Revision 1). \ Clark, W. G. 1991. Groundfish exploitation rates based on life history parameters. Canadian Journal of Fisheries and Aquatic Sciences 48(5):734–750.

Clark, W. G. 1991. Groundfish exploitation rates based on life history parameters. Canadian Journal of Fisheries and Aquatic Sciences 48(5):734–750.

Commission. 2011. Proposal for a Regulation of the European Parliament and of the Council on the Common Fisheries Policy. COM(2011) 425. Brussels.

Fox, W. W. 1970. An exponential yield model for optimizing exploited fish populations. Transactions of the American Fisheries Society 99:80–88.

Gompertz, B. 1825. On the nature of the function expressive of the law of human mortality and on the mode of determining the value of life contingencies. Philosophical Transactions of the Royal Society (London) 115:513–585.

Graham, M. 1943. The fish gate. Faber and Faber, London.

Graham, M. Editor. 1956. Sea fisheries: their investigation in the United Kingdom. Edward Arnold, London.

Holt, S. J. 1949. Biodynamic assessments of fish populations. Annual Report of the Challenger Society 3:10.

Holt, S. J. 1958. The evaluation of fisheries resources by the dynamic analysis of stocks, and notes on the time factors involved. International Commission for the Northwest Atlantic Fisheries, Special Publication 1: 77–97.

Holt, S. J. 2002. ICES involvement in whaling and whale conservation and implications of IWC actions. ICES Marine Science Symposia 215:464–473.

Holt, S. J. 2004. Foreword to the 2004 printing. Pages i–xxiii *in* R. J. H. Beverton, and S. J. Holt. On the dynamics of exploited fish populations. Blackburn Press, New Jersey.

Holt, S. J. 2006. The notion of sustainability. Pages 43–82 *in* D. M. Lavigne, editor. Gaining ground: in pursuit of ecological sustainability. International Fund for Animal Welfare, Guelph, Canada and the University of Limerick, Limerick, Ireland.

Holt, S. J. 2007. New policy objectives and management procedures for EU Fisheries: a commentary and suggestions to The Greens/European Free Alliance in the European Parliament.

Holt, S. J. 2008. Foreword. Pages ix–xxi *in* A. I. L. Payne, J. Cotter, and T. Potter, editors. Advances in fisheries science: 50 years on from Beverton and Holt. Wiley-Blackwell, Oxford.

Holt, S. J. 2009. Sunken billions—but how many? Fisheries Research 97:2–10.

Holt, S. J. 2020. Three lumps of coal: doing fisheries research in Lowestoft in the 1940s. ICES Journal of Marine Science doi:10.1093/icesjms/fsaa199.

Hulme, H. R, R. J. H. Beverton, and S. J. Holt. 1947. Population studies in fisheries biology. Nature 159:714–715.

Mace, P. M. 2001. A new role for MSY in single-species and ecosystem approaches to fisheries stock assessment and management. Fish and Fisheries 2:2–32.

McCall, A. D. 1990. Dynamic geography of marine fish populations. Washington Sea Grant Program, Seattle.

Myers, R. A., K. G. Bowen, and N. J. Barrowman. 1999. Maximum reproductive rate of fish at low population sizes. Canadian Journal of Fisheries and Aquatic Sciences 56:2404–2419.

Payne, A. I. L., J. Cotter, and T. Potter. Editors. 2008. Advances in fisheries science: 50 years on from Beverton and Holt. Wiley-Blackwell, Oxford.

Pella, J. J., and P. K. Tomlinson. 1969. A generalized stock production model. Bulletin of the Inter-American Tropical Tuna Commission 14:421–496.

Quinn, T. J., and R. Deriso. 1999. Quantitative fish dynamics. Oxford University Press, New York.

Ricker, W. E. 1946. Production and utilization of fish populations. Ecological Monographs 16: 375–391.

Ricker, W. E. 1954. Stock and recruitment. Journal of the Fisheries Research Board of Canada 11:559–623.

Schaefer, M. B. 1954. Some aspects of the dynamics of populations important to the management of commercial marine fisheries. Bulletin of the Inter-American Tropical Tuna Commission 1:25–56.

Shepherd, J. G. 1982. A versatile new stock-recruitment relationship for fisheries and the construction of sustainable yield curves. Journal du Conseil International pour l'Exploration de la Mer 40:67–75.

Smith, T. D. 1994. Scaling fisheries: the science of measuring the effects of fishing, 1855–1955. Cambridge University Press, Cambridge, UK.

von Bertalanffy, L. 1938. A quantitative theory of organic growth. (Inquiries on growth laws II.) Human Biology 10(2):181–213.

(The) World Bank and Food and Agriculture Organization (World Bank and FAO). 2009. The sunken billions. The economic justification for fisheries reform. The World Bank, Washington, DC.

Appendix 1: Southern North Sea Plaice Example

This demersal stock was known to be seriously overfished by steam trawlers in the 1930s. It had partially recovered, in accordance with predictions, because of the suspension of most fishing during the Second World War, a period referred to as the Second Great Fishing Experiment (the first being the First World War). Wartime losses had greatly reduced the trawl fleets, and the UK proposal to the new international management authority, created in 1946, the Permanent Commission, was that they should, in total, not be reconstructed to more than about 60% of pre-war tonnage, and also that the codend mesh size be increased above the prevailing 70 mm. There was, post-WWII, no thought of seeking MSY; accepted policy was simply to improve the state of the fishery and of the Plaice stock by regulation. The fishery was a mixed one for Plaice and Sole, but while the price of Sole was relatively high, the quantity and total value of catch of that species was a relatively small proportion of the total. British, French, German, Dutch, Belgian, and Danish trawlers had been engaged, but the English was, by far, the biggest contribution to the total effort.

The pre-war fishing mortality rate of Plaice had been about $F = 0.73$, so this would have been reduced to, and held at, about 0.44. The natural mortality rate M was estimated as 0.1, so $f = F/M$ had been about 7.3, to be reduced to about 4.4. Pre-war, this fishery was thought to have been close to a steady-state, but had become barely profitable. The ratio $m = M/K$ was estimated to be 1.2; w_c was 0.043 of $W = 2.8$ kg. I here round these, respectively, down to 1.0 and up to 0.05, so the results can easily be compared with the tables in this brief.

No evidence was found for changes in the average recruitment, though it might have been higher in the late 19th century and the early years of the 20th before the biomass of the stock had been substantially reduced.

With these parameter values, f_{msy} would be about 1.4. So, the proposed fleet reduction would hardly have "cured" overfishing, but the sustainable catch would have been 18% bigger with a catch rate 86% higher than the pre-war catch and catch rate, so should be profitable. A further fleet reduction to f_{msy} level would have provided a further increase of 29% in the total catch.

Relatively small increases in codend mesh size would have brought worthwhile increases in the sustainable catch of Plaice, if the fishing mortality rate were to be held at $f = 4.4$. The benefit can be appreciated if we were to assume that the mesh would be increased to, say, double w_c, to 0.1. The additional gain in sustainable yield would be 31%, with a corresponding gain in catch rate. Such a mesh change would, however, have led to a significant loss of Sole, which would have discriminated against the countries with a strong market preference for Sole, particularly France.

If the pre-war situation had really been near the point of zero profit, there would have been huge economic advantages from reducing the fishing effort well below that needed to obtain *msy* with the prevailing mesh size and if the ratio of the price of Plaice to the cost of exerting a unit of fishing mortality was unchanged. The *msey* of 78% of the *msy* would be obtained with $f = 0.45$, so with a catch rate more than twice as high.

Appendix 2: North Sea Haddock: Example of Density-dependent Recruitment

The Haddock in the North Sea were known to have been quite seriously affected, already in the last decades of the 19th century, particularly after the introduction of the steam trawlers. Scottish data gave an indication that recruitment to this stock might have declined in the period 1922–1937 when the spawning biomass was declining, after the partial recovery allowed by the drastic reduction of fishing effort in the region during the First World War. But, these data were inadequate for determining the parameter values of a Beverton–Holt stock–recruitment function. Accordingly, a series of five options were calculated in addition to the assumption that recruitment had not changed. These predicted recruitment towards the end of the 1930s decade to be in a range from 0.9 to 0.1 of the recruitment after the First World War. How much less that number was than the steady-state number in the original "virgin" stock is not known, so that, at best, only an overestimate of steepness h can be calculated from these data.

The yield-per-recruit model (see Annex Figure 2, p. 547) gave, for the 70-mm codend mesh being used at the time (implying $w_c = 0.2$), an $f_{msy} = 3.03$. The average values pre-war were $F = 1.0$, $M = 0.2$, so $f = 5.0$. The values for f_{msy} with density-dependence were, as expected, less than with constant recruitment (f about 2.0), but not very much less. The option with the most vigorous density-dependence had an f_{msy} at about 1.8. The *msy* predicted from the extreme option was, however, sevenfold that from constant recruitment. The less extreme options gave *msy*s double or triple those with constant recruits.

Appendix 3: MSY and w_{MSY}

Quinn and Deriso (1999) provide an expression for the time (age of fish) in a cohort—t^*—when the cohort, declining in number exponentially at a rate M and in which the individuals are increasing in body size according to the L. von Bertalanffy equation with parameters K and W, attains its maximal total weight: $\ln[(3 + m)/m]/K$. Substituting and rearranging gives us: $w_{MSY} = [3/(3 + m)]^3$ and $MSY = [m/(3 + m)]^m \times [3/(3 + m)]^3$. Thus, MSY is a simple function only of the fundamental life history parameter $m = M/K$.

A feature of these expressions is that they show that if $m \leq 1.50$, and since fish will always be caught before they reach the maturity size $w_s = 0.3$ $(8/27 = 0.2963)$, **any attempt to reach MSY would quickly exterminate the population since there would be no following cohorts.**

The expression used for the index of biomass of the total unexploited population, derived from the simple sustainable yield-per-recruit model, is:

$$B_{total} = \sum_{n=0}^{3} U_n \,/\, (1 + n/m)$$

where $U_0 = +1$, $U_1 = -3$, $U_2 = +3$, and $U_3 = -1$.

Appendix 4: The Stock–Recruitment Relationship

It has been a basic tenet of this brief that, for the purpose of evaluating alternative management objectives and procedures on the basis of generalized population models, it is useful to express, as far as possible all parameters—constants and variables—as ratios, e.g., $m = M/K$, $f = F/M$, $w = W/W_{inf}$. Following this, I deal as follows with the matter of the relevant consequences of annual recruitment depending on the size or activity of the spawning parent stock. I suggest that the easiest way to think about this question at its root may be as follows.

Consider that the mature females in a parent stock (s) shed N viable eggs in one spawning season. In the period until the age at which the survivors become the number of recruits R to the stock, this new cohort diminishes in number by a prerecruit mortality rate, which can vary with age throughout the prerecruit period. The various exponential rates vary at each phase or moment during the prerecruit period in a manner which is itself linearly density-dependent in relation to the number surviving at each time.

Beverton and Holt (1957) showed that, in this case, the initial magnitude of the spawning is related to the number of recruits R produced by the cohort by a simple equation involving the asymptotic number of recruits, R_∞ that would theoretically be produced by an infinite number of eggs, and a scaling factor b. It is convenient to write the ratio R/R_∞ as r thus:

$$r = s/(s + b).$$

We presume that r tends to unity when s tends to infinity, and $r = 0$ when $s = 0$.

Now, suppose we have two sets of observations for two time-periods, giving us, on average, r_1 from average spawners s_1 and r_2 from spawners s_2. These can, but need not, be absolute numbers; more usually, they would be indices of both quantities that are presumed to be directly proportional to the absolute numbers, most likely from catch-per-unit-effort data; so b can easily be estimated from the two linear simultaneous equations. This procedure is similar to the common procedure for F and M from two datasets providing two values for $Z = F + M$.

Two likely situations where this approach can be followed are:

- An apparently stable situation where fishing has been continuously intense for several years, the stock possibly having been overfished for several years with near-zero gross profit—an example is the point of intersection of all the graphs in Annex Figure 3, p. 548; and
- Data are available pertaining to the original unexploited stock or, more likely, from the very early years of exploitation. In this case, we can refer to the R/R_∞ estimate as r_0, that being the value at which the unexploited stock had settled in a steady-state; this alone sufficiently defines the stock–recruitment relationship for use in calculating the self-regenerating model for that stock from the yield-per-recruit model results.

Further, it is convenient to re-scale s so that its value is 1 when $r = r_0$ so that $r_0 = 1/(1 + b)$ from which $b = (1 - r_0)/r_0$ so that $r = 1/[1 + (1 - r_0)/Sr_0]$. Then, for any value of r, we can write d_r, the recruitment depletion, as $d_r = r/r_0 = 1 / [r_0 + (1 - r_0)/s]$.

Some researchers have defined “steepness” (h) as the relative amount by which recruitment would be reduced if the spawning stock in the unexploited population were to be reduced by fishing to 20% of its original effective size. This specified degree of reduction is an arbitrary choice, but, however defined, steepness is a simple function only of r_0, putting $s = 0.2$.

d_r can be calculated by putting d equal to the biomass of the spawning stock for a given value of $m = M/K$ and selectivity, relative to that biomass when the stock is unexploited. The resulting values are then multiplied by the corresponding steady-state catch-per-unit-effort from the yield-per-recruit expression.

It is of some importance to see to what extent calculations of r from this expression might depend on the precise shape of the convex (from above) curve of r against s for any given value of r_0. An alternative arbitrary function with similar general properties ($d = 0$ for $s = 0$, and $d = 1$ for $s = 1$) could be a simple power function such as: $d = [1 - (1 - r_0^{\,s})]/r_0$.

The curve of this expression diverges considerably from the Beverton–Holt curve for values of r_0 in the range 0.55 to 0.95, but the two nearly coincide outside that range, so theoretical calculations of *steepness* would depend critically on the value of the recruitment into the unexploited stock (relative to the theoretical maximum recruitment). Such discrepancies can distinctly affect calculations of recovery rates of substantially depleted stocks, but would have little importance for self-regenerating, steady-state calculations near to or below fishing rates yielding *msys*, at least for operations of relatively low selectivity.

Annex Figure 1

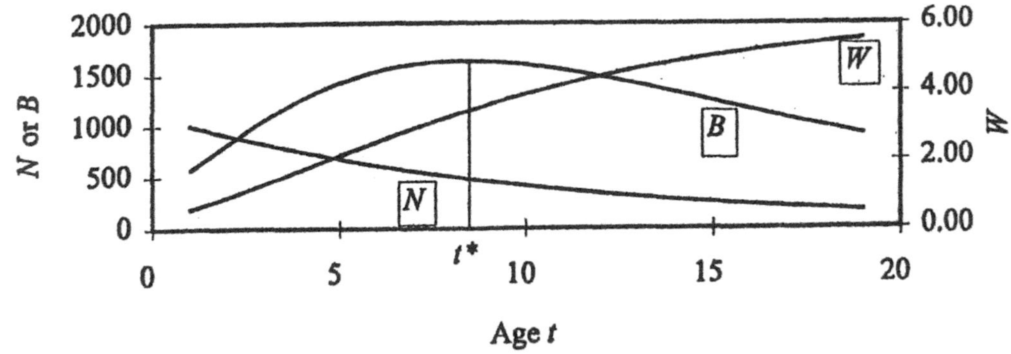

This illustration is from *Quantitative Fish Dynamics* (Quinn and Deriso 1999), Figure 6.1a, p. 241, and is reproduced with permission of Oxford Publishing Limited through PLSclear. It is from data for an unfished population of Sablefish *Anoplopoma fimbria* in the Gulf of Alaska.

The graph labeled N shows the exponential decline in number of a cohort ($M = 0.1$). Graph W shows the average weight of an individual fish as a function of its age, with fitted L. von Bertalanffy function ($K = 0.15$. $W_\infty = 6.20$ kg). Graph B shows the product of graphs N and W, that is, the cohort biomass as a function of age/time, with a maximum (MSY in our terms) at about 8½ years.

So, $m = M/K = 0.67$. The above formulae give us $w_{\mathrm{MSY}} = 0.55$ (54.7%), $\mathrm{MSY}/RW_\infty = 0.175$.

Annex Figure 2

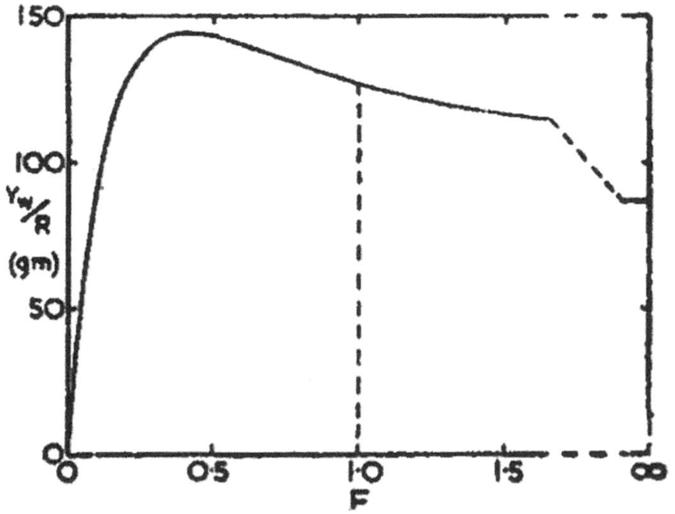

Figure 17.24 taken from Beverton and Holt (1957), p. 325, with the original legend, slightly modied. Haddock: yield against fishing mortality, 70-mm mesh. [Yield-per-recruit, Y_w/R, as a function of F with $t_p' = 1.83$ years. Vertical broken line shows pre-war fishing mortality ($F = 1.0$). The remaining parameters used to construct this and other Haddock diagrams of §17 are as follows: $M = 0.20$, $W_\infty = 1209$ g, $K = 0.20$, $t_0 = -1.066$ years, $t_p = 1$ year, $t_\lambda = 10$ years.] Data from Marine Laboratory, Aberdeen, Scotland, courtesy Basil Parrish and Cyril Lucas.

The above graph is typical of a yield-per-recruit curve with constant parameters and with fairly low selectivity as defined in this brief, so it has a maximum with finite fishing mortality. In terms of definitions in this brief, $m = M/K = 1.0$; $f_{msy} = 2.50$; pre-war relative fishing rate f was 1.0/0.20 = 5.0; t_λ is not relevant in the present context, but tells us that 10 cohorts contribute significantly to the recruited stock, and 9 of them to the exploited phase.

$w_{MSY} = 0.42$, MSY/RW = 0.1055, $w_c = 0.04$, thus fish were liable to capture long before maturity at $w_s = 0.3$.

This graph, produced in 1948, shows that a reduction of the pre-war fishing rate by one-half would lead to a sustainable catch about 15% bigger, thus with a catch rate 120% higher. The scale of the peak of the curve indicates that by "sacrificing" only about 5% of the potential *msy*, 95% of the theoretical *msy* would be secured by exerting only about 80% of the effort required to obtain *msy*.

The geometry of the figure also shows that a sustainable catch of the pre-war size could have been obtained with only about one-quarter of the pre-war effort. Furthermore, if—as was likely—the pre-war situation produced just a bare profit or none, a very large maximum gross profit could have been obtained by exerting less than about one-fifth of the pre-war effort.

These estimates of the benefits of much less fishing effort in terms of yield per constant recruitment, are minimal because, as will be seen in Figure 3, there was evidence that recruitment was density-dependent, and average annual recruitment would have been higher with a larger spawning stock being available as a consequence of the imposition of a much lower exploitation rate.

The adjacent Figure 17.25 in Beverton and Holt (1957) (not reproduced here) shows the benefits that would have been available in the pre-war situation by substantially increasing codend mesh size so that age of first liability to capture would be increased by anything up to 5 years (from 1.8 years), keeping fishing effort unchanged, so that the catch rate would be increased proportionately. The amount of increase in sustainable catch would have been about the same as from reducing the fishing effort to ***msy*** level, but, of course, the gain in catch rate would be only one-half of the catch rate gain from reducing effort by 50%.

Annex Figure 3, Effects of Density-Dependent Recruitment

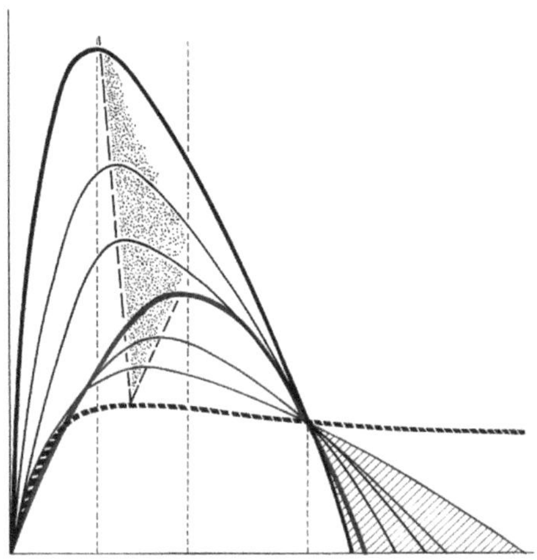

This graph is reproduced from the front cover of the 2nd–1993–edition of Beverton and Holt (1957) and continued on the 3rd edition; it is not in the book itself.

The graph gives results for analyses of data for the Haddock exploited in the North Sea bottom-trawl fishery. The x-axis is the fishing mortality rate; the y-axis is sustainable yield *sy*. The figure was prepared by Beverton, I think, in about 1990.

The thickly dashed curve is the same as that in Annex Figure 1, p. 547 of this brief. Above it are three higher curves (two with thin lines, the third—upper—bolder) whose peaks are indicated by a near-vertical, left-leaning dashed line. These are for three degrees of density-dependence of recruitment on spawning stock abundance as given by the Beverton and Holt self-regenerating model, using the well-known Beverton–Holt stock–recruitment function $R = S/(S + 1)$ (see text). The dataseries available at the time (1949) indicated density-dependence, but with a low degree of statistical significance; the three sample graphs are all compatible with the data. (The three curves—one bold, the others thin—whose peaks are joined by a right-leaning dashed line, are not relevant to this discussion. They are there essentially to illustrate the consequences of possible density-dependence of growth rate.)

The important points here are that:

1. if recruitment is density-dependent on spawning stock size, whether weakly or strongly, the value of f_{msy} is somewhat lower than for constant average recruitment;
2. weak or strong density-dependence of recruitment on spawning stock size has very big implications for predictions of *msy*, but relatively little for f_{msy}. So, efforts to regulate towards *msy* by control of catches are likely—I think virtually certainly—to fail because of the inevitable great uncertainty in estimates of the strength of density-dependence of recruitment on parent stock. Control of fishing effort is far more likely to be successful.

In the present context—discussion of f less that f_{msy}—this matter of recruitment dependence is less important than in broader consideration of rational fishing, because attention is directed mainly to the benefits of maintaining a relatively large stock—closer to the original unexploited size—in which recruitment density-dependence is likely to play a rather small part, at least as far as most of the bony fishes are concerned.

Finally, it is worth noting, from this example, how perilously close to collapse and extinction persistent overfishing can bring the stock and fishery.

Chapter 47

The Benefits of a Fisheries Management Target for an Average Annual Catch Less Than the Theoretical Maximum Sustainable Yield (MSY)[1]

Stock assessments show that catch targets set at significantly less than the theoretical maximum sustainable yield (MSY) would deliver economic and social benefits very much greater than the expected benefits of an MSY management target. Or, to put it another way, would any serious management committed to sustainability, either in terms of quantity or value of output and net profit, seriously consider, in any circumstances, doubling or tripling the production cost to obtain just 10% more output?

It is critically important that high sustainable yields in managed fisheries be profitable for those who catch them, and this requires that they be obtained with an adequate catch rate. Since the catch rate inevitably declines with increasing fishing intensity (because of the reduction in the biomass of the stock), this means that a limit to the desired sustainable yield must be set to ensure profitability.

Several benefits have long been known: (i) precaution against errors in stock assessments and failures of management, including possible errors arising from the use of deterministic rather than stochastic population models; (ii) reduced impact of the fisheries on marine ecosystems; (iii) fewer discards of unwanted catch; (iv) increased average size of the targeted fish species; (v) a lesser effect of natural recruitment variability on catches; (vi) lesser vulnerability to the effects of environmental changes, especially of ocean climate. The most practical important benefit is, however, lessened risk of economic distress to fishers resulting from unnecessarily low catch rates (catch per effort). That is the subject of this memorandum.

A sustainable catch is not necessarily profitable, in the simple sense of its market value exceeding the cost of taking it, primarily vessel operational costs—especially for fuel—and crew. Even exploitation of the resource at MSY level will not necessarily be profitable and, in fact, is likely not to be, considering the common unprofitability of <u>unsustainable</u> fishing. *Any* reduction of fishing intensity somewhat below that required to obtain MSY will generate profit or enhance profitability. The crucial question is *how much reduction*? There is no simple answer to that question, but alternatives can easily be explored with the population models used for making the primary fish stock assessments. It should be noted, however, that there is a lower bound to the desirable level of fishing effort and hence average catch; that is set at the level at which the difference between the value of the catch and the costs becomes less than whatever it would have been at the MSY level. This threshold would, however, generally be much lower than would be contemplated in practice.

[1]Editors' note: written by Sidney Holt as a memo to Michael Earle and dated October 21, 2012.

The possibilities can be explored by inspection of the curves of sustainable average catch as a function either of the fishing mortality rate or of the biomass of the population provided by alternative models and assessment procedures—choice among which depends partly on the available data—and specifically on the shapes of those curves close to and to the left of their maxima. There are three general types of model and procedure:

(1) "Surplus production" models predicting sustainable yield as a simple function of the biomass of either the recruited or the exploited phase of the stock or of that of the female spawners.
(2) Age-structured models in which sustainable catch is assumed to be primarily a function of the natural mortality and growth rate parameters of the fish (or, more generally, the ratio of those two parameters, identified as $m = M/K$).
(3) Multispecies models that combine some features of (1) and (2).

Here, I record the results for a **sample** (that is, a "sacrifice" of just 10% of the theoretical catch) in terms of the long-term benefit as a percentage reduction in fishing effort. (Results of different levels of sacrifice can easily be calculated, as well as the threshold limits.)

Surplus Production Approach

Recent published analyses of existing data for a large number of fish species and stocks have concluded that MSY is to be obtained with a range of biomass between about 40% [Fox's (1970) version of the basic Pella and Tomlinson (1969) model] and 50% (Schaefer's version of the same model; Schaefer 1954) of the pre-exploitation ("virgin") biomass. (In some cases, this pertains to *recruited* biomass and in others to biomass of *spawners*, but this has little effect on the conclusions that are relevant here.) There is an indication that the relative MSY level is rather higher for larger species than for smaller ones.

The shapes of the peaks of the yield–biomass curves are always determined only by a very simple, assumed power function in the models, one parameter of which fully determines the MSY level relative to the unexploited biomass.

The results for 10% catch "sacrifice" are:

a) MSY level at 40% of "virgin": 90% of MSY obtained with 74% of MSY effort
b) MSY level at 50% of "virgin": 90% of MSY obtained with 68% of MSY effort

So, the catch rate (CPUE) would be about 30% higher if the chosen target is 90% of MSY rather than MSY itself.

Age-structured Model Approach

There are two obvious possibilities here. One is the simple, widely used Beverton and Holt (1957) model for sustainable yield-per-recruit. The other is the self-regenerating model by the same authors in which recruitment is represented as a simple function of the biomass of the spawning phase of the population. While this second version of the Beverton and Holt model predicts very widely differing values for sustainable yield from severely depleted population, it makes little difference to predictions of what would be the population level at fishing intensities for MSY or less. Furthermore, it hardly comes into play when fishing rate in kept at or below MSY level, so here, I look only at the results from the yield-per-recruit version.

These results depend somewhat on the value of *m*—generally somewhere between 0.5 and 2.0—but they do depend also on the *selectivity* of the fishing, which is usually expressed as the age and/or size at which fish first become liable to capture (depending both on the fishing gear and the distribution of mobile fish (geographically and by e.g., depth) and the timing of their movements in relation to the fishing operations. This selectivity I refer to, in terms of body weight *w*, is expressed as a percentage of the final theoretical weight of fish of the particular species *W*. Surplus production models take no account of selectivity. A value of *w* = 0.3 means that fish would become liable to capture only after they have reached the average size at maturity.

The results are summarized in the table below for SY = 0.9MSY. In this table, *M/K* is the ratio of the natural mortality rate to the exponent of the growth function, *w* is the selectivity in terms of the ratio of the weight of a fish at first liability to capture to the theoretical asymptotic weight, and F/F_{msy} is the ratio of the fishing intensity to obtain 90% of the MSY to that needed to obtain MSY.

The limitation to possible selectivity in the lower part of the table results from the absence of possible MSYs with higher selectivities when *M/K* is relatively high, that is generally when *M* is high.

Clearly, sustainable catches of 90% of MSY can generally be obtained with less than one-third to about one-half of the fishing effort required to obtain MSY.

(The full Table 4, in Holt (2012, Chapter 46), shows the considerable benefits that could be obtained with a lesser "sacrifice" of 5%, or even of only 1%, of the theoretical MSY.)

m = M/K	*w*	F/F_{msy} %
0.25	0.4	39
	0.3	43
	0.2	46
	0.1	52
0.5	0.4	33
	0.3	40
	0.2	45
	0.1	49
1.0	**0.3**	26
	0.2	38
	0.1	47
1.5	0.2	26
	0.1	42
2.0	0.1	36

Pamela Mace has looked at the same model from another point of view, asking what would the sustainable yield be, as a percentage of MSY, if the fishing effort (mortality rate) were to be held at 75% of that needed to provide MSY (see Restrepo et al. 1998). She used both the Beverton and Holt and the Ricker stock–recruitment relationships in the self-regenerating model, with M/K values of 3.0, 2.0, 1.5, 1.0, and 0.5, focused on the case where selectivity was the same as the age at maturity. Calculated SY/MSY ratios were all in the narrow range 94.4–98.1%; so, a "sacrifice" of from only 2 to 5 or 6% of catch for an effort reduction of 25%.

Mace also pointed out that if a similar stochastic rather than the deterministic model were used, the application of this approach to target-setting would need to be more precautionary.

Generalized Global Multispecies Synthesis (see Annex)

Figure 2 of Worm et al. (2009) suggests that an overall MSY is obtainable with exploitation rate (ratio of sustainable catch to stock biomass) of 0.5 and that 90% of that would be obtained with an exploitation rate of 0.2. The cited exploitation rates are roughly equal to fishing mortality rates, approximately proportional to fishing effort, of 0.69 and 0.22, respectively; thus, the "sacrifice" of 10% of the theoretical MSY would be obtained at 32% of the cost and hence catch-rates 2.8-fold higher than would be obtained with a strict MSY target.

Conclusions

The relatively smaller advantage of taking less than MSY, indicated by the surplus production models, arises from the fact that the sustainable yield–fishing effort curves are rather peaked, and this is solely because of the particular assumption about the shape of the yield–effort curves near to the maxima which is incorporated in these models; it is not at all dependent on data, regardless of data that might be used to estimate MSY and the population biomass from which that could be obtained. Even so, the profitable effort savings from aiming at less than MSY, and the benefits from exerting lower effort, are considerable.

It is, however, obviously better to base management policy on the findings from age-structured and multispecies models, which are based on population data, and provide rather flat-topped yield–fishing effort curves. These confirm that the potential economic and social benefit of aiming somewhat lower than MSY is very great. How much lower is a matter of economic and social policy, guided by scientific analyses.

In sum, 90% of the MSY can be obtained with a catch rate (CPUE) two- to three-fold that available if the management target is MSY. Bigger operational gains and gross profit can be obtained by a bigger "sacrifice," but not proportionately so.

Acknowledgments

I am grateful for comments on and corrections of an earlier draft of this memorandum offered by Pamela Mace and Trevor Branch.

References

Beverton, R. J. H., and S. J. Holt. 1957. On the dynamics of exploited fish populations. Her Majesty's Stationery Office, UK.

Fletcher, R. 1978. On the restructuring of the Pella–Tomlinson system. Fishery Bulletin US 76:515–521.

Fox, W. W. 1970. An exponential yield model for optimizing exploited fish populations. Transactions of the American Fisheries Society 99:80–88.

Hilborn, R., U. Stewart, T. A. Branch, and O. P. Jensen. 2011. Defining trade-offs among conservation, profitability, and food security in the California Current bottom-trawl fishery. Conservation Biology 28(2):257–266.

Holt, S. J. 2006. The notion of sustainability. Pages 43–82 *in* D. M. Lavigne, editor. Gaining ground: in pursuit of ecological sustainability. International Fund for Animal Welfare, Guelph, Canada and the University of Limerick, Limerick, Ireland.

Holt, S. J. 2012. Reform of the EU Common Fisheries Policy: How to achieve sustainable and profitable fishing. The Greens/European Free Alliance in the European Parliament. 63 pp. Available: https://www.greens-efa.eu/files/assets/docs/how_to_achieve_sustainable_and_profitable_fishing_holt.pdf.

Pella, J. J., and P. K. Tomlinson. 1969. A generalized stock production model. Bulletin of the Inter-American Tropical Tuna Commission 14:421–496.

Restrepo, V. R., G. G. Thompson, P. M. Mace, W. L. Gabriel, L. L. Low, A. D. MacCall, R. D. Methot, J. E. Powers, B. L. Taylor, P. R. Wade, and J. F. Witzig. 1998. Technical Guidance on the Use of Precautionary Approaches to Implementing National Standard 1 of the Magnuson-Stevens Fishery Conservation and Management Act. Appendix A. Equilibrium Implications of Fishing at 75% FMSY. NOAA Technical Memorandum NMFS-F/SPO-31.

Schaefer, M. B. 1954. Some aspects of the dynamics of populations, important for the management of the commercial fisheries. Bulletin of the Inter-American Tropical Tuna Commission 1(2):25–56.

Thorson, J. T., J. M. Cope, T. A. Branch, and O. P. Jensen. 2012. Spawning biomass reference points for exploited marine fishes, incorporating taxonomic and body size information. Canadian Journal of Fisheries and Aquatic Sciences 69:1556–1568.

Worm, B., R. Hilborn, J. K. Baum, T. A. Branch, J. S. Collie, C. Costello, M. J. Fogarty, E. A. Fulton, J. A. Hutchings, S. Jennings, O. P. Jensen, H. K. Lotze, P. M. Mace, T. R. McClanahan, C. Minto, S. R. Palumbi, A. M. Parma, D. Ricard, A. A. Rosenberg, R. Watson, and D. Zeller. 2009. Rebuilding global fisheries, Science 325:578–585.

Annex

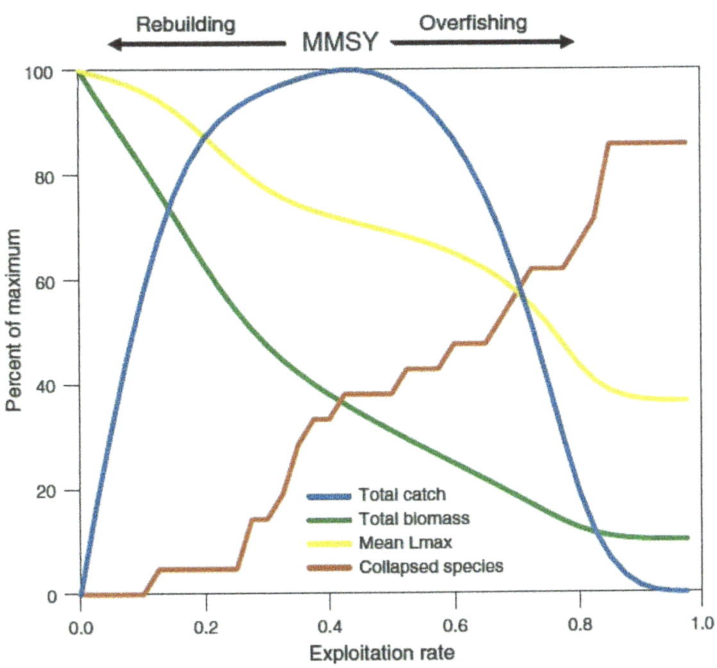

Figure 1. From Worm et al. (2009). Reprinted with permission from AAAS.

In this figure from Worm et al. (2009) illustrating a "model fisheries community" (Georges Bank, Northwest Atlantic) with the MSY at an exploitation rate (roughly the fishing mortality, proportional to a calibrated fishing intensity) of 0.5, the 90% of MSY value, at $F = 0.2$ is at the edge of the slightly sloping upper plateau, where the sustainable catch curve (blue) begins to fall away sharply towards zero, on the left.

The curve for total biomass (green) is the same shape as the catch rate (CPUE) curve would be.

The yellow curve shows how the average size (length) of the fish increases as the exploitation rate decreases, because more fish in each population have a chance to continue their growth.

Chapter 48

Application of "Less than MSY" Management Policy to So-called "Mixed Fisheries"[1]

Introduction

The term "mixed fisheries" is usually used with one of two meanings:

1. a fishery in which two or more species are being caught simultaneously in the same area by the same fishing gear or gears, or
2. a fishery in which two or more species that are known or presumed to be interacting biologically (usually as either competitors or predator–prey pairs) are being caught in the same management area.

These situations are entirely different in their possible consequences and must be dealt with separately.

It is easiest to discuss this in terms of a more or less realistic example. I remember my first experience attending the meeting of the then "Permanent Commission" (later the North-East Atlantic Fisheries Commission) in, I think, 1948. The Commission was discussing proposals for increasing the codend mesh size used in bottom trawls and the related question of minimum legal lengths of landed whole iced fish. The Commissioner for France stood up in plenary session waving two frying pans, one bigger than the other, and saying "The French housewife (all shopping for food and cooking was done in those days by 'wives') prefers this one, in which to cook Sole (*Solea solea*), while the British housewives use larger ones in which to fry Plaice (*Pleuronectes platessa*)." That was the meeting at which my colleague Ray Beverton and I demonstrated our age-specific population model with an analogue computer looking like an upright pinball machine and shedding steel balls.

Here, let us concentrate only on the use of the model to illustrate the mixed trawl fishery, as it then was in the Southern Bight of the North Sea, for the two species, with a particular fixed mesh size. Both were overfished (though not as much as they had been before the war) in the sense of being subject to fishing mortality rates much higher than those required to hold their populations at levels and in conditions (age and size compositions) that could yield continuing average catches by total weight at local *msy* levels. By this term, I mean the peaks of the respective curves of sustainable yield against fishing mortality, corresponding with the ages (size) of first liability to capture of each of the two species.

Now, think of the situation in which regulators are seeking to bring both populations to their respective local *msy* states, so as to maximize the total sustainable catches of both,

[1]Editors' note: a short note written by Sidney Holt to Michael Earle dated December 6, 2012.

combined. They are thinking of maximum weight of catch, though later they might want to think about market value, though this can be complicated by the price ratio being different in the different national markets. So, to keep it simple, we shall continue to focus on weight. It is also worth remembering that the arguments now to be made apply equally to other weight objectives than *msy*, such as "90% of *msy*."

In these circumstances, the fishing mortality rates of the two species' populations will differ because their catchabilities q, the ratio of the fishing mortality rate to the effort exerted, will differ. But this difference may not be large.

The two flatfish species may have different natural mortality rates and different intrinsic growth rates, the K parameter of the L. von Bertalanffy equation for growth in weight most commonly applied in assessing the states of fish stocks. But, the ratio $m = M/K$ varies among species very much less than either of its components, and it is that ratio which mainly determines the dynamics of all populations as far as management of their exploitation is concerned.

The other parameter of the growth equation W_∞, the asymptotic weight individual fishes can be expected to reach if they live long enough, varies greatly between species—so greatly affects the total catches—but does not affect the shape or scale of the curve of sustainable yield against fishing mortality rate or effort.

The differences between the dynamic parameters for the two species means that they will show, for example, different mortality/effort requirements for the *msy* situation. However, the relatively small differences, and the flatness of the yield curves, mean that a single fishing mortality for both species will usually provide a combined sustainable yield rather close to the sum of the optimal single-species values. This suggests an approach to sub-optimal regulation of a mixed two-species fishery.

If there are more species in the fishery, the issue is naturally more complicated, but the same approach could be taken for more numerous species groups. In practice, the best compromise could be found by computer simulation.

Species Interaction

Discussion of this problem usually begins with consideration of the classical Lotka-Volterra models for the interaction between a single predator and its single prey, usually in numerical terms, though for fishery management, it is necessary to use a more complex model in terms of fish weight.

It has long been known that this simple model displays complex behavior, with trajectories of population size ranging from constant steady state through declines that may lead to extinction, increases (that eventually lead to Malthusian breakdowns of the system), oscillations, and limit cycles. More recently, Robert May (1976) showed that the same simple model may also display chaotic behaviour, with unpredictable outcomes, depending on an impossibly precise knowledge of the initial state of the system.

In practice, of course, there is rarely, if ever, any such thing as an isolated two-species system. It is not formally difficult to expand the same approach to three (which allows treatment of competition of two predators for the same prey), four, and more interacting species. The results, however, present worrying problems, quite apart from the increasing difficulty of finding real data adequate for their testing and possible application. We also have to beware of unwarranted assumptions which can appear obvious, but may be quite wrong or misleading. It has, for example, sometimes been assumed that increases in the number of interacting species lead to greater stability of the system and/or less likelihood of chaos. These assumptions have been used to support the idea of greater species biodiversity in an ecosystem ensuring stability. Neither of these assumptions is automatically true; it is not difficult to create plausible/reasonable theoretical models in which the opposite holds.

So, what is to be done about management? What is certain, judging from recent symposia on the subject, is that, in no case, is enough yet understood about the dynamics of complex ecosystems to manage their exploitation adequately and safely by applying multispecies mathematical models. This is true notwithstanding the current proliferation of "ecosystem models" based on simplistic assumptions about the biological productivity of a population being related solely to its biomass, even excluding the role of fish size and age composition.

One approach, in these circumstances, might be to ignore all interactions—presumed or known, identified, and perhaps even measured. Then, later, after perhaps some years of steady management, on this basis to see what unexpected changes the data show and make arbitrary adjustments—"corrections"—to the management details. However, another approach is possible which may provide more stability to the fishery. This might be to treat all the possibly interacting species as if they constitute a mixed, non-interactive set and manage the whole on the basis of an assumption that all of them are being exploited relatively unselectively. A first approximation to the possible general or average level of selectivity might come from observations, if data are available or the degree to which species begin to be exploited just before they become sexually mature (at a weight roughly equal to 30% of their asymptotic weight), or substantially earlier or later in their life cycles. Management on the general assumption of non-selective exploitation would be precautionary, calling for a low, managed fishing mortality rate, and providing relatively high, more profitable, catch rates. It would be useful if simulations of such an approach could be made based of one of the better researched multispecies fishery systems.

I would end this advisory memo with a comment on the kind of regulation needed for managing any multispecies mixed fishery. This is that managing by setting individual species catch limits will always present great problems. From the examples given here, for reasonable approaches to management both for sustainable production and profit, the only way forward has to be by control of fishing mortality rate through control of fishing intensity, appropriately calibrated. It is even possible, I think, that an approach to assessment and management analogous to the well-known (but undesirable) "surplus production" models, focused on the biomass of some defined portion of the entire exploited ecosystem, related to total (weighted by indices of species values) "pooled" catches would provide a first approximation to a desired sustainable exploitation of that system by way of total mixed catch and profitability.

References

May, R. M. 1976. Simple mathematical models with very complicated dynamics. Nature 261:459–467.

Chapter 49

The Some Good and Mostly Bad about Maximum Sustainable Yield as a Management Target[1]

An American historian, Dr. Carmel Finley, has described in detail how maximum sustainable yield (MSY) was used, during the post-World War II negotiation of a US–Japan peace treaty, as a quasi-legal, pseudo-scientific concept to force Japanese fishers to "abstain" from fishing for Halibut *Hippoglossus stenolepis* and salmon *Oncorhynchus* spp. in waters adjacent to the coasts of North America (while not restricting US tuna fishing close to South and Central America) (Finley 2011). It was said that the fish resources in that region were "fully utilized," under joint governmental management, of Canadian and US fishing operations. The claim was based mainly on studies by North American scientists, particularly Milner "Benny" Schaefer, who applied a logistic model to the Yellowfin Tuna *Thunnus albacares* of the Eastern Pacific and thereby launched what we now know as "surplus production" theory (Schaefer 1954). This followed in the footsteps of the Norwegian scientists Johan Hjort, Per Ottestad, and Gunnar Jahn who published, in 1933, their classic study of whaling. Their paper concluded: "*It is clear that a stock of whales cannot replace more than a certain number of individuals, and a wisely conducted whaling industry must act accordingly if the stock is to escape the fate of the stocks off Iceland or the coast of Spain and Portugal. There is, however, a definite level to which a stock can be reduced, while preserving the maximum capacity of regeneration and at the same time securing an optimum catch*" (Hjort et al. 1933).

Hjort et al. (1933) drew a symmetrical curve of steady yield against the number of animals, but did not attribute it to P. F. Verhulst's classic 1838 study that launched "the logistic curve" of numerical population growth.[2] Schaefer's unwarranted shift to consideration only of gross population size by weight (biomass) set us off, I think, down a seriously mistaken path of fishing theory and management.[3] A few years after Hjort et al. (1933), Michael Graham

---

[1]Editors' note: a presentation by Sidney Holt to the September 2012 ICES ASC for the Theme Session "Evolution of management frameworks to prevent overfishing" (Holt 2012a).

[2]Verhulst was a Belgian mathematician who put forward theorems of restraints on human population growth that countered the theory of exponential growth leading to catastrophe propounded by the British Rev. Thomas Malthus, who published six editions of his "Essay on the Principle of Population" between 1798 and 1826. This pre-Darwinian controversy was of profound importance to the new proto-capitalist society needing to believe in "progress."

[3]There are many problems on that path, particularly arising from the absence of empirical evidence regarding its basic assumptions about the forms of density-dependence, but the most damaging of which is that population productivity is a simple function of population biomass. Two populations of a species, having the same biomass, can differ immensely in their biological productivities. The surplus yield model undervalues and impedes consideration of the importance of fishing selectivity in the consideration of sustainable fishing scenarios.

used the "sigmoid curve" as a model for describing what was subsequently labeled "growth overfishing" by David Cushing, but this involved taking account only of growth in body size and the natural and fishing mortality rates (Holt 2006).

The US delegation tried, but failed, to get Schaefer-style MSY and the "Abstention Principle" acknowledged as universal fisheries management objectives at a UN Technical Conference on the Law of the Sea in Rome, held in 1955 at FAO headquarters. A vigorous, but polite, debate there between my mentor, Michael Graham, and Schaefer—on the UK and US delegations, respectively—revealed a disagreement in principle between seeking a physical "ideal" (the latter) and "moving in the right direction" (the former). The following legal-diplomatic UN conference, in Geneva 1958, rejected the abstention "principle," but accepted a qualified MSY-type objective. The Fisheries Convention signed there was never effective, however, mainly because of the failure of the Geneva Conference to agree on the breadth of the Territorial Sea. Eventually, a conditional-MSY objective was written into the well-known UN Law of the Sea Convention of 1982 that came into effect in 1994, for nearly all states, but not the USA.[4]

It having been decided at the beginning of the Third Millennium to revise the Common Fisheries Policy (CFP) of the European Union, the EU Commission put forward a proposal regarding the adoption of an MSY objective as the basis for the revised policy. At the time, I supported that move, mainly on the grounds that in providing for recovery of overfished and depleted stocks, it was better than the then-current policy, essentially of merely seeking stability through a currently sustainable catch, or perhaps somewhat less than that if some recovery was desired. However, the MSY principle applied carelessly has its downside. I recalled when, in 1974–75, the International Whaling Commission (IWC) adopted a <u>numerical</u> MSY objective for all whaling, the Japanese position was that this meant that all "underfished" species and stocks <u>must</u> be reduced to MSY levels, otherwise we were "<u>wasting</u>" a God-given resource. I guess the God was Janus, the two-faced one.

In 2011, the Commission published its proposal for a regulation on the CFP:

1. The regulation should aim at securing the growth of all stocks to levels <u>above</u> those that can produce the MSY <u>as fast as possible</u> and by 2015, where biologically possible. (*Holt's emphasis*)
2. The resulting economic and social hardships must be cushioned financially through the European Maritime and Fisheries Fund.

Such references to "level" are commonly intended to refer to ***biomass***, but biomass *of what* is rarely specified. In fisheries discourse, it practically never means "of the entire population," which would includes the total weights of larvae. "Stock" is usually assumed to mean "recruited animals," but what we regard as recruitment is partially determined by biological events (shift of habitat, metamorphosis in the case of e.g., the flatfishes, Pleuronectiformes) and partially whether the human predator regards the animals collectively as a potential exploitable resource. Sometimes biomass seems to be taken to apply to the exploited segment of the recruited stock, which might be all those fish above a size/age of first liability to capture by trawls or possibly

---

[4] "...measures shall ... be designed to maintain or restore populations of harvested species at levels which can produce the maximum sustainable yield, as qualified by relevant environmental and economic factors, including the economic needs of coastal fishing communities ... and taking into account fishing patterns, the interdependence of stocks and any generally recommended international minimum standards..."

seines, but in the case of gillnet and hook-and-line fisheries, to exclude those above the size/age of non-liability to capture. Yet another possibility, perhaps the most relevant to efforts to manage in order to ensure unintended depletion of the resource does not occur, is spawning stock, or mature stock, the segment of the population comprised of sexually mature animals, which is assumed to be proportional to the total fecundity or reproductive rate of the population. On one side stand such alternative definitions, on the other, the methods of estimating biomass or, more usually, an index of it, by acoustic surveys, experimental fishing, from catch and fishing effort data or by other means. These estimates often do not coincide exactly with any of the definitions. I was invited by The Greens/European Free Alliance in the European Parliament to advise on how far above MSY levels it would be appropriate to aim (Holt 2012b, Chapter 46). It was nice to do the calculations on my laptop rather than with the hand-cranked calculator and slide rule Ray Beverton and I used in the 1940s (Anderson 2011). I interpreted that as: "how far below the value of the fishing mortality rate (F) that would generate MSY should the fishing effort be reduced?" I spent most of the next two years (mid-2010–2012) looking at that question, and I did so mainly by examining age-structured population models based on the equations for steady-state yield-per-recruit with constant growth and natural mortality parameters as a function of the ratio of F to the natural mortality in the exploited phase M ($f = F/M$), of the ratio of M to the growth coefficient K in the von Bertalanffy equation describing fish growth in weight ($m = M/K$), and the size-selectivity of fishing, defined by the lower end of the selective range, by fish weight, $w_c = W_c/W_\infty$.[5] After that, I looked at the consequences of recruitment R being dependent on the size and fecundity of the sexually mature phase of the stock. This can most easily be modeled as the apparently one-, but really two-, parameter equation proposed by Ray Beverton and me in 1957, combined with the yield-per-recruit model to provide what we called a **self-regenerating system**; the single parameter I call r_0, the ratio of the average annual number of recruits to the unexploited ("virgin") stock to the theoretical asymptotic number where the fecundity and/or biomass of spawners were infinite.[6] Throughout this exercise, it seemed to me important to simplify and generalize as far as possible since the Commission was seeking to define a general policy. For teleost fish, r_0 is likely to be close to unity, for skates and rays (Chondrichthyes) substantially less than that, and for cetaceans much less.

At this point, it is worth considering why one would wish to deviate from MSY. There are several reasons, but the two most common are: (i) as a precautionary measure, and (ii) to lighten the fisheries footprint on the ecosystem as a whole, including incidental destruction of non-target species. But there seemed to me to be a third and much stronger reason: (iii) to ensure profitability, because without that there would eventually be no fishing, except perhaps with continued subsidies. I recalled the Great Law of Fishing enunciated by Michael Graham in his classic book *The Fish Gate* (Graham 1943) and quoted at the 1955 conference: "**Fisheries that are unlimited become inefficient and unprofitable**."

It seems to be generally assumed that sustainable exploitation will usually be profitable, and that exploitation at MSY level will always be so, but there is no empirical or theoretical basis for that hope. I have come to the contrary belief that MSY exploitation will commonly not be economically profitable, will certainly be economically inefficient and cannot ever be

[5]See Beverton and Holt (1966). In the tables in this paper, selectivity was given as a length c relative to the theoretical asymptotic length. Here, I use, for the index of selectivity, a fish size at first liability to capture as a fraction of the asymptotic weight $w_c = c^3$.

[6]The two constants in the original version of this expression are, in fact, only scalars. Now, I would set both the spawners and the recruits in the unexploited stock at unity.

economically or socially optimal. I put this conclusion as a possible question: **Is aiming for MSY a recipe for eventual bankruptcy?**

Further, a sort of crude economic optimum—the maximum difference between the market value of a sustainable catch and the cost of taking it— is certain to come from a fishing effort substantially less than that required to secure MSY. I am not suggesting that would be a social or economic optimal target for managing fishing but it would surely be nice to know roughly where it is.

With respect to targets and optimality, the experience of the IWC is worth noting here. In developing a process for revising its management procedure, it noted three objectives: (i) obtaining a high continuing catch over a defined finite period, (ii) having negligible risk of inadvertently depleting any stock to a dangerously low level, and (iii) avoiding, if possible, the need for substantial changes in catch limits from one year to the next. These are in conflict, interactive, and cannot all be simultaneously optimized (maximized or minimized as the case may be). A kind of weighted average among them was considered, but abandoned for want of a methodology for objectively determining appropriate weighting factors. The only practicable solution turned out to be to get agreement on priorities among the various objectives. Fisheries managers have something to learn from this (Papastavrou and Cooke 2006).

So, where is MSY? First, it is well understood that MSY, strictly speaking, comes from catching all the residual members of every cohort (year class) as soon as they reach the age and size at which the total weight of the cohort attains a maximum, that is, just before the cohort weight declines as a result of the specific natural mortality rate overtaking the specific growth rate.[7] But that is impossible except perhaps in a few special cases such as migrating salmon. So we opt for exerting a finite fishing mortality and a selectivity less than the critical value needed for MSY. For each chosen selectivity, we generate a curve of yield-per-recruit against f. For low selectivities, these curves have maxima, which I call *local msy* or just *msy* (lower case); at higher selectivities, they are asymptotic. So, for every stock, with each dynamic parameter m value, there is an infinite number of local *msy*s. It is convenient, for purposes of comparison and generalization, to express these as ratios to the MSY for that value of f, and the local selectivity parameter as a ratio of the selectivity for MSY, expressed as always as fish weights. All the sustainable catch values can be expressed as per the product of the recruitment R and the asymptotic body weight W_∞.

The yield-per-recruit curves that have peaks—that is, mostly the ones for w_c equal to or less than 0.3—are rather flat-topped, more so than any of the corresponding curves that commonly arise in surplus production calculations, which usually assume a parabolic or simple power-function shape. This means that a "sacrifice" of catch by taking somewhat less than *msy* will be "rewarded" by being available with a substantially lower fishing effort and hence a higher catch rate (catch-per-unit-effort), hence potentially more profitable. The threshold selectivity value 0.3 is significant because it is approximately the weight of fishes when they become mature, being also the point of inflexion of the von Bertalanffy curve for growth in body weight. And the higher the selectivity value, the higher is the fishing effort required to obtain the local *msy*.

From the same model, we can easily calculate the biomass of the exploited (selected) stock as a fraction of the stock prior to exploitation, an index of the degree (%) of reduction of catch-per-unit-effort (catch rate, hence profitability) by fishing. This leads to an interesting

[7]The relevant expressions are: $MSY/RW_\infty = [m/(3 + m)]^m \times [3/(3 + m)]^3$; critical weight: $w_{MSY} = [3/(3 + m)]^3$; age at critical weight: $a_{crit} = \ln[(3 + m)/m]/K$; basic von Bertalanffy expression for body weight relative to asymptotic weight: $w = (1 - e^{-aK})^3$.

finding. Look at the last column of Table 1 below. It shows the percentage decline in catch rates resulting when the initial exploited stock abundance is reduced by fishing down to *msy* level—depending on the selectivity and varying rather little with the value of $f = F/M$—the sustainable catch rate would be between about 10 and 20% of what it would have been at the initiation of the fishery. This is, of course, a substantially greater depletion than is commonly assumed when surplus production and similar models are used for stock assessments. The depletion expected when seeking maximum economic sustainable yield would be in the region of half of this, that is, to between about 30 and 50%.[8]

In Table 1, by the way, the range of *m*–values is wider than has been found in field studies; the dynamics of a population of fin whales *Balaenoptera physalus* (apart from the consequences of density-dependent reproduction) turn out to be about the same as those of Haddock *Melanogrammus aeglefinus*. I regard a baleen whale as a very obese Herring *Clupea harengus*!

For selectivities of 0.3 and less, the percentage depletion of the exploited stock segment is the same as that of the mature stock segment. This makes it easy to calculate the corresponding self-generating model.[9]

Table 1. Values for *msy*, f_{msy}, and depletion (CPUE) (*f* is the ratio *F/M*).

$m = M/K$	w_c	*msy/RW*	*msy*/MSY%	*fmsy*	CPUE $(msy/f)/(sy/f_0)$%
0.25	0.4	0.37	88	4.3	14
	0.3	**0.34**	**83**	**3.3**	**17**
	0.2	0.33	80	2.8	18
	0.1	0.29	71	2.1	22
0.5	0.4	0.22	94	5.2	11
	0.3	**0.21**	**89**	**3.4**	**15**
	0.2	0.18	82	2.4	19
	0.1	0.17	73	1.7	23
1.0	0.4	0.11	100	Infinite	
	0.3	**0.10**	**97**	**6.3**	**8**
	0.2	0.09	91	3.0	15
	0.1	0.08	80	1.7	21
1.5	**0.3**	**0.06**	**99**	**Infinite**	
	0.2	0.05	97	5.3	15
	0.1	0.04	87	2.1	18
2.0	**0.3**	**0.03**	**95**	**Infinite**	
	0.2	0.03	100	1.9	5
	0.1	0.03	93	2.8	14

[8]In his perceptive little book, Rogers (1995) wrote that the popular, but entirely arbitrary, reference point for management—$F_{0.1}$—lies between F_{msy} and F_{mesy}. But, it is not self-evident that $F_{0.1}$ will always be higher than F_{mesy}.

[9] The expression from the Beverton and Holt stock–recruitment relationship, for the reduction of recruitment arising from the reduction of the virgin spawning stock by fishing is: $d_r = r/r_0 = 1/[r_0 + (1 - r_0)/s]$, where *s* is a number between unity and zero. If *s* is put equal to 0.2 in this expression, *d* becomes what has been called the "steepness" *h* of the stock–recruitment relationship. It would be reasonable to ask what difference to the self-regenerating equations might result from use of a stock–recruitment relationship other than the inverse hyperbola of Beverton and Holt. An obvious alternative might be a simple power function giving a convex (from above) curve with $r = 0$ when $s = 0$, $r = 1$ when $s =$ infinity, and scaled to $r = r_0$ when $s = 1$, e.g., $r = 1 - a^{-s}$ so $a = 1/(1 - r_0)$ and $r = 1 - 1/(1 - r_0)^{-s}$. [This happens to be exactly the form of a relationship between a linear dimension such as total body length of a fish and its age, discovered by Ford (1933) for North Sea Herring *Clupea harengus*, and rediscovered by Walford (1946) and applied to a wide range of fish species; it is the core of von Bertalanffy's general expression for growth.] Application of this and similar alternative stock–recruitment functions can lead to slightly different predictions of recruitment change with increased fishing effort, but virtually no significant difference to the predicted shift of the *msy* towards a lower *f*-value. A more flexible expression with an additional parameter to the Beverton and Holt expression, as proposed by Shepherd (1982), could give significantly different results quantitatively, especially for greatly depleted stock situations, where depensation (the "Allee effect") may come into play. That should be investigated.

It turns out that the location of *msy*s with respect to the fishing effort (*F/M*) varies rather little with density-dependent recruitment; the fishing mortality rate for *msy* is always somewhat less than that for constant average recruitment. The quantity *msy* itself is, however, immensely variable, highly dependent on the precise value of r_0. This unpredictability, together with the common natural fluctuations in recruitment, led me to the conclusion that the setting of total allowable catches (TACs) is a very inefficient, even hazardous, way of regulating fisheries to meet an MSY-related policy objective. The hazard is enhanced by the fact that, in a stock with many coexistent cohorts, a high fishing mortality rate (relative to a low natural mortality rate), as called for by an *msy* objective, the size and productivity of the stock are strongly conditioned by one or a few of the youngest cohorts.

I have not yet mentioned some major issues. One is the transient forms of the yield equations, needed to assess the problems of moving responsibly from one population state to another—steady—one; that would take longer than we have here. The other arises from the fact that, unlike diamonds, sustainable states are not "forever." That is rarely understood by those who talk so much about the "sustainability" of this or that as if it will be eternal or nothing. For scientific analysis, they—steady-states—must be given fixed terms, possibly arbitrarily, but preferably related at least to the lifespan of the species in question, the durability of human systems for management, and the power of computing systems for conducting the many necessary simulations of fishery histories and prospective management systems. The Scientific Committee of the IWC, which possibly has more experience of dealing with this problem than most, found that, in fact, the choice of time-frame can make important differences to conclusions and management procedures. (I use that term here in the sense of, for example, medical procedures which specify exactly how and in which order steps in the execution of an algorithm must be taken.)

Other important matters are, of course, the estimation of essential parameters and stochastic processes.

I would like to close with two more of Michael Graham's aphorisms. The first, from the 1955 Rome conference: "**The trail of fishery science is strewn with opinions of those who, while partly right, were wholly wrong**." The second, in his address to the United Nations Scientific Conference on the Conservation and Utilization of Resources (UNSCCUR), held at Lake Success, New York, in 1949, the first, post-war UN conference devoted to this subject: "**The World does not stand still while scientists put their minds in order**." Graham's *Bons Mots* may be as valid today as they were then.

I thank Michael Earle for his helpful critical reading of earlier drafts of this presentation.
Thank you.

References

Anderson, E. D. 2011. Standing on the shoulders of giants. ICES Insight 48:4–12.

Beverton, R. J. H., and S. J. Holt. 1957. On the dynamics of exploited fish populations. Her Majesty's Stationery Office, UK.

Beverton, R. J. H., and S. J. Holt. 1966. Manual of methods for fish stock assessment. Part II. Tables of yield functions. FAO Fisheries Technical Paper 38 (Revision 1).

Finley, C. 2011. All the fish in the sea: Maximum sustainable yield and the failure of fisheries management. University of Chicago Press.

Ford, E. 1933. An account of the herring investigations conducted at Plymouth during the years from 1924 to 1933. Journal of the Marine Biological Association of the United Kingdom 19(1):305–384.

Graham, M. 1943. The fish gate. Faber and Faber, London.

Hjort, J., G. Jahn, and P. Ottestad. 1933. The optimum catch. Hvalrådets Skrifter 7:92–127.

Holt, S. J. 2006. The notion of sustainability. Pages 43–82 *in* D. M. Lavigne, editor. Gaining ground: In pursuit of ecological sustainability. International Fund for Animal Welfare, Guelph, Canada and the University of Limerick, Limerick, Ireland.

Holt, S. J. 2012a. The some good and mostly bad about maximum sustainable yield as a management target. ICES Document CM 2012/L:02.

Holt, S. J. 2012b. Brief for the Green Group in the European Parliament on reform of the EU Common Fisheries Policy (CFP): how to achieve sustainable and profitable fishing. Unpublished manuscript dated February 2012.

Papastavrou, V., and J. G. Cooke. 2006. The sustainable use of oceanic wildlife: what lessons can be learned from commercial whaling? Pages 113–128 *in* D. M. Lavigne, editor. Gaining ground: In pursuit of ecological sustainability. International Fund for Animal Welfare, Guelph, Canada and the University of Limerick, Limerick, Ireland.

Rogers, R. A. 1995. The oceans are emptying: Fish wars and sustainability. Black Rose Books, Montreal.

Schaefer, M. B. 1954. Some aspects of the dynamics of populations important to the management of commercial marine fisheries. Bulletin of the Inter-American Tropical Tuna Commission 1:25–56.

Shepherd, J. G. 1982. A versatile new stock-recruitment relationship for fisheries and the construction of sustainable yield curves. Journal du Conseil International pour l'Exploration de la Mer 40:67–75.

Verhulst, P. F. 1838. Notice sur la loi que la population suit dans son accroissement. Correspondance Mathématique et Physique 10:113–121.

Walford, L. A. 1946. A new graphic method of describing the growth of animals. Biological Bulletin 90(3):141–147.

Chapter 50

What is a Fish Stock Assessment?[1]

It was a great and unexpected honor to be the first speaker at this conference. I have returned to fisheries assessment and management only four years ago after several decades of absence while doing other things. Perhaps there is some advantage to come from being out of things for a time; one might be able to avoid the problem of not being able to see the wood for the trees.

Some of you might have heard earlier this year some of what I have to say at one of the several gatherings I have been attending, hawking my wares. Sorry about that. Notwithstanding what is said in the abstract for this opening, I'm going to keep to the narrow agenda of this first session, especially as I remain a bit skeptical towards the "ecosystem approach to management" of fishing.

I should say, too, that I have long been interested, as a man of peace, in military thinking. One particular issue is the relation between strategy and tactics. Sun Tzu, in his *Art of War*, an ancient Chinese treatise, wrote: "*Strategy without tactics is the slowest route to victory. Tactics without strategy is the noise before defeat.*"

To prepare for this event I read—for the first time—the Larkin Lecture due to have been delivered in Vancouver in 1995 by my old colleague Ray Beverton. Ray was unable to do so because of illness, and he passed away soon after. Tony Pitcher wrote a version of what Ray would

Sidney and co-convener Mark Dickie-Collas during the conference panel discussion. Photo by Steven Cadrin.

[1]Editors' note: opening keynote address given by Sidney Holt on July 17, 2013 at the ICES-sponsored World Conference on Stock Assessment Methods held in Boston July 17–19, 2013.

have said (Beverton 1998) from notes passed to him by Ray's wife, Kathy, and published it in *Reviews in Fish Biology and Fisheries* in 1998, entitled: "Fish, fact and fantasy: a long view."

Although Ray and I had taken different paths for half a century, I was delighted to find we both still thought about things in our common field in the same way, though—as usual—he expressed his thoughts politely and carefully; I was inclined to be abrasive. I'll be that now at the risk, I guess, of offending some of you in this audience. So be it.

It seems Ray and I agreed on four things:

1. **Surplus production** assessment methods are fundamentally unscientific and misleading. At best, I'd describe them as *curate's eggs*—mostly bad, but possibly, if one is irrationally benign, good in parts.
2. Setting **catch limits** is about the least efficient way to try to manage fishing. Its worst aspect is the chaotic nature of recruitment and the impossibility of predicting it, but there are other flaws.
3. **MSY** is not a wise target for a rational fisheries policy or even as a valid reference point. More about that later.
4. We were not favorably disposed to the idea of **reference points** as guides for management, at least not those proposed in the literature so far (no names, no pack drill).

So, what can I now say constructively in my remaining 25 minutes? I also read the Report of the ICES Workshop 3 on Implementing the ICES F_{msy} Framework held in Copenhagen in January 2012 (ICES 2012a), a document of the ICES Advisory Committee, and also the Report of the Working Group on Methods of Stock Assessments (WGMG) held in Lisbon in October 2012 (ICES 2012b). I found them extremely complicated and difficult to understand. Almost incomprehensible. That made me think of the trouble we in the Revised Management Procedure (RMP) Development Group of the Scientific Committee of the International Whaling Commission (IWC) had taken to make our new approach to management of whaling understandable to other scientists, politicians, lawyers, diplomats, as well as media persons. Our chosen way was to appoint a chairperson/spokesperson who could and would be able to do that—it was the late Australian scientist, Geoffrey Kirkwood.

So, let me say what I think a fish stock assessment isn't. It's not just the output from a model, nor a review of all we know, or think we know, about a fish population. It should be, I think, a pointer to the options for managing a fishery and their likely consequences, and those options should usually include an existing policy and the procedures and actions related to it, intended to implement it; any other proposed policies; and—possibly—alternative, well specified policies.

Now, a word about "procedures." A while ago, my friend Justin Cooke, author of the IWC's Revised Management Procedure, chided me—quite justly—for continuing to refer to what the IWC used to called its New Management Procedure (NMP) as a "procedure," but which is not a "procedure" at all; rather, simply a poorly defined policy aim and suggestions for how to implement that. On the other hand, the RMP is a "procedure," in the sense used in good surgical practice, consisting of two unambiguous algorithms for executing it—one for calculating a catch limit according to the Commission's specifications, the other for applying the first one to several possible adjacent or spatially overlapping stocks.

With Justin's advice in mind, I have spent the past few years trying to see how to clear a path through the dense thicket of thistles, dead bushes, and fallen branches, illuminated by a few flowers, that now constitutes the "ecosystem" of the science for rationally managing fishing for continuity, sustainability, and profitability. I've suffered some scratches, but so far no broken bones. Among other things, I have concluded that progress has, for several decades, been seriously impeded by the descendants of Benny Schaefer's famous 1954 and 1957 papers applying an aberrant version of P-F. Verhulst's logistic equation of 1838 for human population growth to the Yellowfin Tuna *Thunnus albacares* of the Eastern Tropical Pacific.

Despite Benny's good intentions—arising from the fact that neither he nor anyone else was, at the time, able to determine the age of a tuna—this has gestated the unhelpful phenomenon of world-wide *biomass worship*.

I see that at least one of the contributors to this conference has referred to Benny's model as the "Graham/Schaefer model." The existence of that is a myth spread by Tim Smith (1994) via his book *Scaling Fisheries*. Michael Graham must be writhing in his grave. He and I, with colleagues in FAO and the UK, worked hard together, in 1955, to dissociate us all from Benny's—and the "official" US—approach to assessment and international management. That was during the International Technical Conference on the Conservation of the Living Resources of the Sea, hosted by FAO in Rome in 1955.

Let me now turn to more positive things. Take MSY. We have long known that if MSY exists, it is unreachable. It is the weight of the oldest cohort in a population when it reaches its maximum weight, the so-called *critical mass*. An *infinite* fishing effort would be required to take it. We could try to get close to it by starting to fish the entire stock, of which it is part, very selectively and very intensively when that cohort is approaching its critical mass. Alternatively—and more realistically—we could do our best to begin exploitation earlier in the lives of the fishes and try to optimize that locally, aiming at what I call a *local msy*.

A rough measure of our success would be the ratio of *local msy* to the theoretical MSY. That is defined by the interaction of two other ratios—the ratio of the fish weight at which exploitation begins to the theoretical asymptotic weight of fish in that stock, and the ratio of the natural mortality rate in the exploited stock—M—to a parameter defining the rate of growth towards the asymptote, usually von Bertalanffy's K.

A third, basic ratio is the **exploitation rate**, which I find more useful to define as F/M rather than the more usual $E = F/Z$.

A fourth, much more elusive, parameter, yet another ratio, is what has been called the steepness h of the curve relating the number of annual recruits to the numbers of eggs laid or larvae originally produced by their parents.

You will see that I love ratios even more than some of you might like biomass! Here, I assert that the selectivity of fishing (that is, by size and age of fish) is at least as important for assessment and management as is the fishing intensity. One of the intractable problems of surplus production methodology is that it does not recognize that and is unable to cope with it.

But, anyway, who wants MSY or, for that matter, *any local msy*? Scientists don't. Fishers don't. Some governments do, beginning with the United States (the USA promoted it in 1945 as a way of trying to exclude Japanese fishing vessels from the Northeast Pacific—they called it "full utilization"[2]). MSY is mostly there because administrators, politicians, diplomats, and lawyers haven't yet been able to agree on anything better, and it sounds pretty good to any of the wider public and media persons who might be interested.

---

[2]Fully documented by Carmel Finley in her excellent book *All the Fish in the Sea: Maximum Sustainable Yield and the Failure of Fisheries Management* (Finley 2011).

There is a widespread belief that sustainable use of a renewable resource is a good thing—especially for future human generations.

That's usually accompanied by the unstated assumption that it will be profitable—though surely not as profitable, in the short or medium term, as *unsustainable* exploitation.

The negotiators of the UNCLOS found a way out by making the MSY a universal target of fishing, but taking on a condition that one can depart from it for specified social and economic reasons.[3]

Most people in this business seem to assume that taking a *local msy* will be profitable—that is, that the market value of the product will at least be more than the cost of acquiring it. But, that is not necessarily so—and it is quite likely not to be so in many situations. If there is no surplus of income over market value, there will eventually be no fishery, or it will be maintained only with subsidies which meet the value difference and will be paid for by what Italians call *i solti ignoti*, the usual unknowns, i.e., the taxpayers.

The loophole in UNCLOS was actually put there to placate countries newly engaging in commercial fishing who thought they should be allowed, initially, to deplete fish stocks like everyone else had been doing. But, the derogation is worded so that it can apply both ways. Here, I want to introduce my first quotation from my esteemed mentor, Michael Graham. In his classic 1943 book *The Fish Gate*—and, subsequently, at the UN Conference in 1955—Michael Graham expressed what he called his Great Law of Fishing: "Fisheries that are unlimited become inefficient and unprofitable." (I would add two points: that mostly happens when the market is expanding; and that limited—but badly regulated—fisheries can also become unprofitable.)

So, we come to the recent decision by the organs of the European Union that fishing must be regulated in such a way as to take high, but sustainable, catches from stocks with bigger biomass—there's that damned word again—than is needed to provide MSY—whatever that is. This primary decision intends to mean, in plain English: "take a high but sustainable catch with less intense fishing effort." We are, of course, left with an open end to the question "Which *msy*?" And, I must say, the question of the time-frame for "sustainability"—it cannot be, like diamonds, "*forever*," especially in the light of climate change; the *K*-parameter is temperature-sensitive, for a start. The time-scale is basically a function of the lifespan of the fish species, and the typical duration of human management instruments.

In fact, the European Union's decision contains, in two short paragraphs, two entirely contradictory requirements. The poor, ambiguous drafting originates from confusion about biomasses and fishing rates, as well as from some differences of opinion about what would be a suitable, viable policy. But that's another topic.

An international workshop held in Virginia in 1975, supported by many official and non-governmental organizations, came to that same conclusion 40 years ago about the relation between the value of the gross product and the cost of producing it. Lee Talbot—the organizer—and I wrote the consensus report (Holt and Talbot 1978) which said that high sustainable yields were desirable *provided they did not involve the waste of the other natural resources* needed to take them. We all had in mind such things as the fuel used in fishing. (That was our way of evading economic and consequent political matters.)

What matters most to fishers—given the currently favored sustainability criterion—is what we would call the *catch rate*, and we can express a first approximation to that as the

---

[3]MSY "as qualified by relevant environmental and economic factors" UNCLOS 61.3 (EEZ) and 119.1a (High Seas).

ratio of the sustainable yield to the fishing mortality rate. (There, you see how I love ratios. I'll return to that, but an important new one is the ratio of the weight of a female fish when it begins to lay eggs to the asymptotic weight, which is usually in the range 0.25 to 0.3.)

So, my next question was: **how much less effort** is needed than for any chosen *local msy* catch? I looked at this in terms of apparent *sacrifice*—how much sacrifice of potential yield to get a good catch rate. I tried the consequences of 10%. It turned out that for all values of the various ratios, we would expect to get 90% of any *local msy* with not much more than 50% of the fishing effort. Tell me, what owner of a factory or farm would decide to double his or her production cost just to get a mere 10% more of what they had been producing?

An assessment of this matter needs to estimate the *curvature* of graphs of sustainable yield in the region of their peaks. Surplus production approaches give relatively peaked curves, coming from their usual embedded assumptions that the curves are well described by simple power functions, such as those of Schaefer and Fox (1970) and, in fact, of all the family of Pella and Tomlinson (1969), of which those two are commonly used examples. These all seriously underestimate the potential advantages of "sacrifice." **Age-structured assessment models tend, on the other hand, to generate rather flat-topped (technically speaking, *asymmetrical platykurtic*) curves**.

This particular assessment process had two facets that interested me. One is the value of what Ray Beverton and I called **comparative population dynamics**. One of Ray's last publications (Beverton 1992) returned to that theme, especially the relative invariance of the *M/K* ratio. I have to thank Prof. Eric Charnov for reintroducing me to this subject and the many papers relating to it that he and several others have contributed in recent decades.

The other facet that confronted me was this: I began to think not just about generalizations in stock assessment modeling, but more specifically focused comparative dynamics. That happened when I was invited by the European Parliament to make suggestions about assessing stocks for which there are few of the essential data, and managing fisheries that exploit them, deliberately or incidentally. I see that ICES calls these "data limited and data poor." Personally I like "data sparse" or maybe "data impoverished"; some of my co-conspirators seem to like "data free."

I'll close this diatribe with two more aphorisms from Michael Graham. One of them was in his opening address to the 1955 conference I've mentioned before: "**The trail of fishery science is strewn with the opinions of those who, while partly right, were wholly wrong**." The other was on an earlier occasion—his closing address to the first UN Conference on Natural Resources and their Conservation held at Lake Success, New York in 1947: "**The World does not stand still while scientists put their minds in order**."

Thank you for your patience.

References

Beverton, R. J. H. 1992. Patterns of reproductive strategy parameters in some marine teleost fishes. Journal of Fish Biology 41:137–160.

Beverton, R. 1998. Fish, fact and fantasy: a long view. Reviews in Fish Biology and Fisheries 8:229–249.

Finley, C. 2011. All the fish in the sea: maximum sustainable yield and the failure of fisheries management. University of Chicago Press.

Fox, W. W. 1970. An exponential yield model for optimizing exploited fish populations. Transactions of the American Fisheries Society 99:80–88.

Graham, M. 1943. The fish gate. Faber and Faber, London.
Holt, S. J., and L. M. Talbot. 1978. New principles for the conservation of wild living resources. Wildlife Monographs 59:3–33.
ICES. 2012a. Report of the Workshop 3 on Implementing the ICES Fmsy Framework, 9–13 January 2012, ICES Headquarters. ICES Document CM 2012/ACOM:39.
ICES. 2012b. Working Group on Methods of Fish Stock Assessments (WGMG), 8–12 October 2012, Lisbon, Portugal. ICES Document CM 2012/SSGSUE:09.
Pella, J. J., and P. K. Tomlinson. 1969. A generalized stock production model. Bulletin of the Inter-American Tropical Tuna Commission 14:421–496.
Schaefer, M. B. 1954. Some aspects of the dynamics of populations, important for the management of the commercial fisheries. Bulletin of the Inter-American Tropical Tuna Commission l(2):25–56.
Schaefer, M. B. 1957. Some considerations of population dynamics and economics in relation to the management of commercial marine fisheries. Journal of the Fisheries Research Board of Canada 14:669–681.
Smith, T. D. 1994. Scaling fisheries. The science of measuring the effects of fishing, 1855–1955. Cambridge University Press, Cambridge.
Verhulst, P. F. 1838. Notice sur la loi que la population suit dans son accroissement. Correspondance Mathématique et Physique 10:113–121.

Chapter 51

A Dummy's Guide to Managing Fishing[1]

1. Fish Population Dynamics 101[2]

A fish is a living system. It is born small, increases in size ("grows"), and eventually and inevitably dies ("mortality"). In between, it might reproduce itself, and usually will, if it lives long enough and grows to sufficient size.

A group of fishes will do the same. However, it must reproduce or the population it comprises will no longer exist—it will become ***extinct***. If that extinction is caused by human actions, we say it has been ***exterminated***.

A fish is an open system. It takes in nutrients (**food**) and **assimilates** them to its body ("anabolism") and so grows, but uses part to provide energy to maintain its life ("catabolism"). Together, the processes of anabolism and catabolism make its ***metabolism*** or metabolic state.

Anabolism is primarily a surface process and, as such, its rate is proportional to the square of the fish's size, measured as length. Catabolism is a mass process, so its rate is roughly proportional to the cube of its length. The interaction between the two leads to the fish growing in length along a curve that decreases in steepness as the fish gets older—and bigger—approaching a final maximum size—an asymptote—which it would reach if it lived forever.

Growth in fishes, and in other floating creatures such as whales and dolphins, is in principle **indeterminate**. This is probably because, unlike elephants, dinosaurs, and birds, they don't have to struggle against the force of gravity. Their specific gravity (weight per unit volume) is close to that of the medium in which they live, fresh- or seawater. They have close to neutral buoyancy. They displace an amount of water equal to their own (Archimedes' Principle). When he realized this, Archimedes jumped out of his bathtub, shouting "Eureka," and slipped on the bathroom floor, because it was wet from some of the displaced water slopping over the rim of his bath. Fishes and whales control their buoyancy, and hence their displacement, in various ways: air-filled swimbladders, oil-filled body parts, holding their breath, swimming with fins and/or tails. The control of buoyancy determines their life style. Baby flatfishes float, then turn on their sides, and go to the seabed. Tuna progress by gliding, with slight negative buoyancy, then swimming back up to the sea surface, repeatedly.

---

[1]Editors' note: Sidney Holt's original intent was to write a series of articles for a "Dummy's Guide to Managing Fishing," but he never completed it. His original motivation was Earle's "silly questions" which he found valuable in helping him to clarify his thinking and writing. This chapter, however, has been given the above title and consists of a series of short notes written by Holt to Earle over several years that the editors feel fall into this category. They were conceived independently, hence the lack of flow between the sections.

[2]Editors' note: written November 30, 2011 and intended, by Holt, as the opening for the "Dummy's Guide."

We are much more interested in the weight of a fish than its length. The curve of its weight against its age, being related to the cube, is S-shaped ("sigmoid"). At first, it gets steeper, then gradually bends over and gets ever less steep, so the weight approaches an asymptote that we symbolize as W_∞. Where the curve's bend changes from steeper to less steep, we call its ***inflexion***. In the curve we shall be using, due to Ludwig von Bertalanffy, the weight of the fish varies as the cube of its length, and the inflexion occurs when the fish has reached 30% of its asymptotic weight. In general, the fish reaches sexual maturity when it is close to its inflexion weight. It is convenient to express this as a fraction of the asymptotic weight, and write the symbol in miniscules: $w_{mat} = W_{mat}/W_\infty$. We shall henceforth nearly always designate all weights in this way: w.

The age of the fish at maturity depends on its actual growth rate, more specifically on the value of a coefficient of growth we symbolize as K, which is a measure of the rate at which it grows at first, speeds up, and then slows. But, as we go on, we shall usually find it more convenient to refer to sizes than to ages, even though we need to know something about it in the first place in order to determine the values of parameters such as K and mortality rates.

Beware: *Equations ahead!* Readers who really object to algebra can skip the next paragraphs. An equation is two things—numbers, expressions—joined by an equals sign (=). Where you might occasionally see something that looks like an equals sign, but has three horizontal lines instead of two, you are looking at an identity (≡).

The curve of fish growth in length is defined by the equation:

$$L = L_\infty(1 - 2.718\text{^}K_t).$$

Don't worry about the 2.718; that's a magic number—*e*—that is always appearing in the branch of mathematics called calculus. The symbol ^ stands for "to the power of"; we could also have written it as 2.718^{K_t}, but if you have ever used a spreadsheet such as Microsoft's EXCEL or Apple's NUMBERS, you'll be familiar with my symbol. K is the growth coefficient and t is the age of the fish.

The inflected curve of growth in weight is simply the cube of the above one for length:

$$W = W_\infty(1 - 2.718\text{^}K_t)\text{^}3$$

and we'll write $w = W/W_\infty$.

Now, let's turn to mortality rates. Another state, ***morbidity***—rate of sickness or disability—is also important in considering humans and domesticated livestock and aquaculture, but in wild animals, especially fish, it is rarely considered (sick and disabled wild fish usually die quickly, either from the sickness or enhanced vulnerability to predators). Mortality is commonly expressed as a percentage dying over a certain time, usually a year. This is not convenient for us because we shall need to be able to add different causes of mortality together, and you can't do that with percentages. For example, 10% plus another 5% might be something like 15%. But, what do you do with 60% + 50%? To get around this, and for other reasons, we resort to logarithms and speak of instantaneous coefficients of mortality. In particular, if M is the coefficient of mortality from natural causes and F is that caused by fishing, then the total mortality rate is $Z = F + M$.

The relationship between F, M, and Z can be expressed in various ways. One of them is that $E = F/(F + M)$, where E is called the ***exploitation rate*** and runs from 0 (when there is no fishing at all) to 1 (when fishing mortality overwhelms the natural mortality). Here, we shall use, instead, the ratio F/M for a good practical reason—we want our measure of the fishing mortality rate to bear a simple relationship with the amount of fishing that caused it. That amount we shall call ***deployed fishing power*** or ***exerted fishing effort***, having to remember that they must be ***calibrated*** according to the kinds of vessels and gears being used and calculated according to the geographical and seasonal distribution of their deployment in relation to the presence of fish. We can write the identity $F/M \equiv E/(1 - E)$.

We shall also have to be careful about possible changes in M, but safely use the F/M ratio if M is more or less constant, as we usually assume it to be.

The curve of changing fish size against time, age, or size whose scale and shape are determined by the values of W_∞ (y-scale) and K (x-scale and shape) is a property of an individual, but we can apply it to a group of fish if we assume all fish in the group are more or less the same, or use it to represent average conditions of growth. The curve of a batch fish number against time or age is a collective, a statistic. It can also be drawn as a curve of number against fish size. It is, from above, concave and in form ***exponential***.

Beware, another equation. The number N in a batch at any age or time is related to the starting number N_0 as: $N = N_0 \times 2.718\text{^}{-t(F + M)}$. If there is no fishing, we just put $F = 0$ in this equation.

Now, let's consider the history of a batch of fish all born (eggs laid and hatched at the same time, or at least in the same spawning season): we call such a catch a ***cohort***, and it commonly constitutes an ***age class*** of fish in a population. At first, there are very many small ones and much later fewer larger ones. Multiplying the growth and mortality equations together gives us the total weight of the cohort as it passes through life. Because of the exponential shape of the numbers curve and the sigmoid shape of the growth curve, the curve of total weight has a peak, a maximum, at an intermediate time or age. That peak will occur when all the fish have attained a certain ***optimum size***.

If we want to make the biggest possible catch from that cohort, we wait until the fish have reached the optimum age and then catch them all. That catch would be the *maximum sustainable yield* (MSY) if taken from a population consisting of numerous similar cohorts, spaced from year to year. The series of N_0s we would, in fisheries lingo, probably call the recruitment, since we would start our calculation not at birth, but at a time when they have become "fish" in which we might be interested and could be available to us if we wished.

2. Recruitment[3]

This is the term applied to the number of fish that arrive each year into the harvestable stock of a population. Historically, a common feature of sea fishing has been the great annual variability of total catches, and catch rates—that is, the catch per unit fishing effort (CPUE) exerted. It was realized, towards the end of the 19th century and early in the 20th, when methods were found possible to determine the ages of caught fish—mainly by counting rings in hard parts such as scales and ear bones (*otoliths*)—that the greater part of the variation was often due to variations in the numbers of recruits to each new cohort. In relatively long-lived species, the differences in sizes of age classes could sometimes persist for a few years.

[3]Editors' note: a note written to Michael Earle on November 4, 2014, intended for the "Dummy's Guide."

Generally, researchers originally presumed that large numbers in cohorts came from large numbers of parents, and vice versa. Several important species stocks were observed for many years, particularly during periods when stocks were being reduced by persistent intensifying fishing, but any statistically significant relationships could not be found. An exception was the North American Pacific salmon *Oncorhynchus* spp. that were caught in very large numbers as they ascended their home rivers to spawn. Eventually, however, as the young of some species came to be exploited for industrial production of oil and fishmeal—for a practically unlimited market—it was found that the abundance of mature fish could be drastically reduced by fishing so that the numbers of recruits was sharply reduced, sometimes resulting in the collapse of the stock and of the fishery based on it. A classic example was the collapse, in the 1950s, of the Anchoveta (*Engraulis ringens*) stock that was being exploited along the coasts of Peru and northern Chile and had supported, until then, the world's largest fishery.

Several scientists proposed, in the 1950s–1970s, mathematical expressions that they thought might describe possible relationships between spawning stock size and recruit numbers (e.g., Ricker 1954; Beverton and Holt 1957; Cushing 1971; Shepherd 1982). All of these would predict the virtual absence of a relationship when spawners were abundant, but sharp declines as the spawners were reduced—for whatever reason—the decline being increasingly steep as the numbers of mature fish were reduced, eventually becoming catastrophic as zero recruits and spawners were approached. The expression devised by William Ricker had a difference, however. It predicted that recruitment could also decline after reaching a peak when the spawning stock further increased; there appeared to be some evidence for this phenomenon in the Pacific salmon. The most commonly applied expression is that by Beverton and Holt (1957); it is the simplest, and I shall use it here. It is derived from the idea that throughout the prerecruit phase of life, the mortality rate of the young fish is high and is density-dependent, that is, it is higher the more individuals remain alive in the cohort.

Every mature female bony fish lays a huge number of eggs every spawning period—typically annually. To a first approximation, the number of eggs is proportional to the weight of the animal. So, in an algebraic expression, it may be appropriate to say the fecundity is directly proportional to the total weight (*biomass*) of the mature female part of the stock. There is, however, some evidence that older females can produce more eggs per unit body weight and, further, that their eggs might be more viable than those of younger animals. For practical purposes, it is satisfactory to treat fecundity as proportional to biomass and then make adjustments to that in case of evidence of departure from the norm.

It has been hypothesized that the extremely high fecundity of most bony fishes is an evolutionary response to danger to the very young—their vulnerability to many predators, perhaps competition for food by their siblings, and the dangers of being swept away by currents and rides. In recent years, however, what I think is a more plausible hypothesis has been suggested by several biologists (e.g., Longhurst 2010). Most of the bony fishes are cannibals. The marine species are practically all carnivores, except for a few of the smallest herring-like species—anchovies and sprats *Sprattus* spp., for example—the pelagic larvae of many species that, when the nutrients in the egg yolks have been exhausted, begin to feed on free-living marine bacteria, phytoplankton, and the smaller types of zooplankton, including their own siblings. Their parents then begin to recoup their investment in reproduction by preying on their young. In evolutionary terms, this thus provides a source of food both for juveniles and at least some adults in an environment in which all the photosynthesizing plants are tiny and mobile, as are most of the animals that feed on those plants;

this is in sharp contrast to the terrestrial environment with many large and sessile plants and animal life associated with them.

The fact of extensive cannibalism means that the mortality rate of the young fish is likely to be related more to the number of parents rather than to their own abundance, which could account for the apparent absence of a simple direct relationship between reproduction and recruitment.

The Beverton and Holt expression relating recruitment to reproduction mentioned above is $R = a \times S/(1 + b \times S)$, where R is the number of recruits, S is an appropriate measure of the size of the parent stock, a and b are constants (really only scaling factors), and the ratio a/b is the asymptotic recruit number; let's call that R_{asymp}. It is convenient for us to refer to recruitment in relative terms, that is, the ratio of R to R_{asymp}; call that r. That is equivalent to rescaling R, so that R_{asymp} is 1, meaning $a = b$.

A stock that has not been harvested—in fisheries science jargon called "virgin"—would, ideally, settle into what engineers call a *steady-state* varying around a constant average size and age composition mainly by virtue of varying annual recruitment. The average recruitment would be determined by average reproduction and the average prerecruit rate of natural death. This "virgin" recruitment must, by our model, be less than R_{asymp} since that would require infinite reproduction, but it can, in principle, be quite close to that. Let us call the ratio of the number of recruits to the "virgin stock" to the asymptote as V.

Harvesting will necessarily reduce the overall fecundity of the stock by reducing the numbers of mature females, but by how much will depend on the intensity of fishing and its selectivity. If all mature fish as well as immature fish in the harvestable stock are liable to capture, then the simplest stock models can tell us by how much the biomass of the mature stock is reduced by fishing, in particular when the fishing is sustainable, that is, generating and maintaining a stock in a steady-state, but different from the virgin, assuming the number of recruits is not diminished by the reduction in biomass. Let us call the virgin biomass S_v, the new harvested biomass S, and the ratio of the two s. Similarly, we can define the ratio of the reduced recruitment V to that of the virgin stock v, as h.

The Beverton–Holt stock–recruitment expression can be rearranged to give us the new reduced recruitment—as a fraction of the virgin recruitment—as follows: $h = s/s_v + (1 - v)$. That is sufficient to allow us to convert the yield-per-recruit calculations to yields with new adjusted recruitments in what Beverton and I called a self-regenerating model. Note that this expression is independent of the parameter values of the basic stock–recruitment expression.

As an example of the properties of this expression, consider the case where the virgin recruitment is quite close to the theoretical maximum, say 95%. If the spawning stock biomass is halved, the recruitment is still about 95% of virgin. If, however, the spawning stock is reduced to 10% of virgin, the recruitment is still as high as 65% of the virgin level. With a further reduction of stock to 5% or less, the recruitment collapses rapidly. But, suppose the virgin stock is quite high with respect to the theoretical maximum, and the spawning stock is then reduced to one-half its original biomass, the recruitment is reduced to 80% of virgin, but thereafter the decline is less steep.

If data are available indicating the rough general level of recruitment decline as fishing progresses, then the value of v can be inferred sufficiently well to allow self-regenerating calculations.

This is illustrated in the following graph (Figure 1). The purple line is the yield per recruit (y) as a function of fishing effort expressed as the ratio of fishing mortality rate F to natural mortality rate M (x in the equation). In this example, the selectivity, $c = 0.5$, which means fish

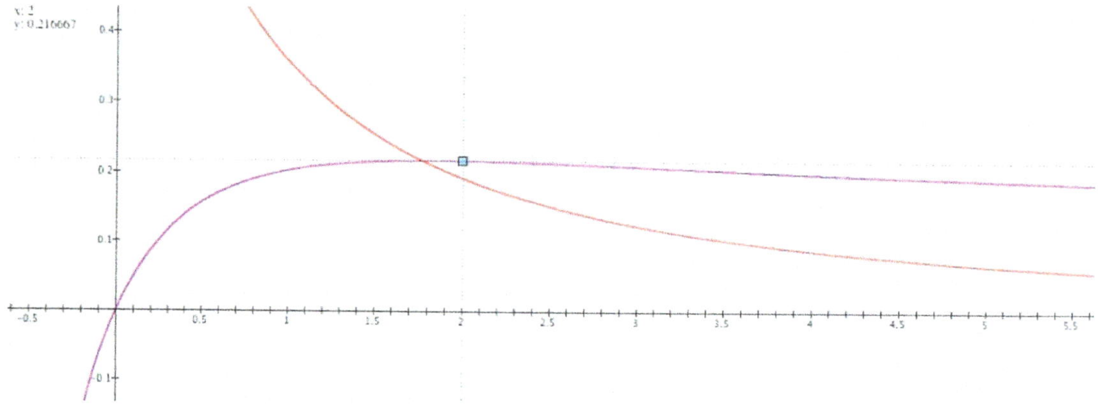

Figure 1. Purple graph: $y = \frac{x}{1+x}\{1 - [\frac{3\,(1-c)}{1+k\,\frac{1}{1+x}}] + [\frac{3(1-c^2)}{1+2k\,\frac{1}{1+x}}] - [\frac{(1-c^3)}{1+3k\,\frac{1}{1+x}}]\}$

Red graph: $$y = \frac{1}{1+x}\{\frac{1 - [\frac{3\,(1-c)}{1+k\,\frac{1}{1+x}}] + [\frac{3(1-c)^2}{1+2k\,\frac{1}{1+x}}] - [\frac{(1-c)^3}{1+3k\,\frac{1}{1+x}}]}{1 - [\frac{3\,(1-c)}{(1+k)}] + [\frac{3(1-c)^2}{(1+2k)}] - [\frac{(1-c)^3}{(1+3k)}]}\}$$

first become liable to capture when they reach 50% of their final (asymptotic) length, which is when they reach *w* at 12.5% of their asymptotic weight. Sexual maturity comes when *w* is 30% or a bit more, so this is a fishery that takes mostly mature animals, but also quite a few immature ones (typical North Sea trawling).

The basic parameter *k*, the ratio of the natural mortality rate *M* to the instantaneous body growth parameter *K* (which defines the speed at which the growing fish approaches its asymptotic weight) is here fixed at 1.5. This value is at about the mid-range of values for bony fishes, which are mostly between 1 and 2. In this example, there is a local *msy* at about *F/M* = 2. The curve is rather flat on both sides of that maximum. You will see that if, for example, you reduce the fishing effort to about one-half of that, which would provide local *msy*, you still get a sustainable catch of more than 90% of the *msy*.

The other graph (red) shows the catch rate (which is proportional to the biomass of the mature part of the stock) as a function of *F/M*. Here, the catch rate is shown not in absolute terms, but as a fraction of the catch rate from the earliest years of exploiting a virgin stock. You see that if you fish with the intensity needed to secure *msy*, you will reduce the exploited stock to about 20% of its virgin biomass. If you decide to take only 90% of *msy*, the stock is reduced only to a bit less than one-half of its virgin abundance.

The graph shows the catch per recruit declining only slowly if the *msy-F/M* value is exceeded, but the catch rate—catch per unit effort—declines rapidly. The decline is much faster in both catch and CPUE if the recruit numbers are reduced as a result of reduction in the size of the mature part of the stock.

With mature stock abundance reduced to such a degree as shown here, it may be likely that the annual recruitment is reduced; this is especially so if fishing is aimed at *msy*. The effects of this are twofold: to reduce the *msy* catch itself somewhat (in proportion to the reduction in recruitment), and to move the *msy* point slightly to the left—that is, the somewhat smaller *msy* becomes obtainable with slightly less fishing effort—but the CPUE changes little.

The scale of these effects can be calculated given some knowledge of the general relationship between the number of annual recruits and the abundance of the parent population.

3. Surplus Production[4]

Introduction

A way—with the above name—of assessing exploited fish stocks and managing their exploitation has distorted and impeded real fisheries science for more than half a century. The method has no scientific credibility and, in my opinion, is based on two absurd premises. Some further absurdities have recently appeared, but we'll come to those later.

The first is that a fish population can be regarded as an undifferentiated, amorphous lump of "protoplasm." From such a lump, a piece can be chopped off, by we humans, and become "catch," after which the remainder of the lump will continue to grow in weight at the same rate as it had been growing. Hence, a year later, we could chop off another lump, possibly the same size as the first chop, and so on, continuing to obtain a "sustainable" catch every year. And, if we do it right, we can "optimize" those catches. In the real world, the residual protoplasmic lump will most likely continue to increase in weight, but certainly at a rate in the coming year different from its growth rate over the preceding year.

The second absurdity is that this marvelous regrowth is just a simple function—or perhaps a not so simple one—of the weight of the lump. To make this look "scientific," we Latinize and Greekify and call the weight "biomass" and the catch "yield." The biomass is not really a property of the actual population, but merely the number of fish in it multiplied by their average weight. And, that "average weight" is not the mean weight of a group of similar animals of sizes varying about some central value, as with, say, a "normal bell-shaped curve," but rather the arithmetical mean of a hugely disparate array, including babies, juveniles, immature grown-ups, sexually mature adults, and the aged.

The term itself—surplus production—is ideologically tainted. It came from the reasonable notion of Karl Marx and Frederick Engels in the 19th century that "surplus value" is the value of product made by a worker in excess of what he needs for himself and his family, while the rest—the surplus—accrues to the owner of the tools and the resources that went into the production. So, the fish are kindly providing, for our benefit, a surplus that they do not need.

Unfortunately, many people associated with sea fisheries—fishers themselves, administrators, lawyers, economists—have come to believe not only that surplus production is a scientifically sound idea, but also that certain numbers associated with it from the years of its invention (late 1940s to early 1950s)—something labeled as *optimum* or "maximum" sustainable yield being obtained when biomass is at 50%, or alternatively 40%, of its size before fishing began ("virgin" size, as is said)—are valid. Neither such figures nor the procedure itself have an empirical or theoretical basis. It is, I think, shameful that respected international organizations, such as FAO of the United Nations and The World Bank could publish—as they did a few years ago—a booklet in which the 50% assumption is made by

---

[4]Editors' note: written on November 24, 2014 and intended, by Holt, as an appendix for the "Dummy's Guide."

their consultants—biologists and economists—in their estimation of the possible futures of global fisheries, both by quantity and monetary value (World Bank and FAO 2009) [*Editors' note: Holt published his rebuttal in Holt (2009)*]. Possibly what could be agreed is that somewhere in the range of possible biomass, from the "virgin" state to extermination, there is an "optimum" state—from our point of view, that is—but even that is misleadingly simplistic. We might agree that the best—a high, profitable, sustainable fishery—lies somewhere not too big, not too small, but just right.

A fish population is *not* an amorphous lump, but rather a structured entity and, furthermore, one that changes its structure as it grows and in response to being exploited by us. So, the potential sustainable catch depends on that changing structure as much as it does on the size of the population, however it is measured, whether in English (or French, Spanish, Russian, or Chinese) or in Greco–Latin and defined alternatively as "biological material derived from living or recently living organisms—animals, plants, fungi, etc." or "as the mass of living organisms in a given area or ecosystem, measured in various ways (depending on the reason for measuring it): live wet weight, dried organic mass, or mass of organically bound carbon." In the fisheries context, it is usually expressed as total wet weight—alive or dead —and may be applied to just a part of the fish population: mature animals (spawning), immatures, females, juveniles, recruited stock (see below), and so on.

To understand the absurdity of the fundamental "surplus production" assumptions (also, subsequently, how biomass is measured in the fisheries context), it is useful to consider the life of a single *cohort* of fish in a population, i.e. all those born—spawned—in a particular year during a single spawning season. They die naturally at a more or less steady rate after passing through the early stages when most of them get eaten by predators (including their siblings and parents). When they have passed through those dangerous early stages, they often change where they live, some metamorphose—change completely their look, shape, habitat, and way of life—and they may become of interest to us, and we then call them *recruits*. The recruits increase in size. Fish, being cold-blooded animals with *indeterminate* growth, increase in weight all their life, at first very rapidly, then later more and more slowly. The weight of the cohort will at first grow—when body growth of individual fish is faster than the natural death rate—then eventually decline when body growth is slow, but natural death continues. So, somewhere along the time–age line, the cohort must reach and pass a maximum weight. We could get a maximum catch from that cohort if we could catch all the remaining fish just as they each that critical age and size. But, of course, we don't have the means to do that.

If we are to treat our chosen cohort in the surplus production manner and take a sustainable catch from it, we need to look not so much at its biomass, but at its *rate of incremental growth* as it get older. This has a different pattern from the growth of the biomass. Because of the S-shapes (Greek: *sigmoid*) of the curve of body growth and the reversed J-shape of the curve of death (exponential), the growth rate of the cohort (weight increment per year) is at first fast, then slows, and then goes into reverse. That is, at a certain age, the weight of the cohort begins to decline; eventually its growth rate goes to zero, then becomes negative, as it passes the critical age and body size. If our cohort really was an amorphous lump, we would get the best catch by chopping off a piece equal to the growth rate and repeating that every year, thereby getting a sustainable catch. Clearly, we could do that to all the cohorts in the population as they reached the size of maximum growth rate.

To give some idea how this works, we may graphically relate the biomass and the incremental growth rate of a cohort. I have selected values of two defining parameters—K, which defines the speed at which a fish approaches its final maximum size ($K = 0.2$) and natural mortality coefficient $M = 0.3$. This gives $M/K = 1.5$, which is the value most commonly found in data for a wide variety of fish species and stocks. With these values, the critical body size is about 30% by weight of the maximum size of the fish, and the critical age is at about 8½ years (which is also when the incremental cohort growth becomes negative). The fastest incremental cohort growth is when fish are about 3 years old—so that's the age to chop them if you want the highest sustainable catch. Getting the best from all the cohorts simultaneously is rather more complicated, but, in principle, the same. It is further complicated in that the fish become mature when they are about 30–40% of their final size and begin to reproduce. So, the population growth is enhanced by the arrival of new recruits a year or two later—depending on how long the eggs and larvae take to reach recruitment size and state.

The potential productivity of the entire recruited population to provide sustainable catch is the sum of the growth rates of all the cohorts in it. At any given time, they are of different ages, and the total depends on the relative participation of all of them in the whole population. So, obviously, the potential surplus production depends on the composition of the population much more than on its total size, however measured. So, equally obviously—I hope—the maximum potential (MSY) cannot be defined in terms of how big the population is in relation to its "virgin" size.

Now, we must look, in light of those considerations, at how proponents of "surplus production" try to determine the parameters of their model. The starting point is always a time-series of catches by the fishing industry and a corresponding series of estimates of the relative sizes of the exploited population, if not of its absolute size. The basic index is the catch per unit fishing effort (CPUE) at a time or during a year, calibrated in some way. But, those catches are not "sustainable" if the stock is not the same size at the end of each year as it was at the beginning of the year. So, the scientist doing this adds to each catch, or subtracts from it, an amount determined by the change in the size of the stock from the beginning of the year to the end of the year. That is more like a sustainable catch that can be plotted in a graph against the indices of total stock weight. Scientists expect that plot to be a smooth curve going from 0–0 to 0–1, a simple polynomial being preferred—that will have a hump at an intermediate stock size—usually assumed to be one-half the virgin size, but from many studies, perhaps closer to 40% of it. So, there is a target for managing the fishing.

However, their composition, by age groups, of the catch or stock from which the sustainable catches were calculated is not that of the "virgin" because it will have been changed by non-sustainable fishing. The "correction" of the actual catch to a theoretically sustainable number is, at best, a rough approximation. In fact, the bias can be devastating. Consider a common case where this assessment is being made, after a period of expanding or at least unregulated fishing, in the expectation that it will give the rationale for regulation of future catches or effort to correct overfishing. The stock in such a case will have an exceptionally large number of faster-growing young, smaller fish which, as we have seen, have a relatively high growth rate. They will, therefore, contribute out of proportion to an estimated high stock growth rate, out of proportion, that is, to what would be expected from a stock that has been fished at a more or less constant level and would have what we call a steady-state age composition. Whatever else happens, we would, in these circumstances, be overestimating the possible sustainable catch, which is not at all what we want to do in such circumstances.

I have, of course, here greatly simplified the assessment process. There might be other sorts of information about indices of stock biomass, for example, from fishing research surveys or echosounding. A. C. Simpson (1959) years ago estimated the size of the Plaice *Pleuronectes platessa* population in the North Sea by sampling, with plankton nets, their pelagic eggs and converting those numbers to total weight of the mature females in the population. Other clever methods have sometimes been used. But, the problem remains how to convert those biomasses to estimates of potential productivity. Naturally, that can be done if we have the means of determining how old the fish are, but that information is precisely what is lacking for those inventing and using surplus yield logic. Milner Schaefer (1954) invented the idea when he was in California studying Yellowfin Tuna *Thunnus albacares* precisely because he had no way to tell their ages.

Another universal and serious problem with using the surplus yield approach is that what is a sustainable catch depends at least as much on the selectivity of the fishery—that is, the size and age at which fish first become liable to capture and retention by the fishing gear used and the way it is used—depth of water, season of fishing, location of fishing, type of seabed, depth of fishing, and so on, as on the *intensity* of fishing which determines what proportion of each cohort is to be caught. The surplus production approach to stock assessment and management simply cannot deal with this matter. Biological productivity of other species and ecosystems that humans can tap comes from the interactions of reproduction, growth, and death. A "biomass" does not grow, reproduce, or die. *Fish* are born, grow, and die, and each of them grows, perhaps reproduces, and certainly dies—with probabilities mostly in accordance with its age.

The New Absurdities

With the expansion of commercial processing of fishes for the production of fish oil and meal, mostly from young fishes and small species, meeting a practically unlimited market, concern has grown about the phenomenon of *recruitment overfishing*, brought about by great reductions in the number of spawning fishes in many stocks by the excessive capture of first-time spawners and, worse, the depletion of the numbers of younger recruited animals that, if allowed to live, would become mature and contribute to the maintenance of the stock and its productivity. In an attempt to define this and find ways of avoiding this situation through appropriate management action, some scientists have invented a quantity called B_{lim} (ICES 1997), a biomass—or a spawning-stock biomass—*below which it is likely that future recruitment will be impaired.* The meaning of "impaired" is generally not defined. The term can apparently also be applied to a biomass below which insufficient information is known about the fish to make reliable calculations. But, this usage is even more vague than the basic one.

There is no evidence, for any fish species or stock, that there is a clear break or kink in the distribution of reproductive potential and biomass, either of the stock as a whole or of the spawning part of it. Some kind of "limit" as a so-called *reference point* (a fashionable term in the fisheries management sphere) is, nevertheless, thought by some to be useful and has been incorporated in, for example, the management processes of the Commission of the European Union. For this purpose, there has to be some way of calculating and estimating B_{lim}, and the following way of measuring this non-existent feature has been used in the work of the International Council for the Exploration of the Sea (ICES).

To understand this, one must know something about scientists' experience in trying to find a secure, reliable, quantitative relationship between the number of eggs and larvae produced in a season and the very much smaller number of juvenile fish that is subsequently recruited. Several algebraic expressions for doing this have been proposed over the past 50 or so years, along with attempts to fit data to them satisfactorily. The trouble is, partly, the great variability in annual recruitment for most species and stocks. All those models indicate a smooth curve relating egg or larvae numbers to recruitment that obviously begins at the origin (no eggs, no recruits) and then proceeds to a maximum (asymptote). In that later phase, the recruitment bears little, if any, relation to the initial number of eggs. An exception is an expression derived by the Canadian scientist William Ricker (1954) in which, when the initial number is very high, the eventual recruitment actually begins to decrease. Ricker found indications of that happening in the Pacific salmon he was studying—that spawn up the river of their birth—but the phenomenon is surely exceptional.

Because recruitment has no identified measurable role in surplus production and biomass theory, scientists using it had to find another way of dealing with this changing recruitment problem. The way found was to suppose that when stock and recruitment levels are high, the recruitment is either unchanging (apart from the usual wild fluctuations) or, if it was changing, this was along a straight line with a slight upward or downward slope. However, if and when the recruitment and spawning stock were small, recruitment would simply be proportional to the biomass of the spawning stock, falling on a steep straight line passing through the origin. So, the overall graph of recruits against stock would be like a hockey stick, which is what its proponents call it (e.g., Barrowman and Myers 2000; Mesnil and Rochet 2010). The place where the change from one straight line to the other was identified as the point where further decline in biomass would "impair" future recruitment is labeled B_{lim}.

The next problem was how to fit the best hockey-stick graph to data and so have a statistically reproducible way of getting a figure for B_{lim}. Such a way was developed in the early 2000s by a number of mathematicians working with ICES, who called the illusory kink, bend, or fold in the hockey stick the "break point," and generally denoted that as B_{lim}. One feature of this method of treating stock and recruitment data is that when, for a stock, there is little or no decline in recruitment with declining stock size, the method necessarily produces a break point just below the data for the smallest observed stock size. But, clearly it could—if it were to exist—be at a smaller stock size, but the stock has not yet been reduced so much by fishing that there are any data to fit.

The absurdity of this rather desperate approach to stock assessment for management purposes is that a quantity called F_{lim} has been defined as the fishing mortality rate that will hold the stock at the B_{lim} level or allow it to attain that level or return to it one day.

Administrators and politicians struggling to bring fish stocks to high, and hopefully, profitable sustainable levels have been led to believe that B_{lim} and F_{lim} are real properties of fish populations and fisheries.

The fear that has led to the B_{lim} nonsense is that a managed fishery might collapse, so a concern is understandable not to allow fishing to continue to be so intense as to lead to a dramatic decline in recruitment. In the ensuing hand-wringing, it is forgotten that whether that might happen depends as much on the selectivity of the fishery—is it focused on catching young to convert to meal and oil, or on catching adults as food for humans, for example—as on the intensity of the fishing effort.

Everyone thinks they know what "collapse" means, but it has not been defined in the fisheries literature (Gibbon, 1776–1788, had a similar problem when writing about the "fall" of the Roman Empire). In some places, it has been defined as the condition in which the catch, or alternatively the stock, has been reduced to 10% of its previously bigger figure. The most common such event is the reduction in the scale of the fishery—and usually also the geographical range of its operations. It may be said "there are no longer any fish," but the reality is that there are not enough of them to be worthwhile for fishers to lose their lives and resources to continue to go out and catch them. The worst consequences of continued intense fishing are usually the loss or extermination of species that are caught incidentally. A well-known example is the virtual extermination of the Common Skate *Dipturus batis* in the North Sea as a result of trawling there for Plaice, Sole *Solea solea*, and Haddock *Melanogrammus aeglefinus*. But, the danger of doing this in a regulated fishery is probably real and could even arise from the fashionable effort to obtain MSY. That is because that effort calls, by definition, for considerable reduction in the size of the stock. It is generally assumed that fishing for MSY would be profitable, but that is just a hope—maybe yes, maybe no.

Avoiding a B_{lim} situation is rather like setting a speed limit for motor vehicles—it might save lives, but will not eliminate the risk of deadly accidents. And even for that, the vehicles must have properly functioning speedometers; there is no such equivalent in the management of fishing or in assessing fish stocks. If we really want to avoid collapse, then a start might be to abandon the drive for maximum sustainable yields.

4. The Two Hoaxes: How the Management of Global Fisheries was Led Astray[5]

This is the story of how a few biologists with mathematical skills perpetrated hoaxes which led authorities with responsibilities for managing sea fishing for high sustainable catches to adopt a destructive overarching policy, thinking it is "scientific," but which is, in fact, very likely to be globally destructive and drive fishing toward insolvency.

The first is that the most desirable policy is one that will maximize long-term catches: its label is maximum sustainable yield (MSY).

The other is more subtle, less well-known, and called "the hockey stick": it is basically the assumption—for which there is no evidence—that for every fish stock, there is a biomass (its total weight called B_{lim}) below which its effective reproduction becomes seriously impaired.

Procedures for determining MSY are old historically and still strongly linked to a concept called "*surplus production*" (Schaeffer 1954; Pella and Tomlinson 1969; Fox 1970), while the "*hockey stick*" (e.g., Barrowman and Myers 2000; Mesnil and Rochet 2010) is quite a new invention. Neither has any scientific basis or credibility.

5. Optimal Biomass[6]

Surplus production modelers define their "MSY" as the largest sustainable catch that can be taken from a population at equilibrium that has been reduced by fishing to a certain

---

[5]Editors' note: written on November 8, 2014 and intended, by Holt, for the "Dummy's Guide."

[6]Editors' note: written on June 27, 2013 following the adoption of the "Basic Regulation" of the CFP that month and intended as part of his critique of surplus production models; see also Chapter 52.

percentage of its original biomass before exploitation. In the original model, due to M. B. Schaefer (1954), this was assumed to be 50%, but recently, it has commonly been assumed to be 37%, based on a model by W. W. Fox (1970).[7] There is no sound empirical or theoretical support for either value. Other models in the same family—due to Pella and Tomlinson (1969) —postulate values between 37% and 80% or more.

It may be of interest to look at the percentages predicted by age-structured population models such as the Beverton and Holt (1957) sustainable yield per recruit and their self-regulating model incorporating a function relating the annual number of recruits to the number and sizes of female parents. Such percentages have no value as a basis for managing fishing, but so many people involved in management cling to the erroneous idea that the biomass is the sole determinant of the productivity of a fish stock and, hence, the sustainable yield that is available that a comparison is instructive. The question is, however, complicated by the fact that sustainable yield, and its possible maximization, depend more on the sizes and age composition of the fishes constituting the biomass than on its gross weight. So, any maximum catch is a function of the age–size selectivity of the fishery.

The biomasses associated with any selective fishery can be deduced from the tables of yield per recruit in Beverton and Holt (1966),[8] but the process is rather complicated. So, I give some sample calculations here.

Beverton (1992) has shown that, for a wide range of marine species of bony fishes, the observed value of the Beverton–Holt invariant $m = M/K$ is mostly between 0.5 and 1.5 and that, within that range, different taxonomic groups tend to be in a more limited range [long-lived Clupeiformes—herrings, etc: 0.35–0.6; short-lived species—sprats, sardines, anchovies, etc: 1.3–1.6; flatfishes such as Plaice: 0.8–1.4, but with differences between the sexes; long-lived Gadiformes—cods, etc: 0.8–1.3; short-lived Gadiformes—Norway Pout *Trisopterus esmarkii*: 1.5–2.5; Redfish *Sebastes fasciatus* in the Atlantic: 0.5–1.0, and in the Pacific: 0.2–0.4].[9]

Here, I look at the calculations for three values of m – 0.5, 1.0, 1.5 – beginning with the central value 1.0.

Remembering that for any m, there is only one MSY, at w_{crit}, but obtainable only with infinite fishing effort, I look at a *near*-MSY with a very high exploitation rate [$F' = 19$, i.e., $E = F/(F + M) = 0.95$] fished highly selectively and, alternatively, a very low selective fishing at $w = 0.001$ and very low exploitation rate of $F' = 0.0256$ ($E = 0.05$).

For this case, MSY' (MSY/RW) = 0.105 and is obtained at $w_{crit} = 0.41$. This means that all the fish in the *near*-MSY catch would be mature. The index of the very lightly exploited recruited stock biomass ($B_v' = B/RW$) is 0.227, and the very highly and selectively exploited biomass B_{MSY}' is 0.173. Thus, the biomass for the *near*-MSY fishery would be 0.173/0.227 = 76% of the biomass of the virtually unexploited stock.

Next, we suppose that the fishery is unselective and at an intensity ($F' = 0.8$) that gives the lowest *local msy*, which is $msy/RW = msy' = 0.057$. In this case, the biomass of the recruited stock exploited to yield *msy* is 31% of the biomass of the unexploited stock. Then, we note that this local *msy* is only 54% of the weight of MSY itself.

---

[7]The Fox model is based on the population and growth function published by B. Gompertz (1832) and is an extreme of the family of surplus production functions proposed by Pella and Tomlinson (1969).

[8]These tables were partially copied—for a few selected values of the parameter M/K—in Annex 1 of Gulland (1983).

[9]If the natural mortality rate in the recruited stock is independent of age, the mean age is $1/M$ years. That is reduced to $1/(F + M)$ when there is fishing.

I have provided these results only for the extremes of selectivity—unselective, to the theoretical maximum at the critical oldest cohort weight. Within that range, there is an infinite number of possible selectivities, with stock depletion percentages at *msys* from 31% to 76%.

Now, it is worth looking at what saving of fishing cost (effort) would be made if we were to be content with a target of some good proportion of MSY, say 90%. To be able to compare the catch rates (*Y'/F'*), we aim at 90% of the *Y'* for *F'* = 19.0 (which is itself 99.9% of MSY), that is 0.095. That is, for the yield at the same selectivity from *F'* = 2.3 and a "sacrifice" of only 10% of MSY, we need to exert only 12% of the effort that would be needed to closely approach *MSY*.

Now, suppose we decide to fish unselectively (w_{crit} = 0.001, *F'* = 0.82) to obtain 90% of the *local msy*, i.e., 0.057 × 0.9 = 0.05. That is obtainable with *F'* = 0.43, which is about half the effort needed to catch *local msy*. The fishing effort savings for intermediate selectivities would be within this wide range.

We now have to see whether these results depend critically on the chosen value of *m* = *M/K*. We look at the range values *m* = 0.5 and 1.5. The ratio of the biomass of the MSY exploited stock to that of the virtually unexploited stock is 77% and 83%, respectively. The ratio of the biomass of the minimally selectively exploited stock at local *msy* to that of the virtually unexploited stock is 31% and 34%, respectively. Clearly, at least over the chosen range *m* = 0.5–1.5, the value of *m* makes little difference to the predicted depletion either by highly selective fishing at *near*-MSY or at practically unselective fishing at *msy*.

We now need to look at the consequences of assuming that there is no trend in the average number of recruits with changes in abundance of parents. If *m* = 1.5, the critical fish weight is about 0.3, that is close to the inflection of the growth curve and to the average size at sexual maturity. So, the "fishing close to MSY" scenario will give catches of almost all mature fish and over the entire lifespan of egg production. But, if *m* = 0.5, *near*-MSY catching takes place close to *w* = 0.64; the catch will be all mature fish—a relatively small number of large spawners—but not affecting most of the reproductive part of the stock, consisting of a large number of relatively smaller spawners. It is not practicable to generalize about the consequences of this, especially since estimation of the parameter r_0 of the Beverton–Holt stock–recruitment relationship—the recruitment into the unexploited stock, expressed as a fraction of the theoretical maximum recruitment with infinite egg number—is generally very uncertain. However, it is known from many trial calculations that the *msys* and the *sys* obtainable by moving from one steady-state to another by changing fishing intensity are highly unpredictable. However, not only is it sure that all *local msys* with reduced fishing will be shifted toward a lower fishing intensity, whatever the value of r_0 as a consequence of some further flattening of the *Y'–F'* curve.

Now we can look at the overall conclusions to be drawn from the study presented here:

1. The value of *m* = *M/K* makes rather little difference to the level of stock depletion caused by intensive fishing up to *msy* levels, but the selectivity of fishing makes a big difference, with—in all cases—depletion ranging from less than 0.3 for unselective fishing up to about 0.8 for highly selective fishing.
2. Unselective *msy'* is about half of MSY' regardless of the value of *m*.
3. MSY', unselective *msy'*, and the catch rate *msy'/F'* are roughly inversely proportional to *M/K*.

References

Barrowman, N. J., and R. A. Myers. 2000. Still more spawner–recruitment curves: the hockey stick and its generalizations. Canadian Journal of Fisheries and Aquatic Sciences 57:665–676.

Beverton, R. J. H. 1992. Patterns of reproductive strategy parameters in some marine teleost fishes. Journal of Fish Biology 41:137–160.

Beverton, R. J. H., and S. J. Holt. 1957. On the dynamics of exploited fish populations. Her Majesty's Stationery Office, UK.

Beverton, R. J. H., and S. J. Holt. 1966. Manual of methods for fish stock assessment. Part II. Tables of yield functions. FAO Fisheries Technical Paper 38 (Revision 1).

Cushing, D. H. 1971. The dependence of recruitment on parent stock in different groups of fishes. Journal du Conseil International pour l'Exploration de la Mer 33:340–362.

Fox, W. W. 1970. An exponential yield model for optimizing exploited fish populations. Transactions of the American Fisheries Society 99:80–88.

Gibbon, E. 1776–1788. The history of the decline and fall of the Roman Empire, Volumes I–VI. J. M. Dent, London.

Gompertz, B. 1832. On the nature of the function expressive of the law of human mortality, and on a new mode of determining the value of life contingencies. Philosophical Transactions of the Royal Society of London 123:513–585.

Gulland, J. A. 1983. Fish stock assessment. A manual of basic methods. FAO/Wiley Series on Food and Agriculture, 1. John Wiley and Sons, Chichester, UK.

Holt, S. J. 2009. Sunken billions—but how many? Fisheries Research 97:2–10.

ICES. 1997. Report of the Study Group on the Precautionary Approach to Fisheries Management, Copenhagen, Denmark, 5–11 February 1997. ICES Document CM 1997/Assess:7.

Longhurst, A. 2010. Mismanagement of marine fisheries. Cambridge University Press, Cambridge.

Mesnil, B., and M-J. Rochet. 2010. A continuous hockey stick stock-recruit model for estimating MSY reference points. ICES Journal of Marine Science 67:1780–1784.

Pella, J. J., and P. K. Tomlinson. 1969. A generalized stock production model. Bulletin of the Inter-American Tropical Tuna Commission 14:421–496.

Ricker, W. E. 1954. Stock and recruitment. Journal of the Fisheries Research Board of Canada 11(5):559–623.

Schaefer, M. B. 1954. Some aspects of the dynamics of populations important to the management of commercial marine fisheries. Bulletin of the Inter-American Tropical Tuna Commission 1:25–56.

Shepherd, J. G. 1982. A versatile new stock–recruitment relationship for fisheries and the construction of sustainable yield curves. Journal du Conseil International pour l'Exploration de la Mer 40: 67–75.

Simpson, A. C. 1959. The spawning of the plaice in the North Sea. Fishery Investigations, London, Series II 22(7):1–111.

World Bank and FAO (The World Bank and Food and Agriculture Organization). 2009. The sunken billions. The economic justification for fisheries reform. The World Bank, Washington, DC.

Chapter 52

Critique of the Objectives of the EU's Common Fisheries Policy[1]

Article 2.2 of Regulation 1380/2013[2] states: "*The CFP shall apply the precautionary approach to fisheries management, and shall aim to ensure that exploitation of living marine biological resources restores and maintains populations of harvested species above levels which can produce the maximum sustainable yield.*" And further: "*... maintaining populations of fish stocks above biomass levels capable of producing maximum sustainable yield, the maximum sustainable yield exploitation rate shall be achieved by 2015 where possible and, on a progressive, incremental basis at the latest by 2020 for all stocks.*"

There are, in my opinion, three fundamental problems with these words that render the regulation impossible to implement consistently with scientific knowledge.

They present two formally incompatible options for the optimum state and/or size of an exploited fish stock. The difference between these is far from trivial, especially in economic terms and the future of managed fisheries. In particular, the option of reaching the MSY exploitation rate by 2015—or any other date—represents the deliberate choice of an economically inefficient management regime.

I present the three problems in—in my judgment—descending order of importance. As to the first problem, MSY is actually only attainable by catching all of the fish that survive to a calculable critical size and age from the oldest cohort ("age group") in a multicohort population by exerting an infinitely high, and extraordinarily selective, fishing rate. Critical size and age are a function only of the natural mortality rate of the fish and the pattern and rate of its growth in weight (and that is always higher/bigger than the median size and age of the attainment of reproductive maturity). So, catching MSY is manifestly impossible; MSY is, in reality, an entirely theoretical, but always unattainable, quantity.

What, as an alternative, less rigorous, but possibly reachable, target can be called a "*local msy*" would be attained with a **finite** exploitation rate in which selective catching begins before fish reach the critical size/age. A *local msy* is always less—and can be much less—than MSY. It is the maximum—peak—of a curve of sustainable ("steady-state") catch against fishing mortality rate—which is a straightforward function of the fishing effort or intensity—for which the fishing selectivity is specified in terms of the age/size at which fish begin to be vulnerable to existing fishing gears and operations (by depth, season, location, etc.). If the latter changes, the *local msy* changes. In general, if selectivity is such that young/small fish are caught—as in fisheries for meal and oil production—the *local msy* will be relatively low, but attained with a relatively mild fishing intensity. If the fishery is essentially for human food,

[1]Editors' note: written by Sidney Holt for The Greens/European Free Alliance in EU Parliament and dated July 12, 2014.

[2]Regulation (EU) No. 1380/2013 of the European Parliament and of the Council on the Common Fisheries Policy.

perhaps concentrating on mature fish, the *local msy* will be larger, but attained only with much more fishing effort. The choice to be made is an economic and social one, not a scientific one.

The second problem is that the regulation focuses on a single desirable or required **biomass** of each fish stock The biological productivity—and hence the potential sustainable catch—depend both on the size of the fish population and its composition by size/age. That composition depends on and reflects the history of the exploitation of the stock and on the annual variation in certain quantities, especially the annual number of "recruits" to the "fishable" stock. Two populations of a species might have the same biomass, but if they differ substantially in their size/age compositions, they will have quite different productivities and hence sustainable yields. The stock composition depends critically on the intensity—and selectivity—of its exploitation, over time, especially recent time. The age/size compositions depend on many factors, but especially on two—the differing numbers of "recruits" (young fish of catchable size) each year and the history of their exploitation—including the selectivity of that exploitation (i.e., at what age/size young fish begin to be vulnerable to the fishing gear being used and the operational practices—such as the location and season of fishing on naturally segregated animals). So, mere specification of a desirable biomass—whether or not it is intended to be of a population in some theoretical "MSY" or "*local msy*" state—is no guarantee at all that the stock will be underfished or overfished if fishing effort and/or catches are adjusted and regulated in the hope of maintaining that state. (Note that this is true whether "desirable" or "target" biomass is specified in absolute quantity or as a percentage of the presumed biomass of the population before its exploitation began, so-called "virgin biomass.")

The third general problem is the vagueness of the provision to maintain populations "above" MSY (or *local msy*) levels and corresponding exploitation rate. (Exploitation rate might mean fishing mortality rate or simply the ratio of catch to average biomass.) How much "above?" Without some specification, it cannot be possible for scientists to advise fully on the consequences of regulations governing either total allowable catches or permissible fishing effort.

That the stock should be kept above an MSY or *local msy* target is commonly regarded as a precautionary measure. In a way, it is, but that is by no means the only way or the best way to ensure precaution. What is far more important comes from the shapes of the curves of sustainable yield against fishing mortality or effort. These are always fairly flat-topped (regardless of the assessment procedure used, either "surplus production" or age-structured dynamic population models) and that is supremely important at or near their peaks, which is what management for *local msy* aims at. This means that the sacrifice of a relatively small fraction of the allowable/sustainable catch means a saving of a very much larger fraction of the fishing effort needed to take it. Hence, the cost of catching vis-à-vis the market value of the catch is much less than used in efforts to attain *local msy*; hence, the rate of profit from fishing is much higher.

In fact, there may even be no "sacrifice" in economic terms. This is because even with a marginally lower fishing intensity, the average size of the fish in the catch will be slightly bigger, and in many fisheries will attract a somewhat higher price per tonne.

Finally, it should be mentioned that the attainment of any given exploitation rate by 2015 depends on whether the stock is currently being exploited above that rate or below it; in colloquial terms, whether it is overfished or not, i.e., "underfished." If it is underfished, it is clearly absurd to allow or encourage fishing effort to increase so far as to reduce it to attain a level or state below the "above" target level.

Chapter 53

Escaping the Stranglehold[1]

Introduction

In a delightful parody of the notion of sustainable development, a young, Spanish "activist" distributed, at a recent conference in Brussels devoted to marine ecosystem conservation, a coffee mug bearing the slogan "To Maximum Sustainable Yield and Beyond." It also carried a linear graph showing total fish biomass rising to the sky! An equivalent idea, posing as a campaign against climate change, is that we can control human-caused global warming by all of us consuming even more energy so long as it is generated by solar capture, wind, tides, damming more rivers, or by nuclear fission, instead of by drastically reducing the *production* of fossil fuels and not merely the consumption of them.

Even the application of an already-agreed fisheries management policy of keeping fishing mortality rates less than what would produce maximum sustainable yield seems now to be extremely difficult for the governing Council of Ministers of the EU, if not impossible. This notwithstanding the undoubted economic and resource conservation benefits of such a policy.

At nearly 90 years of age, I do not have much personal interest left in saving the world. But I shall leave it with the firm belief that science, properly done and used, could do so. I am equally convinced that our political–economic system, based on property, hierarchy, and greed, will not permit that. When the First World War had practically led to the destruction of industrial society, the Russian Bolsheviks proposed an alternative. But Lenin and Trotsky were right and Stalin was wrong; the revolution would fail if it did not become global. It did fail. Pope Francis seems to have similar ideas, as expressed in his amazing recent encyclical *Laudato Si*.

Here, I try to take a rational view of possible futures for sea fishing as a worthwhile pursuit and important contributor to human food supplies. But I won't bet on any of them happening.

---

[1]Editors' note: this was a document submitted by Sidney Holt to the September 2015 ICES Annual Science Conference (ASC) for the Theme Session "Advancement of stock assessment methods for sustainable fisheries" (Holt 2015) and was an update of an earlier contribution by Sidney to the September 2012 ICES ASC for the Theme Session "Evolution of management frameworks to prevent overfishing" (Holt 2012; Chapter 49). The 2015 ICES ASC document was also an expansion of a presentation he gave in February 2015 to a special meeting of the Committee on Fisheries of the European Parliament on Maximum Sustainable Yield in Fisheries Management (Chapter 55). In the process, he considered that it had become a form of manifesto. Instead of orally presenting this document (Holt 2015) at the 2015 ICES ASC, Sidney instead gave a different talk entitled "What is overfishing?" which is reproduced as Chapter 57.

A Manifesto

What is Stock Assessment?

When I decided to contribute to this Theme Session, I thought to begin by defining "Fish Stock Assessment." I had the cozy remembrance that it was a phrase used casually by a group of scientists in late 1949 at the Lowestoft Fisheries Laboratory in England—Michael Graham, its director, David Cushing, Ray Beverton, and myself, with Basil Parrish, our colleague at the Marine Laboratory in Aberdeen, Scotland, and E. S. Russell, Graham's predecessor.

I found no documentary evidence for this, however. An Internet search for use of the term revealed about 50 major published references to it, but with a great diversity of meanings and few attempts at definition. John Gulland, a next-generation scientist at Lowestoft, popularized the phrase, and his first book bore it as its title *Manual of Methods for Fish Stock Assessment. Part 1. Fish Population Analysis* (Gulland 1969), but with no definition. It appears again in Gulland's next book (as editor) *Fish Population Dynamics* (Gulland 1977) and yet again as the title of the opening chapter "Stock assessment methods: the first fifty years" (Smith 1988) in the second edition of the 1977 book *Fish Population Dynamics: The Implications for Management* (Gulland 1988). In what I regard as the best general text on stock assessment and population dynamics—*Quantitative Fish Dynamics*—Terry Quinn and Rick Deriso (1999) frequently use the assessment term, but in varying ways and not defined. They do, however, provide a clear definition of "stock": "the aggregate of a population that can be managed as a discrete unit."

Michael Graham recruited most of the new post-war staff at Lowestoft from among young scientists who had been engaged in operations (British term)/operational (US term) research, groups of mathematicians, physicists, biologists, chemists, and others who had used scientific knowledge and methods to improve strategy and tactics of the Allied military. P. M. S. Blackett, one of the founders of this new discipline, defined it succinctly as "the analysis of data in order to give useful advice" (Nye 2004; see especially Chapter 3: Operational research and atomic weapons 1936–1962). Graham regarded the new fisheries science as a peace-time application of their methods.

"Assessment" was a frequently used word in the literature of operational research; it seemed to be about the transition from data analysis to the offering of advice. So "stock assessment" became a sort of *applied population dynamics*, analogous to the well-known pairs "pure and applied mathematics," "pure and applied physics," etc. Indeed, some of the interests of stock assessment practitioners were close to the military activities, e.g., the theories of search for submarines, detection of enemy aircraft, and the laying of artillery were similar to the theory of efficient search for fish.

The application of stock assessment inevitably became closely associated with the problems of managing fishing in order to correct or avoid "overfishing", but there were evidently other possible applications, including the productive development of new fisheries.

Here it seems pertinent to refer to one of Michael Graham's much-quoted maxims. In his classic 1943 book (Graham 1943) (and, subsequently, in 1955, to a Conference of the United Nations), Michael Graham expressed his Great Law of Fishing: "Fisheries that are unlimited become inefficient and unprofitable." I will add that this can happen also to badly managed fisheries and those for which the market for their products expands geographically or by virtue of a change in their use, such as by a shift from human food to fishmeal and oil.

Articles and books about stock assessment rarely contain clear, unambiguous definitions of "stock" and, indeed, some usage is ambiguous, one being a "management stock," which could be applied to all the fish in a politically designated "management area" or region regardless of whether it contained a single self-contained biological population. Here, I regard the term as generally applying to that *part* of a biologically self-contained population (of a species of fish, crustacean, or mollusk) of interest to humans for exploitation now or in the future. The "part" is usually defined by a lower body size—hence age—boundary, so that eggs and larvae are excluded. From that point of view, the stock is an *open biological system* into which young animals are regularly *recruited* and from which there is continuous loss by death either "natural" or by capture. If the stock is regarded as occupying a certain space, then it may be necessary to provide for consideration of gains and losses by immigration and emigration. Other inputs to and outputs from the open system, that are usually taken to be of second-order interest, are nutrients and excreta.

Recruitment is commonly enumerated as the abundance of the youngest *cohort* in the exploited or exploitable population, i.e., the stock. However, for the management of fishing by some methods—such as by setting maximum catch limits—it is desirable to enumerate the *prerecruits*, that is, the cohort that will become recruits in the next recruitment period, usually a year.

Management Policy

At an early stage in its career, stock assessment was seduced by an evil vampire with the remarkable name *Maximum Sustainable Yield*—nicknamed MSY. That was a political act arising from the negotiation of a peace treaty between the authorities of Japan and the USA and most of the Allied Powers (but excluding the USSR, the People's Republic of China, and a few others) that came into force in 1952. The USA had sought to impose on Japan an *abstention principle* that would prohibit a country's fishers from operating in an area regarding which the coastal state claimed to be already *fully utilizing* the specified fishery resources. Japan's delegation wanted to know the meaning of *full utilization*, to which the US response was "taking MSY." This was the invention of a group of US biologists led by Milner Schaefer, studying the Sardine *Sardinops sagax*, Halibut *Hippoglossus stenolepis*, and Yellowfin Tuna *Thunnus albacares* fisheries of the USA, who had developed a stock assessment procedure that came to be called *Surplus Production* theory. MSY was defined as the peak of a parabolic or similar curve of sustainable catch against the biomass of the stock. Two other US biologists, William Herrington and Wilbur Chapman, active in both the political and commercial spheres, pursued, respectively, the "full utilization" and "abstention" ideas vigorously for several years both in the Americas and in the UN system. But they did succeed in inserting it into a new International North Pacific Fisheries Convention involving the USA, Japan, and Canada, but—interestingly—excluding the USSR.

The USA tried, but failed, to get the abstention principle inserted in the text of the document coming from an International Technical Conference on the Conservation of the Living Resources of the Sea convened in Rome in 1955 at the headquarters of the Food and Agriculture Organization of the UN (United Nations 1955). The USA did, however, succeed in putting into the report the idea that MSY is one possible target for managing fishing. No more was heard about abstention in the UN context (partly because of an emerging consensus that the concept of the Territorial Sea could be extended from three or twelve nautical miles to up to 200 miles). Opponents of the MSY management imperative pointed out—correctly—that, apart from anything else, it would always automatically give an advantage to the strongest contestant in any dispute, particularly those who could make the best scientific case for claiming he was fully utilizing the resource. Notwithstanding that concern, MSY was written into the Convention on Fisheries that emerged from the UN Conferences on the Law of the Sea held in Geneva in 1958 and 1959. That Convention never became operational because the Geneva conferences had failed to resolve the crucial and highly controversial matter of the breadth of the Territorial Sea. That and other matters were eventually settled in the comprehensive UN Convention on the Law of the Sea (UNCLOS) that came into force at the end of 1994, in which an "Exclusive Economic Zone" (EEZ), having some of the features of a Territorial Sea, was defined.

In a remarkable effort to provide a compromise, the UNCLOS awards the coastal state the sovereign right "to maintain or restore populations of harvested species in its EEZ at levels which can produce the maximum sustainable yield, as qualified by relevant environmental and economic factors." However, although the coastal state must determine its capacity to harvest the living resources of the EEZ, where it does not have the capacity to harvest the entire allowable catch (as determined by the MSY criterion), or, presumably, the wish to do so, it must give other states access to the surplus of the allowable catch, presumably taking into account the coastal state's decision as to what exactly constitute relevant environmental and economic factors. (This history is fully documented and analyzed in a book by the American historian, Carmel Finley 2011.)

Relaxing the Conditions of the Target

The UNCLOS enshrines MSY as a principle objective of managing fisheries for sustainable catches, but with a very important condition: "as qualified by relevant environmental and economic factors." This offers us a way of escape from the stranglehold of managing fishing in order to maximize physical yield. There may or may not be environmental or, indeed, social reasons for substantial departure from the MSY objective, but there are certainly always good economic reasons for doing that, principally because the costs involved in taking a high sustainable catch are at least as important as the size and market value of the catch. The European Union has recognized that by adopting as its basic management objective to fish with intensity *less* than would be required to take MSY. How much less is not yet decided, but that decision will have to take into account both economic and social aims and must be made in accord with the best scientific advice.

Discussion of "how much less" has, until now, focused on a target fishing mortality rate F as a percentage of F_{MSY} or, alternatively, as a target sustainable yield (SY) that is some predetermined percentage of MSY. I prefer, here, to refer to the former simply because next year's SY always depends critically on the next incoming recruitment, which can be guessed but is never known and can be extremely variable.

It is unfortunate to find oneself depending on the old, discarded target to define the new one. There are, of course, other maxima in fish stock assessments, one being the **maximum sustainable economic yield** (MSEY), the maximum difference between the market value of the catch and the cost of taking it, discussed by several scientists and economists, beginning more than 60 years ago. One could define an F_{target} somewhere between F_{MSEY} and F_{MSY} (remembering that F_{MSEY} is always less than F_{MSY} and can be much less) as a specified multiple of F_{MSEY} lying within that range. But I think our problem goes deeper than that. Why define a social and economic target simply as a quantity by weight or gross or net value?

F_{MSEY} and MSEY can be easily, though approximately, estimated if the break-even catch-rate is known, from simple economic information—that is the catch-rate the cost of obtaining which is equal to the value of the catch.

There are, naturally, several social reasons why it might be desirable to manage so that the fishing mortality/effort is more than that needed to secure MSEY: maintaining employment, meeting a nation's need for a reserve of experienced seamen, maintain ship-building and repair capacity, and so on.

A Little History

Johan Hjort and Michael Graham both published, in the 1930s (Hjort et al. 1933; Graham 1939), curves of sustainable catch against fishing rate, with peaks they called "optima," but neither of them suggested that these should be regarded as management targets. In fact, their curves were very different from each other. Hjort's was the catch, in numbers, of fin whales *Balaenoptera physalus*, based on mortality and calf birth rates. Graham's was for catch, in weight, of an unspecified fish, based on mortality and body growth rates (Holt 2014).

Not Peaks, but Slopes

We are so habituated to thinking in terms of maxima that it is hard to think outside that box. But things perhaps get easier if we think about *rates* and *slopes* rather than *quantities*. Whether or not they have maxima (depending on the selectivity of the fishery), curves of sustainable yield-per-recruit against fishing mortality always have a gentle slope immediately to the left of the peak or asymptote. This means that the catch rate (Y/F) in that region is much higher than that associated with the peak, while the catch itself is only slightly less than the peak catch. This has enormous economic implications, especially as there is no guarantee that a sustainable fishery at its maximum will be profitable.

The slope of a curve of sustainable yield (by weight or value) is determined by the interaction of a natural mortality rate by which fish die at a relatively unvarying percentage (logarithmic) rate throughout their lives after recruitment, and the curve of increase in individual body weight, which continues throughout their lives. This latter curve always has a high **specific** *growth rate* when the fish is young and small. As it grows, *the absolute rate of growth* reaches a maximum, visible as an inflexion in the growth curve. The inflexion is commonly close to or somewhat lower or higher than the median size at which the fish attains sexual maturity and begins to breed. After that, growth continues, but at an ever-decreasing absolute rate, so that body weight approaches an upper asymptote.

Cohorts

The interaction between two rates of change—growth and death—results in the total weight of a cohort (all the fish born in the same spawning period) increasing, reaching a maximum when the weight of the fish in the cohort reaches a certain *critical size* at a *critical age*, these depending only on the value of the ratio M/K (natural mortality rate to the driving parameter of the von Bertalanffy growth function) for that species–stock. I think it was William Herrington who first pointed this out in the 1930s (Herrington 1935), observing that a maximum catch could be obtained by refraining from fishing until the fish in the oldest cohort had reached critical size and then catching all of the survivors in that cohort. This practice would likely be theoretically sustainable because critical size must always be bigger than the size at the inflexion of the growth curve and will usually be bigger than the size at maturity.

It is worth considering the implications for conclusions about critical ages and sizes, particularly of possible variation in the rate of mortality with age and changes over time in the values of either or both parameters of the body growth function. These will generally affect conclusions quantitatively, but not much in general terms. The critical values will still depend on the equality, but opposite sign, of the absolute mortality rate and the absolute growth rate at a certain point, the latter depending on the changing values of both K and the asymptotic weight W.

Of course, such an MSY is not, in practice, attainable because we do not have the technical means of catching all the surviving fish at one time in one cohort and leaving the other younger cohorts alone. However, the efficiency of any particular selective, but eumetric, fishing for a local MSY (termed *msy*) can be indexed by its ratio to the theoretical MSY.

A fishery at or near local *msy* level causes a substantial reduction in the biomass of the stock from its level in the unexploited ("virgin") state and, depending on the selectivity, this may cause the average recruitment to decrease to somewhat less than that experienced in the virgin or lightly exploited stock. The effect of this is to reduce further the slope of the curve of yield against fishing mortality rate in the vicinity of the peak, and hence enhance the economic advantage of managing the fishery to hold the fishing mortality rate to a level substantially less than that which would provide *msy*. However, because catch and effort change in the same direction the change in CPUE—hence profitability—will be small.

A problem in discussing this further is that the highest SY obtainable from a stock depends critically on the selectivity by which it is exploited. With low selectivity (fish beginning to be caught when they are quite young and small), the peak of the SY curve, what I have called the local *msy*, occurs at a relatively low fishing mortality rate, and is relatively small in weight. If selectivity is high—fish perhaps not beginning to be caught before reaching sexual maturity—then local *msy* will be higher, but obtained only by exerting a substantially higher fishing effort and therefore with diminished catch rate. The actual scale of these quantities—sustainable catches and fishing mortality rates—is determined by the value of the ratio M/K (natural mortality rate to the driving parameter of the expression for growth in individual weight for the particular stock).

More about Selectivity

Published advice about MSY or some other reference point related to it often contains no reference to the selectivity of the fishery being assessed. This is so, for example, for much of the advice offered to the EU Commission by ICES. Advice of this kind, if based on age-structured assessments, is useless or misleading. A minimal requirement is that the assessment model applied to a fishery employing towed nets, such as trawls and seines, should incorporate a function specifying the sigmoid curve of selection against body size or age—perhaps approximated by a knife-edge —or the corresponding bell curve for hooks and line and gillnets, perhaps approximated by two parallel knife edges. (Of course, if the advice comes from a surplus production model, then it cannot carry any such indication because such models assume that age and body size play no part in determining the productivity of the stock.)

There appears to be a common assumption that the selectivity of the fishery cannot be affected by the management regime, and that selectivity is determined entirely by fixed sizes of meshes and hooks. This is both wrong and a dangerous assumption. While a managed change in the fishing mortality rate might not, in itself, cause a change in selectivity, there are several situations in which it can do so. One is in a mixed fishery. If management of two or more of the species affects the relative abundance of the species, then, since virtually all species are partially geographically segregated, the abundance changes will lead fishers to adapt their operations, usually by a geographic or seasonal shift. Similar changes in selectivity can arise for the same reasons if a marine protected area (MPA) that prohibits fishing is declared. Parts of fish stocks inhabiting MPAs practically always have age/size, sex, or reproductive states somewhat different from those of the overall stock composition, so the selectivity of the newly geographically restricted fishery will change. Assessments offering management advice should, therefore, always refer to the original selectivity, whatever that might be, and also suggest what changes might occur under the new regime.

Minimal Economics

If a management target other than some maximum is to be chosen, it seems to me to imply that "stock assessment" has to include at least some economic data and calculations, at least concerning the costs of fishing, the market values of fishes, and, especially, information indicating what would be the threshold catch rate to be ensured in order to avoid multiple bankruptcies of fishing operations. In those circumstances, it surely becomes necessary to review the basic method of management to maintain the desired fishing intensity.

Historically, the advice of scientists has been to manage primarily by regulating the fishing power of the fleet or fleets. However, since the end of World War II, it has proven difficult to do that for political and even constitutional reasons. Therefore, instead of regulating the *input* to the fishing system, it has become almost universal practice to place restrictions only on the *output* in a variety of ways. One of the first restrictions used, in the international management of the Halibut fishery in the eastern North Pacific, was to limit the duration of the legal fishing season by setting opening and closing dates. Elsewhere and later, and now practically worldwide, extremely complex efforts are made to regulate the output by setting catch limits—total allowable catches (TACs) in the current jargon. This has led to many difficult problems. Among them are the following:

- Estimating the desirable catch calls for knowledge of the extremely variable, essentially unpredictable, recruitment in the coming year.
- Since fishing incorporates a random process, it is impossible to avoid unintended excess catch—and also unintended catches of non-target species. This can be partially corrected by annual adjustments to TACs, but this means that the fishing is caused to be less stable than it needs to be. Similar considerations apply to the—usually lesser—consequences of imposing other limitations on landed catches, such as

> setting minimum legal length limits. And the methods to regulate output become even more intractable in mixed fisheries—that is in which several target species are caught simultaneously in the same operations.

Attempting to ensure rational fishing by controlling outputs is equivalent to driving a vehicle using only the brakes instead of adjusting both the brakes and the throttle.

Imposed Inefficiency

In his evidence presented to a special meeting of the Atlantic States Marine Fisheries Commission in Baltimore, Maryland in 1942, the fisheries scientist R. A. Nesbit (1943) well described these forms of regulatory measure as *imposed inefficiency*. He said to Commissioners: "I earnestly recommend that the Commission adopt a policy and program that will protect the fishermen as well as the fish. In one important respect, such a program will be easier to develop than a conventional program based on **a policy of imposed inefficiency**" (*Holt's emphasis*).

Rational, Irrational, and Eumetric Fishing

Then, a year later, a Canadian biologist, A. G. Huntsman, wrote in *Science* (Huntsman 1944) "irrational fishing is where the take in proportion to the effort fails to yield a satisfactory living to the fisherman," so bringing economics into the discussion. Eventually, in 1949, Huntsman told the United Nations Scientific Conference on Conservation and Utilization of Resources (UNSCCUR 1951), held that September at Lake Success, New York, that "the take should be increased only as long as the extra cost is offset by the added revenue from sales" (Huntsman 1951). Huntsman's formulation points us to a way to identify a useful reference point on the curves of catch value against cost related to fishing intensity.

Incidentally, the concept of "rational fishing," much loved by Hjort, Russell, and Graham, as well as by Beverton and myself, appeared in the fisheries literature earlier than usually thought. In his chapter for Gulland's *Fish Population Dynamics: The Implications for Management* (Gulland 1988), Smith (1988) drew attention to the wise words of a German scientist H. M. Kyle, published in 1905: "rational fishing is not subject to natural laws, but rather depends on the monetary side of the matter, the expenditure on boats and gear on the one hand and the gross and net income on the other" (Kyle 1905).

In 1957, Beverton and I added to this the idea of *eumetric* fishing in which selectivity and fishing effort are optimally balanced. The need for such an approach in practice became evident during a workshop (ICNAF 1961) jointly convened in Lisbon, Portugal, by ICNAF, ICES, and FAO in May/June 1957, with the section on the effects of fishing led by Beverton and myself. The purpose of the workshop was to evaluate proposals to increase codend mesh sizes to be used by US and Canadian vessels trawling for Haddock *Melanogrammus aeglefinus* on Georges Bank. We found that the proposed increase in mesh size would result in an increased catch and catch rate if the fishing intensity remained unchanged. However, we advised the Commission that, as fishing effort would not be regulated by ICNAF, this meant that Georges Bank Haddock fishing would become more attractive economically, leading to an increase in the fishing mortality rate, which would negate the gains expected from the mesh size increase (for further analysis of these issues, see Holt 2013). The fact is that consequences of this kind, having knock-on effects throughout the fishing system, have never been recognized and accounted for by regulatory authorities.

Suboptimal Fishing and Sustainability

A balance of intensity and selectivity of fishing (that is, selectivity providing maximum catch for a given intensity, or an intensity giving maximum catch for a given selectivity) is not necessarily optimal or even desirable. In early experiments with bottom trawls, it was discovered that increasing codend mesh size could increase the efficiency of the gear, presumably by allowing less towing power to conduct effective hauls. And, of course, in mixed fisheries, it is not possible to optimize mesh size for two or more species with differing biological parameters, and one might have to consider optimizing for the most valuable species.

Already, at the 1955 UN Conference, Michael Graham had said—especially to the US delegation that was pushing hard for an MSY imperative—that "our objective (*Editor's note: according to Holt, he was speaking for Western European attitudes*) is not to seek maxima, but rather conservation measures to make things better." Without using the word, they were advocating *sustainability*, which became a global buzzword soon after Gro Harlem Brundland introduced us all to her quasi-oxymoronic *sustainable development* (World Commission on Environment and Development 1987).

The closest look by scientists at the sustainability concept has, I believe, been made by the Development Group of its Scientific Committee set up by the International Whaling Commission (IWC) in the 1980s to devise a Revised Management Procedure (RMP) for managing commercial whaling. To begin with, the idea of "a procedure" was taken literally in a fisheries context for the first time, mimicking its use in medical practice, particularly surgery: following in sequence a fixed set of predefined actions, equivalent to an algebraic algorithm. The procedure has to be tested by simulating application, using an appropriate stochastic stock dynamics model. To do that, a *finite* "management period" needs to be selected, and the effectiveness of the procedure judged from the cumulative catch during the period and the state of the stock at the end of the chosen management period. This also involves setting multiple criteria, such as avoiding unintended depletion of the stock to below some critical level, expressing all such events in terms of predefined probabilities.

In the IWC, managers had also determined that the system should be as stable as practicable, not permitting frequent or substantial changes in regulation, except in case of perceived emergency. This is a far cry from the popular myth that sustainability is—like Shirley Bassey's and Ian Fleming's—*Diamonds are Forever*, provided "the environment" does not change.

The key to the IWC success in this endeavor was the decision by the Development Group to go far beyond developing a population model and to simulate the entire management process (e.g., Butterworth and Punt 1999). The credit for this must go to three brilliant members of the Group—Justin Cooke, Bill de la Mare, and Doug Butterworth—and its convener, Geoff Kirkwood, providing a combination of talents in mathematics, statistics, biology, and engineering from three continents. This little collection mimicked the structure and working methods of the best wartime operations research teams that had molded Graham's vision of the future of fisheries research.

Dynamics of Transition

Included in my understanding of "stock assessment" is consideration of options for effecting a transition from a stock that is in a state that is not optimal, and probably not "steady," to the chosen optimally managed state providing high and relatively stable catches. The main consideration is the speed at which the desired transition is to be accomplished, and one measure of performance could be the *cumulative catch during the transition period.* The speed is, of course, determined by the pattern of fishing intensity exercised during the transition as well as by the biologically determined dynamics of the stock. Managers might wish to make the transition with minimum delay, but that could have unforeseen social and economic consequences. Scientific advice based on the population dynamics model plays an important role here.

Monitoring

Monitoring whether the fishery is performing as predicted is evidently an important way of testing the assessment procedures under management, but it is an assessment activity to which far too little attention has been given in practice. Design of a monitoring system must include all such matters as:

- examining whether the population model used to make the assessment is structurally appropriate;
- determining whether the estimates of values of the parameters in the model are correct;
- checking that the agreed management measures are all being taken in a timely manner and being enforced effectively;
- checking the routine procedures for recording catches and their compositions, including unintended catches; and
- checking the validity of assumptions concerning selectivity.

Additionally, the monitoring system must include rules by which regulations are to be changed if significant departures from the intended or expected situation are discovered.

Acknowledgments

I have discussed the opinions expressed in this manifesto during the year with several other scientists, but especially with Michael Earle, Adviser to the Greens/EFA Group in the European Parliament. I am particularly grateful to Dr. Rainer Froese for his many suggestions for improving an earlier version, practically all of which I have included in this revised version. Thanks also to my son Tim for his relentless editing.

References

Butterworth, D. S., and A. E. Punt. 1999. Experiences in the evaluation and implementation of management procedures. ICES Journal of Marine Science 56:985–998.

Finley, C. 2011. All the fish in the sea: maximum sustainable yield and the failure of fisheries management. University of Chicago Press.

Graham, M. 1939. The sigmoid curve and the overfishing problem. Rapports et Procès-Verbaux des Réunions du Conseil Permanent International pour l'Exploration de la Mar 110:17–20.

Graham, M. 1943. The fish gate. Faber and Faber, London.

Gulland, J. A. 1969. Manual of methods for fish stock assessment. Part 1. Fish population analysis. FAO Manual in Fisheries Science No. 4.

Gulland, J. A. Editor. 1977. Fish population dynamics. John Wiley & Sons, New York.

Gulland, J. A. Editor. 1988. Fish population dynamics: the implications for management. John Wiley & Sons, Chichester, UK.

Herrington, W. C. 1935. Modifications in gear to curtail the destruction of undersized fish in otter trawling. United States Bureau of Fisheries, Investigational Report 24.

Hjort, J., G. Jahn, and P. Ottestad. 1933. The optimum catch. Hvalrådets Skrifter 7:92–127.

Holt, S. J. 2012. The some good and mostly bad about maximum sustainable yield as a management target. ICES Document CM 2012/L:02.

Holt, S. J. 2013. The evolution of the objectives, science and procedures of fisheries management since 1945. *In* I. Heidbrink, and M. McCarthy, editors. Fisheries management in a historical perspective. Report of the 12th Conference of the North Atlantic Fisheries History Association, Old Dominion University, Norfolk, Virginia, 19–22 August 2009. North Atlantic Fisheries History Association, Hull, UK.

Holt, S. J. 2014. The graceful sigmoid: Johan Hjort's contribution to the theory of rational fishing. ICES Journal of Marine Science 71:2008–2011.

Holt, S. J. 2015. Escaping the MSY stranglehold. ICES Document CM 2015/A:01.

Huntsman, A. G. 1944. Fishery depletion. Science 99(2583):534–535.

Huntsman, A. G. 1951. Research on use and increase of fish stocks. Pages 169–171 *in* Proceedings of the United Nations Scientific Conference on the Conservation and Utilization of Resources, 17 August – 6 September 1949, Lake Success, New York, Volume VI, Wildlife and Fish Resources, United Nations, New York.

ICNAF. 1961. Fishing effort, the effect of fishing on resources and the selectivity of fishing gear. Proceedings of the Joint Scientific Meeting of ICNAF, ICES and FAO held 27 May – 3 June 1957 in Lisbon. Volume 1 – Reports. FAO, Rome.

Kyle, H. M. 1905. Statistics of the North Sea fisheries. Part II. Summary of the available fishery statistics and their value for the solution of the problems of overfishing. Rapports et Procès-Verbaux des Réunions du Conseil Permanent International pour l'Exploration de la Mer 3.

Nesbit, R. A. 1943. Biological and economic problems of fishery management. Pages 23–53 *in* W. H. Herrington, and R. A. Nesbit, editors. Fishery management. US Fish and Wildlife Service, Special Scientific Report 18.

Nye, M. J. 2004. Blackett: physics, war and politics in the twentieth century. Harvard University Press, Cambridge, Massachusetts.

Quinn, T. J., and R. Deriso. 1999. Quantitative fish dynamics. Oxford University Press, New York.

Smith, T. D. 1988. Stock assessment methods: the first fifty years. Pages 1–33 *in* J. A. Gulland, editor. Fish population dynamics: the implications for management. John Wiley & Sons, Chichester.

United Nations. 1955. Report of the International Technical Conference on the Conservation of the Living Resources of the Sea, 18 April–10 May 1955, Rome. United Nations Publication, Sales No. 1955.II.B.2.

UNSCCUR (United Nations Scientific Conference on the Conservation and Utilization of Resources). 1951. Proceedings of the United Nations Scientific Conference on the Conservation and Utilization of Resources, 17 August – 6 September 1949, Lake Success, New York, Volume VI, Wildlife and Fish Resources. United Nations, New York.

World Commission on Environment and Development. 1987. Our common future. Oxford University Press, Oxford, UK.

Chapter 54

A Comment on Management of European Fisheries for High Sustainable Yields, in Particular in the Baltic Sea[1]

As scientists who have spent most of our working lives conducting studies of fish population dynamics and the application of that discipline for the rational management of sustainable fisheries, it is extremely disheartening to see the authorities of the European Union setting fishing opportunities for the Baltic Sea which will impede a speedy recovery of the mostly depleted stocks, with a high risk of collapse for the western Baltic Cod *Gadus morhua*, which will be fished in 2016 at more than twice the F_{MSY} rate. This decision violates the management policy formally agreed by the basic organs of the Union: the European Parliament, the Council of Ministers, and the Commission. That policy requires that fishing opportunities be derived from establishment of a fishing mortality rate (F) less than that required to provide maximum sustainable yields (F_{MSY}), so that stock sizes are rebuilt to and maintained above levels that can produce the maximum sustainable yield (MSY), as required by the reformed Common Fisheries Policy (CFP).

Further, the Council and Commission have proposed a range of F values for the proposed multiannual plan for Baltic stocks that largely exceed F_{MSY}. There is no provision in the CFP permitting higher fishing mortalities than F_{MSY}, except for narrowly defined cases during a transition period until 2020.

The Commission had asked the International Council for the Exploration of the Sea (ICES) to provide estimates of ranges of plausible values around F_{MSY}, but without indicating the conditions governing those ranges. Many assumed—wrongly—that when the ICES advisers subsequently referred to 95%, that these figures had something to do with statistical confidence limits of the estimates, but that is not so. No such limits have been calculated or approximated. In fact, the ICES suggestions were not for "best" estimates of F_{MSY} values, but for best estimates of the F values resulting in catches no less than 95% of MSY. The selection of the 95% MSY limit is arbitrary (as ICES itself points out) and obscures the fact that the resulting F-ranges are highly asymmetrical, that is, the upper F value resulting in sustainable catches of 95% of MSY may exceed F_{MSY} by far, resulting in drastically increased costs of fishing, reduced catch per effort, and stock sizes well below the level that can produce MSY, thus being incompatible with the CFP. Instead, the lower value of F resulting in 95% of MSY may be a more reasonable and CFP-compliant management target.

The reasons for setting limits of F less than F_{MSY} are twofold. Firstly, it cannot be validly assumed that a fishery managed to provide MSY will be economically worthwhile—that it

[1]Editors' note: this comment was written by Sidney Holt and Rainer Froese and dated November 1, 2015. It was intended to inform the negotiations on the Baltic Sea fisheries management plan.

will be profitable by virtue of the market value of the catch being more than the cost of taking it. An F less than F_{MSY} would enhance the likelihood that the regulated sustainable fishery would be profitable. Secondly, a lower F will have a lesser impact on the productivities of the ecosystems being exploited and so contribute to enhanced conservation status of living marine resources.

Parliament has suggested that F should be set within the range $F = 0$ to F_{MSY}, with a target of $F = 0.8F_{MSY}$. The Commission has to date offered no guidance as to an optimum value within these ranges based on social and economic considerations that would be in conformity with the CFP nor has it asked ICES for scientific advice as to the implications of various options within that range. Instead, ICES advice continues to offer several options for continued overfishing ($F > F_{MSY}$) which are incompatible with the CFP. One of these options, namely fishing western Baltic cod in 2016 at $2.6F_{MSY}$, has just been chosen by the Council.

At the ICES Annual Scientific Conference held in Copenhagen in September 2015, one of the authors of this letter (SJH) presented calculations of the consequences, for sustainable catch, of various fishing mortality rates, and catch rates (catch per unit effort, that determine profitability), specifically for rates at 50% of the F_{MSY} level, for all known sets of values of the basic dynamic population parameters.[2] The calculations show that, with such fishing rates, the sustainable catches would be between about 99% and 90% of the MSYs observed in fishes. This range of results depends mainly on the selectivity of each fishery—i.e., whether fish begin to be caught when they are young and small or much later, as adults and even as mature adults, which would lead to a more natural age structure and higher catches. In all cases, the catch rates and the resulting profits will be nearly double what would be given by fishing at F_{MSY}. (The results of those calculations will be published in the *ICES Journal of Marine Science*). (*Editors' note: they were not published.*)

We suggest that, until such time as comprehensive analyses have been made by ICES and by the Commission of optimum rate of catch per unit effort and gear selectivity resulting in high yields and profits with low impact on the ecosystem, it would be sensible to consider adopting a provisional value of $F = 0.5F_{MSY}$.

We realize that in some stocks, implementation of this policy may lead to strongly reduced catches in the first year, which will, however, be more than compensated by continuous higher catches and profits in subsequent years. Stakeholders should be involved in finding the most acceptable transition path from the current high-cost-low-catch situation to the future low-cost-high-catch situation.

[2]See Chapter 57.

Chapter 55

Oral Presentation to the Committee on Fisheries of the European Parliament during a Public Hearing on Maximum Sustainable Yield in Fisheries Management[1]

I am addressing the matter of escaping from the doctrine, the **stranglehold**, of strict MSY-related targets.

Parliament and Council have adopted the Commission's **historic proposal** of January 2013 that the overall target should be to regulate the fishing mortality rate so that it is less than would be required to obtain MSY.

The agreed text **does not say "How much less."** That makes a practical problem for the Commission's scientific, technical, and economic advisers.

Another formulation was that fishing intensity should be such that the stock **biomass** is always greater than needed to generate MSY.

That raises another and different type of problem because the biomass changes with next year's recruitment, which is highly **variable** and practically **unpredictable**.

It is generally better to **monitor**—and to **control**—systems (like cars, trucks, trains, and machinery) by dealing with the human **input** of energy, if necessary limiting it with a throttle, rather than by simply controlling the **output**—with a brake.

Regulation by fishing effort input greatly improves the **stability** of the industry. Regulation by output adds to natural **instability**.

Sidney Holt speaking to Committee on Fisheries of the European Parliament. Photo by Tim Holt.

[1]Editors' note: presentation by Sidney Holt on February 23, 2015.

Some experts argue that keeping the stock size above the MSY-generating level is a **precautionary measure**, so avoiding the possibility of inadvertently reducing the stock below MSY level and so causing what we call *growth overfishing*.

True, but **avoiding waste** of excessive fishing effort and costs is practically far more important in practice. We can live with less catch than the maximum, but if there is no profit, fishing must either cease or be massively subsidized.

The EU's revised policy should ensure that future catches are sustainable, slightly less than a theoretical maximum, but obtained with much less wasteful fishing effort and, so, at far less cost.

MSY is the main management target in the UNCLOS (United Nations Convention on the Law of the Sea). UNCLOS also says it is OK to deviate from that target for social and economic reasons.

There is no unique MSY; *local msy* varies with selectivity. Selectivity does not depend only, or even mainly, on gear type (mesh, hook size).

Figure 1 illustrates a particular selectivity (capture begins before maturity) and dynamics driver ($M/K = 1.5$—typical for mackerels Scombridae); recruitment is not density-dependent.

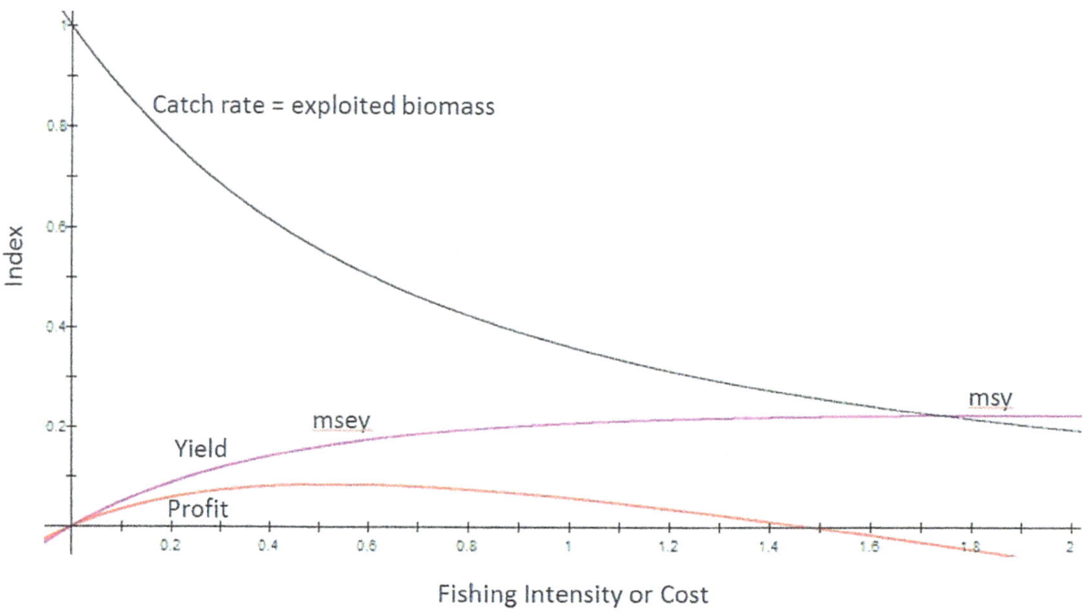

Figure 1. Graphs for a stock of a typical teleost (bony fish) species with dynamic natural mortality—growth rate driver $M/K = 1.5$.

Explanation of graphs

This is near the lower range of this parameter for the bony fishes from about 1.4 to 2.1 (the characteristic value for numerous species of *scombrids*—e.g., mackerels – is 1.6). The M/K value does not concern the general size of the species, which could be anything between that of a sardine Clupeidae and of a Cod *Gadus morhua*.

The curves shown are for a fishery with ***selectivity*** of fish length at first liability to capture of one-half the maximum length that the fish will attain if it lives long enough ($c = 0.5$), which is equivalent to the weight of a fish about 15% of the average weight of a full-grown fish (w/W =

0.15). This means that juvenile fish are not being caught, but near-adults are liable to be caught before they are sexually mature, which occurs generally when w/W = 0.3–0.4.

Here, I refer to maximum sustainable yields of a fishery with an identified selectivity as *local msys*, symbolized with miniscules. The ***absolute MSY***, when selectivity is at the critical age and size of fish, is unobtainable in practice, as it requires extremely selective fishing at infinite intensity.

Note that if the selectivity were to be increased, one way or another, towards the critical values of c and w/W, then both the ***local msy*** and the fishing intensity needed to obtain it also increase. Note also that the general range of values of the ratio M/K does not much affect the qualitative conclusions drawn here, though it does affect the numerical values cited.

The main, domed graph (purple line) is an index of average annual sustainable catch plotted against the fishing intensity. The catch index is an index of catch per average annual number of **recruits**, i.e., of young fish entering the exploitable population and so theoretically liable to capture. But, the value of the catch index depends on the selectivity of the fishing gear and also of the fishing operations (their location, seasonality, depth of capture, and so on).

The catch index must be multiplied by W (in kg)—the body weight to which all the fish would grow if they lived long enough—and by the average annual number of recruits R to give the total catch in kg. For the present purposes, the total market value of the catch can be taken to be directly proportional to its weight index and thus to its total weight, although, in practice, the price of a fish may increase somewhat with its size, and the average size of fish in the catch must decline as fishing intensity increases—this is because the fish will generally be caught when they are younger if the fishing is more intense.

The fishing intensity, on the x-axis, is not the fishing mortality rate F, but the ratio of F to the natural mortality rate or F/M. So, for example, a value F/M = 1.0 means the fishing mortality rate is equal to the natural mortality rate (these rates are not percentages, but are logarithmic—*exponential*). Such a rate would cause the number of fish in the exploited stock to be reduced to about one-half of those originally in the unexploited ("virgin") stock. The biomass of the exploited stock would be reduced to less than one-half that of the unfished stock because fish are being caught earlier in their growth than would be their average age and size in an unexploited stock.

I use here the index F/M rather than F itself to facilitate comparisons between different stocks and species.

To a first approximation, we can assume that the index F/M is directly proportional to the total cost of taking the catch, and the ratio of the catch index to F/M as proportional to the average catch rate (*catch per unit effort*—CPUE) in the fishery.

This graph shows a maximum average sustained catch of index 0.22 being obtained by a fishing intensity of F/M = 1.85.

The graph like a J in reverse (black line) is a plot of the ratio of the sustained catch index to the fishing intensity index. This is a measure of the biomass of the exploited phase of the stock, and equally of the catch rate. The biomass is calibrated to that of the stock before it was exploited as unity. The curve, therefore, shows the ***degree of depletion*** of the exploited phase of the stock as a function of the intensity of fishing. It shows that, in this example, the stock at *msy* level is depleted to about one-fifth of its unexploited biomass (this contrasts with a common assumption that at MSY the biomass is between 50 and 40% of the "virgin" size).

With that degree of depletion, it is quite likely that reduction of spawning females has reduced the annual recruitment. In that case, the F/M for *msy* would be less than that indicated here, and the *msy* itself would also be lower. If the relation between spawning stock size and recruitment has been determined, then these shifts can easily be determined exactly, but for this example, I assume recruitment has not changed. That will not qualitatively affect the conclusions I draw from the example.

The third, humped, curve (red line) at the bottom left is a plot of the economic sustainable yield (***msey***) against the fishing intensity index; *sey* is the difference between the value of the catch and the cost of taking it, in this example being, respectively, proportional to the weight of the catch (no allowance being made for dependence of price on size of fish) minus the cost of taking it, assumed to be proportional to the fishing intensity. This curve, naturally, is zero at the origin (no fishing at all) and zero at an *F/M* value I call ***break-even***. Break-even is a threshold beyond which continued fishing is not worthwhile to the fishers because their expenses are more than the market value of their catches. This *sey* curve, whatever the location, always has a maximum near the middle range, but it is not necessarily symmetrical. The peak I call **maximum sustainable economic yield (*msey*)** is taken with a fishing intensity of *F-msey*.

The *F/M* value at break-even can be higher or lower than that value which generates *msy*. (That taking *msy* will be profitable cannot, contrary to popular opinion, be taken for granted.) For this example, I have assumed that *F/M* is slightly lower than that which generates *msy*, which implies that adjusting intensity to catch *msy* would hardly be worthwhile.

In this example, we see that *msey* is taken by a fishing intensity of 0.5. That is just over 20% of the intensity needed to take *msy*. The sustainable yield index at *msey* is 60% of *msy*. To take it would be profitable, but not very. And to get it, we would have to forego 40% of the potential sustainable catch.

We can do better by aiming at an *F/M* value somewhere between the *msey* and *msy* reference points. We can look, for example, at the consequences of forfeiting—sacrificing—20% of the potential sustainable catch, taking an *sy* of index 0.18 instead of 0.22. This reduced yield could be taken with *F/M* = 0.65, which is at 65% less expense than needed to take *msy*. That would be highly profitable, providing a profit of about 80% over cost. With this intensity of fishing, the exploited stock would be depleted only to a bit less than one-half of its unexploited level, so with a much smaller chance of adverse effect on the recruitment rate.

Reduction of sustained catch—a "**sacrifice**"—of only 10% would result from a fishing effort—and, therefore, at a cost—of about 50% of what is needed to obtain MSY.

That is because models providing curves of sustainable catch against fishing mortality rate are always rather **flat-topped**. The reason for this is that the pattern in the mid-range of the curve is determined almost entirely by the interaction between the natural mortality rate in one direction and the growth rate—increase of body weight—in the other direction.

Potential economic benefits of holding the fishing intensity to less than is needed to take MSY are enormous.

This is **not some new idea**. A. G. Huntsman told the first UN Conference on Conservation and Utilization of Resources in 1949: **"The take should be increased only as long as the extra cost is offset by the added revenue from sales"** (Huntsman 1951).

R. A. Nesbit, in giving evidence in 1942 to a hearing about management, told a US fishery commission, **"the Commission should adopt a policy that will protect the fishermen as well as the fish … such a program will be easier to develop than a conventional program based on a policy of imposed inefficiency"** (Nesbit 1943).

In 1974, a series of international workshops involving 100 scientists, economists, and managers agreed that while MSY had been a useful educational tool, it was quite inadequate as a proper target for management (Holt and Talbot 1978).

We have known for many years that the **profit** from sustainable fishing—i.e., the difference between the market value of the catch and the cost of taking it—has a maximum. We call that **MSEY, maximum sustainable economic yield**. It is somewhere between the MSY level and a much lower fishing intensity.

It is not difficult to calculate if we know just a few things, especially what is the **break-even** catch rate—the minimum catch per unit effort by a fishing unit that makes continued fishing worthwhile.

It is commonly assumed that fishing at MSY level will be profitable, but that is by no means necessarily so.

I am not suggesting that **MSEY**—which some researchers have called the "optimum"—should be the policy target, any more than MSY should be. There are other good reasons for not seeking that particular "optimum." But, the biologically and economically desirable target should be somewhere between MSEY and MSY.

The scientific and economic advisers to the Commission should be asked by Parliament to look at such options. Their efforts should be addressed not merely to the dynamics of the fish stocks, but to the dynamics of the fishing operations and other features of the industry.

Three aims: A—economical, B—sustainable, and C—stable.

Conservative precautionary management, while "saving" the fish in the process, is supposed to be for *our* benefit, and especially to make fishing a worthy occupation again.

Some additional precautions are needed. One concerns the fear that even a well-managed fishery could lead to an occasional **unintentional depletion** of the stock to a level where there are insufficient mature females alive to guarantee normal recruitment. So, a high priority is to maintain fishing effort at a level that will have a negligible probability of ever inadvertently causing what we call **recruitment overfishing**,

Attention must be given to devising optimal patterns for **transition** of current stock states to the agreed sustained states. Such optima would balance the desire for the biggest **cumulative catch** in the circumstances against the need for fishing to be worthwhile during the transition.

Then, we need to design a **monitoring system** to ensure, when the policy is implemented, that it is performing as expected and, if it is not, to make appropriate corrections.

References

Holt, S. J., and L. M. Talbot. 1978. New principles for the conservation of wild living resources. Wildlife Monographs 59.

Huntsman, A. G. 1951. Research on use and increase of fish stocks. Pages 169–171 *in* Proceedings of the United Nations Scientific Conference on the Conservation and Utilization of Resources, 17 August – 6 September 1949, Lake Success, New York, Volume VI, Wildlife and Fish Resources. United Nations, New York.

Nesbit, R. A. 1943. Biological and economic problems of fishery management. Pages 23–53 *in* W. H. Herrington, and R. A. Nesbit, editors. Fishery management. US Fish and Wildlife Service, Special Scientific Report 18.

Chapter 56

Adjunct to Oral Presentation to the Committee on Fisheries of the European Parliament during a Public Hearing on Maximum Sustainable Yield in Fisheries Management[1]

In my presentation to Panel 1 on "**Reliability and Limits of MSY**," I address the matter of departing from strict MSY targets. I shall not be formally addressing Panel 2 on the "Practical Application of MSY in Fisheries Management," but I cannot resist giving my general opinion of this here. It is that managing fishing primarily by setting TACs is about the worst way of securing long-term sustainability. It is like driving using only the brakes instead of the accelerator or its equivalent to control the vehicle, be it a car, truck, or train. It wastes fuel, wears out the brakes and tires, and increases the chance of accidents.

The TAC management system is essentially the imposition of inefficiency on fishing operations. As the biologist R. A. Nesbit told the US Atlantic States Marine Fisheries Commission in 1942 (Nesbit 1943), "I earnestly recommend that the Commission adopt a policy and program that will protect the fishermen as well as the fish. In one important respect such a program will be easier to develop than a conventional program based on a policy of imposed inefficiency."

On this occasion, Nesbit also asserted, correctly, that the absolute MSY can, in theory, be attained only by catching all of the surviving fish in the oldest age class (age group) in the stock when that cohort reaches its maximum total weight at a certain *critical age and size of fish*. That is obviously impossible. The values and properties of MSY depend very much on the selectivity of the particular fishery—that is, at what size and age the fish first become vulnerable to capture. For any given size and age, there can be a maximum sustainable catch, which I refer to as *local MSY* and symbolize as *msy*. I assume that when we mention MSY, we are actually talking about a particular chosen **local** *msy*.

On February 6, 2013,[2] this Parliament opened the door a little for us to escape the half-century stranglehold of maximum sustainable yield doctrine.

Parliament has said the prime objective of managing fisheries in which European nations participate should be to adjust the intensity of fishing in such a way that the fishing mortality rate is less than what would be needed to generate an *msy*. That means keeping fish stocks bigger than they need to be to produce *msy*. Council and the Commission eventually formally agreed with that, but in somewhat garbled and ambiguous language. (They needlessly intro-

[1]Editors' note: prepared by Sidney Holt on February 23, 2015.

[2]Editors' note: this is the date Parliament adopted the Rodust report P7_TA(2013)0040 on the European Commission's proposal for the basic regulation of the Common Fisheries Policy.

duced the undefined term "exploitation rate," but did not define it. It has several very different meanings in the literature of fisheries science and management.)

No one has yet dared to say how far they think the lower fishing mortality rate should be from the *msy* rate. That presents a problem because it limits the possibility for scholars (scientists and economists) to give useful management advice. Here, I make some suggestions for resolving this.

Parliament's idea was not a new one. Throughout the years, from the time of World War II, several people have published valid arguments that an *msy* in itself, while admittedly a useful reference point in certain circumstances, is not a good target for rational, managed fishing. A distinguished Canadian biologist, A. G. Huntsman, told the United Nations Scientific Conference on the Conservation and Utilization of Resources, held at Lake Success, New York in 1949: **"The take should be increased only as long as the extra cost is offset by the added revenue from sales"** (Huntsman 1951). In an earlier statement in 1944 in an article in Science (Huntsman 1944), Huntsman had usefully defined "irrational fishing" as "where the take in proportion to the effort fails to yield a satisfactory living to the fisherman."

One myth, or assumption, to be discarded is that fishing at *msy* would always or automatically be profitable. That is by no means sure.

Several economists and mathematicians, among them H. S. Gordon and C. Clark, as well as Ray Beverton and me (Beverton and Holt 1956), have described the advantages to be had from lower fishing rates. In 1977, Peter Larkin went so far, famously, as to publish "An epitaph for the concept of MSY." A collective, carefully considered, mainly negative, opinion of the MSY history was formulated in a sequence of international meetings of over 100 leading scientists convened at Airlie House, Virginia, in 1974. That opinion is, I think, worth reproducing here (see Annex).

Nevertheless, MSY has had a long life of its own and is still revered—perhaps because, among other things, it is assumed to be "scientific" by many administrators, lawyers, and others engaged in management. It is embedded in the 1982 UN Convention on the Law of the Sea, but with a very important condition, an escape clause: "*...as qualified by relevant environmental and economic factors*."[3] The authorities of the European Union have, wisely and effectively, used that clause—without actually referring to it—in the basic regulation for the revised Common Fisheries Policy. Here, it must be said that there is practically no such thing as a unique MSY from any stock. There is a theoretical one, but it is unobtainable. What we can discuss is what I call **local *msys***, each determined by the selectivity of a particular fishery.

The variation of local *msy* with selectivity is not trivial—*msy* for a particular stock, in which, for example, juveniles are exploited for oil and meal production as compared with exploitation of adult, possibly mature, fish as food for humans, can vary by twofold or more; the fishing intensity required to obtain these *msy*s may well vary by three- or fourfold.

MSYs and *msy*s have sometimes been described as "optima," but usage of this term in resource management is remarkably ambiguous. In an address to the International Whaling Commission (IWC) in 1972, Sir Peter Scott, representing the World Wildlife Fund (WWF) expressed the general bewilderment: "I am astonished by the confusion even here in the Commission which surrounds the phrasing 'maximum sustainable yield,' 'optimum sustainable yield,' 'actual sustainable yield,' and maybe we should be thinking in terms of maximum sus-

[3]This conditional clause was inserted at a late stage in the negotiation of UNCLOS to accommodate "developing countries" that wished to be free to exploit newly identified, previously unexploited stocks and species unsustainably, at least in the early stages of the fishery. The agreed words are, however, such that it can be interpreted either way.

tainable good. The phrase is not mine—it is Victor Sheffer's" (Scheffer was one of the most distinguished and knowledgeable scientists in this field, and long-time chair of the US Marine Mammal Commission).

That year, the United Nations had voted in favor of declaring a ten-year moratorium of all commercial whaling. The IWC tried to avoid that and opted, in 1974, for a management policy based on MSY doctrine. By 1982, that policy had demonstrably failed, an indefinite moratorium was adopted by the Commission, and a process to devise another management procedure launched. The new scheme was adopted in 1994 and the *M* of MSY was abandoned, the *S* was, naturally, retained, but enhanced by the notions of "stability" and "robustness."

A distinguished fisheries scientist Joe Horwood (2008), who had participated in the IWC process, later expressed his opinion broadly as follows: "Fishery managers have not wanted the bed-rock advice on stock (and ecosystem) sustainability confused with economic advice; it is difficult enough as it is. It can be difficult to pin economists down to give the required parameters (values, costs, elasticities needed over the next decade); the task itself is difficult, but so is predicting future fish recruitment. Sustainability is a common currency. However, are we really interested in the optimum? There are many reasons why we are not."

Michael Graham, representing the UK at the UN Technical Conference on the Law of the Sea in Rome in 1955, expressed a similar sentiment: "We in Europe wish not to seek Utopia, but to move in the right direction, to improve our fisheries, combat overfishing, and ensure a good living in future for our fishermen."

All stock assessment models show that curves of sustainable yield (in weight or market value) plotted against fishing intensity (fishing mortality rate *F* or against overall cost) are rather flat-topped, but gently declining on the left-hand side towards lower *F* or cost and bigger total biomass. That means that a **substantial** reduction of fishing intensity, below the *msy* level, produces a sustainable catch only slightly less than the *msy*, at a **higher catch rate (CPUE)** and, therefore, greater **profitability**—a term that I am defining here, crudely, as the difference between the market value of the catch and the cost of taking it. That profit reaches a maximum (*msey*) at a certain fishing intensity. Reduce the intensity further and the profit eventually falls back to zero again. The *percentage* rate of profit, however, changes as the fishing intensity decreases.

(Note that there is no *MSEY*—or, rather, it is always zero—because an infinite selective fishing intensity would be required to capture the theoretical *MSY*.)

Calculations of *sey* curves and *msey* depend, as a first approximation, on estimates of the ***break-even*** fishing intensity in each instance, i.e. where profit from a sustainable catch is zero. If this fishing intensity is **higher** than that needed to generate *msy*—i.e., *MSY* policy would be profitable—the intensity for *msey* would be only a little less than that for *msy*. If sustainable fishing is already **un**profitable—i.e., when the break-even fishing intensity is **below** the level for *msy*—then the intensity for *msey* could be much lower, probably with greater "sacrifice" of potential catch.

When fishing intensity is reduced, the average **size** of fish in the catch increases. This may call for adjustment to the total value of the catch where market price is higher for larger fish than for smaller ones.

The crucial management question is, therefore: where should the optimum be found between *msy* and *msey*? That is for the management authorities to decide on the basis of an assessment of social needs. But, scientists and economists can clarify the options and the probable consequences of adopting any one of them.

I suggest that one parameter to be taken into consideration is the expected sustainable catch as a fraction (%) of the *msy*. That is one measure of the "sacrifice" of potential catch being made for the general economic and social benefit.

I finish with some observations about the transition scenarios from a possible over- (or even under-) fishing situation to rational sustainable fishing. The Council, Parliament, and Commission have decided on two target dates for reaching the first gate to Utopia—2015 and 2020. But, I don't think it asked any of the fishes whether they could make it—i.e. grow and breed fast enough. But, it seems that the scientific advisory groups were not asked for their opinions, either. I think that was an error that should be corrected soon. Several general studies have been made of the feasible options for a satisfactory transitional process—I made one myself in the 1950s—and these should be explored.

Acknowledgement

I wish to thank Michael Earle, Leslie Busby, and Tim Holt for their helpful comments and reminders—especially those beginning "I don't understand..."—during the preparation of this brief.

References

Beckett, C. W., L. Brewer, C. E. Holler, Jr., W. Hubbard, and F. D. Rossini. 1974. Final report. Conference on thermodynamics and national energy problems held at Airlie House, Warrenton, Virginia, June 10–12, 1974. National Academy of Sciences, Washington, DC.

Beverton, R. J. H., and S. J. Holt 1956. A review of methods for estimating mortality rates in exploited fish populations, with special reference to sources of bias in catch sampling. Rapports et Procès-Verbaux des Réunions du Conseil Permanent International pour l'Exploration de la Mer 140:67–83.

Holt, S. J., and L. M. Talbot. 1978. New principles for the conservation of wild living resources. Wildlife Monographs 59.

Horwood, J. 2008. After Beverton and Holt. Pages 49–62 *in* A. Payne, J. Cotter, and T. Potter, editors. Advances in fisheries science: 50 years from Beverton and Holt. Blackwell Publishing, Oxford.

Huntsman, A. G. 1944. Fishery depletion. Science 99(2583):534–535.

Huntsman, A. G. 1951. Research on use and increase of fish stocks. Pages 169–171 in Proceedings of the United Nations Scientific Conference on the Conservation and Utilization of Resources, 17 August – 6 September 1949, Lake Success, New York, Volume VI, Wildlife and Fish Resources. United Nations, New York.

Larkin, P. A. 1977. An epitaph for the concept of MSY. Transactions of the American Fisheries Society 106:1–11.

Nesbit, R. A. 1943. Biological and economic problems of fishery management. Pages 23–53 *in* W. H. Herrington, and R. A. Nesbit, editors. Fishery management. US Fish and Wildlife Service, Special Scientific Report 18.

Annex

Excerpt from the Report of the Conference on Thermodynamics and National Energy Problems, June 10–12, 1974, Airlie House, Warrenton, Virginia (Beckett et al. 1974):

"The concept of maximum sustainable yield as a simple function of stock size was developed to provide a generalized approximate description of the response of a stock or species to exploitation. It has served a useful evolutionary function, as a means of generalized description, and as an elementary teaching aid for students and administrators. It has provided a preliminary conservation goal where it has been used to try to avoid or correct over-exploitation. It has played a significant role in the evolution of understanding of wild populations.

However, like some other simplified concepts, maximum sustainable yield has become institutionalized in a more absolute and precise role than intended by the biologists who were responsible for its original formulation. It is being expected to perform functions for which it was never intended, serving for example, as the sole conceptual basis for, or goal of, management in some cases. Once a concept has been adopted and institutionalized, it is difficult to change it. In this case, because of its institutionalization, the concept of maximum sustainable yield is now an obstacle to the acceptance of concepts that derive from present ecological knowledge, and that would provide a more adequate basis for management. A more adequate basis is especially necessary at this time when the overall impact of man on the biosphere is increasing and diversifying as never before."

This critique deals only with scientific issues, not—as in my presentation—with the fact that the simplistic MSY concept is economically untenable.

The Airlie workshops also formulated some **Principles of Conservation of the Living Resources of the Sea**. The first of these is "*...avoiding wasteful use of other natural resources expended to secure the supply of food and other marine products*." The intention here was primarily to avoid excessive fishing effort, requiring use of more fuel and provision of vessels as well as manpower, and is expressed in the report as: "*(MSY) focuses attention (only) on the output from resource use, without regard to the input of energy, of other natural resources, and of human skill and labor required to secure the output*."

The second principle was a form of what we now call the Precautionary Principle:

"*... providing a margin of safety to allow for unpredicted variations and characteristics of marine resources and their environment, and for the fact that the application of measures may be subject to delay or be otherwise imperfect*."

And, the third Principle was: "*... securing in the first place a supply of food for human consumption*."

Chapter 57

What is Overfishing?[1]

The organizers of this Theme Session of the 2015 Annual Science Conference of the International Council for the Exploration of the Sea (ICES) have, I think, circulated a text of my intended presentation this morning, entitled “Escaping the stranglehold.”[2] It began by my asking “What is an assessment?” The stranglehold I had in mind was the universal acceptance of something called maximum sustainable yield (MSY) as a target for managing sustainable fishing. But, I was also concerned by the continued use in some quarters of a pseudo-scientific and misleading process for estimating MSY called the method of surplus production.

My friend Trevor Branch recently widely circulated on the Web a blog I wrote in October 2011 entitled “Maximum Sustainable Yield: The Worst Idea in Fisheries Management” (Holt 2011). I stand by that opinion, especially as that blog also took a swipe at “surplus production.”

But, I have decided, since writing those pieces, to try to take a more positive view, suitable for this occasion—and limited time—and discuss another popular idea—that of “**overfishing**.”

“**Overfishing**” was a term generated in the late 19th century, but codified in the late 1930s, especially by the English scientists E. S. Russell and Michael Graham. It expressed an opinion about the state of a fishery, especially with respect to the excessive size of the fishing fleets, rather than about the states of the fish stocks that were being sharply affected by the operations of those fleets (Russell 1942; Graham 1951). Another different use of the term, linked to the state of the fish stock itself, describes a level of depletion of the stock by fishing such that future recruitment will be impeded. This has become more fashionable in post-World War II years since, especially, industrial fishing for fishoil and meal production, especially from young fish and small species, has led to collapses both of stocks and of fisheries. The Hilborn and Hilborn (2012) book is, I think, the best popular introduction to this matter.

In 1944, a Canadian biologist, A. G. Huntsman, wrote that “irrational fishing is where the take in proportion to the effort fails to yield a satisfactory living to the fisherman,” so bringing economics into the discussion. Eventually, in 1949, Huntsman (1951) told the United Nations Scientific Conference on Conservation and Utilization of Resources (UNSCCUR), held that September at Lake Success, New York, that “the take should be increased only as long as the extra cost is offset by the added revenue from sales.”

In terms of the well-being of commercial fishing, disobeying that rule is, I think, the most useful definition of overfishing.

---

[1]Editors’ note: oral presentation given to the Theme Session “Advancement of stock assessment methods for sustainable fisheries” at the 2015 ICES Annual Science Conference.

[2]Editors’ note: that text, originally intended for oral presentation, is reproduced as Chapter 53 and is a different story than presented here in Chapter 57.

However, in the decade after World War II, usage changed. This came about mainly as a result of wide adoption of the idea of ensuring the sustainability of fisheries by management procedures that would—directly or indirectly—ensure that stocks were not depleted by fishing to levels or reduced to states insufficient to provide MSYs or, if they had been so damaged, would be allowed to recover to so-called **optimal** levels and states.

The origin of international policies to enforce MSYs lies in the peace treaty negotiations between the Governments of Japan and the US. It was intended to support an "exclusion principle" and related rules that would favor the fishing operations of one state over another, and especially favor the state with the strongest ability to conduct the requisite scientific studies (Finley 2011). MSY was said to be synonymous with "full utilization" of the resource.

Sustainable Yield

There is a common misunderstanding that "sustainability" is some sort of intrinsic property of fish populations and that it arises as a consequence of density-dependence of one or more parameters of a model describing their dynamics. On the contrary, **potential sustainable yields are generated by the act of fishing**. Fishing reduces the numbers of fish alive in the sea, but, more importantly in this context, it alters the age composition of the population. This alteration does not arise because of the selectivity of fishing operations, although that can enhance the composition changes. The basic change in age composition is simply a consequence of the increased mortality rate caused by fishing. This increases the relative abundance of younger, hence smaller, fish compared with the older, bigger ones.

While the natural mortality rate in the exploited phase of the stock may be relatively constant, the pattern of body growth of the fish is quite different. Animals which, like fish, have indeterminate growth patterns—some increase in size continuing through life—increase in weight with age along a sigmoid curve. Their fastest rate of growth is when they are at the inflexion of that curve, which is usually about one-third of their potential final *asymptotic* weight. Their *specific growth rate* (in mathematical terms dW/Wdt) is fastest when they are small and declines continuously throughout life.

If an unfished stock is in a steady-state—neither increasing nor decreasing—the total weight of all fish dying naturally is equal to the increase in total weight of the survivors, plus an annual increment when a new brood of recruits arrives. That weight of growth is the number of fish times the average specific growth rate of all of them. A change in composition of the stock as a consequence of fishing increases that overall growth in the population simply because the specific growth rate of the younger is faster than that of the older animals. If fishing is suddenly stopped, the stock will begin to grow, to increase in total weight. If that increase is regularly removed by humans, we have a sustainable yield or, as some would say, "a surplus." There is no density-dependence, except that which controls the flow of recruits.

I should perhaps explain why graphs of sustainable yield against fishing intensity are inevitably rather flat in the range from f_{msy} to substantially lower fishing intensities than that. Their shape is determined by the interaction between a negative mortality rate and the positive function of growth in body weight. As fishing intensity increases, the number of live fish decreases, but the average specific rate of growth increases because of the change in age composition of the stock towards younger fish. These two rates tend to cancel each other.

This universal feature of fish population dynamics is not captured by the usual application of the assessment method called "surplus production" in which the potential population growth rate (hence sustainable yield) is assumed to be a simple power function of the biomass of the population or some defined part of it. The actual form of the relationship between biomass and growth rate cannot be represented by a simple polynomial with only a few terms.

Management

The UN Convention on the Law of the Sea (UNCLOS 1994) enshrines MSY as a universal management objective, but it also offers a loophole: "… as qualified by relevant environmental and economic factors..." This qualification was inserted during the UN negotiations to satisfy less-developed countries that wished to exploit stocks unsustainably—at least temporarily—when they were establishing new fisheries on previously unexploited stocks. The loophole is, however, as eventually worded, two-sided; that is, it facilitates states managing fishing adopting measures intended to keep stocks *above* "MSY levels" and thus holding catches to something less than MSY.

The European Union (EU) has agreed, by a consensus of its three basic administrative and political components—Parliament, Council, and Commission—that the target for management in its revised Common Fisheries Policy (CFP) is to manage fishing so that the fishing intensity is less than that needed to provide MSY. This is not a new idea; it was long ago advocated by at least two international gatherings of economists and scientists (Holt and Talbot 1978; Mangel et al. 1996). But, the EU has not yet specified how much less? The purposes of "less" are to ensure stability and that each fishery has a sufficiently high catch rate (catch per unit effort; CPUE) to be worthwhile—that is, the market value of the catch being adequately greater than the cost of taking it. The "less than" also has a precautionary element, providing some protection against scientific uncertainty and errors in stock assessment, for which the scientists working in ICES to provide management advice to the EU have traditionally selected confidence levels plus or minus 5%. As we shall see, the application of the "less than" principle for social and economic reasons is likely to call for a "margin" much wider than that.

The decision "how much less?" is evidently an economic and social one. Scientists have, however, an important role to play in advising what are likely to be the effects—on sustainable catches, catch rates, fish stocks, and marine ecosystems—of alternative economic and social options. Over the past 60 years, several economists have suggested as a target, securing maximum sustainable economic yield (see Gordon 1954; Beverton and Holt 1957; Clark 1976). An approximation to this can be calculated fairly approximately if certain economic information is available, most importantly the break-even catch rate—when the market value of the catch is equal to the cost of taking it. This will, however, commonly call for a much reduced fishing intensity and catches well below the supposed MSY level. In my opinion, f_{MSEY} might be considered as a lower bound to possible "less than" decisions.

I shall use a definition of relative fishing intensity f, the ratio of the fishing mortality rate F—an exponential coefficient—to the natural mortality rate M in the exploited phase of the lives of the fishes in the stock; f has a simple relationship with another index of fishing intensity, used in some published tables and expressions: the exploitation rate: $E = F/(F + M)$. Thus, $E = f/(f + 1)$ or $f = E/(1 - E)$ (Beverton and Holt 1957; Gulland 1983).

Here, I think something must be said about MSY. It is a mythical number or, at least, a quantity unobtainable by us. As early as 1943, R.A. Nesbit was perhaps not the first to show that MSY—though he didn't call it that; Hjort later called it the optimum catch—would be obtained by catching all the surviving fish in the oldest cohort (year class) of the stock as soon as they reach a critical size and age, which is when the biomass of that cohort reaches its maximum. That is clearly impossible, in practice, for a wild fish stock, but it can serve as a reference point against which other sustainable catches can be measured.

Graphs of sustainable yield as a function of fishing intensity are sometimes peaked, depending on the *selectivity* of the fishery (basically the size and age at which fish first become liable to capture by the fishing gear used and the pattern of the fishing operations, by location, depth, and season). In hook-and-line and gillnet fisheries, it may be necessary also to specify the age and size above which the fish are not retained. Where such peaks occur, it is mainly in situations where selectivity is low (i.e. small young fish are being caught). I refer to them as local MSYs and indicate them by lower case, i.e., *msy* and the fishing intensities that generate them as f_{msy}. The selectivity of, for example, trawl and seine fisheries can be defined by ratios of the size at first liability to capture to the asymptotic size (weight W; length L) of the fish on the graph of body growth against age, c if growth is specified as body length, w if expressed as body weight.

ICES has, to date, been providing advice to the Commission apparently with the generally unstated presumption that changes in regulations that alter the fishing intensity—either directly or as a result of catch limitations—or the seasonal or geographical distribution of fishing will not alter the selectivity. We must remember that while selective exploitation changes both the abundance and age and size composition of the stock, **unselective** fishing also does that (in fact, it is just such changes in composition that *create* potential sustainable yields, but that is another story).

Given an implausible assumption of no change in selectivity, it seems reasonable to express any "less than" option as a specified percentage of the relevant *f-msy* and the resulting sustainable yield *sy*, as a percentage of the corresponding *msy*. Then, a practical question is this: if we are trying to regulate fisheries for numerous species and stocks using a single general rule, would we have to define such a percentages for each of them or could we opt for the same percentage measure of "less than" for all of them? The rest of this presentation examines that question.

To illustrate this problem, I have chosen a fairly drastic "less than": fishing with an intensity of only 50% of the intensity required to take *msy*. I have looked at a range of selectivities (at levels low enough that there are *msy*s). To deal with differences between species, I have chosen to do a limited number of calculations over a range of values of the basic dynamic parameter $m = M/K$. According to Charnov (1993), a dataset for 175 species and stock analyses extended over an m-range from about 1.0 to 2.1. Charnov calculated mid-values for this range by several methods. Two linear regressions gave central values of 1.6 and 1.89. The simple average of all the data was 2.07, and the geometric mean 1.75. I have looked at the consequences of selected values of m = 1.0, 1.5, 1.75, and 2.0.

Table 1 below shows the chosen values of m, c, and w; $w = 0.3$ is important because that is the body size at which body growth rate is highest—the inflexion of the growth curve. Commonly, sexual maturity occurs just after this. This table also gives the critical sizes, and corresponding MSYs. The latter are indices, MSY divided by the asymptotic weight W and the average annual number of recruits to the stock, assuming there is no time-trend in that number.

Table 1.

$m = M/K$	$k = 1/m$	w_{MSY}	l_{MSY}	MSY/WR
1.0	1.0	0.422	0.75	0.105
1.5	0.67	0.296	0.67	0.057
1.75	0.57	0.252	0.63	0.044
2.0	0.5	0.216	0.60	0.035

Results

Table 2 below summarizes the results of calculations of sustainable yields, *msys*, etc. using the parameter values given in Table 1. They show the well-known properties of the simple Beverton–Holt expression for sustainable yield (the most convenient format of that expression was provided by Gulland 1983, p. 152, Equation 5.11, and repeated in Quinn and Deriso 1999).

I draw your attention to the last column of Table 2. This shows that in the ranges of the chosen m and c values, the sustainable catches at the lower fishing intensity (50% of that required to take *msys*) are rarely less than 90% of *msy* and commonly up to 99%. The catch rates at this lower intensity would be nearly double those expected from an *msy* policy. It is clear that an *msy* policy is crazy economics, whether or not it is profitable. Imagine an entrepreneur producing a desirable object in his factory. Is it likely that he would double his investment and operating costs merely in order to produce at most 10% more such objects, and possibly just 1% more!

Figure 1 shows the graphs of *sys* y/f and dsy/df with the parameter values $m = M/K = 1.75$; $c = 0.36$; $w = 0.05$.

Table 2.

l/L	w/W	msy	f_{msy}	msy/f_{msy}	sy/fsy	sy/msy
$m = 1.0$						
0.35	0.04	0.11	1.22	0.09	0.17	0.94
0.42	0.07	0.14	1.48	0.095	0.18	0.96
0.55	0.17	0.205	2.5	0.082	0.19	0.99
0.67	0.30	0.31	8.2	0.038	0.08	1.00
$m = 1.75$						
0.35	0.04	0.18	0.20	0.133	0.231	0.87
0,42	0.07	0.20	0.22	0.133	0.217	0.90
0.55	0.17	0.27	2.25	0.12	0.240	0.91
0.67	0.30	0.36	3.5	0.10	0.19	0.94
$m = 2.0$						
0.35	0.04	0.20	1.5	0.056	0.941	0.90
0.42	0.07	0.22	1.5	0.011	0.97	0.91
0.55	0.17	0.27	2.2	0.234	0.060	0.87
0.67	0.30	0.37	9.36	0.10	0.19	0.95

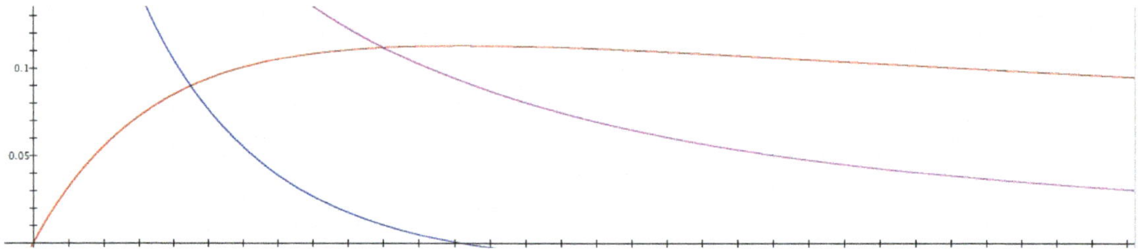

Figure 1. The x-axis is fishing mortality, and the y-axis is sustainable yield (red), stock biomass (purple) and the first derivative of yield as a function of fishing mortality (blue).

Density-dependence

Table 3 below shows the consequences of fishing in producing a smaller exploited stock than existed before fishing began. "Depletion" is the ratio of the $0.5f_{msy}$ exploited stock biomass to the biomass of that part of the stock before fishing began. In general, the exploited part of the stock will be reduced to and held at about one-fifth of its "virgin" size. It is possible that such depletion will bring about some reduction in the annual recruitment resulting from a lower production of eggs and larvae from a reduced number of mature females, and perhaps enhanced by their smaller average size.

It has recently become fashionable to refer to the slope h of the curve of recruitment against mature stock size or some similar measure of reproductive potential of the population. Slope has been defined as the percentage reduction in recruitment caused by a stock reduction to one-fifth—by coincidence about the same as the depletion we predict from a $0.5f_{msy}$ management policy.

The slope h is usually measured on a Beverton–Holt stock–recruitment model which has an upper asymptote for the recruitment from a very high reproduction of eggs and larvae to a practically proportional relationship if reproduction is greatly reduced.

Unfortunately, we rarely know the shape and scale of that curve and, even if we did, the figure for slope depends critically on the location on that curve of the "virgin" starting point; the slope is not a single number derived from the curve's parameters.

Table 3.

$m = M/K$	$c = l/L$	Depletion
1.0	0.35	0.166
	0.42	0.182
	0.55	0.193
	0.67	1.075
1.75	0.35	0.231
	0.42	0.240
	0.55	0.217
	0.67	0.191
2.0	0.35	0.240
	0.42	0.267
	0.55	0.236
	0.67	0.215

We can easily, however, deduce the general effects of such a relationship. Remember we are interested only in the part of a sustainable catch–fishing intensity curve at intensities less than that required to take a local *msy*. The effects of any reduction in recruitment as fishing intensity increases are twofold. The sustainable catch will be slightly bigger than the simple *per-constant average recruitment* calculations tell us, but the local *msy* will be obtained at a slightly lower fishing intensity. The overall situation will be unchanged—an immense economic benefit with a higher catch rate (and larger average size of fish in the catch) and rather little penalty from a somewhat lower total catch).

Other differences from reality of the simple population model should be investigated theoretically, though I do not think any of them would change our general conclusions. The most important of these differences are possible density-dependent changes in the body growth parameters, variation of the natural mortality rate with age of fish within the exploited phase of the stock, and perhaps overall density-dependent changes in either of both of the parameters K and W of the von Bertalanffy growth curve.

References

Beverton, R. J. H., and S. J. Holt. 1957. On the dynamics of exploited fish populations. Her Majesty's Stationery Office, UK.

Charnov, E. L. 1993. Life history invariants. Some explorations of symmetry in evolutionary ecology. Oxford University Press, New York.

Clark, C. W. 1976. Mathematical bioeconomics: the optimal management of renewable resources. John Wiley, New York.

Finley, C. 2011. All the fish in the sea: maximum sustainable yield and the failure of fisheries management. University of Chicago Press.

Gordon, H. S. 1954. The economic theory of a common property resource: the fishery. Journal of Political Economy 62:124–142.

Graham, M. 1951. Overfishing. Pages 20–24 *in* Proceedings of the UN Scientific Conference on the Conservation and Utilization of Resources, 17 August–6 September 1949, Lake Success, New York, Volume VII, Wildlife and Fish Resources. United Nations, Department of Economic Affairs, New York.

Gulland, J. A. 1983. Fish stock assessment. A manual of basic methods. FAO/Wiley Series on Food and Agriculture, 1. John Wiley and Sons, Chichester, UK.

Hilborn, R., and U. Hilborn. 2012. Overfishing: what everyone needs to know. Oxford University Press, New York.

Holt, S. J. 2011. Maximum sustainable yield: The worst idea in fisheries management. Blogpost on Breaching the Blue, 3 October 2011. Available: https://breachingtheblue.com/2011/10/03/maximum-sustainable-yield-the-worst-idea-in-fisheries-management/#comment-940FirefoxHTML\Shell\Open\Command. (March 2021).

Holt, S. J., and L. M. Talbot. 1978. New principles for the conservation of wild living resources. Wildlife Monographs 59:1–33.

Huntsman, A. G. 1944. Fishery depletion. Science 99:534–535.

Huntsman, A. G. 1951. Research on use and increase of fish stocks. Pages 169–171 *in* Proceedings of the UN Scientific Conference on the Conservation and Utilization of Resources, 17 August–6 September 1949, Lake Success, New York, Volume VII, Wildlife and Fish Resources. United Nations, Department of Economic Affairs, New York.

Mangel, M., L. M. Talbot, G. K. Meffe, M. T. Agardy, D. L. Alverson, J. Barlow, D. B. Botkin, G. Budowski, T. Clark, J. Cooke, R. H. Crozier, P. K. Dayton, D. L. Elder, C. W. Fowler, S. Funtowicz,

J. Giske, R. J. Hofman, S. J. Holt, S. R. Kellert, L. A. Kimball, D. Ludwig, K. Magnusson, B. S. Malayang, C. Mann, E. A. Norse, S. P. Northridge, W. F. Perrin, C. Perrings, R. M. Peterman, G. B. Rabb, H. A. Regier, J. E. Reynolds, K. Sherman, M. P. Sissenwine, T. D. Smith, A. Starfield, R. J. Taylor, M. F. Tillman, C. Toft, J. R. Twiss, J. Wilen, and T. P. Young. 1996. Principles for the conservation of wild living resources. Ecological Applications 6(2):338–362.

Nesbit, R. A. 1943. Biological and economic problems of fishery management. Pages 23–53 *in* W. H. Herrington, and R. A. Nesbit, editors. Fishery management. US Fish and Wildlife Service, Special Scientific Report 18.

Quinn II, T. J., and R. B. Deriso. 1999. Quantitative fish dynamics. Oxford University Press, New York.

Russell, E. S. 1942. The overfishing problem. Cambridge University Press, Cambridge, UK.

Schaefer, M. B. 1953. Fisheries dynamics and the concept of maximum equilibrium yield. Pages 53–64 *in* C. P. Idyll, editor. Proceedings of the Gulf and Caribbean Fisheries Institute, Sixth Annual Session, Miami, Florida, November 1953. University of Miami, Coral Gables, Florida.

Postscript

Conversion of an *sy–f* graph to one illustrating economic yield (difference between market value of catch and cost of taking it) is done choosing a value q for the ratio of the market value of a unit of sustainable catch to the cost of exerting a unit of exploitation rate f.

In Figure 2, I illustrate the choice of q such that the break-even point is at the point of *msy*. This reveals an *mey* point at about $f = 0.45$, which is about 32% of f_{msy}.

The value for f_{mey} is sensitive to the value of q. In the example, for instance, if the break-even f is 50% higher than f_{mey}, the f_{mey} is 0.6. If, on the other hand, break-even is at $f_{50\%}$ less than f_{msy}, then the f_{mey} is about 0.35.

The highest rate of profit [(value of catch/cost of taking it)/cost] is highest when f is near zero, and declines steadily—nearly linearly to zero—at the break-even f.

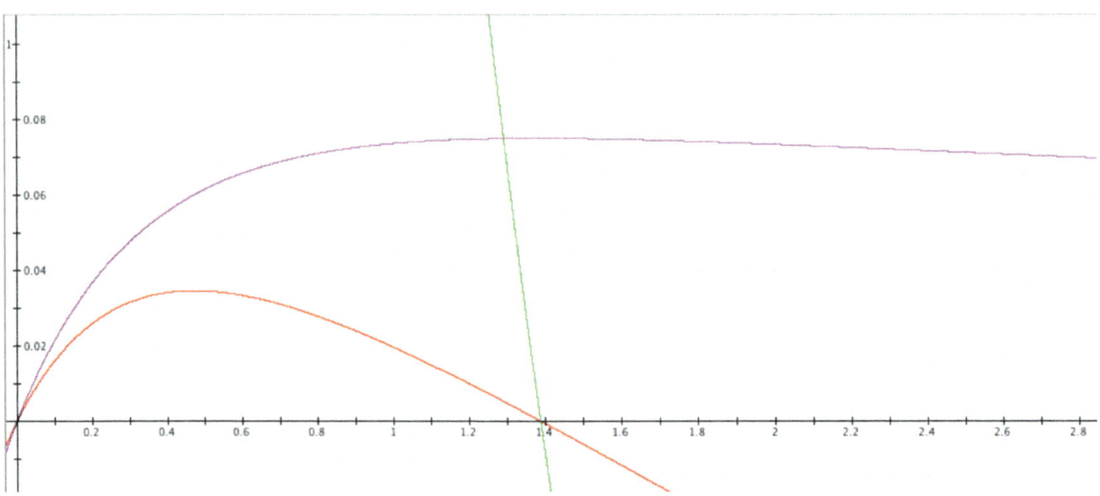

Figure 2. Sustainable economic yield. The x-axis is fishing mortality, and the y-axis is value (revenue, purple; profit, red; profit/revenue, green).

Chapter 58

North Sea Demersal Fisheries Management Consultation—EU[1]

This document is my contribution to the Consultation as a EU citizen. I am British, resident in Italy, date of birth February 28, 1926. I am "retired," but still researching and publishing in scientific journals. I was a consultant to the EU Parliament regarding the Baltic Sea Plan.

I began research in 1946 concerning the North Sea, but a few years later began my long association with the Food and Agriculture Organization of the United Nations in Rome, where I became Director of the Fisheries Resources and Operations Division of the Department of Fisheries. I also worked for a few years as Director of Marine Science in the United Nations Educational, Scientific and Cultural Organization, Paris and Secretary of the Intergovernmental Oceanographic Commission (IOC).

First, I must say that creation of separate management plans for the demersal and—presumably—pelagic fisheries in this area does not make sense. For example, there are certainly interspecific relations between some of the demersal species and pelagic ones. Some of the larger demersal species such as Cod *Gadus morhua* eat smaller pelagic species and the young of the larger pelagic (as well as being cannibalistic). In any case, while Cod is usually grouped among the demersals, it is also quasi-pelagic, unlike e.g., the flatfishes.

The EU institutions have determined that the principle objective for managing sustainable fishing in EU waters is to set regulations to ensure that long-term sustainability is maintained by a fishing effort that will generate fishing mortality rates of all exploited stocks at values less than those needed to produce maximum sustainable yields (MSYs). That was a highly commendable decision.

Provision of useful scientific advice regarding regulation is evidently only possible if targets are clearly defined. So, the first priority must be to decide how much lower the fishing mortality should be than F_{MSY}. Although the "lower than" provision has a precautionary quality, its main purpose is to ensure that fishing remains worthwhile—i.e., profitable in our current economic system. That is entirely in accordance with the United Nations Convention on the Law of the Sea Convention that mandates a basic MSY target, but with provision for some divergence from that to meet economic and environmental considerations.

(It is a commonly held presumption that fishing for MSY will be profitable simply because it aims at high catches. But, MSY generally calls for high, and, therefore, expensive, fishing intensity. The dynamics of fish populations are such that, universally, sustainable

[1]Editors' note: written by Sidney Holt and dated May 1, 2015 as a submission to a consultation by the European Commission concerning the forthcoming multiannual plan for North Sea demersal fisheries. See https://ec.europa.eu/info/sites/info/files/consultation-multi-annual-plan-north-sea-demersal-fisheries-summary_en_0.pdf.

catches not very much less than MSY can be taken with levels of fishing mortality much less than F_{MSY} and so with much higher levels of catch rate—catch per unit effort—and, therefore, more likely to be profitable.)

Determination of an appropriate lower value of F—as a % of the F_{MSY}—calls for extensive simulations to clarify the options for socio-economic decisions. One economic option suggested over the years by scientists and economists is the maximum economic sustainable yield (MESY), obtained when the fishing mortality is considerably less than that required to obtain MSY, and the difference between the market value of the catch and the cost of obtaining it is maximized. This is not difficult to estimate roughly, given appropriate economic data. However, that being done, it remains necessary for authorities to have a policy regarding the desired size of the industry for a variety of social reasons that can well be different from an economic optimum.

Meanwhile, it is surely necessary to decide, rather arbitrarily, on a working value of the "lower than" fishing mortality. It seems to me that in making such a choice, the EU should choose the same value for all regions of the EU waters. Parliament has proposed a very reasonable option $F = 80\%$ of F_{MSY} for the Baltic, and that could be adopted in the interim for the North Sea. It has an advantage of being close to a working value of 75% adopted by the USA and Canada for the Northwest Atlantic with closely related fish stocks such as Cod and Haddock *Melanogrammus aeglefinus* (it would be good also to negotiate compatible rules with Iceland and Norway). The 80% rule would generally provide sustainable catches at 90–95% of MSY.

The second, high priority is to decide on a rational transition process from the current situations to a new sustainable pattern of exploitation providing high (though not quite maximal) yields at relatively low cost and hence reasonable profit. Again, it is essential for scientific advisers to simulate various options for the speed of transition. It is well known that, in terms of maximizing or optimizing the *cumulative* long-term yields, the best is to cease fishing on stocks that are currently at levels and in conditions that are away from the sought eventual optima. But, that is obviously not practicable except perhaps in a few cases where the continued productivity, or even existence, of a fish stock is endangered by continued exploitation. A slower "recovery" would mean sacrifice of some cumulative catch, and the simulations should reveal what would be the social costs and gains of that "sacrifice."

Evidently, it will sometimes be necessary to depart from the F less-than-MSY rule and close a fishery. That derogation can be provided for by a supplementary rule that sets F at zero (or perhaps at a value less than the general level) if scientific data indicate that the size of the spawning stock has become, or is likely to become, less than some threshold value and thus collapse of the stock and of the fishery is threatened. Similarly, closure might be temporarily appropriate if catches unintentionally exceed catch limits or if there are unacceptable unintentional catches of protected species. Clear, unambiguous rules regarding these eventualities must be established.

It is not unreasonable to set dates by which a chosen transition process will be completed, but in doing so, it must be remembered that success in keeping to the preferred timetable depends equally on the chosen levels of fishing intensity and the dynamic biology of the fish stocks.

In the North Sea, there are several multispecies "mixed fisheries" in which several species are caught more or less simultaneously in the same gear, especially in bottom and midwater trawls. Some of these catches are unintended; others have a variety of market values. The proportions of the various species taken are, to a considerable degree, under the control of fishers, taking advantage of the natural partial segregation by species as well as sex and size/age—by location, season, water depth, bottom type, and so on. This means that it is reasonable, even in such cases, to determine catches in the standard way, that is, by treating the species stocks separately. Where selection is not practicable, then it would be reasonable to determine the F-values—if they differ as between species—at an intermediate level, possibly taking into account the different market prices.

In some mixed fisheries, there is evidence or suspicion of biological interactions among some of the species—predator–prey, competition for food or space, etc. However, in very few cases is there yet sufficient knowledge to take fully and quantitatively into account such relationships in the management arena except sometimes in a very general way involving social decisions, such as that fishing for human food is desirable in contrast with catching juveniles and small species for reduction to oil and fishmeal. In general, the practice is necessarily to manage as though such interspecific biological relationships do not exist and monitor over the years whether changes in stocks are or are not in accord with the simple expectations.

I conclude by suggesting that the establishment of a comprehensive monitoring system to check that scientifically based expectations are being met is an essential, if costly, part of a management system that will ensure high sustainable and profitable catches into the future and a good living for fishers.

Chapter 59

How Bad Science Impedes the Good: an Example[1]

In their thoughtful article in the online journal *Frontiers in Marine Sciences* entitled "The good, the bad and the ugly science: examples from the marine science arena," Parsons and Wright (2015) give examples mainly relating to oceanography and climate. My example concerns the management of sea fishing in the expectation of securing high sustainable catches far into the future. My intention is essentially to demonstrate how bad science can, in the political area, impede both the development of good science and its practical application.

In the 1930s, the basic dynamics of exploited fish populations had been elaborated, especially by the research of Johan Hjort and his colleagues in Norway studying especially Cod *Gadus morhua* and Herring *Clupea harengus* (e.g., Hjort et al. 1933) and British naturalists E. S. Russell and Michael Graham in England studying especially Cod and Plaice *Pleuronectes platessa* and also by the work of W. F. Thompson and F. H. Bell (1934) in the Northwest USA on salmon *Oncorhynchus* spp. and Halibut *Hippoglossus stenolepis* and, a little later, W. E. Ricker in Canada. The key to many puzzles about the great annual variability of annual catches was the method of determining the ages of caught fish by counting rings in their scales and later, in the hard objects in their ears—ear stones or otoliths. This revealed that the greater part of the "fluctuations," as they were called, arose from the variation in the numbers of young fish ("recruits") entering the exploited populations from year to year, but also, from these studies, natural mortality and body growth rates as well as information about the onset of sexual maturity. This was supplemented by information from multipeaked distributions of the lengths of landed fish and mark–recapture experiments. There had long been a controversy about the numerical sizes of fish populations. A few scientists—among them the distinguished T. H. Huxley—maintained that they were so numerous that fishing could not much affect them. That was resolved by a Danish biologist C. G. J. Petersen (1891) and in 1904 independently by T. W. Fulton in Scotland who both used the mark–recapture method to count Plaice. They were not infinite in number.

The extensive use of age determinations that allowed the calculation of mortality and growth rates, coupled with the observations that catch rates declined, that increases in total catches slowed with increasing fishing intensity, and that total catches could even decline if effort was not reduced, led to the realization that fishing was substantially affecting the size of the stock and that steady high catches might be obtained by fishing with a medium intensity. This idea was referred to as optimal fishing and any excess beyond that was overfishing, calling for serious ameliorative management measures to be taken.

---

[1]Editors' note: a short note written by Sidney Holt dated July 14, 2015. Though it was intended for publication, it was never submitted and as far as the editors are aware, it was never published.

By the mid-1940s, catch rates in the valuable tuna *Thunnus* spp. fisheries of southwestern North America were showing declines, and the question arose as to whether management to limit catches and fishing effort was needed there. Good statistics of catch and effort were available, but tunas could not be aged, so the new age-structured methods of assessment were not available. The key scientist there was Milner Schaefer working from southern California. His idea (Schaefer 1954) was based on a mathematical model of human population growth published by a Belgian mathematician, P-F. Verhulst, in 1838 in response to the hypothesis of Thomas Malthus (published in the period 1798–1830; e.g., Malthus 1798) that populations increased geometrically (i.e., exponentially) until they ran out of resources (access to which could increase only linearly with population increase) and then collapsed. Verhulst proposed that the geometric rate of increase would decline as the population grew. As an example—he had no relevant data—he assumed that the rate of decline would be linear—mathematically the simplest possible assumption—and this produced a symmetrical S-shaped population growth curve, approaching a stable upper limit—asymptote, with the S-bend (inflexion) at a population size halfway to the upper limit. The hypotheses of Malthus and Verhulst were of great political interest, in the case of the latter especially because the new capitalist entrepreneurs wanted large stable labor forces. In most commentaries, little justice has been awarded to Malthus' idea. This is because the second part of it has been overlooked. Malthus observed that human population consisted of a wealthier few, with good access to resources, and an impoverished many, with little or no access. Malthus reasonably assumed that many of the former group would survive, and they would eventually form another vigorously growing society. In this, he was perhaps ahead of Karl Marx and Fred Engels in giving great influence to the class structure of society. Marx's *Das Kapital* was not published until 1867, but *The Communist Manifesto* appeared in 1848 (Marx and Engels 1848)—the year of the European revolutions. While the economic theory of *surplus value*, which, as we shall see in a moment, is analogous to the *surplus production* (SP) procedure of fish stock assessment, is usually associated with F. Engels; it was apparently first formulated by the French politician—the first self-described "anarchist"—and philosopher P. J. Proudhon in 1840.

Schaefer supposed that the total weight—"biomass"—of a fish population would change with time in accordance with Verhulst's numerical growth model. This, of course, is inherently inconsistent, because if any population were to increase in number according to the simple logistic, its biomass growth pattern could not follow the same kind of trajectory. Ignoring that, Schaefer argued that, at any point in the growth curve, a sustainable catch could be taken of size equal to the slope of the growth curve at that point, so holding the population biomass constant. Furthermore, a maximum sustainable catch (MSY) could be taken by holding the population at its size at the inflexion (this, naturally, would apply to any sigmoid growth curve, not merely to the logistic).

Schaefer's idea came to be known as *surplus production* theory, though I do not think he invented the term. This is analogous with the Marx–Engels surplus value which recognizes that a worker in industrial capitalist society produces more than enough to satisfy his own needs and the "surplus" accrues to the owner of the means of production as a marketable commodity. So, the fish population, in struggling to grow towards its natural equilibrium, produces more biomass than it needs, and so the "surplus" can be taken by we humans.

It was soon realized that the symmetrical logistic—which said that catch could be maximized by holding the stock at 50% of its unfished biomass—was too restrictive, and two of Schaefer's younger colleagues, J. J. Pella and P. K. Tomlinson, published, in 1969, an alternative version with the symmetrical parabola of Schaefer–Verhulst replaced by a polynomial power function. This allows the stock size giving MSY to be anywhere between close to the unexploited biomass and 37% of that.

Unfortunately, Schaefer's theory coincided with the negotiation of a peace treaty be-tween Japan and the USA. The USA was seeking to prohibit Japanese or other "foreign" fishing vessels from operating in the Northeast Pacific, especially for migrating salmon and Halibut, and to that end was pushing an "abstention principle" with the idea that if a coastal state was fully utilizing a resource off its coast, other states should abstain from exploiting it. The Japanese negotiators wanted to know the intended meaning of "full utilization," to which the US response was "taking MSY." The US Administration then spent several years trying—but failing—to get the abstention principle recognized in international law, especially at UN conferences in 1955 in Rome and 1958 in Geneva. However, practically by default, the man-agement objective of MSY was retained as the prime policy, though with various softening conditions attached to it. A residual of the abstention idea remains in the UN text eventually adopted in 1994, but essentially circumvented by the decision to grant coastal states the right to declare Exclusive Economic Zones (EEZs) extending up to 200 nautical miles from their coastlines.

One problem is that MSY can be defined, but is, in fact, impossible to take. This is because it could, in theory, only be obtained by catching *all* the surviving individual fishes in the oldest cohort in the population just as that cohort reaches a maximum total weight at a critical age when the fish all reach a certain critical size. A normal fishery would begin to capture fish in all recruited cohorts (age groups) starting at some lower age; all fisheries are selective in that sense. Depending on the particular selectivity, there can be a local MSY (*msy*), less than the theoretical MSY. If selectivity is low (fish being liable to capture when they are young and small), the local MSY may be obtainable with a fairly low fishing effort and so with the stock biomass not much reduced from the virgin state. That is typical of fisheries for producing fishmeal and oil. Fisheries for human food begin to take fish when they are older and bigger—perhaps already sexually mature—and then the local MSY, if one is available (depending on the particular dynamic properties of the stock), will be somewhat bigger, but produced only by much more intense, and so more costly, fishing effort and reducing the virgin stock to a much lower level—and practically always to considerably less than 37% of the virgin size.

In efforts to manage fishing to gain local MSYs, the surplus production method of stock assessment has been very commonly used, and results from it are generally liked by lawmakers. It looks straightforward and comprehensible and calls for very little scientific in-formation—merely a series of catch statistics and corresponding indices of stock biomass, mostly coming from data for the catch per unit of effort (CPUE). So, it is cheap, there being no need for the difficult studies of fish ages, mortality, growth, and reproductive rates. How-ever, the results of these exercises are pseudo-scientific nonsense and extremely misleading, leading to serious mismanagement of fisheries to which they have been applied. Let us look at the methodology.

We begin with a series of annual catches and corresponding indices of stock biomass from CPUE. SP theory says that sustainable catches (that is, from a stable stock) should bear a simple polynomial/power relationship with the stock biomass indices. So, we need to adjust the observed catch sizes to represent sustainable catches. This is done by adding to, or subtracting from, the recorded catches the difference between the stock biomasses at the beginning and end of the catch year. But, here is the rub—we have only indices of biomasses, not the absolute quantities. So, an estimate of the ratio of these quantities to the indices has to be found, and there are various ways of trying to do that. So, with the adjustments made to the catch data, a standard curve-fitting procedure is used to estimate the power parameter defining the polynomial curve and hence the biomass corresponding to its peak, the local MSY, and also where the stock now is in relation to the "optimal" abundance—even close to a low level where the model would break down (in general, at less than 37% of the virgin size) and the stock perhaps brought towards extermination by continued intense fishing. The process allows labels to be attached to the stock—it is or is not "overfished" or even "endangered."

But, productivity cannot be a simple function of biomass. Two otherwise identical stocks, but with different age compositions, can have very different rates of productivity. Their productivities derive from the interaction of loss in number from death and increase in total body weight. This is because body growth follows a sigmoid trajectory through life. With a fast rate at the inflexion—which is usually at about one-third of the final weight the animal will attain if it lives forever and much slower if it is much older or younger than that (the *specific* growth rate as a proportion of size—dW/Wdt in conventional calculus terms—is highest in the very young animal) —the proportion of older and younger animals is crucial in determining productivity and hence sustainable yield.

If the biomasses had the same age composition in all the observed time-series, then the curve-fitting would make some sense, but would still depend on the actual composition. But, in practice, that cannot be so, anyway, by definition. The time-series must span a range of stock sizes, and the bigger ones will be from earlier years in the history of the fishery when the stock had not been much changed by fishing and so had a relatively high proportion of older, bigger fish. The smaller biomasses come later when fishing has reduced the stock and it has a higher proportion of younger fish than before. The curve-fitting of adjusted, pseudo-sustainable catches to those is, therefore, meaningless and any statistically significant fit will be strongly biased in obvious ways.

The situation is so much worse if there has been a change in the selectivity of the fishery during its history—known or unnoticed, accidental or deliberate. Some approximate corrective adjustment might be made by examining the changes in the body–length compositions of the catches, but that is rarely done.

These problems with the surplus production method are not merely academic ones. As recently as 2009, the World Bank and the Food and Agriculture Organization (FAO) of the UN published a report (World Bank and FAO 2009) in which the economists assumed that maximum sustainable catches would be obtained globally if all fish stocks were held at one-half their original sizes—*à la* Schaefer. The World Bank–FAO conclusions are probably qualitatively correct, but certainly not quantitatively so (Holt 2009).

References

Fulton, T. W. 1904. The rate of growth of fishes. 22nd Annual Report of the Fisheries Board of Scotland Part 111:141–240.

Hjort, J., G. Jahn, and P. Ottestad. 1933. The optimum catch. Hvalrådets Skrifter 7:92–127.

Holt, S. J. 2009. Sunken billions—But how many? Fisheries Research 97:2–10.

Malthus, T. R. 1798. An essay on the principle of population. J. Johnson, London.

Marx, K. 1867. Das Kapital. Kritik der politischen Ökonomie. Verlag von Otto Meissner, Hamburg, Germany.

Marx, K., and F. Engels. 1848. Manifest der Kommunistischen Partei. Kommunistischer Arbeiterbildungsverein, London.

Parsons, E. C. M., and A. J. Wright. 2015. The good, the bad and the ugly science: examples from the marine science arena. Frontiers in Marine Science June 4, 2015, https://doi.org/10.3389/fmars.2015.00033.

Pella, J. J., and P. K. Tomlinson. 1969. A generalized stock production model. Bulletin of the Inter-American Tropical Tuna Commission 14:421–496.

Petersen, C. G. J. 1891. Eine Methode zur Bestimmung des Alters und des Wuchses der Fische. Mitteilungen des Deutschen Seefischerei-Vereins11:226–235.

Schaefer, M. B. 1954. Some aspects of the dynamics of populations, important for the management of the commercial fisheries. Bulletin of the Inter-American Tropical Tuna Commission l(2):25–56.

Thompson, W. F., and F. H. Bell. 1934. Biological statistics of the Pacific halibut fishery. 2. Effect of changes in intensity upon total yield and yield per unit of gear. Report of the International Fisheries (Pacific Halibut) Commission 8.

Verhulst, P. F. 1838. Notice sur la loi que le population suit dans son accroissement. Correspondance Mathématique et Physique 10:113–121.

(The) World Bank and Food and Agriculture Organization (World Bank and FAO). 2009. The sunken billions. The economic justification for fisheries reform. The World Bank, Washington, DC.

Chapter 60

Suggestions Regarding Implementation of the Primary Management Objective of the Common Fisheries Policy of the European Union[1]

Part I. The Basics

The European Parliament, Council of Ministers, and the Commission have agreed to the following text defining the prime management objective of the revised Common Fisheries Policy (CFP) (Reg. 1380/2013, Art. 2.2 first sub-paragraph): "*2. The CFP shall apply the precautionary approach to fisheries management, and shall aim to ensure that exploitation of living marine biological resources restores and maintains populations of harvested species above levels which can produce the maximum sustainable yield*" (EU 2013).

This is entirely compatible with the management principle in the 1982 United Nations Convention on the Law of the Sea (UNCLOS) which requires that [Art. 61.4]: "*... (States) take into consideration the effects (of taking specified management measures) on species associated with or dependent upon harvested species with a view to maintaining or restoring populations of such associated or dependent species above levels at which their reproduction may become seriously threatened.*"

The specified measures are for a conditional MSY policy [Art. 61.3]: "*designed to maintain or restore populations of harvested species at levels which can produce the maximum sustainable yield, as qualified by relevant environmental and economic factors, including the economic needs of coastal fishing communities ... and taking into account fishing patterns, the* ***interdependence of stocks*** *and any generally recommended international minimum standards...*" (*Holt's emphasis*)

Here, the phrase concerning "environmental and economic factors" was inserted during the UNCLOS negotiations mainly to provide a way for "developing" countries to exceed MSY temporarily while establishing new fisheries. The wording adopted does not, however, specify a direction of permissible departure from MSY.

This same EU paragraph continues (second subparagraph in Art. 2.2): "*In order to reach the objective of progressively restoring and maintaining populations of fish stocks above biomass levels capable of producing maximum sustainable yield, the maximum sustainable yield*

[1]Editors' note: written by Sidney Holt and dated February 24, 2016. It was conceived as the first draft of a consensus statement criticizing MSY as a management objective, intended to form the basis for discussion at a workshop of scientists Sydney intended to convene. In the original, the summary was at the beginning of the article; here, we have placed it at the end.

exploitation rate shall be achieved by 2015 where possible and, on a progressive, incremental basis at the latest by 2020 for all stocks."

This introduces two new ideas that are not included in the UNCLOS. First, the idea of a critical or target level for the biomass of a stock, replacing or supplementing the idea of an optimal intensity of fishing; and, second, the reference to an "exploitation rate." That is not defined in the EU text, but it appears, from the context, not to be "the exploitation rate," as formally used in fisheries science, i.e., the ratio of the fishing mortality rate (as an exponential coefficient) to the total mortality rate, so $E = F/(F + M)$, from which we can define a more useful *relative fishing mortality rate* F^*: $F/M = E/(1 - E) = F^*$.

A following paragraph in the EU text is as follows (Art. 2.3): "*3. The CFP shall implement the ecosystem-based approach to fisheries management so as to ensure that negative impacts of fishing activities on the marine ecosystem are minimised, and shall endeavour to ensure that aquaculture and fisheries activities avoid the degradation of the marine environment.*"

So, paragraph 2 above refers to precaution needed because of uncertainties in assessing the state and productivity of a stock, and an error in overestimating MSY is evidently less desirable than an underestimate. That precaution might be obtained by adopting a management rule not very far from the optimal estimate. But, the other objectives—ensuring a good economic state and having low impact on the environment generally (*Editors' note: other objectives relating to these aspects are found in Art. 2.5 of Reg. 1380/2013*)—will possibly call for much bigger departures from an MSY target, and it is the scales of such departures that we need to investigate.

The second part of paragraph 2 introduces the idea of a careful, timed <u>transition</u> from one undesirable state of a fishery to another, desired targeted state. This will most likely call for a temporary departure from the decided target fishing intensity.

We must, from these preliminaries, now turn to a more serious problem with both the UN and EU texts. This is that they assume that MSY is both a measurable and an achievable target (and an economically and socially worthy one). But, it is not achievable by we humans since it could be obtained only by catching all the survivors of the oldest cohort (year class) in the population just as the cohort itself reached its maximum weight, at what is called its *critical age*.

It has been known for more than half a century that MSY is the weight of the oldest cohort (that is, all those born in the same spawning season) in an otherwise unexploited fish population just as the cohort reaches a maximum weight at its critical age, and the fish remaining alive in it reach a critical average weight w_{crit} expressed as a fraction (percentage) of the weight the fish would theoretically attain if it lived to an extreme old age W. [The growth of fishes is what we call *indeterminate*—that is, the fish continues to grow throughout its life, not stopping after sexual maturity as do terrestrial vertebrates, mammals (except, apparently, whales and dolphins), and birds.]

So, MSY could be obtained by catching all the fish in the oldest cohort when they reach critical age and size. But that is physically impossible in a fishery for wild, free-living fish.

Our practical solution is to begin catching from most or all cohorts before the oldest one has reached its maximum total weight. This introduces the idea of the <u>*selectivity*</u> of fishing. I shall refer here to selectivity as the age or size at which catching begins. For several methods of fishing, using, for example, trawls or seines, all or most fish are theoretically catchable after they reach that size, but for some methods—such as gillnets and hooks and lines—there is, in principle, also an upper size limit after which the fish are not retained by the gear.

For the rest of this paper, in order to establish principles, we shall be considering in detail only the fishing operations with a lower size limit. It is usually assumed that it is determined by the size of net mesh or hooks. That is largely true, but another major factor is the behavior of the fish—for example, whether the larger, older fish tend to be found in a different area than the smaller, younger ones, or living at different depths of water, or nearer to or further from coasts or other ocean features. The shapes of histograms of age compositions of fish in sampled catches give us a good idea of the actual selectivity.

It is not hard to write a simple mathematical model from which to calculate what would be sustainable catches in terms of a range of fishing intensities (i.e., F^*) and of selectivity w. [It is sometimes convenient to express the selectivity in terms of the length of the fish relative to its final size—c—rather than its weight (w is roughly equal to the cube of c).] But, these will differ according to a particular parameter for the general dynamics of the species or stock. That is the ratio of the *natural mortality rate* M in the phase of life during the selective exploitation, to a parameter K of the S-shaped (sigmoid) curve of growth in body weight.

This ratio M/K has a fairly narrow range among species of the bony fishes; species or populations with a high M tend to have a high K, and vice versa. The ratio M/K tends to have similar values among related species, such as scombroids (mackerels), cod-like fishes (gadoids), and flounder-like flatfishes (Pleuronectiformes). The mid-range value of all of them is 1.75, with a general range from about 1.5 to 2.1.

Most of the calculations I shall illustrate are of *catches-per-recruit*, the number of recruits being the average number in a cohort when it reaches an age at which the fish are big enough to be worth catching. The "worth" depends, of course, on the "processing" technology and the market. If the fishery is intended to produce biological material to convert industrially into fishmeal and oil, then recruitment might be considered to be at a relative length 0.1 or even less, while if it is an industry for fish for human food, recruitment might be considered to have happened at a relative length of 0.2 or more. To get an idea of this scale, consider that the curve of body weight has an inflexion (where it is most steep) at 0.3, and sexual maturity generally occurs at that size or somewhat bigger. A relative weight of 0.3 (that is, a bit less than one-third of the final weight) is reached when the relative length is 0.67 (that is, two-thirds of the final length).

Most experience has been that light- or medium-intensity fishing does not reduce the spawning stock so much that the recruitment is much affected (reduced, one would expect). Intense, unregulated fishing, especially of young fish, can, however, lead to reduced recruitment, sometimes devastatingly reduced, so we do need to take care. Even if we are unable to predict how much recruitment might change, we can examine feasible or likely options.

Curves of estimated sustainable yield against fishing intensity, for a given selectivity of the fishery, often, but not always, are dome-shaped, with a peak that we can call *local MSY* or *msy*. There is always a peak if selectivity is low, but often not if it is high. However, these curves of sustainable yield are always rather flat-topped at fishing intensities below that for the maximum catch. So, fishing less intensely than is needed to obtain *msy* yields sustainable catches not much less than *msy*. This is an extremely important feature with respect to seeing how to implement the EU management procedure.

Here, it is time to review the origins of sustainable yields.

It is usual to begin with the assumption of an unexploited population in a steady state. That means a kind of equilibrium, at what is rather misleadingly often called "carrying capacity" in which there may be year-to-year changes in abundance, but no trend upward or downward. In such a state, the losses by natural death are equal to the sum of the average growth of body weight and a periodic immigration of young “recruits.” In that state, there is no sustainable yield.

A sustainable yield is created entirely by the onset of fishing, changing, as a result of a generally increased mortality rate, the age composition of the phase of the population called “the exploited stock.” This is a consequence of the pattern of body growth in weight, the *specific* rate of which—as a percentage of its current size—is highest in the youngest fish and diminishes throughout life, so the average growth rate is reduced by the change in the average age of the surviving fish. It should be noted also that the simple annual growth rate (increase per unit time) of a fish is small when it is young, reaches a maximum when it is 30% on its way to full size, and then reduces steadily into old age.

If the natural mortality rate is not changed by fishing, a difference between mortality and growth is thereby created. So, if fishing was to cease, the population would grow in weight so long as the numerical size of the stock is not reduced so much as to cancel out the overall effects of the increase in average growth rate. A sustainable catch in a particular year would be the quantity taken that would exactly impede regrowth. If fishing intensifies, the result will eventually be a reduction in the potential sustainable yield, as the increasing average growth rate will be nullified by further reduced numbers.

Now, we need an overall look at the curves of SY against F^* and particularly how they change with the selectivity of fishing. Considering a stock with a particular value of the population parameter M/K, the local *msy* increases as selectivity increases, as does the value of F^*_{msy}. But, the latter change is more than the former, so a slightly higher *msy* is obtained, but at the cost of a lower catch rate msy/F^*, and hence less economic benefit.

These results are qualitatively the same for all observed values of M/K, but, naturally, will vary quantitatively.

While most attention has usually been given to changes in SY as a function of F^*, selectivity being assumed constant or even unspecified, changes should be examined in the other major dimension: SY with changes in c with a fixed given F^*. There are also maxima in that direction, and the maximum point for a given value of F^* is defined as *eumetric fishing*. (Colin Clark defined fishing away from that point as *cacometric*; Clark 1985.) If we look at both dimensions simultaneously, we see a eumetric fishing curve which reaches a maximum asymptote at MSY.

Unfortunately, the approach to fisheries assessment described here has, for many decades, been corrupted by an unscientific idea of what is called “surplus production,” invented originally by M. B. Schaefer in the 1940s. Schaefer (1954) had a valid reason for his approach —there was, at the time, no way to determine the ages of the tunas he was studying, so, therefore, no good estimates of natural mortality and growth rates. This idea is that biological productivity, leading to a potential sustainable yield, is dependent only on a measure of stock biomass, regardless of the age composition of the stock. This leads to major errors in assessment of the stock and the fishery, compounded by the common absence of a clear definition of which biomass: (i) that of the recruited stock, (ii) the selectively exploited stock, (iii) the part of the stock comprised of mature, spawning fishes, or (iv) just the females. In the administrative and popular documents, and even also in scientific papers, these are described as stock levels, although that term is often considered to be a relative one, for example, as a percentage of some reference biomass, most commonly the biomass before exploitation of the stock in question began.

The reality is that the productivity of a stock, its potential sustainable yield at that time, depends as much on its age/size composition as on its size, and the age composition at a particular time can be very variable, not in a steady state, for two main reasons: (i) the number of recruits varies very much annually, and (ii) because the observed age composition has arisen from a prolonged series of changing exploitation rates.

Schaefer's method was subsequently promulgated vigorously for national political purposes by the US government in the post-World War II years. Catching "MSY" was described as "full utilization of the resource" and used in an attempt to claim exclusive fishing rights by a state in the ocean closest to its shores under what was called "the abstention principle," a "principle" that was decisively rejected by other states and the UN.

Naturally, much more complex population models than the simple age-structured one described above are needed to deal with research findings concerning the simple assumptions of constant natural mortality and body growth (for example, whether M varies with age of the recruited fish, or body growth with food supply, and hence with population abundance or density). Adjustments of this kind make quantitative differences in, for example, calculated sustainable yields, but rarely important qualitative differences. However, the variability that can cause most trouble is undoubtedly that of the number of recruits, and especially if that number is dependent on the number and size of the spawning females.

The general relationship between the size of the population of female spawners and subsequent recruitment has been the subject of numerous models. That most used is the simplest and was formulated by Beverton and Holt (1957). It has two parameters: (i) an asymptotic recruitment arriving from a theoretically infinite spawning stock and (ii) another defining the rate of approach to that asymptote as spawning stock increases. With a very small population size, commensurate with an extreme depletion of the original unexploited stock, the dependence of recruitment on spawning stock size is virtually linear. The shape of its curve is very similar to that of the growth in body size by length of an individual fish, but the algebraic expressions and the theories behind them are quite different. I use the Beverton and Holt model here primarily because our interest is mainly in the mid-range of the curve of SY and F^*, while most of the others differ from Beverton and Holt mainly for very high or very low population size.

The simple Beverton and Holt model can be reduced effectively to one parameter: the recruitment in the unexploited stock expressed as a fraction of the asymptotic number. (For this and other models, it is common in the scientific literature to define the degree of reduction of recruitment that would result from a reduction in the spawning stock as to one-fifth of the unexploited number as the *steepness* of the stock–recruitment curve.) In general, if there is a dependence of recruitment on spawning stock size in the mid-region near *msy*, then the maximum SY is a bit higher than msy and is obtained with a slightly lower F than F^*_{msy}.

Part II. Suggested Management Options

First Option

As the CFP links the management target to "maximum sustainable yield," one might begin by taking that literally—to ensure that catches are less than MSY. The lowest at a very small population size, commensurate with an extreme, is found when selectivity is at its lowest. In the case of a stock with $M/K = 1.75$, the *msy* is lowest when selectivity is theoretically zero, and is almost 50% of MSY and would be taken when F^* is approximately 0.7. So, to ensure that the sustainable catch is less than MSY, it would be necessary to keep below an F^* of 0.7 to ensure the CFP criterion would be met regardless of the selectivity of the fishery.

This would, of course, be less than F^*_{msy} for any realistic selectivity. For example, if c were 0.6, $w = 0.2$, where *msy* is at 83% of MSY, for $F^* = 1.9$, the target would be less than 58% of F^*msy, and the sustainable catch would be 0.0813 per recruit, which is 97% of *msy*. If the selectivity of the fishery were to be starting exploitation at $c = 0.67$, $w = 0.3$—that is, as they are approaching sexual maturity—the target catch would be 0.0903 per recruit, which is still as much as 88% of *msy*. So, it seems that to set a target at or below the lowest possible *msy* with unselective fishing might not be a bad idea.

Remember that the critical selectivity for species stocks in which $M/K = 1.75$, beyond which one should certainly not choose to go, is $c = 0.75$, $w = 0.4$.

Any reader wishing to check these numbers without undertaking tedious calculations can do so by consulting and interpreting the tables of sustainable catch per recruit in Appendix I of Gulland (1983). These are a reduced form of more extensive tables given by Beverton and Holt (1964).

This management target option can be summarized as: **regulate the fishing intensity or the catches such that the fishing mortality rate is never more than about double the natural mortality rate**.

Second Option

Take advantage of the flatness of the curve of sustainable yield against fishing rate. For this, we assume that the selectivity that pertained to the pre-CFP fishing does not change—or, at least, not much—as a result of the new regulations. In fact, it might well change as a result of fishers seeking to optimize their operations to profit from the CFP changes, but managers would need to monitor this and perhaps adjust the regulations accordingly.

It should not be assumed that fishing for any particular *msy* will be profitable—it often would not be—but reducing the fishing mortality to less than F^*_{msy} must increase the probability of profit by increasing the catch rate and also, perhaps, by increasing the average size of the fish in the catches.

There are, in principle, two ways of achieving the desired situation: (i) by directly reducing the fishing intensity to the desired level or (ii) by setting appropriate catch limits. The latter is, however, extremely inefficient both because the change in average total catch will be relatively small and because of its annual variability resulting from the usual large variations in annual recruitment. This has been described by scientists giving evidence in official "hearings" in the USA in 1942 (Nesbit 1943) and in the UNSCCUR Conference in 1949 by A. G. Huntsman (UNSCCUR 1951) as *enforced inefficiency*—that is, requiring a fishing unit to catch less than its capacity.

Here, I summarize the consequences of exploiting an $M/K = 1.5$ stock at a fishing intensity one-half of what would be needed to take *msy* for a series of alternative selectivities: c = 0.35, 0.42, 0.55, and 0.67 by length, and w = 0.04, 0.07, 0,17, and 0.30 by body weight. The highest sustainable yields range from 0.87 to 0.94 of the corresponding *msy*s and the catch rates (sy/F^*_{sy}) are, naturally, nearly double what would be obtained by fishing for *msy* in each case. The *msy* values range from 0.18 to 0.36, with corresponding F^* values from 1.2 to 3.5.

These comparisons well reveal the impossibility of effectively regulating sustainable fishing by setting catch limits. The total catches would vary little over a wide range of fishing intensities and would, in any case, be obscured by the inevitable natural fluctuations in recruitment and other natural variability. There is really no escape from the need to have good data about fishing power and effort and to regulate that directly.

Third Option

Now, we depart from the idea of choosing a management target directly related to *msys* in order to consider economic features of the system. Again, to do this thoroughly would call for a great array of data and complex—and in many ways, highly uncertain—analyses. But a simple approach, not even taking discounting into account, can get us some way toward worthwhile conclusions, I think.

To a first approximation, the market value of a catch is proportional to its quantity by weight. And, correspondingly, the cost of taking that catch is proportional to the fishing intensity exerted, which is proportional to the fishing mortality rate induced. I call the difference between these *gross profit*.

If we scale the two quantities, value and cost, and plot them on the same graph, the point at which the straight diagonal line running up from the origin intersects the catch value curve is the break-even point at which the profit is zero. Another straight line parallel to the cost line and tangential to the value curve indicates, where the two lines touch, a point of *maximum sustainable profit* or *msey*. Is that a potential target meeting the EU specification? It is on the descending limb of the sustainable catch–fishing mortality curve, so it will be at a low F^* value. It points to a simple rule offered by Huntsman in 1949 at the UNSCCUR conference (UNSCCUR 1953): "*The take should be increased only as long as the extra cost is offset by the added revenue from sales.*"

Michael Graham, who was also prominent at that conference—in fact, leading the discussion about fisheries—had, a few years earlier, published his Great Law of Fishing: "*Fisheries that are unlimited become inefficient and unprofitable*" (Graham 1943).

Graham was essentially redefining overfishing in terms of threatened unprofitability. This was not a novel idea. Hjort et al. (1933) and Russell (1942) incorporated it in their view of "optimum fishing." The British naturalist and fisheries inspector Ernest Holt had said, towards the end of the 19th century, that every fish should have a chance to spawn at least once before being destroyed (Holt 1895). All three were essentially saying that most fish should be allowed to live longer in order to grow bigger. The parallel moves to increase the size of the meshes of trawl codends have achieved the same objective: allow fish to grow bigger and hence get a higher catch rate without changing the fishing intensity.

These views of leading scientists and managers clarified at the time that "overfishing" and related problems were not only, or even mainly, about the biological states of the exploited fish populations. But I do not think there are valid social reasons for managing fishing simply to obtain maxima, whether of catch quantity or profit. Rather, they can guide us in seeking not maxima, but *optima*—another useful term much used by the fisheries experts of the 1930s and after. In current terms, the two maxima can give us not targets, but *boundaries* to our choice of targets. Profitable sustainable fishing can evidently be secured by exerting a fishing effort very much less than that needed to attain *msys*, but there is no benefit to be obtained by reducing effort to below that which will provide *mseys* under prevailing conditions of markets and technology.

Fourth Option

Lastly, we must consider the other advantage of aiming at a target somewhat less than *msy* and much less than F^*_{msy}, in terms of minimizing deleterious impacts on the targeted species, on others in the ecosystem of which the target species is part, and on the habitat of all. This is impossible to quantify generally, but I think we can say for sure that in meeting the profit and quantity targets, any such impact will be greatly reduced, including especially unwanted catches of undersized specimens of the target species and its predators and competitors. It is commonly thought that catches of undersized fish are evidence of an inadequate fishing selectivity, but it must be remembered that the increased frequency of undersized individuals in catches is also a consequence of an excessively high fishing rate, hence total mortality rate. The impact on habitat by gears such as bottom trawls will obviously be reduced, if not eliminated.

This is not the place to try to examine the social consequences of regulation to drastically reduce fishing mortality. It has to be said, however, that doing so does not necessarily lead to the employment of much fewer people in fishing. This was examined by several European countries at the end of World War II, when one-half of the pre-war vessels had been destroyed, and the urgent question was whether and how fast to rebuild the fleets. It was then recognized that fishing is an extremely laborious process and that the industry would benefit if the fishers could still earn good livings by working fewer days per year if the fish stocks were kept at a high productive level after their wartime recovery from overfishing before the war.

Summary

Following an explanation of what the notion of sustainability means and implies with respect to managing sea fisheries, this paper offers suggestions for implementing the European Union's policy decision that management—by setting catch limits or by other means—must ensure that the target for the intensity of fishing (hence the rate of fishing mortality) must be less than that which would provide maximum sustainable yield (MSY). The EU has, however, not yet determined "how much less."

Several options are suggested here, with different outcomes, but all ensuring that catch rates, and hence profitability, will be much enhanced, but with very limited "sacrifice" of total sustainable catch. All options would also greatly reduce the amount of unwanted catch that must be discarded—either at sea or on shore—including undersized individuals of the target species, and other undesired, incidentally caught marine life, including vulnerable and threatened species of fish and other organisms. Such a policy would also reduce damage done to marine habitats, especially by mobile bottom-fishing gears.

The benefits of substantially reduced fishing intensity are not trivial. In one example, the limitation of fishing intensity to one-half of that required to obtain MSY would certainly ensure profitability by doubling catch rates. The average size of targeted fish in the catch would also be increased, further enhancing the overall market value of the catch by virtue of increased price. The total catch would, however, remain quite close to the theoretical MSYs, never less than 80%, but usually more than 90% of the theoretical possible maximum.

References

Beverton, R. J. H., and S. J. Holt. 1957. On the dynamics of exploited fish populations. Her Majesty's Stationery Office, UK.

Beverton, R. J. H., and S. J. Holt. 1964. Tables of yield functions for fishery assessment. FAO Fisheries Technical Paper 38.

Clark, C. W. 1985. Bioeconomic modeling and fisheries management. John Wiley & Sons, New York.

EU. 2013. Regulation (EU) No 1380/2013 of the European Parliament and of the Council of 11 December 2013 on the Common Fisheries Policy. Official Journal L354/22 28.12.2013:22–61.

Graham, M. 1943. The fish gate. Faber and Faber, London.

Gulland, J. A. 1983. Fish stock assessment. A manual of basic methods. FAO/Wiley Series on Food and Agriculture, 1. John Wiley and Sons, Chichester, UK.

Hjort, J., G. Jahn, and P. Ottestad. 1933. The optimum catch. Hvalrådets Skrifter 7:92–127.

Holt, E. W. L. 1895. An examination of the present state of the Grimsby trawl fishery, with especial reference to the destruction of immature fish. Journal of the Marine Biological Association of the UK 3:337–448.

Nesbit, R. A. 1943. Biological and economic problems of fishery management. Pages 23–53 *in* W. H. Herrington, and R. A. Nesbit, editors. Fishery management. US Fish and Wildlife Service, Special Scientific Report 18.

Russell, E. S. 1942. The overfishing problem. Cambridge University Press, Cambridge, UK.

Schaefer, M. B. 1954. Some aspects of the dynamics of populations, important for the management of the commercial fisheries. Bulletin of the Inter-American Tropical Tuna Commission l(2):25–56.

United Nations Scientific Conference on the Conservation and Utilization of Resources (UNSCCUR). 1951. Proceedings of the United Nations Scientific Conference on the Conservation and Utilization of Resources, 17 August – 6 September 1949, Lake Success, New York, Volume VI, Wildlife and Fish Resources. United Nations, New York.

Chapter 61

A "Less than F_{msy}" Management Strategy[1]

Today, I finished some simple calculations to illustrate what a "less than" strategy means and the practical importance of taking selectivity of fishing into account.

I look at a typical example of a bony fish in the mid-range of the main driving population parameter, $M/K = 1.75$. I present indices of MSY and local *msy*, as well as SYs, which give the actual values if multiplied by the annual number of recruits and the final maximum body weight of an old individual in kg. These indices are best expressed as percentages of MSY.

The fishing intensity is expressed as the ratio F/M, which is proportional to exerted calibrated fishing effort by the fleets per year.

MSY is a theoretical, but unobtainable, quantity of fish catch. It is the weight of the oldest cohort in the fish population when the cohort arrives at its maximum total weight. The weight of the individual fish at that time is the ***critical weight***, usually expressed as a percentage of the asymptotic weight; its age then is the ***critical age***. We can also look at the ***critical length*** of the fish, which is roughly the cube root of the critical weight. The critical age can only be determined when the value of K is known.

The MSY index for a stock with $M/K = 1.75$ is 0.10535. The critical size is $W = 0.40$. By exerting a fishing mortality rate 20-fold the natural rate, you can get a sustainable catch (*msy*) 99.7% of MSY. However, by exerting a fishing mortality rate only threefold the natural rate, you still get a sustainable catch (*msy*) 91% of MSY. But, to do that involves reducing the selectivity weight from the MSY threshold of 0.4 to 0.2.

Now, let us see what happens if we reduce the fishing effort, say, to one-half that required to obtain *msy*, that is, F/M from 3 to 1.5. We get a sustainable catch 94.5% of the *msy* (same selectivity). Exert a fishing rate of only one-third of F/M_{msy}, i.e., $F/M = 1.0$, and we still catch sustainably 86% of the *msy*. Evidently, such a policy adjustment will enormously increase the profitability of the fishery, but also greatly reduce the impact on the ecosystem in terms of unwanted species catch, catch of undersized fish, impairment of seabed by excessive bottom trawling, and so forth.

Suppose we think we can increase the sustainable catch by increasing the fishing effort—say, by increasing F/M from 3 to 6.0. We would gain only 6% of sustainable catch and have only one-half the catch rate (CPUE), possibly making the fishery unprofitable even if it was profitable with $F/M = 3.0$.

---

[1]Editors' note: Holt wrote this short paper as background for a presentation that Michael Earle, who, at the time, was Adviser for The Greens in the European Parliament, gave about the incorporation of MSY into the Common Fishery Policy at a workshop entitled "Towards Ecosystem Based Fisheries Management in the Baltic Sea," which was held in Stockholm June 16–17, 2016 and was cosponsored by The Fisheries Secretariat, the International Council for the Exploration of the Sea (ICES), and Stockholm University Baltic Sea Centre. Earle's talk was "MSY in the CFP: A Political History," which was the origin of what was later published as Earle (2021).

These figures all come from the basic Beverton and Holt (1957) expression for ***yield-per-recruit*** as a function of fishing effort and selectivity, that is, implying that changes caused by fishing on the biomass of the spawning part of the stock, with fishing intensity in the range of *F/M* from zero to 3, did not cause a reduction in the annual number of recruits. In most closely observed fisheries, such an inevitable reduction is to be considered seriously only if the fishing intensity is considerably higher than *F/M* = 3, with correspondingly greater depletion of the spawning stock. It is, however, not difficult to make adjustments for any slight reduction in recruitment by a less intensive fishing regime. The given numbers change, but the general pattern of stock behavior and yield potential does not. Such a recruitment change leads to slightly lower *msy* catches and slightly lower values of *F/M* needed to take them.

It is of interest to see how the choice of an *M/K* value influences the conclusions. I have looked at the lowest value found among bony fishes: *M/K* = 1.0. The MSY index is much less than for an *M/K* of 1.75—being 0.04 rather than 0.10535. But, the local *msy*, if *F/M* = 3.0, is 93% of MSY rather than 91%, as for *M/K* =1.75. It is obtained by a selectivity of 0.125 rather than 0.2.

More important are the consequences of a “less than” fishing intensity policy of 50% of *F/M-msy*. The sustainable catch is still as much as 87% of what the *msy* policy provides, compared with the 95% obtained by a 50% less than when *M/K* = 1.75, with corresponding benefits to catch rate (CPUE).

The conclusion is, therefore, that the benefits of a drastic “less than” policy do not much depend on the value of the driving parameter of the *SY–F/M* expression. Taking an upper level for *M/K* = 2.0, as observed in the bony fishes, leads to essentially the same conclusion.

Our ability to make predictions of the kind presented here comes principally from (i) the conduct of intensive long-term scientific research by some western European countries—especially England, Scotland, Denmark, Netherlands, and Belgium—and by administrations serving both the eastern and western coasts of USA and Canada; and (ii) the collection of a long series of detailed data about catches and fishing effort and practices.

It remains to look at the scale of depletion of the exploited stock caused by the selected management targets. For this purpose, it is convenient to consider the exploited stock biomass in an initial phase with very small fishing intensity. I take as the example here the sustainable catch rate is *F/M* = 0.053; this happens to be the lowest value given in the tables by Beverton and Holt (1966) or Gulland (1983), *E* = 0.05, so the reader can easily check other options. For *M/K* = 1.75, selectivity *w* = 0.216, an *F/M* value of 3 to provide *msy*, the stock is depleted to one-sixth of its virgin biomass. This is to be compared with the depletion of either one-half or to one-third guessed by those who rely on so-called surplus production assessments. So, the one-third value would actually result from a management target for *F/M* corresponding with about one-half of the effort required to take the local *msy*.

References

Beverton, R. J. H., and S. J. Holt. 1957. On the dynamics of exploited fish populations. Her Majesty’s Stationery Office, UK.

Beverton, R. J. H., and S. J. Holt. 1966. Manual of methods for fish stock assessment. Part II. Tables of yield functions. FAO Fisheries Technical Paper 38 (Revision 1).

Earle, M. 2021. Maximum sustainable yield in the EU’s Common Fisheries Policy—a political history. ICES Journal of Marine Science, doi.org/10.1093/icesjms/fsab037.

Gulland, J. A. 1983. Fish stock assessment. a manual of basic methods. FAO/Wiley Series on Food and Agriculture, 1. John Wiley and Sons, Chichester, UK.

Chapter 62

Comment on an EU Document Entitled: COM Baltic Plan Articles 4–5[1]

Target fishing mortality rates are, in Article 4 Targets, paragraphs 2 and 4, proposed in Annex 1 of the draft regulation. They are said to be in accord with the ranges of upper and lower values offered by ICES. The upper values are defined as fishing mortality rates that will provide sustainable catches of slightly less than MSY values. The selectivity of the fishing operation is not specified, and I would presume it is assumed to be the selectivity pertaining in the data series used to calculate MSYs, whatever that was or perhaps still is. The draft regulation is confusing and ambiguous because it mixes terminology mainly regarding "biomasses" (*B*) relating to the "surplus production" method of calculation and that used in analysis by population models, such as rate of fishing mortality *F*. Another serious source of confusion is the reference to B_{lim}, the spawning-stock biomass below which the stock must not be driven. In fact, B_{lim} is not a property of a fish stock; it is merely an artificial and basically arbitrary limit imposed by some ICES scientists. It is, in any case, not an element in the EU's basic management decision, and it is not considered a valid and useful quantity by fisheries scientists generally.

The B_{lim} discussion in this draft regulation obscures and confuses the consideration of managing fishing to secure catches from stocks maintained above levels that would provide MSY. Maxima in curves of sustainable yield SY against fishing mortality are governed mainly by the relation between natural mortality rates and patterns and rates of growth in fish body weight, not by reproduction. The effect of the reduction of population level by fishing becomes important only when the stock is greatly reduced by overfishing. It is hardly a controlling factor under management conditions designed to keep stocks bigger than are required to provide MSYs.

I decided to calculate one sample case to illustrate the issues. The calculation is for a bony fish species stock, with the value of the main driving parameter, the ratio *M/K* at its median for all fish species so far analyzed: 1.75. The calculation has to be made for a chosen selectivity of the fishery, and I chose a size at first liability to capture of 0.2, that is, the ratio of fish weight to the theoretical maximum weight: *w/W*. That selectivity is about normal for a fishery for direct human consumption. The fish length equivalent is $l/L = 0.6$. I express fishing intensity as the ratio *F/M* symbolized as F^*, *the relative fishing mortality rate*. This is proportional to the fishing effort properly calibrated. It is better to refer to this ratio rather than to *F* itself in order to make comparison across a variety of species easier.

[1]Editors' note: a comment written by Sidney Holt for Michael Earle and dated March 4, 2016. At the time, the Council and Parliament were negotiating the Baltic Sea multiannual plan, and this note comments on a draft text from the Commission that included *F* values above F_{MSY}. The Articles referred to are those in *Regulation (EU) No 2016/1139 of the European Parliament and of the Council establishing a multiannual plan for the stocks of cod, herring, and sprat in the Baltic Sea and the fisheries exploiting those stocks*. Official Journal L, 191:1–15.

This fishery would have its *local MSY* (symbolized as *msy*) at $F^* = 2.0$—that is, when the fishing mortality is twice as much as the natural mortality. The *msy* would be 0.194. This is an *index*; the actual *msy* would be that number multiplied by W and by the annual number of recruits.

A sustainable yield of 95% of *msy* would be obtained with a much lower fishing effort: $F^* = 1.0$. The catch rate (catch per unit effort) would be nearly double what would be obtained if *msy* was the target. So, a decision to forgo only 5% of the *msy* would halve the cost of taking the annual catch.

This perhaps unsurprising result is simply a consequence of the fact that the curves of sustainable catch against fishing effort are always rather flat-topped in the region below the *msy* level.

This calculation was made assuming that fishing intensity leading inevitably to reduced size of the exploited population—and hence the size of the stock of mature spawners—does not affect the annual number of recruits. All our experience is that such reduction has virtually no effect on recruitment until depletion has proceeded as a result of overexpansion of the fishery. If, however, the reduction of the population by fishing <u>does</u> affect the number of future recruits, even with a fishing intensity less than is needed to attain *msy*, then the SY-F^* curve would be even flatter.

It is, I think, immediately obvious that it is practically impossible to distinguish between the *msy* and the 0.95 × *msy* scenarios by setting TACs. With a huge difference in fishing effort and very little difference in the SYs, the effects of such a small change in the catch limit would be completely obscured by the natural variations of the fish population—even by just the annual variations in recruitment alone.

If a bigger "sacrifice" of SY could be contemplated—say 10% or even 20%—the required fishing effort would only be about one-quarter of what is needed to obtain *msy*, and it could probably be monitored reasonably by TACs.

In considering this, it is useful to look at how much the original unexploited stock size would be reduced (depleted) by various fishing intensities. Targeting *msy* would reduce the overall size (biomass) of the stock to about 20% of its unexploited level. Targeting 0.95 × *msy* would reduce it only to 40% of its initial size. Targeting an SY of, say, 15% less than *msy*, would ensure the population was never reduced to less than one-half of its original size.

In thinking about these options, it is important to understand that a sustainable yield from a fish stock is entirely created by human action: the reduction of the population by the act of exploiting it. Fishing changes not only the size of the population, but also its composition—increasing the relative numbers of small young fish. (It is a common error to think that this happens as a consequence of selectivity. Selectivity contributes, but the change in composition results, over time, simply by non-selective fishing reducing the lifespans of the fish.) The younger fish have a higher rate of body growth than older, bigger ones. The young fishes are growing fastest when they reach about 30% of their theoretical maximum weight W, after which growth slows steadily. (Unlike terrestrial vertebrates and birds, fish growth continues until eventual death; they do not have to cope with gravity.) So, fishing changes the delicate balance between body growth and natural death. If average growth rate is caused to increase, exceeding the natural death rate, then the population as a whole is caused to grow, unless it is stopped by fishing. The amount of catch that would stop the population growth is what we call sustainable yield. After a certain continuing reduction in the average growth rate as the fish get older, the relentless, continuing natural death overcomes the increased average growth rate, and the SY begins to decline, passing its maximum *msy*. It should be noted that the rate of *reproduction* in the population plays no part in determining sustainable yields except by ensuring an annual *recruitment* of young fish.

Chronological Bibliography of Sidney J. Holt Publications and Writings[1]

Hulme, H. R, R. J. H. Beverton, and S. J. Holt. 1947. Population studies in fisheries biology. Nature 159:714–715.

Margetts, A. R., and S. J. Holt. 1947. The effect of the 1939–1945 war on the English North Sea Trawl Fisheries. Rapports et Procès-Verbaux des Réunions du Conseil Permanent International pour l'Exploration de la Mer 122:26–46.

Holt, S. J. 1949. Biodynamic assessments of fish populations. Annual Report of the Challenger Society 3:10.

Holt, S. J. 1949. The fishing intensity on the plaice stock of the southern North Sea. Annales Biologiques Conseil Permanent International pour l'Exploration de la Mer 4:111.

Holt, S. J. 1951. Review of W. F. Thompson. "The Effect of Fishing on Stocks of Halibut in the Pacific," and Anon. "Regulation and Investigation of the Pacific Halibut Fishery in 1947 and 1948." Journal du Conseil Permanent International pour l'Exploration de la Mar 17(3):320–322.

Holt, S. J. 1954. Some observations on the concept of "maximum equilibrium catch" in relation to conservation problems. Unpublished.

Holt, S. J. 1955. On the foraging activity of the wood ant. Journal of Animal Ecology 24(1):1–34.

Holt, S. J. 1955. Observations on the concept of maximum sustainable yield as a criterion for the wise exploitation of the living resources of the sea. Paper submitted to the International Technical Conference on the Conservation of the Living Resources of the Sea, Rome, 18 April–10 May 1955. Unpublished.

Kesteven, G. L., and S. J. Holt. 1955. A note on the fisheries resources of the North West Atlantic. FAO Fisheries Papers 7.

Beverton, R. J. H., and S. J. Holt. 1956. A review of methods for estimating mortality rates in exploited fish populations, with special reference to sources of bias in catch sampling. Rapports et Procès-Verbaux des Réunions du Conseil Permanent International pour l'Exploration de la Mer 140:67–83.

Beverton, R. J. H., and S. J. Holt. 1956. The theory of fishing. Pages 372–441 *in* M. Graham, editor. Sea fisheries: their investigation in the United Kingdom. Edward Arnold, London.

Kesteven, G. L., and S. J. Holt. 1956. Classification of international conservation problems. Pages 350–371 *in* Papers Presented at the International Technical Conference on the Conservation of the Living Resources of the Sea: Rome, 18 April to 10 May 1955. United Nations Publication, Sales No: 1956.II.B.1. UN, New York.

Beverton, R. J. H., and S. J. Holt. 1957. On the dynamics of exploited fish populations. Her Majesty's Stationery Office, UK.

Holt, S. J. 1957. Extracts from correspondence concerning the Biarritz Report and the Lisbon meeting. Contribution P.2 to Joint Scientific Meeting of ICNAF, ICES and FAO on Fishing Effort, the Effect of Fishing on Resources and the Selectivity of Fishing Gear, Lisbon, 27 May–3 June 1957.

[1]Editors' note: entries in bold print are the basis for chapters in this book. We are grateful to Vassili Papastavrou for providing the titles of a significant number of papers by Sidney which we had not found. We have included everything that we have encountered, but some may have escaped our attention, particularly those tending to be called Sidney's work in the popular press. Therefore, this bibliography should not be viewed as a definitive record of his writings. A website is being constructed by colleagues of Sidney where many of these works will be available. See http://www.sidneyholt.website/.

Holt, S. J. 1957. A method of determining gear selectivity and its application. Contribution S.15 to Joint Scientific Meeting of ICNAF, ICES and FAO on Fishing Effort, the Effect of Fishing on Resources and the Selectivity of Fishing Gear, Lisbon, 27 May–3 June 1957.

Holt, S. J. 1957. General tables for characteristics of steady-state yields. Contribution P.11 to Joint Scientific Meeting of ICNAF, ICES and FAO on Fishing Effort, the Effect of Fishing on Resources and the Selectivity of Fishing Gear, Lisbon, 27 May–3 June 1957.

Holt, S. J. 1957. A note on the relation between the mortality rate and the duration of life in an exploited fish population. Contribution P.25 to Joint Scientific Meeting of ICNAF, ICES and FAO on Fishing Effort, the Effect of Fishing on Resources and the Selectivity of Fishing Gear, Lisbon, 27 May–3 June 1957.

Holt, S. J. 1957. The relation between size-at-age, annual mean size and the mortality rate of fish. Contribution P.26 to Joint Scientific Meeting of ICNAF, ICES and FAO on Fishing Effort, the Effect of Fishing on Resources and the Selectivity of Fishing Gear, Lisbon, 27 May–3 June 1957.

Holt, S. J., and E. Akyuz. 1957. Exploratory and experimental trawling in the Sea of Marmara. Contribution P.17 to Joint Scientific Meeting of ICNAF, ICES and FAO on Fishing Effort, the Effect of Fishing on Resources and the Selectivity of Fishing Gear, Lisbon, 27 May–3 June 1957.

Holt, S. J., and P. H. Thomas. 1957. A mathematical model for line and gillnet fisheries. Contribution P.19 to Joint Scientific Meeting of ICNAF, ICES and FAO on Fishing Effort, the Effect of Fishing on Resources and the Selectivity of Fishing Gear, Lisbon, 27 May–3 June 1957.

Holt, S. J. 1958. The evaluation of fisheries resources by the dynamic analysis of stocks, and notes on the time factors involved. Pages 77–95 *in* Some Problems for Biological Fishery Survey and Techniques for Their solution (A Symposium held at Biarritz, France, 1–10 March 1956). ICNAF Special Publication No. 1.

Holt, S. J. 1958. Report of the International Training Centre on the Methodology and Techniques of Research on Mackerel (*Rastrelliger*). Bangkok, December 5, 1958. Mimeograph.

Beverton, R. J. H., and S. J. Holt. 1959. A review of the lifespans and mortality rates of fish in nature and their relation to growth and other physiological characteristics. Pages 142–180 *in* G. E. W. Wolstenholme, and M. O'Connor, editors. Ciba Foundation Colloquium on Aging. Volume 5: The Lifespan of Animals. J. & A. Churchill, London.

Gulland, J. A., and S. J. Holt. 1959. Estimation of growth parameters for data at unequal time intervals. Journal du Conseil Permanent International Pour l'Exploration de la Mer 25(1):47–49.

Holt, S. J. 1959. A preliminary comparative study of the growth, maturity and mortality of sardine. Subject Synopsis (4), World Scientific Meeting on the Biology of Sardines and Related Species. Rome.

Holt, S. J. 1959. Letter to the Editor. Water temperature and cod growth-rate. Journal du Conseil Permanent International Pour l'Exploration de la Mer 24(2):374–376.

Holt, S. J., J. A. Gulland, C. Taylor, and S. Kurita. 1959. A standard terminology and notation for fishery dynamics. Journal du Conseil Permanent International Pour l'Exploration de la Mer 24(2):239–242.

Holt, S. J. 1960. Letter to the Editor. Journal du Conseil Permanent International Pour l'Exploration de la Mer 25(2):225–228.

Holt, S. J. 1962. The application of comparative population studies to fishery biology—an exploration. Pages 51–71 *in* E. D. Le Cren, and M. W. Holdgate, editors. The exploitation of natural animal populations; a Symposium of the British Ecological Society, Durham, 28th–31st March 1960. Blackwell Scientific Publications, Oxford.

Holt, S. J. 1963. A method for determining gear selectivity and its application. ICNAF Special Publication 5:106–115.

Chapman, D. G., K. R. Allen, and S. J. Holt. 1964. Appendix V. Reports of the Committee of Three Scientists on the Special Scientific Investigation of the Antarctic Whale Stocks. Reports of the International Whaling Commission 14:32–106.

Holt, S. J., and J. A. Gulland. 1964. Measures of abundance of Antarctic whale stocks. Rapports et Procès-Verbaux des Réunions du Conseil Permanent International pour l'Exploration de la Mer 155:147–151.

Chapman, D. G., K. R. Allen, S. J. Holt, and J. A. Gulland. 1965. Report of the Committee of Four Scientists. Reports of the International Whaling Commission 15:47–63.

Gulland, J. A., and S. J. Holt. 1965. Note on the effect of various catch quotas. Reports of the International Whaling Commission 15:61–63.

Holt, S. J., and J. A, Gulland. 1965. Supplemental note to the Report of the Committee of Four Scientists. Reports of the International Whaling Commission 15:60.

Beverton, R. J. H., and S. J. Holt. 1966. Manual of Methods for Fish Stock Assessment. Part II. Tables of Yield Functions. FAO Fisheries Technical Paper 38 (Revision 1).

Chapman, D. G., K. R. Allen, S. J. Holt, and J. A. Gulland. 1966. Report of the Committee of Four Scientists with two Addenda. Reports of the International Whaling Commission 16:43–49.

Holt, S. J. 1969. The food resources of the ocean. Scientific American 221(3):178–194.

Holt, S. J. 1974. Application of new IWC policy. Paper SC/SP74/Doc 11 presented to the Sei and Bryde's Whale Special Meeting, La Jolla, December 1974. Unpublished. 4 pp. [Available from the IWC Secretariat]

Holt, S. J. 1974. Some implications of present quotas. Paper SC/SP74/Doc 13 presented to the Sei and Bryde's Whale Special Meeting, La Jolla, December 1974. Unpublished. 5 pp. [Available from the IWC Secretariat]

Holt, S. J. 1974. Whales: conserving a resource. Nature 251:366–367.

Holt, S. J. 1974. A pocket epilogue to the history of whaling. Unpublished. 13 pp.

Holt, S. J., et al. 1974. Report of the Working Group on Biometrics and Management. Pages 14–20 *in* W. E. Schevill, editor. The whale problem: a status report. Harvard University Press, Cambridge, Massachusetts.

Holt, S. J., and C. Vanderbilt. 1974. Materials for an appraisal of the dependence of nations on the living resources of the sea. International Ocean Institute, Msida, Malta.

Holt, S. J. 1975. Marine fisheries and world food supplies. Pages 77–96 *in* F. Steele, and A. Bourne, editors. The man/food equation. Academic Press, London.

Holt, S. J. 1975. The concept of maximum sustainable yield (MSY) and its application to whaling. Document FAO/ACMRR/MM/IV/4, Airlie House Workshop, Virginia.

Holt, S. J. 1975. Report of the Scientific Committee, Annex X. Activities of the FAO/ACMRR Working Party on Marine Mammals. Reports of the International Whaling Commission 25:253–260.

Holt, S. J. 1975. The proposition that there may be circumstances when stocks should be harvested at levels below the level for MSY. Paper SC/27/Doc 12 presented to the IWC Scientific Committee, June 1975. Unpublished. 5 pp. [Available from the IWC Secretariat]

Holt, S. J. 1975. Weight versus numbers. Document SC/27/Doc 13 presented to the IWC Scientific Committee, June 1975. Unpublished. 2 pp. [Available from the IWC Secretariat]

Holt, S. J. 1975. Aspects of determining the stock level for maximum sustainable yield. Document ACMRR/MM/EC/29. Advisory Committee on Marine Resources Research, Marine Mammal Symposium, Bergen, Norway.

Holt, S. J. 1975. Aspects of determining the stock level for maximum sustainable yield. Document SC/27/Doc 14 presented to the IWC Scientific Committee, June 1975. Unpublished. 8 pp. [Available from the IWC Secretariat]

Holt, S. J. 1975. Pelagic whaling—Antarctic and Southern Hemisphere. Document SC/27/Doc 15 presented to the IWC Scientific Committee, June 1975. Unpublished. 3 pp. [Available from the IWC Secretariat]

Holt, S. J. 1976. The maximum sustainable yield in weight, and related matters. Reports of the International Whaling Commission 26:203–214.

Holt, S. J. 1976. Stock assessments of minke whales. Reports of the International Whaling Committee 26:392–394.

Holt, S. J. 1976. Criteria for management—weight or numbers? The relevance of whaling effort. Reports of the International Whaling Commission 26:409–413.

Holt, S. J. 1976. Statistics of catches of large whales. Contribution to Scientific Consultation on the Conservation and Management of Marine Mammals and their Environment, Bergen, Norway, September 1976. FAO, Rome.

Holt, S. J. 1976. UNCLOS, MSY, FAO, Regional Fisheries Bodies and Whales. Contribution to joint meeting of the Club of Rome and the International Ocean Institute on 'Reshaping the International Order,' Algiers, 25–28 October 1976.

Bruce, M., and S. J. Holt. 1977. A world guardian for the future. International Ocean Institute, Msida, Malta.

Holt, S. J. 1977. Estimation of sperm whale population sizes from changes in mean size of whales in catches. Reports of the International Whaling Commission 27:363–367.

Holt, S. J. 1977. A contribution to discussion of a new economic order, with reference to the living resources of the ocean. FAO, Rome.

Holt, S. J. 1977. Marine Fisheries. Pages 38–83 *in* E. L. Borgese, and N. Ginsburg, editors. Ocean yearbook I. University of Chicago Press, Chicago.

Holt, S. J. 1977. Whale management policy. Reports of the International Whaling Commission 27:133–137.

Holt, S. J. 1977. Questions about the sex-ratio in catches of rorquals. Reports of the International Whaling Commission 27:138–140.

Holt, S. J. 1977. Does the bottlenose whale necessarily have a sustainable yield, and if so is it worth taking? Reports of the International Whaling Commission 27:206–208.

Holt, S. J. 1977. Assessment of Southern Hemisphere minke whales. Reports of the International Whaling Commission 27:214–215.

Holt, S. J. 1977. Simulation of Southern Hemisphere sei whale stocks. Reports of the International Whaling Commission 27:216–220.

Holt, S. J. 1977. A possible basis for determining limits to the allowable catch when the degree of density dependence of the net recruitment rate is not known. Reports of the International Whaling Commission 27:230–233.

Holt, S. J. 1977. Management of sperm whaling. Memorandum to Commission for information of the Technical Committee. Unpublished. 4 pp.

Holt, S. J. 1977. International cooperation to protect the whales. Oceans 10(4):62–63.

Holt, S. J. 1977. Statement by the Food and Agricultural Organization (FAO) to the Special Meeting of the International Whaling Commission. Tokyo, 6–7 December 1977. Unpublished. 8 pp.

Holt, S. J. 1977. Statement of FAO to the International Whaling Commission. Unpublished. 7 pp.

Holt, S. J. 1978. Commentary: changing attitudes toward marine mammals. Oceanus 21(2):2–8.

Holt, S. J. Editor. 1978. Mammals in the seas, Volume I, Report of the FAO Advisory Committee on Marine Resources Research Working Party on Marine Mammals. FAO, Rome.

Holt, S. J. 1978. A review of current whale harvesting strategies. Whales and Whaling. 2. Australian Government Public Service, Canberra.

Holt, S. J. 1978. Some implications of maximum sustainable net yield as a management objective for whaling. Reports of the International Whaling Commission 28:191–193.

Holt, S. J. 1978. A comment on the assessments for Southern Hemisphere sei whales in the 1976 Report of the Scientific Committee. Reports of the International Whaling Commission 28:389–390.

Holt, S. J. 1978. Letter dated 25 October 1978. Document MP 8 presented to the Special Scientific Working Group on Management Procedures, La Jolla, November 1978 and Lowestoft, June 1979. Unpublished. 2 pp. [Available from the IWC Secretariat]

Holt, S. J. 1978. Reports of attendance at meetings on whales and whaling. Report to FAO. 13 pp.

Holt, S. J., and L. M. Talbot. 1978. New principles for the conservation of wild living resources. Wildlife Monographs 59.

Holt, S. J. 1979. A simple model of pelagic whaling. Reports of the International Whaling Commission 29:155–157.

Holt, S. J. 1979. Modern whaling in the Indian Ocean, particularly in the temperate tropical zone and the subdivision of stocks. Document SC/31/2 presented to the IWC Scientific Committee, June 1979. Unpublished. [Available from the IWC Secretariat]

Holt, S. J. 1979. Whale stocks in the Indian Ocean sector of the Indian Ocean. Document SC/31/01 presented to the IWC Scientific Committee, June 1979, Cambridge, UK. Unpublished. 30 pp. [Available from the IWC Secretariat]

Holt, S. J. 1979. Calculations of the mean age at recruitment from catch curves (Southern Hemisphere minke whales). Reports of the International Whaling Commission 29:365.

Holt, S. J. 1979. A note on the method of estimated net rate of increase from age compositions. Reports of the International Whaling Commission 29:325.

Holt, S. J. 1979. Further considerations on the assessments of Southern Hemisphere minke whales. Reports of the International Whaling Commission 29:451–459.

Holt, S. J. 1979. Proposal for a modified management policy. Reports of the International Whaling Commission 29:327–333.

Holt, S. J. 1979. Notes for the second meeting. Document MP 13 presented to the Special Scientific Working Group on Management Procedures, La Jolla, November 1978 and Lowestoft, June 1979. Unpublished. 5 pp. [Available from the IWC Secretariat]

Holt, S. J. 1979. Statement concerning H.R. 5033 on North Pacific fur seals to the House Sub-committee on Fisheries and Wildlife Conservation and the Environment. Unpublished. 3 pp.

Holt, S. J. 1979. Report of travel. Report to UNEP. 13 pp.

Holt, S. J., and J. G. Clark. 1979. Reclassification of protected baleen whale stocks. Reports of the International Whaling Commission 29:159–162.

Holt, S. J., and B. Grenfell. 1979. Report of the Sub-Committee on Minke Whales, Appendix II. Note on Southern Hemisphere minke whale stocks and quotas. Reports of the International Whaling Commission 29:81–83.

May, R. M., J. R. Beddington, C. W. Clark, S. J. Holt, and R. M. Laws. 1979. Management of multispecies fisheries. Science 205:267–277.

Allen, K. R., and S. J. Holt. 1980. Report of the Scientific Committee, Annex M. Review of research recommendations and the need for future special meetings and studies, Appendix 2. CPUE as an index of population density. Reports of the International Whaling Commission 30:137.

Beddington, J. R., and S. J. Holt. 1980. Report of the sub-committee on minke whales, Appendix 4. Reports of the International Whaling Commission 30:102–103.

Fortom-Gouin, J. P., and S. J. Holt. 1980. Reasons for recommending zero female catch limits for sperm whales. Reports of the International Whaling Commission Special Issue 2:261–262.

Holt, S. J. 1980. On the protection levels for sperm whales. Reports of the International Whaling Commission 30:177–181.

Holt, S. J. 1980. Procedure to eliminate some biases from estimates of pregnancy rates derived from BIWS data. Reports of the International Whaling Commission 30:575–580.

Holt, S. J. 1980. Calculations from Best, SC/30/Doc 33, Table 3—apparent pregnancy rates. Reports of the International Whaling Commission Special Issue 2:255.

Holt, S. J. 1980. Two concerns about the sperm whale model. Reports of the International Whaling Commission Special Issue 2:231–232.

Holt, S. J. 1980. An approach to the question of defining 'humane killing' in the context of the IWC Workshop. Document HK.6 presented to the IWC Workshop on Humane Killing Techniques for Whales, Cambridge, November 1980. Unpublished. [Available from the IWC Secretariat]

Holt, S. J. 1980. Multilingual vocabulary and notation for fishery dynamics/Compiled by S. J. Holt, chief, Biology Branch, Fisheries Division. FAO, Rome.

Holt, S. J. 1980. Southern Hemisphere minke whales: stock assessments, replacement yields and management procedures. Document SC/32/Mi 2 presented to the IWC Scientific Committee. Unpublished. 50 pp. [Available from the IWC Secretariat]

Holt, S. J. 1980. Sharing our planet with wildlife. Oryx: The International Journal of Conservation 15:259–262.

Holt, S. J. 1980. North Atlantic whale stocks with particular reference to Icelandic exploitation of fin and sperm whales. Document SC/32/O 7 presented to the IWC Scientific Committee, June 1980. Unpublished. 67 pp. [Available from the IWC Secretariat]

Holt, S. J. 1980. Note on whaling in the northeast of the Indian Ocean sanctuary. Document SC/32/O 10 presented to the IWC Scientific Committee. Unpublished. 10 pp. [Available from the IWC Secretariat]

Holt, S. J., and Keller, R. 1980. Interim report on Seychelles cetacean survey. Document SC/32/O 9 presented to the IWC Scientific Committee. Unpublished. 3 pp. [Available from the IWC Secretariat]

Holt, S. J., and Vanderbilt, C. 1980. Marine fisheries. Pages 9–56 in Ocean yearbook online, Volume 2, Issue 1. doi: https://doi.org/10.1163/221160080X00028.

Earle, S., and S. J. Holt. 1981. International protected areas. Memorandum to WWF. Unpublished. 2 pp.

Holt, S. J. 1981. Aspects of determining the stock level for maximum sustainable yield. Pages 73–76 in Mammals in the seas, Volume III, General papers and large cetaceans. FAO, Rome.

Holt, S. J. 1981. History of the regulation of southern minke whaling scientific advice and response of industry. Reports of the International Whaling Commission 31:319–322.

Holt, S. J. 1981. The protection of species. Document SC/33/O 13 presented to the IWC Scientific Committee, June 1981, Cambridge, UK. Unpublished. 32 pp. [Available from the IWC Secretariat]

Holt, S. J. 1981. Report of the Scientific Committee, Annex J. Scientific reasons for removal of southern boundary limit at 55°S. Reports of the International Whaling Commission 31:156–157.

Holt, S. J. 1981. Report of the Scientific Committee, Annex K. Minority statement on S. Hemisphere minke whales—catch limits in Areas I, II and V. Reports of the International Whaling Commission 31:158.

Holt, S. J. 1981. Objectives of management, with particular reference to whales. Pages 349–357 *in* Mammals in the seas, Volume III, General papers and large cetaceans. FAO, Rome.

Holt, S. J. 1981. Maximum sustainable yield and its application to whaling. Pages 21–55 *in* Mammals in the seas, Volume III. General papers and large cetaceans. FAO, Rome.

Holt, S. J. 1981. Why Spain should consider cessation of whaling. Memorandum to IUCN & WWF. Unpublished. 4 pp.

Holt, S. J. 1981. Saving whales and other beasts. Memorandum to UNEP. Unpublished. 24 pp.

Holt, S. J., J. Carr, and L. Fraser. 1981. A note on the sex ratios in Southern Hemisphere catches of minke whales. Reports of the International Whaling Commission 31:315–317.

Holt, S. J., A. Hiby, and J. Harwood. 1981. Report of the Sub-Committee on Sperm Whales, Appendix 10. Estimation of N. Atlantic sperm whale stocks. Reports of the International Whaling Commission 31:99–100.

Chapman, D. G., W. K. de la Mare, S. J. Holt, and J. G. van Beek. 1982. Report of the Scientific Committee, Annex M. Statement on minke whale assessments for the Southern Hemisphere and advice to the Commission. Reports of the International Whaling Commission 32:136–138.

Frizell, J., and S. J. Holt. 1982. Organochlorine compounds in cetaceans—implications for population modelling. Document SC/34/O 14 presented to the IWC Scientific Committee, June 1982. Unpublished. 3 pp. [Available from the IWC Secretariat]

Holt, S. J. 1982. A note on the 1981 assessments of the Peruvian stock of Bryde's whales. Document SC/34/Ba11 presented to the IWC Scientific Committee, June 1982. Unpublished. 16 pp. [Available from the IWC Secretariat]

Holt, S. J. 1982. Notes on assessments of Bryde's whales. Document SC/34/Ba15 presented to the IWC Scientific Committee, June 1982. Unpublished. 30 pp. [Available from the IWC Secretariat]

Holt, S. J. 1982. The abundance of fin whales in the Spain–Portugal–British Isles stock. Document SC/34/Ba16 presented to the IWC Scientific Committee, June 1982. Unpublished. 9 pp. [Available from the IWC Secretariat]

Holt, S. J. 1982. Baird's beaked whale. Document SC/34/SM8 presented to the IWC Scientific Committee, June 1982. Unpublished. 10 pp. [Available from the IWC Secretariat]

Holt, S. J. 1982. The stock boundary for North Pacific sperm whales. Document SC/34/Sp7 presented to the IWC Scientific Committee, June 1982. Unpublished. 13 pp. [Available from the IWC Secretariat]

Holt, S. J. 1982. Report of the Scientific Committee, Annex P. North Atlantic sperm/British-Spanish fin whale. Comment on the interpretation of Spanish and Azores CPUE data and the implications for sperm whales. Reports of the International Whaling Commission 32:147–148.

Holt, S. J. 1982. The history of modern whaling, by J. N. Tonnessen and A. O. Johnsen. Book review. Oryx 16(5):452–453.

Holt, S. J. 1982. Report of the Sub-Committee on Sperm Whales, Appendix 7. Adjustment of CPUE for seasonality and catcher performance in Spanish operations. Reports of the International Whaling Commission 32:84–86.

Holt, S. J. 1982. Report of the Special Meeting on Southern Hemisphere Minke Whales, Cambridge, 22–26 June 1981, Appendix 5. Examination of the question whether the concentration of pelagic whaling operations near certain boundaries of management areas is a result of the setting of catch limits by areas. Reports of the International Whaling Commission 32:708–709.

Holt, S. J. 1982. A note on the status of harp and hooded seals. Unpublished manuscript. 9 pp.

Holt, S. J. 1982. A case for the listing of harp and hooded seals on Appendix I of CITES and all Phocid seals on Appendix II. Unpublished. 10 pp.

Holt, S. J., W. K. de la Mare, and J. G. van Beek. 1982. Report of the Special Meeting on Southern Hemisphere Minke Whales, Cambridge, 22–26 June 1981, Appendix 12. A review of information available for the assessment of minke whales in Area II. Reports of the International Whaling Commission 32:717–721.

Holt, S. J., and D. M. Lavigne. 1982. Seals slaughtered—science abused. New Scientist 93:636–639.

Keller, R. W., S. Leatherwood, and S. J. Holt. 1982. Indian Ocean cetacean survey, Seychelles Islands, April through June 1980. Reports of the International Whaling Commission 32:503–513.

van Beek, J. G., S. J. Holt, W. K. de la Mare, J. M. da Rocha, and Y. Shimadzu. 1982. Report of the Special Meeting on Southern Hemisphere Minke Whales, Cambridge, 22–26 June 1981, Appendix 11. Catch per unit effort: an estimation series from Brazilian minke whaling operations. Reports of the International Whaling Commission 32:715–716.

Chapman, D. K., and S. J. Holt. 1983. Report of the Sub-Committee on Other Baleen Whales, Appendix 7. Sightings estimate of Bryde's whale stock in squares H 32-33-34, J 32-33-34. Reports of the International Whaling Commission 33:139.

Chapman, D. K., W. K. de la Mare, S. J. Holt, and F. Pascal. 1983. Report of the Scientific Committee, Annex M. Comments on moratorium proposals (Agenda Item 8.1), Annex M3. Moratorium proposals. Reports of the International Whaling Commission 33:184.

Chapman, D. K., W. K. de la Mare, S. J. Holt, and F. Pascal. 1983. Report of the Scientific Committee, Annex P. Minority statements on Item 9.2, minke whales, Annex P5. Minke whales: North Atlantic Northeastern and Central stocks. Reports of the International Whaling Commission 33:189–190.

Fortom-Gouin, J. P., S. J. Holt, F. Palacio, and C. Roux. 1983. Report of the Special Meeting on Western North Pacific Sperm Whale Assessments, Cambridge, 1982, Annex O. Minority statement: doubts about the validity of projections. Reports of the International Whaling Commission 33:721.

Holt, S. J. 1983. Report of the Scientific Committee, Annex O. Commission provisional Agenda Item 8: Review of regulatory measures other than catch limits. Reports of the International Whaling Commission 33:187.

Holt, S. J. 1983. The Indian Ocean whale sanctuary. Ambio 12(6):345–347.

Holt, S. J. 1983. Report of the Sub-Committee on Other Baleen Whales, Appendix 4. Catch of fin whales off Spain 1957–1981. Reports of the International Whaling Commission 33:136.

Holt, S. J. 1983. Report of the Scientific Committee, Annex J. Reports of the computing sub-group into Scientific Committee Agenda Items 11.1, 11.2, 11.3 and the working group on computing needs, Appendix 1. Examples of some of the problems recently encountered. Reports of the International Whaling Commission 33:175.

Holt, S. J. 1983. Marine animal killing. Nature 304(5922):110.

Holt, S. J. 1983. Report of the Special Meeting on Western North Pacific Sperm Whale Assessments, Cambridge, 1982, Annex F. Stock definition. Reports of the International Whaling Commission 33:706–707.

Holt, S. J. 1983.Who really threatens whales and seals? Oryx 17(2):68–77.

Holt, S. J. 1983. Letter to editor. Oryx 17(2):104–105.

Holt, S. J. 1983. Conserving whales by regulating trade. International Fund for Animal Welfare. Unpublished. 62 pp.

Holt, S. J. 1983. Identification of specific protective measures that require undertaking by governments and organizations for the enhancement of the survival of threatened populations of marine mammals. Working paper prepared for IUCN for UNEP/FAO Consultation on Marine Mammals. Nairobi, Kenya, 11–14 January 1983. Unpublished. 16 pp.

Holt, S. J. 1983. A note on the establishment of the Indian Ocean whale sanctuary. Presented at Symposium on Marine Mammals of the Indian Ocean. Colombo, Sri Lanka, 22–25 February 1983. Unpublished. 13 pp.

Holt, S. J. 1983. Whales are not cetacean resources. Introductory note for Whales Alive Conference. 7 June 1983. Unpublished. 4 pp.

Holt, S. J. 1983. Cooperation in the conservation of marine mammals in the Indian Ocean. Presented at Pacem in Maribus XII. Colombo, Sri Lanka, 25–28 October 1983. Unpublished. 8 pp.

Holt, S. J. 1983. Marine mammals and CITES. Report to The Seychelles. 4 pp.

Holt, S. J., and W. K. de la Mare. 1983. An analysis of recent RII recruitment rates in the north west Pacific stock of sperm whales. Reports of the International Whaling Commission 33:279–281.

Holt, S. J., F. J. Palacio, and L. Busby. 1983. Peru and the International Whaling Commission (IWC). 43 pp. [Paper printed in Lima, Peru for the Tinker Center for Coastal Studies in Latin America, University of Miami. April 1983.]

de la Mare, W. K., J. van Beek, and S. J. Holt. 1984. Report of the Scientific Committee, Annex N. Minority statements, Annex N8. A rebuttal of minority statement N7. Reports of the International Whaling Commission 34:178–179.

Holt, S. J. 1984. Whalemate. BBC Wildlife 1984:348–349.

Holt, S. J. 1984. Report of the Sub-Committee on Other Baleen Whales, Appendix 14. Peruvian Bryde's whale—sightings. Reports of the International Whaling Commission 34:128.

Holt, S. J. 1984. Whaling. BBC Wildlife 1984:444–445.

Holt, S. J. 1984. Some more politics of conservation. Oryx 18(2):102–106.

Holt, S. J. 1984. Aboriginal/Subsistence Whaling (with special reference to the Alaska and Greenland Fisheries)—International Whaling Commission Reports, Special Issue 4, 1982. Book review. Oryx 18(2):119–120.

Holt, S. J. 1984. The Ecology of Whales and Dolphins, by D. E. Gaskin. Book review. Oryx 18(3):176–177.

Holt, S. J. 1984. Indian Ocean whale sanctuary. Whalewatcher Spring 1984. 3 pp.

Holt, S. J. 1984. Whaling: the home stretch? Guest editorial. Oryx 18(4):194.

Holt, S. J. 1984. Aspects of the assessment and regulation of Bryde's whale in the northwestern Pacific. Document SC/36/Ba3 presented to the IWC Scientific Committee, May 1984. Unpublished. 18 pp. [Available from the IWC Secretariat]

Holt, S. J. 1984. Assessment of Area II minke stock. Document SC/36/Mi10 presented to the IWC Scientific Committee, 1984. Unpublished. 44 pp. [Available from the IWC Secretariat]

Holt, S. J. 1984. Catch per unit effort of minke whales in Korean waters. Document SC/36/Mi15 presented to the IWC Scientific Committee, 1984. Unpublished. 15 pp. [Available from the IWC Secretariat]

Holt, S. J. 1984. Whaling in Brazil. Information paper. International League for the Protection of Cetaceans. Unpublished. 6 pp.

Holt, S. J., K. C. Balcomb, W. K. de la Mare, and J. G. van Beek. 1984. Report of the Scientific Committee, Annex N. Minority statements, Annex N9. Statement on North Atlantic humpback whales. Reports of the International Whaling Commission 35:179.

van Beek, J. G., and S. J. Holt. 1984. Report of the Sub-Committee on Other Baleen Whales, Appendix 10. A note on the Peruvian Bryde's whale stock boundaries. Reports of the International Whaling Commission 34:124–125.

Holt, S. J. 1985. Indian Ocean whale sanctuary—background and update. Presented at Symposium on Endangered Marine Animals and Marine Parks, Cochin, India, 12–16 January 1985. Unpublished. 24 pp.

Holt, S. J. 1985. Same old swan song, one more verse. BBC Wildlife Magazine, July 1985. 1 pp.

Holt, S. J. 1985. Let's all go whaling. Ecologist 15(3):113–124.

Holt, S. J. 1985. Whaling news. Oryx 19(1):3.

Holt, S. J. 1985. Boycotts. Oryx 19(1):46.

Holt, S. J. 1985. Norwegian whaling and policy: an appraisal. International League for the Protection of Cetaceans, Rye, E. Sussex.

Holt, S. J. 1985. The classification of whale stocks and the determination of catch limits, under the current IWC management procedure, with limited information. Reports of the International Whaling Commission 35:487–494.

Holt, S. J. 1985. Whale mining, whale saving. Marine Policy 9(3):192–213.

Holt, S. J. 1985. A future for the minke whale? Presentation to Open Symposium on Norwegian Whaling, Oslo, Norway, 17 June 1985. Unpublished. 15 pp.

Holt, S. J. 1985. Classification of North Atlantic minke stocks. Document SC/37/Mi4 presented to the IWC Scientific Committee, June 1985. Unpublished. 60 pp. [Available from the IWC Secretariat]

Holt, S. J. 1985. Classification of the minke whale stock exploited in Southern Hemisphere Area II. Document SC/37/Mi5 presented to the IWC Scientific Committee, June 1985. Unpublished. 8 pp. [Available from the IWC Secretariat]

Holt, S. J. 1985. Criteria for review of special permits. Document SC/37/O 9 presented to the IWC Scientific Committee, June 1985. Unpublished. 7 pp. [Available from the IWC Secretariat]

Holt, S. J. 1985. The gray whale revisited. Document SC/37/PS20 presented to the IWC Scientific Committee, June 1985. Unpublished. 13 pp. [Available from the IWC Secretariat]

Holt, S. J. 1985. Report of the Scientific Committee, Annex Q. Proposals for recommendations for catch limits for Southern Hemisphere minke whales. Reports of the International Whaling Commission 35:151–152.

Holt, S. J. 1985. Report of the Sub-Committee on Southern Hemisphere Minke Whales, Appendix 9. Brazil Area II minke whales. Reports of the International Whaling Commission 35:89.

Holt, S. J. 1985. Report of the Sub-Committee on Southern Hemisphere Minke Whales, Appendix 10. Equations for stock trajectory in SC/36/Mi10. Reports of the International Whaling Commission 35:89.

Holt, S. J. 1985. Report of the Sub-Committee on Other Baleen Whales, Appendix 11. Equations for stock trajectory. Reports of the International Whaling Commission 35:115.

Holt, S. J. 1985. Report of the Sub-Committee on Other Baleen Whales, Appendix 12. Rationale for use of M = 0.070 in assessments. Reports of the International Whaling Commission 35:116.

Holt, S. J. 1985. Conditional withdrawal of objections. International League for the Protection of Cetaceans. Unpublished. 4 pp.

Kasamatsu, F., T. Miyashita, S. J. Holt, and J. Breiwick. 1985. Report of the Sub-Committee on Other Baleen Whales, Appendix 7. Western North Pacific Bryde's whale sightings estimate. Reports of the International Whaling Commission 35:112–113.

Bjørge, A., I. Christensen, J. Cooke, S. J. Holt, K. Lankester, N. Øien, and T. Øritsland. 1986. Report of the Sub-Committee on Northern Hemisphere Minke Whales, Appendix 5. Comments on scientific permits, Appendix 5.4. Comments to the Korean proposal for a special scientific permit by Bjørge, Christensen, Cooke, Holt, Lankester, Øien, and Øritsland. Reports of the International Whaling Commission 36:88.

Holt, S. J. 1986. Reassessment of the Central Atlantic and West Greenland minke whale stocks. Document SC/38/Mi2 presented to the IWC Scientific Committee, May 1986. Unpublished. 19 pp. [Available from the IWC Secretariat]

Holt, S. J. 1986. Aspects of the assessment and regulation of Bryde's whale in the northwest Pacific. Reports of the International Whaling Commission 36:257–262.

Holt, S. J. 1986. Report of the Sub-Committee on Northern Hemisphere Minke Whales, Appendix 2. Explanation of equations and parameter estimation of SC/37/Mi4. Reports of the International Whaling Commission 36:86.

Holt, S. J. 1986. Planning a comprehensive assessment of whale stocks. Document SC/A86/CA3 submitted to the Special Meeting of the Scientific Committee on Planning for a Comprehensive Assessment of Whale Stocks, Cambridge, April 1986. Unpublished. [Available from the IWC Secretariat]

Holt, S. J. 1986. Loopholes and leviathans. BBC Wildlife 4(6):282–289.

Holt, S. J. 1986. Iceland buys some science. BBC Wildlife 1986:564–555.

Holt, S. J. 1986. Whales remain in danger. The Observer, July 13, 1986.

Holt, S. J. 1986. Mammals in the sea. Ambio 15(3):126–133.

Holt, S. J. 1986. Northeast Atlantic minke stock: Reassessment. Document SC/38/Mi3 presented to the IWC Scientific Committee, May 1986. Unpublished. 34 pp. [Available from the IWC Secretariat]

Holt, S. J. 1986. Short briefing on northeast Atlantic minke whales. Unpublished. 1 pp.

Holt, S. J. 1986. Status and productivity of the northeast Atlantic stock of the minke whale. Unpublished. 10 pp.

Holt, S. J. 1986. Letter to the Secretary. Document SC/38/O 17 presented to the IWC Scientific Committee, May 1986. Unpublished. 3 pp. [Available from the IWC Secretariat]

Holt, S. J., D. G. Chapman, K. Lankester, W. K. de la Mare, and J. G. Cooke. 1986. Report of the Scientific Committee, Annex N. Minority statements, N3: Minority statement concerning item 11.2.6 by Holt, Chapman, Lankester, de la Mare and Cooke. Reports of the International Whaling Commission 36:138.

Holt, S. J., and K. Lankester. 1986. Report of the Sub-Committee on Northern Hemisphere minke whales, Appendix 5. Comments on scientific permits, 5.1, comments by Holt and Lankester. Reports of the International Whaling Commission 36:87.

Holt, S. J., N. Øien, T. Øritsland, A. Bjørge, E. Born, L. Larsen, and K. Lankester. 1986. Report of the Sub-Committee on Northern Hemisphere Minke Whales, Appendix 4. Statement by Holt, Øien,

Øritsland, Bjørge, Born, Larsen, and Lankester. Reports of the International Whaling Commission 36:87.
Holt, S. J. 1987. Report of the Scientific Committee, Annex N. Minority statements, N2: statement on Scientific Committee procedures, Item 8.3. Reports of the International Whaling Commission, 37: 137.
Holt, S. J. 1987. Report of the Scientific Committee, Annex N. Minority statements, N4: statement on minke whales in the northeast Atlantic, Item 9.3. Reports of the International Whaling Commission 37:138–140.
Holt, S. J. 1987. Antarctica and the living resources of the Southern Ocean. Presented at Pacem in Maribus XV, Malta, 7–11 September 1987. Unpublished. 49 pp.
Holt, S. J. 1987. Comments submitted by S. J. Holt (Seychelles). Document SC/D87/17 presented to the IWC Scientific Committee Special Meeting to Consider the Japanese Research Permit (Feasibility Study), Cambridge, December 1987. Unpublished. [Available from the IWC Secretariat]
Holt, S. J. Comments submitted by S. J. Holt (Seychelles) on the research plan for the feasibility study on 'The program for research on the Southern Hemisphere minke whale and for preliminary research on the marine ecosystem in the Antarctic' by the Government of Japan. Document SC/D87/23 presented to the IWC Scientific Committee Special Meeting to Consider the Japanese Research Permit (Feasibility Study), Cambridge, December 1987. Unpublished. [Available from the IWC Secretariat]
Holt, S. J. 1987. Report of the Sub-Committee on Northern Hemisphere Minke and Other Baleen Whales, Appendix 2. Overlap of seasons in Norwegian bottlenose and minke whaling in central Atlantic. Reports of the International Whaling Commission 37:99.
Holt, S. J. 1987. The status of the East Greenland–Iceland fin whale stock. Document SC/39/Ba2 presented to the IWC Scientific Committee, June 1987. Unpublished. 14 pp. [Available from the IWC Secretariat]
Holt, S. J. 1987. Comments on Japanese proposal to catch Southern Hemisphere minke whales. Document SC/39/Mi10 presented to the IWC Scientific Committee, June 1987. Unpublished. 12 pp. [Available from the IWC Secretariat]
Holt, S. J. 1987. Carry on scientist. BBC Wildlife Magazine July 1987:332–333.
Holt, S. J. 1987. Proposed culling of grey seals in eastern Canada. Report to IFAW. 27 pp.
Holt, S. J. 1988. The whale war. Marine Policy 12(4):410–411.
Holt, S. J. 1988. The planet gets a fever. BBC Wildlife 1988:553–555.
Holt, S. J. 1988. Scientific whaling. Marine Pollution Bulletin 19(4):184–185.
Holt, S. J. 1988. Nineteenth century minke whaling in the North Atlantic. Document SC/40/Mi24 presented to the IWC Scientific Committee, May 19887. Unpublished. 2 pp. [Available from the IWC Secretariat]
Holt, S. J. 1988. Report of the Sub-Committee on North Atlantic minke whales, Appendix 5. Comments on research proposals by Institute of Marine Research, Bergen, Norway (SC/39/O 11). Reports of the International Whaling Commission 38:94–95.
Holt, S. J. 1988. Report of the Scientific Committee, Annex R. The Japanese scientific permit proposal for Southern Hemisphere minke whales: background and comments, R3: extracts from SC/39/Mi10. 'Comments on Japanese proposal to catch Southern Hemisphere minke whales under special permit'. Reports of the International Whaling Commission 38:142–144.
Holt, S. J. 1988. Effects of a pause in whaling on fish catches. Unpublished. 2 pp.
Holt, S. J. 1988. The whaling game. For The Mammal Society. Unpublished. 3 pp.
Holt, S. J. 1989. Briefing on Icelandic claims about whales and fish stocks. Unpublished. 4 pp.
Holt, S. J. 1989. Whaling catch. New Scientist 122(1661):83.
Holt, S. J. 1989. Peacetime for Moby Dick. BBC Wildlife 1989:294–296.
Holt, S. J. 1989. Even pessimists rub eyes over new whale figures. BBC Wildlife 1989:473.

Holt, S. J. 1989. 'The state of the Northeast Atlantic minke whale stock'—a critique. Reports of the International Whaling Commission 39:213–218.

Holt, S. J. 1989. Some considerations regarding second-stage testing of management procedures. Document SC/41/O 9 presented to the IWC Scientific Committee, May 1989. Unpublished. 3 pp. [Available from the IWC Secretariat]

Holt, S. J., and S. N. Sexton. 1989, Monitoring trends in dolphin abundance in the eastern tropical Pacific using research vessels over a long sampling period: analyses of 1987 data. Reports of the International Whaling Commission 39:347–351.

Holt, S. J. 1990. Sensitivity of management procedures to discrepancies between catch limits and catches. Document SC/J90/Mg 4 presented to the IWC Scientific Committee, June 1990. Unpublished. 2 pp. [Available from the IWC Secretariat]

Holt, S. J. 1990. Contribution to comprehensive assessment of minke whales in the North Atlantic. Document SC/42/NHMi21 presented to the IWC Scientific Committee, June 1990. Unpublished. 62 pp. [Available from the IWC Secretariat]

Holt, S. J. 1990. Notes on super-compensation and MSYR. Document SC/42/O 13 presented to the IWC Scientific Committee, June 1990. Unpublished. 6 pp. [Available from the IWC Secretariat]

Holt, S. J. 1990. Are measurements of increase rates of depleted stocks useful in estimating MSY rate? Reports of the International Whaling Commission 40:517–524.

Holt, S. J. 1990. Minke myth. Letter to the editor. BBC Wildlife 1990:201.

Holt, S. J. 1990. Norway's harpoon figure gets that old itch. BBC Wildlife 1990:332–333.

Holt, S. J. 1990. Norway's final solution to whales' problems. BBC Wildlife 1990:409.

Holt, S. J. 1990. A global convention for the protection of cetaceans from hazards posed by human activities. Submitted to meeting of the UNEP Marine Mammal Action Plan. Bonn, Germany, 15–16 October 1990. Unpublished. 7 pp.

Holt, S. J., and S. N. Sexton. 1990, Monitoring trends in dolphin abundance in the eastern tropical Pacific using research vessels over a long sampling period: analyses of 1988 data. Reports of the International Whaling Commission 40:471–476.

Holt, S. J., and N. M. Young. 1990. Guide to review of the management of whaling. Center for Marine Conservation, Washington, DC.

Cooke, J. G., and S. J. Holt. 1991. Report of the Sub-Committee on North Atlantic Minke Whales, Appendix 8. Regression analysis of greater Barents Sea CPUE data by area and month, 1951–83. Reports of the International Whaling Commission 41:154–157.

Holt, S. J. 1991. The moratorium must stay. Sonar 5:12–13.

Holt, S. J. 1991. Abstract to SC/42/NHMi21. Contribution to comprehensive assessment of minke whales in the North Atlantic. Reports of the International Whaling Commission 41:587–588.

Holt, S. J. 1991. Abstract to SC/42/O 13. Notes on super-compensation and MSYR. Reports of the International Whaling Commission 41:592.

Holt, S. J. 1991. Possible bias in Icelandic catch of fin whales per unit effort. Document SC/43/Ba1 presented to the IWC Scientific Committee, May 1991. Unpublished. 10 pp. [Available from the IWC Secretariat]

Holt, S. J. 1991. Fin whales in the North Atlantic, review of a comprehensive assessment (author's summary). Document SC/43/Ba5 presented to the IWC Scientific Committee, May 1991. Unpublished. 9 pp.

Holt, S. J. 1991. Problems in the classification of stocks, with particular reference to northeast Atlantic minke whales. Document SC/43/Mi9 presented to the IWC Scientific Committee, May 1991. Unpublished. 21 pp. [Available from the IWC Secretariat]

Holt, S. J. 1991. The un-ethics of whaling. Why whales? Whale and Dolphin Conservation Society Special Publication 8–16.

Holt, S. J. 1991. Commercial whaling humane? International Fund for Animal Welfare. 54 pp.

Holt, S. J. 1991. The forty-third international whaling games. BBC Wildlife 1991:52–53.

Holt, S. J. 1991. Whales and wherefores. BBC Wildlife 1991:313–314.

Holt, S. J. 1991. Fin whales in the North Atlantic: a comprehensive assessment. World Wide Fund for Nature.

Holt, S. J. 1991. They're not saved yet. Notes on the themes of the Fourth Biennial Conference of the American Cetacean Society, Monterey, California, November 1990. Unpublished. 10 pp.

Holt, S. J. 1991. Some lessons from twenty years of efforts to manage commercial whaling and to save whales. Prepared on behalf of IFAW for the Planning and Coordinating Committee of the Marine Mammals Action Plan. Unpublished. 24 pp.

Holt, S. J., and C. Carlson. 1991. Implementation of a revised management procedure for commercial whaling. International Fund for Animal Welfare.

Holt, S. J. and W. K. de la Mare. 1991. Report of the Scientific Committee, Annex S. Minority statements, statement by Holt and de la Mare on adoption of agenda Re: management advice. Reports of the International Whaling Commission 41:219.

Holt, S. J., and N. M. Young. 1991. Guide to review of the management of whaling, 2nd edition. Center for Marine Conservation, Washington, DC.

Cooke, J. G., W. K. de la Mare, S. J. Holt, K. Lankester, and T. Lyrholm. 1992. Minority statement advice on selection of a revised management procedure for commercial baleen whaling. Reports of the International Whaling Commission 42:267–269.

Holt, S. J. 1992. The debate on whaling. Nature 359(6390):9.

Holt, S. J. 1992. No room for whalers in the EC. The Independent. 27 November 1992.

Holt, S. J. 1992. Whalers lose. New Scientist 135(1834):46.

Holt, S. J. 1992. Iceland does the deed. BBC Wildlife 1992:59.

Holt, S. J. 1992. Whales for the grandchildren. BBC Wildlife 1992:61–62.

Holt, S. J. 1992. Abstract to SC/43/Ba1. Possible bias in Icelandic catch of fin whales per unit effort. Reports of the International Whaling Commission 42:757.

Holt, S. J. 1992. Selection of an RMP. Reports of the International Whaling Commission 42:134–135.

Holt, S. J. 1992. Abstract to SC/43/Mi9. Problems in the classification of stocks, with particular reference to northeast Atlantic minke whales. Reports of the International Whaling Commission 42:754.

Holt, S. J. 1992. Abstract to SC/43/Ba5. Fin whales in the North Atlantic, review of a comprehensive assessment (author's summary). Reports of the International Whaling Commission 42:757.

Holt, S. J. 1992. Comments on SC/F92/Mg2, SC/43/O15+O8 and SC/F92/Mg3. Document SC/F92/Mg11 presented to the Special Meeting of the Scientific Committee on the Revised Management Procedure, Copenhagen, 2–6 March 1992. Unpublished. 9 pp. [Available from the IWC Secretariat]

Holt, S. J. 1992. Scientific advice on the management of commercial whaling. Document SC/44/O 6 presented at the 1992 meeting of the IWC Scientific Committee. Unpublished. 2 pp. [Available from the IWC Secretariat]

Holt, S. J. 1992. Thoughts about MSY rates and levels. Document SC/44/O 3 presented at the 1992 meeting of the IWC Scientific Committee. Unpublished. 43 pp. [Available from the IWC Secretariat]

Holt, S. J. 1992. Report of the Third Comprehensive Assessment Workshop on Management Procedures, Annex N. Suggestions for further testing procedures, N2: catch data considerations for the 'grand unified procedure'. Reports of the International Whaling Commission 42:303.

Holt, S. J. 1992. Deferring the final farewell. Global 500 1992(4):12–15.

Holt, S. J. 1992. Wildlife in the ocean commons: whales and other beasts. Inequality and the Commons, the Third Biennial Conference of the International Association for the Study of Common Property, September 17–19, 1992, Washington, DC.

Holt, S. J. 1992. Seychelles. Progress report on cetacean research, May 1990 to April 1991. Reports of the International Whaling Commission 42:364–362.

Holt, S. J. 1992. Report of the Comprehensive Assessment Special Meeting on North Atlantic Fin Whales, Annex K. Minority statements, Statement by Holt on adoption of agenda. Reports of the International Whaling Commission 42:644.

Holt, S. J. 1992. Being economical with the truth. Sonar 7:8–10.

Holt, S. J. 1992. The management of whaling. With an appendix on economies with the truth. International Fund for Animal Welfare. 56 pp.

Holt, S. J. 1992. Fin whaling in the Denmark Strait: Part 1—exploratory analysis. Document SC/44/NAB1 presented to the IWC Scientific Committee, June 1992. Unpublished. 29 pp. [Available from the IWC Secretariat]

Holt, S. J. 1992. Increase in blue whales in the Denmark Strait. Document SC/44/O 28 presented to the IWC Scientific Committee, June 1992. Unpublished. 7 pp. [Available from the IWC Secretariat]

Holt, S. J. 1992. Dispersal of baleen whales over Antarctic feeding grounds. Document SC/44/SHB1 presented to the the IWC Scientific Committee, June 1992. Unpublished. 10 pp. [Available from the IWC Secretariat]

Holt, S. J. 1992. Do the great whales have a future? Presentation to "Talking Whales" Whale and Dolphin Conservation Society Annual Conference, London, UK, 31 October 1992. Unpublished. 13 pp.

Holt, S. J. 1992. Animal protection and environmentalism: science. *In* R. D. Ryder, editor. Animal welfare and the environment. Gerald Duckworth and Company, Ltd., London.

Beverton, R. J. H., and S. J. Holt. 1993. On the dynamics of exploited fish populations. Chapman & Hall, New York. (Re-issue of 1957 paper)

Holt, S. J. 1993. Some questions about the implementation of the RMP, with particular reference to the North Atlantic minke whales. Document SC/45/Mg2 presented to the IWC Scientific Committee, April 1992. Unpublished. 52 pp. [Available from the IWC Secretariat]

Holt, S. J. 1993. Abstract to SC/44/O 6. Scientific advice on the management of commercial whaling. Reports of the International Whaling Commission 43:534.

Holt, S. J. 1993. Abstract to SC/44/NAB1. Fin whaling in the Denmark Strait: Part 1—exploratory analysis. Reports of the International Whaling Commission 43:525.

Holt, S. J. 1993. Abstract to SC/44/O 28. Increase in blue whales in the Denmark Strait. Reports of the International Whaling Commission 43:538.

Holt, S. J. 1993. Harvesting of whales. Nature 361(6411):391.

Holt, S. J. 1993. Abstract to SC/44/O 6. Scientific advice on the management of commercial whaling. Reports of the International Whaling Commission 43:534.

Holt, S. J. 1993. Abstract to SC/44/O 3. Thoughts about MSY rates and levels. Reports of the International Whaling Commission 43:534.

Holt, S. J. 1993. Abstract to SC/44/SHB1. Dispersal of baleen whales over Antarctic feeding grounds. Reports of the International Whaling Commission 43:527.

Holt, S. J. 1993. Report of the Scientific Committee, Annex L. Sanctuaries. L2: A response to Annex L1. Reports of the International Whaling Commission 43:206–207.

Holt, S. J. 1993. Sustainable whaling. Marine Pollution Bulletin 26(4):223–225.

Holt, S. J. 1993. Notes on maximum sustainable yield rates. Document SC/45/O 3 presented to the IWC Scientific Committee, April 1993. Unpublished. 5 pp. [Available from the IWC Secretariat]

Holt, S. J. 1993. Protection levels and the sex of whales: some questions about the implementation of the RMP, with some particular reference to North Atlantic minke whales. International Fund for Animal Welfare Technical Briefing 93(5):12.

Holt, S. J. 1993. Minkes' saga, Part I or the North Atlantic massacre. International Fund for Animal Welfare Technical Briefing 93(6):7.

Holt, S. J. 1993. Minkes' saga, Part II or whatever happened to Vesterfjord? International Fund for Animal Welfare Technical Briefing 93(7):9.
Holt, S. J. 1993. Numbers games. International Fund for Animal Welfare Technical Briefing 93(8):5.
Holt, S. J. 1993. Analysis of comments by the government of Japan on the French proposal for sanctuary in the southern hemisphere. International Fund for Animal Welfare Technical Briefing 93(1):23.
Holt, S. J. 1993. Comments on opposition to the French proposal for a Southern Ocean sanctuary. International Fund for Animal Welfare Technical Briefing 93(2):30.
Holt, S. J. 1993. The 95% confidence trick. Why the pause in commercial whaling must continue. International Fund for Animal Welfare Technical Briefing 93(3):36.
Holt, S. J. 1993. Whale sanctuary in the Southern Ocean. The question of its northern boundary. International Fund for Animal Welfare Technical Briefing 93(4):25.
Holt, S. J. 1993. The conservation of whales in the southern hemisphere: Political and practical reasons for supporting the French proposal for the creation of a circumpolar whale sanctuary in the Southern Ocean. International Fund for Animal Welfare Technical Briefing 93(10):13.
Holt, S. J. 1993. The regulation of commercial whaling. Consideration of the commercial moratorium and Norway's resumption of minke whaling in the northeast Atlantic. International Fund for Animal Welfare Technical Briefing 93(11):33.
Holt, S. J. 1993. Resignation of the Chairman of the Scientific Committee of the IWC: a commentary. International Fund for Animal Welfare Technical Briefing 93(9):19.
Holt, S. J. 1993. "Scientific whaling" Mark II. International Fund for Animal Welfare Technical Briefing 93(13):14.
Holt, S. J. 1993. Some comments on scientific aspects of the proposal by the Government of France for a whale sanctuary in the Southern Ocean, including its relationship with the question of implementing the Revised Management Scheme/Procedure. Document SC/45/SBHa3 presented to the IWC Scientific Committee, April 1992. Unpublished. 57 pp. [Available from the IWC Secretariat]
Holt, S. J. 1993. The regulation of commercial whaling. Wilderness 1993:29.
Holt, S. J. 1993. On loving whales, krill and marine ecosystems. Conservation Biology 7(3):451–452.
Holt, S. J. 1993. What is sustainable development? The test case over whaling. Our Planet 5(5):10–13.
Lankester, K., E. Slooten, and S. J. Holt. 1993. Report of the Scientific Committee, Annex O. Minority statements. M3: Minority statement on Item 8.3—northeastern Atlantic minke whales. Reports of the International Whaling Commission 43:218.
Holt, S. J. 1994. Depensatory net recruitment functions. Document SC/46/Mg2 presented to the IWC Scientific Committee, May 1994. Unpublished. 4 pp. [Available from the IWC Secretariat]
Holt, S. J. 1994. Abstract to SC/45/Mg2. Some questions about the implementation of the RMP, with particular reference to the North Atlantic minke whales. Reports of the International Whaling Commission 44:517.
Holt, S. J. 1994. Abstract to SC/45/O 3. Notes on maximum sustainable yield rates. Reports of the International Whaling Commission 44:521.
Holt, S. J. 1994. Abstract to SC/45/SBHa3. Some comments on scientific aspects of the proposal by the Government of France for a whale sanctuary in the Southern Ocean, including its relationship with the question of implementing the Revised Management Scheme/Procedure. Reports of the International Whaling Commission 44:511.
Holt, S. J. 1994. Ahab's return. BBC Wildlife 12:54–58.
Holt, S. J. 1994. *Apropos* a sanctuary for whales in the Southern Hemisphere. International Fund for Animal Welfare Technical Briefing 94(1):16.
Holt, S. J. 1994. The Indian Ocean Sanctuary: a model for the Southern Ocean? International Fund for Animal Welfare Technical Briefing 94(2):16.
Holt, S. J. 1994. Norway's war on the little leviathan of the North Atlantic. International Fund for Animal Welfare Technical Briefing 94(3):26.

Holt, S. J. 1994. The World Conservation Union supports the proposed circumpolar whale santuary in the southern ocean. International Fund for Animal Welfare Technical Briefing 94(5):11.

Holt, S. J. 1994. AIDS debate continues. Nature 367:212.

Holt, S. J. 1994. Commercial whaling: a corrupt industry. International Fund for Animal Welfare Technical Briefing 94(6):17.

Holt, S. J. 1994. Implications of revelations of Soviet under-reporting of Antarctic whale catches. International Fund for Animal Welfare Technical Briefing 94(7):31.

Holt, S. J. 1994. Comment on a proposal by the government of Norway to down-list the minke whales of the northeast Atlantic from CITES Appendix I to Appendix II. International Fund for Animal Welfare Technical Briefing 94(9):9.

Holt, S. J. 1994. Note on the northern boundary of the proposed circumpolar sanctuary for whales in the Southern Ocean. Unpublished. 1 pp.

Holt, S. J. 1994. Some legal aspects of the perceived role of a guardian for future generations in monitoring 'sustainability' of any use of marine living resources. Unpublished. 10 pp.

Holt, S. J. 1994. Actions for the protection of whales in Puerta Vallarta and beyond. Opening statement by the International Fund for Animal Welfare. Meeting of the International Whaling Commission, Mexico, May 1994. Unpublished. 10 pp.

Holt, S. J. 1994. Sustainable development of living marine resources: a torrent of words and little action. Presented at Pacem in Maribus XXII. Madras, India, 4–8 December 1994. 8 pp.

Holt, S. J. 1994. The social fabric: the current generation, future generations and externalities. Paper prepared for Workshop on New Principles of Living Resources Conservation. Washington, USA, 6–9 March 1994. 10 pp.

Holt, S. J. 1994. Whaling. Report of the independent scientific peer review of the catch limit algorithm of the International Whaling Commission's Revised Management Procedure. Letter to National Marine Fisheries Service, USA. 9 pp.

Frizell, J., S. J. Holt, and C. Phillips. 1995. Report of a Workshop to Outline a Programme of Non-Lethal Whale Research in the Southern Ocean Sanctuary. Document SC/47/O19 presented to the IWC Scientific Committee, June 1995. Unpublished. 31 pp. [Available from the IWC Secretariat]

Holt, S. J. 1995. Report of the Scientific Committee, Annex E. Report of the Sub-Committee on Southern Hemisphere Baleen Whales, Appendix 5. Agenda Items 7.2 estimates of abundance of humpback whales, and 8.2 abundance estimates of other species. A comment on SC/46/SH24, Parts 1 and 2. Reports of the International Whaling Commission 45:138.

Holt, S. J. 1995. Italy. Progress report on cetacean research, May 1995–April 1995. Document SC/47/O 24 presented to the IWC Scientific Committee, May 1995. Unpublished. 2 pp. [Available from the IWC Secretariat]

Holt, S. J. 1995. Obituary of Professor Raymond (John Heaphy) Beverton. Reviews in Fish Biology and Fisheries 5:396.

Holt, S. J. 1995. Counting the shadows of whales: the GeeNought [g(0)] saga. International Fund for Animal Welfare Technical Briefing 95(1):9.

Holt, S. J. 1995. The (dis)information age: a reply. Oryx 29(4):222–223.

Holt, S. J. 1995. Viewpoint: Creating confidence. Marine Pollution Bulletin 30(9):583–585.

Holt, S. J. 1996. Response to Hutton. Oryx 30(3):161–162.

Holt, S. J. 1996. The ways of the whales. Obituary for Doug Chapman. The Guardian. July 16 1996.

Holt, S. J. 1996. Sea fisheries and living marine resources. Memorandum to the Executive Committee of the Independent World Commission on the Oceans. Unpublished. 4 pp.

Mangel, M., L. M. Talbot, G. K. Meffe, M. T. Agardy, D. L. Alverson, J. Barlow, D. B. Botkin, G. Budowski, T. Clark, J. Cooke, R. H. Crozier, P. K. Dayton, D. L. Elder, C. W. Fowler, S. Funtowicz, J. Giske, R. J. Hofman, S. J. Holt, S. R. Kellert, L. A. Kimball, D. Ludwig, K. Magnusson, B. S.

Malayang, C. Mann, E. A. Norse, S. P. Northridge, W. F. Perrin, C. Perrings, R. M. Peterman, G. B. Rabb, H. A. Regier, J. E. Reynolds, K. Sherman, M. P. Sissenwine, T. D. Smith, A. Starfield, R. J. Taylor, M. F. Tillman, C. Toft, J. R. Twiss, J. Wilen, and T. P. Young. 1996. Principles for the conservation of wild living resources. Ecological Applications 6(2):338–362.

Holt, S. J. 1997. Managing commercial whaling. Paper presented at International Fund for Animal Welfare and Whale and Dolphin Conservation Society Conference. London, UK, 16 September 1997. 18 pp.

Holt, S. J., and K. Lankester. 1997. Science and the management of marine fisheries on wild stocks. International World Commission on the Oceans Working paper. Unpublished. 10 pp.

Holt, S. J., and P. Sands. 1997. Our ocean. Enhancing global consciousness of the role of the oceans in human affairs, and participation in decision-making. Revised working paper prepared following International World Commission on the Oceans Study Group on Public Awareness and Participation. Unpublished. 27 pp.

Holt, S. J. 1998. Fifty years on. Reviews in Fish Biology and Fisheries 8:357–366.

Holt, S. J. 1998. Brave initiative. New Scientist 2125:58–59.

Holt, S. J. 1998. Counting to kill. New Scientist 2141:52.

Holt, S, J. 1998. Review of “The International Politics of Whaling” by Peter Stoett. Environmental Values 7(3):372–374.

Holt, S. J. 1998. Whaling and the ‘Irish proposals’. Oryx 32(2):84.

Holt, S. J. 1998. Sustainable use of whales. Oryx 32(3):166.

Holt, S. J. 1998. Loopholes exploited in whaling regulation. Nature 393:206.

Anonymous. 1999. Protocol for the scientific evaluation of proposals to cull marine mammals. Report of the Scientific Advisory Committee of the Marine Mammals Action Plan. October 1999. Available: http://www.cull.org.

Holt, S. J. 1999. The ocean… our future. Book review. Marine Pollution Bulletin 38(6):514–517.

Holt, S. J. 1999. Whaling and international law and order. Marine Pollution Bulletin 38(7):531–534.

Holt, S. J. 2000. Whales and whaling. Pages 73–88 *in* C. R. C. Sheppard, editor. Seas at the millennium: an environmental evaluation. Volume III. Global Issues and Processes. Elsevier Science, Amsterdam.

Holt, S. J. 2000. Whales and whaling at the third millennium. Expanded version of Whales and whaling chapter. *In* C. R. C. Sheppard, editor. 2000. Seas at the millennium: an environmental evaluation. Unpublished. 64 pp.

Holt, S. J. 2000. Still in danger. New Scientist 2239:50.

Holt, S. J. 2000. Response to Dan Goodman. Marine Pollution Bulletin 40(1):91–92.

Holt, S. J. 2000. Moratoria and bans. International League for the Protection of Cetaceans Occasional Paper 2000-1. Unpublished. 42 pp.

Holt, S. J. 2000. International aspects of whale-watching. Working paper for IFAW/ICRAM Wrap-up Workshop on Whale Watching. Tuscany, Italy, 6–10 February 2000. 58 pp.

Holt, S. J. 2000. Introduction—setting the stage. Presentation at Symposium on Culling Mammals organized by The Mammal Society and the International Fund for Animal Welfare. London, UK, 24–25 November 2000. 8 pp.

Holt, S. J. 2000. Ethical aspects of the sustainable use of marine living resources, especially of whales. Contribution to the Fifth World Congress on Bioethics.

Holt, S. J. 2001. A comment on Tore Schweder’s ‘Protecting whales by distorting uncertainty: non-precautionary mismanagement’. Fisheries Research 52:227–230.

Holt, S. J. 2001. Sharing the catches of whales in the southern hemisphere. Pages 322–373 *in* R. Shotton, editor. Case studies on the allocation of transferable quota rights in fisheries. FAO Fisheries Technical Paper 411.

Holt, S. J. 2001. Opening statement by the International Ocean Institute. Meeting of the International Whaling Commission in London, UK, July 2001. Unpublished. 2 pp.

Holt, S. J. 2001. Opening statement by the International League for the Protection of Cetaceans. Meeting of the International Whaling Commission in London, UK, July 2001. Unpublished. 2 pp.

Holt, S. J. 2001. The determination and use of possible upper limits for targets for short- or medium-term reduction of bycatches of harbor porpoises in the EU EEZ. Unpublished. 6 pp.

Holt, S. J., V. Papastavrou, and C. Phillips. 2001. Response to report on the CITES conference in Nairobi, insofar as it refers to whales. Oryx 35(1):84–85.

Holt, S. J. 2002. The whaling controversy. Fisheries Research 54:145–151.

Holt, S. J. 2002. Whaling and whale conservation. Marine Pollution Bulletin 44:715–717.

Holt, S. J. 2002. On defending the whaling moratoria. Unpublished. 25 pp.

Holt, S. J. 2002. ICES involvement in whaling and whale conservation, and implications of IWC actions. ICES Marine Science Symposia 215:464–473.

Holt, S. J. 2003. Is the IWC finished as an instrument for the conservation of whales and the regulation of whaling? Marine Pollution Bulletin 46:924–926.

Holt, S. J. 2003. The tortuous history of 'scientific' Japanese whaling. BioScience 53(3):204–206.

Holt, S. J. 2003. Don't fix it. New Scientist 2380:34.

Beverton, R. J. H., and S. J. Holt. 2004. On the dynamics of exploited fish populations. Blackburn Press. Caldwell, New Jersey. (Re-issue of the 1957 paper).

Holt, S. J. 2004. Counting whales in the North Atlantic. Science 303(2):39.

Holt, S. J. 2004. Foreword to the 2004 printing. Pages i–xxiii in Beverton, R. J. H., and S. J. Holt. 1957. On the dynamics of exploited fish populations. Blackburn Press, New Jersey.

Holt, S. J. 2004. The experience of Antarctic whaling. Pages 131–150 *in* A. I. L. Payne, C. M. O'Brien, and S. L. Rogers, editors. Management of shared fish stocks. Blackwell Publishing, Oxford.

Holt, S. J. 2004. Preliminary thoughts. Pages 21–28 *in* F. Briand, editor. Investigating the roles of cetaceans in marine ecosystems. CIESM Workshop Monograph No. 25. CIESM, Monaco.

Holt, S. J. 2004. In defence of the Southern Ocean whale sanctuary. International Fund for Animal Welfare. 25 pp.

Holt, S. J. 2004. Whales and ice. Common themes of Antarctica, whaling and the law of the sea. Ocean Yearbook 18:364–382.

Holt, S. J. 2005. Are whales enemies of humans? Memorandum for the International Fund for Animal Welfare, Third Millennium Foundation, International League for the Protection of Cetaceans and Global Ocean. Unpublished. 6 pp.

Holt, S. J. 2005. The loss of whales and fish. Pseudo-science 2: Whales are eating 'our' fish. Memorandum prepared for Committee on Fisheries meeting of the Food and Agriculture Organization. Unpublished. 4 pp.

Holt, S. J. 2006. The notion of sustainability. Pages 43–82 *in* D. M. Lavigne, editor. Gaining ground: in pursuit of ecological sustainability. International Fund for Animal Welfare, Guelph, Canada and the University of Limerick, Limerick, Ireland.

Holt, S. J. 2006. Science is rooted in conversations. Pages 137–145 *in* K. L. Anderson, and C. Thiery, editors. Information for responsible fisheries: libraries as mediators: Proceedings of the 31st Annual Conference: Rome, Italy, October 10–14, 2005. International Association of Aquatic and Marine Science Libraries and Information Centers, Fort Pierce, Florida. Available: https://hdl.handle.net/1912/1480.

Holt, S. J. 2006. Propaganda and pretext: Japan's strategy to gain control of the International Whaling Commission (IWC). Marine Pollution Bulletin 52:363–366.

Holt, S. J. 2006. Commentary on "Implementing sustainability in EU fisheries through maximum sustainable yield". The Greens/European Free Alliance in the European Parliament. Unpublished. 5 pp.

Holt, S. J. 2006. Commentary on Commission Staff Working Document. The Greens/European Free Alliance in the European Parliament. Unpublished. 5 pp.

Holt, S. J. 2006. Whales competing? An analysis of the claim that some whales eat so much that they threaten fisheries and the survival of other whales. International League for the Protection of Cetaceans. Available: https://wwfeu.awsassets.panda.org/downloads/iwc_holt_report_eng.pdf.

Holt, S. J. 2007. New policy objectives and management procedures for EU fisheries. A commentary and suggestions. A briefing paper prepared for the European Policy Office of the WWF. Unpublished. 67 pp.

Holt, S. J. 2007. Whale tale. New Scientist 2611:20.

Holt, S. J. 2007. Whaling: will the phoenix rise again? Marine Pollution Bulletin 54:1081–1086.

Holt, S. J. 2007. New policy objectives and management procedures for EU Fisheries. A commentary and suggestions to The Greens/European Free Alliance in the European Parliament. A study sponsored by Ms Marie-Hélène Aubert, MEP. Unpublished. 23 pp.

Holt, S. J. 2007. Statement to Fisheries Committee of EuroParliament. Unpublished. 4 pp.

Busby, L., and S. J. Holt. 2008. Annotated Guide to the Schedule of the International Convention for the Regulation of Whaling. International Fund for Animal Welfare, London.

Holt, S. J. 2008. So, farewell, fishes and whales! Telos II:3–20.

Holt, S. J. 2008. Does history serve conservation? Pages 3–14 *in* K. L. Anderson, editor. Changes on the horizon. Proceedings of the 33rd Annual Conference of the International Association of Aquatic and Marine Science Libraries and Information Centers, October 7–11, 2007, Sarasota, Florida. IAMSLIC, Fort Pierce, Florida. Available: https://hdl.handle.net/1912/2946.

Holt, S. J. 2008. Three lumps of coal: doing fisheries research in Lowestoft in the 1940s. A talk by Sidney Holt to a CEFAS seminar in Lowestoft, Suffolk, UK, 25 April 2008. Unpublished. 20 pp.

Holt, S. J. 2008. Foreword. Pages ix–xxi *in* A. Payne, J. Cotter, and T. Potter, editors. Advances in fisheries science. 50 years on from Beverton and Holt. Blackwell Publishing, Oxford.

Holt, S. J. 2008. Sustainable Use of Wild Marine Living Resources: Notion or Myth? Pages 185–201 *in* L. L. Rockwood, R. E. Stewart, and T. Dietz, editors. Foundations of Environmental Sustainability. The Coevolution of Science and Policy. Oxford University Press, New York.

Holt, S. J. 2008. Greed enthroned: Shall future generations eat fish or whales? A lecture at Gresham College, London, November 19, 2008. Unpublished.

Holt, S. J. 2008. Game-play: Moratorium and management. Calypso Log Magazine, (issue?): 10–13.

Holt, S. J. 2009. Sunken billions—but how many? Fisheries Research, 97: 2–10.

Holt, S. J. 2009. Comments by Sidney J. Holt to the Commission of the European Communities on the Green Paper: Reform of the Common Fisheries Policy. Unpublished. 33 pp.

Holt, S. J. 2009. Opportunity: Needs meet possibilities. Negotiating sanctuaries for whales. A contribution to the First International Conference on Marine Mammal Protected Areas. Maui, Hawaii, 30 March–3 April 2009.

Holt, S. J. 2009. Sanctuary for whales: an unfinished history. Report to International Fund for Animal Welfare.

Holt, S. J. 2009. Opening statement by Dr Sidney Holt on behalf of NGOs attending the Intersessional meeting of the IWC in Rome in March 2009 and as the observer for the Antarctic and Southern Ocean Coalition (ASOC). Unpublished.

Holt, S. J. 2009. Wasteful whaling: a memorandum. Unpublished. 13 pp.

Holt, S. J. 2009. Hurrah for our brave whalers! Unpublished. 4 pp.

Holt, S. J. 2010. Sanctuary: 'Saving the Whale', in Six Movements. IFAW and Global Ocean.

Holt, S. J. 2010. Some simple thoughts about the science of fisheries management. Unpublished manuscript dated September 16, 2010. 2 pp.

Holt, S. J. 2010. Dealing in whales: an aide-memoire about restoring the functionality of the IWC. Unpublished manuscript dated February 18, 2010. 10 pp.

Holt, S. J. 2010. Wherever is MSY? Unpublished manuscript dated November 15, 2010. 27 pp.

Holt, S. J. 2010. Norwegian minke whaling in the North Atlantic: a memoir. Unpublished manuscript dated February 13, 2010. 14 pp.

Jacquet, J., D. Pauly, D. Ainley, S. Holt, P. Dayton, and J. Jackson. 2010. Seafood stewardship in crisis. Nature 467:28–29.

Holt, S. J. 2011. What is the theory of fishing? Draft chapter submitted to ISPRA, June 9, 2011. Unpublished.

Holt, S. J. 2011. Science, politics and economics in the International Whaling Commission. Aquatic Mammals 37(3):420–430.

Holt, S. J. 2011. Unfinished business. Pages 73–80 *in* A contribution to Telos Volume V—Special Edition in Honour of Guido de Marco. Fondation de Malte, Valletta.

Holt, S. J. 2011. Statement by the Antarctic and Southern Ocean Coalition (ASOC) to the 63rd meeting of the International Whaling Commission in Jersey, Channel Islands, July 2011. ASOC Secretariat, Washington, DC.

Holt, S. J. 2011. Maximum sustainable yield: the worst idea in fisheries management. Published on October 3, 2011. Available: https://breachingtheblue.com/2011/10/03/maximum-sustainable-yield-the-worst-idea-in-fisheries-management/.

Holt, S. J. 2011. An overhaul for MSY. Published on October 10, 2011. Available: https://breachingtheblue.com/2011/10/10/an-overhaul-for-msy/.

Holt, S. J. 2011. Fish population dynamics 101. Unpublished manuscript dated November 30, 2011. 4 pp.

Holt, S. J. 2012. Brief for The Green Group in the European Parliament on reform of the EU Common Fisheries Policy (CFP): how to achieve sustainable and profitable fishing. Unpublished manuscript dated February 2012. 54 pp.

Holt, S. J. 2012. Global fisheries crisis: whose fault: fishing industries, administrations, scientists, the fish? Presentation at conference on Planet Under Pressure 2012, New Knowledge Towards Solutions held March 26–29, 2012 in London. Unpublished.

Holt, S. J. 2012. Negotiating the Indian Ocean Sanctuary. Journal of Cetacean Research and Management 12(2):145–150.

Holt, S. J. 2012. The some good and mostly bad about maximum sustainable yield as a management target. ICES Document CM 2012/L:02.

Holt, S. J. 2012. The benefits of a fisheries management target for an average annual catch less than the theoretical maximum sustainable yield (MSY). Unpublished manuscript dated October 21, 2012. 9 pp.

Holt, S. J. 2012. Application of "less than MSY" management policy to so-called "mixed fisheries". Unpublished manuscript dated December 6, 2012. 3 pp.

Holt, S. J. 2013. Two approaches to the problem of making assessments for management purposes of fish stocks for which data are deficient. Unpublished manuscript dated April 21, 2013. 2 pp.

Holt, S. J. 2013. What is a fish stock assessment? Keynote address to the opening session of the World Conference on Stock Assessment Methods. Unpublished manuscript of a presentation given July 17, 2013. 7 pp.

Holt, S. J. 2013. The evolution of the objectives, science and procedures of fisheries management since 1945. *In* Heidbrink, and M. McCarthy, editors. Fisheries management in a historical perspective. Report of the 12th Conference of the North Atlantic Fisheries History Association, Old Dominion University, Norfolk, Virginia, 19–22 August 2009. North Atlantic Fisheries History Association, Hull, UK.

Holt, S. J. 2013. Il Massimo Rendimento Sostenibile. La peggiore idea nella gestione delle attività di pesca. Pages 263–281 *in* G. Bulian, and S. Raicevich, editors. In mare altrui. Pesca e territorialità in ambito interdisciplinare. Aracne Editrice, Rome.

Holt, S. J. 2013. Thoughts on the case about 'scientific whaling' brought to the International Court of Justice by the Government of Australia against the Government of Japan. Whale and Dolphin

Conservation blog. Available: https://us.whales.org/2013/07/04/a-guest-blog-on-the-issues-at-the-icj-by-dr-sidney-holt/.

Holt, S. J. 2014. The graceful sigmoid: Johan Hjort's contribution to the theory of rational fishing. ICES Journal of Marine Science 71:2008–2011.

Holt, S. J. 2014. Critique of Regulation No. 1380/2013 of the European Parliament and of the Council of 11 December 2013 on the Common Fisheries Policy. Unpublished manuscript dated July 12, 2014. 4 pp.

Holt, S. J. 2014. Recruitment. Unpublished manuscript written in 2014. 5 pp.

Holt, S. J. 2014. Surplus production. Unpublished manuscript written in 2014. 8 pp.

Holt, S. J. 2014. The two hoaxes: how the management of global fisheries was led astray. Unpublished manuscript written in 2014. 1 p.

Holt, S. J. 2014. Optimal biomass. Unpublished manuscript written in 2014. 8 pp.

Froese, R., C. Walters, D. Pauly, H. Winker, O. L. F. Weyl, N. Demirel, A. C. Tsikliras, and S. J. Holt. 2015. A critique of the balanced harvesting approach to fishing. ICES Journal of Marine Science73(6):1640–1650.

Holt, S. J. 2015. Oral presentation to the Committee on Fisheries of the European Parliament during a public hearing on maximum sustainable yield in fisheries management. Unpublished manuscript of a presentation given February 23, 2015. 7 pp.

Holt, S. J. 2015. An Adjunct to the Presentation to the Committee on Fisheries of the European Parliament. Unpublished manuscript of a presentation given February 23, 2015 to a public hearing on Maximum Sustainable Yield in Fisheries Management. 6 pp.

Holt, S. J. 2015. North Sea Demersal Fisheries Management Consultation—EU. Unpublished manuscript dated May 1, 2015. 9 pp.

Holt, S. J. 2015. How bad science impedes the good: an example. Unpublished manuscript dated July 14, 2015. 13 pp.

Holt, S. J. 2015. Escaping the MSY stranglehold. ICES Document CM 2015/A:01.

Holt, S. J. 2015. What is overfishing? Presentation to the ICES Annual Science Conference 2015 Theme Session on Advancement of Stock Assessment Methods for Sustainable Fisheries dated September 2015. 10 pp. This was a revision of his original written submission (ICES Document CM 2015/A:01) to the Theme Session.

Holt, S. J., and Froese, R. 2015. A comment on management of European fisheries for high sustainable yields, in particular in the Baltic Sea. Unpublished manuscript dated November 1, 2015. 4 pp.

Engelhard, G. H., R. H. Thurstan, B. R. MacKenzie, H. K. Alleway, R. C. A. Bannister, M. Cardinale, M. W. Clarke, J. C. Currie, T. Fortibuoni, P. Holm, S. J. Holt, C. Mazzoldi, J. K. Pinnegar, S. Raicevich, F. A. M. Volckaert, E. S. Klein, and A-K. Lescrauwaet. 2016. ICES meets marine historical ecology: placing the history of fish and fisheries in current policy context. ICES Journal of Marine Science 73(5):1386–1403.

Froese, R., C. Walters, D. Pauly, H. Winker, O. L. F. Weyl, N. Demirel, A. C. Tsikliras, and S. J. Holt. 2016. Reply to Andersen et al. 2016. "Assumptions behind size-based ecosystem models are realistic". ICES Journal of Marine Science 73:1656–1658.

Holt, S. J. 2016. Suggestions regarding implementation of the primary management objective of the Common Fisheries Policy of the European Union. Unpublished manuscript dated February 24, 2016. 9 pp.

Holt, S. J. 2016. Comment on an EU Document entitled: COM Baltic Plan Articles 4–5. Unpublished manuscript dated March 4, 2016. 3 pp.

Holt, S. J 2016. Evolution of the theory of rational fishing. Keynote lecture at the Conference "Historical ecology of semi-enclosed basins: past, present and future of seas at risk," held at the Department of Biology, University of Padua, Chioggia, Italy, 4 October 2016. Unpublished.

Holt, S. J. 2016. A "less than *F*-msy" management strategy. Unpublished manuscript written in June 2016. 2 pp.

Holt, S. J. 2016. Save the whale! Memoirs of a whale hugger. International Fund for Animal Welfare. 48 Hour Books, Akron, Ohio.

Pauly, D., R. Froese, and S. J. Holt. 2016. Balanced harvesting: the institutional incompatibilities. Marine Policy 69:121–123.

Holt, S. J., and S. Raicevich. 2018. Evolution of the theory of rational fishing. The case study of the North Sea. Regional Studies in Marine Science 21:74–78.

Holt, S. J. 2020. Becoming a marine scientist: helped by a daily quota of three lumps of coal. ICES Journal of Marine Science 77:463–468, doi:10.1093/icesjms/fsaa009.

Holt, S. J. 2020. Three lumps of coal: doing fisheries research in Lowestoft in the 1940s, ICES Journal of Marine Science, doi: 10.1093/icesjms/fsaa199.

Holt, S. J. Undated. Interactions between marine mammals and fisheries. Unpublished. 3 pp.

Holt, S. J. Undated. On the ethics of continued commercial whaling. Unpublished. 16 pp.

Holt, S. J. Undated. Recommendations on technology assessment. Unpublished. 1 pp.

Holt, S. J. Undated. Suggestions for further actions by IWC. Unpublished. 2 pp.

Index

A

B

C

D

E

F

G

H

J

N

T

Y

Z